KV-118-178

SOURCE BOOK ON THE FERRITIC STAINLESS STEELS

Other Source Books in This Series:

Applications of the Laser in Metalworking
Innovative Welding Processes
Electron Beam and Laser Welding
Brazing and Brazing Technology
Gear Design, Technology and Performance
Copper and Copper Alloys
Materials for Elevated-Temperature Applications
Maraging Steels
Powder Metallurgy
Wear Control Technology
Selection and Fabrication of Aluminum Alloys
Materials Selection (Volumes I and II)
Nitriding
Ductile Iron

Metals Science Source Book Series:

The Metallurgical Evolution of Stainless Steels
Hydrogen Damage

SOURCE BOOK ON THE FERRITIC STAINLESS STEELS

A comprehensive collection of outstanding articles from the periodical and reference literature

Compiled by
Consulting Editor

R. A. LULA

Consulting, Inc.
Natrona Heights, Pa.

American Society for Metals
Metals Park, Ohio 44073

Library of Congress Card No.: 81-71377

ISBN: 0-87170-120-0

PRINTED IN THE UNITED STATES OF AMERICA

Contributors to This Source Book

G. AGGEN
Allegheny Ludlum Steel Corp.

A. PAUL BOND
Climax Molybdenum Co.

K. G. CARROLL
United States Steel Corp.

RENE CASTRO
Société d'Electrochimie, d'Electrométallurgie et des Aciéries d'Ugine Savoy

HUNG-CHI CHAO
United States Steel Corp.

A. DATTA
Allied Chemicals Corp.

J. A. DAVIS
Allegheny Ludlum Steel Corp.

RALPH M. DAVISON
Climax Molybdenum Co.

J. J. DEMO
E.I. du Pont de Nemours & Co., Inc.

E. J. DULIS
United States Steel Corp.

R. M. FISHER
United States Steel Corp.

I. A. FRANSON
Allegheny Ludlum Steel Corp.

R. H. KALTENHAUSER
Allegheny Ludlum Industries, Inc.

N. KINOSHITA
Kawasaki Steel Corp.

C. W. KOVACH
Colt Industries

H. D. KURTZ
Colt Industries

PETER LILLYS
Crucible Steel Co. of America

REMUS A. LULA
Allegheny Ludlum Industries, Inc.

G. E. MOLLER
Consultant

T. NAKAZAWA
Nippon Steel Corp.

A. E. NEHRENBERG
Crucible Steel Co. of America

T. J. NICHOL
Allegheny Ludlum Steel Corp.

N. OHASHI
Kawasaki Steel Corp.

Y. ONO
Kawasaki Steel Corp.

H. W. PAXTON
Carnegie Institute of Technology

L. S. REDMERSKI
Colt Industries

Y. SOGO
Nippon Steel Corp.

ROBERT F. STEIGERWALD
Climax Molybdenum Co.

MICHAEL A. STREICHER
E.I. du Pont de Nemours & Co., Inc.

T. SUNAMI
Nippon Steel Corp.

S. SUZUKI
Nippon Steel Corp.

HELMUT THIELSCH
Welding Research Council
New York, N.Y.

ROLAND TRICOT
Société d'Electrochimie, d'Electrométallurgie et des Aciéries d'Ugine Savoy

R. O. WILLIAMS
Cincinnati Milling Machine Co.

K. YOSHIOKA
Kawasaki Steel Corp.

NOTE: Affiliations given were applicable at date of contribution.

CONTENTS

INTRODUCTION

Ferritic stainless steels were developed in Europe just before the start of World War I, about the same time as the austenitic steels. Large-scale industrial use started with the 12% Cr martensitic, the 17% Cr ferritic, and the 18Cr-8Ni austenitic steels. As production methods progressed, as fabrication techniques improved, and as the demand for these steels expanded to increasingly diverse fields of application, modified versions of the three basic compositions were developed. Gradually, the popularity of the austenitic stainless steels increased until, at the present time, the production volume of austenitic steels greatly exceeds the combined volume of the ferritic and martensitic steels. Therefore, more research and development have been concentrated on the austenitic steels.

There is good reason for the widespread use of the austenitic steels: they have excellent fabricating properties, can be readily welded, have good mechanical properties over a broad temperature range, and possess a resistance to corrosion that satisfies the needs of most industries. The martensitic and ferritic stainless steels, on the other hand, have acquired a more restricted field of application. The martensitic steels are on the lower scale of corrosion resistance, because they contain only 11-15% Cr, but they do exhibit a useful combination of strength, ductility, and toughness. For this reason, these steels are used primarily for structural components in power plants, steam and gas turbines, and the petrochemical industry, and for cutlery and tools.

The ferritic stainless steels, which contain 11-30% Cr, can exhibit corrosion resistance equivalent to that of the austenitic steels when molybdenum is added. Oxidation resistance of the ferritic steels is also very good, especially for the high-chromium grades such as Type 446, or when aluminum and rare earths are added.

In spite of equivalent corrosion resistance and lower-cost alloying elements, the ferritic steels have been surpassed by the austenitic steels primarily because of their lower formability and toughness, and reduced weldability. These drawbacks are minimized in light gages, so the bulk of current use of ferritic steels is in light-gage sheet form.

Structure of the FeCr Ferritic Stainless Steels

The basis for understanding the constitution of the ferritic stainless steels can be found in the Fe-Cr phase diagram shown on page 2 of this Source Book. In the 15-30% Cr range, where most of the commercial alloys are found, the structure is a body-centered cubic alpha ferrite from room temperature to the solidus temperature. The solubility for the interstitial elements carbon and nitrogen is very low, and therefore they appear as carbides and nitrides of chromium.

There are two very important features of the Fe-Cr phase diagram. First is the gamma-loop, which extends to 12% Cr in the binary system, but is extended to much higher chromium levels in commercial alloys because of the influence of carbon, nitrogen, and other austenitizing elements. Titanium, columbium, aluminum, and silicon have the reverse effect of contracting the gamma-loop to lower chromium levels. The second significant feature is the area in the center of the diagram that shows a pure sigma and duplex sigma-alpha zone. This portion of the phase diagram has been modified to include the presence of alpha prime, a fine precipitate of high-chromium solid solution that is responsible for the well-known "885 °F embrittlement" phenomenon. The domains of sigma and alpha prime can be extended substantially by addition of several alloying elements or by cold working. Chi phase forms in the higher-chromium steels containing molybdenum. References to the

Table 1. Typical Compositions of the Ferritic Stainless Steels

Designation	C	N_2	Si	Mn	Cr	Other
AISI 405	0.06	0.02	0.30	0.50	13.0	0.25Al
409	0.05	0.01	0.30	0.50	11.0	0.40Ti
429	0.05	0.02	0.30	0.50	15.0	—
430	0.05	0.03	0.30	0.50	16.50	—
430F	0.06	0.03	0.30	0.50	16.50	0.3S
434	0.04	0.03	0.30	0.50	16.50	1.0Mo
436	0.04	0.01	0.30	0.50	16.50	1.0Mo, 0.40Cb
439(HWT)	0.05	0.02	0.30	0.50	18.00	0.50Ti
442	0.08	0.04	0.30	0.50	21.00	—
446	0.08	0.08	0.30	0.50	26.00	(N_2 optional)
			New Developments			
18-SR	0.05	0.02	1.10	0.20	18.00	2.0Al
18Cr-2Mo	0.02	0.02	0.30	0.50	18.00	2.0Mo, 0.25Ti, 0.3Cb
E-Brite® 26-1	0.003	0.008	0.20	0.10	26.00	1.0Mo, 0.01Cb
SEA-CURE®	0.01	0.025	0.20	0.20	26.00	3.0Mo, 2.5Ni, 0.4Ti
AL 29-4	0.005	0.01	0.10	0.10	29.00	4.0Mo
AL 29-4-2	0.005	0.01	0.10	0.10	29.00	4.0Mo, 2.0Ni
Fe-28Cr-2Mo	$0.01x$	$0.01x$	$0.1x$	$0.1x$	28.00	2.0Mo

NOTE. E-Brite® 26-1, AL 29-4 and AL 29-4-2: Allegheny Ludlum Steel Co.; SEA-CURE®: Crucible Steel Co.; 18-SR: Armco Steel Co.; Fe-28Cr-2Mo: THYSSEN EDELSTAHLWERKE.

original and the modified phase diagrams are included in two of the papers presented in this book.

An important facet of the ferritic stainless steels that can be explained by the phase diagram is the absence of a complete phase transformation, which could be exploited for grain refinement. As a result, the ferritic stainless steels have an inherent tendency to grain coarsening when exposed to high temperature. A large-grained structure is undesirable, however, because it impairs room-temperature toughness and formability. The only methods of refining the grain structure are (1) to finish hot rolling at a temperature below that at which recrystallization occurs, and (2) to cold work and follow by a recrystallization anneal. This thermomechanical process for grain refinement requires heavy dimensional reduction, so most ferritic stainless steels are used in light-gage sheet form.

"New" and "Old" Ferritic Stainless Steels

When reviewing the various commercial stainless steels, it is appropriate to distinguish between the "old" and the "new" steels. The old ferritic stainless steels are represented by the standard AISI grades shown in Table 1. The bulk of current production is in these grades. The new ferritic steels, also shown in Table 1, are primarily proprietary grades only now starting to penetrate the market. The origin of the two families of ferritic stainless steels is to be found in melting technology. Until the development of the oxygen decarburization (AOD) and vacuum decarburization (VOD) processes and the availability of large vacuum induction furnaces, the ferritic stainless steels were produced only in electric furnaces. Carbon and nitrogen cannot be reduced economically in an electric furnace under about 0.05% C+N. The AOD, VOD, and vacuum induction processes, on the other hand, can economically produce steels with lower carbon and nitrogen content (C+N $\leq$ 0.05%). Because low carbon and nitrogen are mandatory for improving the toughness and weldability of the ferritic steels, the advent of the AOD and VOD processes prompted the development of new higher-alloyed ferritic stainless steels, which not only have superior mechanical properties but also have better corrosion resistance as a result of alloying with higher amounts of molybdenum and chromium. At one time, all AISI steels were melted in the electric furnace; now, most producers use the economically advantageous AOD process. Among the new steels, 18Cr-2Mo and SEA-

CURE® are melted by the AOD process. E-Brite® 26-1, AL 29-4-2, and Fe-28Cr-2Mo are vacuum induction melted.

Technical Characteristics of the Ferritic Stainless Steels

The ferritic stainless steels have several important metallurgical characteristics that must be taken into consideration during processing, fabrication, and application. These characteristics have to be reviewed in comparison with those of the austenitic steels, which would be the alternative materials for similar applications.

Tensile Properties. There is an appreciable difference between the mechanical properties of the ferritic and the austenitic steels. The room-temperature yield strength of the ferritic steels in the annealed condition is 20 to 40 per cent higher than that of the austenitic steels. The rate of work hardening of the ferritic steels is lower, however, and hence they cannot attain the same strength as the austenitic steels through cold working. The lower rate of work hardening accounts at least in part for the lower elongation of the ferritic steels. At high temperature, the ferritic steels have lower strength than the austenitic steels, but their oxidation resistance is very good.

Toughness and Embrittling Phenomena. Similar to all steels with a body-centered cubic crystal structure, the ferritic stainless steels exhibit a ductile-to-brittle transition temperature that for many grades is close to or even above room temperature. The interstitial elements carbon and nitrogen have a potent effect on the ductile-to-brittle transition temperature; a low interstitial elements content is essential for a low transition temperature and hence good toughness at room temperature. For this reason, all the newly developed ferritic steels have very low carbon and nitrogen content, obtained by special melting techniques.

Small grain size also has a favorable influence on toughness. Because grain refinement is feasible only through severe thermomechanical working, the best toughness is attained in thin gages, sheet, or wire, rather than in heavy sections such as plate, bar, or forgings.

Toughness of the ferritic stainless steels is influenced by three embrittling mechanisms: (1) "885 °F embrittlement," (2) sigma-phase precipitation, and (3) high-temperature embrittlement.

"885 °F Embrittlement." All ferritic stainless steels, with the possible exception of Type 409, are susceptible to this phenomenon when exposed in the 720-950 °F temperature range. Peak hardening and embrittlement occur at approximately 885 °F as a result of the precipitation of alpha prime, a coherent chromium-rich particle. This embrittlement becomes more pronounced with increasing chromium content. In order to avoid this form of embrittlement during processing, fast cooling is recommended through the 700-950 °F range. Exposure to this temperature in service should be avoided.

Sigma-Phase Precipitation. Sigma phase, an intermetallic compound, may form in the higher-chromium ferritic steels when they are exposed for prolonged periods at 1100 to 1500 °F. Sigma formation is enhanced by molybdenum, silicon, titanium, and columbium. This precipitation results in room-temperature embrittlement as well as loss of corrosion resistance, and hence should be avoided both in processing and in service.

High-Temperature Embrittlement. Carbide and nitride precipitation subsequent to heating at very high temperatures can result in embrittlement as well as susceptibility to intergranular corrosion, especially in alloys containing high interstitial elements. Both consequences are significant because they occur in the weld heat-affected zone of these steels.

Intergranular Corrosion. The ferritic stainless steels are susceptible to intergranular corrosion after being heated above about 1800 °F. The heat-affected zone of weldments is the more likely candidate for this form of corrosion, which is caused by the precipitation of chromium carbides at the grain boundaries. The only practical method of preventing intergranular corrosion is by alloying with titanium and/or columbium to form the carbides of these elements. Lowering the carbon content below 0.03%, as in austenitic stainless steels, is not effective. Even with 0.01% maximum carbon content, it is necessary to add carbide-stabilizing elements.

Corrosion and Stress Corrosion. Because the chromium content of the ferritic stainless steels ranges from 11 to 29 per cent, the general corrosion resistance can vary from moderate to excellent. Additions of molybdenum increase the resistance to corrosion by producing a more stable passive film.

The 11-12% Cr steels are strictly for structural use in components requiring a minimum corrosion or oxidation resistance at reasonable cost. Automotive exhaust systems are undoubtedly the largest single application for Type 409. The intermediate-chromium steels, represented by Types 430 and 434 and forming the bulk of the ferritic steel production, are used for their decorative-functional attributes in automotive trim, appliances, cooking utensils, and tableware. The higher-chromium steels, usually alloyed with molybdenum, have general corrosion and pitting corrosion resistance equal to or actually exceeding that of the austenitic steels. Types 18Cr-2Mo, E-Brite® 26-1, SEA-CURE®, and AL 29-4 are used in the petrochemical and chemical industries and in seawater applications. Type 446 is used primarily for oxidation resistance.

One of the most important features of the ferritic stainless steels is their resistance to stress-corrosion cracking in chloride solutions. Because the austenitic steels are highly susceptible to stress corrosion, the ferritic steels provide a more economical solution than the nickel-base alloys for applications requiring this property.

Banding Deformation. Types 430 and 434 steels plastically formed by stretching operations exhibit a nonuniform flow pattern termed "ridging" and "roping," caused by rolling textures formed during processing. This pattern is undesirable, especially since these steels are used primarily in applications requiring good appearance. Although controlled thermomechanical processing can eliminate the hot roll texture and minimize this problem, the best results are obtained by using columbium-stabilized steel, Type 436.

* * * * *

The papers presented in this Source Book have been carefully selected to provide basic as well as applied information about all of the above properties of the ferritic stainless steels.

The American Society for Metals extends most grateful acknowledgment to the many authors whose work is presented in this book, and to their publishers.

R. A. Lula
Consulting Editor

Cover photo, which shows a heat exchanger with E-Brite® 26-1 alloy (XM-27) tubes and clad tube sheets being fabricated for organic acid service, is courtesy of Mastercraftsmen.

SOURCE BOOK ON THE FERRITIC STAINLESS STEELS

J. J. Demo[1]

STRUCTURE, CONSTITUTION, AND GENERAL CHARACTERISTICS OF WROUGHT FERRITIC STAINLESS STEELS

REFERENCE: Demo, J. J., *Structure, Constitution, and General Characteristics of Wrought Ferritic Stainless Steels, ASTM STP 619,* American Society for Testing and Materials, 1977.

ABSTRACT: High chromium-ferritic stainless steels have good general corrosion and pitting resistance and are resistant to stress-corrosion cracking. Despite these desirable properties, the alloys have found little use as materials of construction. This lack of use is a result of significant losses in ductility, toughness, and corrosion resistance when these alloys are subjected to moderate or high temperatures. Names given to the phenomena causing loss in properties include 475°C, sigma phase, and high-temperature embrittlement. This publication summarizes the literature describing the causes, the cures, and the limitations imposed on alloys when these problems occur. The most seriously limiting problem—high temperature embrittlement and loss or corrosion resistance—is discussed in considerable detail. The key role that interstitial carbon and nitrogen play on notch sensitivity and loss of ductility and corrosion resistance following a high-temperature exposure as in welding is defined. Good as-welded properties, the absence of which has severely restricted the use of ferritic stainless steels, depend on controlling interstitial carbon and nitrogen. The publication describes three methods that are being used for interstitial control. It is now possible to produce ferritic stainless steels which are tough and which have excellent corrosion resistance and ductility in the as-welded conditions. Several new high-chromium ferritic alloys with these desirable properties are being produced commercially.

KEY WORDS: ferritic stainless steels, embrittlement, sigma phase, properties, corrosion resistance, notch sensitivity, interstitial, stabilization

The high chromium-iron steels represent the fourth class of alloys in the family of stainless steels, the other three classes being austenitic, martensitic, and precipitation-hardening stainless and heat-resisting steels. Ferritic stainless steels are iron-based alloys containing from about 12 to 16

[1]Senior consultant, Materials Engineering, Engineering Service Division, Engineering Dept., E. I. du Pont de Nemours and Co., Inc., Wilmington, Del. 19898.

and 30 percent chromium. The high chromium limit is arbitrary and is meant simply to include all commercially produced alloys. Ferritic stainless steels, though known for more than 40 years, have had more restricted utility and less wide use than the austenitic stainless steels. Reasons for this include: the lack of the ductility characteristic of austenitic stainless steels, their susceptibility to embrittlement, notch sensitivity, and poor weldability, all factors contributing to poor fabricability. However, with the increasing cost of nickel, the high resistance of the ferritic steels to stress-corrosion cracking, and their excellent corrosion and oxidation resistance, intensive research over the decade of 1960's has resulted in ferritic alloy compositions which have good weldability and fabricability.

Structure and Constitution

In theory, the ferritic stainless steels are structurally simple. At room temperature, they consist of chromium-iron alpha (α) solid solution having a body-center-cubic crystal (bcc) structure. The alloys contain very little dissolved carbon; the majority of the carbon present appears in the form of more or less finely divided chromium carbide precipitates. They remain essentially ferritic or bcc up to the melting point. A typical constitution diagram, as published by the American Society for Metals [*1*][2] is reproduced in Fig. 1. Attention is directed to the lower chromium end of the phase diagram at the region of intermediate temperatures, about which several points can be stated.

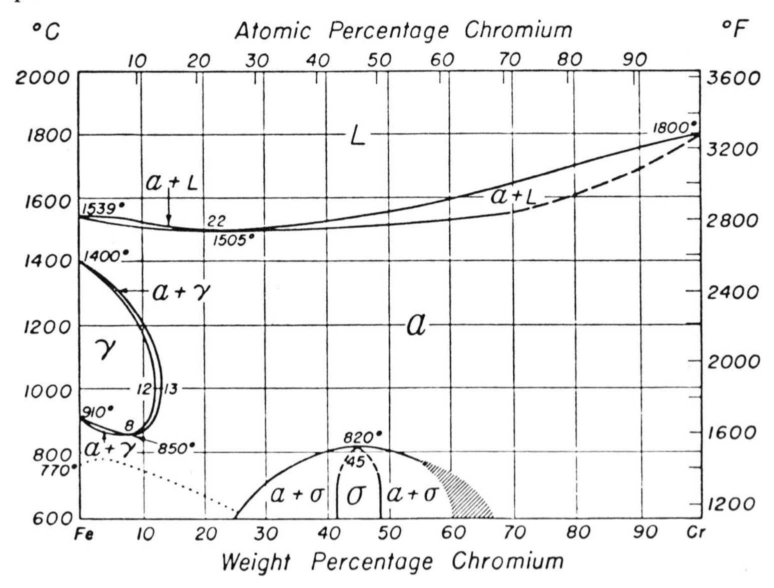

FIG. 1—*Chromium-iron phase diagram* [1].

[2] The italic numbers in brackets refer to the list of references appended to this paper.

1. Chromium is a member of a group of elements called ferrite formers which extend the α-phase field and suppress the gamma (γ)-phase field. This property results in the so-called γ loop extending in a temperature range from 850 to 1400°C.

2. As shown in Fig. 1, the transformation in chromium-free iron from α to γ phase occurs at about 910°C. As chromium is added, the transformation temperature is depressed to about 850°C at 8 percent chromium and then rapidly increased so that, at 12 to 13 percent chromium, the transformation temperature is about 1000°C.

3. The transformation from γ to α which occurs at about 1400°C for pure iron is depressed with increasing chromium to about 1000°C at 12 to 13 percent chromium. At this point, the upper and lower temperature curves for transformation join up to close off and form the γ loop. Beyond 12 to 13 percent chromium, transformation to γ is no longer possible, and an alloy would remain ferritic or bcc all the way from below room temperature to the melting point. It will be seen that this maximum limit for the existence of γ phase is very much a function of austenitizing elements and can be moved to higher chromium levels in the presence of these elements, particularly the interstitials carbon and nitrogen. For alloys with less than about 12 percent chromium, an $\alpha \rightarrow \gamma$ transformation occurs on heating into the γ range just as in pure iron. Ferritic stainless steels cannot normally undergo this transformation upon heating and cooling.

4. Between the γ loop and α-phase field, there is a narrow transition region where an alloy at temperature will have both α and γ phases which, depending on the quench rate, may or may not be retained at room temperature.

Effect of Carbon and Nitrogen

The location of the γ loop in the iron-chromium phase diagram has been carefully studied recently by Baerlecken, Fischer, and Lorenz [*2*]. These investigators used magnetic measurements at elevated temperatures to determine the effect of carbon and nitrogen (and nickel) on the formation of austenite. The detailed α and γ regions these workers developed for pure iron and chromium alloys are shown in Fig. 2. The lowest point in the γ loop occurs at 840°C and 6.5 percent chromium. The greatest width of the two-phase field is at about 1075°C, and the complete enclosure is reached at about 11.5 percent chromium. Since the greatest expansion of the γ region is to about 10.6 percent chromium, the two-phase $\gamma + \alpha$ region is very narrow in this high-purity alloy.

Additions of austenitizing elements, particularly carbon and nitrogen, cause the outside boundary of the $\gamma + \alpha$ two-phase field to shift to higher chromium levels. The powerful effects of carbon and nitrogen in this

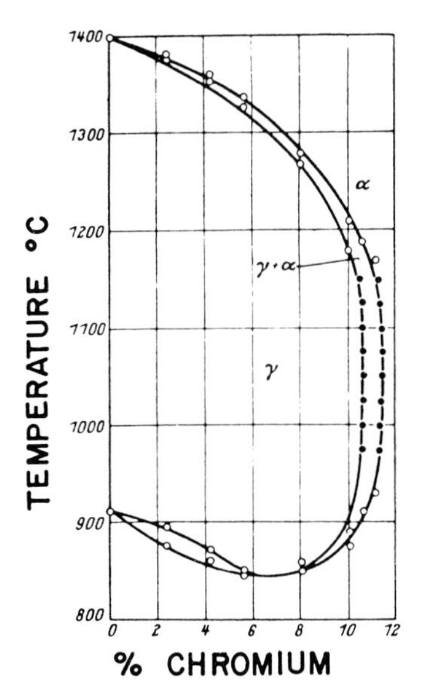

FIG. 2—*Iron loop of the iron-chromium phase diagram for alloys with about 0.004 percent carbon and 0.002 percent nitrogen* [2].

regard are shown in Figs. 3*a*,*b* taken from work by Baerlecken et al [*2*]. Two effects are observed: namely, an expansion of the two-phase region to higher chromium contents and the shifting of the maximum extension of the $\gamma + \alpha$ phase field to higher temperatures. For example, it is seen in Fig. 3*a* that 0.013 percent carbon and 0.015 percent nitrogen shift the maximum expansion of the $\gamma + \alpha$ from 11.5 to 17.0 percent chromium, while 0.04 percent carbon and 0.03 percent nitrogen shift it to about 21 percent chromium. At still higher carbon levels, for example, 0.2 percent,

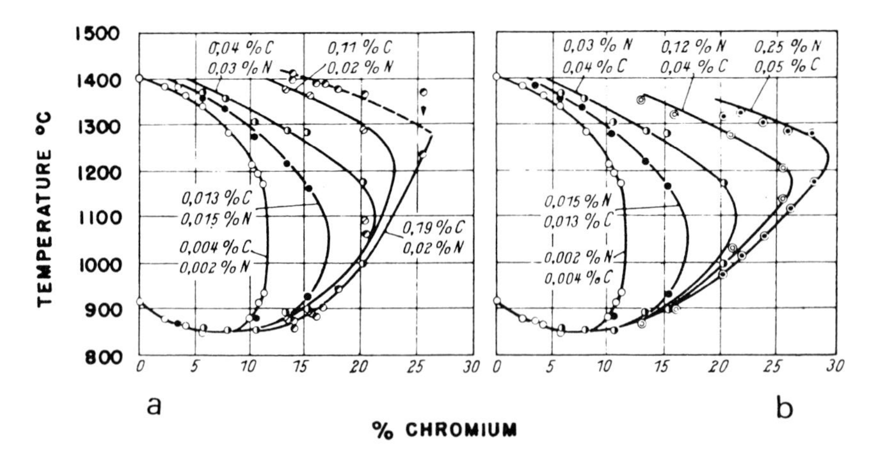

FIG. 3—*Shifting of the boundary line* ($\gamma + \alpha$)/α *in the system iron-chromium through increasing additions of carbon or nitrogen* [2].

an expansion to 26 percent chromium is observed. In addition, the point of greatest expansion of the two-phase region is shifted to higher temperatures, from about 1075°C for the pure alloy to about 1300°C for an alloy containing 0.2 percent carbon. The expansion of the two-phase region to even higher temperatures is limited by the solidus temperatures of the alloys. As evident from Fig. 3*b*, nitrogen acts similar to carbon as a powerful expander of the two-phase $\gamma + \alpha$ field. In an alloy containing 0.25 percent nitrogen, the point of greatest width of the two-phase field is shifted from 11.5 percent chromium for the pure alloy to 28 percent chromium and to higher temperatures, from 1075 to about 1250°C.

Besides its effect on extending the two-phase $\gamma + \alpha$ region, carbon, because of its low solubility in the α matrix, is rejected from solid solution as complex carbides, $(Cr,Fe)_7C_3$ and $(Cr,Fe)_{23}C_6$ which precipitate predominantly at the grain boundaries [*3,4*]. When an alloy is heated to temperatures of 1100 to 1200°C and the carbon level exceeds about 0.01 percent [*2*], the carbon cannot be held in solid solution even with rapid quenching, and the complex carbides are formed, which drastically affect the properties of the alloys.

It is at once obvious from this that whether or not an alloy remains completely bcc depends very much on the chromium and the interstitial levels. At low-chromium levels, even alloys with relatively low interstitial levels may have a duplex structure when heated to temperatures around 1100°C. Alloys at the higher chromium levels can only have a duplex structure if they also contain high interstitial levels; otherwise, these alloys are fully ferritic at all temperatures. In general, beyond about a 13 percent chromium content, heating to high temperatures no longer produces massive transformation of α to γ; therefore, grain refinement and hardening by heat treatment and quenching are no longer possible.

Strengthening Mechanisms

As just described, the ferritic stainless steels are characterized by the essential absence of the α to γ transition upon heating to high temperatures. Consequently, hardening by the γ to martensite transition upon cooling will not normally occur. This transformation mechanism is utilized in carbon and alloy steels and martensitic chromium-iron alloys for achieving high hardness and strengths. The influence of heat treatment temperature on hardness for two ferritic stainless steels and a martensitic stainless steel are depicted in Fig. 4, based on data from Refs *5* and *6*. In contrast to the 13 percent martensitic steel which undergoes the $\alpha \rightarrow \gamma \rightarrow$ martensite transition, the hardness on an 18 and 20 percent chromium steel varies little with temperature. The fact that there is some hardening occurring is due to the small amount of austenite formed at temperature which is dependent upon the chromium content and inter-

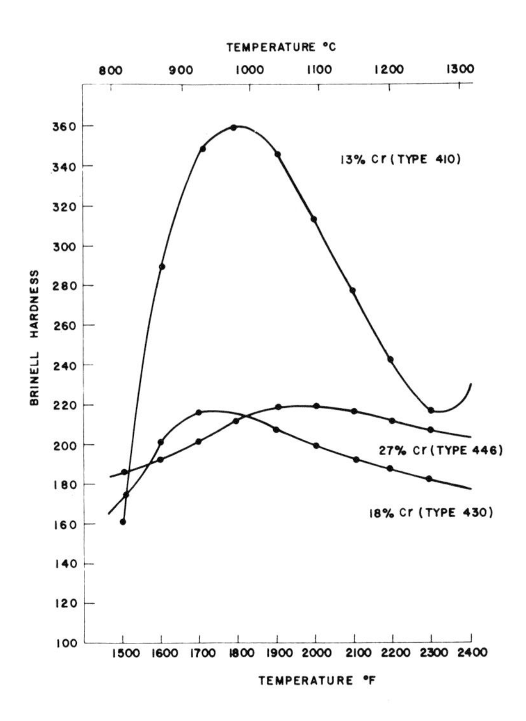

FIG. 4—*Relative hardening of ferritic and martensitic chromium-iron alloys after water quenching from indicated temperature* [5,6].

stitial levels. For ferritic stainless steels, the hardening effect is strongest at the lower chromium ranges but can occur even at higher chromium levels if the carbon content is increased so as to expand the α to γ, two-phase region as shown in Figs. 3*a*,*b*.

In summary, it is not possible to harden or strengthen ferritic stainless steels significantly by heating them to high temperatures and cooling rapidly. In these alloys, the slight increases in hardness or strength when the alloys are heated above 850°C are related to increases in grain size (through grain growth) and the presence of small volumes of austenite which reverts to martensite upon cooling.

Strengthening by Heat Treatment

Though a ferritic iron-chromium stainless steel cannot be strengthened or hardened by the classical $\alpha \rightarrow \gamma \rightarrow$ martensite mechanism, the alloys are susceptible to significant strengthening by heat treatment. In general, the strengthening mechanisms that do occur are not desirable because the alloys are embrittled, and ductility and toughness at room temperature are severely reduced. Distinction will be made between three separate and distinct forms of embrittling mechanisms: (*a*) sigma (σ)-phase precipitation, (*b*) 475°C embrittlement, and (*c*) high-temperature embrittlement.

Surveys by Thielsch [*6*], Rajkay [*7*], Kaltenhauser [*8*], and Demo and Bond [*9*] have summarized the extensive literature, describing these embrittling phenomena. These effects have been major deterrents to the use of ferritic stainless steels as engineering materials. Because of their importance to engineering properties, a summary of each of these embrittling effects is given here with details as to the nature and cure of each.

Sigma Phase—This occurs in iron-chromium alloys containing between 15 to 20 and 70 percent chromium exposed to temperatures from about 500 to 800°C.

475°C Embrittlement—This embrittlement occurs when a ferritic iron-chromium stainless steel is heated between 400 and 540°C. An increase in hardness and tensile strength is observed, concurrent with a substantial decrease in ductility and impact resistance.

High-Temperature Embrittlement—This embrittlement derives its name from the fact that ferritic alloys with moderate to high carbon and nitrogen levels are brittle at room temperature following exposure of the alloys to temperatures above about 1000°C.

While all these embrittling phenomena can affect severely the mechanical properties of a ferritic chromium-iron stainless steel in an engineering application, one is particularly serious, namely, high-temperature embrittlement. This effect is serious because a useful engineering material must be capable of being welded and heat treated while maintaining ductility, toughness, and corrosion resistance.

Sigma Phase

Examination of the phase diagram in Fig. 1 shows a second zone at lower temperatures centered about 45 percent chromium in addition to the γ loop. A detailed part of the σ portion of the chromium-iron phase diagram by Cook and Jones [*10*] is shown in Fig. 5. Pure σ forms between 42 and 50 percent chromium, while a duplex structure of both α and σ phases has been found to form in alloys with as little as 20 and as much as 70 percent chromium when they are exposed to temperatures between 500 and 800°C. The existence of a compound in the iron-chromium system at about 50 percent chromium was suggested as early as 1927 [*11,12*]. It was not until 1936 that the intermetallic compound, iron-chromium, was definitely identified as σ phase [*13,14*]. σ phase is an intermetallic compound containing one atom of iron with one atom of chromium which is hard, nonmagnetic, and consists of a tetragonal unit cell. Extensive research has been reported describing the effort to establish the structure and transformation characteristic of this iron-chromium compound [*15–20*]. σ phase forms in other alloy systems when two metals with a bcc and face-centered-cubic (fcc) structure are alloyed together

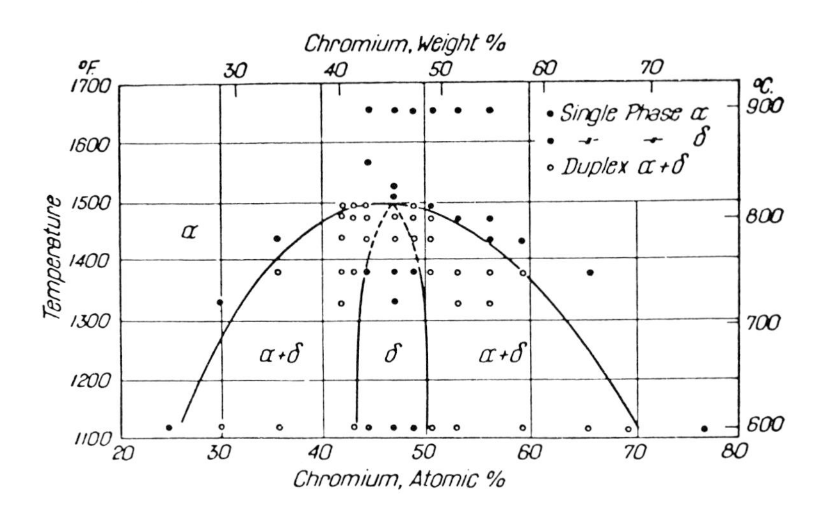

FIG. 5—*Iron-chromium phase diagram in binary high-purity 25 to 76Cr alloys. This diagram shows pure σ phase from 44 to 50 percent chromium and mixed α and σ phase from about 25 to 44 and from 50 to 70 percent chromium with phase boundaries for temperature interval of 1100 to 1500°F (595 to 815°C)* [10].

and have atomic radii not differing by more than 8 percent [*3*]. Elements like molybdenum, silicon, nickel, and manganese shift the σ-forming range to lower chromium content [*3,13,14,20,21*]. σ phase forms readily on heating alloys containing 25 to 30 percent chromium to 600°C but only after a relatively long-time exposure. In alloys containing less than about 20 percent chromium, σ phase is difficult to form [*3*]. Cold work enhances the rate of σ-phase precipitation [*14,19,21–24*]. The formation of σ phase is accompanied by an increase in hardness and a severe reduction in ductility and toughness [*25*], especially when these properties are measured at ambient temperatures. An important consideration is that, in most chromium-iron alloys, σ forms very slowly, requiring hundreds of hours. This is shown by the data in Fig. 6 from Shortsleeve and Nicholson [*20*] describing the threshold times of σ formation at 595 and 650°C as a function of chromium content. Based on these data, weld deposits and casting normally would not have sufficient time in the appropriate temperature ranges for σ to form in alloys, especially those containing 15 to 33 percent chromium [*6*]. Only with long isothermal holds can σ phase form to severely reduce the ductility and toughness of chromium-iron alloys. A representative microstructure [*25*] of a 27Cr alloy exposed for 3144 h at about 540 to 565°C shows a structure composed of ferrite, spheroidized carbides, and σ phase (Fig. 7). The σ phase has been precipitated as an essentially continuous series of islands around the ferrite grain boundaries. Under some exposure conditions, these islands are preferentially attacked (Fig. 7*c*) indicating that the presence of σ phase is also detrimental to the corrosion resistance of

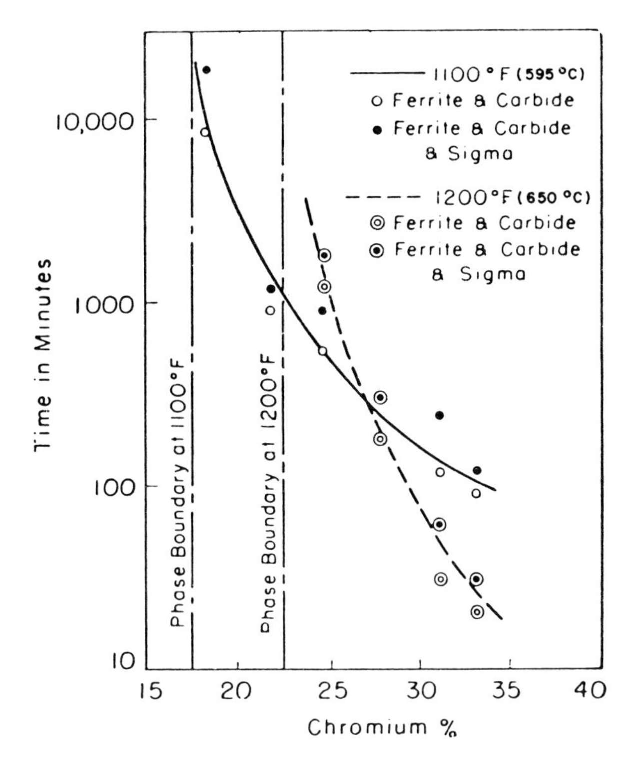

FIG. 6—*Effect of chromium content on the threshold times of σ formation at 1100 and 1200°F (595 and 650°C)* [20].

chromium-iron alloys. Fortunately, σ phase developed in an alloy may be brought into solution with relatively short holding periods of an hour or more by heating at a temperature over 800°C [*6*]. Alloys containing nickel, molybdenum, and manganese may require longer holding periods or higher temperatures to dissolve σ phase.

475°C Embrittlement

When chromium-iron alloys containing 15 to 70 percent chromium are subjected to prolonged heating at temperatures between 400 and 540°C, the alloys harden, and a drastic loss in ductility is observed. The hardening phenomenon is referred to as 475°C (885°F) embrittlement because peak hardness on aging occurs at this temperature, as illustrated in Fig. 8 [*6,25*] for chromium-iron stainless steels heated for long times at selected temperatures. The effect of prolonged heating at 475°C on increasing the strength and decreasing the ductility in an Fe-27Cr alloy is shown in Fig. 9 from the work of Newell [*25*]. As suggested by these data, hours of exposure at 475°C are required before noticeable changes in hardness and tensile properties are observed. However, notched specimens may reveal this embrittlement in a much shorter time. Zapffe et al

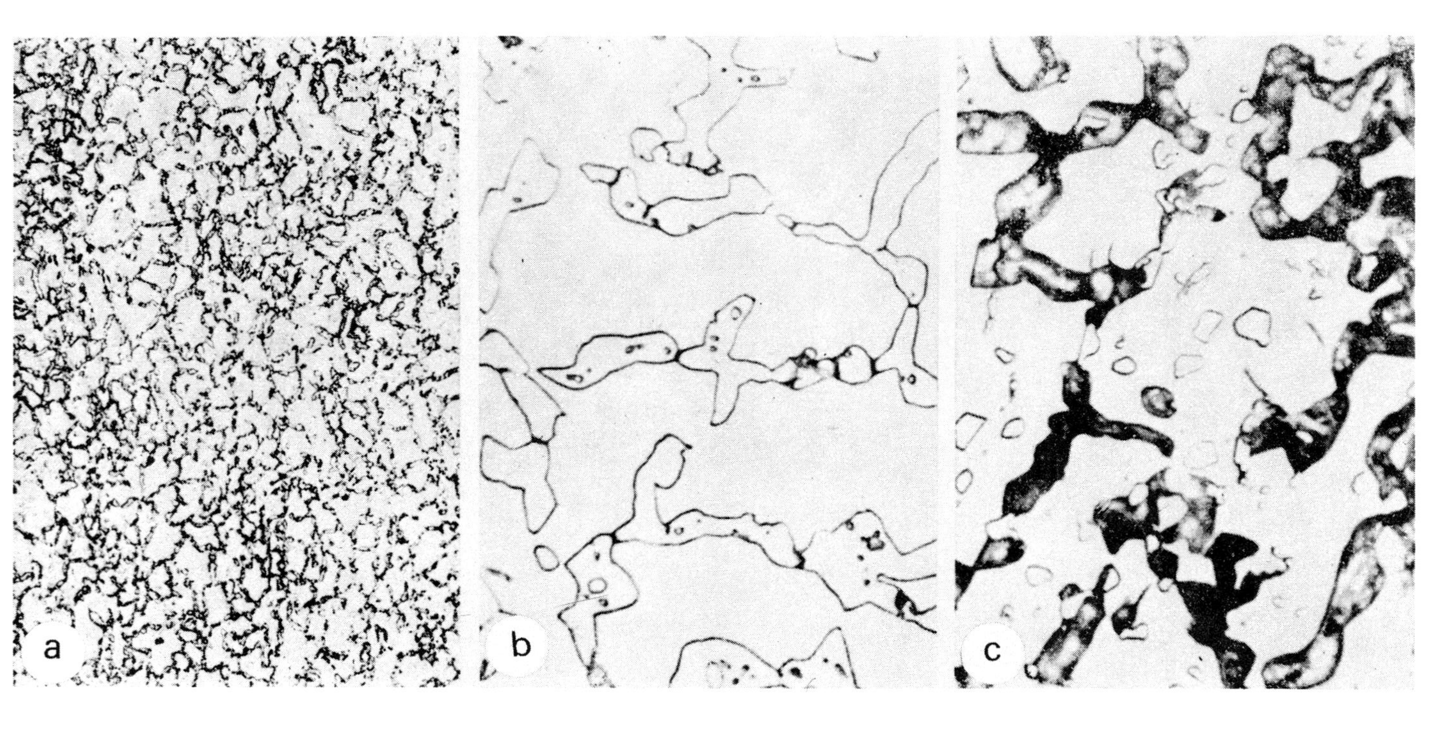

FIG. 7—(a) *27Cr-Fe alloy (air melted) showing ferrite, carbides, and intergranular σ-phase constituent formed after heating 131 days at approximately 1050°F (565°C)* (∽ ×200). (b) *σ phase with some small spheroids of carbides around which σ phase has been formed* (∽ ×1000). (c) *Etched 10 s in aqua regia during which σ phase is blackened and eaten out. Carbides appear mainly in center of ferrite grains* (∽ ×1000) [25].

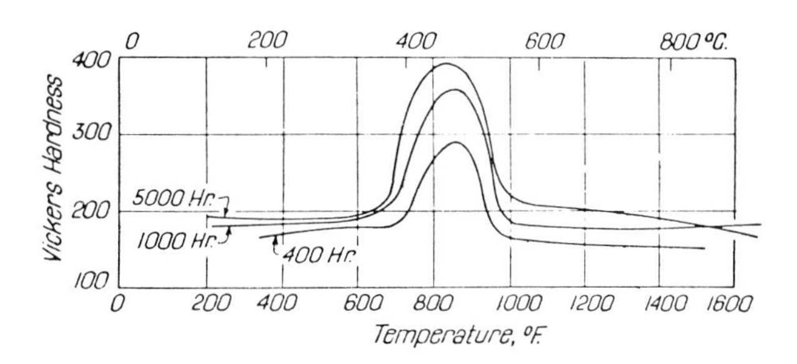

FIG. 8—*Hardness surveys on bars after prolonged heating in a temperature gradient. Analysis: 0.20 percent maximum carbon, 1.50 percent maximum manganese, 0.025 percent maximum sulfur and phosphorus, 0.75 percent maximum silicon, 26 to 30 percent chromium, 1.00 percent maximum nickel, 0.12 to 0.25 percent nitrogen* [25].

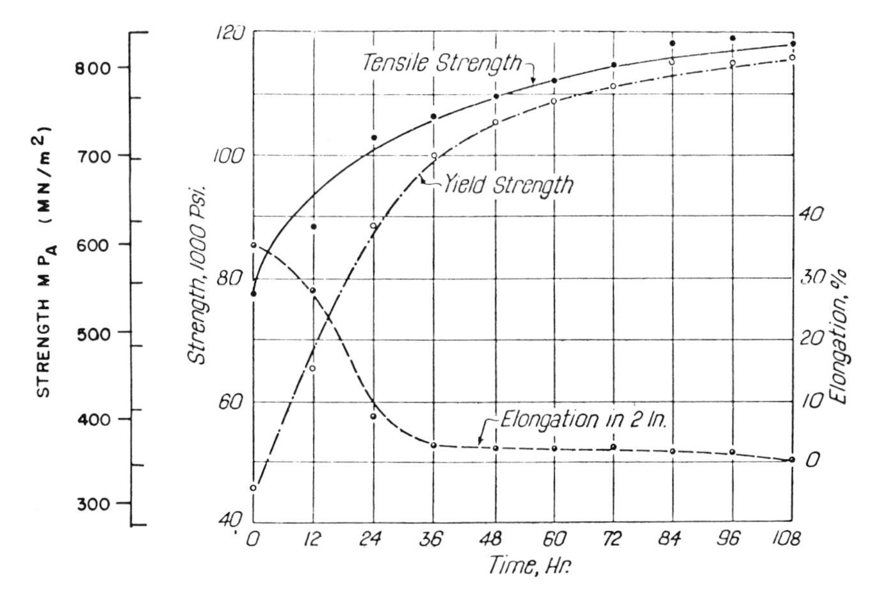

FIG. 9—*Graph showing effect of aging time at 885°F (475°C) on room-temperature tensile properties of 27Cr-Fe alloy, air melted* [25].

[*24,26*], using a 26Cr stainless steel, were able to show embrittlement on notch specimens in a bend test after only the first half hour of exposure at 475°C.

Large reductions in impact strength may also be noticed after only short-time exposure to embrittling temperature. Data taken from Colombier's book [*3*] are summarized in the following table to illustrate this point for a steel with 0.08 percent carbon, 0.4 percent silicon, and 16.9 percent chromium. The mechanical properties (tensile strength, yield strength, and elongation) show essentially no effect of the embrittling treatment, but a drastic loss in impact strength is recorded. As noted by

	Annealed	Heated 4 h at 450°C
Tensile, tsi[a]	37.0	38.0
Yield point, tsi	21.0	25.4
Elongation, %	22.8	23.6
Impact values, kg/cm^2	12.0	1.4

[a]tsi = tons per square inch.

Colombier, after repeated heating in the range of 450 to 500°C, the elongation starts to decrease and, after a few hundred hours, may be virtually zero, while an increase in tension of some 7 to 8 tsi occurs.

The cause for the increase in strength and hardness and drastic reduction in ductility was unexplained for a long time. X-ray diffraction analysis by Becket [*27*], Riedrich and Loib [*28*], and Bandel and Tofaute [*29*] on specimens embrittled by heating in the neighborhood of 475°C showed no changes in lattice parameter. Newell [*25*], while reporting no changes in lattice parameters, did note that an atomic disturbance was occurring because the diffraction lines in back-reflection spectra of an embrittled specimen were diffuse and broad. Early investigators could detect no significant change in metallographic features, although Riedrich and Loib [*28*] and Newell [*25*] did report a grain boundary widening. Based on this observation, Riedrich and Loib concluded that the embrittlement was caused by a precipitate along the grain boundary. In separate work, Bandel and Tofaute [*29*] discovered that an embrittled 18Cr alloy had a lamellar precipitate. Prior to 1951, 475°C embrittlement was considered to be related to σ-phase formation and received the attention of many studies. Newell [*25*] in 1946 showed that the embrittlement would occur in practically carbon-free alloys as well as in those of moderate carbon content. Bandel and Tofaute [*29*], Riedrich and Loib [*28*], and Newell [*5*] showed that the degree of hardening as a function of temperature exposure was proportional to chromium content. The data from Riedrich's and Loib's work using experimental low-carbon alloys are produced in Table 1. The alloys in the annealed condition at the start

TABLE 1—*Hardnesses[a] of chromium-iron alloys after 1000 h exposure at 885°F (475°C) as a function of chromium content.*

Chromium, % by weight	Brinell Hardness	Chromium, % by weight	Brinell Hardness
14.5	150	20.1	260
15.9	195	21.9	270
17.0	223	23.7	290
19.4	252	28.7	320

NOTE—Data by Newell [*5*] and Riedrich and Loib [*28*].
[a]The hardness of annealed alloys at the start was about 145 Brinell.

all measured about 145 Brinell. The hardnesses after 1000-h exposure at 475 °C are tabulated in the table. N. D. Newell's [5] data reproduced in Fig. 10 also show the relationship of chromium content in commercial al-

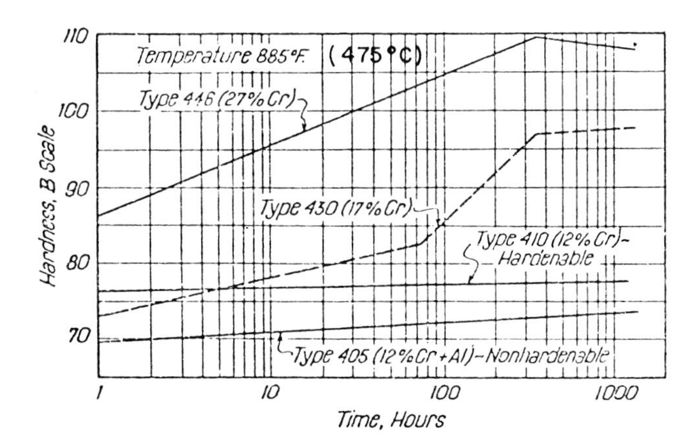

FIG. 10—*Age hardening tendencies at 885 °F (475 °C) of 12, 17, and 27Cr-Fe alloys* [5].

loys to hardness increases. Rapid increases in hardness with exposure times were noted for Types 446 and 430 stainless steel, with little change noted for Types 410 and 405 stainless steel (12 percent chromium). These data indicate that the hardening occurs in alloys containing over about 15 percent chromium, increases with chromium content, and is particularly pronounced for a commercial Type 446 stainless steel containing 26 percent chromium.

In 1951, Heger [*30*] emphasized the suggestion first made by Bandel and Tofaute [*29*] and Newell [*5*] that the embrittlement was due to a precipitation hardening process involved in the early stages of formation of σ phase. This was in contrast to the theory of a minor impurity precipitation which was rendered untenable because alloy additions [*30*] and prior heat treatment [*29*] did not improve resistance to the embrittlement, and even the purest chromium-iron alloys embrittled [*27*]. The cause, as put forth by Heger [*30*] and Newell [*5,25*], was precipitation of some phase which is inherent in the chromium-iron alloy system itself and not an impurity. Heger postulated that 475 °C embrittlement occurred when a chromium-iron alloy heated in the embrittling range undergoes the reaction

$$\alpha \text{ phase} \rightarrow \text{transition phase} \rightarrow \sigma \text{ phase}$$

The transitory phase that proceeds σ formation at higher temperature was said to cause 475 °C embrittlement. Newell described the embrittlement phenomenon as being related first to an initial lattice change or distortion in the α matrix leading subsequently to gross precipitation of σ phase when the atomic mobility is increased at but slightly higher temperatures. As Heger pointed up, the transition phase is intermediate in structure between α and σ and is coherent with the matrix α phase. The presence of this coherency between the two different structures causes large resistance to dislocation motion typical of a precipitation-hardening mechanism [*31*]. The generated stresses then increase the hardness and cause other property changes. While both Heger and Newell felt 475 °C embrittlement was related to a transition state before σ phase is formed, Heger pointed up a very serious contradiction of this hypothesis. Houdremont's [*6,32*] data reproduced in Fig. 11 illustrate this problem. The alloys containing 17 and 28.4 percent chromium did not reveal any hardening because of σ-phase formation (600 to 900 °C). However, significant hardening in the 475 °C embrittlement range (400 to 550 °C) did occur. It thus appeared that 475 °C embrittlement phenomenon is definitely inherent to the chromium-iron binary system, as suggested by Heger and Newell, but it is not related directly to the formation of σ phase as postulated.

Based on extensive research work from 1951 to 1964, two hypotheses have been suggested to describe the mechanism of 475 °C embrittlement. One school—Masumota, Saito, and Sugihara [*33*] and Porney and Bastien [*34*]—have attributed the changes in physical properties of alloys with aging at about 475 °C to atomic ordering. In fact, Takeda and Nagai [*35*] claimed to have found X-ray verification for super lattices corresponding to Fe_3Cr, FeCr, and $FeCr_3$. However, all published attempts to observe super lattices in chromium-iron α alloys using more sensitive neutron diffraction have been unsuccessful [*36–38*]. The second and more prevalent school of thought for the occurrence of 475 °C embrittlement is the formation of a coherent precipitate due to the presence of a miscibility gap in the chromium-iron system below about 550 °C, in a chromium range where σ phase can form at higher temperatures. The data published by Fisher, Dulis, and Carroll in 1953 [*39*] gave the first indication of the presence of a coherent precipitate. These investigations were able to extract fine particles about 200 Å in diameter from a 28.5Cr alloy aged for 1 to 3 years at 475 °C. The material was found to be nonmagnetic, to have a bcc structure with a lattice parameter (a = 2.877 Å) which is between that of iron and chromium, and to contain chromium in the range from 61 to 83 percent. Williams and Paxton [*40*] and Williams [*37*] confirmed these results and were the first to propose explicitly the existence of a miscibility gap below the σ-forming region in the equilibrium diagram of Fig. 1. The extent of the miscibility gap is shown in Fig. 12 as published by Williams [*37*]. Alloys aged within the gap would separate into chromium-rich ferrite (α^1) and iron-rich ferrite (α). Reversion to the unaged condition occurs when the specimens are heated above

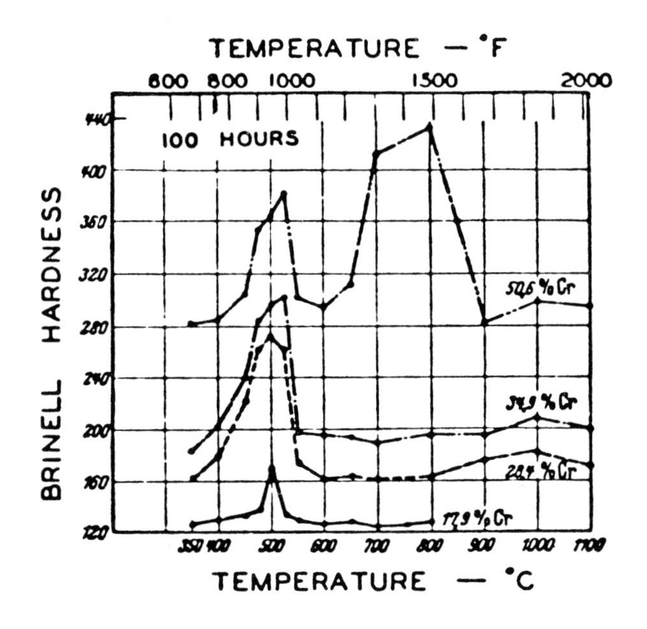

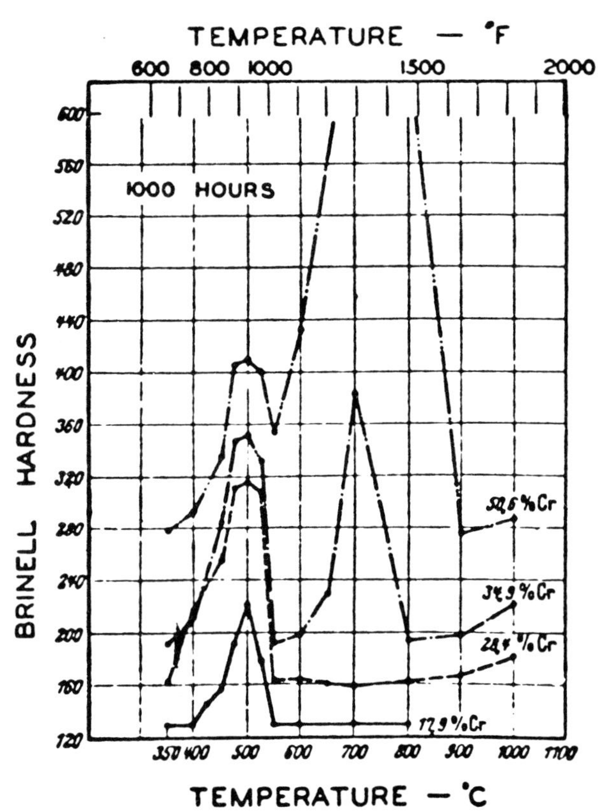

FIG. 11—*Effects of temperature and aging time on hardness increase caused by 885°F (475°C) brittleness and σ-phase precipitation in several chromium-iron alloys* [6,32].

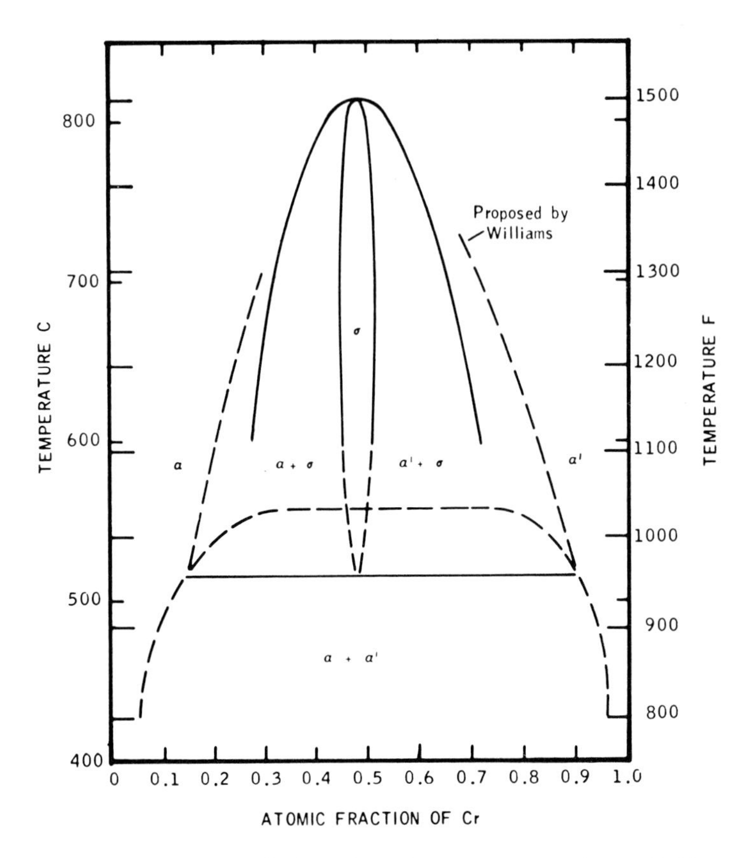

FIG. 12—*Phase diagram of the iron-chromium system according to Williams* [37].

550°C in agreement with all experimental data. The mechanism of age hardening by α–α^1 precipitation in the chromium-iron alloy system has been studied in detail by Marcinkowski, Fisher, and Szirmae [*41*] who analyzed the influence of the formation of a chromium-rich precipitate (α^1) on the deformation markings in the vicinity of hardness impressions. In addition, using extraction-replica and thin-foil transmission electron microscopy, these workers were able to show the existence of a coherent chromium-rich precipitate in an iron-rich matrix in a 46 percent by weight chromium-iron alloy heated for thousands of hours at 500°C (Fig. 13).

In more recent work on lower chromium content alloys, Grobner [*42*] established the critical temperature range as well as the kinetics of the embrittling process in chromium-iron alloys containing 14 and 18 percent chromium. He showed that, for an 18Cr alloy, 475°C embrittlement occurs in a temperature range from 400 to 500°C for both a commercial purity alloy and a low interstitial alloy. However, the vacuum-melted alloy of higher purity requires a longer exposure time to embrittle, as shown by Grobner's data on impact strength reproduced in Fig. 14. In addition to drastic changes in toughness, embrittled alloys, whether high

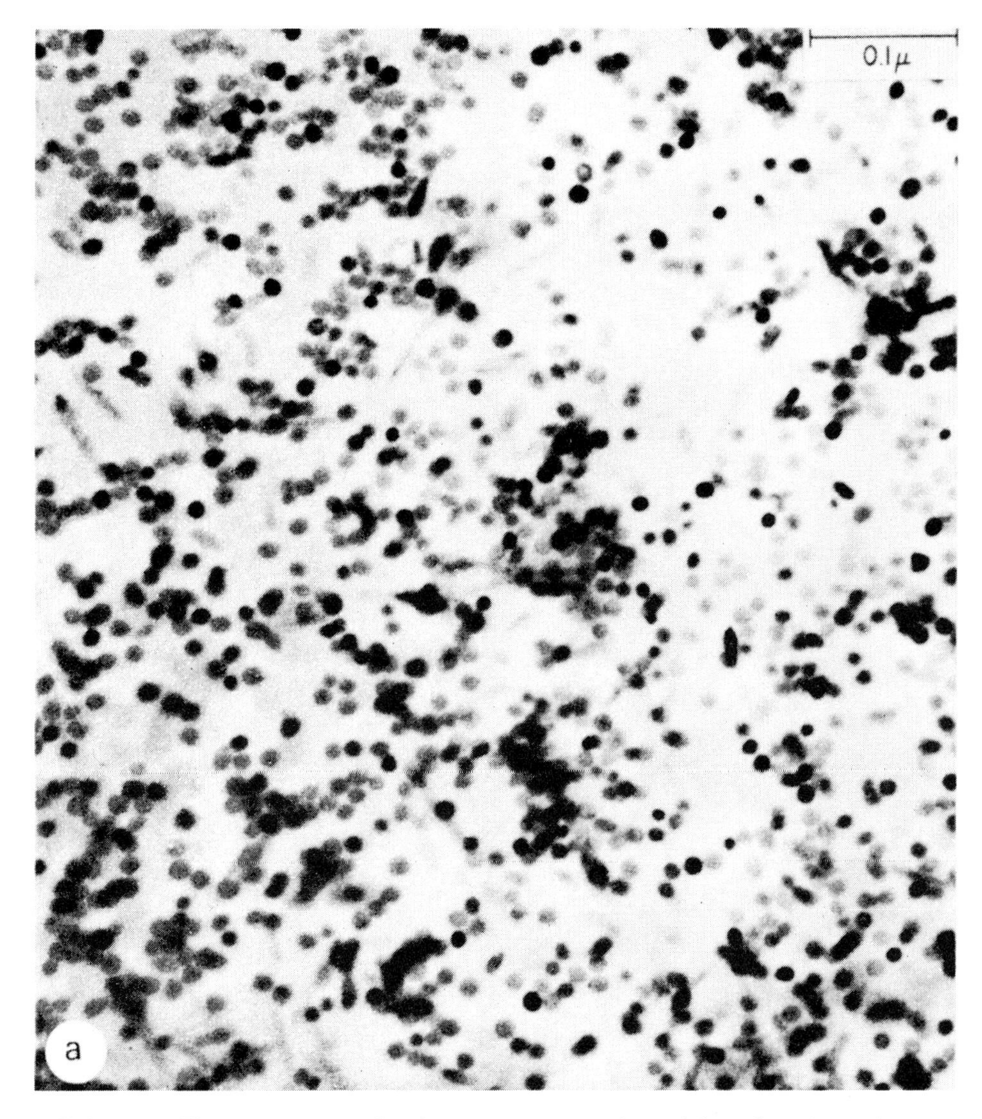

FIG. 13*a*—*Electron micrograph of an extraction replica of bcc chromium-rich zones existing in a 47.8 atomic percent chromium-iron alloy after annealing 9650 h at 500°C.*

purity and vacuum melted or air melted, show increases in yield and tensile strength and drastic reduction in elongation, as shown in Table 2. Finally, Grobner showed that even steels with chromium contents as low as 14 percent show embrittlement when exposed in the temperature range of 370 to 485°C, but only after much longer exposure times as compared to the 18Cr steels.

Many investigators have studied the effect of additives on 475°C embrittlement in chromium-iron steels [*3,6,27,29,30,42*]. Heger [*30*] noted that, in general, the addition of alloying elements offered little or no improvement in preventing the embrittlement. The effects of additives

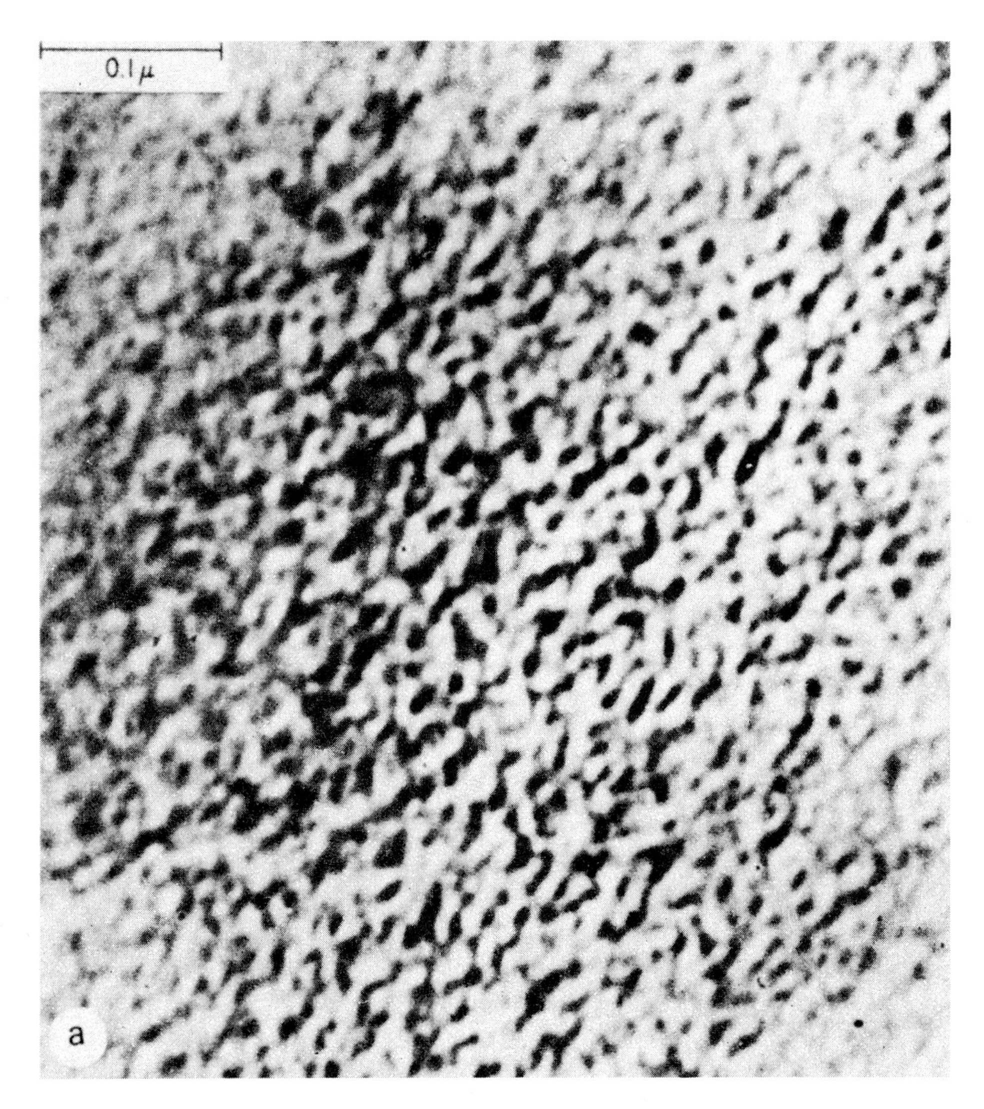

FIG. 13*b*—*Transmission electron micrograph showing structure existing in a 47.8 atomic percent chromium-iron alloy after annealing 3743 h at 500°C* [41].

on 475°C embrittlement taken from Heger [*30*] are summarized in Table 3. While not an additive, cold work [*6*] intensifies the rate of 475°C embrittlement. In addition to drastic reduction in toughness and ductility, embrittled chromium-iron alloys show a severe reduction in corrosion resistance [*24,29*]. Data from Newell's article [*24*], tabulated in Table 4, show that an embrittled alloy corrodes about four to twelve times more rapidly in boiling 65 percent nitric acid than does an annealed specimen. The accelerated corrosion on embrittled alloys is probably due to the selective corrosion of the iron-rich ferrite formed because of the miscibility gap or, as suggested by Hodges [*43*], due to formation of chromium-rich carbide precipitates.

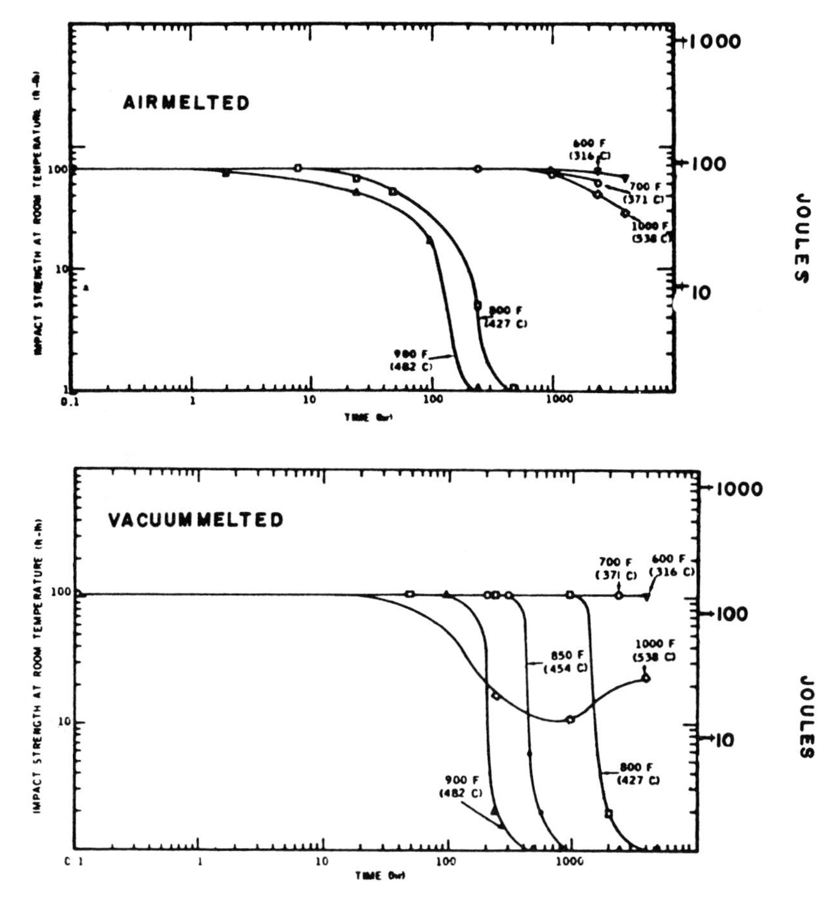

FIG. 14—*Impact strength at room temperature of half-size Charpy V-notch specimens after aging at indicated times and temperatures*; *air-melted versus vacuum-melted 18Cr alloys* [42].

The 475°C embrittlement may be alleviated and toughness and corrosion resistance restored by heating embrittled alloys at temperatures of 550°C and higher for a long enough time. As pointed up by Newell [*24*], 1 h suffices to remove the embrittlement at 593°C or higher, while 5 h are required at 582°C, and over 1000 h at 538°C. For compositions susceptible to σ-phase precipitation in a relatively short time, as indicated in Fig. 12, the 475°C embrittled alloy can be heated to temperatures above about 800°C (that is, above the σ-forming region) and rapidly cooled to remove the embrittlement [*6*].

Summary 475°C Embrittlement

Holding chromium-iron alloy in the temperature range of 400 to 540°C causes the alloys to become brittle and lose corrosion resistance. The

TABLE 2—*Tensile properties at room temperature of a vacuum-melted and air-melted alloy after indicated aging treatments.*

Alloy Type	Aged Temperature, °C	Aged Time, h	0.2% Offset Yield Strength, ksi (MPa)	Tensile Strength, ksi (MPa)	Elongation, % (50.8-mm gage)
1		0	30.9 (213)	49.9 (344)	32.5
1	316	4000	31.6 (217)	50.8 (350)	38.3
1	427	4800	69.8 (481)	84.1 (579)	13.0
1	482	2400	73.4 (505)	100.3 (690)	11.5
2		0	43.0 (296)	69.8 (481)	22.5
2	316	4000	38.5 (265)	69.1 (476)	25.5
2	427	960	73.7 (508)	95.8 (660)	17.0
2	482	2400	83.4 (540)	104.0 (717)	12.5

Alloy	Cr	C, % by weight	N
1 vacuum melted	17.4	0.002	0.003
2 air melted	18.0	0.044	0.091

NOTE—Data by Grobner [*42*].

TABLE 3—*Effects of composition on 475°C embrittlement in chromium-iron alloys.*

Element	Effect on 475°C Embrittlement[a]
Cr	intensifies
C	no effect[a]; intensifies[b]
Ti, Cb	intensifies
Mn	lowers slightly
Si	intensifies
Al	intensifies
Ni	low amounts; intensify large amounts; decrease
N	very slight[a]; intensifies[b]
P	intensifies
Mo	intensifies
Severe cold work	intensifies[c]

[a] Heger [*30*].
[b] Grobner [*42*].
[c] Thielsch [*6*].

TABLE 4—*Average rate of corrosion in boiling 65% nitric acid for 27Cr alloy after aging at 475°C.*

Condition	Corrosion Rates, mil/year
Annealed	8.9
Embrittled 500 h at 475°C	31
Embrittled 6000 h at 475°C	109

NOTE—Data by Newell [*25*].

embrittlement is caused by formation of coherent chromium-rich precipitate in the iron-rich matrix as a result of a miscibility gap in the chromium-iron phase diagram below about 550°C in the chromium range from about 15 to 70 percent. The 475°C embrittlement phenomenon and σ-phase embrittlement are not related metallurgically. Significantly shorter exposure times can cause 475°C embrittlement, as compared to σ-phase embrittlement. As with σ phase, it is unlikely that welding or a high-temperature heat treatment can produce 475°C embrittlement even if the alloy experiences a relatively slow cool through the embrittling temperature ranges. A very heavy section may suffer 475°C embrittlement during cooling. The fact that 475°C embrittlement or σ-phase embrittlement will not normally occur in alloys welded or annealed at high temperatures and then cooled makes these embrittling problems less serious. However, alloys with more than about 16 percent chromium should not be used for extended service between 370 and 540°C, especially if the alloy is cycled from room temperature to the operating temperature during process shutdowns or excursions.

High-Temperature Embrittlement and Loss of Corrosion

When high-chromium stainless steels of intermediate and high interstitial content are heated above about 950°C and cooled to room temperature, they may show a severe embrittlement [*31*] and loss of corrosion resistance [*44*]. The effects can occur during welding, isothermal heat treatments above 950°C, and casting operations. The effect on ductility and corrosion resistance is shown in Fig. 15 [*45*] for a commercial AISI

a

b

FIG. 15—*Effect of welding on* (a) *ductility and* (b) *corrosion resistance of AISI Type 446 steel; 26 percent chromium, 0.095 percent carbon, 0.077 percent nitrogen; corrosion tested in boiling ferric sulfate-50 percent sulfuric acid solution* [45].

Type 446 stainless steel after it is welded. Of all the detrimental effects which can occur in high-chromium ferritic stainless steels following heat treatment, the so-called *high-temperature embrittlement* is most damaging because operations such as welding, heat treatment, and casting—all operations necessary for a material of construction—can cause serious loss in ductility and corrosion resistance. Not surprisingly, this problem has been a severe deterrent, limiting extensive use of air-melted ferritic stainless for commercial construction.

Background

Exceptionally thorough survey articles by Thielsch [*6*] and Rajkay [*7*] summarize all the research done up to about 1966 on the causes for the severe loss of corrosion resistance and ductility when high-chromium steels of moderate interstitial contents are heated to high temperatures. Most early investigators studied either the embrittlement phenomenon or the corrosion loss phenomenon, although, as latter investigators showed, the property losses are related to a single mechanism. Up to the 1960's, two theories were offered to explain the severe embrittlement.

1. The segregation or coherent state theory [*6*] postulated that embrittlement resulted from a clustering or segregation of carbon atoms in the ferrite matrix. During rapid cooling, most of the dissolved carbon in solid solution does not reprecipitate as carbides. Instead, the carbon atoms in the supersaturated ferrite phase group as coherent clusters which harden (that is, embrittle) the matrix, much in the manner of certain age-hardening alloys. Annealing affected alloys between 700 and 800°C causes the carbon to precipitate as carbides, thereby removing the carbon atom clusters and the embrittlement.

2. In the martensitic mechanism first described by Pruger [*46*], regions in the alloy of relatively high-carbon content transform to austenite at elevated temperature. During subsequent cooling, these regions transform to brittle martensite. Annealing in a temperature range of 700 to 800°C removes the embrittlement by transforming the martensite to ferrite and chromium carbides.

The theories proposed up to about 1960 to explain the severe intergranular attack on high-chromium ferritic alloys following high-temperature exposure are summarized next.

1. Houdremont and Tofaute [*47*] postulated that a carbon-rich austenite forms at the sensitizing temperature. When cooled, easily dissolved iron carbides precipitate at the grain boundaries between the austenite and ferrite phases. By annealing at about 750°C, the iron carbides are converted to chromium carbides which resist chemical dissolution, and, therefore, the material becomes resistant to intergranular attack.

2. Hochmann [*44*] also proposed the need for austenite formation at temperature but suggested that intergranular corrosion occurs by preferential attack on the grain boundary austenite phase itself because of its low-chromium and high-carbon content.

3. Lula, Lena, and Kiefer [*48*] reject any mechanism of intergranular attack that requires the formation of austenite. As they pointed up, operations aimed at preventing austenite formation following high-temperature exposure did not prevent intergranular corrosion. These workers proposed that the stress surrounding the carbide or nitride precipitates formed during cooling are the cause of rapid corrosion on the matrix adjacent to the precipitates. Annealing between 650 and 815°C annealed out the stresses caused by the precipitated phases, thus restoring corrosion resistance.

An interesting point is the lack of any suggestion by the early investigators that the high-temperature loss in corrosion resistance was due to chromium depletion adjacent to chromium-carbide precipitates. This mechanism proposed by Bain et al [*49*] in 1933 and detailed by Ebling et al [*50*] has been generally accepted as the cause of intergranular attack in 18Cr-8Ni austenitic stainless steels. Application of this phenomenon to explain intergranular corrosion in a ferritic stainless steel was difficult because the effects of heat treatment on the austenitic and the ferritic stainless steel were opposite. It was known that an austenitic stainless steel would sensitize by holding in a temperature range from 400 to 800°C and that the corrosion resistance of a sensitized specimen could be restored by heating to temperatures above about 950°C which dissolved the chromium carbide precipitates. In direct contrast, an air-melted ferritic stainless steel is sensitized whenever it is heated above about 950°C. Further, the corrosion resistance of a sensitized ferritic specimen can be recovered by annealing in a temperature range from 700 to 850°C which corresponds to a part of the range where an austenitic stainless steel is sensitized. In light of these contrasting effects, attribution of the chromium depletion theory as the mechanism for corrosion loss in ferritic stainless steel did not appear possible.

High-Temperature Loss of Corrosion Resistance

In the early and mid-1960's, a renewed interest in ferritic stainless steels resulted in research studies into the mechanism for the *high-temperature embrittlement and corrosion* loss phenomenon. It was Baumel in 1963 [*51*] who discussed the application of the chromium depletion theory to ferritic stainless steels. In the ten years between 1960 and 1969, investigations by Bond and Lizlovs [*52*], Bond [*53*], Baerlecken et al [*2*], Demo [*45,54*], Hodge [*43,55*], and Streicher [*56*] confirmed that the chromium depletion theory is the most plausible explanation for the severe intergranular attack which occurs when alloys of moderate carbon and nitrogen content are heated above about 950°C and then cooled to room temperature. Henthorne [*57*] has compiled an extensive summary on the factors causing intergranular attack in iron- and nickel-base alloys.

Demo [*54*] showed, in 1968, curious effects of heat treatment on the corrosion resistance of a commercial AISI Type 446 stainless steel exposed to Streicher's ferric sulfate-sulfuric acid test (M. A. Streicher, ASTM Bulletin, No. A229-58, April 1958, pp. 77–86, ASTM A262-70, Part 3, and Ref *56*). In a series of tests summarized in Table 5, he showed the

TABLE 5—*Effects of thermal treatment on the corrosion resistance of AISI Type 446 stainless steel.* [a]

Specimen[b] Designation	Condition	Corrosion Rate,[a] mil/year (exposure h)
1	as-received	30 (120)
2	30 min, 1100°C, water quench	780 (24)
3	30 min, 1100°C, air cool	800 (24)
5	30 min, 1100°C, water quench + 30 min, 850°C, water quench	42 (120)
4	30 min, 1100°C, slow cool to:[c]	
	1000°C, water quench	767 (120)
	900°C, water quench	27 (120)
	800°C, water quench	20 (120)
	700°C, water quench	18 (120)
	600°C, water quench	25 (120)

NOTE—Data by Demo [*54*].

[a]Exposed to boiling ferric sulfate-50% sulfuric acid solution (ASTM Recommended Practice A 262-70, Part 3).

[b]See Fig. 6 for microstructure of designated specimens.

[c]2.5°C/min in furnace.

poor corrosion resistance of specimens water quenched or air cooled from 1100°C and the recovery when the sensitized specimens were reheated to 850°C. However, specimens which were cooled slowly in the furnace from 1100°C to temperatures below 1000°C and then quenched displayed corrosion resistance equivalent to the annealed specimen. Based on these data and metallographic examination, Demo proposed two hypotheses but favored the one describing the existence of a chromium carbide-nitride precipitation range in the temperature region from 500 to 900°C similar to that observed for an austenitic stainless steel. Bond [*53*] in 1968 described the effects of carbon and nitrogen level on the sensitization of ferritic stainless steels containing 17 percent chromium. He concluded that intergranular corrosion of ferritic stainless steel is caused by the depletion of chromium in areas adjacent to where chromium-rich carbides and nitrides precipitate. A portion of Bond's data is tabulated in Table 6 and show the effects of interstitial levels on the intergranular corrosion re-

TABLE 6—*Results of the boiling 65% nitric acid test on selected 17Cr alloys containing carbon and nitrogen.*

		Corrosion Rate, mm/dm² day, for the average of 5 Successive 48-h Periods After 1 h of Heat Treatment at Indicated Temperature, water quench			
Carbon, %	Nitrogen, %	1450°F (788°C)	1700°F (926°C)	1900°F (1038°C)	2100°F (1150°C)
0.0021	0.0095	167	185	148	326
0.0025	0.022	164	944	...	761
0.0044	0.057	186	...	577	357
0.012	0.0089	126	824	1817	...
0.061	0.0071	147	619	1574	...

NOTE—Data by Bond [*53*].

sistance of a series of 17Cr alloys in boiling 65 percent nitric acid. Based on these and other data, Bond concluded that a 17Cr-Fe alloy containing 0.0095 percent nitrogen and 0.0021 percent carbon was resistant to intergranular corrosion after sensitizing heat treatment in the range of 900 to 1150°C. He also pointed up that alloys containing more than 0.022 percent nitrogen and more than 0.012 percent carbon were quite susceptible to intergranular corrosion after sensitizing heat treatments at temperatures higher than about 926°C (1700°F). The low interstitial requirement for resisting sensitization needs to be emphasized in contrast to the interstitial levels of about 0.06 percent carbon and 0.03 percent nitrogen in a typical air-melted 17Cr Type 430 stainless steel. Bond also showed, through electromicroscopic examination of the alloys susceptible to intergranular corrosion, grain boundary precipitates which were absent in alloys not susceptible to such corrosion.

Demo [*45*] in 1971 showed that a 26Cr alloy containing 0.014 percent carbon and 0.004 percent nitrogen after heating to 1000°C had excellent intergranular corrosion resistance if quenched and poor resistance if air (or more slowly) cooled. These results by Bond and Demo showed that ferritic stainless steel could be subjected to high temperature without resulting loss of corrosion resistance, providing low maximum interstitial levels were maintained in the alloys, obviously much below levels for air-melted alloys. For these low interstitial ferritic alloys, the response to heat treatment is remarkably similar to that observed for austenitic chromium-nickel stainless steels. If heated above 1000°C and water quenched, the intergranular corrosion resistance is excellent, and no grain boundary precipitates of chromium carbides are observed; upon slow cooling, severe intergranular attack may be observed along with grain boundary precipitates of chromium carbides [*49,58*]. A clear relationship between heat treatment, the presence of an intergranular precipitate, and corrosion resistance exists for a low interstitial ferritic alloy.

For a commercial air-melted Type 446 (26 percent chromium) stainless steel, Demo [*45*] showed there is not a one-to-one correlation between loss of corrosion resistance and the presence of a continuous grain boundary precipitate in contrast to austenitic stainless steels and low interstitial ferritic stainless steels. These results are shown in Fig. 16 for specimens whose corrosion resistance is summarized in Table 5. For example, specimens with microstructures numbered 2 (water quenched from 1100°C) and 3 (air cooled) had poor intergranular corrosion resistance consistent with the continuous chromium-rich precipitates observed in the grain boundaries. Specimens 1 (as-received) and 4 (cooled slowly from 1100°C) showed excellent intergranular corrosion resistance consistent with the essential lack of any grain boundary precipitate. However, Specimen 5 (reheated at 850°C) showed excellent intergranular corrosion resistance but also a heavy grain boundary precipitate usually indicative of poor corrosion resistance. It is lack of correlation between corrosion resistance and grain boundary precipitate and the fact that commercial alloys become sensitized when heated above about 1000°C that have held back the acceptance of the theory that the *high-temperature sensitization* problem in ferritic stainless steel is caused by chromium-rich precipitates formed at grain boundaries when the alloys are cooled through a temperature from about 400 to 900°C.

Using selected Fe-26Cr alloys of reduced interstitial levels, Demo [*45*] defined the sensitization range for 26Cr-Fe alloys (0.1 in. thick). To do this, he used alloys which had excellent corrosion resistance when quenched from 1100°C but poor corrosion resistance if air cooled such that the quenched alloys were supersaturated in carbon and nitrogen at room temperature. The water-quenched alloys were then subjected to a second heat treatment at selected times and temperatures between 400 and 1000°C. The alloys were corrosion tested and microstructurally examined. From the corrosion data, Demo constructed a time-temperature-sensitization envelope (TTS) for 26Cr-Fe ferritic stainless steel as shown in Fig. 17. The microstructure of selected examples (whose positions are marked on Fig. 17) and their corrosion resistance are tabulated in Fig. 18. Specimen 1 heated at 1100°C and water quenched had good corrosion resistance consistent with the absence of grain boundary precipitates. Specimen 2, on the other hand, air cooled from 1100°C, had poor corrosion resistance consistent with the grain boundary precipitates of chromium-rich carbides and nitrides. Specimen 3, water quenched from 1100°C, then reheated at 900°C, had good corrosion resistance and a harmless discontinuous intergranular precipitate. Comparing micrographs of Specimens 3 to 1 offers very strong evidence that chromium-rich materials precipitate even at 900°C, followed by agglomeration and growth of the precipitates. Specimen 4, reheated for a short time at

700°C, exhibited good corrosion resistance, but the presence of a continuous intergranular precipitate is evident. At this temperature, the rapid precipitation of chromium-rich carbides and nitrides to relieve supersaturation was compensated for by almost simultaneous diffusion of chromium which healed the initial chromium-depleted areas adjacent to the precipitate. On the other hand, Specimen 5, heated for the same time as Specimen 4 but at 600°C, showed severe intergranular corrosion as well as a continuous grain boundary precipitate. Under these conditions of time and temperature, the rapid precipitation of chromium-rich precipitates occurred to relieve supersaturation; but, with reduced diffusion rates, chromium atoms did not have time to diffuse into the depleted areas. This fact is demonstrated by Specimen 6, also exposed at 600°C, but for a long time (6 h). The specimen had a continuous grain boundary precipitate, but, unlike Specimen 5, it had good corrosion resistance. The longer hold time allowed the chromium atoms to diffuse into the chromium-depleted areas adjacent to the grain boundary precipitate restoring the alloys corrosion resistance. In consideration of Fig. 17, the line marking the low temperature side of the TTS band reflects the minimum time and temperature required for chromium-rich carbides and nitrides to precipitate from a supersaturated solid solution and to form chromium-depleted zones. The line marking the high-temperature side of the sensitization zone represents for 0.1-in. material, minimum conditions of time and temperature necessary for chromium diffusion to occur and heal the depleted areas after the matrix supersaturation in interstitial levels has been relieved by precipitation of chromium-rich carbides and nitrides.

These data are consistent with the chromium depletion theory for intergranular corrosion. When ferritic stainless steels are heated above about 950°C, the chromium-rich carbide and nitride precipitates are dissolved in solid solution. If the interstitial levels are very low, they will be maintained in solid solution if the specimen is cooled rapidly. For these quenched alloy compositions, the corrosion resistance is good. If these alloys now supersaturated in carbon and nitrogen are subjected to temperatures from about 500 to 950°C, chromium-rich carbides and nitrides precipitate rapidly to relieve supersaturation. In the temperature range of about 700 to 950°C, almost as soon as precipitation occurs, chromium diffuses into the chromium-depleted areas formed when precipitation first occurred, increasing the chromium content of these areas. Consequently, intergranular corrosion resistance is good, despite the observed presence of an intergranular precipitate. In the temperature range of 500 to 700°C where chromium-rich carbides and nitrides rapidly precipitate, the resulting chromium-depleted zones are not healed except during long hold times because the diffusion rate of chromium is markedly lower at these lower temperatures.

In light of this description, the chromium depletion theory can also

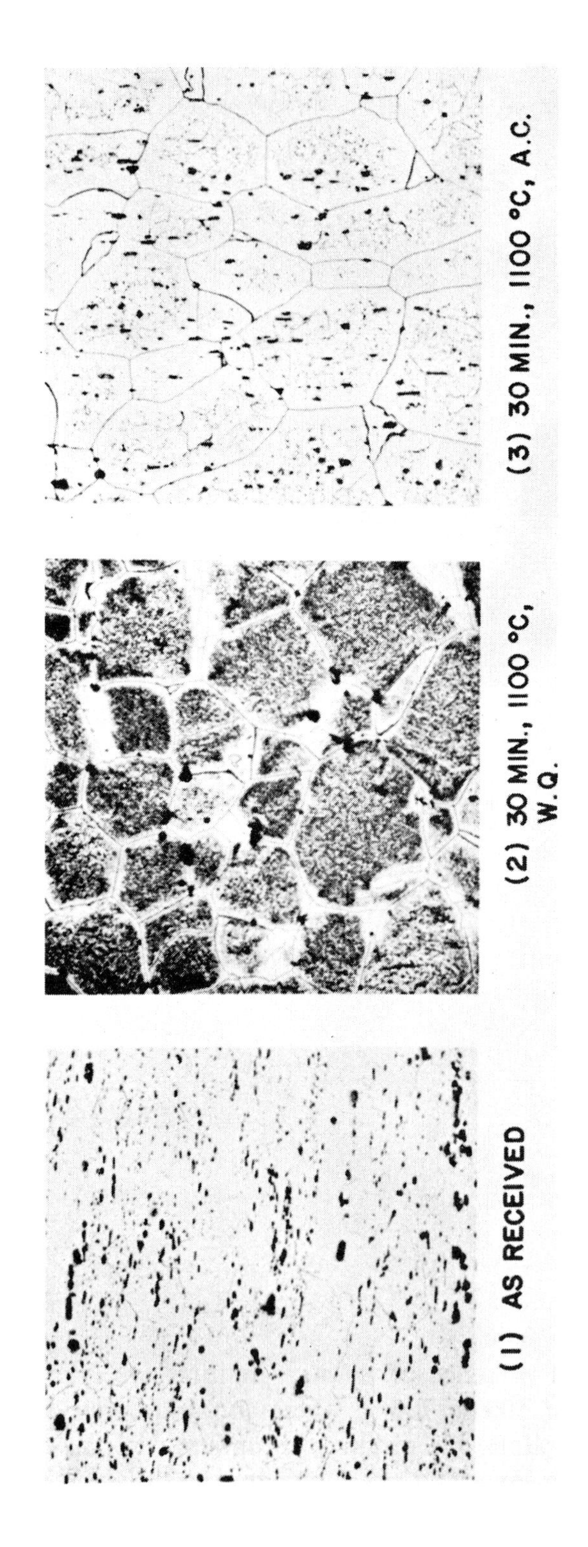

(1) AS RECEIVED

(2) 30 MIN., 1100 °C, W.Q.

(3) 30 MIN., 1100 °C, A.C.

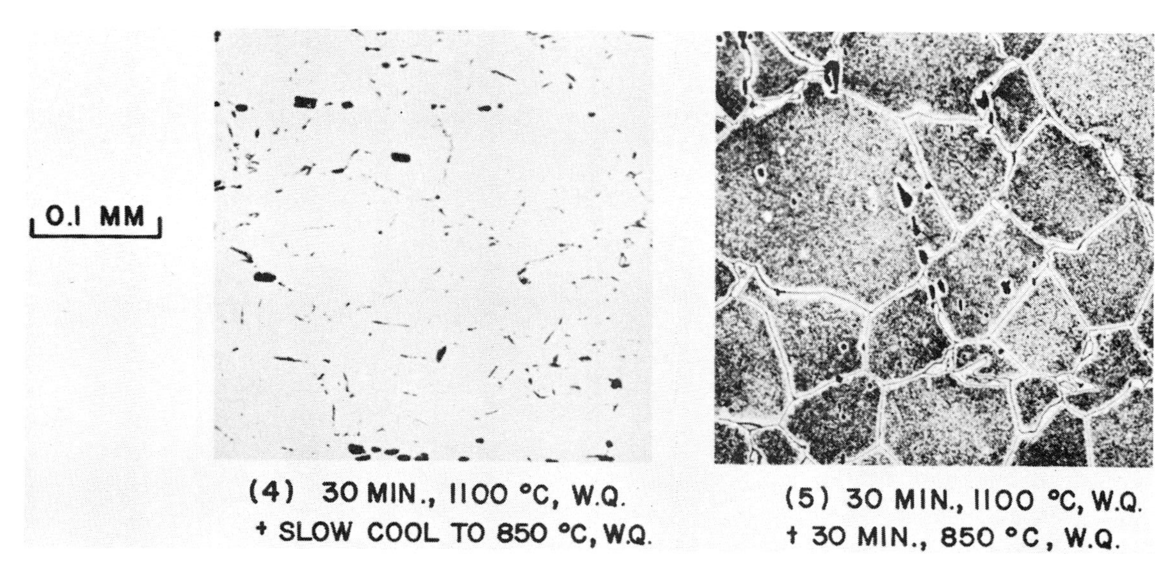

FIG. 16—*Relationship between thermal treatments and microstructure on an AISI Type 446 steel; 26 percent chromium, 0.095 percent carbon, 0.077 percent nitrogen; see corrosion rates for specimens in Table 5* [45].

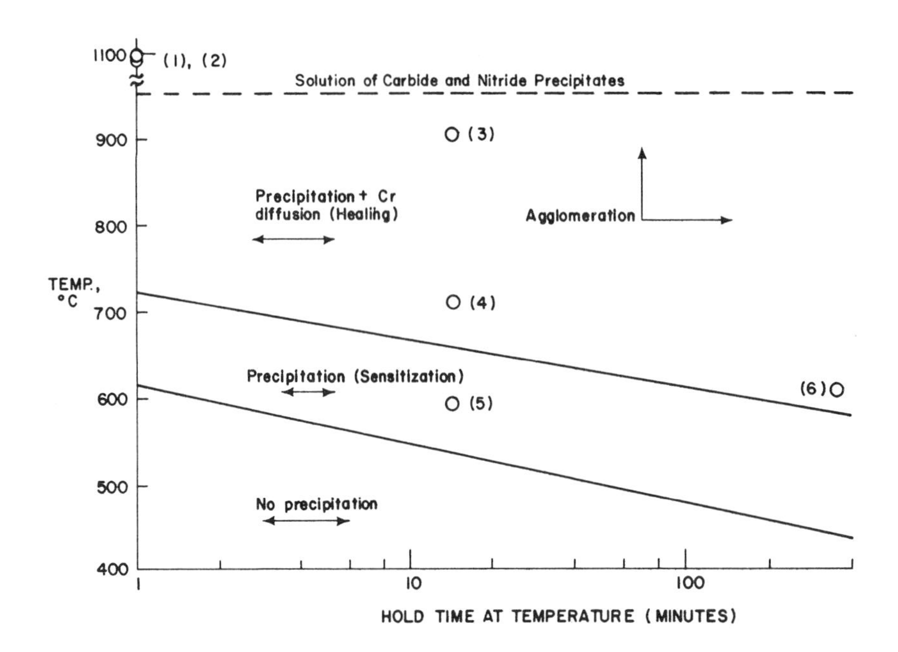

FIG. 17—*TTS diagram for Fe-26Cr alloys treated initially at 1100°C, 30 min, water quenched, then reheated at selected lower temperatures for different times; corrosion tested in boiling ferric sulfate-50 percent sulfuric acid solution; (carbon + nitrogen) = 180 ppm* [45].

explain the effects of heat treatment on the corrosion resistance and microstructure on the high interstitial alloys like AISI Types 430 and 446 stainless steel. For these alloys, in contrast to the low interstitial alloys, even rapid quenching from temperatures above 950°C is not fast enough to prevent precipitation of chromium-rich carbides and nitrides and chromium depletion results. This behavior is produced by the high driving force for precipitation in the intermediate temperature ranges caused by a combination of high interstitial supersaturation, rapid diffusion of carbon and nitrogen as compared to chromium, and high rates of precipitation in the ferritic matrix. The corrosion resistance of a sensitized high interstitial alloy can also be restored by heating in the temperature range of about 700 to 950°C. This heat treatment permits chromium to diffuse into the depleted areas, raising the chromium content and restoring corrosion resistance, even though a grain boundary precipitate may still be observed in the microstructure (Specimen 5, Fig. 16). A long-time heat treatment in this temperature range, as used in commercial annealing practices, may even allow sufficient time for the fine grain boundary precipitate to agglomerate into large, discontinuous precipitates which are thermodynamically more stable and often observed for long-time heat treatments (Specimens 1 and 4, Fig. 16).

In 1970, Hodges [*43,55*] demonstrated that high-purity alloys from 17 to 26 percent chromium were highly resistant to intergranular cor-

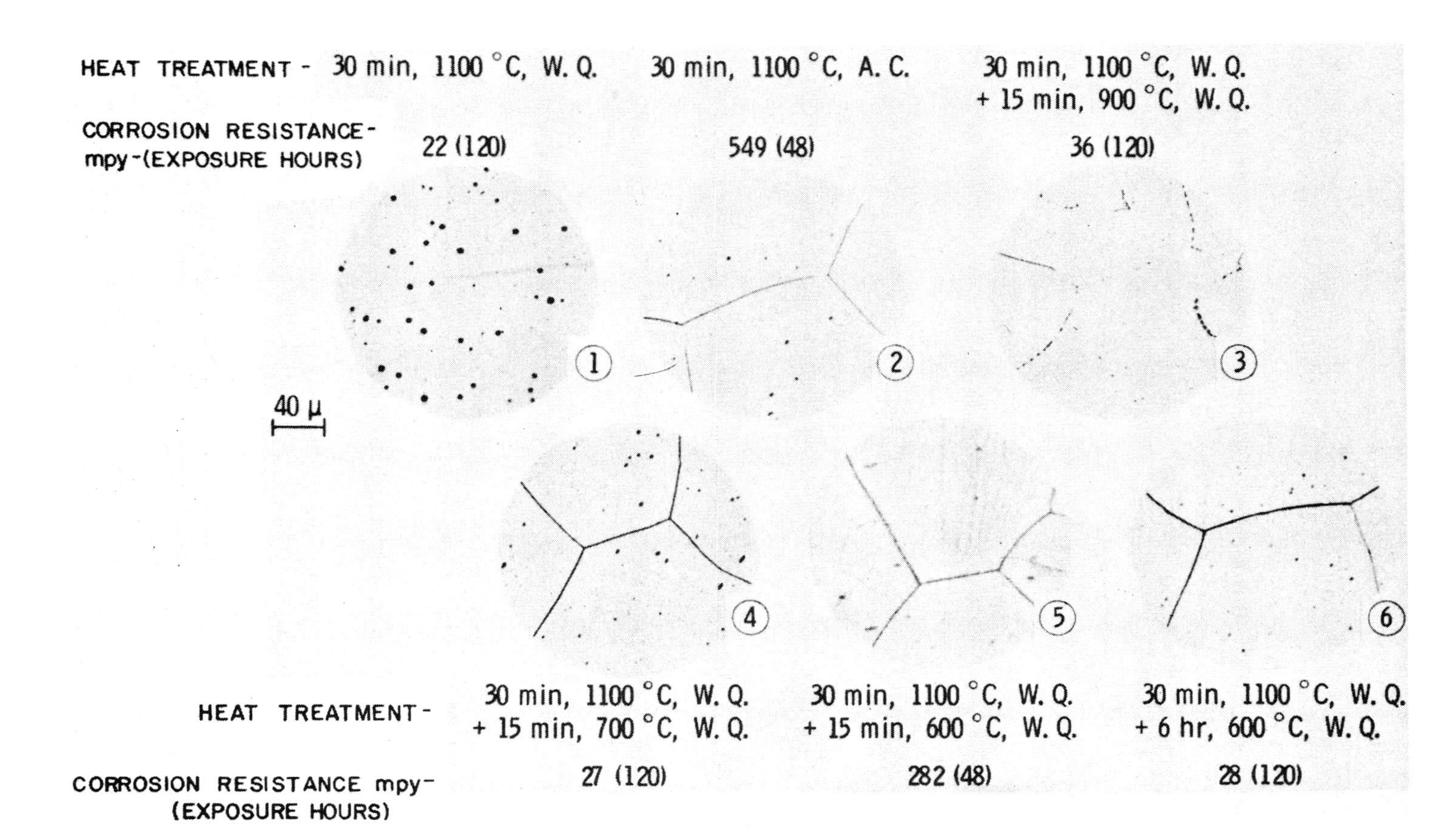

FIG. 18—*Microstructure and corrosion resistance of alloys treated as indicated by numbered positions on TTS diagram in Fig. 17; corrosion tested in boiling ferric sulfate-50 percent sulfuric acid solution* [45].

rosion following exposures to temperatures above about 950°C. Based on his work, he concluded that sensitization in ferritic stainless steels could be explained best on the basis of the chromium depletion theory. Hodges noted that the sensitization-desensitization behavior of ferritic stainless steels containing moderately low interstitial levels could be described by the same type of TTS curve as reported for austenitic stainless steels, the difference being that the time sequence for sensitization-desensitization for austenitic stainless steel is minutes and hours, but only seconds and minutes for ferritic stainless steels. Hodges also showed that molybdenum as an alloying element in high-purity ferritic stainless steels shifts the sensitizing envelope to longer times, thereby delaying sensitization and loss of intergranular corrosion resistance. Streicher [*59*] showed that, for high-purity ferritic alloys, the interstitial levels which are tolerable without sensitization occurring is increased when molybdenum additions up to 6 and preferably 4 percent are made.

Streicher [*56*] has studied in detail the effects of carbon and nitrogen level and heat treatment on the microstructure and corrosion resistance of chromium-iron alloys. In particular, he has documented carefully the regions of attack in specimens which are fully ferritic or a mixture of ferrite and austenite, depending on where the alloy composition was located according to the phase diagrams in Fig. 3 when it was exposed to high temperatures. As an example taken from his work, a commercial Type 446 steel (26 percent chromium) containing 0.098 percent carbon and 0.21 percent nitrogen contains up to about 40 percent austenite when water quenched from a heat treatment at 1150°C. As shown in Fig. 19*a* from Streicher's work, no chromium-rich precipitates of carbon and nitrogen are formed at the boundary between austenite grains or in the austenite grains themselves because of the high solubility of carbon and nitrogen in austenite. When this specimen was subjected to the ferric sulfate-sulfuric acid corrosion test, (Fig. 19*b*), severe intergranular attack occurred at the ferrite-ferrite and ferrite-austenite boundaries because carbon and nitrogen were rejected due to the low interstitial solubility in ferrite. However, no attack is observed at the austenite-austenite boundaries because the interstitials are maintained in solid solution due to their high solubility in austenite. Streicher [*56*] also showed that heat treatments have no detrimental effects on the intergranular corrosion susceptibility of high-purity 16 and 25Cr-Fe stainless steels (carbon $\leqslant$ 0.0086 percent, nitrogen $\leqslant$ 0.0025 percent). From his work, he also concludes that the chromium depletion theory can adequately explain intergranular corrosion and other forms of localized attack occurring in chromium-iron stainless steels.

Based on the work of a number of investigators, the *high-temperature loss of corrosion resistance* observed when ferritic alloys of high interstitial contents are heated above about 950°C may be explained by

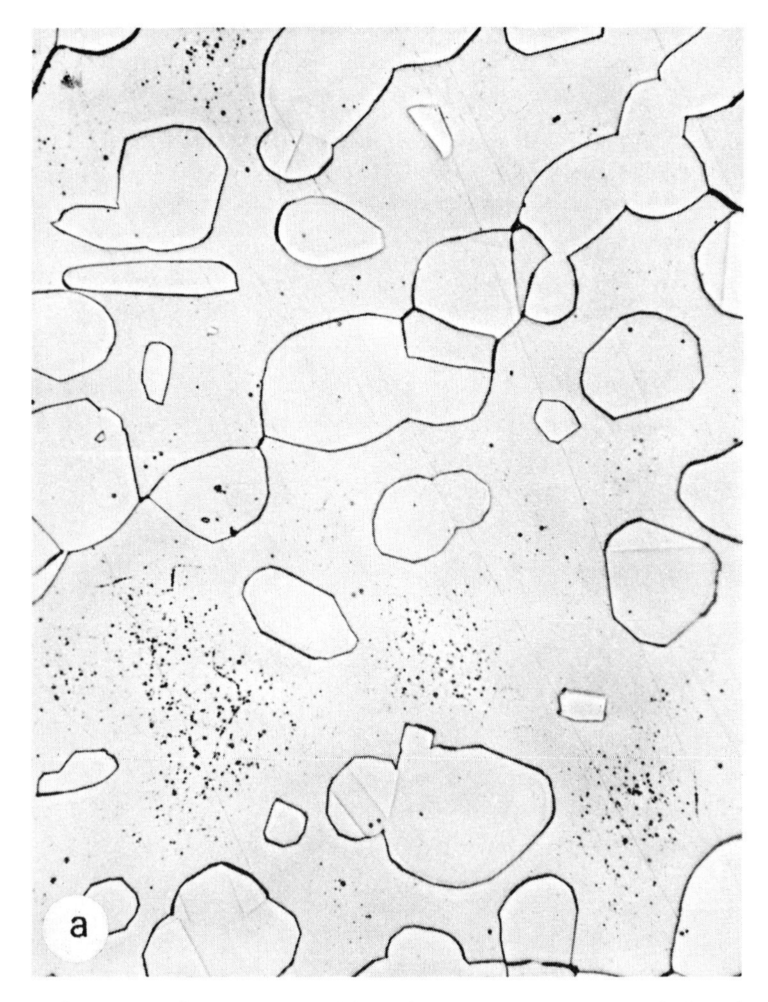

FIG. 19*a*—*Structure of Type 446 steel heated at 1150°C* (×*500*). *Etched surface shows "dot" precipitate in ferric matrix, but not near austenite grains, which have absorbed carbon and nitrogen from adjacent ferrite.*

the chromium depletion theory which is accepted widely as the cause for sensitization in austenitic stainless steels. No support for alternate theories proposed to explain the *high-temperature corrosion loss* phenomenon can be found. These alternate theories were based on (*a*) preferential dissolution of iron carbides, (*b*) accelerated corrosion due to stresses surrounding chromium rich precipitates, (*c*) presence of austenite or its decomposition products at grain boundaries, and (*d*) galvanic action between the precipitate and the surrounding metal matrix.

High-Temperature Embrittlement

Whenever commercial high chromium-iron ferritic stainless steels con-

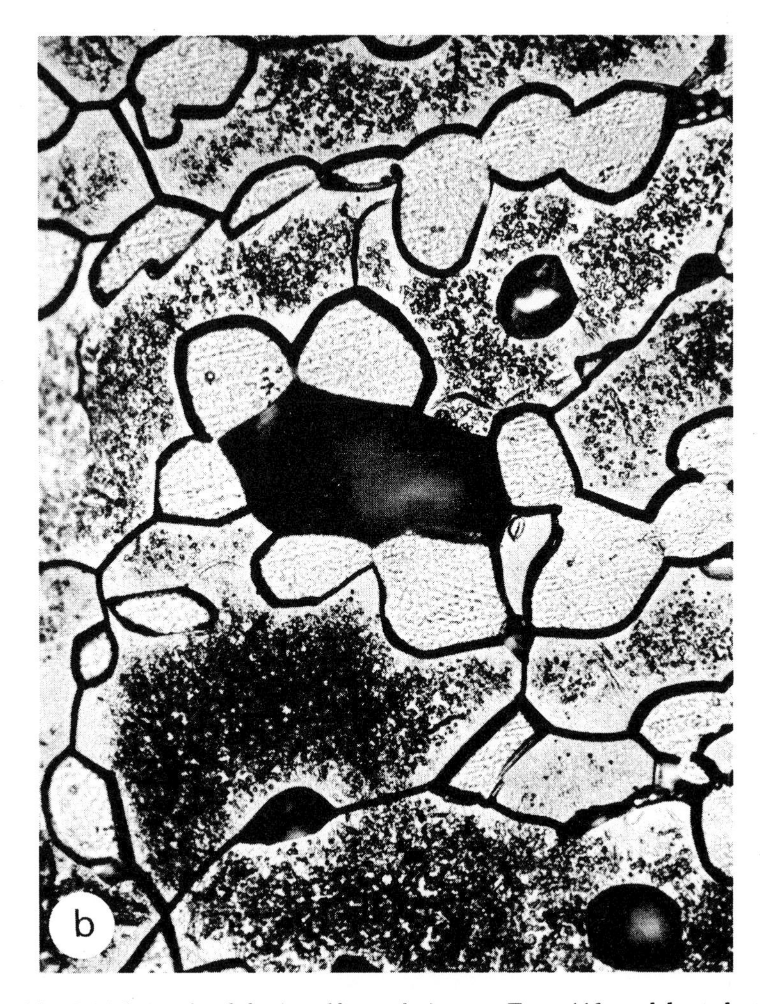

FIG. 19*b*—*Initial attack of ferric sulfate solution on Type 446 steel heated at 1150°C, water quenched (×500). Intergranular attack on ferrite-ferrite and ferrite-austentite boundaries. No intergranular attack on austenite-austenite boundaries. Localized attack on dot precipitates which are absent in zones near austenite grains but not at ferrite-ferrite boundaries; exposed 14 h in boiling ferric sulfate-50 percent sulfuric acid solution* [56].

taining moderate to high interstitial levels are heated above about 1000°C, the alloys at room temperature show an extreme loss in toughness and ductility. If an embrittled alloy is reheated in a temperature range of 750 to 850°C, the ductility of the alloy is restored. Two early theories, coherent state and the martensitic mechanism, proposed to explain this high-temperature embrittling phenomenon were described earlier. Baerlacken et al [*2*], Demo [*45*], Semchyshen et al [*60*], and Plumtree et al [*61*] have shown through their work that *high-temperature embrittlement* is related to interstitial levels in the alloy just as is loss of intergranular corrosion resistance.

Baerlecken et al [2] studied the effects of heat treatment on the toughness (impact strength) of air-melted steels containing moderate levels of carbon and nitrogen and vacuum-melted steel containing very low interstitial levels. Using air-melted steel, these workers concluded that the variable toughness as a function of heat treatment and chromium content was connected with the structure and morphology of the carbide precipitates. Using high-purity, vacuum-melted alloys given a two-stage heat treatment (30 min at 1050°C cooled to 800°C for 25 h) these workers showed that the impact transition temperature was shifted to higher temperatures as the chromium content increased. However, if these same low interstitial alloys were given a second high-temperature heat treatment by reheating to 1050°C which would normally embrittle a high interstitial alloy, they discovered that excellent low-temperature impact values were observed for alloys at all chromium levels. These results are shown in Fig. 20 from their work. They explained this effect of heat treatment on toughness by the fact that the solubility of the interstitials in the α solid solution decreases with increasing chromium content. There-

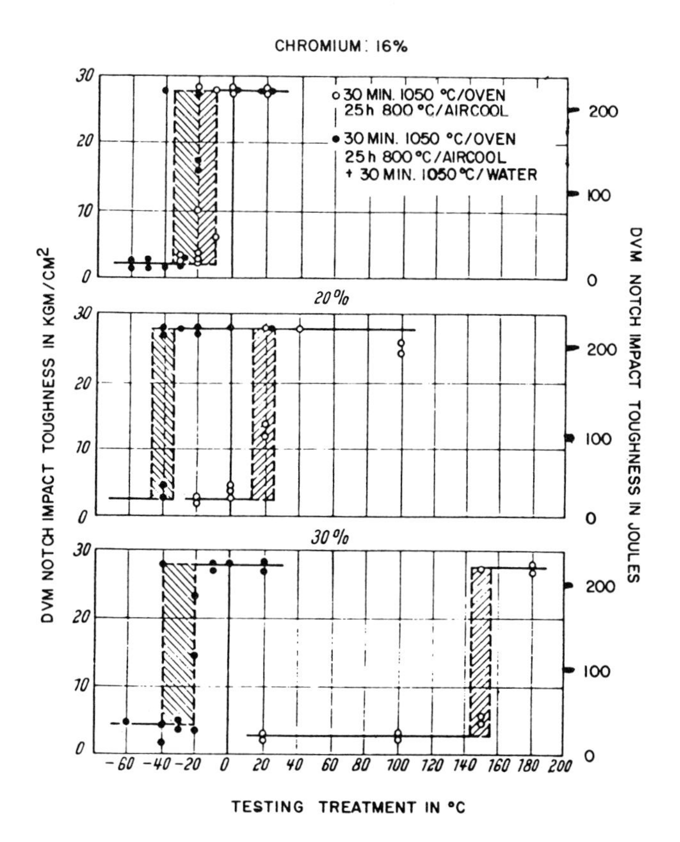

FIG. 20—*Effect of heat treatment on position of the sharp drop of the notch impact toughness-temperature curve of vacuum-melted chromium steels with 16 to 30 percent chromium* [2].

fore, the initial two-stage heat treatment on specimens of the same interstitial levels produces more precipitates of carbides and nitrides the higher the chromium content of the alloy is, with resulting decrease in toughness (that is, increase in the transition temperature). On the other hand, by a second solution heat treatment and rapid quenching, it is possible even with steels of high-chromium content (and reduced carbon and nitrogen solubility) to keep the very low interstitial levels of vacuum-melted alloys in solid solution. Therefore, carbides and nitrides do not precipitate, and their deleterious effects on notch impact toughness do not occur. Another important conclusion from their work is that grain size has little effect on notch impact behavior because high-chromium, vacuum-melted alloys even with grain size ASTM 1-3 displayed transformation temperature of about −40°C.

Demo [*45*] studied the effect of thermal treatment on the ductility of high and low interstitial purity alloys containing 26 percent chromium. As shown in Table 7, loss of ductility is observed only when the commercial

TABLE 7—*Effect of thermal treatments on the ductility of stainless alloys.*

	Elongation (in 25.4 mm) for the Indicated Stainless Steels After Heat Treatment		
Condition	AISI Type 446	High Purity[a] 26Cr	AISI Type 304
Annealed	25	30	78
30 min, 1100°C, water quench	2	30	84
30 min, 1100°C, air cool	27	32	85
30 min, 1100°C, slow cool[b] to 850°C, water quench	33	30	...
30 min, 1100°C, water quench + 30 min, 850°C, water quench	27	29	...
120 min, 677°C, air cool[c]	...	...	84

NOTE—Data by Demo [*45*].
[a]0.014% carbon, 0.004% nitrogen.
[b]Cooled in furnace at 2.5°C/min.
[c]A heat treatment known to cause sensitization in AISI Type 304.

AISI Type 446 stainless steel containing high interstitial levels is heated to 1100°C and water quenched. Importantly, the same alloy cooled more slowly from the high-temperature exposure (that is, air or slow cooled) shows excellent ductility, as does the embrittled alloy when it is annealed by a second heat treatment at 850°C. In contrast, the ductility of the low interstitial 26 percent chromium steel is not adversely affected by heat treatment. These observations suggested to Demo that loss of ductility (that is, tensile elongation) in chromium-iron alloys when heated to high temperature, as in welding or isothermal heat treatment, was also related to interstitial content, as suggested by Baerlacken et al, and cooling rate. Demo also examined the fractured edges of the heat-treated, air-melted Type 446 stainless steels in cross section, as shown in Fig. 21. The water-quenched specimen (brittle) shows intragranular cleavage. The air-cooled specimen (ductile) shows intergranular cleavage with some localized deformation and elongation of the grains. The slowly cooled specimen (ductile) shows a fibrous shear structure with considerable localized deformation and elongation of the grains. Based on these observations, Demo concluded that precipitation of carbides and nitrides in the grain boundaries does not grossly affect ductility because both air-cooled and water-quenched specimens had intergranular precipitates, but the failure mode in the brittle specimen was predominantly intragranular. Moreover, an embrittled specimen reheated to 850°C shows restored ductility despite the presence of a heavy intergranular precipitate, as shown in Fig. 16 (Specimen 5). Finally, as shown in Table 7, an 18Cr-8Ni austenitic stainless steel, which has been sensitized by heating at 677°C to cause grain boundary precipitation of chromium carbides, retains excellent ductility.

Using thin-film electron microscopy examination, Demo has shown structural difference between embrittled and ductile alloys. Transmission micrographs of the water-quenched specimen (brittle) and the air-cooled specimen (ductile) are shown in Fig. 22. For the water-quenched specimens, precipitates are noted, not only in the grain boundaries as expected from the optical micrographs, but also on nearly all dislocations. On the other hand, no precipitates are observed on the dislocations of the air-cooled specimens, although grain boundary precipitation is shown in the optical micrograph. Unrestrained motion of the dislocations in the water-quenched specimen is blocked by the precipitate, with a resulting increase in strength and reduction in ductility and toughness.

Demo proposed that precipitation of chromium-rich carbides and nitrides on dislocations in the grain body and not on grain boundary surfaces is responsible for the severe loss in ductility when ferritic alloys are heated to high temperature. Demo suggests two possibilities to explain why chromium-rich precipitates form on dislocations in rapidly cooled specimens and not on more slowly cooled specimens, even though both contain chromium-rich precipitates in the grain boundaries. First, the high interstitial contents of the air-melted alloys, combined with the high interstitial supersaturation, serves as a strong driving force during quenching for rapid precipitation on all high-energy surfaces such as grain boundaries and dislocations. During a slower cool, the longer relative time available to relieve supersaturation may allow for diffusion of carbon and nitrogen to grain boundary areas where the supersaturation is relieved by precipitation on the more preferred higher energy surfaces of the grain boundary. An alternate explanation is that dislocation nucleation during the rapid quench occurs simultaneously with rejection of carbon and nitrogen as

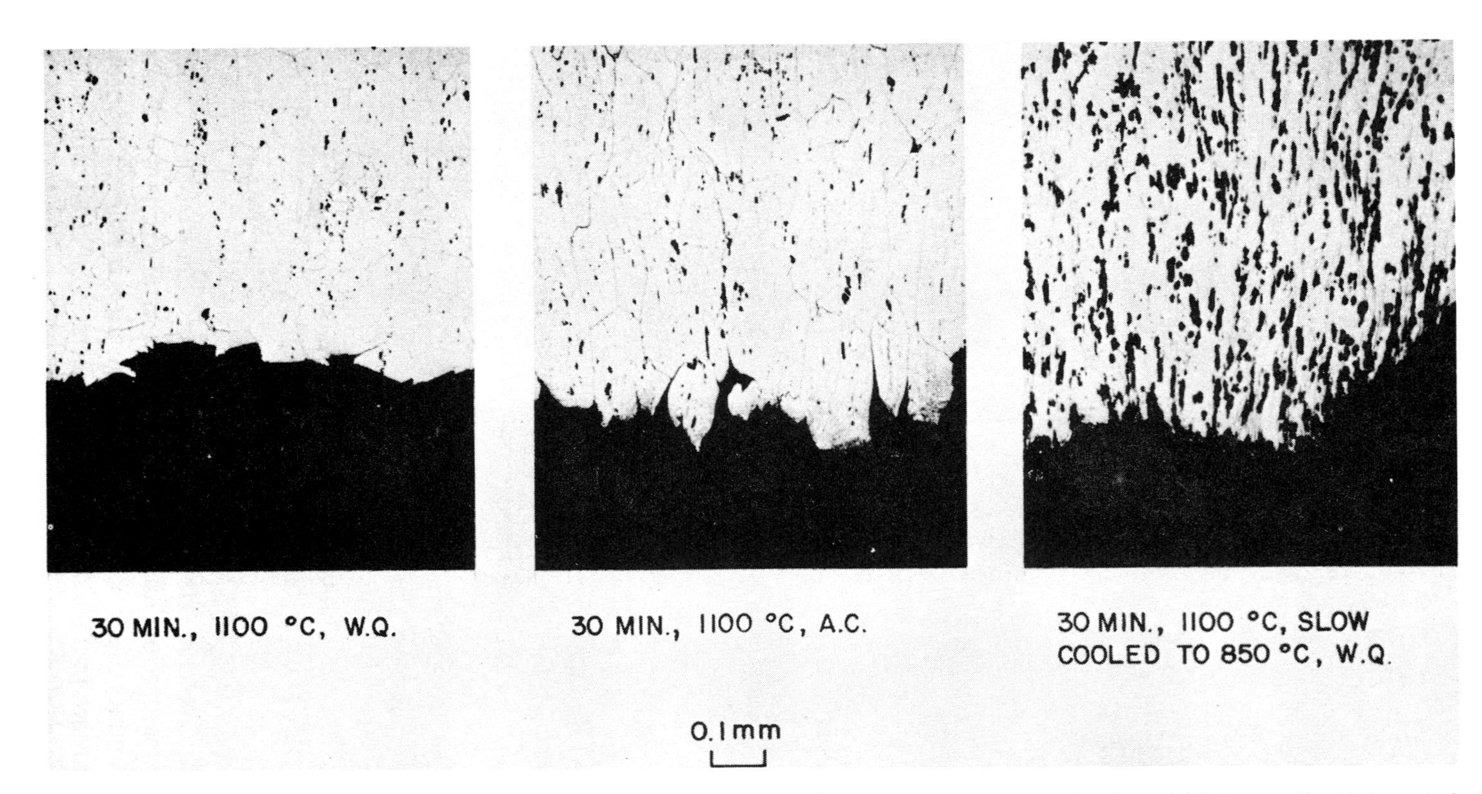

FIG. 21—*Effect of cooling rate on morphology of the fracture edges of a tension specimen made of an AISI Type 446 stainless steel heated to 1100°C; 0.095 percent carbon, 0.077 percent nitrogen. Furnace slow cool ~2.5°C/min* [45].

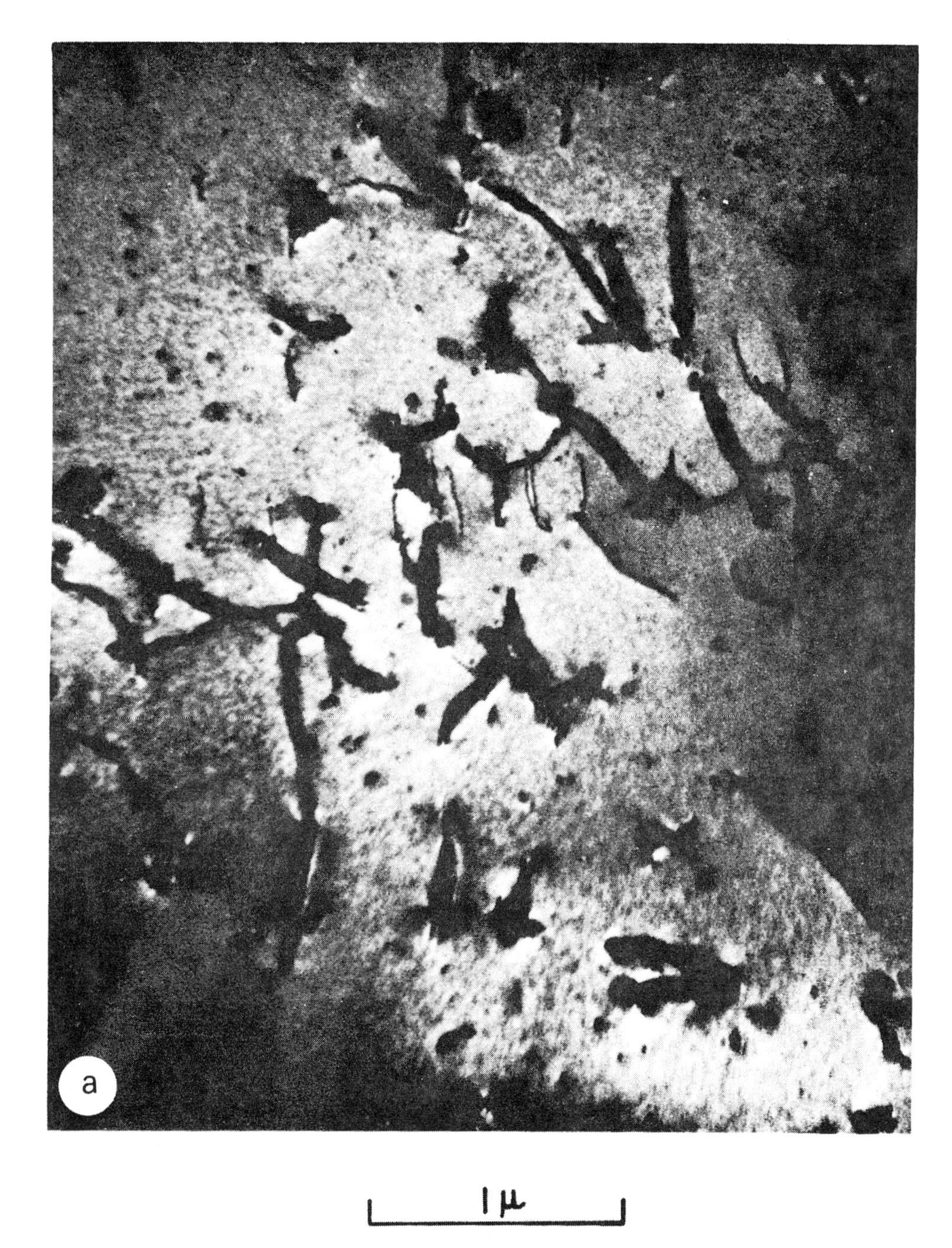

(*a*) Water quenched.

FIG. 22—*Transmission electron micrographs of an AISI Type 446 stainless steel heated to 1100°C; 0.095 percent carbon, 0.077 percent nitrogen* [54].

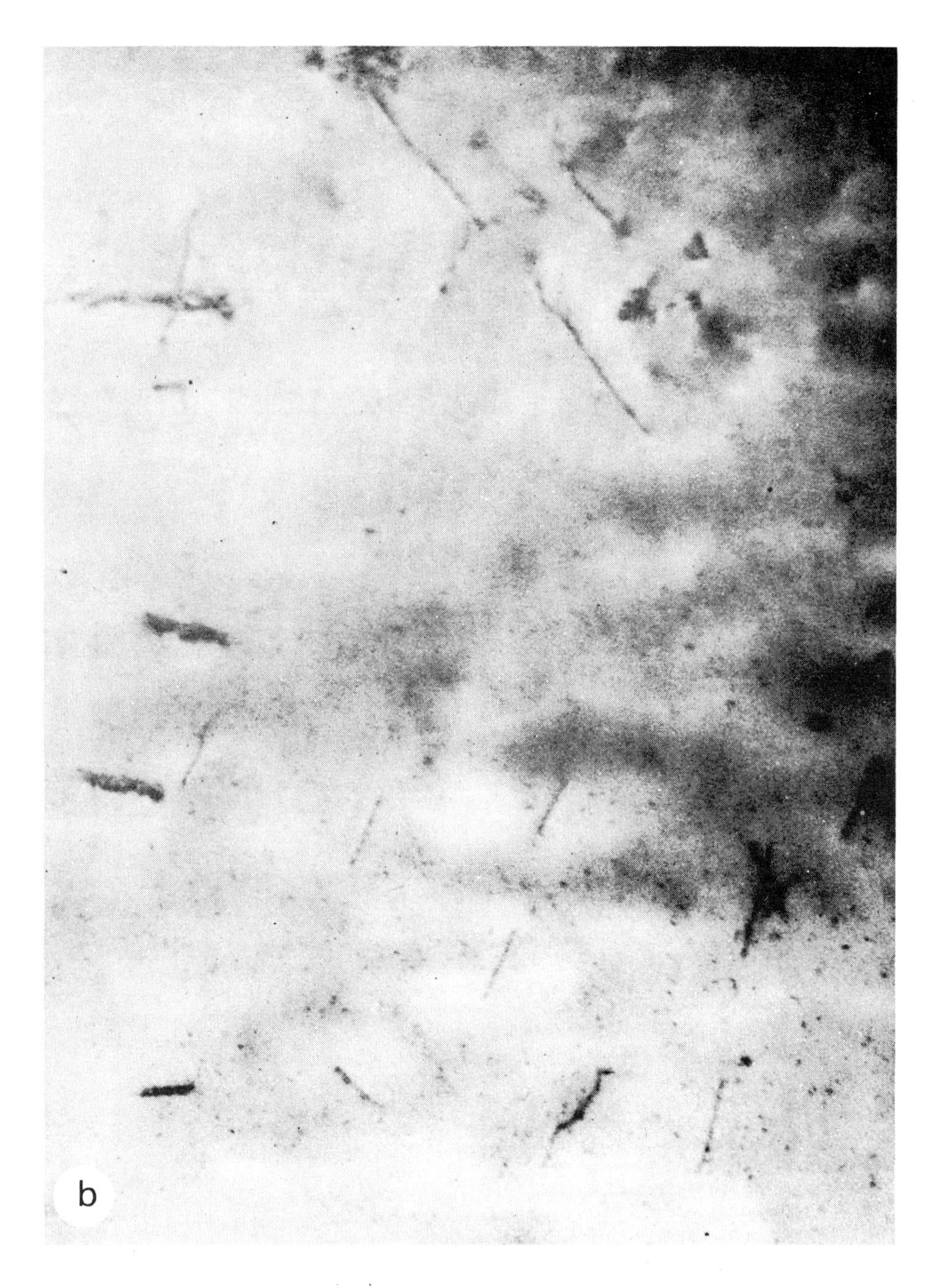

(*b*) Air cooled.

FIG. 22—*Continued.*

chromium-rich carbides and nitrides precipitate to relieve the supersaturation. With a slower cool and, therefore, reduced thermal stresses, relief of supersaturation by precipitation on high-energy surfaces may occur before dislocations are nucleated by the effects of thermal stresses. Demo associates the poor ductility of chromium-iron alloys subjected to high-temperature exposure to fine, dispersed precipitates in the matrix preventing easy movement of dislocation similar to a hardening mechanism in a precipitation-hardened alloy [*62*]. An increase in strength is observed with a concomitant tremendous reduction in ductility and toughness.

Semchyshen, Bond, and Dundas [*60*] have reported the effects of chromium, carbon, and nitrogen levels and heat treatment on the toughness (impact resistance) of chromium-iron alloys containing from about 14 to 28 percent chromium. The impact resistance of a 17Cr steel in the annealed condition as a function of carbon content is shown in Fig. 23*a*. Note the relatively low transition temperatures even at carbon levels of 0.061 percent. These data may be contrasted with impact energy absorbed for the same specimens sensitized by heating to 1150°C as shown in Fig. 23*b*. When the carbon content of the alloys exceeds about 0.018 percent carbon, a large increase in transition temperature is observed. As noted, these alloys contain low levels of nitrogen (<0.0010 percent). These same workers repeated this work on a series of alloys containing increasing levels of nitrogen, with carbon being held to low levels below 0.004 percent. The impact data for the annealed specimens and those annealed, as well as heated at 1150°C are shown respectively in Figs. 24*a,b*. The annealed specimens containing nitrogen up to 0.057 percent displayed excellent impact resistance. However, when the nitrogen content in the 17Cr alloys exceeded about 0.022 percent nitrogen, the impact transition temperature was shifted to high temperatures following a thermal treatment at 1150°C. The same effects between carbon, nitrogen, heat treatment, and impact resistance were found for alloys containing 26 percent chromium. Semchyshen et al also showed that a 17Cr alloy containing 0.01 percent carbon or 0.02 percent nitrogen contained grain-boundary precipitates, while alloys of higher purity remained free of grain boundary precipitate after quenching from temperatures above about 925°C. Based on these data and observations, they conclude that *high-temperature embrittlement* (that is, the observed increase in transition temperature when ferritic alloys are quenched from temperatures above about 1000°C) is caused by precipitates of chromium-rich carbonitrides mainly on the grain boundaries.

Plumtree et al [*61*] showed in recent work that the impact transition temperature increased linearly with the total interstitial content for 25Cr-Fe alloys annealed and water quenched. The data show a change in transition temperature from below room temperature to above in the range from 350 to 450 ppm total carbon plus nitrogen plus oxygen. These workers also showed that the impact transition temperature increases with the second

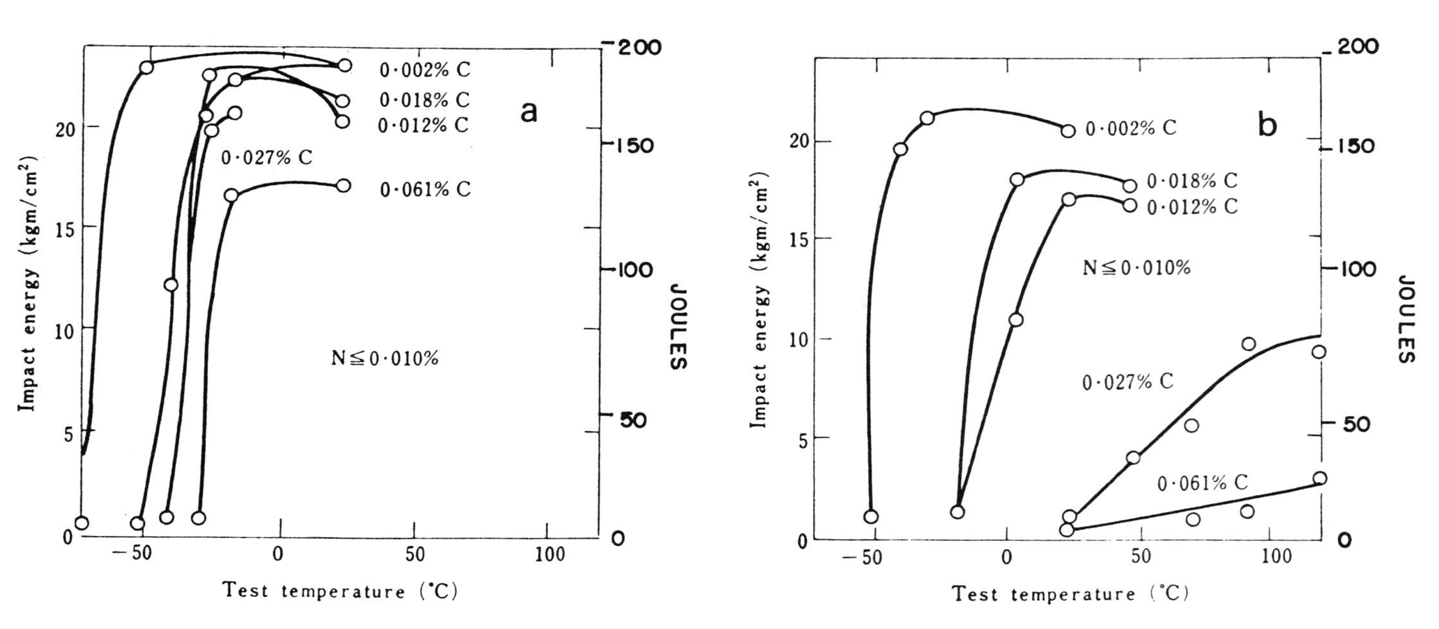

FIG. 23—*Transition curves for quarter-size Charpy V-notch impact specimens of 17Cr-0.002 to 0.61C. Ferritic stainless steels heat treated at*: (a) *815°C for 1 h and water quenched or* (b) *815 + 1150°C for 1 h and water quenched* [60].

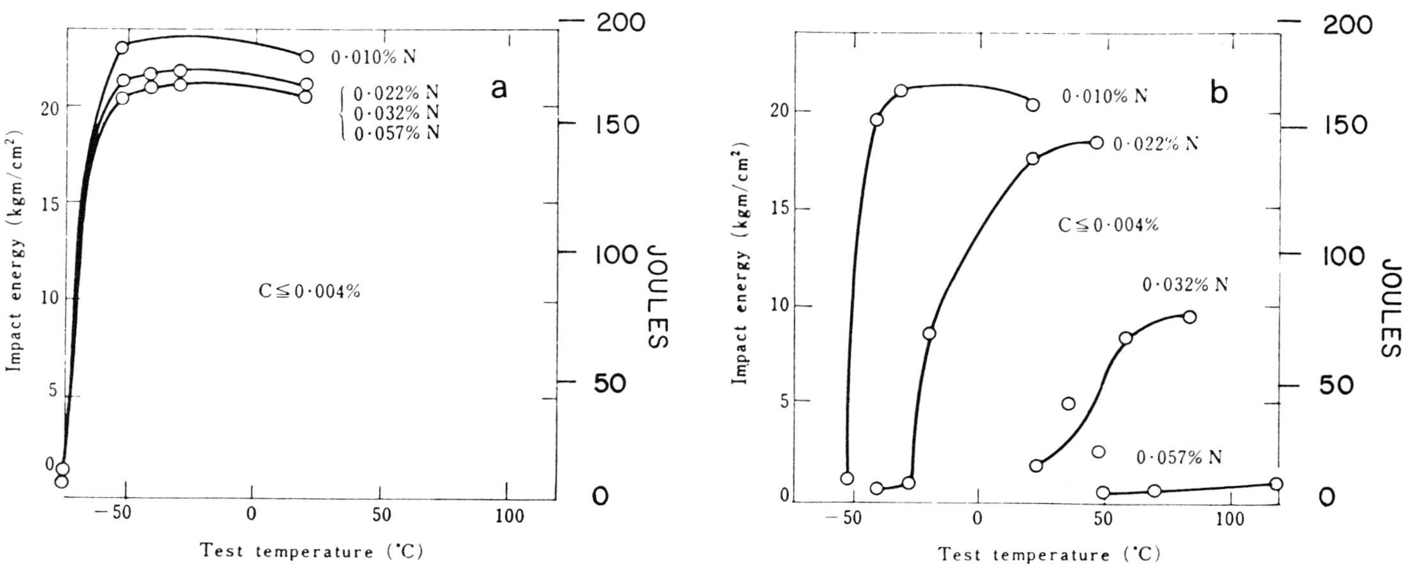

FIG. 24—*Transition curves for quarter-size Charpy V-notch impact specimens of 17Cr-0.010 to 0.057N ferritic stainless steels heat treated at*: (a) *815°C for 1 h and water quenched or* (b) *815 + 1150°C for 1 h and water quenched* [60].

phase content, and they proposed that increasing amounts of second phase inhomogeneously distributed throughout the matrix, particularly at the grain boundaries, lower the effective surface energy of a crack to promote cleavage failure and brittleness.

Baerlecken et al [2], Demo [54], Semchyshen et al [60], and Plumtree et al [61] all conclude from their data that the so-called *high-temperature embrittlement* phenomenon of high-chromium ferritic stainless steels is due to precipitation of chromium-rich carbides and nitrides caused by relief of supersaturation when the alloys are exposed to high temperatures. Baerlacken et al and Semchyshen et al find the embrittlement occurs because the chromium-rich precipitate forms on grain boundaries, while Demo and Plumtree claim the embrittlement occurs because a finely dispersed precipitate in the grain matrix hinders dislocation motion, and precipitates in the grain boundaries do not necessarily indicate alloy embrittlement. Perhaps the difference in thought resides in testing severity. A grain boundary precipitate may be detrimental to the high-rate energy absorbing requirement of the impact test but perhaps not detrimental to the slow-rate energy absorbing requirement of the tensile or slow bend test. The important point, however, is the general agreement that the *high-temperature embrittlement* phenomenon observed in ferritic stainless steels is dependent upon the levels of carbon and nitrogen in the alloys. The embrittlement problem in chromium-iron alloys will manifest itself whenever alloys containing moderate or high interstitial levels are heated to temperatures above about 950°C. The embrittlement is caused by precipitation of chromium-rich carbides and nitrides on grain boundaries or dislocations or both. It is at once apparent that the same precipitation mechanism causing embrittlement also produces the serious loss of corrosion resistance when ferritic alloys are heated to high temperatures.

Notch Sensitivity in Annealed Alloys

Unlike the austenitic stainless steels, annealed chromium-iron stainless steels are highly notch sensitive in a manner similar to mild and low alloy steels [25,63,64]. In 1935, Krivobok [65] showed that the room-temperature impact resistance of chromium-iron alloys in the presence of a notch declined sharply when the chromium content exceeded about 15 percent, and this decline in properties was independent of carbon content in the range of 0.01 to 0.20 percent. These data are shown in Fig. 25. Heger [66] did a comprehensive study of the effects of notching on the tensile and impact strength of annealed air-melted 27Cr-Fe alloys measured at room temperature and at elevated temperatures. These data were reported by Newell in 1946 [24]. Looking first at the effect of a notch on tensile properties, Heger showed that the tensile strength increased and the elongation decreased in the presence of a notch. These effects are shown in Fig. 26 for

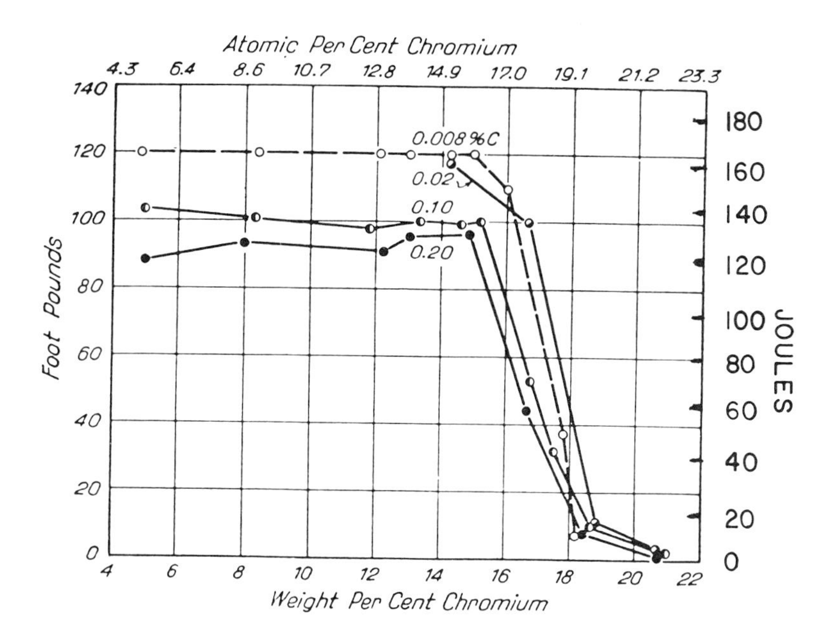

FIG. 25—*Effect of variation in chromium and carbon content on notch impact toughness of commercial chromium stainless steels* [65].

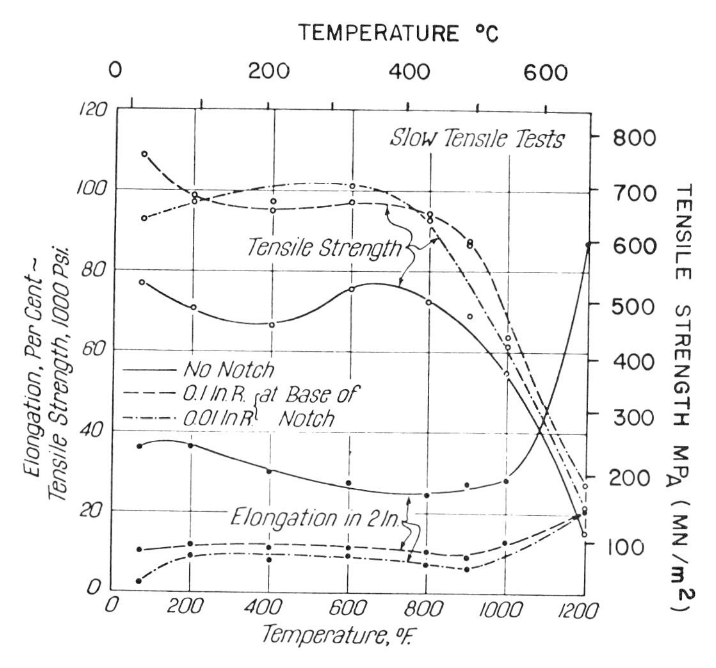

FIG. 26—*Effect of a notch on the short-time, high-temperature tensile properties of 27Cr-Fe air-melted alloy* [25,66].

short time tests run from room to elevated temperature. The difference in tensile and elongation properties between notched and unnotched specimens remained, even up to temperatures of 426 to 538°C. When Heger considered the effect of a notch on the impact resistance of a 27Cr-Fe alloy, he found a surprising difference in the impact energy absorbed between the notched and unnotched specimens. Similar to Krivobok's data, the notched specimen showed low impact energy absorption, but the unnotched specimen had high impact energy absorption. These data from Heger's work are shown in Fig. 27 along with the effect of testing tem-

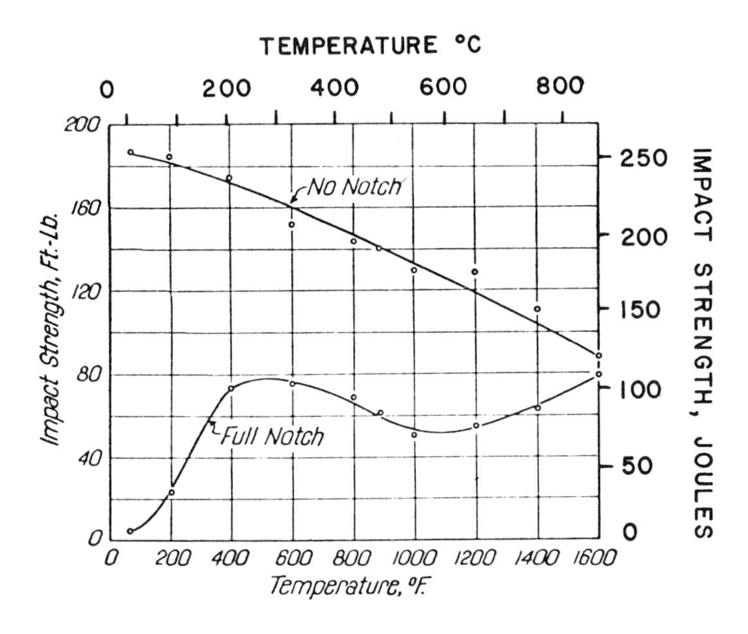

FIG. 27—*Effect of temperature on impact properties of 27Cr-Fe, air-melted alloy. No notch versus full notch* [25,66].

perature. There is a wide difference in the energy absorbed between the unnotched and notched specimen at room temperature which persists up to about 870°C testing temperature. However, the magnitude of the difference in impact absorption energy between the two types of specimens decreases as temperature increases because the impact absorption energy for the unnotched specimen declines and increases for the notched specimen. At temperatures of 870°C and above, the alloy was no longer notch sensitive. These data indicate the value of preheating air-melted chromium-iron alloys to reduce notch sensitivity in severe forming operations.

Based on data from Krivobok and Heger, it was generally believed into the late 1940's that notch sensitivity occurred independent of carbon and nitrogen content [*63,67*] and, in fact, was associated with the high-chromium content of the alloys *per se*. Later work by Hochmann [*68*] and Binder and Spendelow [*69*] showed that notch sensitivity in annealed chromium-iron alloys with high-chromium levels was again basically caused by the presence of critical levels of interstitial elements, particularly carbon, nitrogen, and oxygen. The dramatic effects of interstitial level on the impact behavior of annealed chromium-iron alloys were discovered when vacuum melting techniques were developed and used to produce alloys with extremely low interstitial levels.

The impact strength measured at room temperature of vacuum-melted chromium-iron alloys as a function of chromium content is shown in Fig. 28 from Binder et al [*69*]. The maximum interstitial content of the alloys

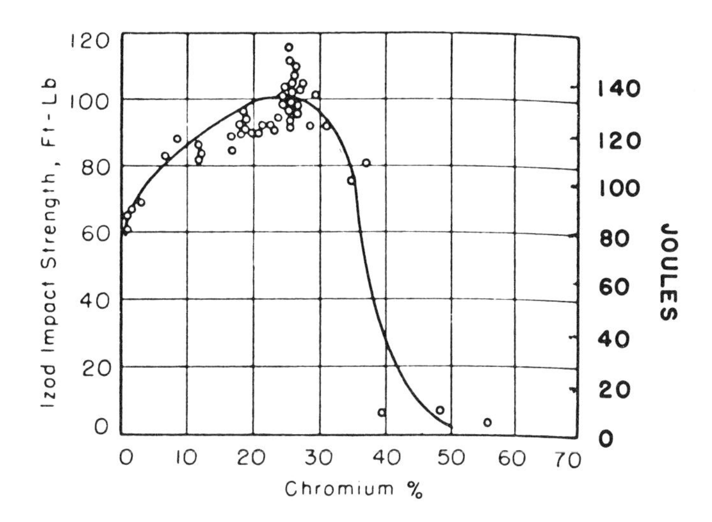

FIG. 28—*Impact strength of vacuum-melted chromium-iron alloys* [69].

used to generate these data include carbon 0.015 percent, nitrogen 0.01 percent, and oxygen 0.04 percent. As shown in Fig. 28, the toughness of these high-purity alloys increases as chromium content is raised, reaching a maximum at about 26 percent chromium, where unheard-of impact strength values of 100 ft·lb (136 J) are apparent. In contrast to Krivobok's data, [*65*] (Fig. 25) which showed a sharp drop in impact strength for alloys containing more than about 16 percent chromium, use of the vacuum melting process has raised the level of chromium, for which excellent impact toughness may be obtained in the presence of a notch from 16 percent to somewhat greater than 35 percent. The key to this difference in impact performance between normal air-melted steels and vacuum-melted alloys is primarily in their carbon and nitrogen content. The relationship between interstitial content and chromium content on toughness was determined in a very comprehensive investigation by Binder and Spendelow [*69*], as shown in Fig. 29. At chromium levels above about 15 to 18 percent, there is a drastic decrease in the carbon and nitrogen levels tolerable in a high-chromium alloy for high room-temperature impact resistance. For an air-melted steel with carbon and nitrogen contents up to about

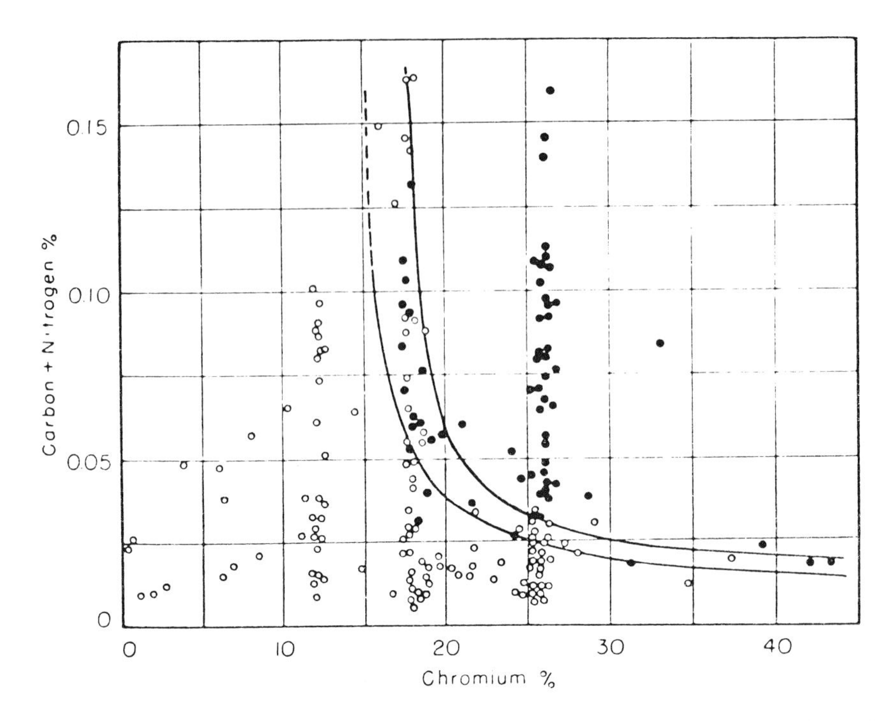

FIG. 29—*Influence of carbon and nitrogen on toughness of chromium-iron alloys. Open circles: high-impact strength alloys; solid circles: low-impact strength alloys* [69].

0.12 and 0.05 percent [*69*], respectively, these data would predict good room-temperature impact resistance is possible only if the alloys contain chromium levels below 15 to 18 percent. The agreement with Krivobok's early data [*65*] on air-melted alloys is excellent, but the conclusion that poor impact resistance was due to chromium content *per se* was incorrect. The low impact resistance of Krivobok's alloys containing more than 15 to 18 percent chromium was not caused by the chromium level but was due to the high interstitial levels characteristic of air-melted steels.

Binder and Spendelow, in their comprehensive investigation, also determined the individual effects of carbon and nitrogen on impact resistance of annealed alloys at two levels of chromium, 18 and 25 percent. These data are tabulated in Figs. 30 and 31, respectively. Both figures show: (*a*) that a straight line of a 45-deg slope may be drawn to separate the areas of high and low toughness, (*b*) that there is an equivalency in the effect by carbon and nitrogen on toughness, (*c*) that the carbon plus nitrogen sum is critical rather than the absolute value of each separately, and (*d*) that the maximum sum of carbon plus nitrogen tolerable for good room-temperature toughness is 0.055 percent for 18 percent chromium and 0.035 percent for 25 percent chromium.

Until the late 1960's, achieving levels of 0.035 percent and lower was possible only in the laboratory. With the development of economical vacuum melting and vacuum and gas refining techniques, commercial

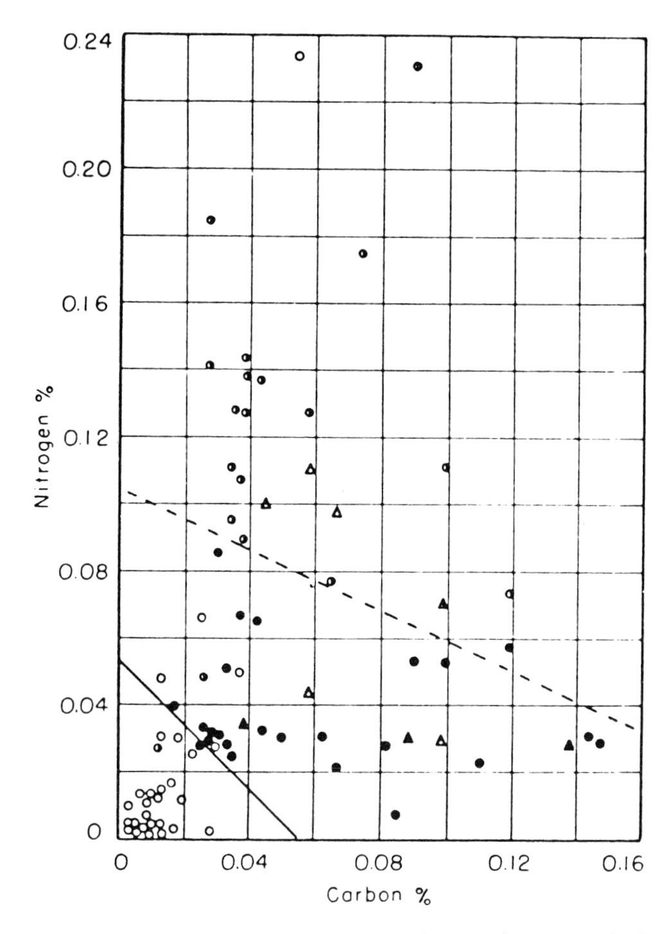

FIG. 30—*Effects of carbon and nitrogen on toughness of 17 to 19Cr-Fe alloys. Open circles: high-impact strength alloys; solid circles: low-impact strength alloys; semi-solid circles: intermediate impact strength alloys; triangles represent commercial arc-melted steels* [69].

production of high chromium-iron alloys became feasible and a reality. These alloys, to be described later, show excellent toughness at room temperature in the presence of notches. However, for available air-melted alloys such as AISI Type 430 (18 percent chromium), Type 442 (20 percent chromium), and Type 446 (26 percent chromium), notch sensitivity is a factor of major importance which must be considered carefully in application of these alloys. Good engineering practices require avoiding surface scratches and reentrant angles, notches, or other forms of stress risers when these air-melted alloys are used under shock loading conditions.

Weldable, Corrosion-Resistant, Ductile Ferritic Stainless Steels

The largest single drawback to the use of ferritic stainless steels has been the loss of corrosion resistance and ductility following exposures to high temperatures, as in welding and isothermal heat treatments. The

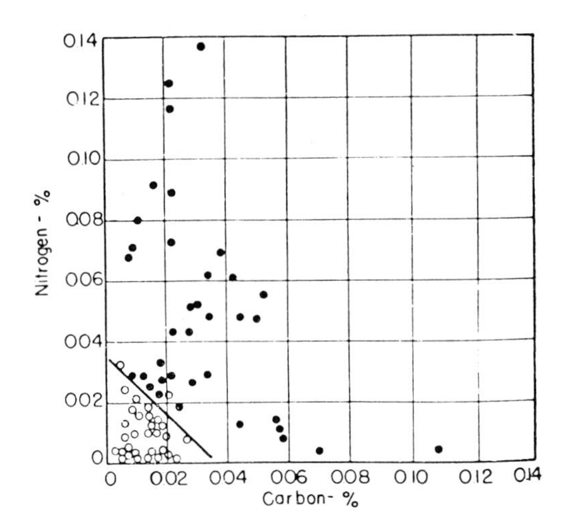

FIG. 31—*Effects of carbon and nitrogen on toughness of 24 to 26Cr-Fe alloys. Open circles: high-impact strength alloys; solid circles: low-impact strength alloys* [69].

other problems with ferritic stainless steel (475°C and σ-phase embrittlements) can be tolerated because relatively long exposure times at moderate temperatures are required to cause these other embrittling problems. However, unless a material has good corrosion resistance and ductility in the as-welded condition, its usefulness as a material of construction is limited severely. As just described, the cause for the serious loss in ductility and corrosion resistance when ferritic stainless steels are exposed to high temperatures and for the notch sensitivity of annealed alloys is related to the interstitial content of the alloys. After research in the early and mid-1960's had shown this fact, the research effort in the late 1960's saw development and commercialization of weldable, corrosion-resistant chromium-iron alloys. To reach this position, three routes to achieve interstitial control were researched and developed. Demo [*70*] has summarized and described these methods used to achieve interstitial controls in ferritic stainless steels.

Low Interstitials

By reducing the interstitial levels below certain minimum values, weldable and corrosion resistance can be produced, as shown in the works of Hochmann [*68*], Demo [*70*], Bond [*53*], Streicher [*56*], and Hodges [*43,55*]. Binder's [*69*] (Fig. 29) early work showed how the impact resistance of annealed ferritic stainless steels varied with interstitial and chromium content. A similar type of study by Demo [*70*] showed the relationship of interstitial level and chromium content on the ductility and corrosion resistance of as-welded alloys. These data are summarized in Fig. 32 and include a comparison to Binder's data for impact resistance on annealed specimens. The carbon and nitrogen levels which can be present in an alloy without affecting the weldability of a chromium-iron alloy are low and decrease rapidly with increasing chromium content. As can be seen from the data in Fig. 32, the interstitial tolerance level for a "good" as-

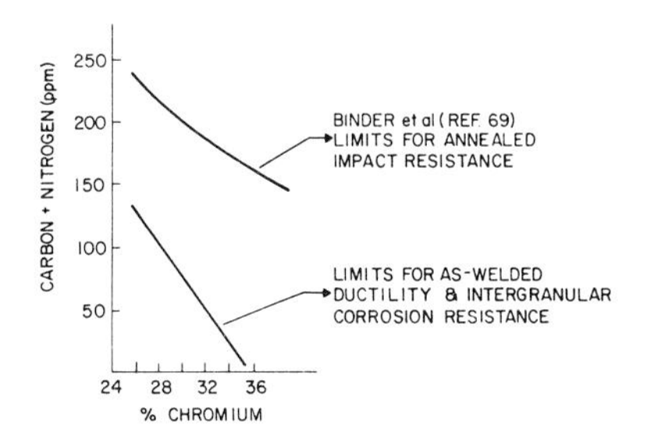

FIG. 32—*Effects of carbon and nitrogen level and chromium content on as-welded ductility and intergranular corrosion resistance of chromium-iron ferritic stainless steels. Comparison to Binder's limit for impact resistance of annealed specimens* [70].

welded specimen at a given chromium level is lower than that needed for impact resistance of an annealed specimen. Demo further studied the variation in the properties of weld ductility and weld corrosion resistance as functions of chromium content and interstitial sum level. These data are tabulated in Table 8. As chromium content increases from 19 to 35

TABLE 8—*Carbon plus nitrogen limits for as-welded properties of intergranular corrosion resistance and ductility as a function of chromium content.*

	Limit for Sum of C + N (ppm) to Have the Indicated Property in an As-Welded Alloy[a]	
Chromium Level, % by weight	Intergranular[b] Corrosion Resistance	Ductility[c]
19	60 to 80	>700
26	100 to 130	200 to 500
30	130 to 200	80 to 100
35	~250	<20

NOTE—Data by Demo [*70*].
[a]Sample thickness: 0.1 in. (2.54 mm) thick.
[b]Intergranular corrosion resistance in boiling ferric sulfate-50% sulfuric acid solution.
[c]No cracks, as determined by bending around a 0.2-in. mandrel.

percent, the amount of carbon plus nitrogen that can be tolerated for intergranular corrosion resistance increases somewhat. Conversely, for as-welded ductility, the sum of tolerable interstitials is reduced drastically.

At low chromium levels, as-welded corrosion resistance is the factor controlling whether an alloy has good weldability; at high-chromium levels, as-welded ductility is the limiting factor. At the 26 percent chromium level, intergranular corrosion resistance is more sensitive to the interstitial sum level than is ductility while at 35 percent chromium; the as-welded ductility is more critically dependent on the interstitial sum than is corrosion resistance.

To produce weldable and corrosion-resistant chromium-iron alloys by this route, it is evident that very low levels of carbon and nitrogen are needed. Until recently, such low levels could only be produced in a laboratory or by using high-purity raw materials. However, technological advances in steel-making practices have made the concept of low interstitial ferritic alloys possible through the development of such techniques as oxygen-argon melting, vacuum refining, and electron-beam refining. Of particular note is the electron-beam continuous hearth refining technique developed by Airco Vacuum Metals and described by Knoth [*71*]. This process has the advantage of achieving the lowest carbon and nitrogen levels by exposing a high surface to volume ratio of molten metal to a high vacuum for extended periods of time. As the molten metal flows down a series of water-cooled copper hearths, electron beam heat sources provide localized regions of intense heat in the molten metal, causing volatilization and removal of tramp impurity elements. The process is being used currently to produce commercially a high-purity ferritic stainless steel containing nominally 26 percent chromium, 1 percent molybdenum, and balance iron. By maintaining the carbon plus nitrogen level below 250 ppm, it is reported [*72,73*] that this commercially available alloy is ductile and corrosion resistant following welding, has good toughness, and combines resistance to stress-corrosion cracking with good general corrosion and pitting resistance. However, for a weldment to be resistant to intergranular attack, a carbon plus nitrogen sum level at or near 250 ppm is too high. Demo [*45*] reports intergranular attack on a high-purity 26Cr alloy containing 180-ppm carbon plus nitrogen, while Streicher [*59*] (Table 2) shows grain dropping in the weld and heat-affected zone of two high-purity 26Cr-1Mo alloys containing 105 and 230-ppm carbon plus nitrogen, respectively. For complete resistance to intergranular attack following welding or isothermal heat treatment, it appears that the carbon plus nitrogen levels in 26Cr and 26Cr-1Mo high-purity alloy systems must be maintained below about 100 to 120 ppm [*70*] with nitrogen [*59*] less than 90 ppm. The high-impact values at and below room temperature for an electron beam refined 26Cr-1Mo alloy as compared to a 26Cr-1Mo alloy containing 0.08 percent carbon is remarkable, as shown in Fig. 33 [*60, 72*]. The scatterband in the E-Brite 26-1 alloy data is the result of specimen orientation, variations in thermal treatments, and cooling rates.

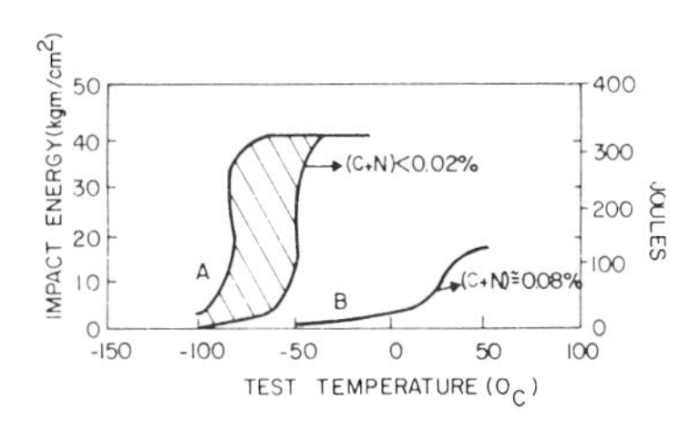

FIG. 33—(a) *Charpy V-notch transition temperature range for commercially produced electron-beam-melted ferritic steel containing 26 percent chromium and 1 percent molybdenum* (*E-Brite 26-1*) [72]. (b) *Transition curve for quarter-size V-notch impact specimens of an air-melted steel containing 26 percent chromium and 1 percent molybdenum* [60].

Interstitial Stabilization

A second means to control interstitials is to add elements to the alloy which form stronger carbides and nitrides than does chromium. Such elements include titanium, columbium, zirconium, and tantalum. The early work by Lula, Lena, and Kiefer [*48*] describes a comprehensive effort to study the intergranular corrosion behavior of ferritic stainless steel, including the effects of titanium and columbium additions. These investigators showed that titanium and columbium additions were not completely effective in preventing sensitization when the alloys were subjected to high temperature. This result was caused by not considering the need to tie up nitrogen as well as carbon and also by the unknown fact at the time that titanium carbide itself is dissolved in highly oxidizing solutions such as the boiling nitric acid solution used in the study.

More recent work by Baumel [*74*], Bond and Lizlovs [*52*], and Demo [*75,76*] have shown that columbium and titanium additions were effective in preventing intergranular corrosion following exposure of ferritic stainless steels to high temperatures such as isothermal heat treatments and welding. To resist intergranular corrosion, titanium additions of about six to ten times the combined carbon and nitrogen level are necessary; for columbium, additions of eight to eleven times are required. The relationship of interstitial content, chromium level, and titanium level for intergranular corrosion resistance and ductility after welding has been studied extensively by Demo [*75,76*]. Bond et al [*52*], Lula et al [*48*], Herbsleb [*77*], Baumel [*74*], and Cowling et al [*78*] have shown that titanium-stabilized alloys may show intergranular attack when exposed to a highly oxidizing solution such as boiling nitric acid due to dissolution of titanium carbonitrides; however, columbium-stabilized alloys resist intergranular attack even in highly oxidizing solutions.

Demo [*75,76*], Semchyshen et al [*60*], Wright [*79*], and Pollard [*80*] have reported the effects of stabilizing additions on the weld ductility of ferritic stainless steels. By introducing titanium or columbium in the ferritic alloy, the level of interstitial which can be present in the matrix without adversely affecting the room-temperature ductility after welding is increased significantly. These data are shown in Table 9 by the tensile ductility measurements on welded 18Cr-2Mo specimens [*60*] and in Table 10 by the slow bend tests on welded 26 to 30Cr alloys [*75,76,79*]. With

TABLE 9—*Effect of stabilizer additions on the tensile ductility of annealed versus welded specimens containing 18% chromium-2% molybdenum.*

C + N, % by weight	Ti or Cb, % by weight	Elongation in 50 mm, %	
		Annealed	As Welded
0.005	0	33	31
0.03	0	31	8
0.07	0.5	34	30
0.06	0.6	28	21

NOTE—Data by Semchyshen et al [*60*].

TABLE 10—*Effect of titanium on the as-welded bend ductility for chromium-iron ferritic stainless steels containing 26 to 30% chromium.*

C + N, ppm	Ti, % by weight	Bend Test Ductility As-Welded
113	0	passed 180 deg, 2t[a]
310	0	passed 180 deg, 1/2t[b]
362	0	failed 90 deg, 2t[a]
450	0	failed 90 deg, 2t[a]
900	0	failed 135 deg, 1t[b]
300	0.22	passed 180 deg, 1/2t[a]
387	0.24	passed 180 deg, 2t[a]
488	0.47	passed 180 deg, 2t[a]
850	0.45	passed 180 deg, 1/2t[b]

[a] Data by Demo [*75,76*], 0.1-in.-thick specimens. t = specimen thickness.
[b] Data by Wright [*79*], 0.06-in.-thick specimens.

stabilizer additions, the interstitial elements are effectively tied up as stable carbides and nitrides such that their effective level in solid solution is reduced. Consequently, stabilized alloys at relatively high levels of carbon and nitrogen act similarly to the very low interstitial alloys just described in having excellent corrosion resistance and ductility (tensile or bend) following exposure to high temperatures which, without stabilization, would cause loss of corrosion resistance and ductility.

The effects of stabilizing additions on the impact properties of chromium iron alloys in comparison to the impact properties of low interstitial alloys, however, presents another story. The effects of stabilizing additives on impact properties have been studied and described by Semchyshen et al

[60] and Wright [79]. Two aspects have been studied, namely, the effect of titanium content and interstitial level on the impact resistance of (a) annealed specimens and (b) specimens heated to high temperatures by welding or isothermal heat treatments. In the annealed condition, the titanium-modified steels exhibit transition temperatures commensurate with their interstitial levels; that is, whatever the impact transition temperature is for the unstabilized, annealed alloy as a function of interstitial level (see Fig. 29), remains about the same or is slightly reduced when the alloy is stabilized.

Stabilizing additions of titanium, however, are useful in reducing the detrimental effects of high-temperature treatments on the impact resistance of high interstitial alloys. These effects of stabilizing additions taken from Semchyshen et al [60] are shown in Fig. 34 for air-melted commercial

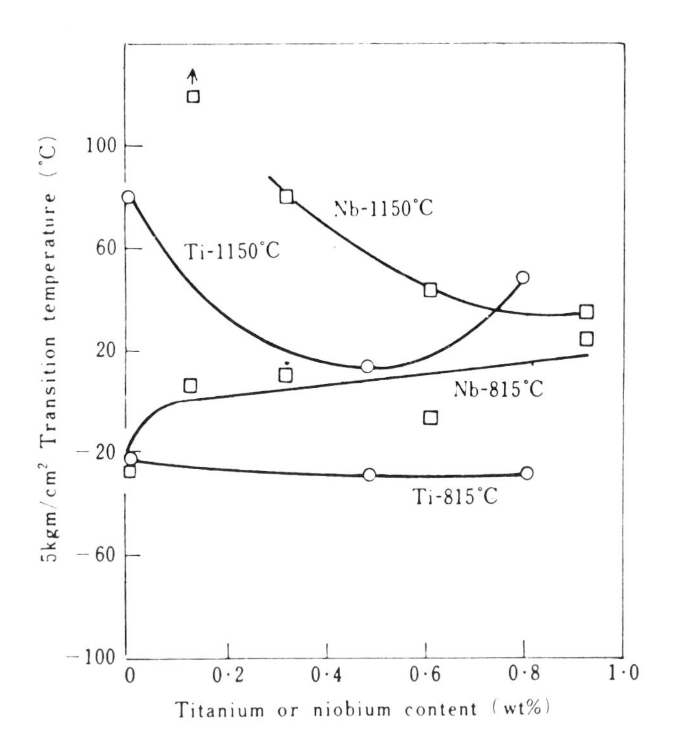

FIG. 34—*Transition temperatures for quarter-size Charpy V-notch specimens air-melted commercial-purity 18Cr-2Mo steels water quenched from 1150°C (sensitized) and 815°C (annealed) as a function of titanium or columbium content* [60].

purity (0.07 percent carbon plus nitrogen) 18Cr-2Mo alloy. Increases in titanium content from 0 to 0.8 percent show little effect on the impact transition temperature of annealed (815°C) specimens. However, increases in titanium content from 0 up to about 0.5 percent improve (lower) the transition temperature when the alloys are subjected to a high-temperature treatment (1150°C). These data also show that columbium additions, though effective in lowering the transition temperature for alloys subjected to a high temperature, were somewhat harmful to the impact resistance of the alloys in the annealed condition. Semchyshen et al [*60*] also showed that titanium additions beyond about ten times the combined carbon and nitrogen content could affect (increase) the impact transition temperature of annealed 18Cr-2Mo alloys (0.07 percent carbon plus nitrogen) as shown in Fig. 35. A precipitation of an intermetallic phase markedly increased impact transition temperatures.

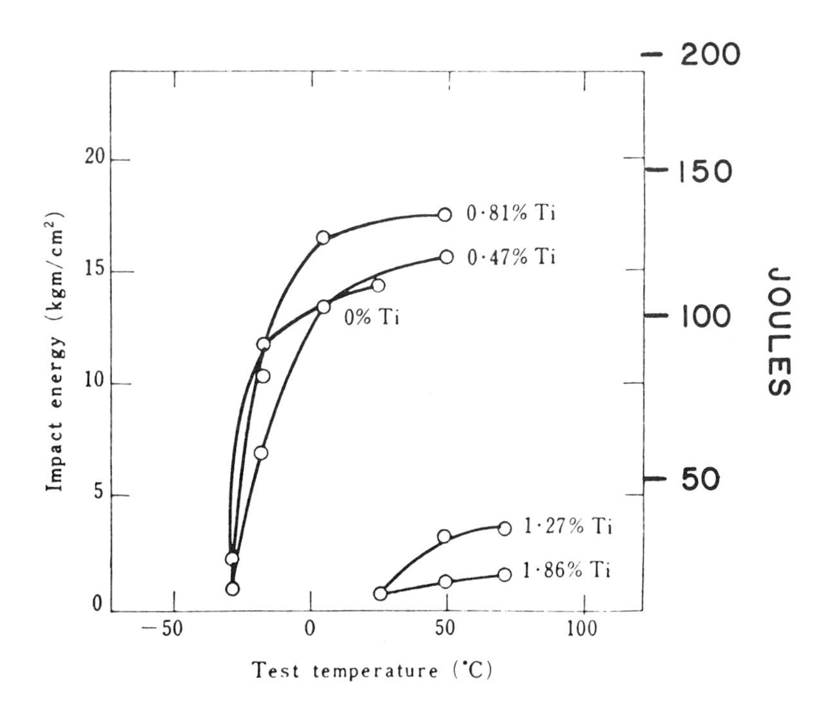

FIG. 35—*Transition curves for quarter-size Charpy V-notch impact specimens of air-melted 18Cr-2Mo; 0 to 1.86Ti ferritic stainless steels heat treated at 815°C for 1 h and water quenched* [60].

Superimposed on the effects of titanium and interstitial content on the ductility and impact resistance of chromium-iron alloys is the marked effect of thickness on these properties, particularly impact resistance. This point is shown by the data in Table 11 taken from Wright's work [*79*]. In order to have acceptable properties (weldability and touchness) at plate gages as heavy as 0.5 in., interstitial sum level in a 26Cr alloy must be as low as or lower than about 100 to 125 ppm. For alloys containing about 300 to 900 ppm total carbon and nitrogen, toughness may be poor at gages above about 0.13 in., and weldability is poor at all gage thicknesses. The addition of titanium increases the thickness or gage level at which

TABLE 11—*Impact transition temperatures as a function of gage and added titanium for annealed 26Cr-1Mo alloys.*

C + N, % by weight	Gage Thickness, in. (mm) Charpy V-Notch, Ductile to Brittle Temperature, °C 0.5 (12.7)[a]	0.12 to 0.14 (3.05 to 3.56)[a]	0.06 (1.52)[b]
0.0065	−57	...	−73
0.0310	149	38	−73
0.0900	162	38	−18
0.0300 + 0.22 Ti	121	−1	−46
0.0850 + 0.45 Ti	107	38	−46

NOTE—Data by Wright [79].
[a] Water quenched after anneal.
[b] Air cooled after anneal.

an alloy will have good weldability. For example, as Wright notes, a 26Cr-1Mo alloy containing 300 ppm total carbon and nitrogen and 0.22 percent titanium, has excellent toughness and weldability (that is, as-welded ductility and corrosion resistance) up to a gage thickness of about 0.13 in.

In summary, adding a stabilizer to an alloy of moderate carbon and nitrogen will not improve the annealed impact behavior significantly but may improve the as-welded ductility and corrosion resistance tremendously. This point is a most important difference between stabilized alloys and low interstitial alloys. The high-purity material will have excellent toughness and as-welded corrosion resistance and ductility. The stabilized material will also have excellent as-welded corrosion resistance and ductility but may not have high room-temperature impact resistance. These effects as shown by Wright [79] are particularly magnified at section thicknesses greater than about ⅛ in. Therefore, for thick plate sections where high room-temperature toughness is an absolute requirement, the low interstitial alloys would be acceptable, but the stabilized grades would not. On the other hand, for thin material, as required in heat exchanger tubing, the titanium stabilized alloys and the low interstitial alloys have similar toughness and weldability properties so that either alloy system may be used.

Weld Ductilizing Additions

A third method which produces high chromium-iron ferritic stainless steels with good as-welded ductility is to add low concentrations of selected elements with atomic radius within 15 percent of the α matrix. This method was developed and investigated extensively by Steigerwald et al [81]. As noted in Fig. 32, the carbon and nitrogen level for as-welded ductility is reduced drastically as chromium content increases. At 35 percent chromium, an impossibly low interstitial level of about 10 to 15 ppm is necessary for as-welded ductility. Steigerwald et al found that, in the presence of low amounts of copper, aluminum, vanadium, and combinations of these elements, alloys with good as-welded ductility could be produced at high interstitial levels. A summary of this work is shown in Fig. 36. With selective additives, an alloy containing 35 percent chromium can

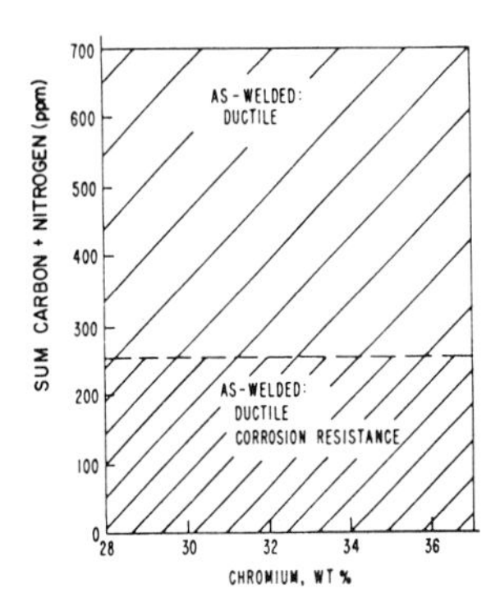

FIG. 36—*Effect of weld ductilizing additives on as-welded ductility and corrosion resistance of high chromium-iron stainless steels; additives, singly or in combination, include aluminum, copper, vanadium, platinum, palladium, and silver in a range 0.1 to 1.3 percent* [81].

be produced with both as-welded ductility and corrosion resistance at interstitial levels of 250 ppm versus the 10 to 15 ppm level needed at this chromium level when the additives are absent. It is remarkable that a ferritic stainless steel containing 35 percent chromium can be made with as-welded ductility at intermediate interstitial levels.

Sigma Phase and 475°C Embrittlement Susceptibility

By the techniques of interstitial control, chromium-iron alloys can be produced which resist the damaging loss of corrosion resistance and ductility following high-temperature exposures, as in welding or isothermal heat treatments. In addition, some alloys can be produced to also have high room-temperature impact toughness in thick sections. Therefore, by interstitial control methods, the *high temperature* exposure and notch sensitivity problems which have limited severely the usefulness of ferritic stainless steels before have been removed. However, though not as serious as the embrittling problems described earlier, the 475°C and σ-phase embrittling phenomenon will still occur in alloy composition made resistant to high-temperature exposure effects. This point has been shown by works of Grobner [*42*], Hochmann [*44*], and in reported data for E-Brite 26-1 [*72*].

Molybdenum Additions

For improving the general corrosion and pitting resistance of chromium-iron stainless steels, the effects of molybdenum additions have been extensively studied and reported on by Bond [*82*], Demo [*70*], Streicher [*59*], and Steigerwald [*83–85*]. Of particular note here is the effect of molybdenum additions on embrittling a chromium-iron stainless steel. Semchyshen et al [*60*] have summarized the data describing the effect molybdenum level has on the toughness of 18 and 25Cr alloys. The effects of molybdenum level on the impact transition temperature of an annealed 25Cr alloy containing high and low interstitials, respectively, are shown in Fig. 37. There is considerable difference in the impact transition temperature of the high-purity alloys (−50°C) and the high interstitial alloys (+50°C) without molybdenum addition, as described earlier. The point, however, is that molybdenum additions up to about 2 to 3 percent have little effect on toughness but an adverse effect when the molybdenum level exceeds about 3 to 4 percent. This decrease in toughness associated with high-molybdenum contents is caused by the formation of chi (χ) phase, a brittle intermetallic compound of iron, chromium, and molybdenum.

Streicher [*86*] has reported on a study of the effects of heat treatment on the microstructure and formation of χ and σ phase in two low interstitial alloys containing 28 percent chromium-4 percent molybdenum and 28 percent chromium-4 percent molybdenum-2 percent nickel. One-hour heat treatment in the range of 700 to 925°C produced only a small amount of σ at the grain boundaries. The largest amounts of σ are formed by heating at 815°C. Heating for 100 h at 815°C caused σ and large amounts of χ phase to form a grain boundaries and within the grains. The complex relationship between chromium, molybdenum, and iron levels on the formation of χ and σ phases has been described in detail by the work of McMullen et al [*87*]. The ternary phase diagram for the iron-chromium-molybdenum system at 898°C (1650°F) isotherm, as defined by McMullen et al, is shown in Fig. 38 [*87*]. χ phase is stable over a wider temperature range than σ [*87*], so annealing above 980°C is required to eliminate it in steels with 18 percent chromium and over 3.5 percent molybdenum [*60*] and in steels containing 28 percent chromium-4 percent molybdenum [*86*].

For 26Cr alloys containing 1 percent molybdenum and stabilized with titanium, Aggen [*88*] and Demo [*89*] have shown that a corrosion-damaging second phase occurs in the alloy when a slight excess of titanium beyond that needed to tie up the interstitials is present. This phase, believed also to be χ, is richer than the matrix in titanium, molybdenum, and silicon and forms by isothermal holds for extended periods in the temperature range from 595 to 850°C. Before it has grown large enough to be seen by an optical microscope, the intergranular corrosion resistance

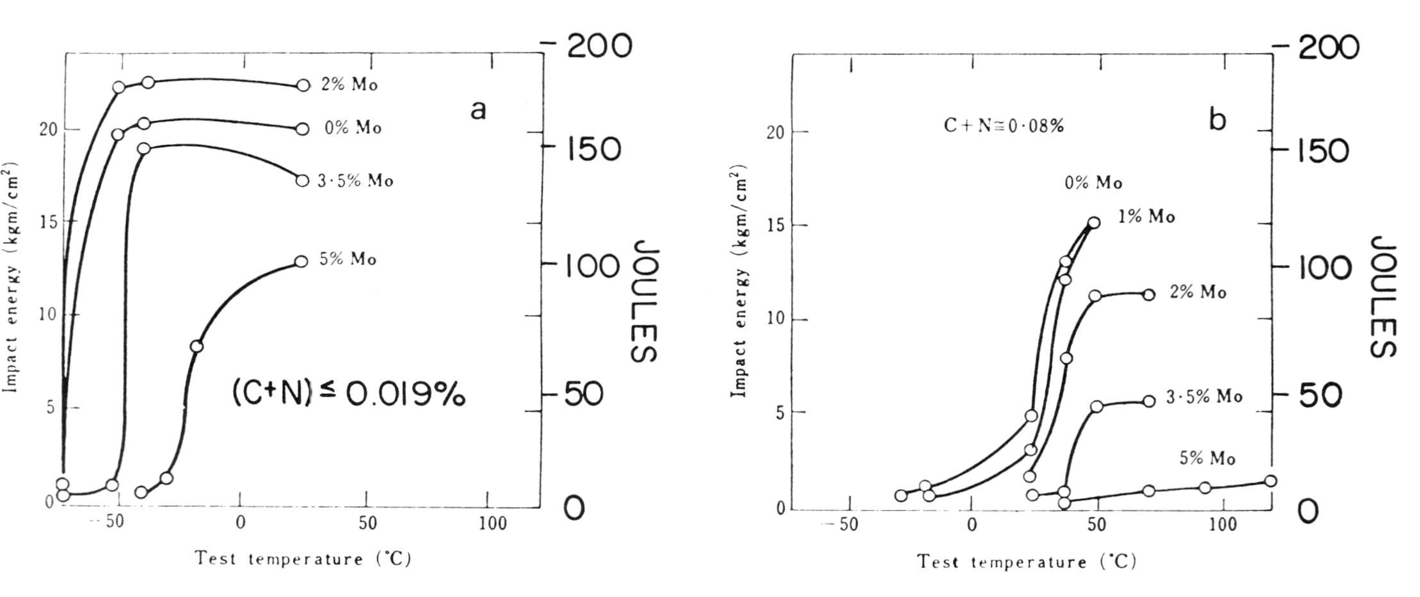

FIG. 37—*Effects of molybdenum additions on impact transition temperatures of annealed 25Cr-Fe stainless steels*: (a) *vacuum-melted stainless steels heat treated at 980°C for 1 h and water quenched or* (b) *air-melted stainless steels heat treated at 815°C for 1 h and water quenched. Quarter-size Charpy V-notch impact specimens* [60].

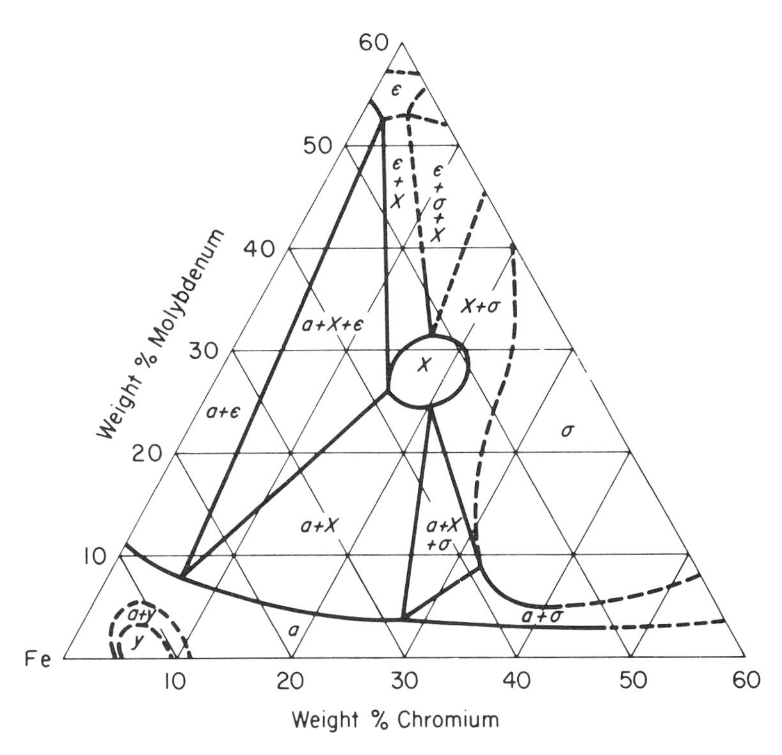

FIG. 38—*Constitution of iron-chromium-molybdenum alloys at 898°C (1650°F).* σ = *sigma*; χ = *chi*; *and* ϵ = Fe_7Mo_6 [87].

of a susceptible alloy will be affected deleteriously when it is exposed to highly oxidizing solutions. The effect only appears in alloys containing molybdenum and excess titanium and can be removed by short annealing treatments at temperatures above about 900°C.

Summary

Chromium-iron alloys are bcc up to the melting point. Therefore, hardening by the ferrite-austenite mechanism upon heating and quenching cannot occur. Ferritic chromium iron alloys can be embrittled by the phenomenon of 475°C embrittlement, σ-phase embrittlement or χ-phase embrittlement for those alloys containing molybdenum. These embrittling mechanisms generally require a long-time isothermal treatment or very slow cool through the intermediate temperature regions such that they do not normally constitute a threat to the weldability and processing of ferritic stainless steels. The serious loss of ductility and corrosion resistance when ferritic stainless steels are subjected to high-temperature exposures as in welding or isothermal treatment and the problem of notch sensitivity have been shown to be related to interstitial levels in the alloys. By development of interstitial control techniques, it is now possible to produce high-chromium ferritic stainless steels which are corrosion resistant and ductile following a high-temperature exposure, as in welding or an isothermal heat treatment. Further, depending on the interstitial levels, high chromium-iron alloys can be produced, having excellent toughness and impact transition temperatures below room temperature. By making chromium-iron stainless steels weldable and tough, and combining this property with stress-corrosion resistance and good general corrosion resistance, attractive new materials of construction are and will be available to compete with austenitic 18Cr-8Ni-type stainless steels.

Several organizations throughout the world are engaged in commercial development of ferritic stainless steels. An 18Cr-2Mo alloy stabilized with columbium or titanium is being produced in Europe. A comprehensive review article on the properties of 18Cr-type ferritic alloys has been published by Schmidt and Jarleborg [*90*]. Where greater corrosion resistance is desired, several new compositions are either in commercial production or are being actively developed. These include a 26Cr-1Mo alloy produced either with extremely low interstitial levels by electron-beam melting techniques or with moderate interstitial levels stabilized with titanium and, to a lesser extent, with columbium. In commercial development are low interstitial alloys of the composition 26 percent chromium-2 percent molybdenum, 21 percent chromium-3 percent molybdenum, 28 percent chromium-2 percent molybdenum, 28 percent chromium-4 percent molybdenum, 28 percent chromium-4 percent molybdenum-2 percent nickel. These alloys have better general corrosion resistance than Type 316 stainless steel. Those like the 28Cr-4Mo and 28Cr-4Mo-2Ni alloys have been shown by Streicher [*59*] to have corrosion resistance in a variety of aggressive environments comparable or better than the highly alloyed nickel-based alloys. The complex relationship between chromium and molybdenum content on pitting resistance, ductility, and weld ductility, as studied by Streicher [*59*] is shown in Fig. 39. As shown by

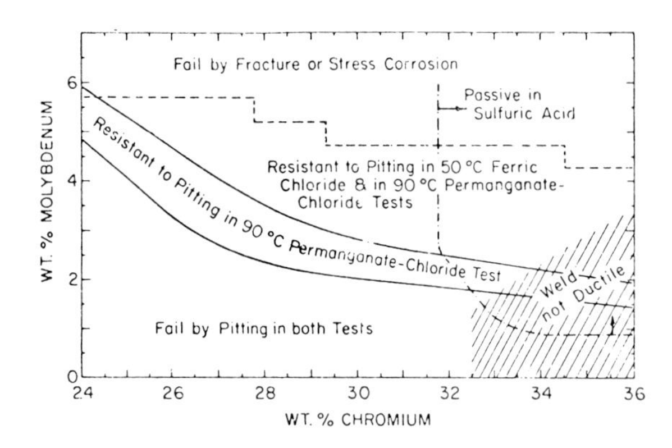

FIG. 39—*Properties of iron-chromium-molybdenum alloys. Pitting tests: 10% FeCl·6H_2O at 50°C (with crevices), 2 percent potassium manganate-2 percent sodium chloride at 90°C. Stress corrosion test: 45 percent boiling (155°C) magnesium chloride on welded U-bend specimens. Sulfuric acid: boiling 10 percent by weight. Carbon: 10 ppm; nitrogen: 200 ppm; carbon plus nitrogen: 250 ppm* [59].

these data, the 28Cr-4Mo-type alloys have better resistance to pitting than any other types of chromium-molybdenum-iron alloys now under development. With a variety of desirable properties available, including good weldability and resistance to stress-corrosion cracking, a family of ferritic stainless steel compositions is becoming available as materials of construction which will replace the austenitic and nickel-based alloys in many applications.

References

[1] Bain, E. C. and Aborn, R. H. in *Metals Handbook,* 1948 ed., American Society for Metals, p. 1194.

[2] Lecken, E. B., Fischer, W. A., and Lorenz, K., *Stahl und Eisen,* Vol. 81, No. 12, 1961, pp. 768–778.

[3] Colombier, L. and Hochmann, J., *Stainless and Heat-Resisting Steels,* Edward Arnolds Ltd. Publishers, England, 1967.

[4] Nehrenberg, A. E. and Lillys, P., *Transactions,* American Society for Metals, No. 46, 1954, pp. 1177–1213.

[5] Newell, H. D., *Metal Progress,* April 1947, pp. 617–626.

[6] Thielsch, H., *Welding Journal,* Research Supplement, Vol. 30, 1951, pp. 209s–250s.

[7] Rajkay, L., *Proceedings,* American Society for Testing and Materials, Vol. 67, 1967, pp. 158–169.

[8] Kaltenhauser, R. H., *Metals Engineering Quarterly,* Vol. 11, May 1971, pp. 41–47.

[9] Demo, J. J. and Bond, A. P., *Corrosion,* Jan. 1975.

[10] Cook, A. J. and Jones, F. W., *Journal of the Iron and Steel Institute,* No. 148, 1943, pp. 217–223.

[11] Chevenard, P., *Travaux et Memoires Du Bureau International Des Poids et Mesures,* Vol. 17, 1927, p. 90.

[12] Bain, E. C. and Griffiths, W. E., *Transactions of the American Institute of Mining and Metallurgical Engineers,* Vol. 75, 1927, pp. 166–213.

[13] Anderson, A. G. H. and Jette, E. R., *Transactions,* American Society for Metals, Vol. 24, 1936, pp. 375–419.

[14] Jette, E. R. and Foote, F., *Metals and Alloys,* Vol. 7, 1936, pp. 207–210.

[15] Smith, G. V., *Iron Age,* Vol. 166, No. 22, 1950, pp. 63–68; No. 23, 1950, pp. 127–132.

[16] Duwez, P. and Baen, S. R. in *Symposium on the Nature, Occurrence, and Effects of Sigma Phase, ASTM STP 110,* American Society for Testing and Materials, 1951, pp. 48–54.

[17] Menezes, L., Roros, J. K., and Read, T. A. in *Symposium on the Nature, Occurrence, and Effects of Sigma Phase, ASTM STP 110,* American Society for Testing and Materials, 1951, pp. 71–74.

[18] Dickens, G. J., Douglas, A. M. B., and Taylor, W. H., *Nature,* No. 167, 1951, p. 192.

[19] Heger, J. J., *Metal Progress,* Vol. 49, 1946, p. 976B.

[20] Shortsleeve, F. J. and Nicholson, M. E., *Transactions,* American Society for Metals, Vol. 43, 1951, pp. 142–156.

[21] Olzak, Z. E., "The Effect of Alloy Composition on Sigma Phase Precipitation in 27% Chromium-Iron Catalyst Tubes, First Step Dehydogenation, Plains Plant," Report No. PR-4-2, The Babcock and Wilcox Tube Co., 25 Aug. 1944.

[22] Heger, J. J. in *Symposium on the Nature, Occurrence, and Effects of Sigma Phase, ASTM STP 110,* American Society for Testing and Materials, 1951, pp. 75–78.

[23] Gilman, J. J., *Transactions,* American Society for Metals, 1951, pp. 101–187.

[24] Zapffe, C. A. and Worden, C. O., *Welding Journal,* Research Supplement, Vol. 30, 1951, pp. 47s–54s.

[25] Newell, H. D., *Metal Progress,* May 1946, pp. 977–1028.

[26] Zapffe, C. A., Worden, C. O., and Phebus, R. L., *Stahl und Eisen,* Vol. 71, 1951, pp. 109–119.

[27] Becket, F. M., *Transactions,* American Institute of Mining, Metallurgy, and Petroleum Engineers, No. 131, 1938, pp. 15–36.

[28] Riedrich, G. and Loib, F., *Archiv fur das Eisenhuttenwesen,* Vol. 15, No. 7, 1941-1942, pp. 175-182.
[29] Bandel, G. and Tofaute, W., *Archiv fur das Eisenhuttenwesen,* Vol. 15, No. 7, 1941-1942, pp. 307-319, (Brutcher Translation No. 1893).
[30] Heger, J. J., *Metal Progress,* Aug. 1951, pp. 55-61.
[31] Thielsch, H., *Welding Journal,* Research Supplement, Vol. 29, 1950, pp. 126s-132s.
[32] Houdremont, E., *Hanbuch der Sonderstahlkune,* Springer Verlag, Berlin, Germany, 1943.
[33] Masumoto, H., Saito, H., and Sugihara, M., *Scientific Reports of Tohoku University,* Series A, Vol. 5, 1953, p. 203.
[34] Pomey, G. and Bastien, P., *Revue de Metallurgie,* Vol. 53, 1956, pp. 147-160.
[35] Takeda, S. and Nagai, N., *Memoirs of the Faculty of Engineering,* Nagoya University, Vol. 8, 1956, p. 1.
[36] Shull, C. G., Wollan, E. O., Koehler, W. C., and Strauser, W. A., American Engineering Council Publication, ORNL-728, Oak Ridge National Laboratory, 1950.
[37] Williams, R. O., *Transactions,* Metallurgical Society of the American Institute of Mining, Metallurgical, and Petroleum Engineers, No. 212, 1958, pp. 497-502.
[38] Tisinai, G. F. and Samans, C. H., *Journal of Metals,* Oct. 1957, pp. 1221-1226.
[39] Fisher, R. M., Dulis, E. J., and Carroll, K. G., *Transactions,* American Institute of Mining, Metallurgical, and Petroleum Engineers, No. 185, 1957, pp. 358-374.
[40] Williams, R. O. and Patton, H. W., *Journal of the Iron and Steel Institute,* No. 185, 1957, pp. 358-374.
[41] Marcinkowski, M. J., Fisher, R. M., and Szirmae, A., *Transactions,* American Institute of Mining, Metallurgical, and Petroleum Engineers, 1964, pp. 676-689.
[42] Grobner, P. J., *Metallurgical Transactions,* Vol. 4, 1973, pp. 251-260.
[43] Hodges, R. J., *Corrosion,* Vol. 27, 1971, pp. 164-167.
[44] Hochmann, J., *Revue de Metallurgie,* Vol. 48, 1951, pp. 734-758, (Brutcher Translation No. 2981).
[45] Demo, J. J., *Corrosion,* Vol. 27, 1971, pp. 531-544.
[46] Pruger, T. A., *Steel Horizons,* Vol. 13, 1951, pp. 10-12.
[47] Houdremont, E. and Tofaute, W., *Stahl und Eisen,* Vol. 72, 1952, pp. 539-545.
[48] Lula, R. A., Lena, A. J., and Kiefer, G. C., *Transactions,* American Society for Metals, Vol. 46, 1954, pp. 197-230.
[49] Bain, E. C., Aborn, R. H., and Rutherford, J. J. B., *Transactions,* American Society for Steel Treating, Vol. 21, 1933, pp. 481-509.
[50] Ebling, H. F. and Scheil, M. A. in *Advances in the Technology of Stainless Steels and Related Alloys, ASTM STP 369,* American Society for Testing and Materials, 1965, p. 275.
[51] Baumel, A., *Archiv für das Eisenhüttenwesen,* Vol. 34, 1963, pp. 135-149; British Iron and Steel Industry Translation No. 3287, London, 1963.
[52] Bond, A. P. and Lizlove, E. A., *Journal of the Electrochemical Society,* No. 116, 1969, pp. 1305-1311.
[53] Bond, A. P., *Transactions,* Metallurgical Society of the American Institute of Mining, Metallurgical, and Petroleum Engineers, No. 245, 1969, pp. 2127-2134.
[54] Demo, J. J., "Effect of High Temperature Exposure on the Corrosion Resistance and Ductility of AISI 446 Stainless Steel," NACE Preprint No. 23, 1968 Conference, Cleveland, National Association of Corrosion Engineers, 1968.
[55] Hodges, R. J., *Corrosion,* Vol. 27, 1971, pp. 119-127.
[56] Streicher, M. A., *Corrosion,* Vol. 29, 1973, pp. 337-360.
[57] Henthorne, M., *Localized Corrosion—Cause of Metal Failure, ASTM STP 516,* American Society for Testing and Materials, 1972, pp. 66-119.
[58] Mahla, E. M. and Neilsen, N. A., *Transactions,* American Society for Metals, Vol. 43, 1951, pp. 290-322.
[59] Streicher, M. A., *Corrosion,* Vol. 30, 1974, pp. 77-82.
[60] Semchyshen, M., Bond, A. P., and Dundas, H. J., *Toward Improved Ductility and Toughness,* Kyoto, Japan, 1971, pp. 239-253.
[61] Plumtree, A. and Gullberg, R., *Journal of Testing and Evaluation,* Vol. 2, No. 5, 1974, pp. 331-336.

[62] Gleiter, H. and Hornbogen, E., *Materials Science and Engineering,* Vol. 2, 1967/1968, pp. 285–302.
[63] Kinzel, A. B. and Franks, R., *The Alloys of Iron and Chromium, Vol. II—High Chromium Alloys,* McGraw Hill, New York, 1940.
[64] Legat, H., *Metallwirtschaft,* Vol. 17, 1938, pp. 509-513, (Brutcher Translation No. 654).
[65] Krivobok, V. N., *Transactions,* American Society for Metals, 1935, pp. 1-56.
[66] Heger, J. J., "The Effect of Notching on the Tensile and Impact Strength of Annealed 27% Chromium-Iron Alloy at Room and Elevated Temperatures," Report No. RR-3-6, The Babcock and Wilcox Tube Co., 28 March 1945.
[67] Lincoln, R. A., dissertation, Carnegie Institute of Technology, Pittsburgh, Pa., 1935.
[68] Hochmann, J., *Comtes Rendus,* No. 226, 1948, pp. 2150-2151.
[69] Binder, W. O. and Spendelow, H. R., *Transactions,* American Society for Metals, Vol. 43, 1951, pp. 759–777.
[70] Demo, J. J., *Transactions,* Metallurgical Society of the American Society of Mining, Metallurgical, and Petroleum Engineers, Vol. 5, 1974, pp. 2253–2256.
[71] Knoth, R. J., "Electron Beam Continuous Hearth Refining and Its Place in the Specialty Steel Industry," presented at the Specialty Steel Seminar, Pittsburgh, Pa., 1969.
[72] "E-Brite 26-1," Brochure No. AV110-3-71, Airco Vacuum Metals, Division of Air Reduction Co., Inc., Berkeley, Calif.
[73] Knoth, R. J., Lakso, G. E., and Matejka, W. A., *Chemical Engineering,* Vol. 77, No. 11, May 1970, pp. 170–176.
[74] Baumel, A., *Stahl und Eisen,* Vol. 84, 1964, pp. 798-802.
[75] Demo, J. J., "Ferritic Iron-Chromium Alloys," Canadian Patent No. 939,936, 15 Jan. 1974.
[76] Demo, J. J., "Ductile Chromium-Containing Ferritic Alloy," Canadian Patent No. 952,741, 13 Aug. 1974; patent applied for in the United States.
[77] Hersleb, G., *Werkstoffe und Korrosion,* Vol. 19, No. 5, 1968, pp. 406–412.
[78] Cowling, R. D. and Hintermann, H. E., *Journal of the Electrochemical Society,* Vol. 117, No. 11, 1970, pp. 1447-1449.
[79] Wright, R. N., *Welding Journal,* Research Supplement, Vol. 50, Oct. 1971, pp. 434s–440s.
[80] Pollard, B., *Metals Technology,* Jan. 1974, pp. 31-36.
[81] Sipos, D. J., Steigerwald, R. F., and Whitcomb, N. E., "Ductile Corrosion-Resistant Ferrous Alloys Containing Chromium," U.S. Patent No. 3 672 876, 27 June 1972.
[82] Bond, A. P., *Journal of the Electrochemical Society,* Vol. 120, No. 5, 1973, pp. 603–613.
[83] Steigerwald, R. F., "Low Interstitial Fe-Cr-Mo Ferritic Stainless Steels," Developments in the Field of Molybdenum-Alloyed Cast Iron and Steel, Soviet-American Symposium, Moscow, Jan. 1973.
[84] Steigerwald, R. F., *Tappi,* Vol. 56, No. 4, 1973, pp. 129–133.
[85] Steigerwald, R. F., "New Molybdenum Steels and Alloys for Corrosion Resistance," NACE Preprint No. 44, 1974 Conference, Chicago, National Association of Corrosion Engineers, 1974.
[86] Streicher, M. A., *Corrosion,* Vol. 30, No. 4, 1974, pp. 115–124.
[87] McMullin, J. G., Reiter, S. F., and Ebeling, D. G., *Transactions,* American Society for Metals, Vol. 46, 1954, pp. 799–811.
[88] Aggen, G., unpublished work, Allegheny Ludlum Industries, Inc., Pittsburgh, Pa.
[89] Demo, J. J., unpublished work, E. I. du Pont de Nemours and Co., Inc., Wilmington, Del.
[90] Schmidt, W. and Jarleborg, O., *Ferritic Stainless Steels with 17% Cr,* Climax Molybdenum GMBH, Dusseldorf, Germany, an affiliate of American Metal Climax, Inc., 1974.

Study of the isothermal transformations in 17% Cr stainless steels

Part I. Nature and morphology of the transformation products

RENÉ CASTRO and ROLAND TRICOT*

THE MOST COMMON of the industrial stainless steels is undoubtedly that containing 15–17% chromium and 0·05–0·1% carbon. Nevertheless, not a great deal has been done in regard to the investigation of the metallography and transformations of this steel, the best-known work being that of Nehrenberg and Lillys.[1]

It is known that this steel consists at room temperature and in the equilibrium state, of ferrite and carbides, but when heated to 850–900°C it undergoes a partial, austenitic transformation. This transformation develops progressively up to a maximum at about 1,100°C, after which it regresses and at about 1,300°C a single, homogeneous phase is again obtained, which remains stable into the solidus. These high-temperature transformations are the cause of the double loop separating the gamma and alpha (or delta) regions in the Fe-Cr. state diagram, the characteristic point of the composition in question passing, depending on the presence of carbon, through a two-phase region of ferrite + austenite, between the two loops.

We have investigated, in this first part of the paper, the isothermal decompositions appearing at decreasing holding temperature: firstly starting at the maximum-austenite temperature of about 1,100°C; and secondly, from the temperature of pure ferrite, *i.e.* at or above 1,300°C.

For this purpose, two melts were prepared, having the following analysis:

	C	Cr	Si	Mn	Ni	N	S	P
Melt A	0·045	17·3	0·45	0·61	0·15	0·028	0·011	0·017
Melt B	0·080	15·7	0·45	0·64	0·12	0·025	0·010	0·016

It will be seen that these lie towards the opposite ends of the range of industrial specifications, as regards the balance of the alpha-promoting (chromium) and gamma-promoting (carbon) elements, this was done in order to obtain at 1,100°C the widest possible difference in the austenite content; the other additives and impurities are practically identical.

These steels were investigated by a number of different experimental techniques:

(*a*) Conventional and phase-contrast micrography;

(*b*) Monocular and stereoscopic electron micrography;

(*c*) Dilatometry;

(*d*) Thermo-magnetometry;

(*e*) Thermo-resistometry;

(*f*) X-ray or electron diffraction;

(*g*) Microanalysis by Castaing probe;

(*h*) Deformation tests by hot-drawing[2];

(*i*) Mechanical strength and hardness tests.

Temperature-development of the austenitic-ferritic equilibrium structure

Cross-checking by different measuring techniques has enabled the A_1 transformation temperature to be determined for heating from:

*Société d'Electrochimie, d'Electrométallurgie et des Aciéries électriques d'Ugine Savoy. Communication to the Autumn Meeting of the Société Française de Metallurgie, Paris, 1961. *Memoires Scientifiques*, **59** (9), 1962.

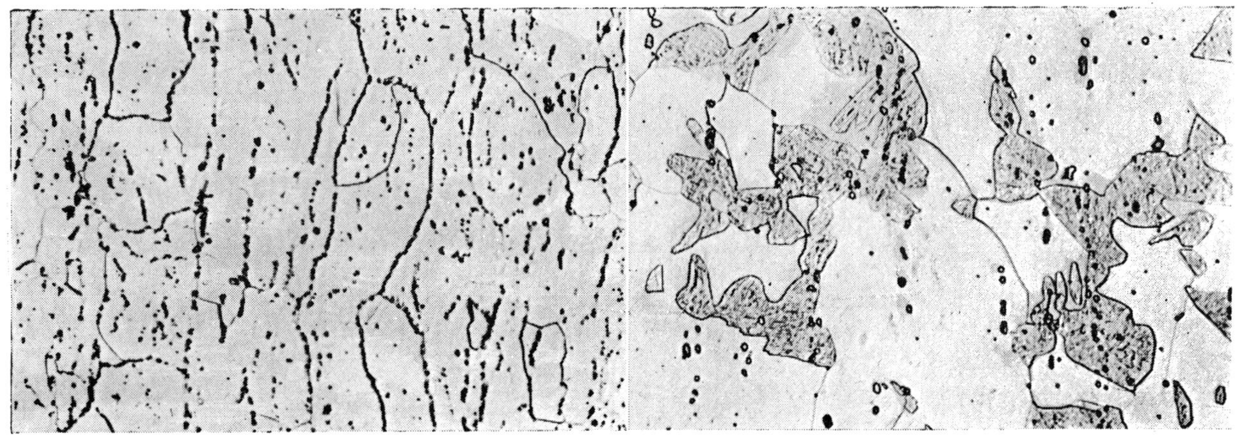

1 *Ferrite + carbide structure after annealing at 800°C Melt A* × 500

2 *Structure after water-quenching at 900°C, 2 h. Undissolved carbides in α and γ* × 500

875°C for Melt A, with the highest ferritic tendency;

840°C for Melt B, with the least pronounced ferritic tendency.

The Curie point of the ferrite was found at 675°C and 695°C for melts A and B respectively.

Fig. 1 shows the starting structure (annealed at 800°C), consisting of a ferritic matrix and coalescent carbides. Beyond the A_1 point, austenite appears, and up to 50° above the A_1 point, relatively little carbide is dissolved (fig. 2); this primary austenite is 0·7 to 1·2% lower in chromium compared with the contiguous ferrite (table 1), as determined by microanalysis.

Fig. 3 shows, for the two melts A and B, the variations in the percentages of the phases in equilibrium against the temperatures. The quantitative evaluation of the phases has been performed by the method of linear analysis described by Howard and Cohen.[3] The quantity of austenite increases progressively from the A_1 point towards 1,050–1,100°C, where it reaches its peak of about 27% for Melt A, to 60% for Melt B. Between 950 and 1,000°C, more rapid dissolution of the carbides in the ferrite is observed (fig. 4), probably owing to the fact that, in spite of reduced solubility of the carbon in this phase, the speed of diffusion of this element is relatively much greater in this case than in austenite. It is necessary to heat practically to 1,050–1,100°C in order to obtain complete solution of the carbides.

As the temperatures increase above 1,100°C, the quantity of austenite decreases progressively (fig. 3). At such high tempering heats, it is impossible to prevent a partial regression of the ferrite to austenite, but the latter is easily differentiated from the equilibrium austenite by its acicular aspect.

Finally, the austenite disappears completely, at a temperature which we have designated the A_5 point, located respectively at 1,250 and 1,350°C for melts A and B. The delta ferrite is the only stable phase between A_5 and the solidus (about 1,450°C for the two melts). Quenching from pure ferrite, causes the appearance of inter- and intragranular austenite in Widmanstätten patterns (fig. 6) which is unavoidable even by energetic quenching of small samples. The ferrite is heavily

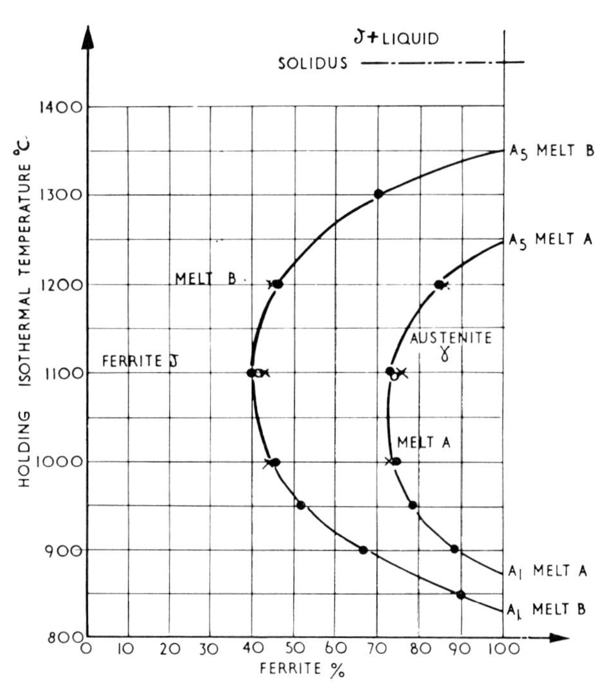

3 *Ferrite/austenite percentages in equilibrium against temperature*
● *experimental points by heating and holding*
× *experimental points by isothermal decomposition. Melt A: 1,300° 5 min→1,000 or 1,100 or 1,200° 1 h, water. Melt B: 1,400° 5 min →1,000 or 1,100 or 1,200° 1 h, water*
○ *experimental points by isothermal decomposition*
Melt A: 1,225° 10 min →1,100° 1 h, water
Melt B: 1,275 10 min → 1,000° 1 h, water

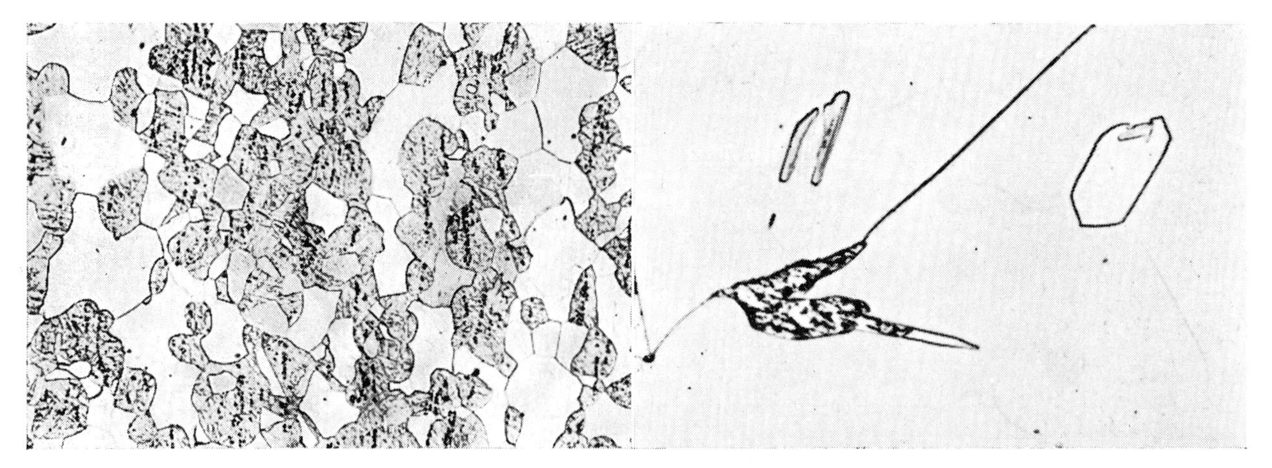

4 *Undissolved carbides in austenite after water-quenching at 950°C, 2 h, Melt B* × 500

5 *TE* 1,225°*C* 30 *min Melt A* × 1,000

veined in the vicinity of the austenite (transformed into martensite with residual austenite at the edges (fig. 7), and in its largest areas, copious precipitation (fig. 8) of carbides, $M_{23}C_6$ and nitrides Cr_2N in needles is observed identifiable in extraction prints. This precipitation disappears in the vicinity of the austenite; this phenomenon of the so-called ' white zone ' is probably due to the very rapid diffusion of the C and N atoms towards the austenite, during cooling.

So long as the proportion of austenite remains appreciable, the size of the ferrite grains only changes slowly with increasing holding temperature; but, in the neighbourhood of the A_5 point and particularly when this point is passed, grain-growth becomes very rapid. This is particularly apparent along any decarburization gradient.

The chromium and iron contents of the two phases—ferrite and austenite—in equilibrium at different temperatures—have been determined by microanalysis (table 1). The ferrite is higher in chromium than the austenite, the difference attaining 2% and more. The equilibrium diagrams of the Fe-Cr-C ternary system[4] similarly show that the ferrite is lower in carbon than the austenite (nearly 0·02% C for the ferrite, against 0·12% for the austenite). In tempering from temperatures above the A_5 point, the composition of the austenite formed during cooling, is substantially the same as the ferrite (table 1).

6 *TE* 1,400°*C* 5 *min Melt B* × 200 **7** *TE* 1,400°*C* 5 *min Melt B* × 1,000 **8** *TE* 1,300°*C* 5 *min Melt A* × 500

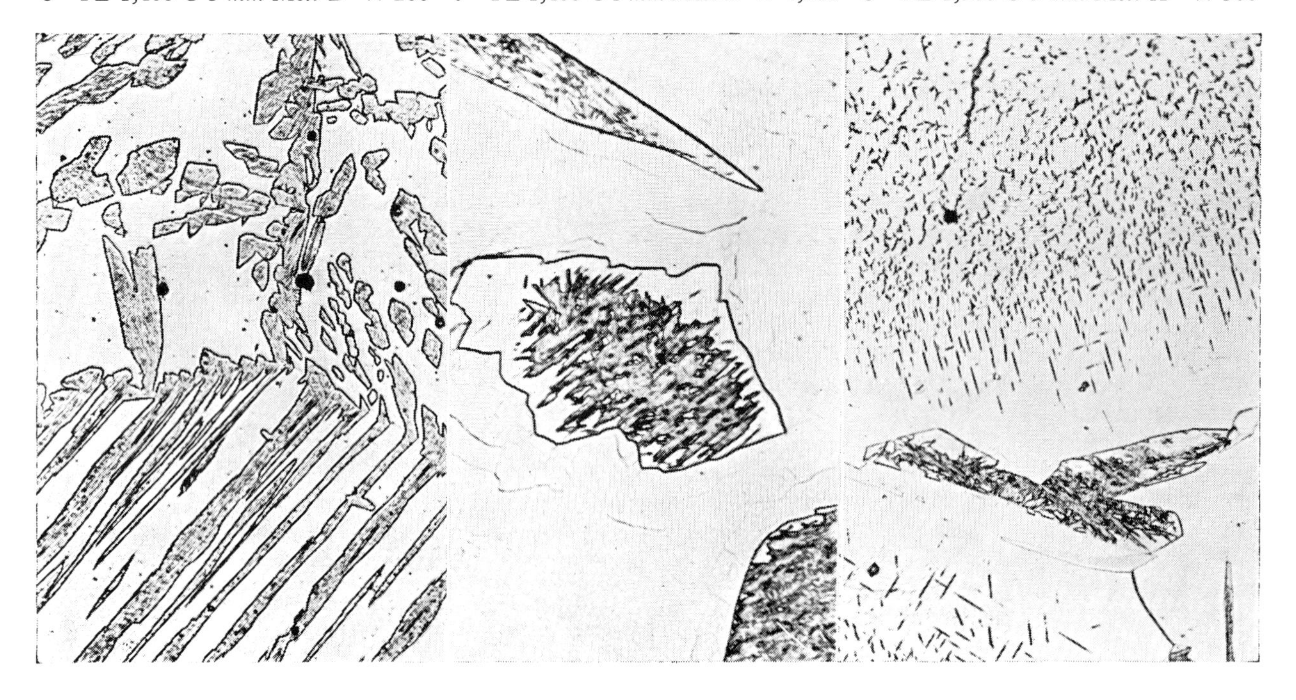

TABLE 1 *Distribution of the alloy constituents in the ferrite and austenite*

Holding temperature (°C)	Melt A				Melt B			
	Austenite		Ferrite		Austenite		Ferrite	
	% Fe	% Cr	% Cr	% Fe	% Fe	% Cr	% Cr	% Fe
875	—	—	—	—	84·3	14·5	15·25	83·55
900	83·9	15·9	17·1	81·7	83·8	15·0	15·7	83·1
1,000	82·2	16·6	18·0	80·8	83·8	15·0	17·2	81·6
1,100	82·4	16·4	18·0	80·8	83·5	15·3	16·8	82·0
1,200	82·4	16·4	17·8	81·0	83·7	15·1	16·7	82·1
1,300	structure 100%				84·1	14·7	15·7	83·1
1,400	ferrite δ				83·4*	15·4*	15·5*	83·3*

Mean of 10 points per result, accuracy ± 2%
*Structure entirely ferritic at 1,400°C: the austenite formed during cooling has the same composition as the delta-ferrite.

The curves of deformability in hot drawing, are shown in fig. 9. This shows the effects of:

(*a*) The appearance and subsequent disappearance of the austenite in the ferrite at rising temperatures; Melt B, which forms more austenite at rising temperature than Melt A, has a more pronounced forgeability minimum.[5]

(*b*) Precipitation embrittlement in the ferrite and in the Widmanstätten-shaped austenite at falling temperatures.[6, 7]* The shift between the curves obtained at rising temperature and those obtained with a previous holding at high temperature, partly corresponds to the thermomechanical embrittlement produced by the precipitation of carbides and nitrides, during the actual process of tensile deformation.

(*c*) The grain-growth embrittlement at falling temperature. Preceding holding at 1,250 or 1,350°C obviously increases the grain size of the delta ferrite and is a supplementary embrittlement factor, causing a shift between the curves at rising and falling temperatures respectively.

Structural components in the isothermal decomposition of an initial two-phase, ferrite + max. austenite, obtained at 1,100°C

TTT curves

Fig. 10 shows the diagram of isothermal decomposition for Melt A, plotted for a starting temperature of 1,100°C, at which there is a maximum proportion of austenite in the two phase, gamma + delta structure, with complete dissolution of the carbides. The TTT curves have the form of a simple C. They have been essentially determined by micrography, dilatometry, and hardness measurement. The following curves have been plotted:

(1) For the start of carbide precipitation;

(2) For the start and finish of the austenite-ferrite + carbide transformation, with curves of the intermediate development.

(3) From the start of the appearance of the D-aggregate and the pearlite (the nature of these constituents will be discussed presently).

The B melt furnishes approximately similar curves (fig. 11). To define their differences, it may be indicated that the peak of the end of the $\gamma \rightarrow \alpha + C$ transformation is found at 690°C for Melt A, with complete transformation in 50 min; and at 720°C for Melt B, with completion in 25 min.

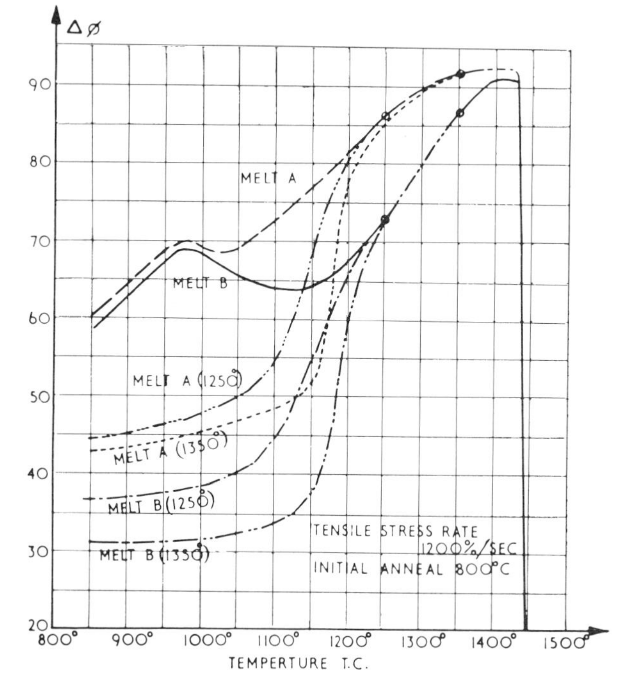

9 *Forgeability curves*
θ *rising* θ *falling after* 1 *min. at*

	θ rising	1,250°	1,350°
Melt A	— — —	—···—···—	------------
Melt B	———	— · — · —	—··—··—

Micrographic investigation

The micrographic aspects described here, have been observed equally in each of the melts investigated, for which they are qualitatively identical.

(a) *Decomposition at* 1,000°*C*. The proportion of austenite varies little, even over long holding periods. Some coarse carbides, in small number, appear at the gamma-delta joints and in the austenite (figs. 12 and 13). The start of regression of the austenite to alpha ferrite in the form of intrusions originating at the gamma-delta boundaries (figs. 12 and 13) is

*It should be remembered that the forgeability curves 'at rising temperature' are obtained by heating the sample from room temperature to the test temperature θ and then applying the tension. The curves 'at falling temperature' are on the contrary obtained by first bringing the sample to a temperature H > θ and then cooling from H to θ before applying tension. In the examples given, H has been chosen at 1,250 or 1,350°C.

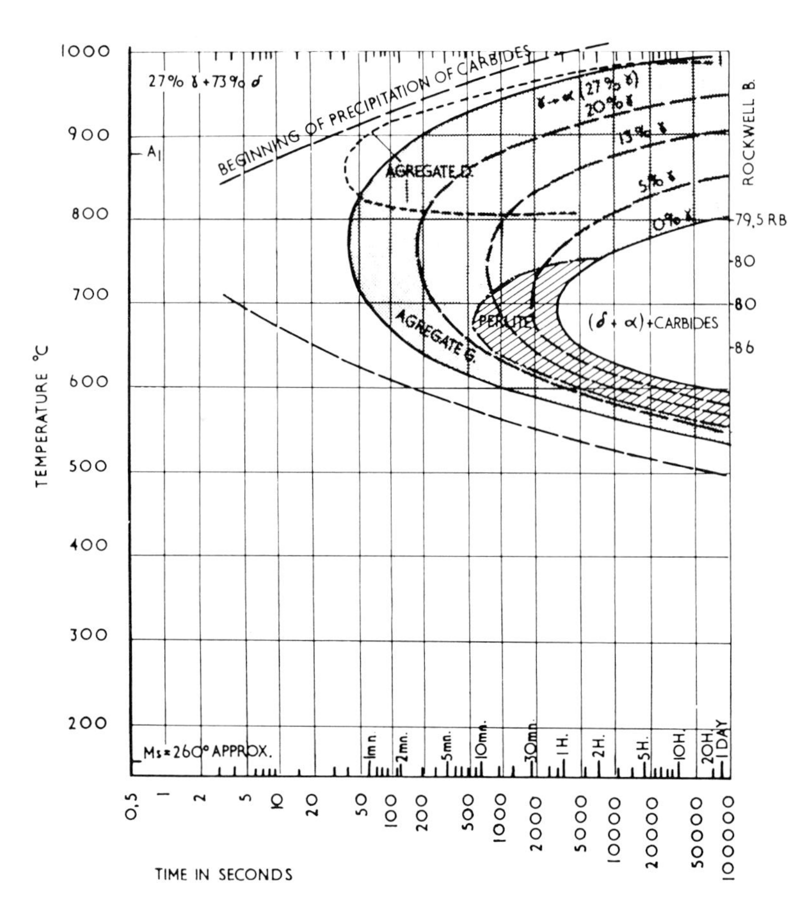

10 *TTT curves for Melt A*

Analysis	C	Si	Mn	Ni	Cr
	0·045	0·455	0·610	0·116	17·3
	N	O	S	P	
	0·028	0·016	0·011	0·017	

Austenitized at: 1,100°, 30 min.

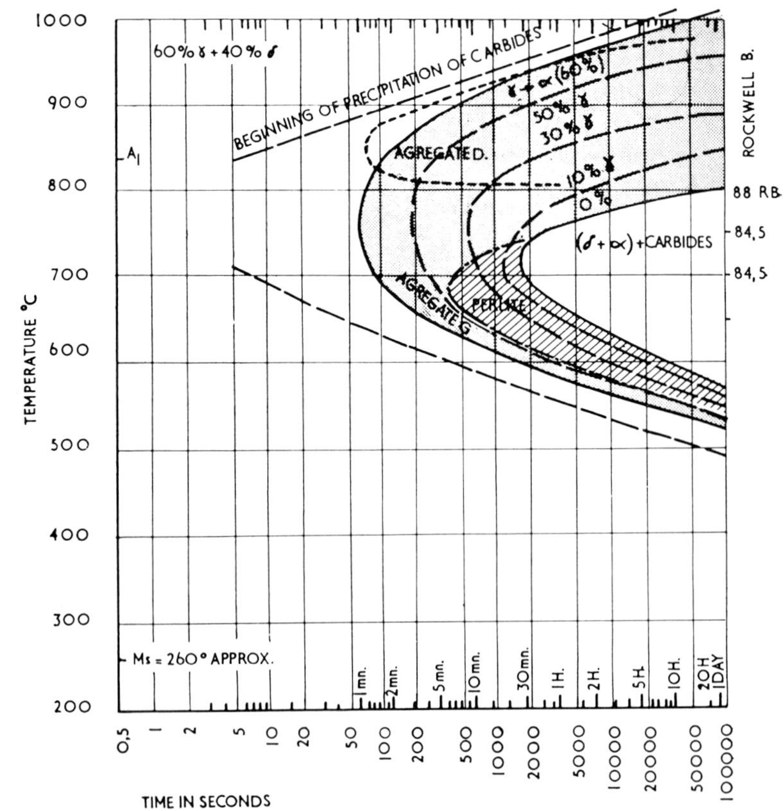

11 *TTT curves for Melt B*

Analysis	C	Si	Mn	Ni	Cr
	0·080	0·45	0·64	0·120	15·7
	N	O	S	P	
	0·025	0·020	0·010	0·016	

Austenitized at: 1,100°, 30 min.

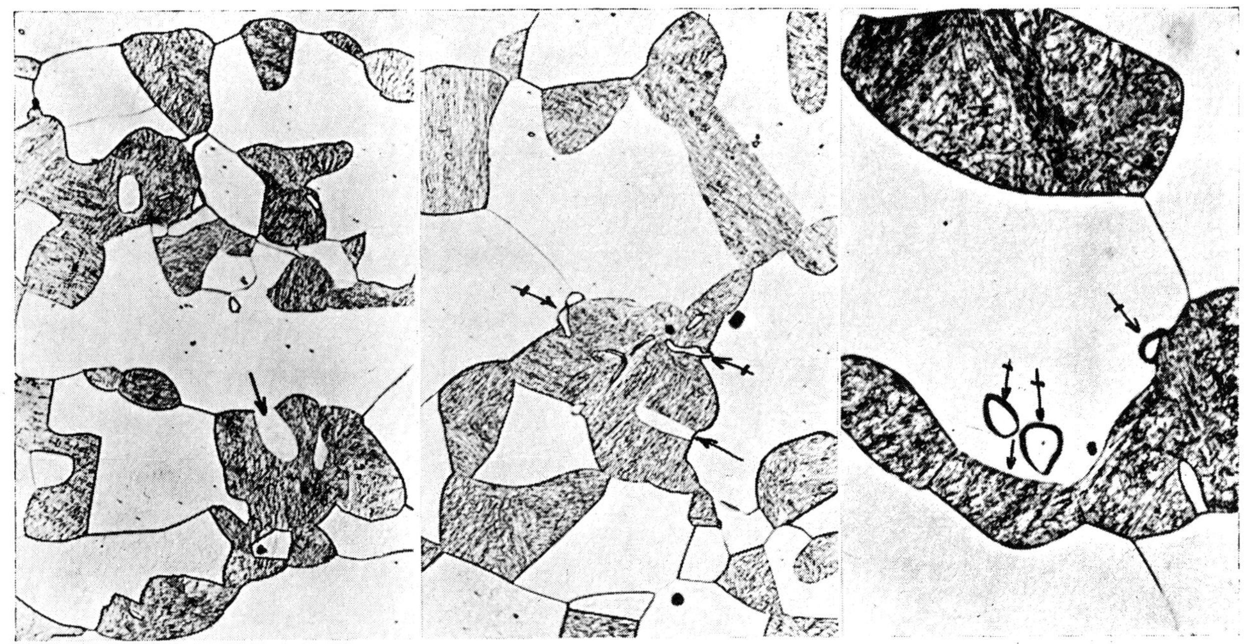

12 1,100°→1,000°*C* 4 *h, Melt A* × 500

13 1,100°→1,000°*C* 24 *h, Melt A* *Carbide regression* × 500

14 1,100°→1,000°*C* 24 *h, Melt A* *Carbides in ferrite by regression* ×1,000

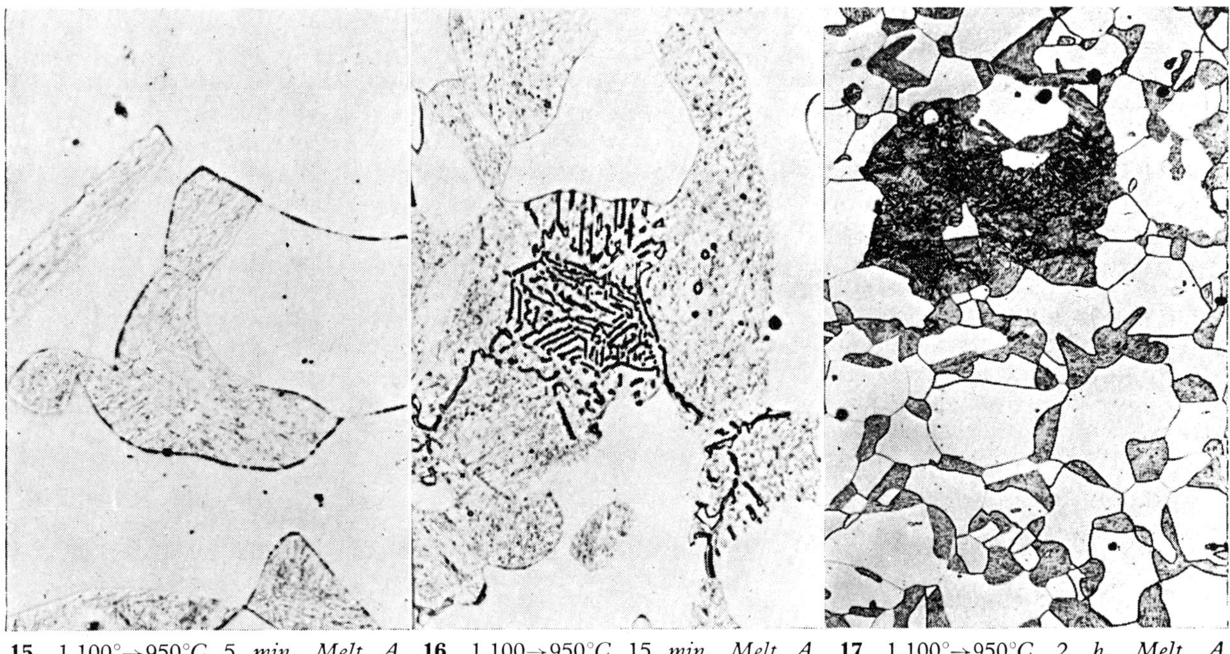

15 1,100°→950°*C* 5 *min, Melt A* *Intergranular carbides* × 1,000

16 1,100→950°*C* 15 *min, Melt A* *D-aggregate* × 1,000

17 1,100°→950°*C* 2 *h, Melt A* *Nodular D-aggregate* × 200

observed, the grain boundaries between the gamma and the alpha being little marked by attack. With progress of the gamma-alpha transformation, the carbides coalesced by long holding, are then found in the alpha-ferrite (fig. 14).

(b) *Decomposition at* 950°*C.* A fine, intergranular, carbide precipitate is observed after 5 min holding at the gamma-delta and delta-delta interfaces (fig. 15), while the quantity of austenite has not yet changed.

After about 15 min, a lamellar or rod-shaped aggregate appears in the delta-ferrite at the boundary of a gamma-delta interface, materialized by intergranular carbides (fig. 16). It consists of

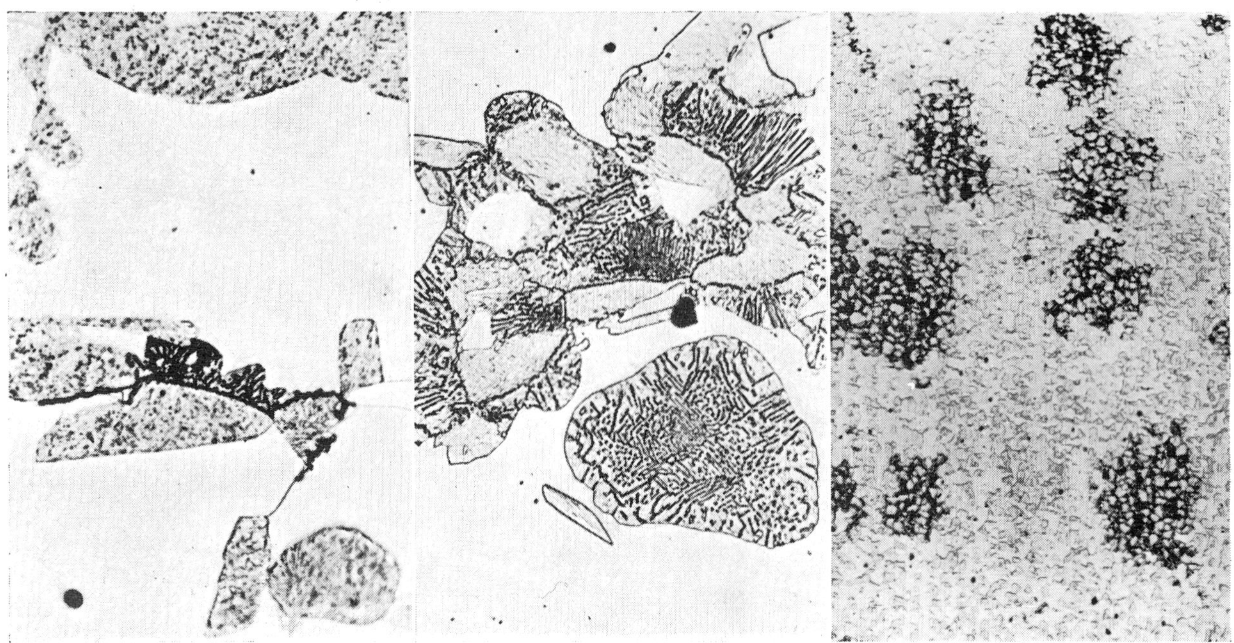

18 1,100°→900°C 1 *min, Melt A* D-aggregate × 1,000 **19** 1,100°→900°C 1 *h, Melt A* D-aggregate × 500 **20** 1,100°→900°C 4 *h, Melt B* Distribution of nodules of D-aggregate × 50

lamellar $M_{23}C_6$ carbides and γ' austenite, which becomes transformed into martensite during cooling (the designation as γ' austenite* is here applied exclusively in order to distinguish it from the initial, gamma-austenite).

ferrite $\delta \rightarrow M_{23}C_6$ carbides + γ' austenite . . .(1)

This eutectoid aggregate principally investigated by K. Kuo[8–12] has been called by the latter, the *delta-eutectoid.* For convenience of notation, we have preferred the term ' D-eutectoid ' or *D aggregate* to prevent the same letter from having to designate, simultaneously, the phase and the association of the decomposition products thereof.

Simultaneously, the reaction

gamma-austenite → alpha-ferrite . . . (2)

commences, the alpha-ferrite patches grow at the expense of the gamma, in acicular form or as gulfs.

The formation of the D-aggregate takes place in nodules distributed fairly regularly at a distance of about one millimetre (fig. 17). After two hours of holding the temperature, the volume of the D-aggregate has hardly changed, amounting to 3–4% of the structure, while outside the nodules, the quantity of austenite has regressed in accordance with reaction (2), and then becomes stable at about 18%. There is, thus, a *double reversion* of the crystal structures, with disappearance of the delta-ferrite and the appearance of alpha-ferrite, reaction (1) being accompanied by a heterogeneous distribution of the carbon.

(c) *Decomposition at* 900°*C.* The phenomena are, as a whole, identical. The D-aggregate, appearing earlier and with a smaller inter-lamella spacing than at 950°C, is preferentially formed on the intergranular carbides, gamma and delta, of the first deposit (fig. 18)† which causes the carbon distribution to become very markedly heterogeneous, the areas of the D-aggregate being spaced between 2 and 0·4 mm apart (fig. 20).

Outside these areas, the regression of the austenite to alpha-ferrite, which starts after 10–15 min, continues and contributes to the development of a heterogeneous structure by the increase in size of certain of the ferrite grains (fig. 21). This regression which never affects the γ' ferrite of the D-aggregate, becomes arrested and after 60 h there remained about 10–12% of $\gamma + \gamma'$ austenite, in Melt A.

Figs. 22 and 23 show under the electron microscope, the structure of the D-aggregate. The carbides, in the form of flat, and often hollow, small rods, appear first in the γ' during the lateral spread of the D-aggregate (fig. 22). This development, already mentioned by Kuo, shows that the carbide is the so-called ' governing constituent ' in reaction (1).

(d) *Decomposition at* 850°*C.* The phenomena are similar. However, the areas of D-aggregate are smaller and more numerous, with shorter and more closely packed lamellae, and no longer form after

*According to K. Kuo.
†Later developing into areas joining the austenite nodules which have remained intact (fig. 19).

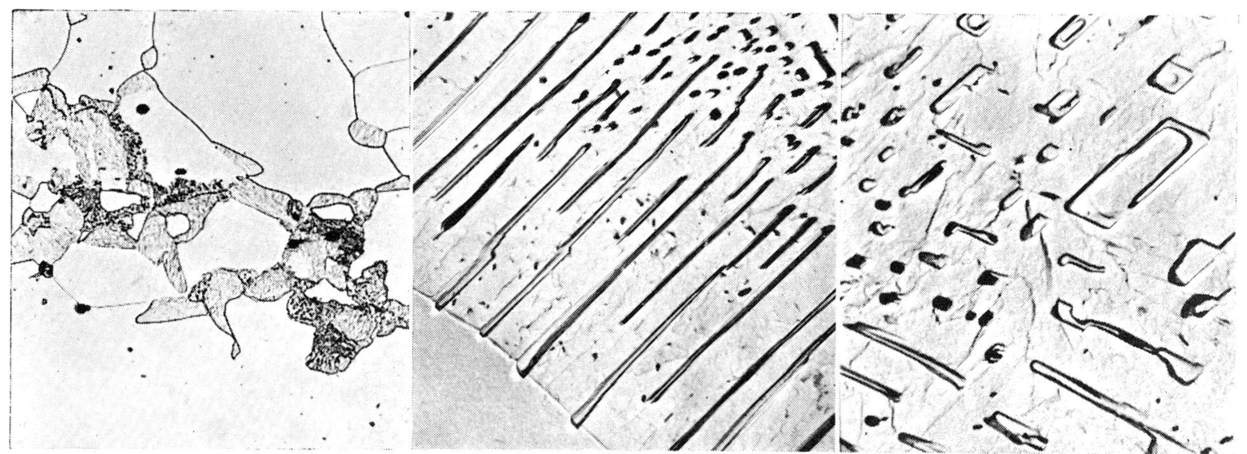

21 1,100°→900°*C* 24 *h*, *Melt A Nodular D-aggregate; elsewhere regression* × 200

22 1,100°→900°*C* 4 *h*, *Melt A D-aggregate* × 6,000

23 1,100°→900°*C* 4 *h*, *Melt A D-aggregate* × 6,000

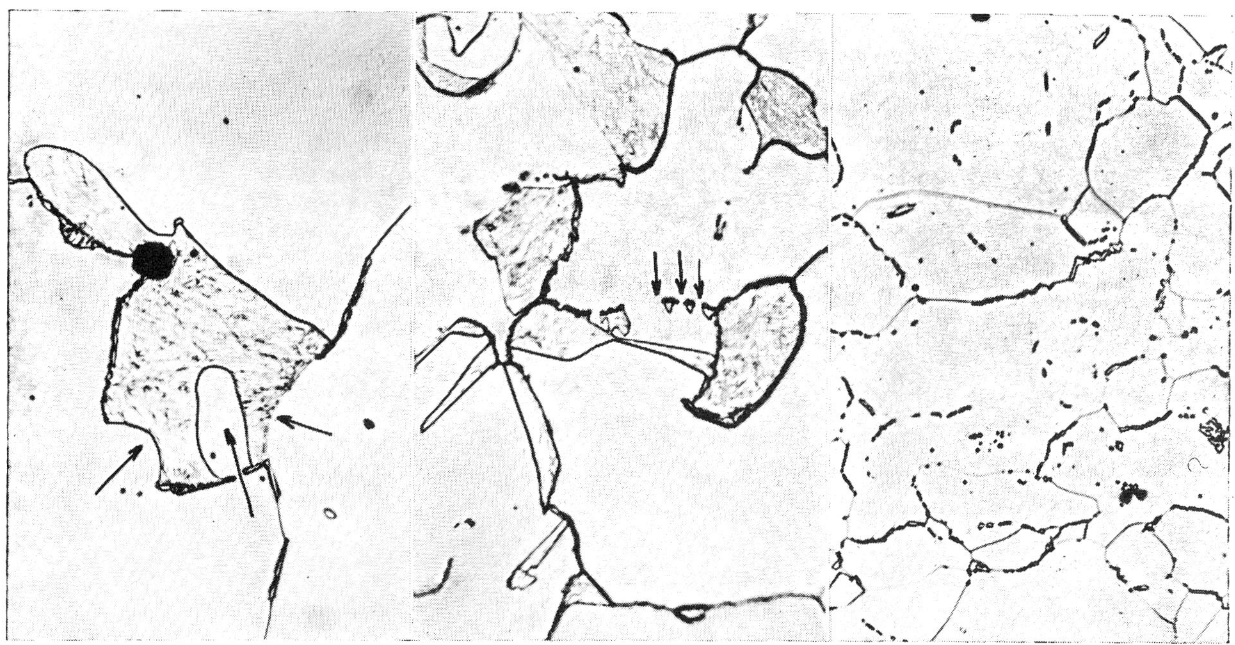

24 1,100°→850°*C* 4 *h*, *Melt A Aspect of regression impeded by intergranular carbides* × 1,000

25 1,100°→800°*C* 10 *min*, *Melt A Intergranular carbides and regression* × 1,000

26 1,100°→800°*C* 24 *h*, *Melt A Regression completed* × 500

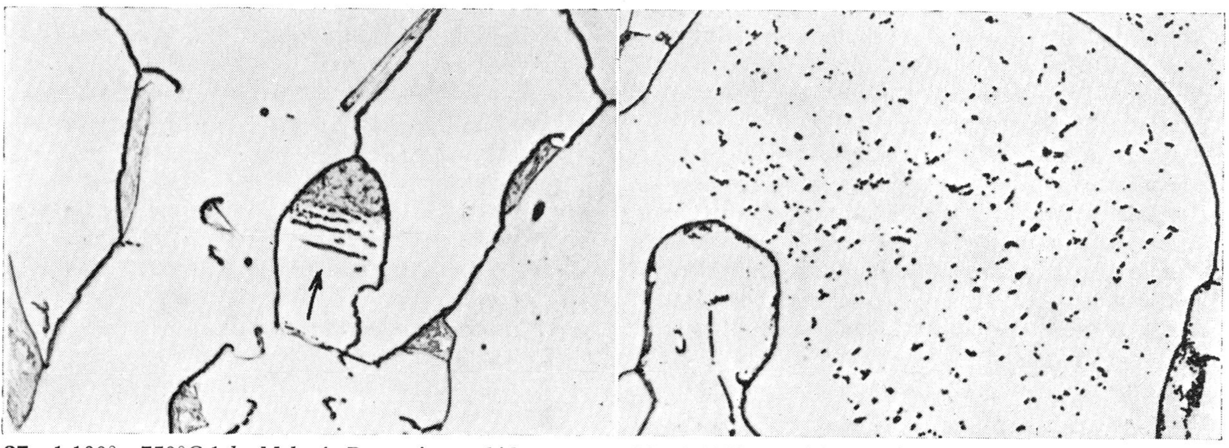

27 1,100°→750°*C* 1 *h*, *Melt A. Regression carbides* × 1,000

28 1,100°→700°*C* 10 *min*, *Melt A. Precipitation in the delta-ferrite* × 1,000

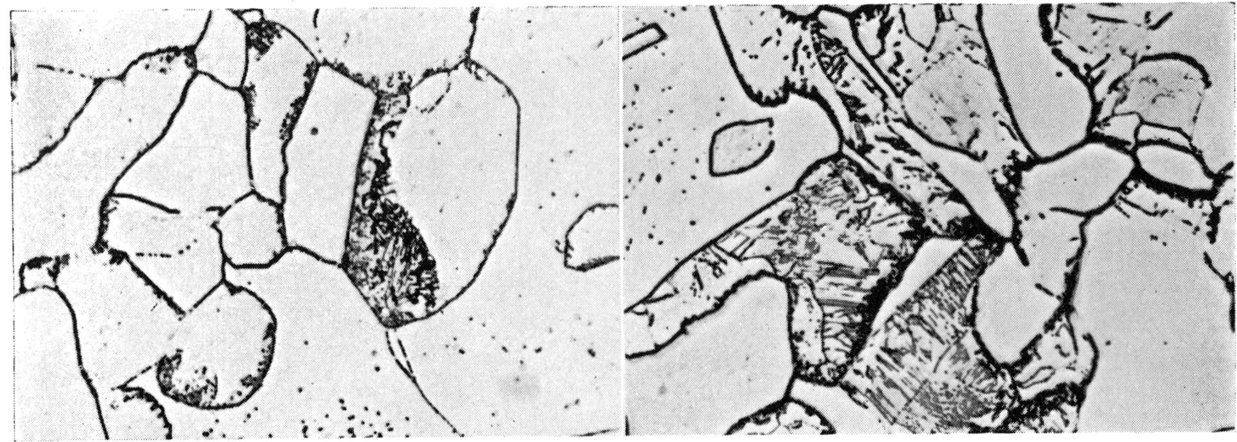

29 1,100°→700°*C* 10 *min, Melt A. G-aggregate and pearlite (etchant electrolytic chromic acid)* × 1,000

30 1,100°→700°*C* 30 *min, Melt A. G-aggregate and pearlite (etchant Vilella)* × 1,000

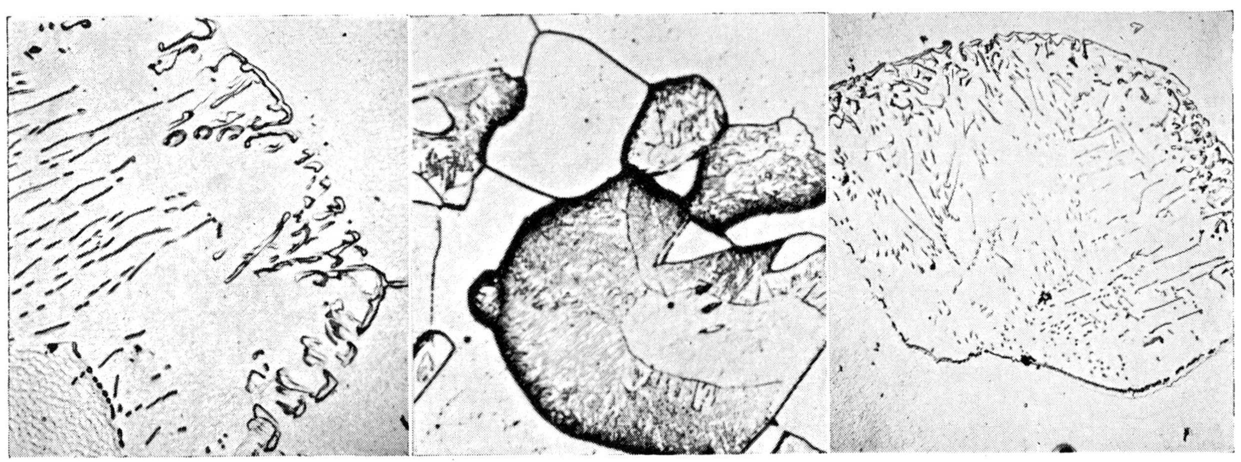

31 1,100°→700°*C* 30 *min, Melt A G-aggregate, pearlite, and regression* × 6,000

32 1,100°→650°*C* 30 *min, Melt A G-aggregate and pearlite (etchant Vilella)* × 1,000

33 1,100°→650°*C* 30 *min, Melt A G-aggregate and pearlite* × 6,000

about 20 min. The gamma-alpha regression, similarly starting earlier, most usually proceeds by shifting of the gamma-delta interface; it becomes retarded when the interface is defined by a line of carbide deposits (fig. 24) and is most frequently observed in the form of gaps or gulfs starting from the breaks in this deposit.

(e) *Decomposition at* 800°*C*. At this temperature, precipitation of carbides at the gamma-delta and delta-delta interfaces, in the form of a continuous network, is exceedingly rapid. While at 825°C the D-aggregate still forms, and is prematurely checked, this is no longer observed at 800°C: on the contrary, the regression phenomena begin earliest, and attain their maximum speed (fig. 25). After 24 h, the transformation to alpha-ferrite and carbides, is complete.

(f) *Decomposition at* 750°*C*. After a very rapid and continuous, initial precipitation of intergranular carbides, only a gamma-alpha regression is observed. However, the rate of diffusion of the carbon and its limited solubility in austenite and ferrite, begin to weaken this regression which is now accompanied by a periodical deposition of carbides appearing in different places along the front of the gamma-alpha transformation, and producing a pseudo-pearlitic appearance (fig. 27).

(g) *Decomposition at* 700°*C*. In addition to the, always early, intergranular precipitation of carbides, there begin to appear fine, intragranular precipitates of carbides and nitrides in the coarse grains of delta-ferrite (fig. 28)—the appearance does not change by long holding periods.

In addition, on the boundaries of the austenitic areas, the development is observed, during short (1–3 min) holding periods, of a very fine and granular aggregate which we shall term the G-aggregate. The growth of this aggregate, of limited

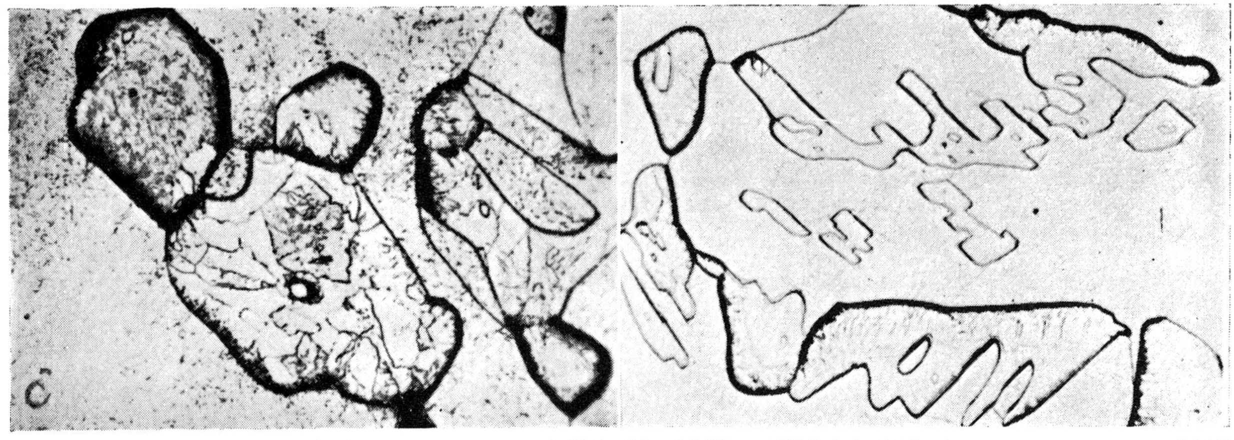

34 1,100°→600°*C* 10 *h*, *Melt A* × 1,000 **35** 1,100°→600°*C* 4 *h*, *Melt A* × 1,000

extent, ceases after about 10 min while the remaining austenite begins to transorm either into alpha-ferrite by regression, or into an aggregate of a pearlitic nature (figs. 29, 30). Decomposition by these two processes in combination is completed within an hour. It should be noted that at 725°C no G-aggregate forms in either melt, while the pearlite begins to appear after about 15 min holding time.

The G-aggregate does not stain so well as the pearlite by electrolytic reagents (chromic acid, fig. 29), while acid reagents (' Vilella,' fluopicrate) stain it more heavily than the pearlite (fig. 30). Fig. 31 shows the characteristic appearance of the G-aggregate under the electron microscope, with its barbed elements as well as the fine lamellae of pearlite.

The G-aggregate has been identified as consisting of $M_{23}C_6$ carbides and ferrite. Without going into detail, it may be indicated that pearlite, as here defined in accordance with authors like Nehrenberg and Lillys,[1] is in reality a lamellar aggregate of alpha-ferrite and M_2N nitride: in other words, a ' nitrogen-pearlite.' This will be investigated in more detail in a second paper.

(h) *Decomposition at* 650°*C.* The phenomena are analogous to those developing at 700°C, with a more continuous and larger border of G-aggregate, and a finer pearlite structure (figs. 32 and 33).

(i) *Decomposition at* 600°*C.* The kinetic mechanism is in this case slower and the structures more confused by reason of the extensive precipitation of carbides in the delta- and alpha-ferrites (fig. 34). The residual austenite transformed into martensite, becomes less and less susceptible to staining and the gamma-alpha regression can develop more particularly in preferential directions (fig. 35).

(j) *Decomposition below* 600°*C.* Below 600°C, decomposition is very slow, and much restricted. At 550°C, 24 h holding time, only a fringe of G-aggregate and incipient regression, are observed. On the contrary, the fine precipitation of carbides and nitrides in the delta-ferrite, continues to be visible, at 550, 500 or 450°C.

The M_S temperature is found at 155°C for Melt A, and 260°C for Melt B; but, if the austenite has been impoverished in carbon by precipitation of carbides, particularly prolific at low holding temperatures, the M_S point may rise considerably and exceed 400°C for Melt B.

(Continued on the next page)

Study of the isothermal transformations in 17% Cr stainless steels

Nature and morphology of the transformation products

Continued

RENÉ CASTRO and ROLAND TRICOT*

Structural constituents of isothermal decomposition of pure delta ferrite obtained at 1,300–1,400°C

FOR the study of the isothermal decomposition of a single and homogeneous phase, delta-ferrite, the starting temperature has been selected as $A_5 +$ 50°C, *i.e.* 1,300°C for Melt A, and 1,400°C for Melt B.

The structures obtained by quenching from the austenite forming during the rapid passage through A_5–A_1 of the gamma-loop (figs. 6, 7, 8), have already been shown earlier in the paper.

Decomposition at 1,200, 1,100 or 1,000°C causes rapid formation of austenite, which after 20 min to 1 h holding time attains the same percentage as in the equilibrium curve (fig. 3) obtained by heating. The precipitation of carbides and nitrides in the delta-ferrite with appearance of the ' white zone ' (fig. 8) during cooling between 1,300 and 1,400°C and the isothermal holding temperature, and is visible after holding for 5 min at the latter temperature, disappears at longer holding times, the carbon then diffusing towards the austenite areas then forming.

The aggregates obtained by decomposition, are qualitatively of the same character. Their arrangement is, however, influenced by the considerable initial growth of the ferritic grain and the rapid formation of inter- and intragranular austenite in Widmanstätten patterns, during cooling.

Curiously, the formation of the D-aggregate is found to take place essentially at contact of the austenite in the intergranular regions, at only one side of the initial grain boundaries of the delta-ferrite (figs. 36*a* and *b*).

The regression phenomena $\gamma \rightarrow \alpha$, which as a whole appear to develop more slowly, proceed in general along oriented fronts (fig. 37). Intergranular precipitation of the carbides is more marked and rapid at the gamma-delta interfaces of the intergranular austenite.

Since the delta grain is very coarse, the precipitates of carbides and nitrides in the delta-ferrite, at some distance from the grain boundaries, are very abundant for holding treatments up to 750°C (fig. 38). Visible for short holding times at temperatures exceeding 750°C; they do not form at such temperatures with holding times of 1–24 h, in consequence of the diffusion of carbon and nitrogen towards the austenite.

The distribution of the G-aggregate appears to follow the same rules as for the D-aggregate (figs. 39*a* and *b*). After formation of the G-aggregate, the austenite becomes in turn transformed into alpha-ferrite with eventual liberation of carbides marking the regression front; or into pearlite. Finally, at the lowest holding temperatures ($\leq$ 650°C), marked veining of the delta-ferrite, such as follows on direct quenching, is again observed (fig. 40).

---

*Société d'Electrochimie, d'Electrométallurgie et des Aciéries électriques d'Ugine/Savoy. Communication to the Autumn Meeting of the Société Française de Metallurgie, Paris, 1961. *Memoires Scientifiques*, **59** (9), 1962.

36a 1,300°C 5 *min* → 900°C 5 *min. Melt A. Distribution of D-aggregate.* × 100 (*reduction of* ⅓ *in reproduction*)

39a 1,400 → 700°C 5 *min. Melt B. Distribution of G-aggregate.* × 100 (*reduction of* ⅓ *in reproduction*)

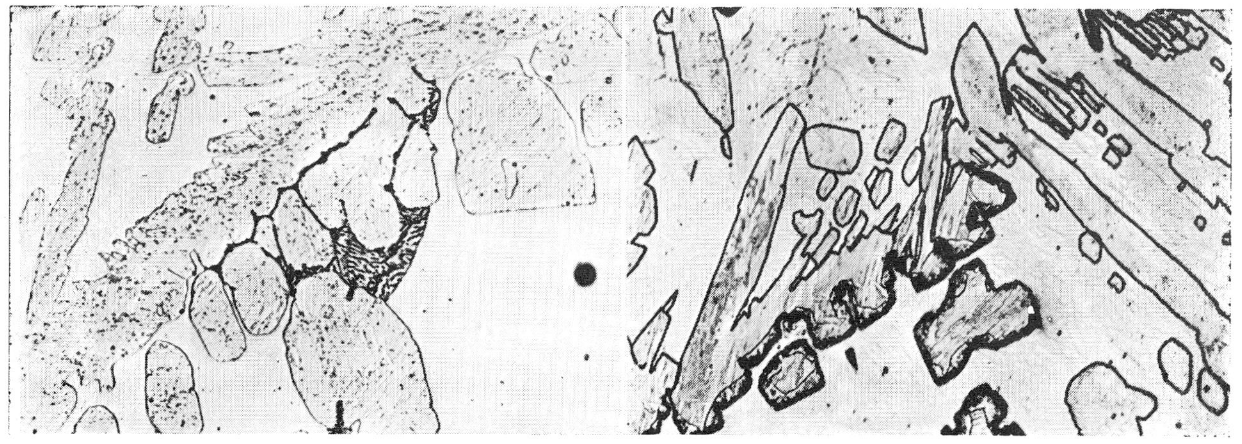

36b 1,400°C 5 *min* → 900°C 5 *min. Melt B. D-aggregate on intergranular austenite.* × 500

39b 1,400 → 700°C 5 *min. Melt B. G-aggregate.* × 500

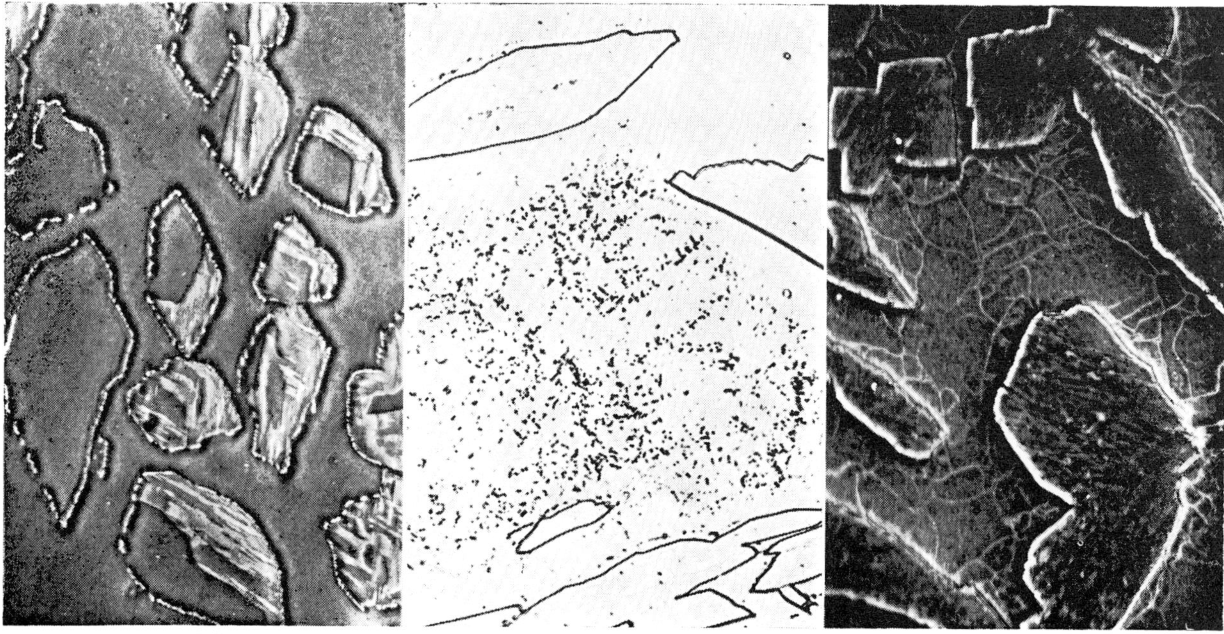

37 1,400 → 800°C 4 *h. Melt B. Regression (phase contrast)*

38 1,400 → 750°C 1 *h. Melt B. Precipitation in delta-ferrite.* × 500

40 1,400 → 650°C 1 *min. Melt B. Veining in delta-ferrite.* (*Phase-contrast*) × 500

DISCUSSION

Precipitation of carbides

(a) *Intergranular carbides.* The precipitation of intergranular carbides which precedes the transformations at all temperatures is so rapid in the range of 800–750°C that it cannot be avoided even by water-quenching small samples from 1,100°C. Their examination shows these to be $(Cr, Fe)_{23}C_6$ carbides with a Cr:Fe ratio of about 2 at 800°C, as determined by microanalysis with extraction replicas according to the IRSID method.[13] The distribution and morphology of these carbides must thus be governed by the rate of nucleus formation on the one hand, the rates of diffusion of C and Cr, and the degree of carbon supersaturation in the austenite and ferrite, on the other hand. It would thus appear that the delta-delta, and particularly the gamma-delta interfaces, are the favourable locations for such precipitation.

(b) *Intragranular precipitation in delta-ferrite.* The intragranular precipitation of carbides and nitrides in the delta-ferrite, observed in isothermal decomposition at under 750°C and in quenching from 1,300–1,400°C, is similarly rapid and depends little on the holding time. This aspect is characteristic for rapid precipitation *in situ* by lowered solubility of carbon and nitrogen in the delta-ferrite; the disappearance of this precipitation in contiguity with the austenite disclosing the limited diffusion of these elements towards the austenite.

Formation of the D-aggregate

Transformation of delta-ferrite into D-aggregate

$$\delta \rightarrow \text{carbides } M_{23}C_6 + \gamma' \ldots (1)$$

develops in the 950–825°C range from the $\gamma - \delta$ interface. The identification of lamellar carbides shows that these are $(Cr, Fe)_{23}C_6$ with a Cr : Fe ratio of about 2·3. The γ' austenite is transformed into martensite during the cooling. This reaction shows many points in common with the pearlitic transformation of Fe-C alloys:

(*a*) The incubation period is very short—less than 1 min at 900–850°C. The curve of the start of the appearance of the D-aggregate forms a simple 'C' the two arms of which appear to have an asymptote at about 1,000–1,050°C for the upper, and 800°C for the lower arm.

(*b*) The curve showing the proportion of delta-ferrite transformed into D-aggregate against the holding time, has the classical, exponential form characteristic for transformations proceeding by nucleus formation and growth (fig. 41 for Melt B at 900°C). This reaction ceases well before complete transformation of the delta-ferrite.

(*c*) The interlamellar spacing of the carbides decreases with the isothermal holding temperature. This distance, which is about $5 \cdot 10^3$ Å in the D-aggregate forming at 850°C is entirely comparable with that of pearlite forming at about 650°C in Fe-C alloys.[14]

(*d*) Nucleus formation starts at the grain boundaries and the number of nuclei increases as the temperature falls. The areas of D-aggregate become more numerous and smaller with decreasing temperature.

(*e*) The lamellar carbides continuing from the preceding in the delta-ferrite, are the governing phase.

Mechanism of formation of the D-aggregate

This reaction has already been observed in tungsten steels,[15] molybdenum steels,[8] high-speed steels of the 18-4-1 type,[9] vanadium steels,[12] Cr-Ni-Mo stainless steels,[10, 11] Cr-Mn steels.[16] K. Kuo,[8–12] and Nehrenberg and Lillys,[1] were the first to suggest a mechanism of formation for this aggregate.

Since the governing phase is the carbide $(Cr, Fe)_{23}C_6$, consideration must be given to the rates of diffusion of C and Cr in the different phases. Comparison of the coefficients of diffusion of carbon in Fe-C alloys[17, 18] shows that carbon diffuses about 50 times quicker in ferrite than in austenite at about 850–900°C (the ratio is about 350 at 650°C, and 20 at 1,100°C). Indications in the literature of the diffusion rate of chromium[19–21] show in first approximation that: chromium diffuses 50–100 times quicker in ferrite than in austenite; carbon diffuses in austenite or ferrite, about 10^5 times quicker than chromium at about 850°C (this ratio is about 10^6 at 700°C, and 10^4 at 1,050–1,100°C).

The formation of the carbide $(Cr, Fe)_{23}C_6$ appears thus to be governed by the diffusion of chromium in the ferrite.

In view of the distribution of chromium and carbon in the delta and gamma phases, and that the surface energy required for the formation of a nucleus is smaller at an interface, a nucleus of carbide $M_{23}C_6$ will form more readily at the gamma-delta interface. Since the delta-ferrite has a higher chromium content, and chromium diffuses more rapidly in the delta than in the gamma phase, the carbide nucleus grows in the delta-ferrite rather than in the gamma-phase. This causes an impoverishment both of the carbon and of the chromium in the immediate vicinity of the carbide rod, and diffusion gradients become established which cause a migration of carbon and chromium towards the impoverished zones.

At the temperatures in question, the local depletion of carbon is easily compensated by the carbon in the delta and gamma grains. This, however, is not the case for the chromium, with its far slower diffusion rate. The delta-ferrite enclosing the carbide rod, being lower in chromium which is

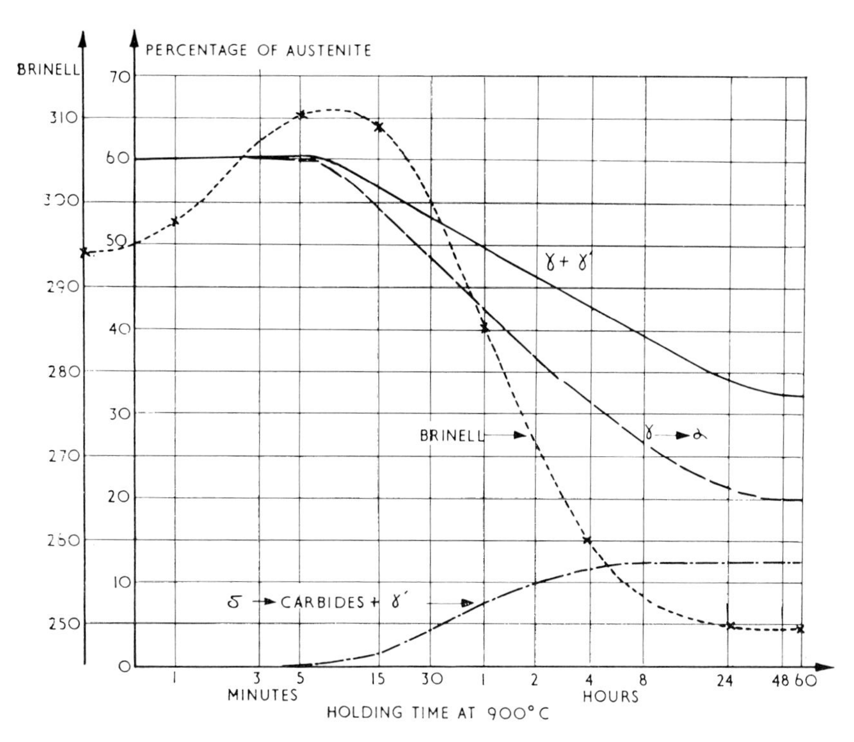

41 *Transformation speeds for isothermal holding at 900°C. Melt B.*
— · — · *Reaction* $\delta \rightarrow$ *carbiaes* $+ \gamma'$ *(D-aggregate)*
— — — *Reaction* $\gamma \rightarrow \alpha$
——— $\gamma + \gamma'$
- - - - - *Brinell hardness*

a ferrite-promoting element, becomes unstable, and transforms into γ' austenite, which is a stable phase at this temperature, in the Fe-Cr-C diagram. The Cr-gradient being thus interrupted, the formation of the γ' phase leads to a local concentration of chromium in the surrounding delta-ferrite, and a new carbon nucleus can be formed. This process of nucleus-formation and lateral as well as longitudinal growth of the carbide rods, will continue until the area (patch) of D-aggregate encounters other patches of D-aggregate or grains of the gamma phase, or until it is arrested by the exhaustion of available carbon, the latter being due both to the lowering of solubility in the gamma and delta phases and the diminishing percentage of austenite in equilibrium.

The upper asymptote of the reaction (1) should theoretically correspond to the temperature of separation of the two-phase domain $\gamma + \delta$ from the three-phase domain $\gamma + \delta +$ carbides $M_{23}C_6$. However, the conditions necessary for the formation of the D-aggregate are not only the primary nucleus-formation of the carbide, but also the secondary nucleus formation of the γ' phase: the upper temperature limit for the formation of the D-aggregate can thus be depressed, either by the higher rate of diffusion of the chromium which prevents the formation of γ', or by the deficiency of nuclei and very rapid coalescence of the carbides at these temperatures. Finally, the temperature of the lower asymptote corresponds very closely with the A_1 point of the γ', depleted in chromium. Contrary to Kuo, no transformation of γ' into ferrite is observed between the A1 and A1' temperatures; the rate of diffusion of the chromium would be too low, even for long holding periods.

It should finally be noted that the mean distribution of the patches of D-aggregate in the structure, gives an idea of the mean free path length of the carbon atoms.

The parallel development of the formation of pearlite in carbon steels, and the D-aggregate, has induced Kuo to call the latter a 'delta eutectoid.'[8] The principal difference between these two transformations consists in that the factor governing the reaction is the rate of diffusion of the chromium in the delta-ferrite, for the D-aggregate, and the rate of diffusion of the carbon in the austenite, for the formation of pearlite.

The $\gamma \rightarrow \alpha$ regression of the austenite

In all cases, the allotropic transformation $\gamma \rightarrow \alpha$ commences after the start of intergranular precipitation of the carbides and the formation of the D-aggregate. This regression most frequently proceeds by shifting of the gamma-delta interface towards the interior of the austenite; it becomes retarded if this interface is established by a continuous deposit of carbides.

At high temperatures, the deposition of the carbides (intergranular precipitation and, eventually, formation of D-aggregate) which precedes and accompanies this transformation, should sufficiently deplete the carbon in the austenite, which is a gamma-promoting element, so as to transform it directly into alpha-ferrite. At temperatures above the A_1 point, this regression is also restricted by the quantity of austenite in stable equilibrium ($\gamma + \gamma'$, figs. 41 and 3). The upper limit of this transformation will thus be the temperature separating the two-phase domain $\gamma + \delta$ from the three-phase domain $\gamma + \delta +$ carbides.

At lower temperatures, the diffusion of carbon towards the grain boundaries is retarded, and the regression $\gamma \rightarrow \alpha$ is accompanied by a periodical release of carbides which determines the regression interface at different instants.

The curves giving the quantity of the gamma phase transformed into the alpha phase against time, have not the characteristic form of reactions by nucleus-formation and growth. In actual fact, the speed of the transformation is not directly proportional to the quantity of remaining austenite, the instantaneous speed appearing to decrease more rapidly than the quantity of austenite. It is probable that the rate of growth is related to the rate of precipitation of the carbides, which should decrease with approach to equilibrium conditions.

G-aggregate and pearlite

The transformation producing the G-aggregate (carbide in barbed form + ferrite) first appears in the austenite on the edge of the gamma-delta grain boundaries, for decomposition between 700–550°C. The carbides are of the type $(Cr, Fe)_{23}C_6$ with a Cr : Fe ratio of about 1·6 at 650°C. Their morphology and the position of the G-aggregate indicate the limited diffusion rate of carbon at these temperatures. This transformation is followed, either by a $\gamma \rightarrow \delta$ regression or by a pearlitic transformation.

The pearlitic transformation exhibits all the characteristics of reactions by nucleus-formation and growth: isothermal diagram, reaction speed, interlamellar spacing decreasing with temperature. Analysis of this aggregate shows the actual presence of small rods of chromium nitride Cr_2N + ferrite. It is, as a matter of fact, intended to study this point in more detail.

Conclusions

The isothermal decomposition of industrial stainless steels with 17% chromium having, at high temperature, either a structure of pure delta-ferrite, or a mixed, ferrite + austenite structure, discloses a sequence of separate transformation stages at regular temperature intervals, and partially overlapping. These can be enumerated as follows for descending temperatures:

(*a*) Formation of gamma austenite from delta ferrite: $\delta \rightarrow \gamma$;

(*b*) Intergranular precipitation of carbides $(Cr, Fe)_{23}C_6$ at the gamma-delta and delta-delta grain boundaries;

(*c*) Decomposition of the delta-ferrite to the D-aggregate: lamellar carbides $(Cr, Fe)_{23}C_6 + \gamma'$ austenite;

(*d*) Regression of the gamma-austenite to alpha-ferrite with possible periodical discharge, at low temperatures, of carbides at the regression interface;

(*e*) Granular or acicular precipitation of carbides $M_{23}C_6$ and nitrides Cr_2N in the delta ferrite;

(*f*) Decomposition of the austenite to G-aggregate: barbed carbides $(Cr, Fe)_{23}C_6$ + alpha-ferrite;

(*g*) Decomposition of the austenite into nitrided pearlite (nitrides Cr_2N + alpha-ferrite);

(*h*) Martensitic transformation of the residual γ and γ' phases.

It is understood that all these structures, which may coexist in a sample subjected to continuous cooling, for instance, may produce extremely complex micrographic structures, which, however, only show definitely the presence of ferrite and carbides $M_{23}C_6$ and nitrides Cr_2N; particularly if it is further considered that a partial coalescence accompanying an extended thermal cycle may yet further distort the original aspects of each of these.

References

(1) A. E. Nehrenberg and P. Lillys, *Trans. A.S.M.*, 1954 **46**, 1176.
(2) A. Gueussier and R. Castro, *Rev. Mettalurg.*, 1958, **55**, 1023.
(3) R. T. Howard and M. Cohen, *Trans. A.I.M.E.*, 1947, **172**, 413.
(4) K. Bungardt, E. Kunze and E. Horn, *Arch. Eisenhüttenwes*, 1958 **29**, 193.
(5) F. K. Bloom, W. C. Clarke and P. A. Jennings, *Metal Progress*, Feb., 1951, 250.
(6) R. Castro and A. Gueussier, *Rev. Metallurg.*, 1958, **55**, 107.
(7) R. Castro and A. Gueussier, *Ibid.*, 1960, **57**, 715.
(8) K. Kuo, *J. Iron St. Inst.*, 1954, **176**, 433.
(9) K. Kuo, *Ibid.*, 1955, **181**, 128.
(10) K. Kuo, *Ibid.*, 1955, **181**, 134.
(11) K. Kuo, *Ibid.*, 1955, **181**, 213.
(12) K. Kuo, *Ibid.*, 1955, **181**, 218.
(13) J. Philibert, G. Henry, M. Robert and J. Plateau, *C.R. Acad. Sci. France*, 1961, **252**, 1320.
(14) R. F. Mehl and A. Dube', 'Phase Transformations in Solids, 1951, pp. 545.
(15) A. E. Nehrenberg and J. G. Y. Chow, *Trans. A.S.M.*, 1953, **45**, 675.
(16) H. Jolivet, unpublished.
(17) J. K. Stanley, *Trans. A.I.M.F.*, 1949, **185**, 752.
(18) C. Wells, W. Batz and R. F. Mehl, *Ibid.*, 1950, **188**, 553.
(19) T. Heumann and H. Boehmer, *Archiv. Eisenhüttenwes.*, Dec., 1960 **31**, 749.
(20) P. L. Gruzin, *Doklady AN U.S.S.R.*, 1954, **94**, 681. (Transl. Brutcher No. 3,331).
(21) S. Gertsriken and I. Dekhtyar, *Zh. tekhn. Fiz.*, 1950, **20**, 1005.

Study of the isothermal transformations in 17% Cr stainless steels

Part II. Influence of carbon and nitrogen

RENÉ CASTRO and ROLAND TRICOT*

FOLLOWING THE PRECEDING communication[1] we considered it would be interesting to pursue the investigation of the decomposition processes in 17% Cr stainless steels by endeavouring to define the influence respectively of carbon and nitrogen, as well as subsidiary alloying constituents, on the one hand in the equilibrium constituents, and on the other hand in the structural constituents of the isothermal decompositions.

For this, we have used principally:

A series of melts, designated C 1 to C 9, low in nitrogen (0·007–0·013%), and with carbon contents varying between 0·010 and 0·400% (25-kg, high-frequency melts in vacuum);

A series of melts with variable carbon content, from 0·010 to 0·100 %, and high nitrogen content:

0·065–0·085% for melts designated N 1 to N 4;

0·150–0·200% for melts designated N 5 and N 6.

(High-frequency melts in air, 10-kg each.)

TABLE 1 *Analysis of experimental melts with various C and N contents*

	% C	% Si	% Mn	% Ni	% Cr	% S	% P	% N
C 1	0·012	0·420	0·590	0·150	17·3	0·015	0·010	0·016
C 2	0·049	0·420	0·385	0·110	17·2	0·015	0·010	0·013
C 3	0·066	0·450	0·420	0·105	17·3	0·015	0·010	0·013
C 4	0·091	0·420	0·460	0·085	17·3	0·015	0·010	0·013
C 5	0·111	0·430	0·450	0·090	17·3	0·017	0·010	0·013
C 6	0·155	0·365	0·595	0·140	17·3	0·008	0·010	0·007
C 7	0·261	0·400	0·455	0·110	17·5	0·009	0·010	0·007
C 8	0·348	0·385	0·545	0·105	17·3	0·010	0·010	0·009
C 9	0·375	0·410	0·550	0·170	17·1	0·010	0·010	0·011
N 1	0·013	0·365	0·520	0·190	17·0	0·007	0·010	0·064
N 2	0·032	0·315	0·430	0·170	16·8	0·006	0·014	0·083
N 3	0·055	0·360	0·470	0·200	17·5	0·007	0·017	0·086
N 4	0·077	0·350	0·510	0·205	17·3	0·005	0·016	0·085
N 5	0·017	0·410	0·380	0·140	17·6	0·009	0·012	0·204
N 6	0·100	0·465	0·470	0·125	17·3	0·008	0·011	0·145

The analyses of these melts are given in table 1.

The same experimental techniques have been used, as previously described.[1]

*Société d'Electrochimie, d'Electrométallurgie et des Aciéries électriques d'Ugine Savoy. Communication to the Autumn Meeting of the Société Française de Metallurgie, Paris, 1961. *Memoires Scientifiques*, **59** (9), 1962.

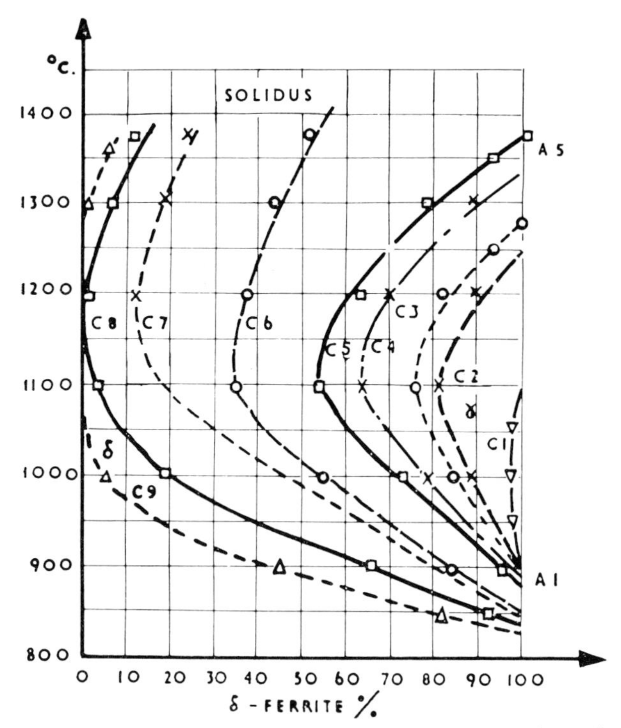

1a *Steels with 17·5% Cr. Percentages of delta ferrite and austenite in equilibrium, against temperature. Influence of carbon (low nitrogen)*

Line	Melt	%Cr	%C	%N
∇ ————————	C1	17·3	0·012	0·016
× — — — — —	C2	17·2	0·049	0·013
○ -------------	C3	17·3	0·066	0·013
× —·—·—·—	C4	17·3	0·091	0·013
□ ————————	C5	17·3	0·111	0·013
○ — — — — —	C6	17·3	0·155	0·007
× -------------	C7	17·5	0·261	0·007
□ ————————	C8	17·3	0·348	0·009
Δ -------------	C9	17·1	0·375	0·011

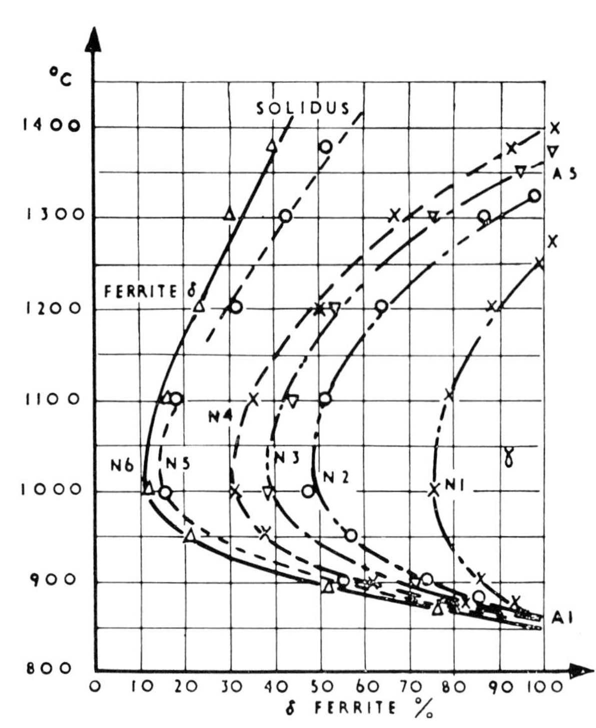

1b *Steels with 17·5% Cr. Percentages of delta ferrite and austenite in equilibrium, against temperature: melts with high nitrogen content*

Line	Melt	%Cr	%C	%N
× —·—·—·—	N1	17·0	0·013	0·064
○ —··—··—	N2	16·8	0·032	0·083
∇ —···—···—	N3	17·5	0·055	0·086
× — — — — —	N4	17·3	0·077	0·085
○ -------------	N5	17·6	0·017	0·204
Δ ————————	N6	17·3	0·100	0·145

Influence of carbon and nitrogen on the development of austenite-ferrite equilibrium at temperatures above the A_1 point

The A_1 temperature of the occurrence of austenite on heating

In sum, the determinations of the A_1 point, as made principally by micrography, thermoresistivity measurements and dilatometry, result in the following, linear formula:

$$\theta_{A_1} = 30(\%Cr) + 73(\%Si) - 250(\%C) - 280(\%N) - 66(\%Mn) - 115(\%Ni) + 405$$

This formula is of course approximate and only valid within the limits of composition not greatly differing from those of industrial, 17% Cr steels. The coefficients of the different elements vary slightly from those given by Post and Eberly[2] for higher-chromium steels (16–30%), as well as by Irvine, Crowe and Pickering[3] for 13% Cr steels.

Equilibrium percentages of ferrite and of austenite

The proportions of delta ferrite and austenite in the equilibrium state, determined by linear analysis against the temperatures, from A_1 to 1,400°C, are shown in fig. 1*a* for melts C 1 to C 9, for constant chromium and nitrogen and increasing carbon content. The curves show a general configuration indicating an increase in the proportion of austenite from A_1 up to a maximum value at a temperature θ_m, followed by a progressive decrease and finally disappearance of the austenite at a temperature which we have designated A_5 while, below approx. 0·120% carbon, the structure may entirely consist of delta ferrite at the very high temperatures between the A_5 point and the solidus, it is not possible to have an entirely ferritic structure with more than approx. 0·012%C, before the point of incipient fusion (fig. 1*a*). The θ_m temperature corresponding to the maximum of austenite, progressively rises from 1,100 to 1,200°C, with increasing carbon percentages. The structure becomes entirely austenitic at 1,200°C for a carbon content higher than 0·35%.

On the contrary, in the high-nitrogen melts (fig. 1*b*), the θ_m temperature is reduced from 1,100 to about 1,000°C, and the increase in the quantity of austenite is thus more rapid between A_1 and θ_m, than in the low-nitrogen melts. These observations may be associated with the development of the micrographic aspect. In fact, while the complete dissolution of the chromium carbides is only observed in the neighbourhood of the θ_m point (*e.g.* 1,050–1,100°C for melts C 2 to C 4, and 1,150–1,200°C for the melts C 6 to C 8), the dissolution of the nitrides starts far earlier and is more rapidly completed (at about 950–1,000°C), than that of the carbides (figs. 2 to 4). Thus, the gamma-promoting action of the nitrogen is more pronounced than that of the carbon, at temperatures between A_1 and 1,000°C, and the shape of the lower branch of the percentage curves is to be attributed principally to the dissolution of the carbides and nitrides.

Gamma-promoting action of carbon and of nitrogen

Seeking a simple relationship between the maximum quantity of austenite at the θ_m temperature on the one hand, and the analysis on the other hand, the following approximate expression can be set up:

$$\gamma\,\%\ \text{max. at } 1{,}100°\text{C} = 470(\%\text{N}) + 420(\%\text{C}) + 30(\%\text{Ni}) + 7(\%\text{Mn}) - 11{\cdot}5(\%\text{Cr}) - 11{\cdot}5(\%\text{Si}) + 186$$

(elements in per cent. by weight).

It will be seen that the gamma-promoting influence of carbon and nitrogen is practically equivalent. On the other hand, if the same formula is applied to temperatures closer to the A_1 point, according to what has already been said, the coefficient for the carbon will be markedly lower than for the nitrogen (*e.g.* at 1,000°C—250 for C, and 470 for N).

The evaluation of the influence of the usual alloying constituents has been obtained from other experimental melts. This formula agrees sufficiently well with those which are found in the literature, for similar alloys (3, 4, 5).

A_5 temperatures

We have plotted in fig. 5, the A_5 temperatures against a parameter (%C) + m(%N) and have found that, putting $m = 0{\cdot}75$, the A_5 temperatures for all the melts fall satisfactorily on one curve. From this, it may be concluded that, in melts with 17% Cr:

(a) The carbon and nitrogen raise the temperature of the A_5 point; but the carbon is about 1·3 times more effective than the nitrogen;

(b) At the same temperature (above θ_m) the proportion of nitrogen corresponding to a completely ferritic structure is higher than the percentage of carbon for the same purpose. Thus, for example, at 1,300°C, pure delta ferrite is obtained with maximum contents of about 0·083% or about 0·117% of nitrogen. Fig. 5 at the same time shows the approximate location of the separation boundaries of the delta and delta + gamma domains, for chromium levels varying from 12 to 22·5%. These curves have been plotted from a differ-

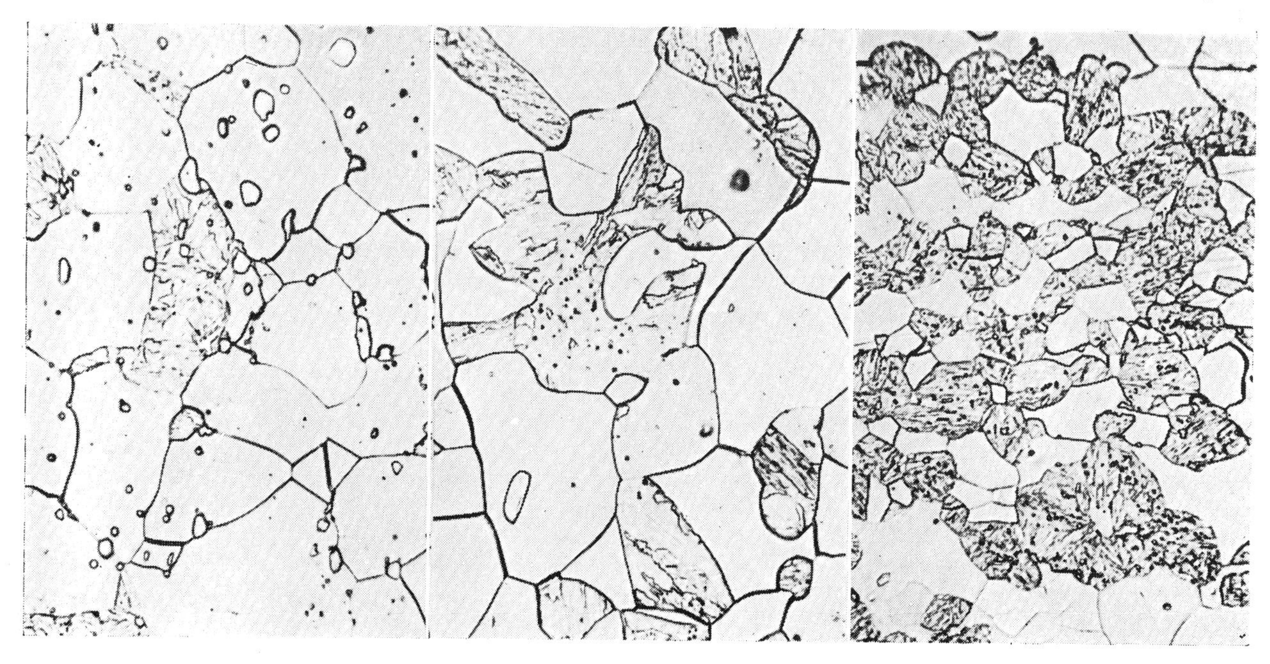

2 *Undissolved carbides after water-quenching at 900°C—2 h, in melt C5 (C=0·111, N=0·013)* × 1,000

3 *Nearly complete dissolution of nitrides after quenching at 900°C, 2 h, in melt N1 (C=0·013, N=0·064)* × 1,000

4 *Nitrides and part of the carbides dissolved after quenching at 900°C, 2 h, in melt N4 (C=0·077, N=0·085)* × 1,000

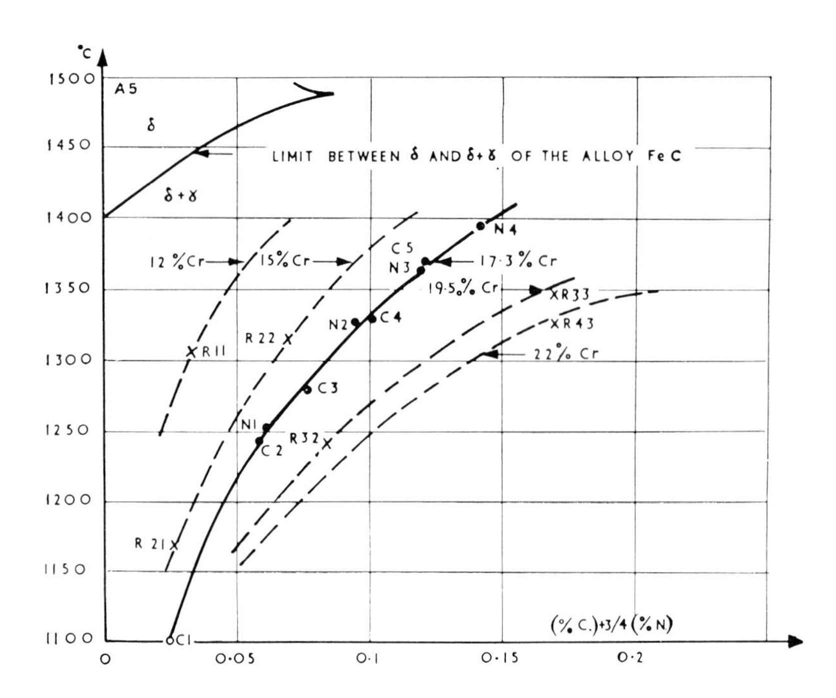

5 *Temperature A5 against carbon and nitrogen contents*

ent series of melts with variable chromium and carbon, and low nitrogen, which will be the subject of a subsequent paper.

Distribution of the alloying constituents in austenite and ferrite: the Fe-Cr-C diagram

Knowledge of the temperatures of the A_1 and A_5 points, the presence or absence of undissolved carbides at each temperature, the respective percentages of austenite and ferrite in the different melts with varying carbon and chromium content, and low in nitrogen, enable the equilibrium diagram to be set up, for industrial Fe-Cr-C alloys with the usual percentages of associated constituents. In view of the fact that the so-called 'tic-line' connecting the characteristic points of the delta

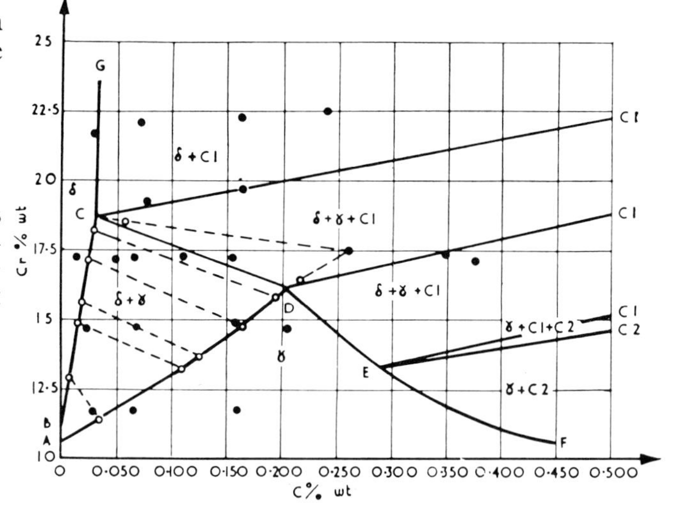

6a ABOVE *Isothermal section at 1,100°C of the ternary Fe-Cr-C diagram.* $N \leqslant 0{\cdot}013$, $Ni \leqslant 0{\cdot}12$, $Si \simeq 0{\cdot}4$, $Mn \simeq 0{\cdot}5$, $C1 = (Cr, Fe)_{23}C_6$; $C2 = (Cr, Fe)_7C_3$

●—points representing the experimental melts:
○—chromium content of ferrite or austenite by electron microsonde titration

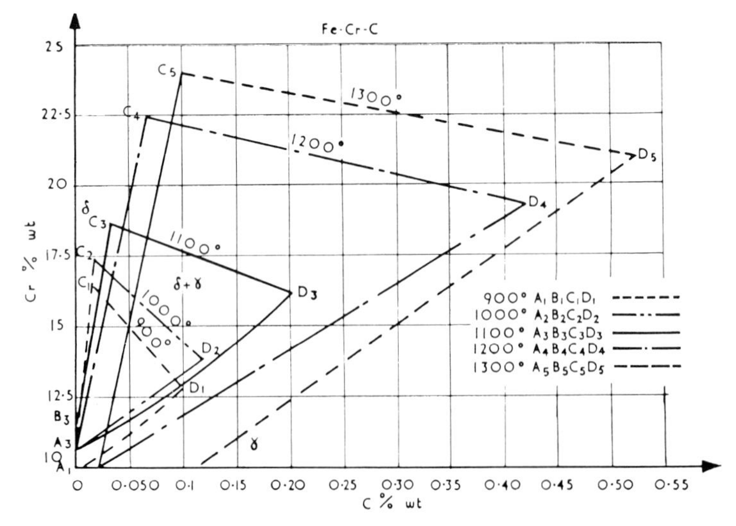

6b LEFT *Position of the two-phase domain* ($\gamma + \delta$) *in the isothermal sections at 900, 1,000, 1,200 and 1,300°C, in the ternary diagram Fe-Cr-C*

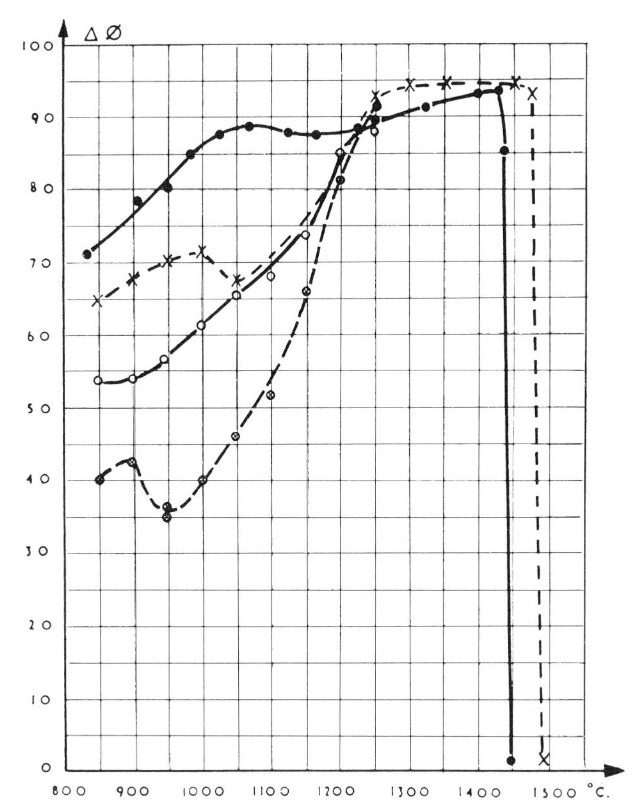

7a *Deformability curves of melts C3 and N1 at rising temperature and descending temperature, from previous holding at 1,250°C*

Melt	% Cr	% C	% N	Temp. Rising	Temp. Descending
C3	17·3	0·066	0·012	——●——	——○——
N1	17·0	0·014	0·065	- - - -×- - - -	- - - -×- - - -

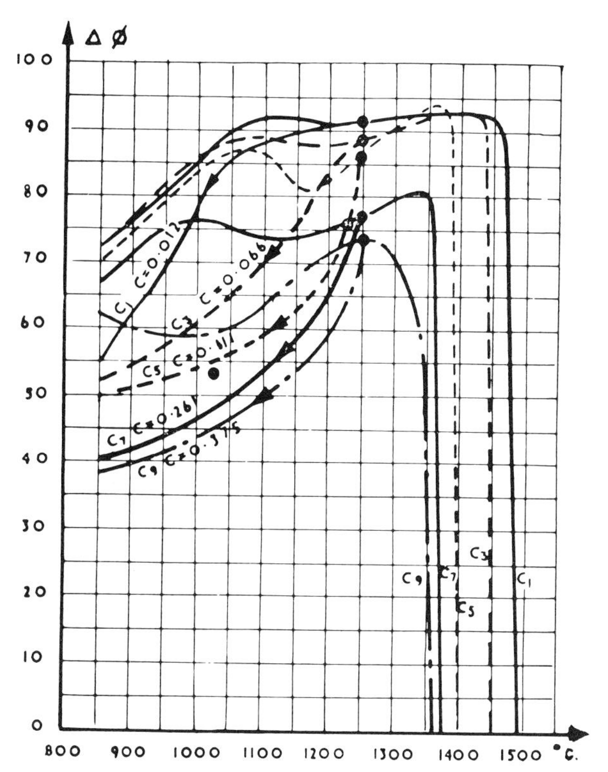

7b *Deformability curves of melts C1 to C9 at rising and falling temperatures from holding at 1,250°C*

Melt	% Cr	% C	% N	
C_1	17·3	0·012	0·016	————————
C_3	17·3	0·066	0·013	— — — —
C_5	17·3	0·111	0·013	- - - - - - - - - - -
C_7	17·5	0·261	0·007	————————
C_9	17·1	0·375	0·011	———·———

ferrite and the austenite in a two-phase, Fe-Cr-C alloy, is not parallel to the vertical planes of the binary alloys, Fe-Cr or Fe-C, it is of greater interest to examine the isothermal sections of the ternary diagram.

By way of example, fig. 6*a* shows the isothermal section at 1,100°C, of such a diagram. Determination of the chromium and iron percentages in the ferrite and austenite, by means of electron-probe microanalysis, in great measure confirms the boundaries of the diagram, and gives the location of the tie-line in the two-phase, gamma-delta domain. The results furnished by microprobe are, on the contrary, far more scattered in the three-phase domain, gamma-delta-$M_{23}C_6$ carbides: probably owing to the difficulty of point analysis in regions free from carbides—particularly in the austenite.

As a whole, the equilibrium diagrams obtained, differ but little from those recently published [6,7]; however, it may be noted that the straight line CD (fig. 6*a*) separating the domain of gamma + delta from delta + gamma + $M_{23}C_6$ carbides, is closer to the iron corner. In general, it may be mentioned, the difference between the chromium contents of the ferrite and the austenite, decreases in the measure that the equilibrium temperature rises (position of the CD line) (fig. 6*b*). The ratio of the chromium in the ferrite to the chromium in the austenite varies, for the 17% Cr steels with low nitrogen, from 1·2 at 1,100°C, to 1·11 at 1,300°C. The nitrogen, which extends the domain of the austenite, decreases the difference between the chromium percentages in the gamma and the delta, since the above ratio becomes respectively 1·08 at 1,100°C and 1·04 at 1,300°C.

Influence of nitrogen on grain size

It has long been known that intentional addition of nitrogen to chromium steels restricts their tendency to coarsening of the ferritic grain at high temperatures. From the preceding it would appear that this influence is to be attributed to the presence of relatively greater quantities of austenite at such high temperatures. Moreover, at the same temperature and with the same proportion of austenite, the marked refining of the grain appears to be due in the first instance to the alpha-gamma transformation

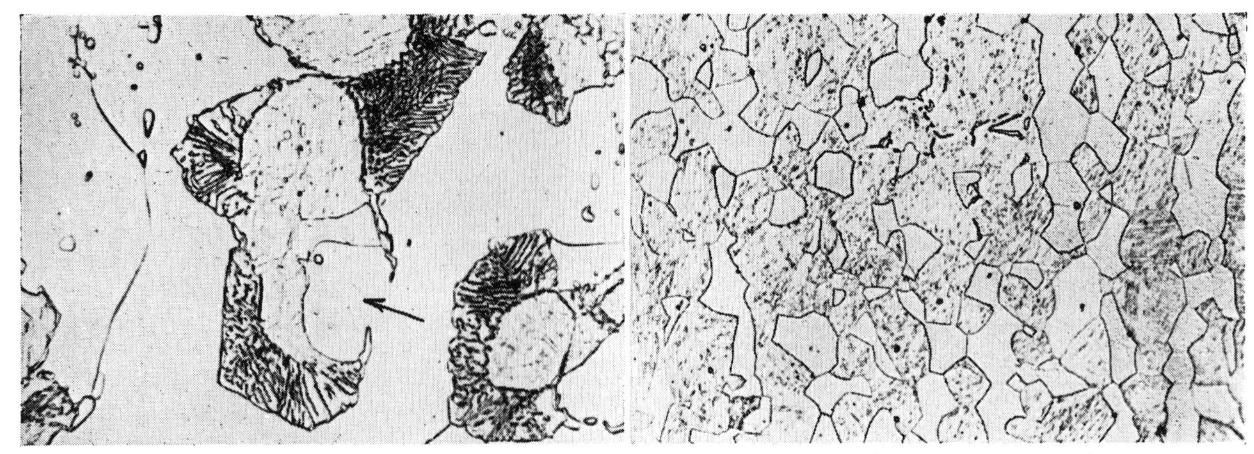

8 *Isothermal decomposition at* 950-800°*C. Aggregate D and regression (Melt C6,* 1,100°*C,* 1*h* → 900°*C,* 2 *h, water)* × 1,000

9 *Isothermal decomposition at* 950 → 800°*C. D-aggregate absent and sparse carbides at* γ-γ *interfaces. Melt N4,* 1,100°*C,* 1 *h* → 900°*C,* 2*h, water* × 500

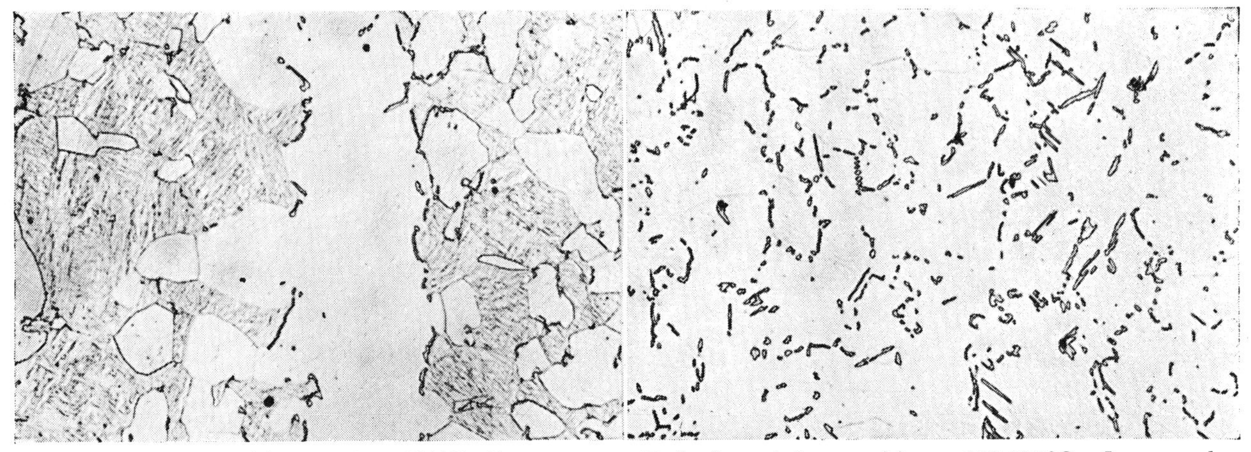

10 *Isothermal decomposition at* 950 → 800°*C. D-aggregate absent, intergranular precipitation and regression. Melt N6* 1,100°*C,* 1 *h* → 850°*C,* 2*h, water* × 500

11 *Isothermal decomposition at* 950-800°*C. Intergranular nitrides and complete regression. Melt N5,* 1,100°*C,* 1 *h* → 800°*C,* 2 *h, water* × 500

during heating, which develops from numerous small areas of austenite uniformly distributed through the ferritic matrix of high-nitrogen melts. This is possibly a consequence of the kind of distribution of the chromium nitrides.

Solidus and hot workability

In addition to what has already been said in the previous paper,[1] comparison of the hot-tensile behaviour of the experimental melts shows that:

(a) The deformability at rising temperatures of the mixed gamma + delta structures, for equal respective proportions, is lowered by increasing nitrogen content (fig. 7*a*). The minimum forgeability is equally lower on the temperature scale, probably as a consequence of the earlier appearance of austenite in high-nitrogen melts, as has been explained under the heading 'Equilibrium percentages.'

(b) The burning temperature (temperature of the solidus) decreases by about 7° when the carbon content is raised by 0·010%. The nitrogen appears to have no effect, at least within the limits investigated.

(c) The observed shift of the curves obtained at rising temperature and those obtained at falling temperature[8] after previous holding at high temperature (1,250°C), is that much more marked, for low-nitrogen melts, the higher the carbon content (fig. 7*b*). This phenomenon partly corresponds to the thermomechanical embrittlement caused by carbon precipitation during elongation. The same applies for melts with increasing nitrogen content; but the harmful influence of nitrogen is at falling temperatures, greater than that of carbon, for identical percentages of austenite, as is shown by fig. 7*a*, where the carbon and nitrogen contents of the melts C 3 and N 1 are practically equal, but transposed. It would appear, therefore, that for reasons still to be discovered, the process of precipitation of nitrides during plastic deformation, has a higher embrittling effect, than that of carbides.

Influence of carbon and nitrogen on the structural constituents of isothermal decompositions.

The preceding paper[1] disclosed a number of transformations in 17% Cr steels. We shall confine ourselves here, to examining the parts played respectively by carbon and nitrogen, in the products of isothermal decomposition, and in particular, in the formation of the D- and G-aggregates, and nitrogen-pearlite.

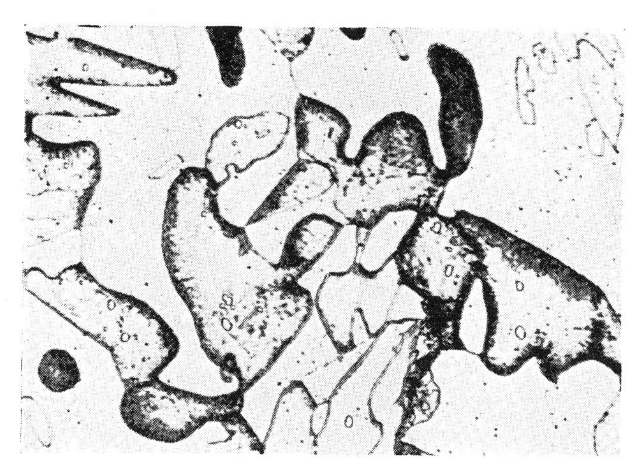

12 *Melt C5 with 17% Cr and low nitrogen. G-aggregate and regression,* 1,100°*C* → 650°*C*, 1 *h, water* × 500

Formation of the D-aggregate (*isothermal decomposition in the* 950–825°*C range*).

It has already been seen, that D-aggregate is formed of γ-austenite and lamellar carbides $(Cr, Fe)_{23}C_6$ with a Cr/Fe ratio of about 2·3 (fig. 8).

In the low-nitrogen melts with 17% Cr, the transformation of the delta ferrite into the D-aggregate is more extensive, with higher carbon content. However, this aggregate is no longer formed in melts with over 0·3% carbon, owing to the absence of delta ferrite in the initial structure at 1,100°C. The gamma-alpha regression is likewise more pronounced, with increase in the quantity of precipitated carbides (the intergranular carbides and the lamellar carbides of the D-aggregate), and the closer the temperature of the isothermal decomposition, approaches the A_1 point.

On the contrary, no D-aggregate is formed from the delta ferrite in high-nitrogen melts with 17% Cr (0·06 to 0·20 N), irrespective of the actual carbon content between 0·01 and 0·1%. For decomposition temperatures above 875°C, intergranular precipitation is slight, and the regression of the austenite to alpha ferrite exceedingly retarded and incomplete (fig. 9). At lower temperatures (850–800°C), an abundant precipitation of intergranular nitrides and carbides results in a marked acceleration of the regression from gamma to alpha (figs. 10 and 11). It thus appears that, before the nitrides have become precipitated, the presence of nitrogen greatly stabilizes the austenitic structure.

It is known, from the researches of Kuo[9] on ferrites of different compositions, that the mechanism of the formation of D-aggregate is controlled

13 *Melt C5 with 17% Cr and low nitrogen. G-aggregate* 1,100°*C* → 675°*C*, 4 *h, water* × 9,000

14 *Melt C5 with 17% Cr and low nitrogen. G-aggregate* 1,100°*C*→675°*C*, 4 *h, water* × 10,000

by the primary formation of nuclei of $M_{23}C_6$ carbide, producing local depletion of the carbon and chromium surrounding the carbide. This depletion of the chromium content not being compensated owing to a lower rate of diffusion of the chromium in the delta ferrite, the latter, becoming unstable, is correlatively transformed into austenite γ^1.

The inhibition of the formation of the D-aggregate in the presence of high nitrogen concentrations may be associated with the fact that in the domain centring on 875°C, where it is formed in the 17% Cr steels, high nitrogen contents cause the subsistence of high proportions of austenite, retaining a great part of the carbon in solution. Thus, the carbon available for this transformation, which is ordinarily provided by the decreased solubility of the carbon in delta ferrite and austenite and the decreasing proportion of austenite, becomes insufficient to release the mechanism of formation of the D-aggregate.

The high nitrogen percentages may equally act by lowering the temperature of the A_1 point, increasing the range of stability of the austenite (the rate of diffusion of the chromium in the ferrite may then be too low at such low temperatures), and by the refining of the grain structure which increases the total free surface at the grain boundaries, and thus

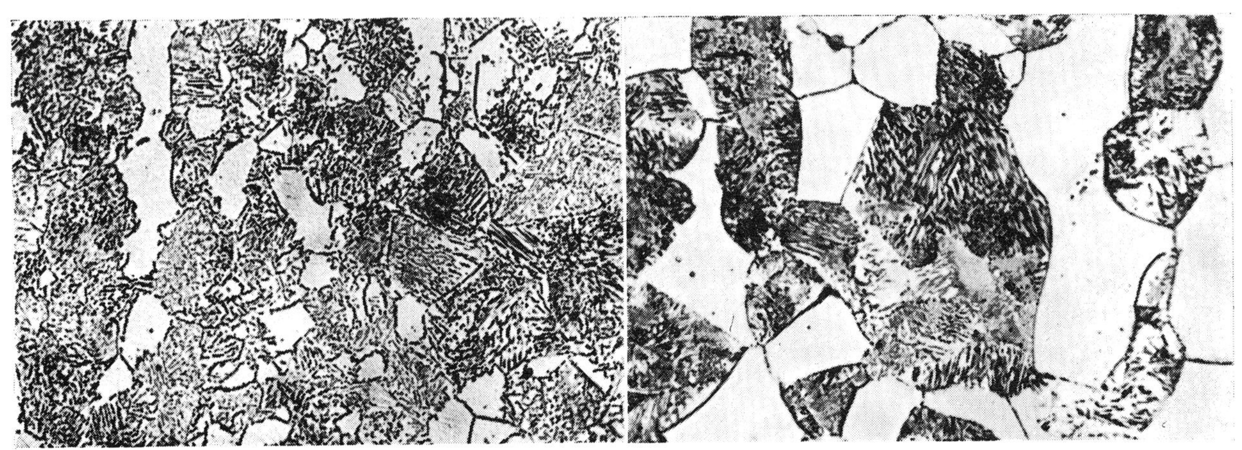

15 *Melt N5 17% Cr, high nitrogen and low carbon. B-aggregate,* 1,100°*C*→750°*C,* 2 *h water* × 500

16 *Melt N5 with 17% Cr, high nitrogen and low carbon. B-aggregate,* 1,100°*C*→650°*C,* 2 *h, water* × 1,000

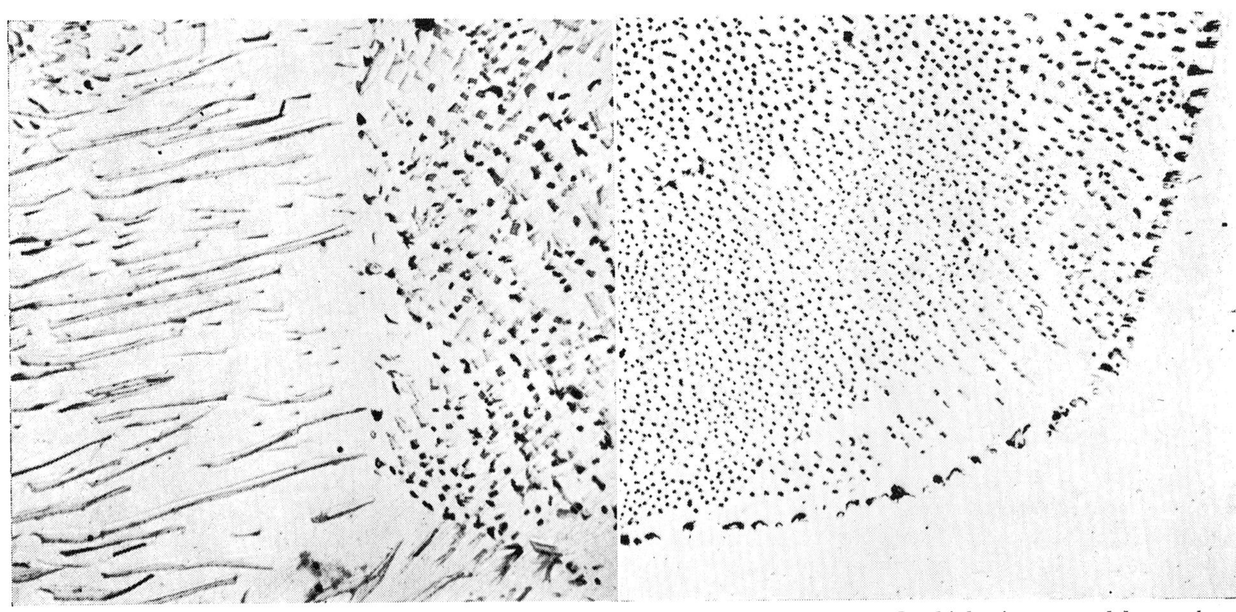

17 *Melt N2 with 17% Cr, high nitrogen and low carbon. B-aggregate.* 1,100°*C*→650°*C,* 1 *h, water* × 6,000

18 *Melt N2 with 17% Cr, high nitrogen and low carbon. B-aggregate.* 1,100°*C*→650°*C,* 1 *h, water* × 6,000

the possibility of intergranular deposition of carbides and nitrides. It should be noted that Nehrenberg and Lillys[10] have observed the formation of the D-aggregate in melts with 21 and 25% chromium and high concentrations of carbon and nitrogen, in which the A_1 temperature is higher according to the higher chromium content.

19 *Melt N2 with 17% Cr, high nitrogen and low carbon. Extraction replica of chromium nitrides.* 1,100°C→650°C, 1 *h, water* × 6,000

The G-aggregate and the nitrogen-pearlite (B-aggregate) (isothermal decomposition at 750–600°C)

It has been seen that the intergranular precipitation of carbides tends at low temperatures (725–600°C) to the formation of a very fine G-aggregate highly susceptible to carbide attack in the form of small rods or barbs, and of alpha ferrite; this aggregate occupying a narrow band bordering the gamma-delta interface within the austenite.

In low-nitrogen melts with 17% Cr and variable carbon content, the width of this band of G-aggregate increases with the carbon content (fig. 12). There will be seen in fig. 13, and particularly in fig. 14, the typical aspect of the carbides of this aggregate, under the electron microscope. The analysis of replicas by electron diffraction shows that these are $(Cr, Fe)_{23}C_6$ carbides with a Cr/Fe ratio of about 1·6, determined by electron-probe microanalysis of extraction replicas prepared by the IRSID method.[11] After the rapid formation of this aggregate, the gamma-alpha regression takes place, with possible localized precipitation of carbides; but, in these, low-nitrogen melts, practically no lamellar aggregate of the pearlitic type is observed.

On the contrary, the high-nitrogen, low-carbon (0·015%) melts show an intensive formation of a pearlitic aggregate here called the B-aggregate, and

20 *Melt N6 with 17% Cr, high carbon and nitrogen. Intergranular carbides and nitrogen-pearlite.* 1,100°C 1 *h*→750°C, 2 *h* × 500

21 *Melt N6 with 17% Cr, high carbon and nitrogen. G-aggregate disclosed by shadow with light Vilella etchant and B-aggregate.* 1,100°C 1 *h*→650°C, 2 *h* × 500

22 *Melt N6 with 17% Cr, high carbon and nitrogen. G-aggregate at edges, regression and B-aggregate.* 1,100°C 1 *h* →650°C, 2 *h* × 1,000

practically no G-aggregate is found (figs. 15 and 16). Examined under the electron microscope, this aggregate B is found to have the shape of small rods of considerable length, and sometimes flattened (figs. 17, 18). The extraction of these rods by separation with bromine (fig. 19) enables identification of a M_2N nitride with a hexagonal lattice, having the parameters $a = 2{\cdot}77$; and $c/a = 1{\cdot}62$. Since the nitrides Cr_2N, Fe_2N have very similar parameters, the indeterminacy was resolved by electron micro-analysis of extraction replicas: this disclosed a Cr/Fe ratio of about 32; this is, consequently, chromium nitride, Cr_2N. The B-aggregate is also called nitrogen-pearlite.

Finally, high-nitrogen and high-carbon melts show, at average decomposition temperatures, *e.g.* 750°C, a decomposition into isolated intergranular carbides and the abundant formation of nitrogenous pearlite (fig. 20). At lower temperatures, for instance 650°C, the G-aggregate is again found, in restricted proportion, and is more readily etched than the nitrogen-pearlite. (Figs. 21 and 22.)

We have further verified these results by nitriding or carburizing melts with very low carbon or nitrogen respectively, and observing the decomposition at 650–700°C. The micrographic examination confirmed the parts respectively played by the carbon (G-aggregate, no B-aggregate) and the nitrogen (B- aggregate, no G-aggregate). Similarly, the formation of B-aggregate and practically no G-aggregate, is again found when observing isothermal decomposition at low austenitizing temperature—900°C for instance—when a large part of the nitrides is already in solution, while a considerable quantity of undissolved carbides still remains.

Decomposition to the rod-shaped B-aggregate, $\gamma \rightarrow Cr\ N + \alpha$ exhibits all the characteristics of a eutectoid transformation with nucleus-formation and grain-growth, C-shaped profile of the TTT curve, rate of transformation and interlamellar spacing decreasing with temperature.

The designation 'nitrogen-pearlite' thus appears to be justified. The TTT curve at the start of transformation appears to approach the A_1 point asymptotically. The growth of this aggregate is probably principally governed by the low rate of diffusion of the chromium of the austenite; since it can be assumed in first approximation, that the rate of nitrogen diffusion in austenite is comparable or slightly lower than that of carbon at the same temperatures.[12]

While the start of formation of the G-aggregate appears to take place sufficiently rapidly, the form and final distribution of the carbides $(Cr, Fe)_{23}C_6$ of this aggregate suggest that the resulting local attenuation of the chromium content is the principal cause of their growth. In fact, while the mobility of carbon in the austenite is still considerable at 650–700°C, the rate of diffusion of chromium is considerably less at such low temperatures, particularly in austenite.[13] Thus, while the sufficient mobility of chromium in delta ferrite at 825–950°C enables the formation of lamellar, D-aggregate from the ferrite, the insufficiently high rate of diffusion of the chromium in the austenite, prevents the lateral and longitudinal growth of a lamellar aggregate of ferrite and carbide $(Cr, Fe)_{23}C_6$ in steels with 17% chromium. It has unfortunately not been possible to determine by microanalysis, the chromium content of the ferrite in aggregate G, in view of its extreme fineness.

Finally, it should be noted that the M_S temperatures of the start of martensitic transformation by cooling quickly from about 1,100°C, vary between about 240 and 150°C, as the carbon content increases from 0·01 to 0·35%. High nitrogen content does not appear niticeably to reduce these temperatures.

Conclusions

It has been shown, on the one hand in the equilibrium constituents, and on the other hand in the products of isothermal decomposition, of steels with 17% Cr, that the carbon and nitrogen have distinct and opposing effects, which may be summarized as follows:

1. At high temperatures, the gamma-promoting influence of nitrogen is comparable with that of carbon, but is exercised earlier, owing to the more rapid dissolution of the nitrides. Analysis by electron probe of the chromium percentages in the ferrite and the austenite respectively, has enabled certain limits to be defined in the ternary, Fe-Cr-C equilibrium diagram. The harmful effect of nitrogen on hot-working properties, is comparatively greater than that of carbon.

2. At isothermal decomposition temperatures between 950 and 825°C, the transformation of the delta ferrite into the D-aggregate (lamellar carbides $(Cr, Fe)_{23}C_6$ + austenite) is inhibited by the presence of high nitrogen contents. In the region of 750–600°C, carbon and nitrogen respectively lead to the formation of austenitic decomposition aggregates of different aspects: the G-aggregate formed of barbed carbides $(Cr, Fe)_{23}C_6$ and alpha ferrite (absence of lamellar ferrite); and the B-aggregate, consisting of rods of chromium nitride Cr_2N and alpha ferrite (nitrogen-pearlite) in lamellar form.

The nitrogen percentages found in commercial alloys of this type are thus by no means negligible from the aspect of their influence on structural transformations, as would be the case in carbon steels.

Acknowledgments

The optical and electron microscopy was carried out with the help of Mr. J. Roche and Mr. R. Devin respectively, to whom our thanks are due.

References

(1) R. Castro and R. Tricot, *Mem. Scient. Rev. Metallurg.*, 1962, **59** (9), 571; *Metal Treatment*, 1964, **31** (229-30), 401, 436.
(2) C. B. Post and W. S. Eberly, *Trans. A.S.M.*, 1951, **43**, 243.
(3) K. J. Irvine, D. J. Crowe and F. B. Pickering, *J. Iron St. Inst.*, 1960, **195**, 386.
(4) J. H. Waxweiler, U.S.A. Patent 2,851,384 of September 9, 1958.
(5) K. J. Irvine, D. T. Llewellyn and F. B. Pickering, *J. Iron St. Inst.*, 1959, **192**, 218.
(6) K. Bungardt, E. Kunze and E. Horn, *Archiv. Eisenhuettenwes*, 1958, **29**, 193 and 26.
(7) E. Baerlecken, W. A. Fischer and K. Lorenz, *Stahl u. Eisen*, 1961, **81**, 768.

(8) R. Castro and R. Poussardin, *C.I.T. of C.D.S.*, February, 1962, p. 505.
(9) K. Kuo, *J. Iron St. Inst.*, 1954, **176**, 433, and 1955, **181**, 128 and 213.
(10) A. E. Nehrenberg and P. Lillys, *Trans. A.S.M.*, 1954, **46**, 1176.
(11) J. Philibert, G. Henry, M. Robert and J. Plateau, *C.R. Acad. Sci. France*, 1961, **252**, 1320.
(12) P. E. Busby, D. P. Hart and C. Wells, *J. Inst. Metals*, 1956, p. 686.
(13) T. Heumann and H. Bohmer, *Archiv. Eisenhuettenwes.*, 1960, **31**, 749.

Study of the isothermal transformations in 17% Cr stainless steels

Part III. Influence of alloying elements other than carbon and nitrogen

R. TRICOT and R. CASTRO*

DISCUSSION

Evolution of the ferrite-austenite equilibrium at temperatures above the A_1 point

Within the limits of the compositions investigated in this paper, the evolution of the austenite-ferrite distribution at temperatures above the A_1 point can be summarized as follows:

(*a*) From A_1 to the θ_m temperature the increase in the austenite component appears to be governed to a great extent by the comparatively slow rate of dissolution of the chromium carbides, and the relatively more rapid dissolution of the chromium nitrides. The dissolution is favoured by the addition of γ-phase promoters, lowering the temperature of the A_1 point and increasing the proportion of austenite.

Conversely the α-promoting elements or both α-phase and carbide-promoting elements, with carbides of greater stability than those of chromium, reduce the proportion of austenite which is formed —other things being equal, of course.

(*b*) From θ_m to the A_f temperature, the progressive diminution of the proportion of austenite appears to be associated in part with the appreciable increase with the temperature of the solubility of C and M in the δ-ferrite. Similarly, the segregation coefficient of the α-promoting and γ-promoting elements in the ferrite and the austenite, progressively tends to unity with increasing temperature.

*Société d'Electro-Chimie, d'Electro-Métallurgie et des Aciéries Electriques d'Ugine, at Ugine (Savoy). Communication to the 'Journée des Aciers. Spéciaux' of the Société Française de Métallurgie and the Cercle d'Etudes des Métaux, Grenoble, June 1963.

Formation of the D-aggregate

The transformation of the δ-ferrite into the D-aggregate, consisting of lamellar carbides $(Cr,Fe)_{23}C_6$ and γ'-austenite, takes place in most of the two-phase alloys at temperatures of isothermal decomposition on either side of the A_1 point. Kuo[17] in his work on the ferrites of varying composition, and Nehrenberg and Lillys[6] for alloys containing 17–25% Cr, were the first to suggest a mechanism for the formation of this aggregate.

Rates of diffusion of carbon and chromium

In view of the nature of this aggregate, the mechanism of its formation is governed by the following considerations:

(*a*) The difference in the distribution of the alloying constituents, Cr and C, particularly in the austenite and the ferrite. Examination of the ternary state diagram Fe-Cr-C shows that the δ-ferrite is higher in chromium and lower in carbon, than the austenite.

(*b*) The limits of solubility of carbon in austenite and in ferrite respectively. In actual fact, the carbon available for the reaction will be furnished both by the drop in solubility and by the diminishing proportion of austenite in equilibrium, between the θ_d and θ_i temperatures.

(*c*) The difference in the rates of diffusion of C and Cr within the constituents. The coefficients of diffusion of C in ferrite and in austenite, have been determined by Stanley[26] and Wells, Batz and Mehl[27] for Fe-C alloys. The values for chromium have been measured most recently by Heumann and Boehmer[28] who have shown that for atomic concentrations exceeding 10% Cr, the diffusion coefficient no longer depends on the concentration. The values of the coefficient D are given in table 6, plotted against temperature.

Extrapolating the values of D to regions not covered by the experimental compositions and the temperature ranges which have been investigated, is not to be recommended. Disregarding this in a qualitative approach, we find that, in first approximation:

The carbon and the chromium diffuse more rapidly in the ferrite than in the austenite, in a ratio of about:

	600°C	900°C	1,200°C
C ..	600	50	10
Cr ..	125	100	50

The carbon diffuses in the ferrite or the austenite about 10^5 times more quickly than chromium at 800–900°C (this ratio becomes 10^6 at 600–700°C, and 10^4 at 1,100°C). It will be noted that the mean distribution of the patches of D-aggregate in the structure (fig. 15*a*) gives the impression of the mean free path of the carbon atoms.

Since the lamellar carbide $(Cr,Fe)_{23}C_6$ is the governing phase of the D-aggregate, the formation of this carbide will consequently be determined by the rate of diffusion of Cr in the δ-ferrite.

Mechanism of formation of D-aggregate

Given the distribution of chromium and carbon in the ferrite and austenite and taking account of the fact that the surface energy necessary for the formation of a nucleus is less at the grain boundary, the interfaces δ–γ are the suitable sites for the precipitation of the carbides. Since the δ-ferrite has a higher chromium concentration, and this element diffuses more rapidly through the ferrite than the austenite, the chromium carbide grows more readily in the ferrite than in the austenite, starting from the intergranular carbide which acts as nucleus.

The apparition of a rod of carbide in the δ-ferrite gives rise to simultaneous impoverishment both of the carbon and the chromium in the immediate vicinity, and diffusion gradients become established. At these temperatures the local depletion of the carbon is relatively compensated by the diffusion of this element. This is not the case for the chro

TABLE 6 *Diffusion coefficients of carbon and chromium in austenite and ferrite*

	Carbon in		Chromium in	
	austenite	ferrite	austenite	ferrite
Region of determination:				
Composition	Fe-C–0·2% C	0–0·68% C	10–22% Cr	10–22% Cr
Temperature	850–1,300°C	515–785°C	950–1,150°C	820–1,440°C
D_0 (cm²/s)	0·365	0·0079	7·1 10^{-5}	1·48
Q (cal/mole)	35,900	18,100	40,600	54,900
D (cm²/s) at				
600°C	4·3 × 10^{-10}	2·5 × 10^{-7}	2 × 10^{-16}	2·5 × 10^{-14}
700°C	3·5 × 10^{-9}	7·2 × 10^{-7}	7 × 10^{-15}	7 × 10^{-13}
800°C	1·7 × 10^{-8}	1·7 × 10^{-6}	1 × 10^{-13}	9 × 10^{-12}
900°C	7·8 × 10^{-8}	4 × 10^{-6}	1·1 × 10^{-12}	8 × 10^{-11}
1,000°C	2·8 × 10^{-7}	6·4 × 10^{-6}	9 × 10^{-12}	5 × 10^{-10}
1,100°C	7·8 × 10^{-7}	1·7 × 10^{-5}	5 × 10^{-11}	2·5 × 10^{-9}
1,200°C	1·8 × 10^{-6}	2 × 10^{-5}	2 × 10^{-10}	1 × 10^{-8}

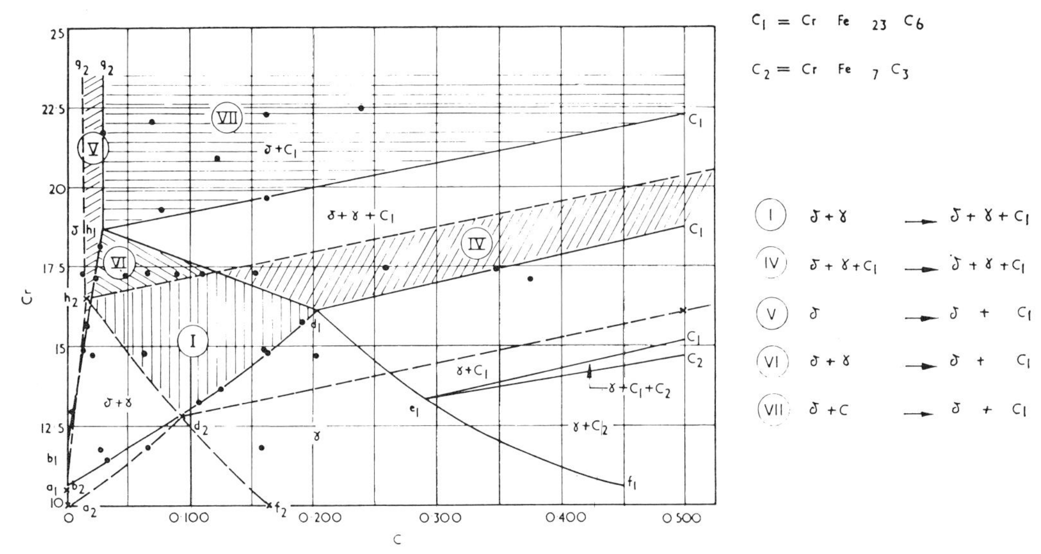

40 *Possibility of formation of D-aggregate by superposition of isothermal sections at 1,100°C and 900°C*

mium, due to the fact that the rate of diffusion of the latter is slower than that of the carbon.

The δ-ferrite which surrounds the carbide rod, impoverished in chromium, which is a ferritic promoter, becomes unstable and is transformed into γ′-austenite, which is a stable phase at this temperature in the Fe-Cr-C diagram. The diffusion of the chromium is consequently impeded and a new nucleus of the carbide can form at the interface γ′–δ.

This process of nucleation and lateral growth, with the longitudinal growth of the rods, will continue until the available carbon is exhausted or other similar patches are encountered.

Conditions for the formation of D-aggregate

In the mechanism described above it will be seen that the conditions necessary for the formation of this D-aggregate are, not only the primary nucleation of the $M_{23}C_6$ carbide, but also a secondary nucleation of the γ′-austenite.

The D-aggregate cannot form in steels with an entirely austenitic, initial structure, and the first condition for its formation is, of course, at least the presence of a mixed structure $\delta + \gamma$ at the starting temperature θ_d. The second condition will be, that at the decomposition temperature θ_i, the $M_{23}C_6$ carbide is formed.

Superposing the isothermal sections in the Fe-Cr-C diagram, for instance for $\theta_d = 1{,}100°$, and $\theta_i = 900°$ (fig. 40), it will be seen that the two-phase steels with 12% Cr or 15% Cr and C below about 0·05%, cannot exhibit the D-aggregate owing to the fact of the absence of the carbides $M_{23}C_6$, in structural equilibrium at 900°.

The upper limit for the formation of D-aggregate lies between about $A_1 + 100°$ and $A_1 + 150°$, for the compositions investigated in this paper. It corresponds practically to the temperature dividing the region of $\gamma + \delta$ from the region $\gamma + \delta + C_1$, or, again, δ from the region $\delta + C_1$.

However, at high temperatures and high chromium concentrations the rate of diffusion of the chromium in the ferrite may become sufficient, either to cause the regression of the γ′ austenite to ferrite (involving the coalescence of the lamellar carbides of the D-aggregate), or prevent the formation of γ′.

This condition, of the possibility of secondary nucleation of the γ′ also arises in the melts with 17% Cr, where it has been seen that the addition of γ-promoting elements favours the transformation of the δ-ferrite, low in chromium, into γ′-austenite. On the contrary, the addition of α-promoting elements may make the secondary nucleation of γ′ impossible in the case of highly ferritic melts.

Finally, the lower limit for the formation of the D-aggregate, which is located between A_1–25° and A_1–350°C (melts with high chromium), practically corresponds to the A'_1 temperature of the γ′-austenite, low in chromium and carbon; secondary nucleation of the γ′ being no longer possible below this temperature. The absence of D-aggregate in the isothermal decompositions towards 800° of the two-phase melts with 12% Cr, is probably due to this fact since, according to the Fe-Cr-C diagram, the A'_1 and A_1 temperatures are merged at about 800°.

The parallel development subsisting between the pearlitic transformation of the Fe-C alloys and the formation of D-aggregate, has led Kuo to call this latter, the δ-eutectoid.[17] The principal difference between these two transformations consists in the fact that the factor governing the reaction is the rate of diffusion of chromium in δ-ferrite in the case of the D-aggregate, and the rate of diffusion of the carbon in the austenite, for the pearlitic transformation.

Regression of the austenite

In all cases the regression of the austenite to α-ferrite at falling temperatures starts after the commencing precipitation of intergranular carbides and eventually after the commencement of formation of the D-aggregate. This regression occurs the more frequently by displacement of the δ–γ interface towards the inside of the austenite.

At temperatures above A_1 this results principally from the impoverishment of the carbon in the austenite and is arrested when the proportion of austenite stable at this temperature, has been attained. At lower temperatures, the diffusion of the carbon towards the grain boundaries is retarded and the regression $\gamma \rightarrow \alpha$ is accompanied by a periodical deposition of carbides $M_{23}C_6$.

This regression, which is very slight in steels with 12% Cr, becomes increasingly important with rising chromium percentage. This phenomenon appears to be directly related to the shape of the lower branch (A_1–θ_m) of the equilibrium curves (figs. 1 and 2). Similarly, this regression is enhanced by the addition of α-promoting elements while γ-promoters depress it.

Decomposition into G-aggregate and pearlite

Melts with variable Cr and C

In the steels with 12–25% Cr and variable carbon content, low in nitrogen, experience has shown that the decomposition of the austenite proceeds differently, depending on whether the initial structure is a mixed $\gamma + \delta$, or wholly austenitic.

Mixed $\gamma + \delta$ initial structure. In this first case, after formation of the intergranular carbides, the decomposition of the austenite starts with the formation of G-aggregate consisting of α-ferrite and $M_{23}C_6$ carbides of globular or short, bacillar shape. This transformation produces local carbon impoverishment, which restricts the development of the G-aggregate and the decomposition of the austenite proceeds by regression $\gamma \rightarrow \alpha$ with possible precipitation of isolated carbides $M_{23}C_6$.

Wholly austenitic starting structure. After intergranular precipitation of the carbides the austenite decomposes successively into two aggregates which can be distinguished by the form of their carbides.

First of all, a lamellar eutectoid starts forming from the grain boundaries, which is easily stained by etching and resembles the fine pearlite, troostite or granulite of the low-alloyed steels at falling decomposition temperatures. The formation of this pearlitic aggregate causes local carbon impoverishment in the neighbouring austenite, which can be detected either by the rise in the M_s–M_f temperatures in melts containing residual austenite at room temperature, or by etching with controlled potential.

The remaining austenite, low in carbon due to the formation of the pearlitic aggregate, then decomposes into α-ferrite and isolated carbides of globular or short, bacillar shape.

These aggregates differ, at least in the initial stages of their formation, by the nature of their carbides. In fact, in the following sequence of carbides: $(Fe,Cr)_3C$ – $(Cr,Fe)_7C_3$ – $(Cr,Fe)_{23}C_6$, an increasing alloyed carbon content will generally be observed and in general, the alloyed carbides increase, *e.g.* $M_{23}C_6$ depending on:

Increasing chromium content.
Decreasing carbon content.
Rising decomposition temperature.

Moreover, the pearlitic eutectoid appears to be principally based on the M_7C_3 carbides, while in the subsequent decomposition of the austenite to ferrite + isolated carbides, the $M_{23}C_6$ carbide predominates. Finally, as in the different stages of the martensitic annealing of 12% Cr steels, it is observed that any carbide can evolve into one more highly alloyed with increasing duration of the isothermal holding period: thus, for example, the carbide M_3C formed in the first few minutes of decomposition at 600–650°C in melts with 12% Cr and high carbon, which eventually disappears.

The following novel conclusion thus appears: the pearlitic aggregate—ferrite-carbides is easily formed with M_3C and M_7C_3 carbides; while $M_{23}C_6$ appears to form with greater difficulty a lamellar aggregate with the ferrite—whether in steels with a mixed initial structure (G-aggregate), or in wholly austenitic steels.

It is possible that the depletion in chromium content caused by the formation of the $M_{23}C_6$ carbide, associated with the low rate of diffusion of chromium in austenite at these temperatures, inhibits any longitudinal growth of the lamellae, as well as the lateral development of new lamellae. On the contrary, it has been found that the $M_{23}C_6$ carbide assumes the lamellar form in the high-temperature transformations of δ-ferrite to D-aggregate.

It may finally be remarked that by comparison with pearlite in the Fe-C alloys and at equal decomposition temperatures, the interlamellar spacing is closer in the pearlitic aggregate, which may be due to the fact that lateral growth is controlled by the rate of diffusion of the chromium in the austenite, which is far less than that of the carbon.

We have not found pro-eutectoid ferrite in 12% Cr steels, but only an intergranular precipitate of carbides, whether in a two-phase or an entirely austenitic initial structure.

17% Cr melts with additives

In these alloys, the γ-phase-promoting elements increase hardenability by retarding or suppressing both the $\gamma \rightarrow \alpha$ regression and the formation of nitrogen-pearlite. There remains, the G-aggregate in melts with a mixed, initial structure. On the other hand, the α-phase-promoting constituents favour the $\gamma \rightarrow \alpha$ regression and the formation of nitrogen pearlite, which is always formed after the G-aggregate.

IN CONCLUSION, these different observations show the continuity in the appearance and nature of the decomposition products in steels with chromium varying between 12 and 25%; as well as the antagonistic influence of α-promoting and γ-promoting alloying constituents.

References

(1) R. Castro, and R. Tricot, *Mém. Scient. Rev. Métallurg.*, 1962, **59**, 571; *Metal Treatment*, 1964, **31** (229), 401; (230), 436.
(2) R. Tricot, and R. Castro, *Mém. Scient. Rev. Métallurg.*, 1962, **59**, 587; *Metal Treatment*, 1964, **31** (231), 469.
(3) C. B. Post, and W. S. Eberly, *Trans. ASM*, 1951, **43**, 243.
(4) K. J. Irvine, D. J. Crowe, and F. B. Pickering, *J. Iron Steel Inst.*, 1960, **195**, 386.
(5) R. T. Howard, and M. Cohen, *Trans. AIME*, 1947, **172**, 413.
(6) A. E. Nehrenberg, and P. Lillys, *Trans. ASM*, 1954, **46**, 1176.
(7) J. Bourrat, L. Colombier, J. Hochmann, and J. Philibert, *Comptes Rendus Acad. Sciences*, 1957, **244**, 1197.
(8) J. H. Waxweiler, U.S.A. Patent No. 2,851,384, September 9, 1958.
(9) K. J. Irvine, D. T. Llewellyn, and F. B. Pickering, *J. Iron Steel Inst.*, 1959, **192**, 218.
(10) R. H. Thielemann, Proced. ASTM, 1940, **40**, 788.
(11) W. Hume-Rothery, and G. V. Raynor, 'The structure of metals and alloys', Inst. of Metals, 1954.
(12) W. Tofaute, A. Sponheuer, and H. Bennek, *Archiv. Eisenhüttenwes.*, May 1936, **8**, 499.
(13) W. Tofaute, C. Küttner, and A. Buettinghaus, *Ibid.*, June 1936, **9**, 607.
(14) N. R. Griffing, W. D. Forgeng, and G. H. Healy, *Trans. AIME*, February 1962, **224**, 148.
(15) K. Bungardt, E. Kunz, and E. Horn, *Arch. Eisenhüttenwes.*, March 1958, **29**, 193, 261.
(16) E. Baerlecken, W. A. Fischer, and K. Lorenz, *Stahl. u. Eisen.*, June 1961, **81**, 763.
(17) K. Kuo, *J. Iron Steel Inst.*, 1954, **176**, 433; 1955, **181**, 128, 213.
(18) A. E. Nehrenberg, *Metal Progress*, November 1951, **60**, 64.
(19) R. L. Rickett, W. F. White, C. S. Walton, and J. C. Butler, *Trans. ASM*, 1952, **44**, 138.
(20) W. Peter, and W. Matz, *Archiv. Eisenhüttenwes*, 1957, **28**, 807.
(21) S. Drapal, *Hutnické Listy*, 1960, **15**, 961.
(22) H. Jolivet, *J. Iron Steel Inst.*, 1939, **140**, 95.
(23) K. Kuo, *Ibid.*, 1953, **173**, 363.
(24) S. Heiskanen, *Jernkontor. Annal.*, 1955, **139**, 361.
(25) J. Koutsky, and J. Jezek, *Hutnické Listy*, 1958, **13**, 1098.
(26) J. K. Stanley, *Trans. AIME*, 1949, **185**, 752.
(27) C. Wells, W. Batz, and R. F. Mehl, *Ibid.*, 1950, **188**, 553.
(28) T. Heumann, and H. Boehmer, *Archiv. Eisenhüttenwes.*, 1960, **31**, 749.

HIGH TEMPERATURE TRANSFORMATIONS IN FERRITIC STAINLESS STEELS CONTAINING 17 TO 25% CHROMIUM

By A. E. Nehrenberg and Peter Lillys

Abstract

The rather complex isothermal transformations which occur in both the austenite and delta ferrite of commercial 17, 21 and 25% chromium steels heated sufficiently for complete solution of carbides (and nitrides, when present) were studied.

Precipitation of the $M_{23}C_6$ *carbide precedes the transformations at all temperature levels. At temperatures below the* A_1 *there is next a separation of ferrite containing little or no precipitated carbide while the earlier grain boundary precipitation reaction continues. The pearlite reaction follows the precipitation and ferrite separation reactions. The amount of ferrite formed from austenite decreases and the amount of pearlite increases with decreasing temperature. At the lowest temperatures for austenite transformation, formation of a dark-etching aggregate precedes the pearlite reaction.*

At temperatures ranging from about 50 °F below the A_1 *temperature to about 200 °F above, a transformation involving the formation of an aggregate of lamellar carbides in austenite from delta ferrite was observed. The transformation exhibits the characteristics of the pearlite reaction and stops when equilibrium is attained among the participating phases, austenite, delta ferrite and carbide. This transformation is not peculiar to high chromium steels, but occurs in other steels which contain ferrite (delta) at high temperatures.*

Some observations are reported in connection with the formation of austenite from delta ferrite during a quench from very high temperatures.

Evidence is presented which indicates that the decomposition of austenite at temperatures slightly below the A_1 *is the result of growth of residual (delta) ferrite grains. There is virtually no nucleation of ferrite at such temperatures. Similarly, at higher temperatures, the changes in proportions of ferrite (delta) and austenite with variations in temperature dictated by equilibrium considerations result from the growth of one phase into the other.*

A paper presented before the Thirty-fifth Annual Convention of the Society, held in Cleveland, October 17 to 23, 1953. Of the authors, A. E. Nehrenberg is supervisor of the Research Laboratory, and Peter Lillys is research metallurgist, Crucible Steel Company of America, Harrison, N. J. Manuscript received April 21, 1953.

Source: *Transactions of the American Society for Metals*, 46, 1954, 1177-1213, © American Society for Metals

DURING an experimental program which involved the study of a number of modified chromium steels possessing mixed microstructures of austenite and ferrite at high temperatures, the authors observed evidence of an unusual kind of transformation which seemingly involved the formation at temperatures well above the A_1* of an agregate structure consisting of lamellar carbides in austenite from high temperature (or delta) ferrite. The lamellar microconstituent resembled pearlite, but consisted at room temperature of lamellar carbides in martensite rather than in ferrite. Some additional work soon showed that the transformation of delta ferrite at high temperatures to an aggregate of lamellar carbides in austenite was by no means peculiar to the experimental compositions being studied. Rather, it was also found in the commercial "nonhardenable" straight chromium stainless steels, in tungsten steels (1)[1], and in a Ni-Cr stainless steel (Type 329) as well.

No indication has been found in the literature that such a transformation had previously been described. As a matter of fact, relatively little has been published concerning the metallography and transformation behavior of the straight chromium steels of higher chromium content.

Although Types 430, 442 and 446 stainless are generally considered to be nonhardenable, they always contain some austenite if heated to appropriate temperatures. In view of the increasing commercial importance of such steels, it seemed reasonable that a knowledge of their transformation characteristics would be of some practical as well as metallurgical interest. Accordingly, these steels were selected for study, not only for the purpose of arriving at an understanding of the unusual delta ferrite transformation to austenite containing lamellar carbides but for the purpose of ascertaining the transformation behavior of the austenite as well.

A single commercial heat of each of the grades Types 430, 442 and 446 was selected at random from warehouse stock for study. The data pertaining to the transformation of austenite to its various products are summarized in the form of conventional TTT-diagrams which are discussed in the first part of the paper. Following this there is a description of the transformations which involve the delta ferrite.

Material

The compositions of the steels studied are shown in Table I. The material was in the form of hot-rolled, commercially annealed and pickled plates. The Type 446 plates were $\frac{3}{16}$ inch thick and the

*The A_1 temperature is defined as the lowest temperature at which austenite forms on heating.

[1]The figures appearing in parentheses pertain to the references appended to this paper.

Source: *Transactions of the American Society for Metals*, 46, 1954, 1177-1213

Table I
Compositions of Steels

Type	C	Mn	Si	Ni	Cr	Mo	Al	N
430	0.09	0.40	0.33	0.34	17.20	0.06	0.010	0.03
442	0.17	0.56	0.46	0.35	20.96	0.04	0.013	0.12
446	0.24	0.46	0.42	0.26	24.85	0.02	0.010	0.17

Types 430 and 442 plates were ¼ inch thick. For convenience the Types 430, 442 and 446 steels will be designated 17, 21 and 25% chromium steels, respectively, in the remainder of this paper.

Procedure

The transformation behavior of the three steels was studied by means of the well-known conventional metallographic procedures which do not require detailed description. The austenitizing temperatures employed were selected to assure complete solution of the carbides in short heating times. Short heating times were used to minimize carburization of the specimens during the austenitizing treatments for which a Sentry furnace equipped with a carbon muffle was used. The optimum austenitizing treatments consisted of a 15-minute heating at 2000 °F (1095 °C) for the 17% chromium steel, and a 5-minute heating at 2300 °F (1260 °C) for the 21 and 25% chromium steels.

Lead baths were used for the transformation studies at temperatures in the range from 900 to 1900 °F (480 to 1040 °C). In those instances where temperatures of 2000 and 2100 °F (1095 and 1150 °C) were required, small laboratory electric muffle furnaces were used.

The etchants consisted of various mixtures of picric and hydrochloric acids dissolved in alcohol, the proportions being varied according to the requirements for a particular specimen (2). For example, for the samples used to establish the time required for the start of carbide precipitation at the various temperatures, the optimum etchant consisted of an alcoholic 5% picric acid solution to which 0.1 to 0.2% HCl was added. When such small concentrations of HCl are present, the etching solutions are particularly sensitive to carbide precipitation and have little or no effect on samples containing no precipitated carbides. When, on the other hand, it was desired to develop details of the microstructure with a minimum amount of attack of the carbides, or of the depleted grain boundaries, HCl concentrations of the order of 2 to 5% in 5% picral generally gave best results.

The lineal analysis procedure described by Howard and Cohen was employed for obtaining some quantitative data (3).

Some X-ray diffraction work was done for the purpose of identifying the intermetallic compounds present in the isothermally trans-

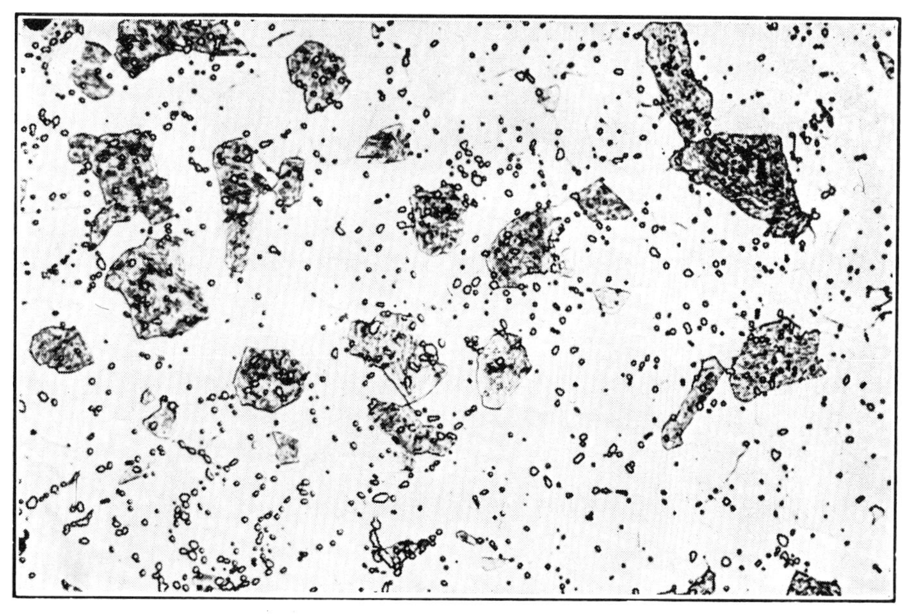

Fig. 1—Residual Intermetallic Compounds in Martensite Formed by Heating 21% Chromium Steel to 1740 °F (40 °F Above A_1 Temperature) and Water Quenching. Tempered 6 minutes at 1050 °F. Etch—5% picral + 2% HCl. × 500.

formed samples. The procedures employed have been described previously (4).

Results and Discussion

A_1 Temperatures

The term A_1 temperature as applied to the 17, 21 and 25% chromium steels of this investigation has the same meaning as in the iron-carbon system. That is, the A_1 temperature is the lowest temperature at which austenite forms on heating.

In this investigation the A_1 temperatures were determined by metallographic examination of samples heated for ½ hour at various temperatures, water-quenched and tempered for 6 minutes at 1050 °F (565 °C) to darken the martensite, when present. The metallographic examination established that the A_1 temperatures for the 17, 21 and 25% chromium steels are 1610, 1700 and 1920 °F (875, 925 and 1050 °C), respectively.

The carbides and nitrides present dissolve very slowly in the austenite with increasing temperature above the A_1. Temperatures 400 to 600 °F (205 to 315 °C) above the A_1 were found to be required for their complete solution.

Fig. 1 shows that the residual carbides and nitrides in martensite, produced by the transformation during cooling of the austenite formed in the 21% chromium steel at a temperature 40 °F above the A_1, are essentially of the same size as those in the areas which were not transformed to austenite. Since the carbon, nitrogen and chromium

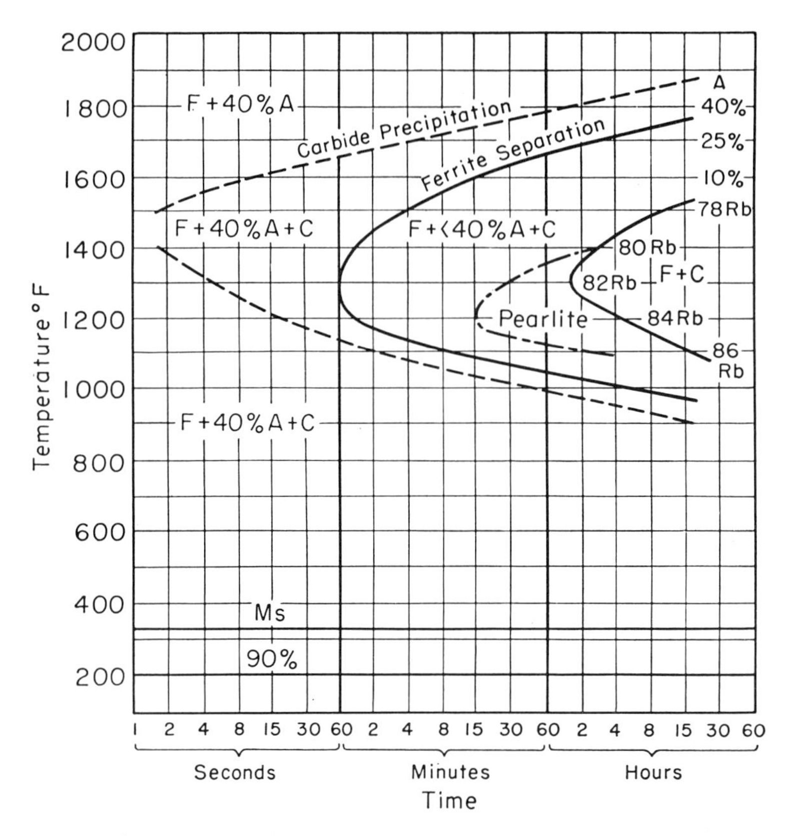

Fig. 2—TTT Characteristics of 17% Chromium Stainless Austenitized at 2000 °F 15 Minutes. A_1 temperature 1610 °F. Hardness as quenched—Rockwell B-101.

are largely concentrated in the carbides and nitrides, and since there has been little solution of these compounds in the austenite formed somewhat above the A_1 temperature, it can be concluded that such austenite is relatively low in carbon, nitrogen and chromium.

Isothermal Austenite Transformation Data

The isothermal transformation data for the austenites of the 17, 21 and 25% chromium steels are summarized in the TTT-curves shown in Figs. 2, 3 and 4. The austenite constitutes 40 to 50% of the microstructures of these steels at the various austenitizing temperatures. The curves are all of the single C type and indicate that these steels do not have a bainite reaction. In this respect the curves are similar to those which have been published for other low carbon steels of lower chromium content (5, 6, 7, 8). As little as 5% chromium together with 0.5% molybdenum and 0.1% carbon seems adequate to suppress completely the bainite reaction (6).

The 17, 21 and 25% chromium steels differ rather widely in the temperature of most rapid transformation. With increasing chromium in this range, the temperature corresponding to the minimum trans-

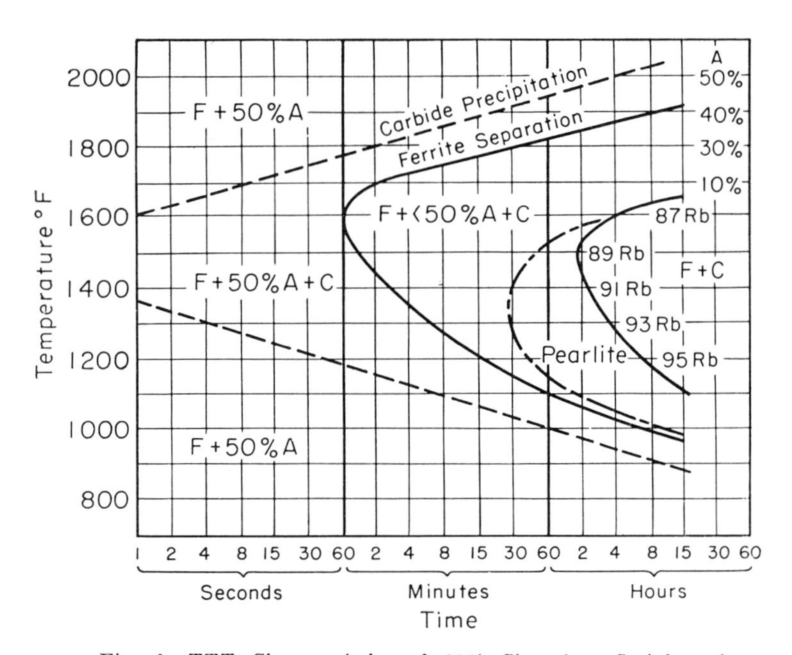

Fig. 3—TTT Characteristics of 21% Chromium Stainless Austenitized 2300 °F 5 Minutes. M_s temperature approximately minus 320 °F. A_1 temperature 1700 °F. Hardness as quenched—Rockwell B-95.

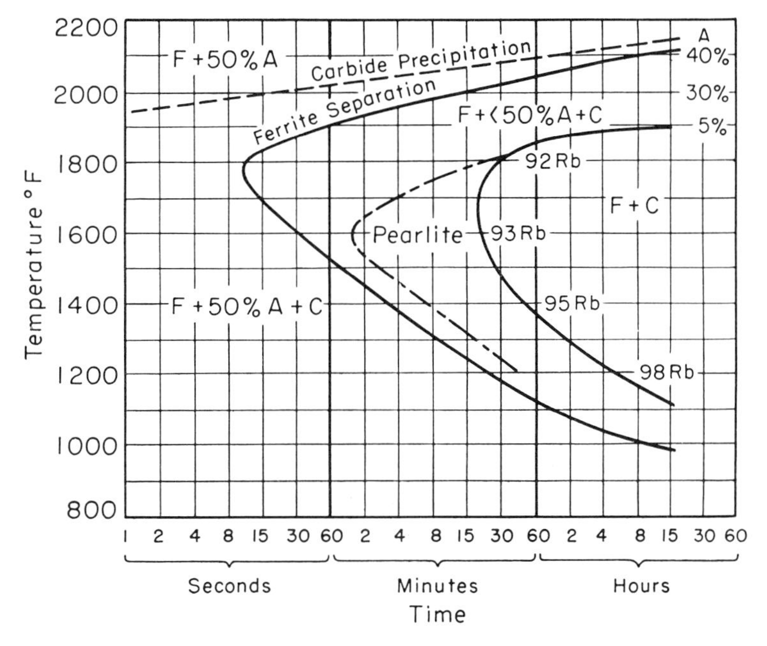

Fig. 4—TTT Characteristics of 25% Chromium Stainless Austenitized 2300 °F 5 Minutes. M_s temperature below minus 320 °F. A_1 temperature 1920 °F. Hardness as quenched—Rockwell B-98.

formation time increases from 1300 to 1600 or 1700 °F (705 to 870 or 925 °C). This increase in temperature of most rapid transformation with increase in chromium evidently is related to the fact that the A_1 temperature also is increased by a corresponding amount.

The steel of highest carbon and chromium content, the 25% chromium steel, can be completely transformed at temperatures above about 1300 °F (705 °C) more rapidly than either the 17% or the 21% chromium steel. In the temperature region of most rapid transformation, 1600 to 1700 °F (870 to 925 °C), the austenite of the 25% chromium steel can be transformed completely in about 20 minutes. About 1½ hours is required to transform completely the austenite of the 17 and 21% chromium steels at the temperatures where the transformation time is a minimum.

Carbide Precipitation

The precipitation of the $M_{23}C_6$ carbide precedes the transformation of austenite to its various decomposition products at all temperatures in each of the steels. This precipitation is so rapid in the 25% chromium steel studied that it could not be completely avoided by a water quench to room temperature. In the 17 and 21% chromium steels, on the other hand, it was possible to avoid precipitation by a direct quench to room temperature and by a quench into molten lead baths at temperatures of 1200 to 1300 °F (650 to 705 °C) and below.

Nature of Intermetallic Compounds Associated With Austenite Decomposition

In the case of the 17% chromium steel, X-ray diffraction analysis of extracted residues showed only the $M_{23}C_6$ carbide to be present in samples completely transformed at high temperatures. Extracted residues from the 21 and 25% chromium steels of high nitrogen content completely transformed at temperatures in the range 1500 to 1900 °F (815 to 1040 °C) for the 25% chromium steel, and 1300 to 1600 °F (705 to 870 °C) for the 21% chromium steel, were all found by X-ray diffraction to contain both the carbides $M_{23}C_6$ and M_7C_3 together with a nitride of the type M_2N.

Preliminary attempts to identify all of these constituents metallographically by the use of staining reagents were unfruitful and this phase of the work was temporarily abandoned. Metallographic identification of the various compounds seemed not to be required in the present investigation, for it was evident by a comparison of the 17% chromium steel which contained only the $M_{23}C_6$ carbide with the more complex 21 and 25% chromium steels that the basic transformation behavior seemed to be the same whether a single carbide or different carbides and the nitride were involved.

In general, it was established that in the 21 and 25% chromium steels, the $M_{23}C_6$ carbide formed first and that the M_7C_3 carbide and the nitride formed considerably later after the transformation of austenite was well advanced.

In view of the difficulty of distinguishing among the various compounds by metallographic procedures, it will be convenient to use the general term "carbide" when reference is made to the intermetallic compounds in the balance of the discussion.

Martensite Reaction

It was established by means of the conventional quench-temper procedure (9) that the austenite of the 17% chromium steel has an M_s temperature of 320 °F and that the martensite reaction is about 90% complete at 200 °F. Small amounts of austenite are retained when this steel is quenched from 2000 °F (1095 °C) to room temperature.

The M_s temperature of the austenite in the 21% chromium steel was found to be about —320 °F. In the case of the 25% chromium steel, no martensite could be formed by a quench and refrigeration treatment at —320 °F.

These data are applicable only to small samples rapidly quenched to avoid as much precipitation during cooling as possible. It will be shown later that the precipitation reaction, if permitted to occur, markedly raises the M_s temperatures of the impoverished austenites.

Microstructures Resulting From Isothermal Transformation of Austenite

Figs. 5, 6 and 7 illustrate the transformation products formed from austenite in the 25% chromium steel at various temperatures in the range 1900 down to 1100 °F (1040 to 595 °C). The microstructures of the 17 and 21% chromium steels go through the same series of changes with decreasing transformation temperature, but the temperature at which a given microstructure is produced varies from one steel to the other in a manner which seems to bear some relationship to the A_1 temperature.

Fig. 5a is typical of the transformation product which forms in the 17, 21 and 25% chromium steels at temperatures close to the A_1. The microstructure resulting from complete transformation at such temperatures consists of rather massive carbides located primarily in what formerly were ferrite: ferrite or ferrite: austenite grain boundaries. The ferrite which existed prior to the start of transformation (delta ferrite) is virtually indistinguishable from the ferrite resulting from the decomposition of the austenite.

Fig. 5b illustrates the microstructure formed at a slightly lower temperature. There is an increase in the amount of carbide precipitated within former austenite grains and in the delta ferrite as well. The amount of carbide in the delta ferrite areas is about the same as in the areas which were formerly austenite.

When the transformation temperature is lowered still further

Fig. 5—Transformation Products Formed in 25% Chromium Steel at 1900, 1800 and 1700 °F. Etch—5% picral + 5% HCl. × 500. (a) 16 hours at 1900 °F; (b) 30 minutes at 1800 °F; (c) 30 minutes at 1700 °F.

the areas which had been austenite contain two types of products as Fig. 5c shows. One is ferrite containing few precipitated carbides and the other is pearlite.

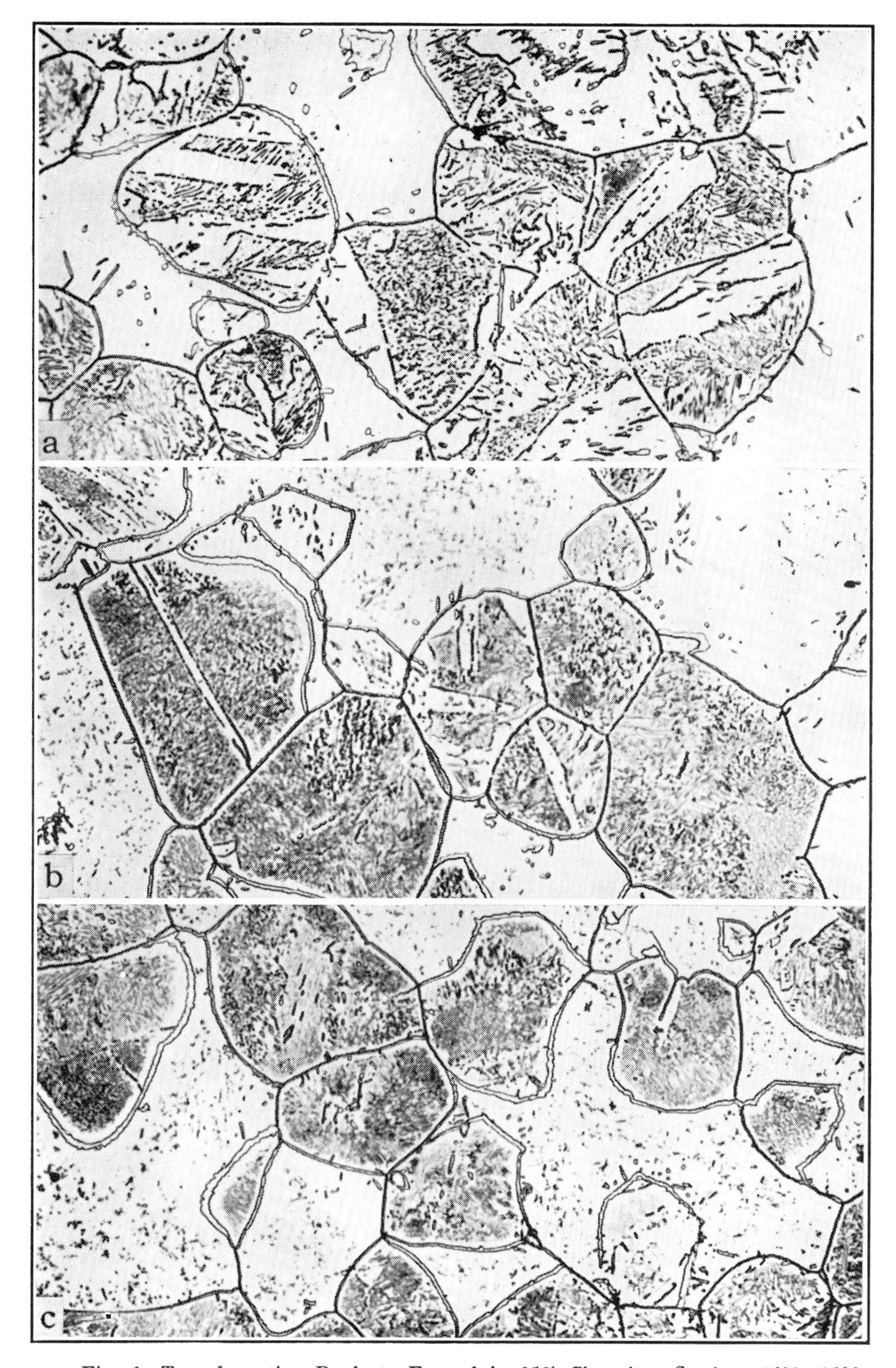

Fig. 6—Transformation Products Formed in 25% Chromium Steel at 1600, 1500 and 1400 °F. Etch—5% picral + 5% HCl. × 500. (a) 30 minutes at 1600 °F; (b) 30 minutes at 1500 °F; (c) 1 hour at 1400 °F.

Fig. 6 shows that with continued lowering of the transformation temperature there is a decrease in the amount of the ferrite and an increase in the amount of pearlite in the areas that were formerly

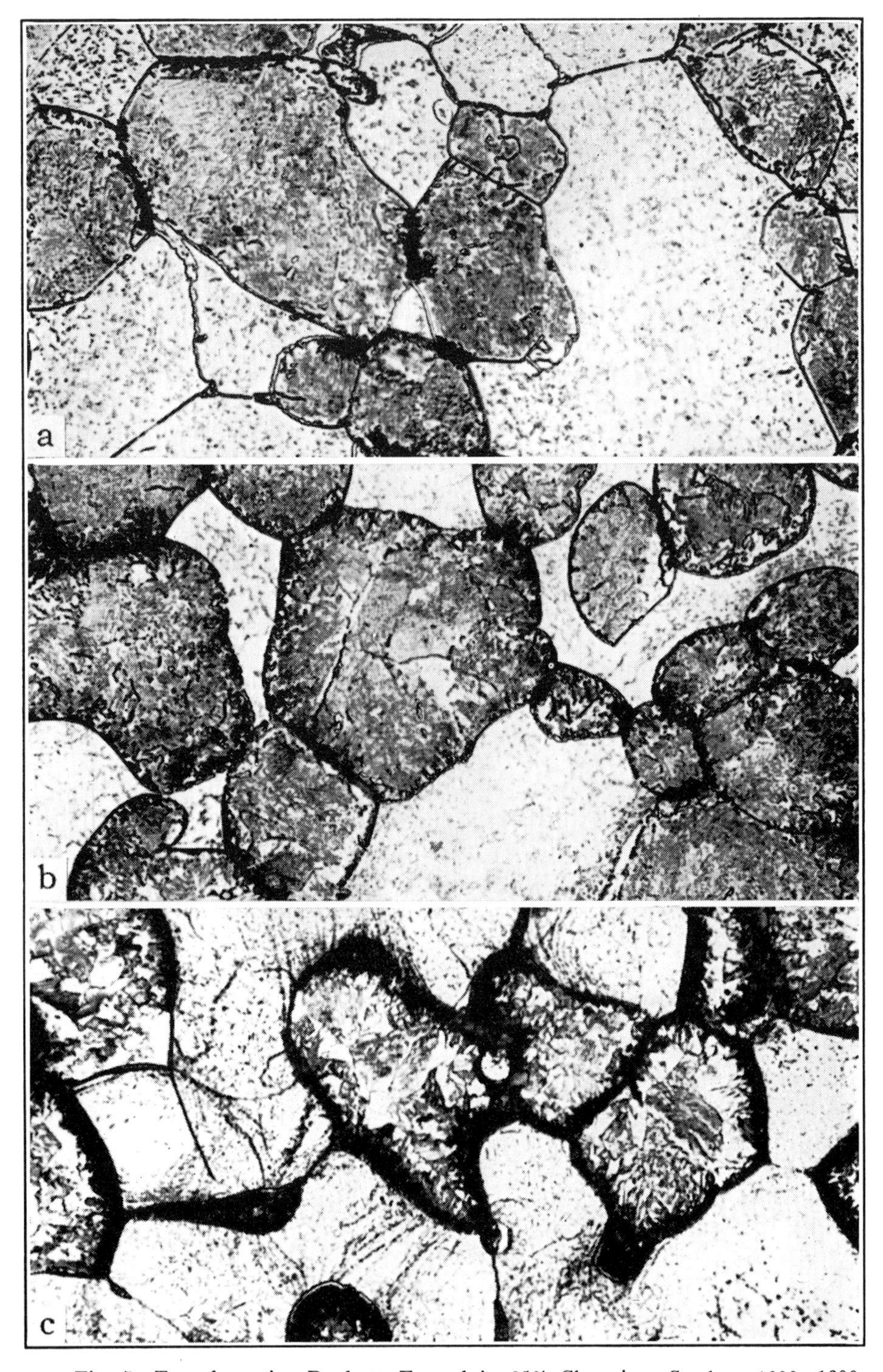

Fig. 7—Transformation Products Formed in 25% Chromium Steel at 1300, 1200 and 1100 °F. Etch—5% picral + 5% HCl. × 500. (a) 4 hours at 1300 °F; (b) 16 hours at 1200 °F; (c) 16 hours at 1100 °F.

austenite. The pearlite and the precipitation in the delta ferrite both become finer with decreasing temperature.

At temperatures of 1300 °F (705 °C) and below there is no

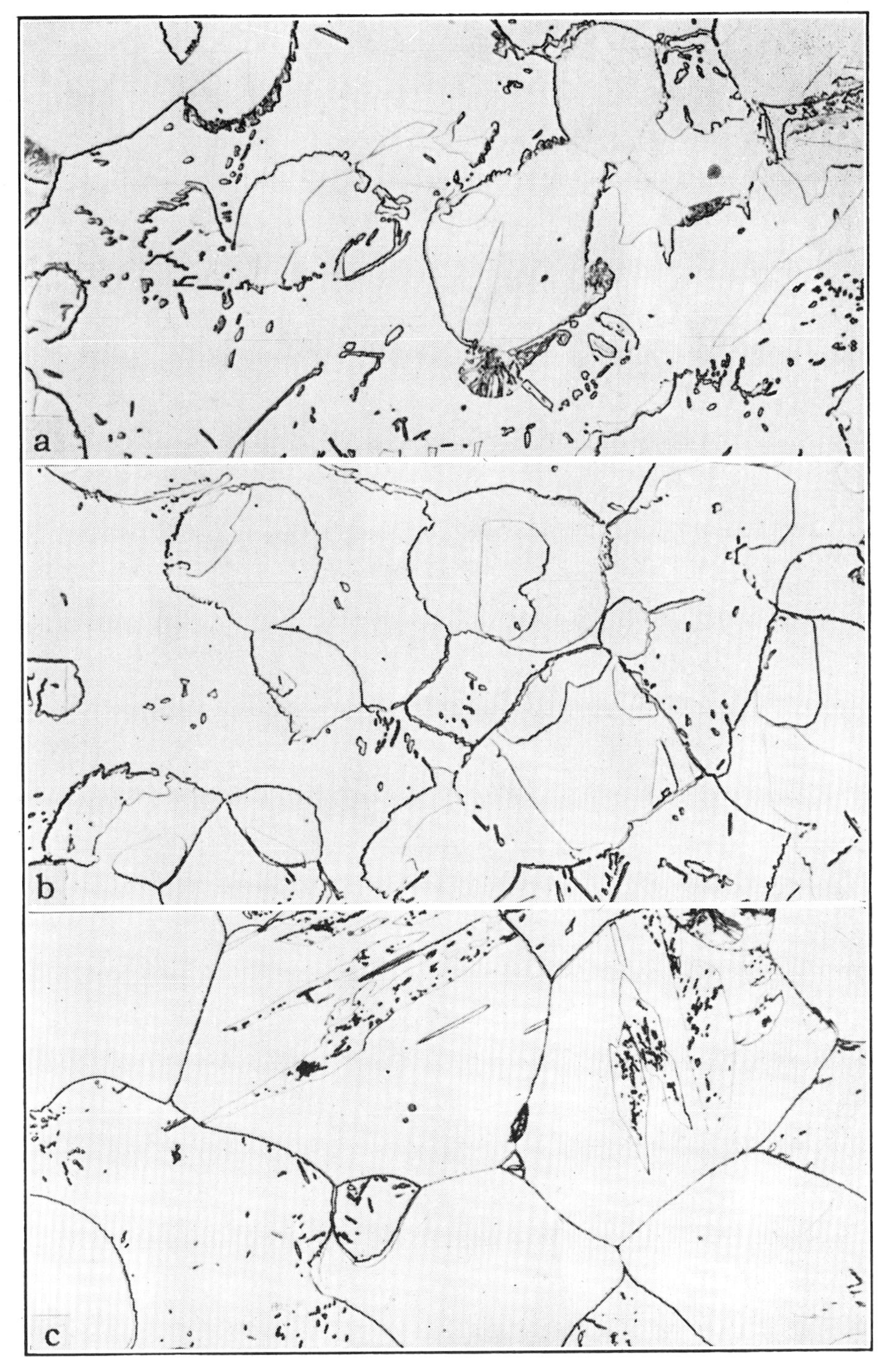

Fig. 8—Intermediate Stages in the Transformation of 25% Chromium Steel. Etch —5% picral + 5% HCl. × 500. (a) 6 minutes at 1800 °F; (b) 10 minutes at 1800 °F; (c) 4 minutes at 1600 °F.

longer any separation of ferrite in the 25% chromium steel as Fig. 7 shows. Instead, there is a fine, dark-etching aggregate which appears in the austenite adjacent to the grain boundaries. The amount of this

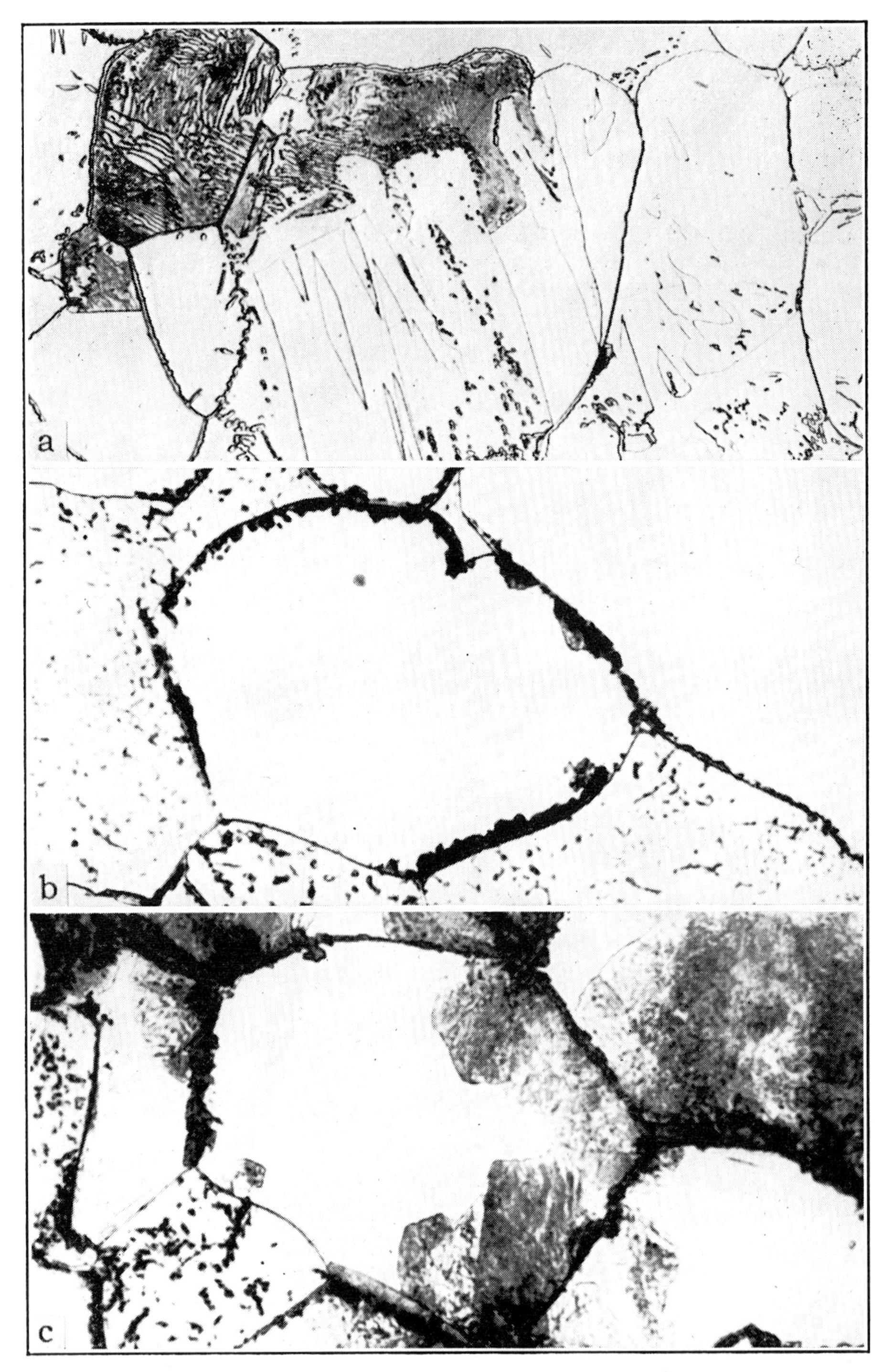

Fig. 9—Intermediate Stages in the Transformation of 25% Chromium Steel. Etch —5% picral + 5% HCl. (a) 6 minutes at 1600 °F. × 750; (b) 30 minutes at 1200 °F. × 1000; (c) 1 hour at 1200 °F. × 1000.

aggregate increases to some extent with decreasing transformation temperature. The transformation product which occupies most of the prior austenite areas is fine pearlite.

Figs. 8 and 9 illustrate intermediate stages in the transformation of the 25% chromium steel at various temperatures. Figs. 8a and 8b illustrate the manner in which the austenite of the 25% chromium steel transforms at 1800 °F (980 °C). The first evidence of transformation appears within the austenite adjacent to the ferrite: austenite grain boundaries where small areas of ferrite are observed. The areas of ferrite grow with time from the boundaries to the interior of the austenite and thus consume the austenite. The ferrite areas which form at 1800 °F (980 °C) are generally irregular in shape and tend to be blocky although acicular areas are occasionally observed as Fig. 8a indicates.

It is of interest that the ferrite, although it occupies areas which were formerly austenite with relatively high carbon content, is quite free from carbide. Since the precipitation of carbide has preceded the transformation of austenite, the austenite presumably is saturated with carbon and cannot absorb carbon from those areas which are being transformed to ferrite. Evidently, then, most of the carbon in those areas diffuses to the grain boundaries where it is precipitated as chromium carbide as the ferrite grows.

Fig. 8c shows that the ferrite which forms from austenite becomes more acicular with decreasing transformation temperature, and that it contains carbide. There is a decrease in the amount of this ferrite with decreasing temperature, with none being observed below 1400 °F (760 °C) in the 25% chromium steel and below 1200 and 1300 °F (650 and 705 °C) in the 17 and 21% chromium steels, respectively.

The effect of temperature on the morphology and quantity of the ferrite formed by the transformation of austenite is similar to that which has been described by Mehl for hypoeutectoid carbon and low alloy steels (10).

Following partial transformation of the austenite to ferrite containing little or no precipitated carbide, pearlite begins to form in the 25% chromium steel, provided the temperature is below 1800 °F (980 °C). In the 17 and 21% chromium steels, the maximum temperatures at which pearlite can form are 1350 and 1550 °F (730 and 845 °C), respectively. Fig. 9a illustrates partial transformation to pearlite in the 25% chromium steel at 1600 °F (870 °C) following the formation of acicular ferrite.

The lower temperature transformation products are illustrated by Figs. 9b and 9c. The first product which forms at 1200 °F (650 °C) in the 25% chromium steel, and at the lower temperatures in the other steels as well, is a dark-etching aggregate which appears at the grain boundaries and is illustrated by Fig. 9b. At a somewhat later time, transformation to fine pearlite begins, as Fig. 9c shows.

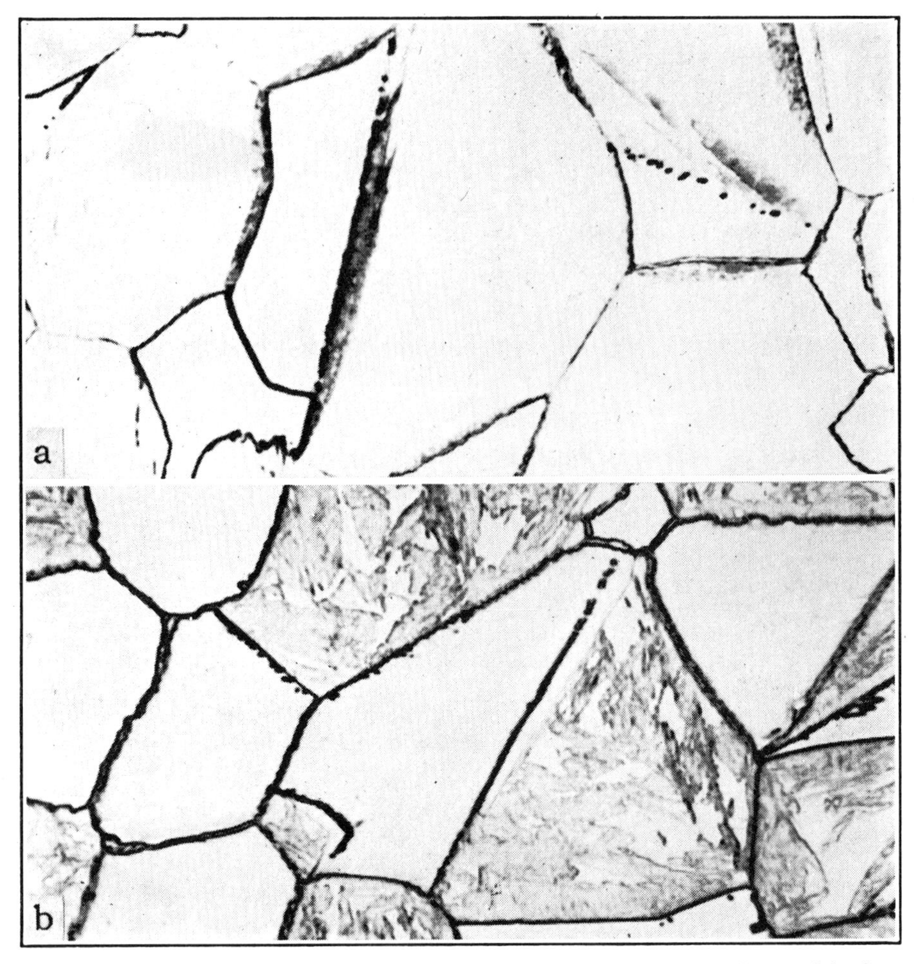

Fig. 10—Depletion of Austenite in 21% Chromium Steel by Carbide Precipitation Prior to Austenite Transformation. × 1500. (a) 2 minutes at 1400 °F. Etch—5% picral + 0.8% HCl. (b) 6 minutes at 1400 °F. Etch—5% picral + 2% HCl.

Depletion of Austenite by Precipitation

The precipitation reaction which precedes transformation at the various temperatures markedly raises the M_s temperatures by depletion of the austenite in carbon and chromium. Metallographic evidence of this depletion is contained in Fig. 10. Although the 21% chromium steel has an M_s temperature of about —320 °F (—160 °C) when small samples are quenched to avoid precipitation, Fig. 10a indicates that a few needles of martensite formed during cooling to room temperature in austenite adjacent to grain boundaries containing a precipitate which was formed during a 2-minute holding at 1400 °F (760 °C). A longer holding time serves to decrease the alloy content of the austenite still more. As the result of this, by the time transformation of the austenite begins, the austenite has become so depleted that much of it transforms to martensite during cooling to room temperature. This is illustrated by Fig. 10b.

The concept of austenite depletion during transformation in the high temperature region is not a new one. Several years ago Lyman and Troiano presented metallographic and X-ray evidence that the austenite in high carbon 6 and 9% chromium steels became depleted during transformation (11).

Essentially, the pearlite reaction in the 17, 21 and 25% chromium steels is one involving the transformation of austenite lowered in carbon and chromium by prior precipitation. Thus, the composition of the austenite at the time of the pearlite reaction approaches that of the austenite formed at temperatures just above the A_1 before it is enriched by the solution of the intermetallic compounds at higher temperatures.

The lines denoting the start of the pearlite reaction in the TTT-curves in Figs. 2, 3 and 4 join the line for complete transformation at higher temperatures. It is to be noted that this composite line approaches the A_1 temperature asymptotically in much the same manner as the single line denoting the start of transformation in low alloy steels approaches the A_1 temperature. This is probably related to the fact that the composition of the depleted austenite just prior to the pearlite reaction is not markedly different from the composition of the austenite formed just above the A_1 temperature.

Reaction Rate Curves

Reaction rate curves for the 25% chromium steel transformed at temperatures of 1900, 1700 and 1500 °F (1040, 925 and 815 °C) are shown in Fig. 11. The rate curve for the 1500 °F (815 °C) temperature level is of the shape characteristic of that for a nucleation and growth process (10). The 1500 °F (815 °C) transformation product of austenite (Fig. 6b) is predominantly pearlite, and it is generally agreed that pearlite forms by a process involving nucleation and growth.

The reaction rate curve for 1900 °F (1040 °C), on the other hand, is of a different type and suggests that some other process is operative. Actually, the metallographic work indicated that at 1900 °F (1040 °C) the austenite was gradually consumed primarily by growth of the existing (delta) ferrite. There appeared to be little or no evidence that nucleation was also involved in the decomposition of the austenite at this temperature.

Fig. 12 illustrates how the 1900 °F (1040 °C) transformation proceeds by growth of the residual ferrite. The area which occupies the center and bottom of the micrograph is austenite. The areas on either side are ferrite. The grain boundary at the upper left containing precipitated carbides is a ferrite: ferrite boundary. The austenite is gradually consumed by movement of the ferrite: austenite interfaces into the austenite. These interfaces are incoherent. Smith

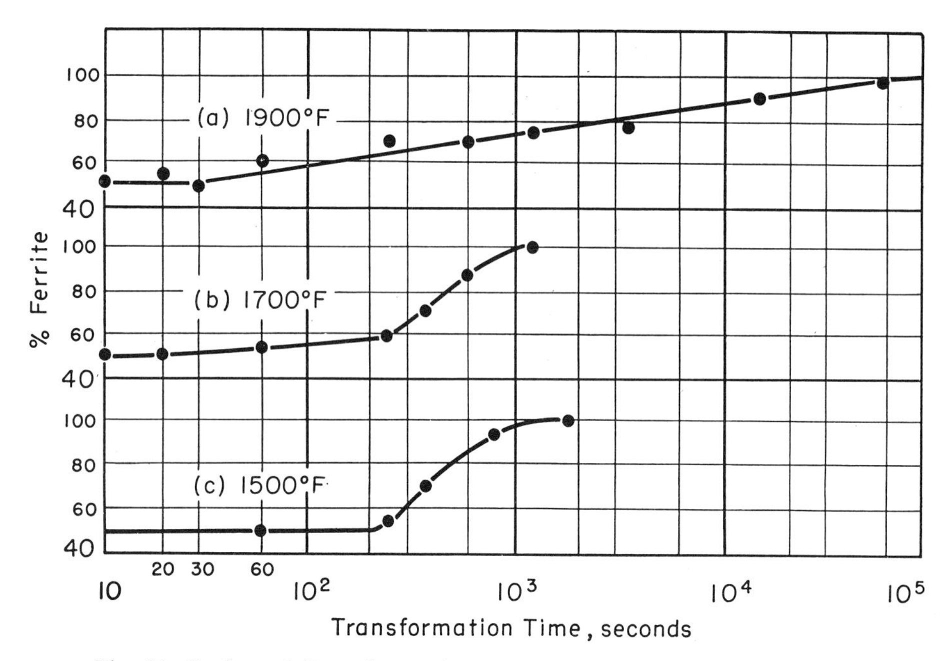

Fig. 11—Isothermal Rate Curves for 25% Chromium Steel Austenitized at 2300 °F 5 Minutes and Transformed at the Indicated Temperatures.

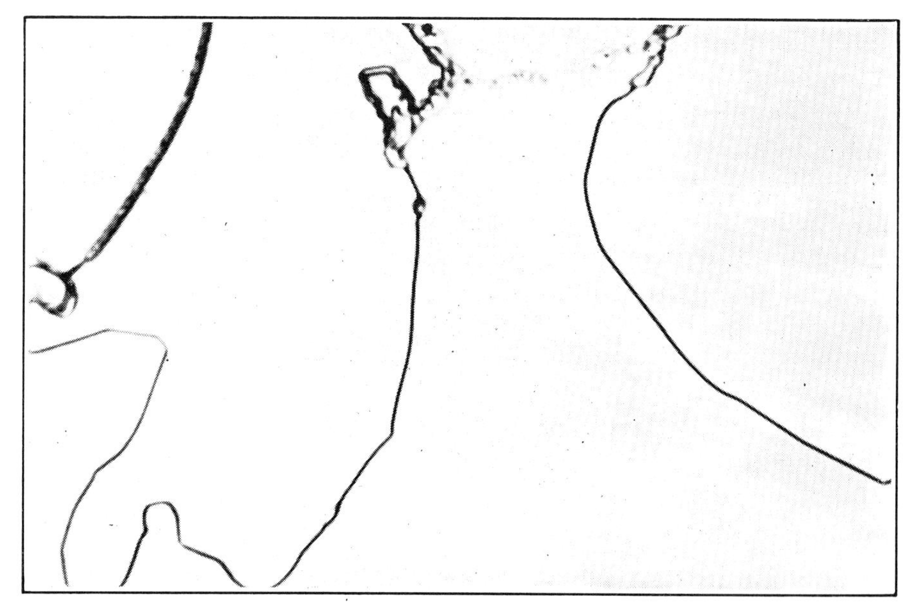

Fig. 12—Transformation of Austenite of 25% Chromium Steel by Growth of Existing Delta Ferrite. Austenite occupies center and bottom of micrograph. Sample austenitized 5 minutes at 2300 °F and transformed 4 minutes at 1900 °F. Electrolytically polished, etched in 5% HCl – 1% picral (Vilella's reagent). × 1500.

recently described how such incoherent interfaces can move during a phase transformation with a minimum expenditure of free energy (12). In the mechanism discussed by Smith it was necessary to assume nucleation of the stable phase in one grain with growth into an adjacent grain proceeding by movement of a disordered interface.

In the present case, grains of the stable phase (ferrite) already exist at the transformation temperature and the transformation proceeds by movement of the incoherent ferrite: austenite boundaries into the austenite.

Contrary to normal growth processes the rate of the 1900 °F (1040 °C) reaction at any instant is not directly proportional to the amount of the remaining austenite. Rather, it was found when an attempt was made to fit the data to the equation for a first-order reaction that the instantaneous rate decreased more rapidly than the amount of austenite decreased. Precipitation of carbide precedes and accompanies the 1900 °F (1040 °C) transformation. It is probable that the rate of growth is related to the rate of the precipitation reaction which would be expected to decrease as the equilibrium conditions for the system are approached.

The rate curve for 1700 °F (925 °C) exhibits characteristics of both the 1900 and 1500 °F (1040 and 815 °C) curves. That is, during the first 4 minutes at 1700 °F (925 °C) the rate curve is like that for the 1900 °F (1040 °C) transformation, whereas after 4 minutes the curve resembles the 1500 °F (815 °C) rate curve. The transformation diagram for this steel (Fig. 4) shows that the abrupt change in slope of the 1700 °F (925 °C) rate curve which occurs at about 4 minutes corresponds with the start of the pearlite reaction.

High Temperature Transformations Involving Delta Ferrite

It is well known that when the high chromium steels and other steels high in elements which restrict the gamma loop are heated to increasingly higher temperatures, austenite begins to form at the A_1 temperature and the amount of austenite increases with increasing temperature to some maximum value which may be 100% or less, depending upon the particular balance of composition. In the steels of the present investigation this maximum amount of austenite has been indicated to be 40 to 50%. With further increase in temperature, the amount of austenite decreases while the amount of ferrite increases. The ferrite which exists with austenite at very high temperatures will be defined as delta ferrite.

The present discussion will be concerned with the transformations which occur during the cooling of delta ferrite from very high temperatures into or through the range of temperatures at which the amount of austenite is a maximum. As this range is approached on cooling from some higher temperature, there is a decrease in the amount of delta ferrite with a corresponding increase in the amount of austenite. If there is residual austenite present at the maximum temperature attained, the increase in amount of austenite during cooling is accomplished simply by the growth of the existing grains of austenite into the delta ferrite.

Source: *Transactions of the American Society for Metals*, 46, 1954, 1177-1213

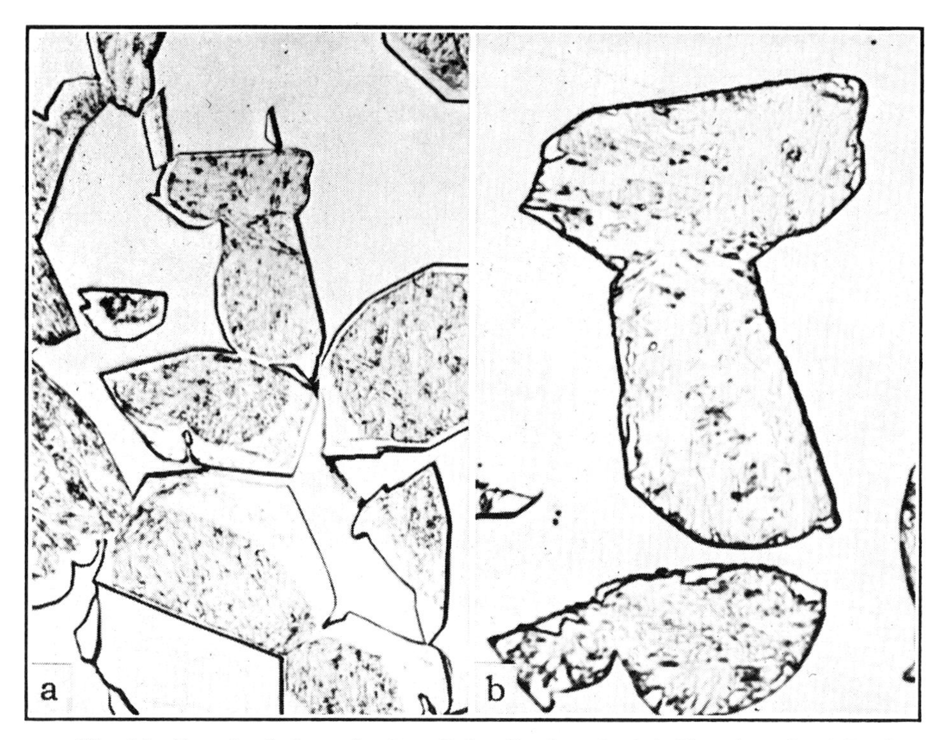

Fig. 13—Growth of Austenite Into Delta Ferrite of 17% Chromium Steel During Water Quench From 2300 °F. In (a) darker-etching core areas were austenite at 2300 °F in matrix of delta ferrite. The light-etching rims surrounding the darker-etching cores are the result of growth of austenite during the quench. (b) Same field as (a) but at higher magnification etched to develop details of grain structure. Specimen was refrigerated at minus 320 °F and tempered 5 minutes at 1050 °F. Electrolytically polished, etched (a) 10 minutes in 5% picral + 0.2% HCl, (b) 20% HCl + 1% picral. (a) × 750; (b) × 1500.

Fig. 13 illustrates this growth of austenite in the 17% chromium steel during a quench from 2300 °F (1260 °C). The growth of austenite occurs so rapidly that it cannot be entirely suppressed during cooling through the temperature region for maximum amount of austenite, even when small specimens are drastically quenched. The light-etching rims around the darker-etching grains in Fig. 13a denote areas formerly delta ferrite into which the austenite grew during the quench. The areas which were transformed to austenite during the cooling are distinguishable by a selective etching procedure from those which were austenite at the maximum temperature because they are of a different composition. Note that there are no well-defined boundaries separating the rim from the core. Fig. 13b shows several grains from the same field as that used for Fig. 13a photographed at a higher magnification after repolishing and re-etching to develop the martensitic structure. Again, there are no boundaries separating the rim from the core and the martensitic markings extend from the core into the rim portion without interruption. Fig. 13b also illustrates that the composition of the rim is different from that of the interior of the grains, for there is more retained austenite in the rim than in the core.

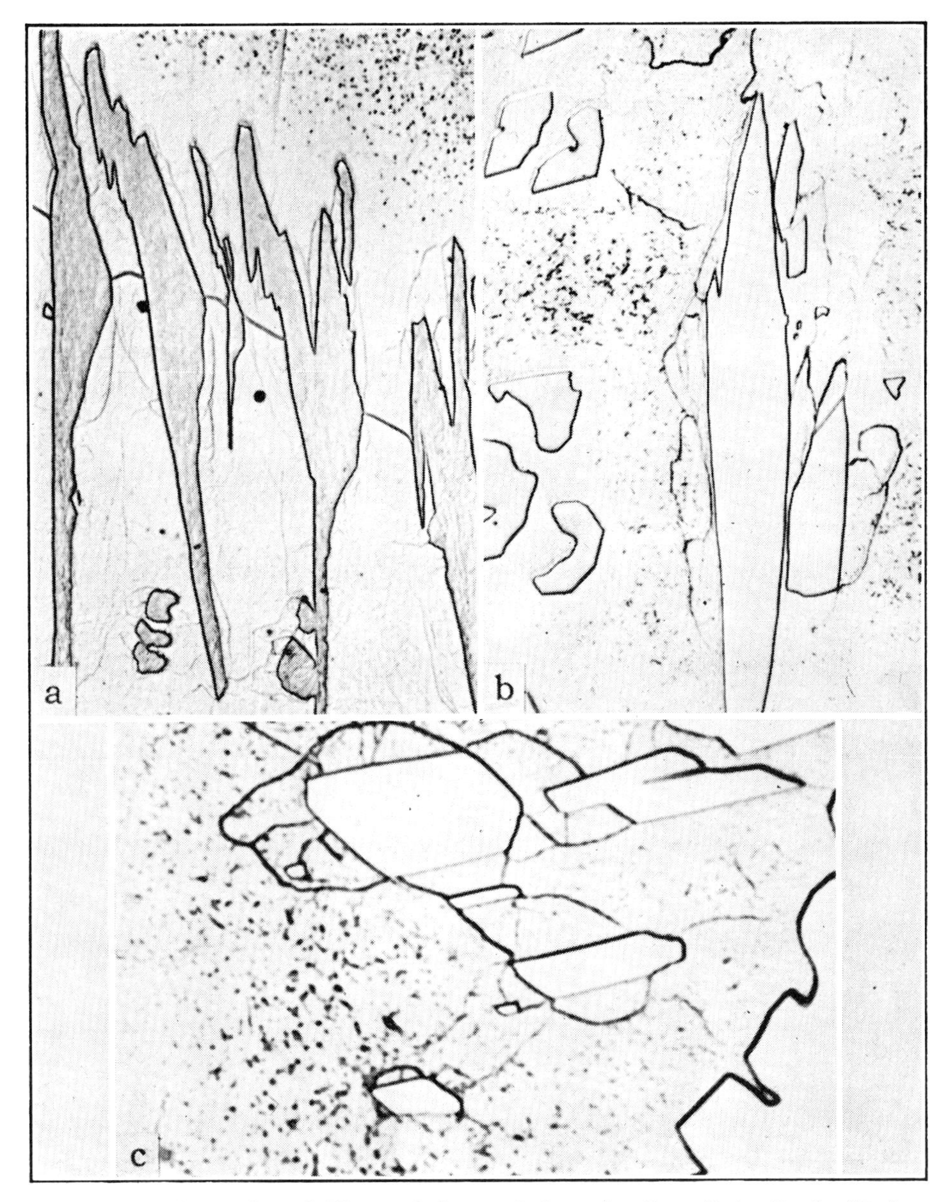

Fig. 14—Formation of Elongated Areas of Austenite From Delta Ferrite During Water Quench From 2500 °F. (a) 0.06% carbon – 12.5% chromium steel, tempered 1 minute at 1050 °F; (b) and (c) 25% chromium steel. Electrolytically polished, etched (a) 5% picral + 0.4% HCl; (b) and (c) 1% picral + 20% HCl. (a) × 500; (b) × 750; (c) × 1500.

In steel which is completely delta ferrite at the maximum heating temperature, the austenite which forms from the ferrite grows in elongated areas along definite planes of the ferrite and forms a Widmanstätten pattern. The literature contains a number of examples of such a distribution of austenite (8, 13, 14, 15).

Substructures which are commonly termed veining are associated with the rapid formation of austenite from delta ferrite during the quench. Veining occurs whether the austenite forms by growth of the existing austenite, or by nucleation and growth along certain planes in the ferrite.

Examples of veining surrounding the irregularly shaped areas resulting from austenite formation during the quench of a 12.5% chromium, and the 25% chromium steel, from 2500 °F (1370 °C) are illustrated in Fig. 14. The literature on veining in metals is voluminous. It has been clearly shown that veining can be produced by deformation at elevated temperatures which involves either externally applied mechanical deformation or deformation arising internally from the volume change which accompanies an allotropic transformation (16, 17, 18). The veining shown in Fig. 14 is of the latter type.

In a recent discussion of substructures in crystals, Guinier (18) pointed out that the factors involved in the formation of veining are the same as those which produce polygonization. Also, that similar subgrain boundaries are involved. In view of this, Guinier suggested that veining is a manifestation of polygonization, that is, the formation of slightly disoriented subgrains within a grain during an annealing treatment which does not produce recrystallization. In this connection, Greninger observed more than 15 years ago that ferrite grains with veining gave rise to broadened diffraction spots corresponding to an orientation dispersion of 1 to 10 degrees (19).

Veining, then, is a consequence of deformation and the pattern of veining can serve the useful purpose of providing an indication of the distribution of the strains resulting from the deformation. In Fig. 14a, for example, note particularly the acicular area of martensite at the extreme left of the micrograph. Veining is absent in the lower grain of ferrite at the left of the straight martensite boundary, but is present on the right. Evidently, the interface on the right was moving into the ferrite during the quench, whereas the interface on the left did not move laterally.

It seems clear that the process involved in the formation of the elongated areas of austenite from delta ferrite during the quench is one of nucleation and growth. The nuclei form at a grain boundary and grow at a much more rapid rate along definite planes in the ferrite than in directions normal to these planes. Further, the lateral growth may be blocked by certain planes in the ferrite so that growth is restricted. The small area at the center of Fig. 14a illustrates an early stage in the growth of a nucleus and the other larger areas represent a more advanced stage.

The distortion of the delta ferrite grain boundary by the delta ferrite → austenite transformation is of particular interest. The distortion is seemingly the result of variable rates of growth in different directions. This produces what amounts to a sudden pulsation or surge which can deform the grain boundary without actual longitudinal motion, or shear, being involved.

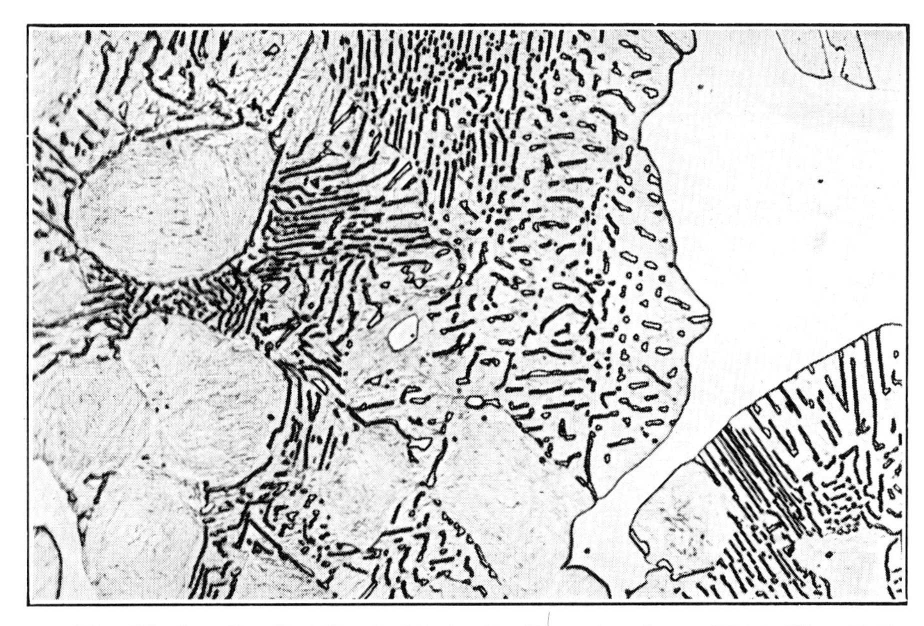

Fig. 15—Lamellar Carbides in Martensite Occupying Areas Which Were Delta Ferrite at 2000 °F. 17% chromium steel heated 15 minutes at 2000 °F, transformed 4 hours at 1750 °F, water-quenched, tempered 5 minutes at 1050 °F. Etch—5% picral + 2% HCl. × 500.

Transformation of Delta Ferrite to an Aggregate of Austenite and Carbides

The transformation of delta ferrite to austenite free from precipitated carbides, which has just been discussed, occurs very rapidly and at very high temperatures. Another transformation product of delta ferrite consisting of an aggregate of carbides in austenite was observed to form in the 17 to 25% chromium steels at a lower range of temperatures.

Fig. 15 illustrates this transformation product. The microstructure shown was produced by a 15-minute heating of the 17% chromium steel at 2000 °F (1095 °C) followed by a 4-hour transformation at 1750 °F (955 °C), water quench. A short tempering treatment was employed to darken the martensite. Prior to the start of this transformation at 1750 °F (955 °C) (which required about a 20-minute holding) the austenite which was present occupied the areas of martensite which do not contain carbides. The lamellar microstructure formed isothermally by the growth of carbide and austenite simultaneously into delta ferrite. All the areas which were austenite at the end of the 4-hour holding temperature at 1750 °F (955 °C) transformed to martensite during the quench to room temperature.

The transformation of delta ferrite to austenite containing lamellar carbides occurs in the 17, 21 and 25% chromium steels in the temperature interval extending from about 50 °F below the A_1 temperature to an upper limit 200 to 300 °F above the A_1. The transformation of delta ferrite to austenite without carbides, previously discussed, occurs at higher temperatures than these and could not be suppressed during cooling to the temperatures at which the austenite + carbide aggregate was observed to form isothermally.

The literature contains many examples of lamellar microstructures in a wide variety of ferrous and nonferrous alloys. Generally speaking, these microstructures are produced by a recrystallization reaction (20) or are the result of an allotropic transformation. Perhaps the most widely known example of a lamellar microstructure of the latter type is the steel structure, pearlite.

In both the pearlite and recrystallization reactions, the lamellar structure is developed by an edgewise growth of the lamellae in phase with a moving interface (20, 21). The transformation of delta ferrite to an aggregate of austenite and lamellar carbides proceeds by exactly the same mechanism. This is illustrated by the micrographs in Figs. 16a and 16b.

Like the pearlite reaction, the delta ferrite → austenite + carbide reaction is nucleated by carbide precipitation. Prior to the formation of the lamellar aggregate, carbide precipitation is observed either at austenite: ferrite interfaces, or at ferrite: ferrite interfaces. Then, there is growth of austenite from the boundary into the delta ferrite. Generally there is simultaneous edgewise growth of carbide lamellae, but in some cases there is growth of austenite into ferrite without edgewise growth of carbide. Fig. 16d is such an example. In this case precipitation in a ferrite: ferrite boundary nucleated the reaction, and it is to be noted that an aggregate of austenite and carbide had formed in the grain at the right, whereas little carbide seems to be associated with the growth into the grain at the left. The growth into the grain at the right is presumed to be of the same type as that illustrated by Figs. 16a and 16b, the essential difference being that a higher temperature is involved, with the result that spheroidization occurs almost as rapidly as the aggregate is formed. Well-defined lamellar structures were found to be quite rare when the austenite + carbide aggregate formed from delta ferrite at the higher temperatures involved in this reaction. Instead, partially or completely spheroidized carbides in austenite were observed.

Fig. 16c shows that the austenite of the 21% chromium steel containing the lamellar carbides has an M_s temperature above room temperature so that a great deal of martensite forms during the cool to room temperature, and there is no martensite in the austenite which existed previously except in the vicinity of the old boundaries where impoverishment due to precipitation has occurred. Since the residual austenite of the steel of Fig. 16c has an M_s temperature of about —320 °F when it is quenched rapidly from 2300 °F (1260 °C) to avoid precipitation, it is clear that the new austenite of the aggregate

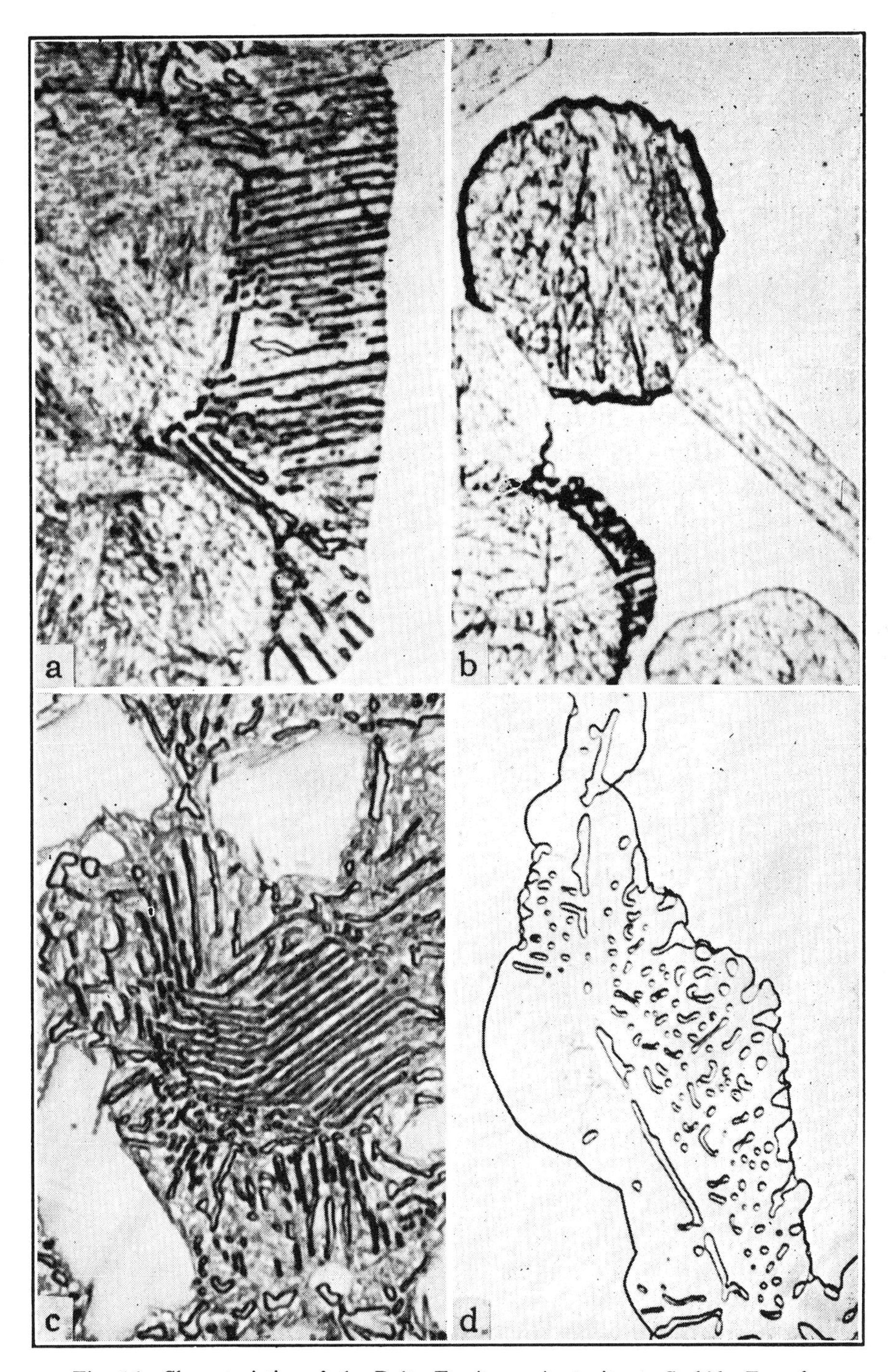

Fig. 16—Characteristics of the Delta Ferrite → Austenite + Carbide Transformation. (a) 17% chromium steel heated 15 minutes at 2000 °F and transformed 30 minutes at 1750 °F. Etch—5% picral + 2% HCl. × 2000. (b) 17% chromium steel heated 15 minutes at 2000 °F and transformed 2 minutes at 1600 °F. Etch—5% picral + 5% HCl. × 2000. (c) 21% chromium steel heated 5 minutes at 2300 °F and transformed 1 hour at 1800 °F. Etch—5% picral + 2% HCl. × 1200. (d) 25% chromium steel heated 5 minutes at 2300 °F and transformed 30 minutes at 2000 °F. Etch—5% picral + 5% HCl. × 1000.

structure formed by the transformation of delta ferrite has a much lower alloy content.

Fig. 16b shows that there has been a great deal of carbide precipitation at certain of the ferrite: austenite interfaces and virtually none at any of the austenite: austenite interfaces. This is a typical observation, for it was consistently observed in this investigation that the most favorable sites for precipitation were the ferrite: ferrite and ferrite: austenite boundaries. Precipitation was observed to occur much more slowly at austenite: austenite interfaces. This would seem to suggest that the diffusion required in the formation of the precipitate can take place more rapidly in the ferrite than in the austenite, but there are no data available to indicate whether or not the elements involved do actually diffuse more rapidly in ferrite than in austenite at the temperatures involved. Further, the chromium content of the ferrite is higher than that of the austenite and this also is a contributing factor.

Mechanism Involved in the Formation of the Aggregate of Lamellar Carbide in Austenite From Delta Ferrite

The transformation involving the formation of an aggregate of lamellar carbide in austenite from delta ferrite, like the pearlite reaction, is nucleated by precipitation of carbide. The areas surrounding the carbide are depleted in carbon and chromium by the precipitation of the $M_{23}C_6$ carbide and this sets up diffusion gradients which cause carbon and chromium to migrate toward the impoverished areas. Carbon diffuses more rapidly and minimizes the carbon concentration gradient. Chromium, on the other hand, diffuses much more slowly so a steeper chromium concentration gradient would exist. The delta ferrite depleted in chromium is unstable at the temperature involved and reverts to the stable phase austenite. In effect, the removal of chromium has served to extend the limits of the gamma field over a broader range of temperature.

As a result of the diffusion which accompanies the precipitation, there is more carbon as carbide in the area now occupied by the transformation product than there was in the delta ferrite from which the austenite-carbide aggregate formed. There are two kinds of evidence that this is the case. In the first place, a sample quenched after partial transformation to the aggregate, and subsequently tempered to precipitate and agglomerate the carbides, shows a smaller volume of carbides in the untransformed delta ferrite than that in the aggregate structure. Secondly, after the formation of the new austenite starts, the quantity of the residual austenite decreases. Much of the carbon involved in the formation of the aggregate comes from the areas of residual austenite which simultaneously transform to ferrite. The net effect of the overlapping reactions is that by the time the delta ferrite → austenite + carbide transformation has stopped there is less austenite in the sample (including that which formed from delta ferrite) than there was initially. The delta ferrite → austenite + carbide reaction does not go to completion, but stops when equilibrium is established among the phases austenite, ferrite and carbide (nitride).

The authors visualize that concentration gradients much like those which have been described for pearlite (21) exist at the moving interface between the lamellar structure and delta ferrite. In other words, at the ends of the lamellae the delta ferrite is impoverished in carbon and chromium and this provides the impetus for the diffusion required for the transformation.

Summary and Conclusions

The present paper describes the rather complex isothermal transformations which occur in both the austenite and delta ferrite of commercial 17, 21 and 25% chromium steels. The heating temperatures and times employed in this work were such that there was complete solution of the carbides (and nitrides, when present).

The transformation of austenite at all temperatures below the A_1 is preceded by carbide precipitation. Next, there is a separation of ferrite containing little or no precipitated carbides while the precipitation reaction continues at the grain boundaries. At temperatures just below the A_1 the austenite can be transformed completely by a combination of the precipitation and ferrite separation reactions.

No pearlite could be formed at temperatures close to the A_1, but at somewhat lower temperatures the pearlite reaction follows partial separation of ferrite. The amount of ferrite in completely transformed specimens decreases with decreasing temperature and the amount of pearlite increases correspondingly. At the lowest temperatures involved in the transformation of austenite, the formation of small amounts of a dark-etching aggregate structure precedes the pearlite reaction.

No bainite reaction was observed in the 17, 21 and 25% chromium steels. Martensite forms only in the 17% chromium steel during a quench from the heating temperatures employed.

At temperatures ranging from about 50 °F below the A_1 temperature to about 200 °F above the A_1, part of the delta ferrite present can be transformed isothermally to an aggregate of lamellar carbides in austenite. The transformation exhibits the characteristics of the pearlite reaction. That is, it is nucleated by carbide precipitation and proceeds by the edgewise growth of carbides in phase with a moving interface which is the result of an allotropic transformation. The matrix transformation is the reverse of that in the pearlite reaction, however. The transformation of delta ferrite to austenite containing lamellar carbides stops when equilibrium is established among the participating phases.

Some observations were reported in connection with the formation of austenite from delta ferrite during a quench from very high temperatures.

Evidence is presented which indicates that the decomposition of austenite at temperatures slightly below the A_1 is the result of growth of residual (delta) ferrite grains. There is virtually no nucleation of ferrite at such temperatures. Similarly, at higher temperatures, the changes in proportions of ferrite (delta) and austenite with variations in temperature dictated by equilibrium considerations result from the growth of one phase into the other.

Acknowledgment

The authors are grateful to the management of Crucible Steel Company of America for giving us permission to publish the results of the work described in the paper and to Mr. P. Payson, Assistant Director of Research, for his encouragement, constructive criticisms and advice. The authors are also indebted to their colleagues in the Research Laboratory, G. A. Stefanelli, A. G. Allten, A. Simon, F. Baureis and Mrs. A. Holcomb, who assisted us in various phases of the work.

References

1. A. E. Nehrenberg and J. G. Y. Chow, Discussion of paper by A. H. Grobe and G. A. Roberts, "Effect of Carbon Content on 18-4-1 High Speed Steel", TRANSACTIONS, American Society for Metals, Vol. 45, 1953, p. 475.
2. A. E. Nehrenberg, "Modified Picral for Higher Alloy Steels", METAL PROGRESS, Vol. 62, July, 1952, p. 91.
3. R. T. Howard and M. Cohen, "Quantitative Metallography by Point-Counting and Lineal Analysis", *Transactions,* American Institute of Mining and Metallurgical Engineers, Vol. 172, 1947, p. 413.
4. A. G. Allten and P. Payson, "The Effect of Silicon on the Tempering of Martensite", TRANSACTIONS, American Society for Metals, Vol. 45, 1953, p. 498.
5. R. L. Rickett et al, "Isothermal Transformation, Hardening, and Tempering of 12% Chromium Steel", TRANSACTIONS, American Society for Metals, Vol. 44, 1952, p. 138.
6. U. S. Steel Corp., "Atlas of Isothermal Transformation Diagrams", 1951.
7. Republic Steel Corporation, "S-Curves for Chromium Stainless Steels", METAL PROGRESS, Vol. 61, April, 1952, p. 96B.
8. A. E. Nehrenberg, "Transformation of Low-Carbon, 12% Chromium Stainless Steels", METAL PROGRESS, Vol. 60, November, 1951, p. 64.
9. A. B. Greninger and A. R. Troiano, "Kinetics of the Austenite to Martensite Transformation in Steel", TRANSACTIONS, American Society for Metals, Vol. 28, 1940, p. 537.
10. R. F. Mehl, "The Decomposition of Austenite by Nucleation and Growth Processes", *Journal,* Iron and Steel Institute, Vol. 159, June 1948, p. 113.
11. T. Lyman and A. R. Troiano, "Isothermal Transformation of Austenite

in One Per Cent Carbon, High-Chromium Steels", *Transactions,* American Institute of Mining and Metallurgical Engineers, Vol. 162, 1945, p. 196.
12. C. S. Smith, "Microstructure", TRANSACTIONS, American Society for Metals, Vol. 45, 1953, p. 533.
13. F. K. Bloom, W. C. Clarke, Jr., and P. A. Jennings, "Relation of Structure of Stainless Steel to Hot Ductility", METAL PROGRESS, Vol. 59, February 1951, p. 250.
14. J. H. G. Monypenny, "Stainless Iron and Steel", John Wiley & Sons, Inc., 2nd Edition, 1931.
15. V. N. Krivobok and M. A. Grossmann, "A Study of the Iron-Chromium-Carbon Constitutional Diagram", TRANSACTIONS, American Society for Steel Treating, Vol. 18, 1930, p. 760.
16. N. P. Goss, "Subboundary Structure of Recrystallized Iron", *Transactions,* American Institute of Mining and Metallurgical Engineers, Vol. 145, 1941, p. 272.
17. A. Hultgren and B. Herrlander, "Hot Deformation Structures, Veining and Red-Shortness Cracks in Iron and Steel", *Transactions,* American Institute of Mining and Metallurgical Engineers, Vol. 172, 1947, p. 493.
18. A. Guinier, "Substructures in Crystals", Imperfections in Nearly Perfect Crystals, John Wiley & Sons, 1952, p. 402.
19. A. B. Greninger, "Transformation Twinning of Alpha Iron", *Transactions,* American Institute of Mining and Metallurgical Engineers, Vol. 120, 1936, p. 293.
20. A. H. Geisler, "Precipitation From Solid Solutions of Metals", Phase Transformations in Solids, John Wiley & Sons, Inc., 1951.
21. F. C. Hull and R. F. Mehl, "The Structure of Pearlite", TRANSACTIONS, American Society for Metals, Vol. 30, 1952, p. 381.

DISCUSSION

Written Discussion: By A. J. Lena, R. A. Lula and G. C. Kiefer, Research Laboratory, Allegheny Ludlum Steel Corp., Brackenridge, Pa.

The authors are to be complimented on their presentation of a very interesting and valuable paper. It was particularly gratifying to us to see that their observations on the rapid rate of carbide precipitation were in general agreement with our work on intergranular corrosion[2] and we believe that their paper provides evidence in support of our proposed theory of intergranular corrosion.

There is one minor point of disagreement and that is in regard to the suppression of precipitation in Types 430 and 442 by water quenching. Unlike Nehrenberg and Lillys, we have never been able to prevent the carbide or nitride precipitation by water quenching of sheet samples of Types 430 and 442 as well as Type 446. This discrepancy may be due to a difference in etching technique for the detection of the grain boundary precipitate. In order to study the effect of etchants, we have etched 0.04-inch sheet samples of an 0.07% carbon Type 430 steel which had been water-quenched after 15 minutes at 2000 °F (1095 °C) in the 5% picric acid – 0.1 to 0.2% hydrochloric acid solution used by Nehrenberg and Lillys and electrolytically in a 10% ammonium persulphate solution which we have favored in our work. A 2-second etch in the ammonium persulphate

[2]R. A. Lula, A. J. Lena and G. C. Kiefer, "Intergranular Corrosion of Ferritic Stainless Steels", see this volume of TRANSACTIONS.

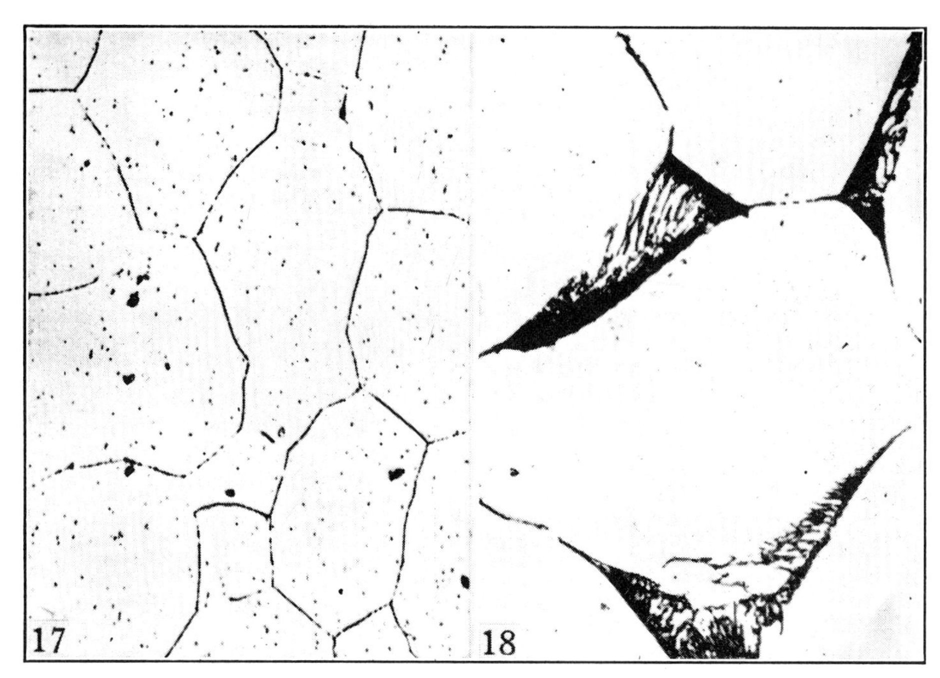

Fig. 17—Intergranular Precipitate in Type 430 Water-Quenched From 2000 °F as Revealed by a 2-Second Electrolytic Etch in Ammonium Persulphate. × 500.

Fig. 18—Photomicrograph of an Austenitic Chromium-Manganese Steel Showing a Lamellar Structure Consisting of Chromium Carbide ($Cr_{23}C_6$) and Austenite Which Developed in Areas Which Were Originally Delta Ferrite as a Result of Aging for 4 Hours at 1400 °F (760 °C). Etchant is electrolytic 10% chromic acid. × 1000.

solution was successful in showing a grain boundary precipitate in these samples, as can be seen in Fig. 17, whereas a 3-minute etch in the picric-hydrochloric acid solution gave no evidence of any precipitation.

The M_s temperature data in this paper provide a basis for explaining the difference of the effect of welding and post weld annealing on the ductility of Types 430 and 442. These differences can be seen in Table II where the tensile properties of a Type 430 and Type 442 steel are given. In the case of Type 430 where the M_s temperature of the austenite which formed during welding is above room temperature, as-welded samples are brittle, due to the presence of grain boundary martensite, whereas in Type 442 where the M_s temperature is below room temperature, the austenite is retained and the as-welded samples remain ductile. That the austenite is retained in this steel can be shown either metallographically or by magnetic measurements, in which case it was found that the weld metal and heat-affected zones were less magnetic than the base metal. Both of these steels are susceptible to intergranular corrosion, and in order to restore immunity it is necessary to anneal at temperatures between 1200 and 1500 °F (650 and 815 °C). This annealing treatment effectively tempers the martensite of the Type 430 so that the ductility as well as corrosion resistance is restored. With Type 442, however, this annealing treatment conditions the retained austenite by the precipitation of chromium carbides with a resultant raising of the M_s temperature so that the transformation to martensite can occur on cooling and the ductility is reduced. A second annealing treatment is necessary to temper the martensite and restore ductility.

Table II
Effect of Welding and Post Weld Annealing on the Corrosion and Tensile Properties of Type 430 and Type 442 Stainless Steel

	Type 430 (0.065-Inch Sheet)				Type 442 (0.046-Inch Sheet)			
Treatment	Y. S.	T. S.	% El.	Inter-granular Cor-rosion	Y. S.	T. S.	% El.	Inter-granular Cor-rosion
As-received	50,880	72,560	27	No	47,870	66,310	21	No
As-welded	50,860	77,500	7	Yes	54,400	74,160	20	Yes
As-welded + 5 min. at 1400 °F	45,300	71,200	25	No	48,890	64,450	6	No
As-welded + 5 min. at 1400 °F + 5 min. at 1400 °F	......	......	..	...	46,450	76,120	18.5	No

We were very much interested in the transformation of delta ferrite to carbide and austenite, for we have observed an identical transformation in the delta ferrite of austenitic chromium-manganese steels. An example of this transformation to a lamellar aggregate in this type of steel is shown in Fig. 18. Although not related to this paper, it might be mentioned here that when this transformation goes to completion in austenitic steels, the presence of initial delta ferrite is ineffective in promoting the formation of sigma in steels capable of sigma formation.

Finally, we should like to ask the authors if they have obtained any quantitative data on the individual effects of carbon and chromium on the positions of the isothermal transformation curves? In this respect we wonder if the more rapid rate of transformation in the Type 446 steel is not due to the high carbon content alone, rather than a high chromium content. We also believe that the phase diagrams available in the literature provide a more suitable means of determining the extent of carbon and chromium partitioning in the ferrite and the austenite of these steels, rather than relying on the use of photomicrographs.

Written Discussion: By R. A. Perkins, senior research assistant, Metals Research Laboratories, Electro Metallurgical Co., Niagara Falls, N. Y.

High temperature transformations in ferritic chromium steels have received little attention in the past few years, and the authors have made an important contribution to this rather complex subject. The effect of small residual amounts of austenite or its decomposition products on the properties of chromium stainless steels is ill-defined, and an improved understanding of the time-temperature transformation characteristics of these steels may lead to a better understanding of observed property changes.

The authors' observation that carbide precipitation precedes austenite decomposition is interesting and significant. We have had occasion to investigate high temperature transformations of austenite and ferrite in 21% chromium steels modified with small amounts of nickel, copper, and molybdenum to produce a duplex structure of ferrite plus 25 to 40% austenite in the annealed condition. Observations on this steel indicated the formation of an unstable austenite which decomposed to a martensitic structure and chromium carbide on heating at 1650 °F (900 °C), followed by rapid cooling, as shown in Fig. 19. The distribution of the coagulated

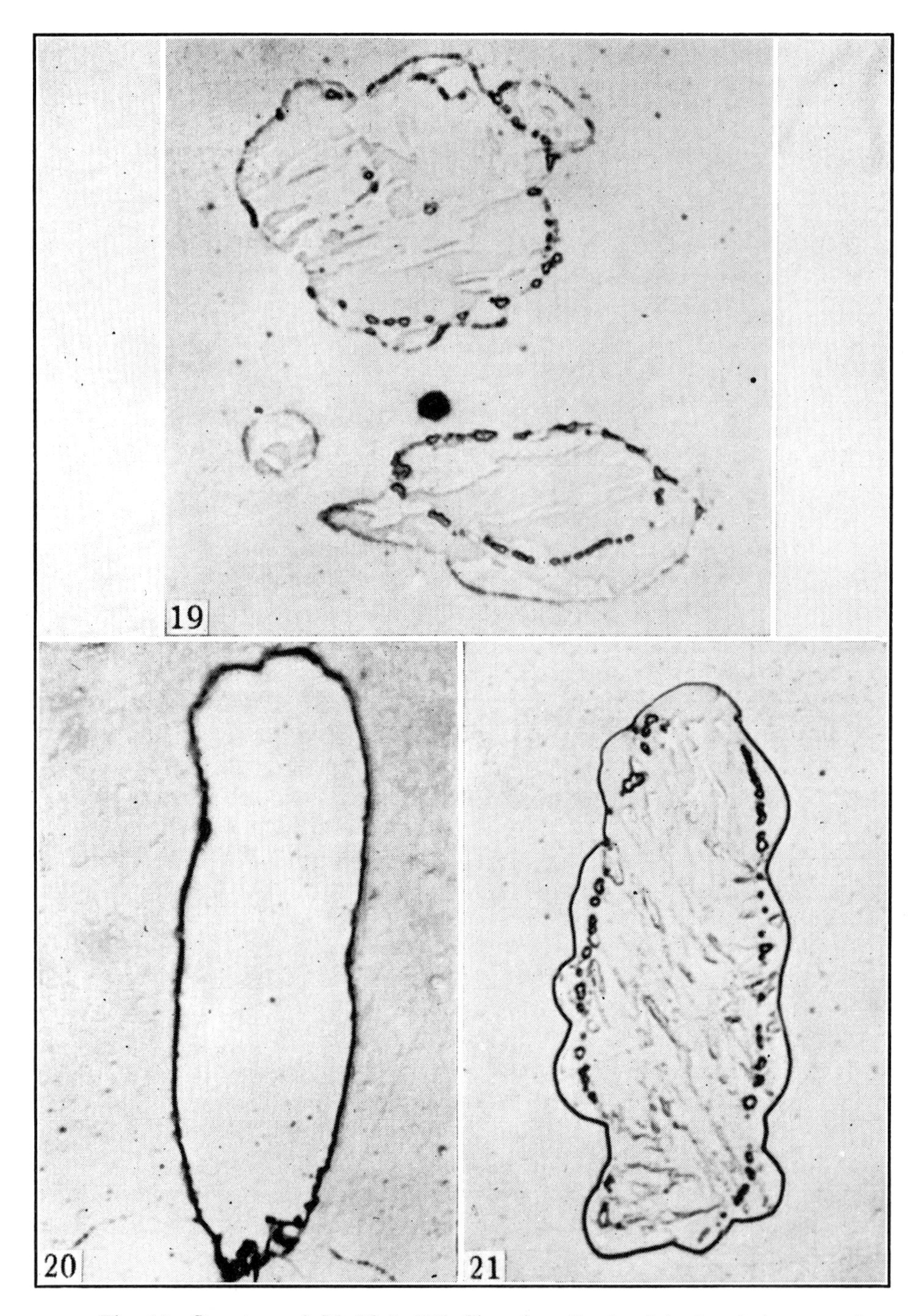

Fig. 19—Structure of Modified 21% Chromium Steel. Islands of decomposed austenite showing coagulated carbides. × 2000.

Fig. 20—Structure of Modified 21% Chromium Steel. × 2000. Island of stable austenite with finely divided carbides precipitated at austenite-ferrite boundary. Air-cooled from about 1470 °F (800 °C) after rolling.

Fig. 21—Structure of Modified 21% Chromium Steel. × 2000. Coagulation of carbides at prior austenite-ferrite boundary with growth and decomposition of austenite island on heating same field 1 hour at 1650 °F (900 °C), helium-cooled.

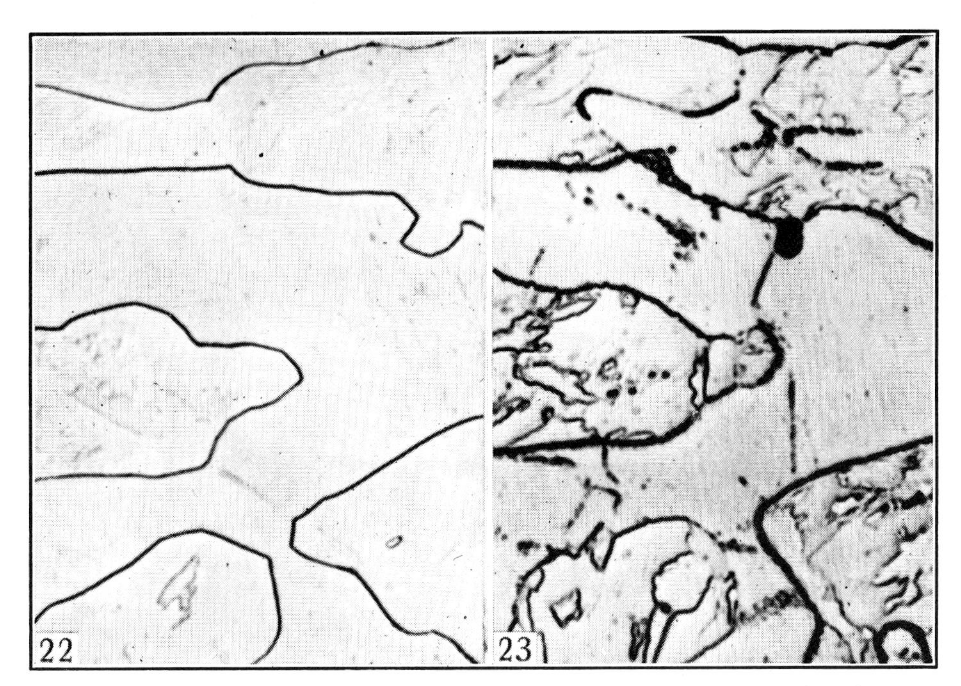

Fig. 22—Structure of Modified 21% Chromium Steel. × 2000. Solution of carbides accomplished by annealing 45 minutes at 1880 °F (1025 °C), helium-cooled.

Fig. 23—Structure of Modified 21% Chromium Steel. × 2000. Subsequent precipitation of finely divided carbides and decomposition of austenite in same field on reheating 1 hour at 1290 °F (700 °C), helium-cooled.

carbide particles suggests formation around prior austenite-ferrite boundaries, and subsequent investigations tended to support this conclusion, as shown in Figs. 20 and 21. It is seen that finely divided carbides precipitated at austenite-ferrite boundaries in the as-rolled structure tended to agglomerate on annealing at 1650 °F (900 °C), while the original austenite islands grew into the ferrite matrix and decomposed to give a martensitic structure. This sample was annealed in helium and lightly polished and re-etched to reveal these changes occurring in the selected field. It would seem that growth occurred on holding and not on cooling, since 1650 °F (900 °C) is in the range for maximum austenite formation. Decomposition probably occurred on cooling, due to an increase in the M_s temperature resulting from precipitation of chromium carbide. These observations check those of the authors in that the carbides apparently precipitate prior to any transformation of the austenite. Annealing these steels at a higher temperature (1880 °F or 1025 °C) effected a solution of the carbides and rendered the austenite more stable on cooling, as shown in Figs. 22 and 23. Subsequent heating at 1290 °F (700 °C) precipitated fine carbides in and around the austenite islands with accompanying decomposition of this phase.

The occurrence of carbide precipitation prior to austenite decomposition appears to be well-defined and is an important concept which may aid in better understanding the behavior of stainless steels at elevated temperatures.

Written Discussion: By Axel Hultgren, Djursholm, Sweden.

After the high temperature treatment, each steel consisted of an equilibrium mixture of ferrite, of higher chromium and lower carbon contents, and austenite, of lower chromium and higher carbon contents. At temperatures below A_1, as defined, both are supersaturated and the isothermal transformation reactions are governed by temperature. That is, of course, always true for austenite, but in this case the temperature dependence makes itself felt also in that, at low temperatures, ferrite and austenite transform more or less independently, whereas with increasing temperature more interaction occurs between the two.

In the low temperature range, carbide particles are precipitated within the ferrite grains and austenite transforms by ferrite separation and eutectoid decomposition, just as hypoeutectoid austenite generally does in the pearlite range. It is not to be expected, however, that equilibrium is reached at apparent completion of transformation. As transformation temperature is increased, a tendency for localized carbide precipitation at ferrite grain boundaries and at phase boundaries becomes evident (Figs. 5 to 7).

The authors attribute the preferred selection of those sites for carbide precipitation tentatively to the diffusion in ferrite being faster than in austenite. The writer would rather suggest, as the main cause, that nucleation of carbide, under the conditions in question, is for some reason facilitated in ferrite[3].

The fact that, in the higher transformation range, no lamellar eutectoid is formed may be explained as follows: After carbide has been precipitated at ferrite grain and phase boundaries and adjoining ferrite has become depleted in carbon and chromium, the austenite is greatly supersaturated with respect to the depleted ferrite, and rapid diffusion, particularly of carbon, will set in, enabling the ferrite to precipitate more carbide and relieving the ausenite of its supersaturation with respect to carbide to such an extent that no carbide is nucleated in it and thus no eutectoid formed. Instead, austenite becomes increasingly supersaturated with respect to ferrite, and may be invaded by ferrite without any carbide forming in the areas previously austenitic. If this interpretation is correct, the carbide particles in ferrite, as seen in Figs. 8c and 9a, were precipitated after, and not simultaneously with, the ferrite separation.

When ferrite is hot-deformed, and veining produced, grain boundaries have been observed to migrate[4]. The "deformed" grain boundary in Fig. 14a probably also resulted from migration, under the influence of transformation stress. Possibly that is what the authors meant.

The formation of an apparent austenite-carbide eutectoid in supersaturated ferrite is an intriguing phenomenon. The authors' suggested explanation, which probably is substantially correct, is based on a number of important observations of structural changes during the process. The short-distance diffusion of chromium is reflected in the spacing of the carbide lamellae, and the long-distance diffusion of carbon in the con-

[3]Axel Hultgren and Collaborators, "Isothermal Transformation of Austenite and Partitioning of Alloying Elements in Low Alloy Steels", *Kungliga Svenska Vetenskapsakademiens Handlingar,* Fourth Series, Vol. 4, No. 3, 1953, p. 20, 38-39.

[4]Axel Hultgren and B. Herrlander, "Hot Deformation Structures, Veining and Red-Shortness Cracks in Iron and Steel", *Transactions,* American Institute of Mining and Metallurgical Engineers, Vol. 172, 1947, p. 493, Fig. 14.

sumption of adjacent austenite and the concurrent retardation of the reaction.

The fact that this transformation was also found to occur in a temperature range down to 50 °F below A_1 where, at equilibrium, no austenite exists can only be explained as a result of the lower chromium content of the original austenite, assisted by the ready diffusion of its carbon, bringing the A_1 of localized areas at the austenite-ferrite boundaries down below the A_1 of the steel as a whole. This effect can, of course, only be temporary. On continued holding below the true A_1 the austenite formed will disappear.

In conclusion, the writer would like to acknowledge his great pleasure in reading this interesting paper.

Written Discussion: By Robert C. Downey, Metallurgical Development Unit, Materials Laboratory, Aircraft Gas Turbine, General Electric Co., Cincinnati, Ohio.

The authors are to be commended for a substantial addition to the present knowledge of transformation characteristics of the straight chromium stainless steels.

It is to be noted that aside from the purely theoretical considerations involved, practical aspects concerning certain behavior phenomena of these steels suggest themselves. Two of these will be mentioned.

1. It has been previously found that the so-called ferritic stainless steels, i.e., those containing 17% chromium or more, can be made subject to intergranular corrosion (sensitized) by heating them into the temperature range for austenite formation, above approximately 1700 °F (925 °C), followed by air cooling. This phenomenon is explainable by virtue of the authors' observation that the austenite-to-ferrite transformation at all temperatures below the A_1, while preceded by harmless intragranular carbide precipitation, is accompanied by intergranular carbide precipitation. This, of course, renders the steel susceptible to intergranular corrosion. It follows that the nature of susceptibility to intergranular corrosion is the same for these straight chromium stainless steels as for the nonstabilized 18-8 types, though the mechanism of sensitization differs. It is fortunate, from the standpoint of harmful carbide precipitation, that the straight chromium stainless steels are not used in service within the sensitization temperature range, i.e., within the temperature range of austenite formation. It is apparent, moreover, that when these materials have been subjected during processing to thermal cycles which involve cooling through the austenite transformation range (e.g., forging, welding) they should not be put into service under severe corrosive condition without a prior stabilizing heat treatment.

2. Another practical aspect concerns the welding of Type 430 stainless steel (17% chromium). It has been noted on occasion that Type 430 stainless steel weldments in the as-welded condition are susceptible to intergranular corrosion in the weld heat-affected zone. In view of the authors' findings this behavior can probably be attributed to harmful grain boundary carbide precipitation, since some precipitation should certainly occur in the heat-affected zone of the weld during cooling. At the same time it is known that martensite exists in the weld zone grain boundaries

of as-welded 430 material. If the supposition of carbide precipitation be correct, then it must be assumed that these grain boundary carbides coexist with the martensite just mentioned. Now it is known that a 1400 °F (760 °C) post-anneal of such weldments removes the susceptibility to intergranular corrosion. It can logically be asked what mechanism results in recovery of stability of the tempered material. Certainly, the martensite will be fully tempered to restore ductility to the weld, but it is wondered what becomes of the grain boundary carbides during this treatment, since it is certain that their harmful concentration is in some manner dissipated.

In austenitic stainless steels which have been sensitized by carbide precipitation, stability is restored by resolution of the carbides on heating to an appropriate temperature. In the case of 430 material, however, it would not seem that resolutioning of the carbides would take place at 1400 °F (760 °C), since the authors have shown that carbides remain largely undissolved, even when heated to much higher temperatures. The authors' comments as to what probable mechanism is involved in restoration of stability by the 1400 °F (760 °C) treatment would be appreciated.

Authors' Reply

The contributions and comments made by each of the discussers are very much appreciated.

We are pleased to know that our work is in general agreement with the work of Lula, Lena and Kiefer[5]. The one point on which we do not appear to agree concerns the suppression of precipitation in Types 430 and 442 stainless by water quenching small samples. We stated in our paper that we were able to suppress the carbide precipitation in Types 430 and 442 stainless, whereas Lula, Lena and Kiefer could not. The discussers have demonstrated that this difference is the result of differences in etching techniques. In our work we adjusted the concentration of HCl in 5% picral until the reagent became sufficiently selective so that it would show an increase in amount of precipitated carbide with time during our isothermal studies and would not show excessive grain boundary attack. Our diagrams, therefore, are intended to show the time required for the start of the carbide precipitation reaction at the various temperature levels under the particular etching conditions employed in our work. In the case of Types 430 and 442 stainless, it was necessary to hold samples for definite times at temperatures of 1200 to 1300 °F (650 to 705 °C) and below for isothermal carbide precipitation, and our etching technique suggested that little or no precipitation had occurred during the quench into the molten lead baths at these temperatures. There was a great deal of precipitation in all of the Type 446 samples quenched to the various bath temperatures and etched in our HCl – picral reagent.

The start of carbide precipitation is certainly difficult to evaluate, for it is possible to produce severe grain boundary attack, particularly by employing electrolytic etching reagents in the absence of discrete precipitated particles. Perryman[6], for example, observed severe grain bound-

[5]R. A. Lula, A. J. Lena and G. C. Kiefer, "Intergranular Corrosion of Ferritic Stainless Steels", see this volume of TRANSACTIONS.

[6]E. C. W. Perryman, "Grain Boundary Attack on Aluminum in Hydrochloric Acid and Sodium Hydroxide", *Journal of Metals,* American Institute of Mining and Metallurgical Engineers, July 1953, p. 911.

ary attack in high-purity aluminum containing less than 0.055% iron, and attributed this attack to a concentration of iron at the grain boundaries, making them susceptible to electrochemical attack.

The data on the effect of post-weld heat treatments submitted by Messrs. Lena, Lula and Kiefer provide an excellent example of how a knowledge of the transformation behavior of the ferritic stainless steels may be put to practical use, and we are very grateful for this contribution to our paper. Also, we were interested in knowing that the lamellar aggregate resulting from the decomposition of delta ferrite was observed by these authors in a chromium-manganese steel.

We have no quantitative data on the effects of carbon and chromium on the transformation characteristics of these steels, but are inclined to agree with Lena, Lula and Kiefer that the more rapid rate of transformation in our Type 446 steel is primarily the consequence of the high carbon content. As the result of the high carbon content, there is a high rate of chromium carbide precipitation, and this impoverishes the austenite, causing it to transform more rapidly than it would if the rate of carbide precipitation were lower because of a lower carbon content.

We are grateful to Mr. Perkins for favoring us with his observations. The micrographs illustrate very clearly that the stability of the austenite is greatly modified as the result of the impoverishment resulting from the precipitation of chromium carbide which precedes the austenite decomposition. The isothermal formation of some austenite at 1650 °F (900 °C) from delta ferrite following and accompanying carbide precipitation and agglomeration at the former ferrite : austenite grain boundaries is also nicely shown by Mr. Perkins' photomicrographs.

We are particularly pleased that Prof. Axel Hultgren was sufficiently interested in our paper to offer his interpretation of some of our observations. It is gratifying to know that he concurs with us in our explanation of the mechanism involved in the formation of the lamellar aggregate from delta ferrite.

There is only one point on which our interpretations differ. In the case of our observation that carbide precipitation occurred preferentially at ferrite : ferrite or ferrite : austenite grain boundaries we suggested that this observation may indicate that the diffusion required in the growth of the carbides occurs faster in the ferrite than in the austenite. Professor Hultgren, on the other hand, prefers to believe that nucleation of the carbide can occur more readily at such sites than at austenite : austenite grain boundaries. There is no way of knowing at present which viewpoint is correct.

In the absence of pertinent diffusion data, our explanation is purely speculative. We were influenced in arriving at our explanation by the observation described in an earlier paper by one of the present authors[7] that the carbides in low alloy hypoeutectoid steels disappear first in the ferrite areas on heating into the $Ac_1 - Ac_3$ temperature region.

Thus, on heating, carbides dissolve more rapidly in ferrite, whereas on cooling, carbides precipitate more rapidly in ferrite. We have been inclined to associate these observations and have tried to arrive at an

[7]A. E. Nehrenberg, "The Growth of Austenite as Related to Prior Structure", *Transactions,* American Institute of Mining and Metallurgical Engineers, Vol. 188, 1950, p. 162.

explanation which would be applicable to both. We know that at the temperatures involved in austenite formation in carbon and low alloy steels the diffusion coefficient for carbon in ferrite is greater than that for carbon in austenite. We have adopted the viewpoint that this accounts for the earlier disappearance of the carbides in ferrite on heating in the $Ac_1 - Ac_3$ range. Conversely, we reasoned that the earlier precipitation of carbide in ferrite during cooling might be accounted for in a similar manner. In the case of the high chromium steels of our recent paper, it is undoubtedly the rate of diffusion of chromium that controls the rate of the precipitation reaction, and data on rates of diffusion of chromium in ferrite and in austenite at the temperatures involved in the precipitation reaction would be pertinent. To the best of our knowledge such data are not available.

In connection with Professor Hultgren's comments on our Fig. 14a, we are afraid we did not adequately develop the point we were trying to make. We agree that the deformed grain boundary is a consequence of the transformation stress. Since the shifting of polishing scratches or other fiducial markings have been used to study shear transformations, i.e. martensite transformations, we thought some metallographers might not be aware that a nucleation and anisotropic growth process might produce a similar shifting of fiducial markings. We classified the delta ferrite → austenite transformation illustrated by Fig. 14a as a nucleation-and-growth transformation because we observed variations in the lengths and widths of the acicular areas of austenite throughout this and other specimens.

Professor Hultgren commented on the fact that the transformation of delta ferrite to austenite plus carbides was observed in a temperature range down to 50 °F below A_1, where, at equilibrium, no austenite exists. It was consistently observed in this investigation that at all temperatures at which this transformation was observed, there was less austenite present at equilibrium than there was at the time the delta ferrite → austenite + carbide reaction began. Thus, although austenite formed from delta ferrite at localized areas in the vicinity of the carbide precipitate, the reverse transformation, i.e. austenite to ferrite, also occurred at the same temperature in other areas. The latter transformation was found to more than offset the former transformation.

Such a result can be rationalized as follows: The precipitation of the $Cr_{23}C_6$ carbide removes carbon and chromium from solid solution in the approximate proportions 1 to 20, by weight. From the standpoint of their relative effects on the gamma loop, about 30 parts by weight of chromium are required to compensate for 1 part by weight of carbon[8]. Since the carbon and chromium are not removed from solid solution in the proportion 1 to 30, there will be an adjustment in the proportions of austenite and ferrite as precipitation of the chromium carbide proceeds.

In the carbide being precipitated, the proportion of carbon, which broadens the gamma loop, to chromium, which restricts it, is greater than the proportion 1 to 30 required to maintain the existing proportion of austenite and ferrite. Thus, the adjustment will be in the direction of less austenite.

[8]Unpublished data, Crucible Steel Company of America.

We appreciate Mr. Downey's comments concerning some practical implications of our work. It is certainly true that post-welding heat treatments are required to eliminate the tendency for intergranular corrosion resulting from carbide precipitation during welding or certain hot working operations on the straight chromium steels. The data submitted by Messrs. Lena, Lula and Kiefer in their discussion of our paper illustrate this point very nicely.

Mr. Downey asks us to offer an explanation for the fact that a 1400 °F (760 °C) post-weld heat treatment will eliminate the susceptibility to intergranular corrosion in Type 430 stainless. It is our opinion that the chromium impoverishment hypothesis which has been offered as an explanation for susceptibility to intergranular corrosion in austenitic steels may apply to straight chromium steels like Type 430 as well. We like this hypothesis because we were able to demonstrate by metallographic procedures that impoverishment of adjacent areas accompanies carbide precipitation in grain boundaries. In our opinion, then, a short reheating at 1400 °F (760 °C) permits the diffusion of chromium into the impoverished areas and thereby eliminates the tendency for intergranular corrosion.

Lula, Lena and Kiefer[5], on the other hand, do not subscribe to the chromium impoverishment hypothesis as applied to the ferritic stainless steels. Instead, they suggest that the grain boundary precipitation of carbide or nitride strains the matrix adjacent to the precipitate and thus makes the grain boundary area susceptible to intergranular attack. The elimination of susceptibility to intergranular attack by a short reheating at 1400 °F (760 °C) is attributed by these authors to the relief of these stresses.

Finally, in closing we wish to again convey our sincere appreciation to the discussers for their stimulating comments. We are glad to have had the opportunity to benefit from this exchange of viewpoints.

Identification of the Precipitate Accompanying 885°F Embrittlement in Chromium Steels

by R. M. Fisher, E. J. Dulis, and K. G. Carroll

IT is well known that ferritic steels containing more than 15 pct Cr when subjected to temperatures in the range of 700° to 1000°F exhibit increasing hardness and decreasing ductility. The phenomenon has been widely termed the "885°F embrittlement," after the temperature of most marked effect.[1,2] In view of the excellent review articles available in the literature[3-6] only a brief account of experimentally established facts need be given here.

The extent of changes in physical characteristics during embrittlement depends on chromium concentration and time at temperature, higher alloy content and longer time both promoting more rapid and extensive changes. In a 27 pct Cr steel, changes in impact strength and in angle of fracture in bending can be detected after only a 1 hr exposure at 885°F; after 50 hr this steel becomes quite brittle. Hardness increases slowly with time during thousands of hours exposure and may attain a maximum hardness number twice as large as that of the unexposed steel.

Microstructural changes accompanying embrittlement have been described as an initial widening of grain boundaries followed by eventual darkening of ferrite grains. Embrittled steels etch more readily, e.g., the weight loss of a 27 pct Cr steel in acid solution may occur at a rate one hundred fold greater following exposure at 885°F. Marked changes which accompany embrittlement have been observed in electrical resistivity, specific gravity, and magnetic coercive force. Changes in physical properties may be readily removed by heating at temperatures above the embrittling range, such as a treatment at 1100°F for 1 hr.

It has frequently been noted that the 885°F embrittlement suggests precipitation on a submicroscopic scale of a chromium-rich constituent, the nature of which has not been revealed by X-ray diffraction. Progressive broadening of the body-centered cubic diffraction lines during embrittlement has been observed,[6] and recent observations by Lena and Hawkes[7] upon single crystals have shown early asterism in X-ray photographs, disappearing within an hour at 900°F. Many workers have ascribed[8-13] the phenomenon to a precipitation of σ phase (FeCr), which is known to cause embrittling effects at temperatures much higher than 885°F. Two general observations, however, suggest that σ precipitation cannot be responsible for the 885°F phenomenon: 1—prior cold work greatly enhances σ formation, whereas it scarcely affects the 885°F embrittlement, and 2—the presence of an alloying element such as nickel or manganese may have an effect on the 885°F embrittlement which is opposed to its effect upon σ formation.

R. M. FISHER, E. J. DULIS, and K. G. CARROLL are associated with the Research Laboratory, United States Steel Corp., Kearny, N. J.

Table I. Chemical Composition of Steels Used

Sample	Aging Time at 900°F, Hr	Chemical Composition, Wt Pct							
		C	Mn	P	S	Si	Ni	Cr	N
A	10,000	0.032	3.13	0.016	0.010	0.35	0.08	28.14	0.084
B	10,000	0.12	0.63	0.014	0.010	0.06	0.10	27.48	0.071
C	0	0.24	0.89	0.020	0.003	0.55	0.48	26.45	0.222
D	10,000	0.24	0.89	0.020	0.003	0.55	0.48	26.45	0.222
E	34,000	0.24	0.89	0.020	0.003	0.55	0.48	26.45	0.222
F*	5,000	0.08	0.55	0.014	0.008	0.05	0.10	16.54	—
G*	10,000	0.08	0.55	0.008	0.027	0.43	0.16	15.80	—
H*	10,000	0.08	0.55	0.009	0.022	0.64	0.11	15.05	—
I*	10,000	0.07	0.55	0.010	0.024	0.56	0.12	14.14	—

* These samples, obtained from Heger[12] and Link and Marshall,[13] were cold-reduced 95 pct before exposure.

The slight enhancement of σ formation and 885°F embrittlement observed in the presence of elements with strong carbide and nitride forming tendencies[2] is probably a consequence of lessened chromium depletion of the matrix. The bar graph in Fig. 1 shows a typical example, taken from two 27 pct Cr steels used in this work, of the hardness after exposure for 10,000 hr at 900°, 1050°, and 1200°F. Steel A (0.03 pct C, 3.13 pct Mn) showed marked hardening at 900° and 1200°F, whereas steel B (0.12 pct C, 0.63 pct Mn) exhibited only the 900°F hardening. The σ phase was found in steel A at the higher temperatures but not in steel B. Presumably σ formation is enhanced by the low-carbon and high-manganese concentrations in A.[14] Thus there are two distinctly different hardening phenomena present which cannot both be ascribed to σ precipitation without invoking a transition phase possessing remarkable properties.

Materials

A number of chromium steels exposed for long periods (5000 to 34,000 hr) at 900°F, as well as unexposed samples of one of the steels, were available for this investigation. Table I gives the chemical compositions and aging treatments of these steels. In addition to these steels exposed in the elevated temperature test furnaces of the National Tube Division, a number of high-chromium steels were heated for short periods in small laboratory air furnaces and lead baths. Supplementing these commercial steels, a sample of high-purity (0.018 pct C, 0.002 pct N) 28 pct Cr iron, exposed 1000 hr at 887°F, was furnished by the Union Carbide and Carbon Corp. In addition, an alloy of iron and chromium of high purity containing 46 pct Cr was used. This

Reprinted with permission from *Journal of Metals*, 197, May 1953, 690-695, © 1953 AIME material, made by vacuum fusion, was obtained from the National Research Corp., Cambridge, Mass.

Optical Microscope Examination of Embrittled Steels

The micrographs of Fig. 2 illustrate the phenomenon of 885°F embrittlement in a 27 pct Cr steel. Ferrite grains in the unexposed steel (sample C) are clear, the outlined or dark etching constituents being chromium carbide and nitride. After an exposure of 34,000 hr at 900°F (sample E), the ferrite appears darkened by a precipitate which is too fine to be resolved by the optical microscope, and a veining or subgrain structure can be seen in the darkened ferrite. This precipitate dissolves rapidly in the ferrite matrix upon heating to 1200°F, as shown in Fig. 2c, leaving the microstructure much as it appeared in the unexposed steel of Fig. 2a.

These changes in microstructure are reflected in hardness of 198, 369, and 202 diamond pyramid hardness for the specimens of Figs. 2a, 2b, and 2c, respectively. These results indicate that a definite microstructural change accompanies the hardness increase during the long time exposure at 900°F, and that the embrittled steel quickly returns to its initial hardness and to a structure similar to the initial microstructure upon heating at 1200°F. The hardness of sample E (369 DPH) was not significantly different from that of sample D (373 DPH).

Electron Microscope Examination

Fig. 3 shows electron micrographs of the samples shown in Fig. 2. These are photographs of uranium-shadowcast plastic replicas stripped from the polished and etched surfaces of the steel before and after exposure at 34,000 hr at 900°F. This one-step plastic replica technique results in negative replicas, in that unetched projections on the original surface appear as depressions in the micrographs. The smooth, unetched regions in both figures correspond to the carbides and nitrides in the steel. The rough background in Fig. 3b indicates the presence of a precipitate in the ferrite matrix which is too fine to be resolved by this replica technique. This is analogous to the ferrite darkening shown in the optical micrograph of Fig. 2b; however, the vein structure was not observed.

A new "Extraction Replica" technique that is applicable to these samples has recently been developed at this Laboratory. This technique will soon be reported elsewhere and will be described here only briefly. It has been found possible to free from the metal matrix the precipitate particles that are partially embedded in the replica plastic by etching the sample a second time through the plastic. The replica contains the actual precipitate particles, thus avoiding the loss in resolution inherent in any replica technique. By suitable manipulation of the electron beam, it is possible to select areas of the replica containing separated particles for diffraction patterns by transmission. An alcoholic solution of picric and HCl acids (grain size reagent) has been found most effective for the stripping technique on chromium steels, although solutions of bromine or iodine in alcohol may also be used.

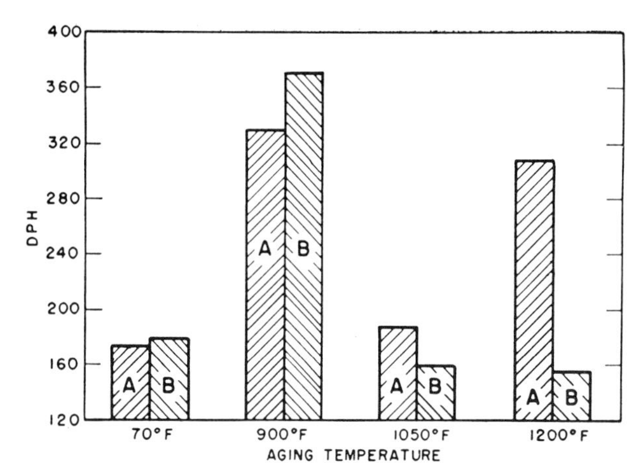

Fig. 1—Effect of aging temperature on hardness of two 27 pct Cr steels after 10,000 hr exposure.

Fig. 4 shows particles of the precipitate which are responsible for the ferrite roughening of Fig. 3b. Fig. 4a is an extraction replica of sample D (10,000 hr, 900°F) and Fig. 4b shows the precipitate in sample E (34,000 hr, 900°F). The precipitate particles seem to be spherical as they always appear circular in the micrographs and never rodlike as would disks on edge. The particles are quite uniform in size, varying from about 150 to 300Å but averaging 200 in Fig. 4a and 230 in Fig. 4b. Resolution of the micrographs here limits accuracy to about ±30Å.

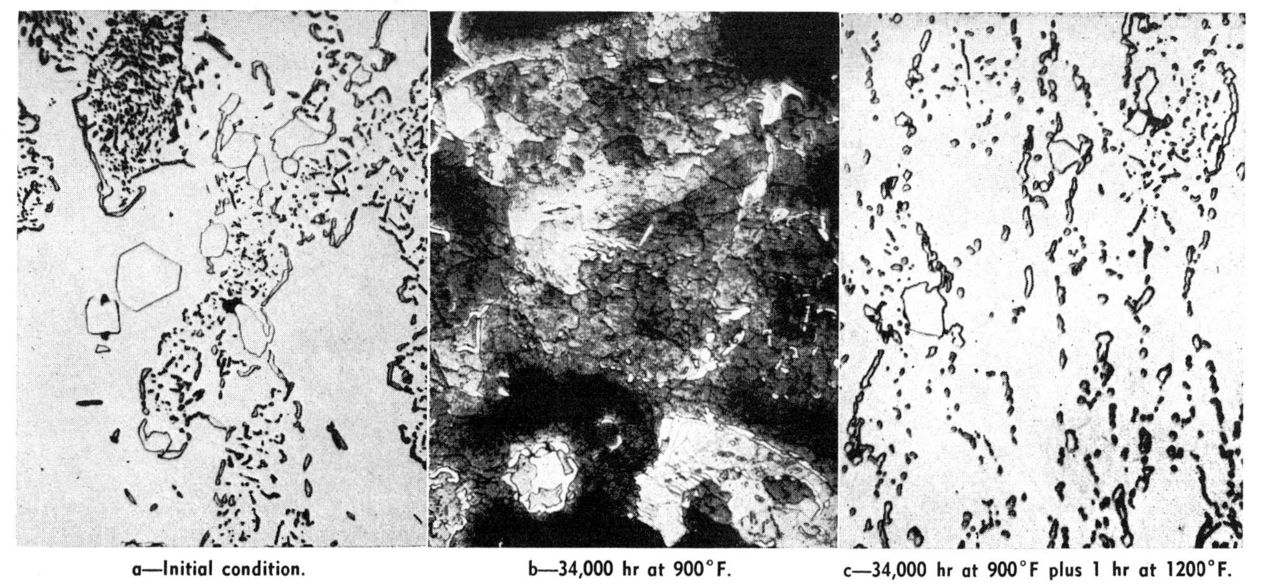

a—Initial condition. b—34,000 hr at 900°F. c—34,000 hr at 900°F plus 1 hr at 1200°F.

Fig. 2—Microstructures of steel C (27 pct Cr). Picric-HCl etch. X1000. Area reduced approximately 50 pct for reproduction.

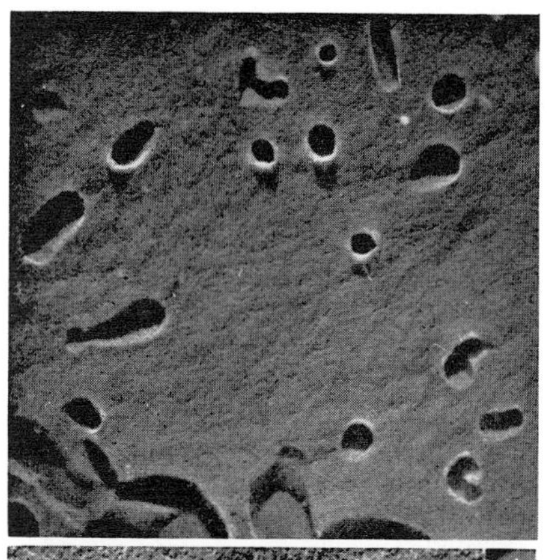

Fig. 3 — Uranium shadowed replicas of 27 pct Cr steel. HCl-picric etch. X15,000. Area reduced approximately 50 pct for reproduction.

a (upper) — Before exposure.

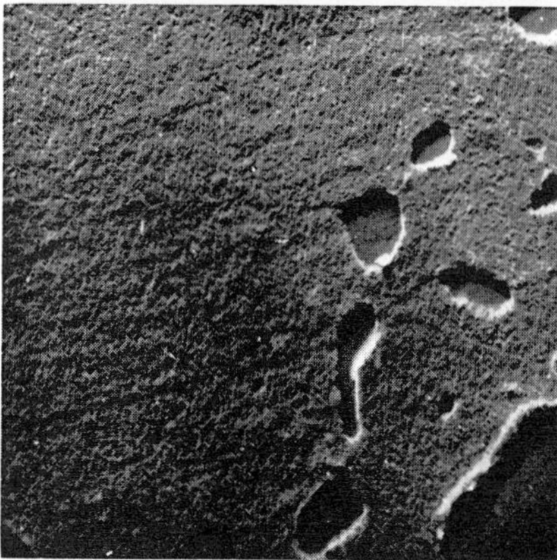

b (lower) — 34,000 hr at 900°F.

Fig. 4a and b also illustrates quantitatively the increase in numbers of particles from 10,000 to 34,000 hr exposure although there is little change in particle size. Some clustering of particles in Fig. 4b results from slightly over etching the second time so that some previously unexposed particles that are not fixed to the plastic are free to cluster together. This is difficult to avoid with such small precipitate particles.

Transmission electron diffraction from the same area of the replica shown in Fig. 4b gives the pattern, reproduced in Fig. 5, of a simple body-centered cubic phase of lattice parameter similar to that of iron or chromium. Because pure iron and pure chromium differ in lattice parameter by only 0.6 pct, it is not possible to distinguish between these two elements, or their solid solution, by means of electron diffraction alone. It is interesting to note the evidence of preferred orientation in the diffraction pattern of Fig. 5, which indicates that the replica technique removed the particles from the steel without disturbing their orientation. The six-fold symmetry in the inmost {110} ring suggests that the precipitate formed with its {110} planes coincident with the six {110} planes of the ferrite matrix.

Electron diffraction examination by means of the same replica technique described above showed the presence of the body-centered cubic precipitate in sample F (17 pct Cr, 5000 hr, 900°F), sample G (16 pct Cr, 10,000 hr), and sample H (15 pct Cr, 10,000 hr). The patterns from these samples were considerably weaker than from the 27 pct Cr steels and the particle diameter of the precipitate was about 125Å in H to 175Å in F. No body-centered cubic pattern was obtained from sample I (14 pct Cr, 10,000 hr) and no precipitate particles were observed.

Weak patterns were obtained for the precipitate in samples of the 46 pct Cr alloy after 700 hr at 900°, 950°, and 1000°F. The strongest pattern was obtained from the 950°F sample and the {200} diffraction line was unusually weak in all these cases.

The patterns from the high-purity Union Carbide and Carbon specimens (28 pct Cr, 1000 hr, 887°F, hardness 275 DPH) were rather weak and here also the {200} line was relatively weaker than in the 10,000 hr samples. The particles were too small to be clearly resolved by the electron microscope and from this fact, and from measurements of line broadening effects, they are estimated to be less than 50Å in diameter.

No precipitate could be observed in extraction replicas of the sample of Fig. 2c (sample E reheated 1 hr at 1200°F) nor was a body-centered cubic electron diffraction pattern obtained from these replicas. A strong body-centered cubic pattern was obtained from an extraction replica of another small piece of sample E (27 pct Cr, 34,000 hr, 900°F) reheated ¼ hr at 1100°F. The precipitate particles were about 120Å in diameter and the hardness of this sample was 239 DPH.

X-Ray Examination

X-ray examination of the etched surface of sample C (before) and samples D and E (after embrittlement) showed no new diffraction lines but only a slight broadening of the original body-centered cubic ferrite lines after exposure. Fig. 6 shows the diffraction pattern of the residue obtained by dissolving a small sample of embrittled steel (sample E, 34,000 hr, 900°F) in grain size reagent (HCl-picric acid in alcohol). This pattern was obtained on a Philips 90° spectrometer with filtered chromium radiation. It shows the presence of Cr_2N and $Cr_{23}C_6$ as minor constituents, along with strong {110} diffraction from a body-centered cubic lattice. This material showed no evidence of ferromagnetism when tested with a small hand magnet and in a magnetic susceptibility balance. Similar patterns were obtained from the surface of the solid

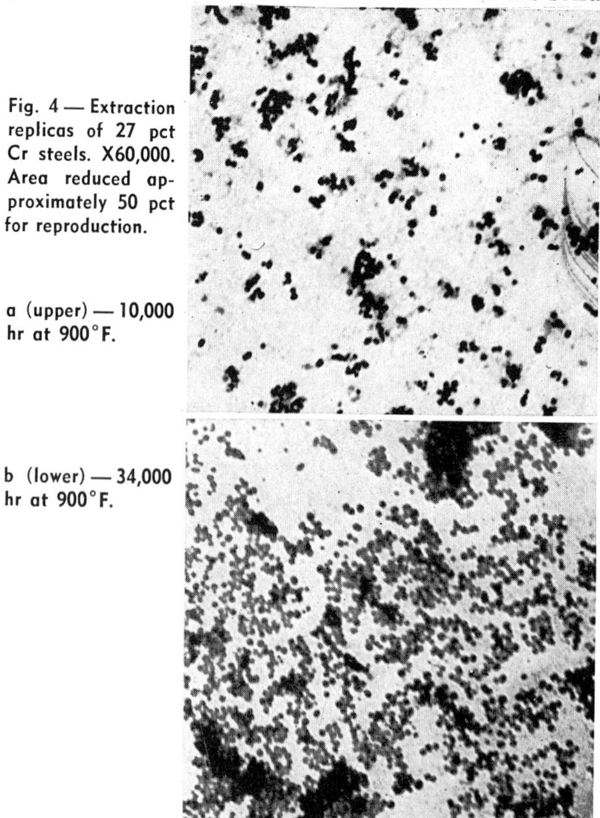

Fig. 4 — Extraction replicas of 27 pct Cr steels. X60,000. Area reduced approximately 50 pct for reproduction.

a (upper) — 10,000 hr at 900°F.

b (lower) — 34,000 hr at 900°F.

steel specimen after it was etched deeply, except that the {110} diffraction line was masked by the ferrite {110} line.

X-ray fluorescence was used to measure the relative Fe-Cr composition of the extract from sample E. A copper target tube was used for excitation, and a sodium chloride crystal as monochromator. For calibration purposes a pure alloy of iron and chromium containing 46 pct Cr was used. Repeated observations on this and other similar extracted materials from samples D and E yielded a result of approximately 85 pct Cr-15 pct Fe. This, of course, includes the chromium in the carbide and nitride as well as the precipitate.

The lattice parameter of the body-centered cubic constituent was measured in the spectrometer modified for back reflection, by measuring the diffraction angle of the {211} reflection, under chromium radiation, in comparison with that from spectroscopically pure iron (a = 2.8665Å). Values were obtained for extracts from steels D and E after 10,000 and 34,000 hr exposures at 900°F of 2.877 and 2.878Å, respectively. Precision of these measurements is limited by particle broadening which prevented accurate resolution of the α doublet. The lattice parameter measurements would imply a chromium content of about 70 pct;[15] however, influence of residual elements as well as particle broadening prevents accurate estimates of composition from these measurements.

X-ray examination of the etched surfaces of samples F and G (heavily cold-worked before exposure at 900°F) showed the presence of σ phase (FeCr) as well as Cr_2N and $Cr_{23}C_6$. The strength of the electron diffraction pattern for the precipitate from these samples indicated that extraction of it for X-ray examination was not feasible. The σ phase was found in sample A after aging at 1050° and 1200°F but not in steels B, D, or E after aging at these temperatures. X-ray observations of σ phase in these samples verifies previously reported identification by metallographic means.[12, 13]

A small amount of residue extracted from the high-purity 28 pct Cr sample was examined by X-ray diffraction. No diffraction lines were obtained from the specimen due probably to the extremely small size of the particles. X-ray fluorescence analysis of the residue gave a chromium to iron ratio of about ten or twelve to one.

Chemical Composition of Extracted Materials

Table II lists the results of chemical analyses of the residues obtained by dissolving embrittled samples in the grain size reagent used previously as an etchant. Samples which were first investigated (D and E) were simply left in the etching solution until considerable residue had collected on the surfaces of the specimens and then removed from the solution and dried. After drying, the surfaces were carefully brushed and the residue collected. To obtain quantitative data on the amount of residue in an embrittled steel, samples A and B were crushed in a mortar, weighed, dissolved in grain size reagent, and the residue washed by decanting in alcohol. After washing, the residue was collected, dried, and weighed. The amount of residue available for analysis varied between 0.2 and 0.5 g. Rather more reliance is placed upon the chemical analysis of steels A and B because more extract was available for analysis and because of improved techniques in handling.

From Table II it is seen that the Cr-Fe ratio agrees

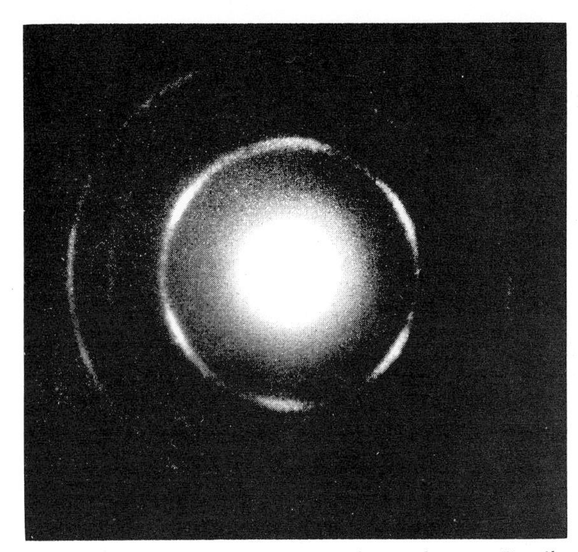

Fig. 5—Electron diffraction pattern of area shown in Fig. 4b.

moderately well with the ratio determined by X-ray fluorescence analysis. While the reported nitrogen contents of the extracted materials from samples A and B are quite consistent with the yield factors and the nitrogen contents of the initial steels (Table I), the carbon values are several times larger than would be expected from the yield factors and the carbon content of the steels. It is probable that these discrepancies are due to organic contamination both from the etchant and from etching products not removed during washing. The estimated carbon contents listed in Table II were obtained by multiplying the original carbon content of the steels by the concentration factor in the extraction. This assumes that all the carbon in the steels was combined as carbide and was recovered in the extraction.

Since the X-ray results showed that the extract consists only of $Cr_{23}C_6$, Cr_2N, and the body-centered cubic precipitate, it is possible to compute the relative proportions of these three phases in the extract, and also the composition of the precipitate. Table III lists the results of these computations, using the nitrogen values from the analyses and the estimated

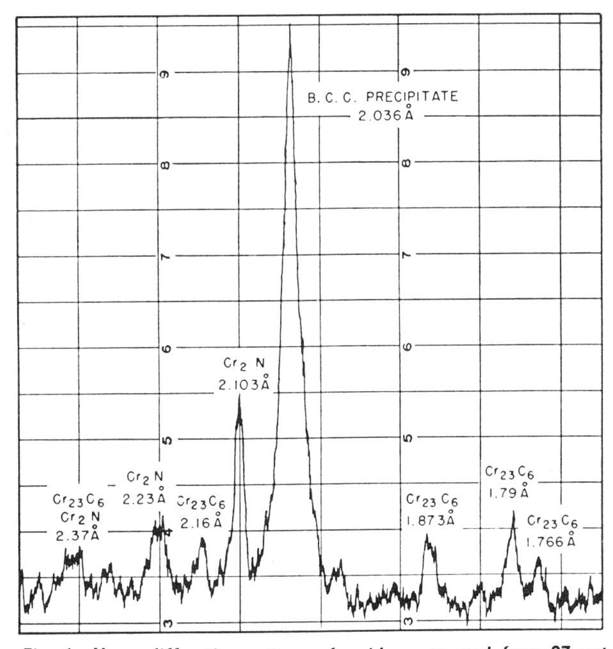

Fig. 6—X-ray diffraction pattern of residue extracted from 27 pct Cr steel exposed 10,000 hr at 900°F.

Table II. Composition of Extracted Material, Wt Pct

Specimen	Yield	Cr	Fe	N	C	Total	Estimated C
A	11.5	64.0	11.3	0.84	1.69	77.8	0.28
B	9.9	66.4	11.1	0.77	2.75	81.0	1.2
D	13.1*	61.5	12.2	1.7	—	75.4	1.8
E	18.5*	70.9	11.6	1.2	1.9	85.6	1.3

* Estimated from nitrogen values of steel and extract.

values for carbon. Agreement among the samples in the composition of the body-centered cubic precipitate is quite good except for sample D which was the smallest sample and yielded the bare minimum of residue that could be analyzed.

The proportions by weight of the constituents in the extracts can be converted into weight proportions in the embrittled steels by using the yield values of Table II. Table IV shows the results of calculating the composition of the steels in this way. Values in parentheses are the values for the carbide and nitride that would be expected from the carbon and nitrogen contents of the steels. The values computed from the analyses of the extract are all higher than the values computed from the steel analyses. This error, whatever its origin, tends to decrease both the apparent chromium content of the body-centered cubic precipitate and the apparent amount of precipitate formed. The chromium content of the remaining ferrite is computed by subtracting the chromium combined as nitride, carbide, and body-centered cubic precipitate from the original steel chromium analysis and calculating its weight percent in the remaining ferrite matrix.

Magnetic Measurements

Preliminary qualitative observations with a small Alnico permanent magnet revealed that the extractions from embrittled steels were either nonmagnetic, or only weakly magnetic. Quantitative measurements were made in a magnetic susceptibility balance. In this instrument a magnetizing field of about 400 oersteds was provided, and the tractive force in an inhomogeneous field was measured on a standard chemical balance. Extracts from specimens D and E revealed no detectable ferromagnetism; those from A and B gave tractive forces equal to about 10 pct of the specimen weights. For comparison, under identical conditions, steel C gave a tractive force equal to twice its weight. Electron diffraction examination of one of the magnetic residues indicated the presence of magnetite, which may account for the slight ferromagnetism.

Small pieces of embrittled steels D and E were investigated over a range of temperatures up to 670°C (1238°F) in the same instrument. No evidence was found for any Curie temperature, other than the principal one at 600° ±5°C (1112° ±9°F) which, according to Adcock,[15] is appropriate to a 27 pct chromium ferrite. No hysteresis effects were found upon heating above and cooling below the Curie temperature; hence no chromium depletion of the ferrite was evidenced. In view of other findings this may be ascribed to the rapidity of solution of the precipitate in the ferrite matrix above 475°C (887°F), because during the course of the measurements at least 30 min was spent in increasing the specimen temperature from 475° to 600°C. It would appear that direct observation of a chromium-depleted ferrite is impossible by magnetic means. The nonmagnetic residue of sample E was examined for a low-temperature Curie point down to —195°C (—319°F). A slight tractive force was observed at —195°C for this sample which might have been due to paramagnetism just above the Curie point. From the work of Adcock,[15] a Curie point below —195°C indicates a chromium content in excess of 78 pct.

Experimental Results

The experimental results described in the preceding sections indicate the formation of a submicroscopic precipitate during the embrittlement of ferritic chromium alloy steels at 885°F.

The electron and X-ray diffraction measurements show that the precipitate has the body-centered cubic structure with lattice parameter between that of iron and chromium. This lattice parameter of 2.878Å is equal to that of a pure alloy of 70 pct Cr-30 pct Fe. Chemical analyses of the residues extracted from embrittled steels indicate that the precipitate is between 78 to 82 pct Cr and that the precipitate can amount to as much as 10 pct by weight in a 27 pct Cr steel exposed for four years. X-ray fluorescence analysis of the extract from a high-purity 27 pct Cr iron aged 1000 hr at 900°F showed a chromium content of 92 to 94 pct Cr.

Magnetic measurements on the precipitate show that it is not ferromagnetic above —195°C, which is the Curie point for a 78 pct Cr-22 pct Fe alloy. Electron microscope observations reveal that the diameter of the precipitate particles depends both on time of exposure and on chromium content of the steel. In 27 pct Cr steels the precipitate diameter varied from less than 50Å after 1000 hr exposure to about 225Å after 34,000 hr exposure. For 10,000 hr exposure, diameter varied from about 125Å in a 15 pct Cr steel to about 200Å in a 27 pct Cr steel. The precipitate particles formed during 34,000 hr at 900°F completely dissolved during 1 hr reheating at 1200°F and shrank from 225 to 120Å in diameter during ¼ hr at 1100°F.

The precipitate was found in steels with chromium contents above 14 pct and was found to coexist with σ (FeCr) in steels which had been severely cold-worked before exposure at 885°F. In well annealed 27 pct Cr steels, σ phase was found after exposure at 1050° and 1200°F in samples that increased in hardness during aging and was not found in samples that did not harden at these temperatures. No precipitate was found to form above 1000°F.

Discussion

As stated in the introduction, changes in properties of ferritic chromium steels during embrittle-

Table III. Computed Composition of Extract, Wt Pct

Specimen	Cr_2N	$Cr_{23}C_6$	Body-Centered Cubic Precipitate
A	9.3	6.5	84.2 (82.5% Cr, 17.5% Fe)
B	8.2	26.5	65.3 (78.6% Cr, 21.4% Fe)
D	18.6	41.0	40.4 (61.0% Cr, 39.0% Fe)
E	11.9	26.9	61.2 (77.5% Cr, 22.5% Fe)

Table IV. Computed Composition of Embrittled Steel, Wt Pct

Specimen	Cr_2N	$Cr_{23}C_6$	Body-Centered Cubic	Cr Content of Ferrite
A	1.07 (0.7)	0.75 (0.56)	9.7	21
B	0.81 (0.69)	2.62 (2.1)	6.5	21
D	2.42 (1.85)	5.37 (4.2)	5.3	19
E	2.20 (1.85)	4.98 (4.2)	11.3	14

ment at 885°F are accompanied by precipitation of a chromium-rich phase. The small particles of the embrittling phase have the body-centered cubic structure of the matrix with a lattice parameter only 0.2 pct larger and are probably coherent with the matrix, thus greatly increasing its hardness due to internal strains. The formation of a fine precipitate also explains the increase in the coercive force of the steel. Chromium depletion of the matrix during embrittlement lowers the corrosion resistance of the steel and also lowers its electrical resistivity. Data in the literature[16] suggest that the rate of corrosion for embrittled steels is several times greater than for a steel of chromium content equal to the composition of the depleted matrix (as calculated in this report). This effect could well be due to the presence of the precipitate particles in the steel.

Formation of the precipitate is difficult to explain. The Fe-Cr phase diagram indicates that ferrite and σ phase should be the stable phases in an equilibrated 27 pct Cr alloy; nevertheless during exposure at 885°F the alloy seems to prefer to separate into iron-rich and chromium-rich ferrites. However, in severely cold-worked steels the precipitate can coexist with σ at 885°F. Rate of growth of the precipitate is unusually slow. From diffusion considerations, the precipitate particles could grow to the observed 200Å diameter in less than 2 hr or, alternatively, in 34,000 hr the diffusion-limited diameter is more than 5000Å. In the cold-worked samples aged at 900°F the largest σ particles were within 20 pct of the calculated diffusion-limited size.

The size and structure of the precipitate is reminiscent of the "clusters," "knots," or "complexes"[17] proposed to account for the changes in properties during age-hardening of nonferrous metals that occur before any microstructural changes can be discerned. Various transitional structures preceding formation of the equilibrium precipitate have been observed[18, 19] by X-ray scattering at small angles from nonferrous metals undergoing age-hardening. In most cases the data have been interpreted as scattering from a two dimensional platelet but occasionally as from a spherical precipitate. The theory of these coherent transitional precipitates, as developed so far,[19, 20] implies that the thickness of the platelet depends on the mismatch between the atom-atom distances in the precipitate and in the matrix, and also on the relative rigidity of the precipitate and the matrix. According to this theory, the small amount of mismatch between the chromium-rich body-centered cubic precipitate and the matrix, and the greater rigidity of the precipitate, could result in very thick platelets or essentially spheres.

If 885°F embrittlement is an age-hardening phenomenon it should exhibit overaging. This has not been observed for 27 pct Cr steels even after four years of aging. However, it is possible that overaging will be observed after further exposure, since the matrix has been depleted in chromium almost to the threshold composition of embrittlement and little more precipitate can form. The fact that the hardness did not increase during heating from 10,000 to 34,000 hr, while the amount of precipitate more than doubled with no change in particle size, may mean that the particles formed first lost coherency at about the same rate as new coherent particles formed. This idea is supported by the observations that the precipitate size, in a fully embrittled sample held a short time in the softening temperature range, is much larger than in a sample embrittled for just a short period although the reheated sample is considerably softer. The resolution treatment may cause the precipitate particles to lose coherence while dissolving and thus soften the steel.

Acknowledgments

The authors wish to thank J. J. Heger, H. S. Link, and P. W. Marshall, of the Research and Development Laboratory, United States Steel Corp., together with A. B. Wilder, of National Tube Div., United States Steel Corp., and W. O. Binder, of Union Carbide and Carbon Corp., for supplying samples used in this investigation.

It is also a pleasure to acknowledge the assistance of many members of the Research Laboratory, including D. S. Miller for assistance in the magnetic measurements and L. S. Darken and G. V. Smith for valuable suggestions.

References

[1] F. M. Becket: On the Allotropy of Stainless Steel. *Trans.* AIME (1938) **131**, p. 15.

[2] G. Riedrich and F. Loib: Embrittlement of High Chromium Steels Within Temperature Range of 570-1100°F. *Archiv. Eisenhuttenwesen* (October 1941) **15**, pp. 175-182.

[3] G. Bandel and W. Tofaute: Brittleness of Chromium Rich Steels at Temperatures Around 930°F. *Archiv. Eisenhuttenwesen* (1942) **15**, No. 7, pp. 307-319.

[4] W. Dannohl, W. Hessenbruck, and E. Hengler: Brittleness Phenomena of High Chromium Steels in Temperature Range of 930°F. *Archiv. Eisenhuttenwesen* (1942) **15**, No. 7, pp. 319-320 (discussion of refs. 2 and 3).

[5] J. J. Heger: 885°F Embrittlement of the Ferritic Chromium-Iron Alloys. *Metal Progress* (August 1951) pp. 55-61.

[6] H. D. Newell: Properties and Characteristics of 27% Chromium Iron. *Metal Progress* (May 1946) pp. 977-1028.

[7] A. J. Lena and M. F. Hawkes: Embrittlement of High Chromium Ferritic Stainless Steels at 475°C. Journal of Metals (February 1952) p. 146.

[8] E. Houdremont: *Einfuhrung in die Sonderstahlkunde* (1935) p. 183. Berlin. Julius Springer.

[9] V. N. Krivobok: Alloys of Iron and Chromium. *Trans.* ASM (1935) **23**, pp. 1-60.

[10] E. C. Bain and R. H. Aborn: The Iron-Nickel-Chromium System. *Metals Handbook* (1939) pp. 418-422. Cleveland. ASM.

[11] G. V. Smith, W. B. Seens, H. S. Link, and P. R. Malenock: Microstructural Instability of Steels for Elevated Temperature Service. *Proc.* ASTM (1951) **51**, pp. 895-917.

[12] J. J. Heger: The Formation of Sigma Phase in 17 Per Cent Chromium Steels. ASTM Special Technical Publication No. 110 (1950) pp. 75-81.

[13] H. S. Link and P. W. Marshall: The Formation of Sigma Phase in 13-16% Chromium Steels. *Trans.* ASM (1952) **44**, p. 549.

[14] C. O. Burgess and W. D. Forgeng: Constitution of Fe-Cr-Mn Alloys. *Trans.* AIME (1938) **131**, p. 272.

[15] F. A. Adcock: The Cr-Fe Constitutional Diagram. *Journal* ISI (1931) **124**, pp. 99-146.

[16] Owen K. Parmiter: Wrought Stainless Steels. *Metals Handbook* (1948) p. 553. Cleveland. ASM.

[17] A. H. Geisler: Precipitation from Solid Solutions of Metals. *Phase Transformations in Solids.* (1951) pp. 493-496. New York. John Wiley and Sons.

[18] J. Calvet, P. Jacquet, and A. Guinier: Age Hardening of a Copper-Aluminum Alloy. *Journal* Inst. Metals (1939) **65**, p. 121.

[19] F. R. N. Nabarro: Diffusion and Precipitation in Alloys. *Symposium on Internal Stresses in Metals and Alloys.* (1948) pp. 237-249. Inst. of Metals.

[20] J. C. Fisher, J. H. Hollomon, and J. G. Leschen: Precipitation from Solid Solution. *Industrial and Engineering Chemistry* (June 1952) **44**, pp. 1324-1327.

The Nature of Ageing of Binary Iron–Chromium Alloys Around 500°C

By R. O. Williams, M.Sc., Ph.D., and H. W. Paxton, M.Sc., Ph.D.

SYNOPSIS

An investigation of ageing (475° C embrittlement) of Fe–Cr alloys has been carried out using electrical resistance, hardness, magnetic measurements, and X-ray diffraction. On the basis of this work it is concluded that a miscibility gap exists below 600° C and is joined by one or two eutectoid reactions to the sigma-forming regions. Chemical and/or magnetic energies may be responsible for the gap and perhaps account for the odd shape.

The gap results in the precipitation of a Cr-rich phase which is always coherent with the matrix. As a result of this coherency, both the matrix and precipitate are isotropically strained to a common lattice parameter. Since no loss of coherency occurs, the alloys do not over-age and particles larger than 200 Å were not found. Because of the nature of the strains, it is proposed that the particles interact with the hydrostatic component of edge dislocations in contrast to shear interaction. 1267

AN EXPERIMENTAL INVESTIGATION has been made of the nature of the ageing of Fe–Cr alloys in the temperature range of 400–550° C. This ageing phenomenon has been known as '475° C embrittlement' of ferritic stainless steels. Such a name arose, no doubt, from the very brittle nature of such materials for no apparent reason when aged in this temperature range. This study has shown that the phenomenon is one of ageing or age-hardening in more or less the usual sense. This problem is of interest for two reasons: this ageing imposes severe limitations on the practical utilization of such alloys and the basic nature of this ageing has not been clearly resolved in spite of extensive investigations.

The change which these materials undergo on ageing in this temperature range can be described briefly. The room-temperature ductility and impact strength decrease to essentially zero and the hardness and strength increase markedly. The increases in physical properties are, however, useless because of the extreme brittleness. The ageing is accompanied by decreases in electrical resistivity and corrosion resistance and an increase in Curie temperature. A significant quantity of heat is also released during ageing. The changes which can be shown by means of X-ray diffraction or microstructural methods are somewhat controversial, but are slight in any case. Ageing effects can be removed completely by holding the materials at about 600° C for a short time. The most important new phase identification was a chromium-rich precipitate resulting from the ageing (Fisher *et al.*[1]). The more important investigations and reviews of the subject are given below:

Bandel and Tofaute,[2] Becket,[3] Fisher, Dulis, and Carroll,[1] Heger,[4] Hochmann,[5] Imai and Kumada,[6] Krivobok,[7] Lena and Hawkes,[8] MacQuigg,[9] Masumoto, Sainto, and Sugihara,[10] Newell,[11] Riedrich and Loib,[12] and Tagaya *et al.*[13]

Three theories have been advanced to explain this ageing. It has been supposed that the presence of some minor phase as an oxide, nitride, phosphide, or carbide is responsible. Other investigators have believed that ordering is the process responsible for the changes. The most popular theory, however, is the one which supposes that the ageing is connected with the tendency to form σ-phase: more explicitly, it is supposed that the ageing is a result of the formation of a transition lattice prior to the formation of equilibrium sigma. There is almost no direct evidence in support of the first two theories, while certain similarities between this system and the classical system (Al–Cu), give considerable support to the last theory. However, it is necessary to suppose that the high-chromium phase is the transition structure and this is the interpretation given by its discoverers and by Lena and Hawkes.[8]

This study was undertaken to provide additional information on this ageing process. Many new details have been discovered and an entirely new interpretation of the phenomenon is given. A previously unrecognized type of age hardening is shown to occur, and it suggested the development of certain new ideas of the mechanism of hardening.

EXPERIMENTAL PROCEDURES

In this work a wide range of purities and compositions was used; these were given several different mechanical and thermal treatments prior to ageing. The processes of ageing and recovery were followed by hardness, resistivity, and magnetic measurements, and considerable X-ray diffraction and microstructural work was done. Normally, isothermal ageing was used, followed by recovery during slow heating which caused the precipitate to go back into solution.

The alloys which were used in this study are listed in Table I. The first seven pairs were laboratory induction melts made by Allegheny Ludlum Research

---

Manuscript received 10th January, 1956.

Dr. Williams formerly Allegheny Ludlum Research Fellow of the Carnegie Institute of Technology, Pittsburgh, Pa., is now Senior Research Metallurgist at the Cincinnati Milling Machine Co., Cincinnati, Ohio, U.S.A.

Dr. Paxton is Assistant Professor of Metallurgy at the Carnegie Institute of Technology.

This work was done as partial fulfilment of requirements for Ph.D. Degree for the first author.

Table I

ANALYSES OF ALLOYS: VALUES IN wt-%

Alloy No.	Cr	Mn	Si	N_2	C	P	S
83	17·74	0·49	0·65	0·021	0·033	...	0·011
83*V*	17·37	0·23	0·60	0·0015	0·016	...	0·006
24	30·24	0·45	0·22	0·19	0·028	...	0·011
24*V*	29·92	0·20	0·23	0·002	0·019	...	0·004
79	30·64	0·50	0·50	0·38	0·024	...	0·013
79*V*	28·55	0·23	...	0·005	0·012	...	...
82	30·74	0·54	0·29	0·29	0·020	...	0·014
82*V*	30·05	0·26	...	0·0015	0·016	...	0·005
90	30·10	0·62	0·61	0·039	0·025	...	0·012
90*V*	29·46	0·41	0·61	0·003	0·012	...	0·006
30	41·27	0·38	0·27	0·11	0·021	...	0·015
30*V*	40·12	0·25	...	0·0025	0·029	...	...
41	52·72	0·30	0·23	0·081	0·041	...	0·011
41*V*	50·58	0·18	0·19	0·0025	0·014	...	0·005
46	46·24	...	0·085	0·004	0·015	...	...
46*V*	...	...	0·094	0·011	0·024	...	0·004
*K*314	14·81	Nil	Nil	0·0022	0·014	About 0·008	About 0·016
*L*681	20·55	Nil	Nil	0·003	0·013	About 0·008	About 0·016
*L*684	23·04	Nil	Nil	0·005	0·008	About 0·008	About 0·016
*K*396	37·74	Nil	Nil	0·0039	0·016	About 0·008	About 0·016
*K*590	55·70	0·01	Nil	0·0034	0·020	About 0·008	About 0·016
70	68·66	...	...	...	0·06	...	...
80	79·50	...	...	...	0·06	...	...

Laboratory for this study. Alloy 46 was supplied by E. J. Dulis of United States Steel, the next five were furnished by W. O. Binder of Electro Metallurgical Company, and the final two came from A. H. Sully of Fulmer Research Institute. The last eight alloys were vacuum-melted alloys made from pure stock, and are generally equivalent to the best used by previous investigators. The second listing of an alloy followed by a '*V*' refers to a special vacuum treatment which is discussed below.

All hardness measurements were made using a Vickers Diamond Pyramid Hardness Tester and a 10-kg load. Normally, the same sample was used for a complete series of tests where the sample was re-aged for an additional increment of time. At least five readings were made after each interval, and this would normally give indications of changes as small as five to ten numbers (2%), unless copious mechanical twinning occurred, as normally happened with higher-Cr specimens.

During fabrication, most of the materials were heated in air. Ageing was carried out either in lead pots or in an inert atmosphere. The only important thermal treatment so far as composition goes was the vacuum annealing homogenization, as denoted by '*V*' in Table I. This process consisted of electrically heating the material in the form of $\frac{1}{8}$-in. rods in vacuum for several hours at 1050° C, in argon at 1200° C for 24 h and finishing in vacuum at 1050° C for 20 h. As shown by the analysis, this treatment gave a great reduction of nitrogen and a small reduction of carbon, manganese and sulphur, and probably of phosphorus and oxygen. The reduction of minor phases in the microstructure was very great. However, some minor phase or phases were put into solution, as evidenced by later microscopic examination.

The resistivity measurements were made by determining the voltage drop across specimens of known size when a current of 1·000 A (accurate to about 0·1%, constant to 0·001%), was passed through them. Readings could be made such as to detect changes of 0·002 Ω-cm. The experimental points, however, varied from the most probable value by about three times this figure for reasons not completely known, but the results were accurate enough for the purpose. Because of the large temperature coefficient of resistance of these alloys, the furnace used for the ageing had to be controlled closely. By suitable design, it was possible to obtain fluctuations less than 0·1° C and drifts of about 1°, which proved satisfactory.

The magnetic measurements were performed by the circuits and arrangement illustrated in Fig. 1. The field could be determined by either going above the Curie temperature or by removing the sample. Apparently, based on experimental results, the small gaps at the ends of the sample introduce no detectable demagnetization force. Measurements were made with fields of 650, 1400, and 2200 oersteds, but as the smallest field saturated the sample only these results

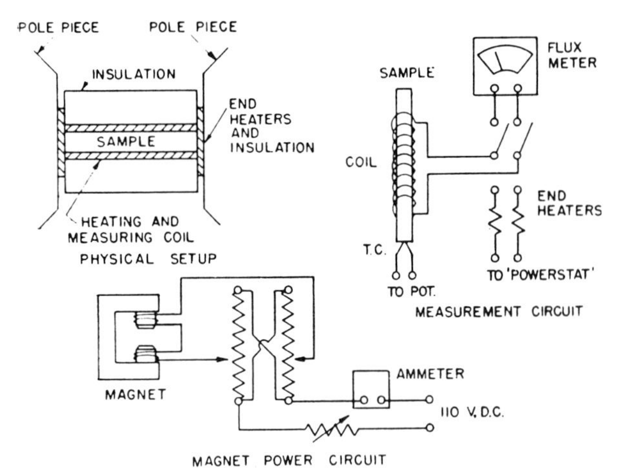

Fig. 1—Apparatus and circuits for magnetic measurements

have been included. Ageing was accomplished by removing the sample and placing it in a lead pot.

EXPERIMENTAL RESULTS

Isothermal Ageing

The changes in physical properties which result on isothermal ageing of these Fe–Cr alloys are given in Figs. 2 to 13 and in Table II. In most of these figures the time scale is modified by adding 1 h to the actual time, so that 0–999 h covers three cycles of log paper.

Figure 2 shows the hardness changes of a series of high-purity alloys which has been received from the Electro Metallurgical Company. To reduce the scatter in the hardness readings, these alloys had been hot-rolled at 875°C by about 50%. This caused about 25% of the material to recrystallize, but there was no evidence during this study that deformed and recrystallized material behaved differently. Figure 3 gives data from certain other alloys which were less pure, and it will be noted that they fit the same pattern. It might be noted that Alloy 41 is the one which was used to make magnetic measurements. Alloy 90 had been hot-rolled and Alloy 79 was aged as a single crystal and in a magnetic field, but no significant differences were found. The smaller scatter from hot-deformed alloys as compared to others is apparent.

While no detailed study has been made of the hardening of cold-worked materials, it has been established that the numerical increase in D.P.N. for such materials is significantly less than for recrystallized materials after 1000 h at 475° C. This was due in part to mechanical recovery during ageing, as clearly demonstrated by the decrease in line widths in Debye–Scherrer patterns. Such alloys had still not recrystallized at 550° C when given the treatment noted in Fig. 8. Also, Alloy *L*681 (23% Cr) failed to recrystallize in 90 days at 500° C after a 95% cold reduction. From this it can be concluded that the presence of the precipitate does not promote recrystallization and may, in fact, retard it.

Becket[3] and Lena and Hawkes[8] have reported high-purity alloys which showed a very slight tendency to age. Lena and Hawkes have concluded

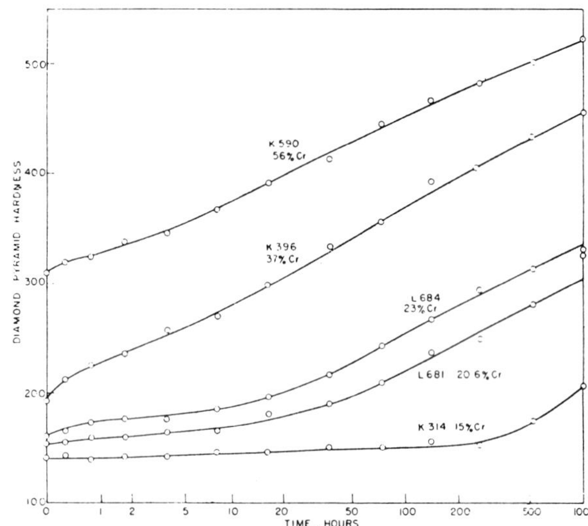

Fig. 2—Hardening characteristics of high-purity Fe-Cr alloys during ageing at 475° C

that either nitrogen must be present and precipitate out or else the alloys have to be deformed. While there may have been justification for this conclusion from their work, the present work does not support this conclusion. An alternate explanation which is yet unproved is that thermal treatment prior to ageing is very important. That prior treatment has an effect has been stated by Samans[14] without details and shown by Lena.[15] To check this possibility a series of samples of Alloy 82 after vacuum annealing were held for 1 h at temperatures between 530° and 800° C, quenched in oil, and then aged at 475° C. The data are plotted in Fig. 4. It should be noted that some slight amount of a minor phase may have precipitated and given rise to some of the effects shown, but this is not likely in that all the samples were held 1 h at 800° C after the vacuum treatment, and this should have precipitated this

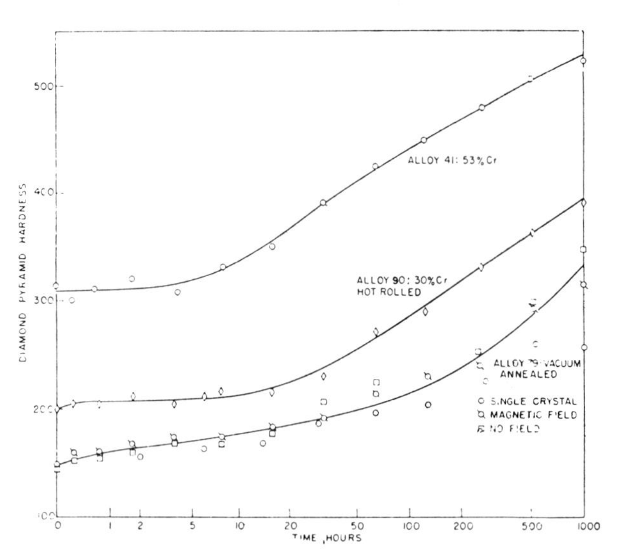

Fig 3—Hardening characteristics of Fe-Cr alloys during ageing at 475° C

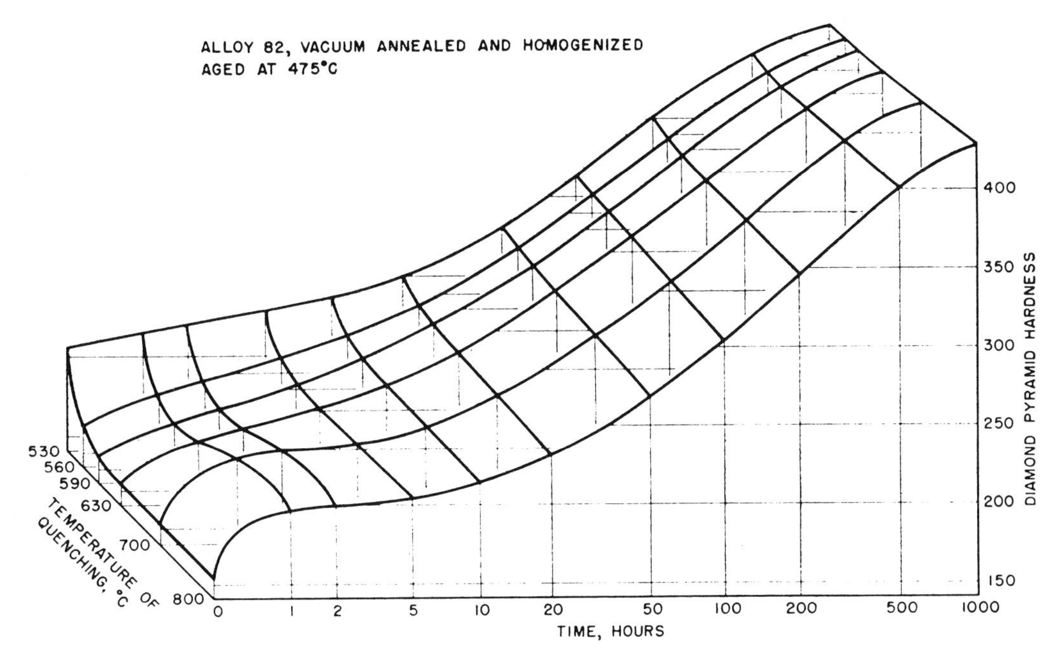

Fig. 4—Effect of quenching temperature on the hardening characteristics of a 30% Cr–Fe alloy during 475° C ageing

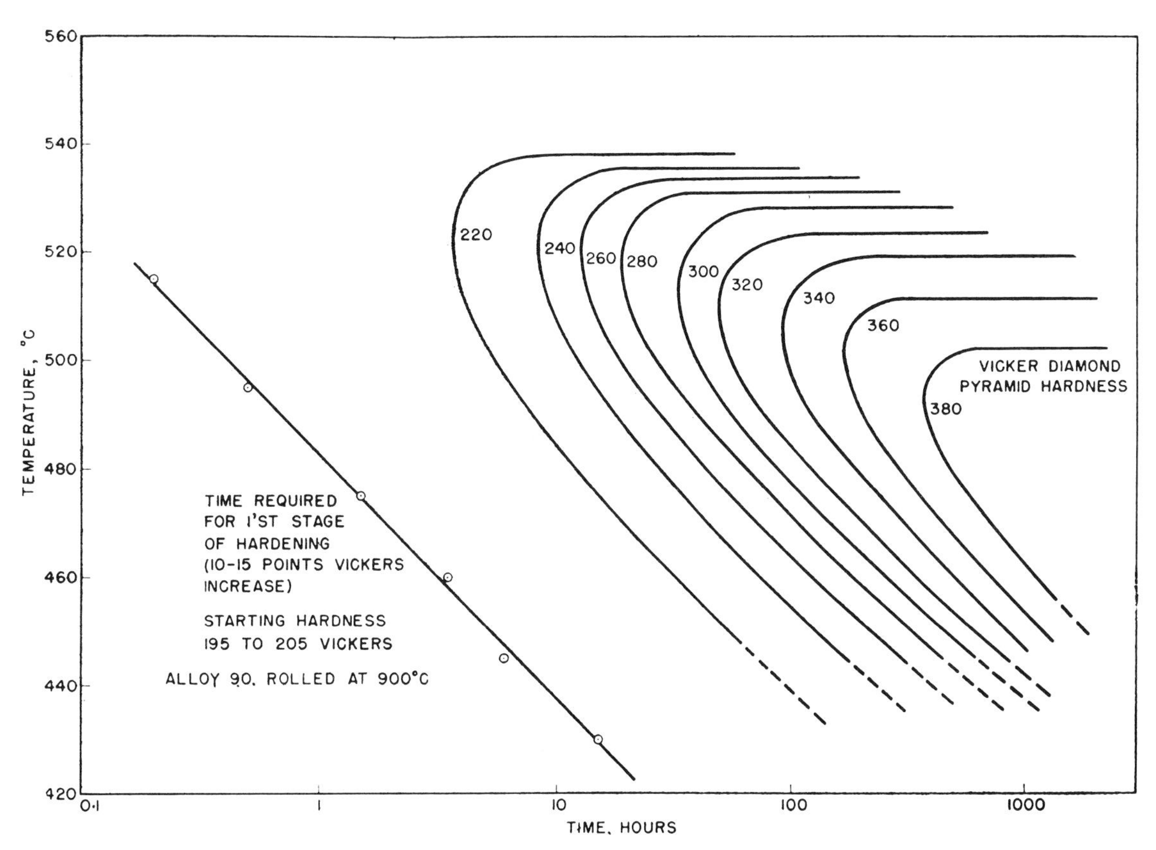

Fig. 5—Time/temperature dependence of hardness of a 30% Cr–Fe alloy during ageing

Table II

SUMMARY OF RESISTIVITY CHANGES OF Fe–Cr ALLOYS RESULTING FROM AGEING 1000 h AT 475° C

Starting State	Alloy No.	Cr content, wt-%	Initial 25° C Resist.	Final 25° C Resist.	Initial 475° C Resist.	Final 475° C Resist.	Dec. % in 475° C Resist.	Initial Temp. coeff. at 25° C, % per ° C	Final Temp. coeff. at 25° C, % per ° C	Re-solution Temp. with Slow Heating, ° C
Cold worked	83	17·7	62·1	59·4	99·8	95·5	4·3	0·178	0·146	...
	24	30·2	57·5	54·7	100·0	90·6	9·4	0·205	0·170	...
	79	30·6	60·4	59·5	102·5	94·1	8·2	0·201	0·144	...
	82	30·7	60·2	57·5	103·4	93·7	9·4	0·185	0·146	...
	90	30·1	61·1	57·9	97·5	89·1	8·6	0·164	0·139	...
	30	41·3	61·7	55·6	103·6	90·0	13·1	0·196	0·152	...
	46*	46·3	55·9	52·5	99·3	86·0	13·5	0·246	0·169	...
	41	52·7	66·0	58·1	103·6	90·5	12·6	0·223	0·145	...
Recrystallized	83	17·7	60·8	...	100·4	96·8	3·6	0·160	...	527
	24	30·2	56·6	...	100·0	91·8	8·2	0·171	...	541–58
	79	30·6	58·6	...	102·0	94·6	7·3	0·179	...	541–55
	82	30·7	59·5	...	103·1	95·0	7·8	0·193	...	541–65
	90	30·1	59·6	...	97·1	90·0	7·3	0·163	...	541–62
	30	41·3	59·5	...	102·0	91·3	10·4	0·201	...	553
	46	46·3	55·5	...	99·0	86·5	12·6	0·229	...	558
	41	52·7	62·7	...	104·4	92·0	11·9	0·196	...	553
Vacuum annealed	83	17·4	62·0	61·5	105·2	103·0	2·2	0·161	0·150	518
	24	29·9	54·8	54·3	100·4	92·5	7·9	0·229	0·170	544–53
	79	28·6	61·5	60·3	104·0	96·7	7·0	0·188	0·151	539–53
	82	30·0	57·7	55·2	104·8	89·5	14·6	0·224	0·167	546–53
	90	29·5	64·8	62·7	106·2	97·2	8·5	0·197	0·135	537–53
	30	40·1	58·0	57·8	100·6	93·0	7·5	0·213	0·162	544–51
	46	46·2	57·8	56·6	114·1	102·5	10·2	0·248	0·176	560
	41	50·6	54·7	53·2	86·0	79·0	8·2	0·226	0·177	546

*Recrystallized

phase. The quenching rate was fairly fast, although oil was used, since the samples were $\frac{1}{8}$-in. rods.

In spite of rather extensive data by previous workers, it was not possible to construct a transformation/temperature curve in terms of physical parameters such as hardness. For this reason, samples of hot-rolled Alloy 90 were aged at different temperatures and the results are assembled in Fig. 5. Two points should be noted. The hardness does not decrease at any time for isothermal ageing, even when the total time is ten times that required for maximum hardness; the final hardness is reached sooner the higher the temperature. The completion of the first stage of ageing (*see* Fig. 13) is shown to give a straight line parallel to the underside of the knee. It corresponds to an apparent activation energy of 55,000 cal/mol.

Extensive work was done on the resistivity changes of the alloys from Allegheny Ludlum in the cold-worked (60% reduction), recrystallized, and vacuum-annealed conditions. Alloy 46 was also used in the last two conditions. Because of the marked similarity of the changes for all alloys, the results have been condensed into Table II. The principal points which are shown are that the decrease in room-temperature resistance is less than that at high temperatures, that the decrease increases with chromium content within the range studied, and that the change in temperature coefficient is small at high temperatures and larger at lower temperatures as a result of ageing.

An example of how the resistance changes as a function of time is shown in Fig. 6 for Alloy 79. The results for the other alloys were very similar. The small vertical displacements result from different specimens and are not significant. Except for the difference in the initial part, the three curves are virtually parallel. The difference in the initial part is a result of the differences in cooling rate prior to ageing. The cold-worked sample was cooled fastest and shows the most pronounced initial effect, while the recrystallized sample was cooled from 1000° to 300° C in about 1 h and shows the least effect. The vacuum-annealed sample had an intermediate cooling rate. This type of behaviour is general and has been studied in some detail.

It has been shown in Table II that the change in resistance at elevated temperatures is greater

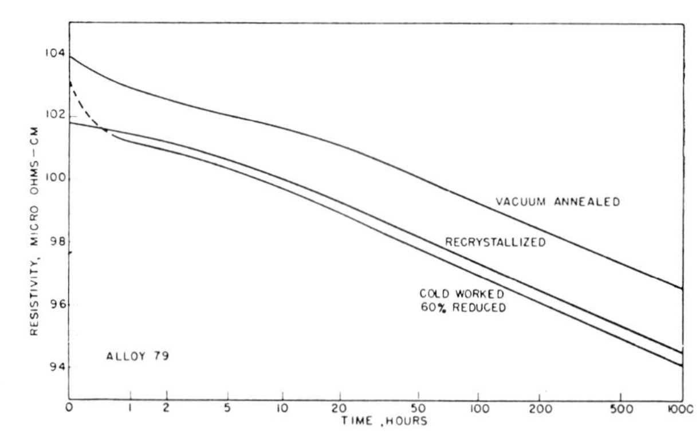

Fig. 6—Resistivity changes for a 30% Cr–Fe alloy aged at 475° C with different initial treatments

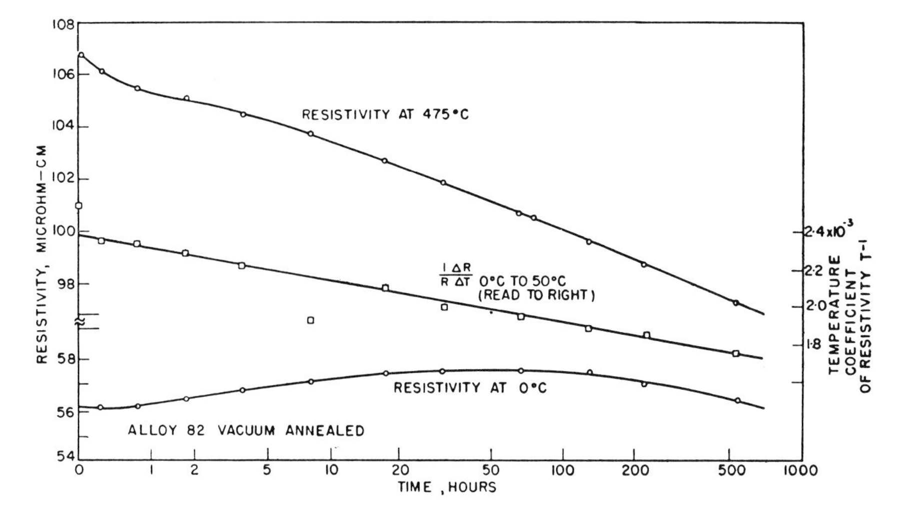

Fig. 7—Changes in resistivity at 0° C and 475° C of a 30% Cr–Fe alloy during ageing at 475° C

than that at low temperatures. To study this effect as a function of ageing time, a sample of Alloy 82 which had been given a vacuum anneal was placed in a ' Vycor ' tube sealed at one end. The necessary connections were made to measure the electrical resistance and the thermocouple was peened into the sample. Helium was used inside to promote heat transfer and protect the sample. This assembly was placed alternately in warm water, an ice bath, and a lead pit at 475° C, and the necessary readings were made to obtain Fig. 7. It should be noted that, had the resistance been plotted for room temperature rather than 0° C, no increase would have been found. These results suggest that there would be a rather large increase in resistance at lower temperatures.

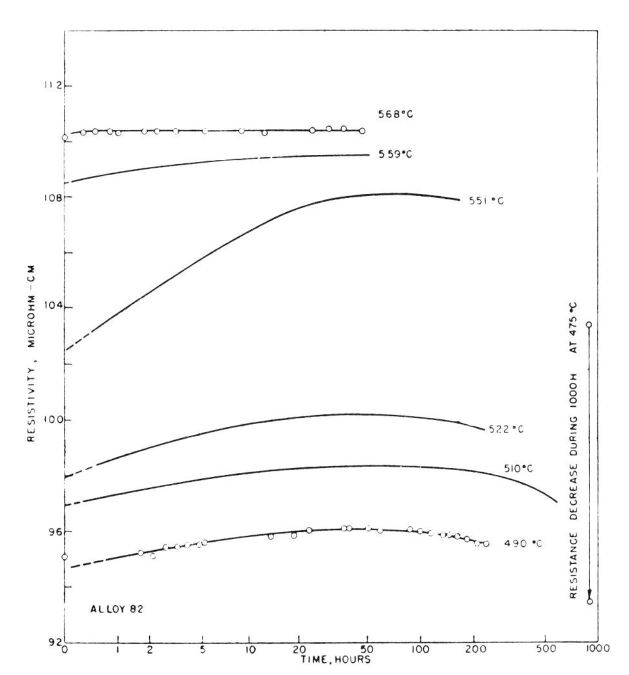

Fig. 8—Resistivity changes with time at successively higher temperature of a 30% Cr–Fe alloy following ageing at 475° C.

It was supposed that the temperature required for complete re-solution of the precipitate could be determined accurately by a series of isothermal ageings at increasing temperatures after the resistance had fallen during precipitation. While this method did not prove to be as satisfactory as continuous heating, it did provide certain information. Typical results are shown in Fig. 8. Since the furnace required about one hour to reach each new temperature the early parts could not be followed accurately. Nevertheless, the results show that the changes are characterized by at least two processes. Initially, the resistance increases due supposedly to changes in composition of the precipitate and matrix. With longer times at the three lower temperatures the resistance starts decreasing eventually, and this is considered a result of additional precipitation. An accurate re-solution temperature is not obtained, but it is apparent that 551° C is roughly the temperature required.

In obtaining the data for Fig. 8, it was apparent that a fairly large resistance change was time dependent, but that most of it could not be followed in that set-up. For this reason, the apparatus used to

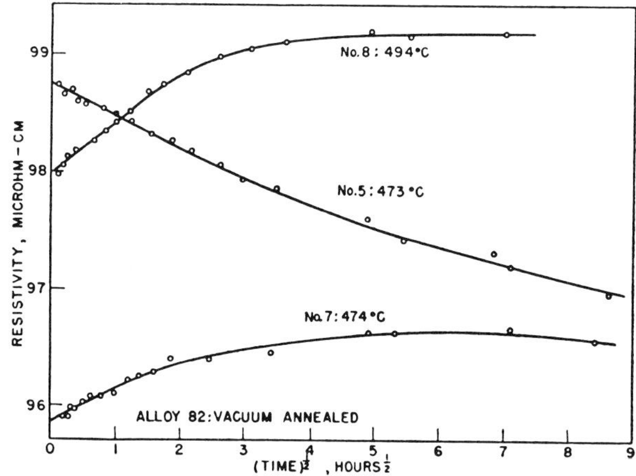

Fig. 9—Time dependence of resistivity of a 30% Cr–Fe alloy after a temperature change of 25° C

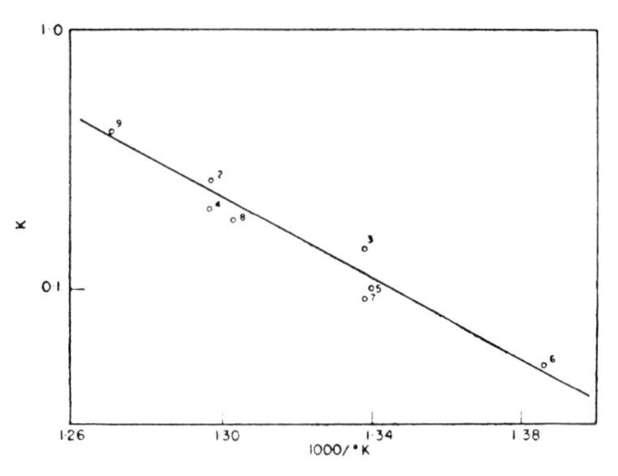

Fig. 10 Log of rate constant for time dependence of resistivity versus reciprocal of absolute temperature

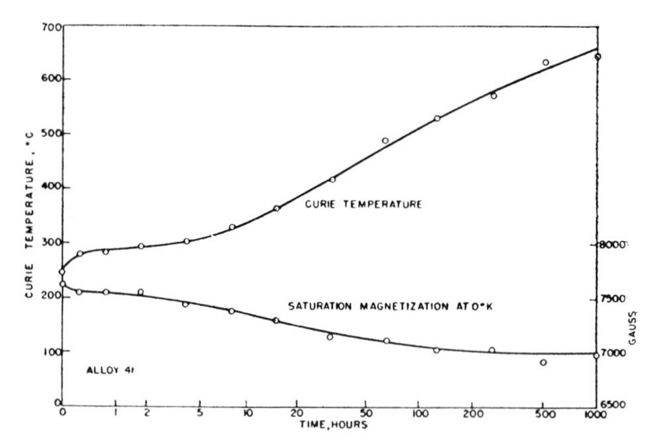

Fig. 12—Curie temperature and saturation magnetization of a 53% Cr–Fe alloy aged at 475° C

obtain rapid temperature changes was used to follow these changes. The sample was aged at 475° C for 70 h and then switched to a lead pot about 20° to 25° higher and the resistance followed. The temperature response was sufficiently rapid that the first reading could be obtained in less than one minute. This switching was continued up and down until about nine steps had been performed. It was possible to represent the initial resistance change as $Kt^{1/2}$ or $Kt^{1/3}$ with approximately the same accuracy. Typical results are plotted in Fig. 9 for three steps. This initial slope K, was corrected for a temperature change of 25° on the basis that its magnitude was proportional to the temperature change. These values are plotted in Fig. 10, log K versus the reciprocal of absolute temperature. The results can be represented adequately by a straight line corresponding to an activation energy of 36,000 cal/mol.

For the magnetic work a sample containing 53% Cr was chosen so that the initial Curie temperature would fall appreciably below 450° C. The results for Alloy 41 are shown in Figs. 11 and 12. In Fig. 11 the magnetization is given as a function of ageing time and temperature for progressive ageing times. The curves are similar at all times to those for pure metals, although the magnetization decreases for this alloy somewhat more rapidly at lower temperatures and less rapidly as the Curie temperature is reached than for pure metals. It is evident that the Curie temperature cannot be directly obtained for times of ageing in excess of 64 h. To obtain the Curie temperature for the longer times and to estimate the saturation magnetization at absolute zero, it was assumed that the data would fit an equation of the form

$$I/I_0 = F(T/\theta),$$

where I is the magnetization at some temperature T, and I_0 is the magnetization at absolute zero. F is some function of the ratio T/θ, where θ is the Curie temperature. By the graphical use of this expression it was possible to obtain reasonably good

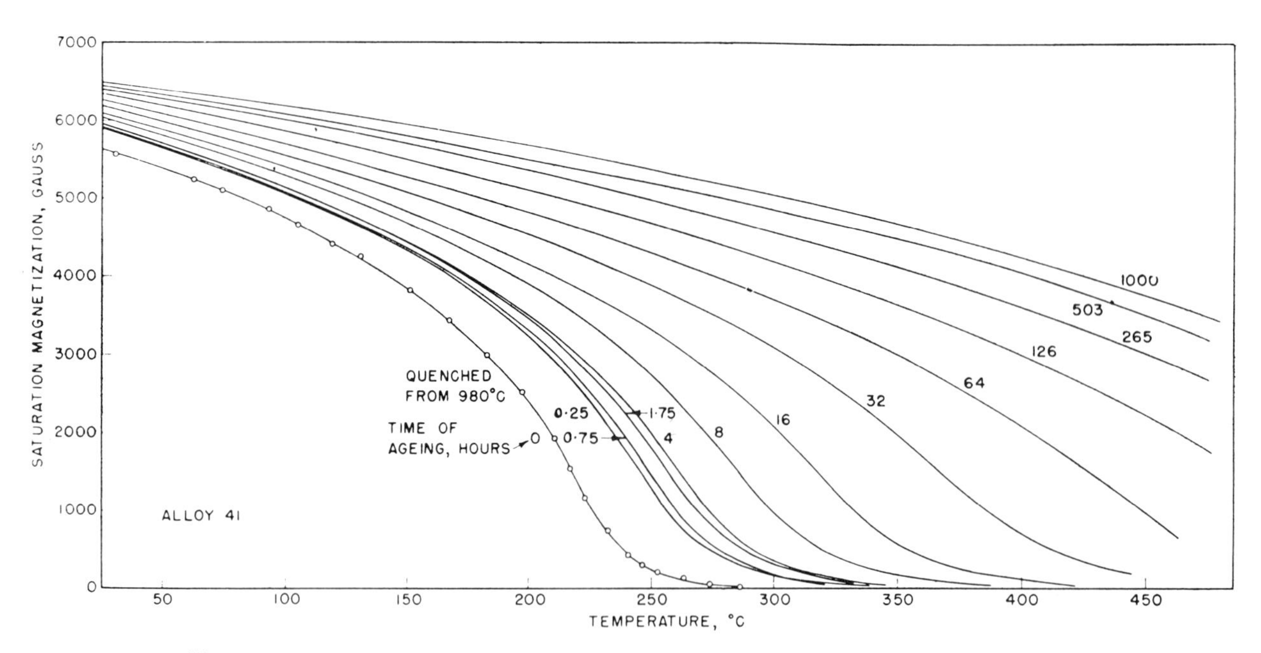

Fig. 11—Magnetization/temperature curves for a 53% Cr–Fe alloy aged at 475° C

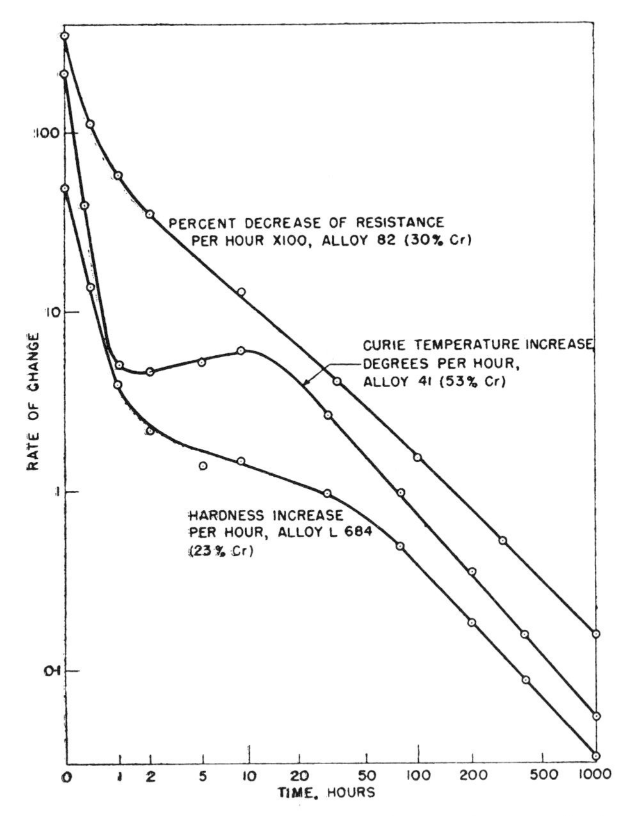

Fig. 13—Rates of change of electrical resistance, Curie temperature, and hardness of Fe–Cr alloys during 475° C ageing

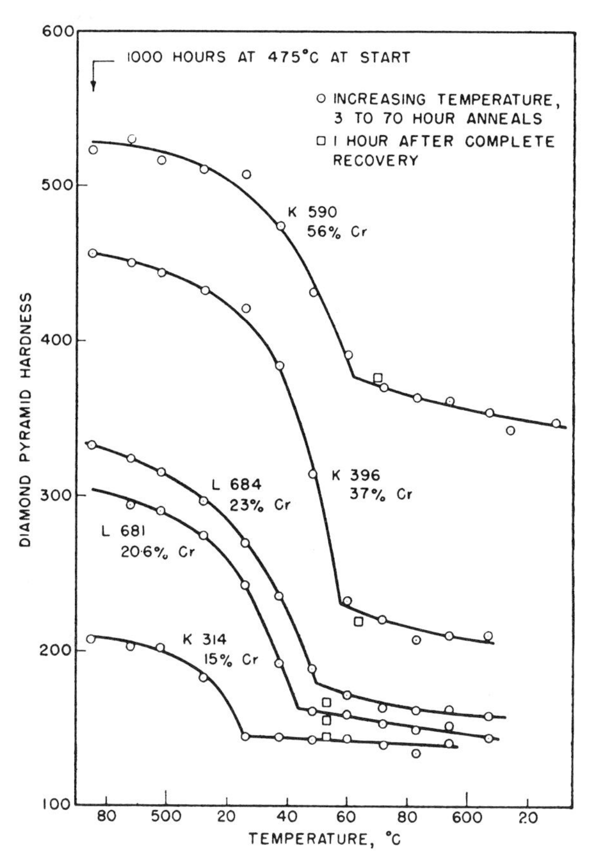

Fig. 14—Hardness recovery of aged high-purity Fe–Cr alloys annealed at successively higher temperatures

values of Curie temperature and saturation magnetization at absolute zero, although this expression does not hold exactly. These quantities are plotted in Fig. 12 as a function of time. The marked similarity

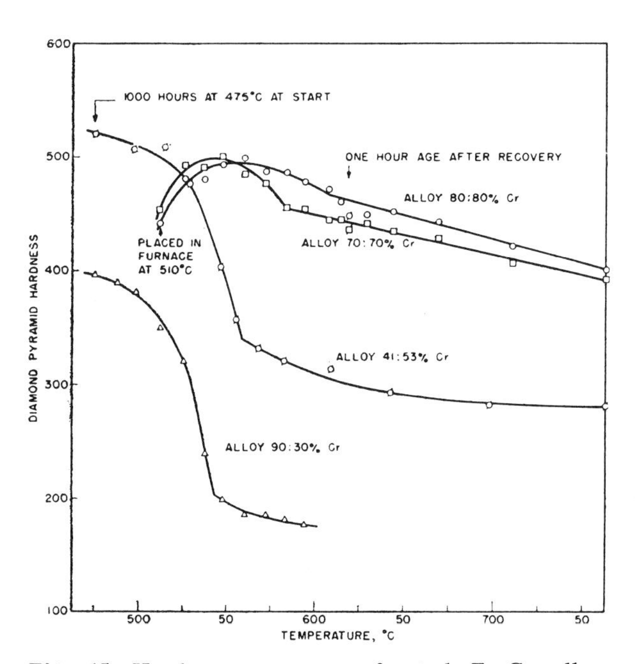

Fig. 15—Hardness recovery of aged Fe–Cr alloys annealed at successively higher temperatures

between hardness, resistivity, and Curie temperature should be noted.

While the method of plotting the time scale was convenient for representing the data, neither this nor any other method is capable of showing well the rates of change when they cover such a wide range. For this reason actual rates for hardness, resistivity, and Curie temperature change have been extracted from the graphs already presented and are given in Fig. 13. It is unfortunate that data adequate for the determination of all three rates from a single alloy did not exist. For this reason, it is not known if the second ' break ' would coincide for a given alloy. It should be noted that the rates change by a factor in excess of a thousand during the ageing. It is convenient to speak of the ageing as occurring in three stages as defined by the two more-or-less sharp breaks in this figure.

Effects of Elevation of Temperature

So far, all the results have been obtained from isothermal treatments, and for hardness and resistivity it has been demonstrated that essentially constant values are reached with sufficient holding times. This immediately suggests that following these properties with continuous heating could be used to study the effects of the re-solution of the precipitate. It has been found that step heating is entirely satisfactory for hardness and magnetic measurements, but continuous heating is preferable for resistivity.

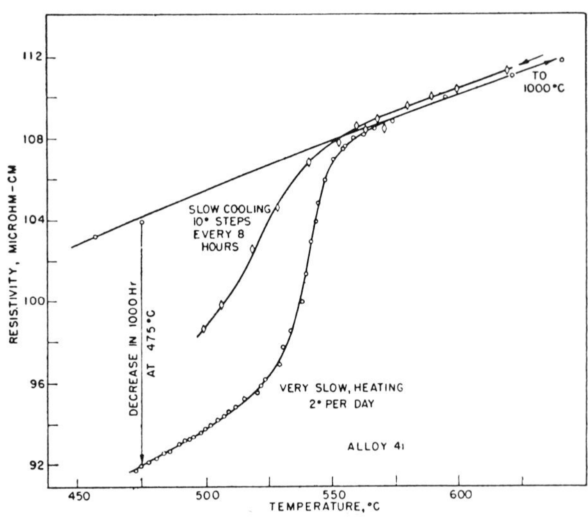

Fig. 16—Recovery of resistance of an aged 53% Cr–Fe alloy during slow heating

The principal results for hardness are shown in Figs. 14 and 15. It will be noted that the series of alloys in Fig. 14 is the same as given in Fig. 2 for isothermal ageing. It has been shown during this work that about three days are sufficient to establish the new hardness after a temperature change for the lower temperatures and less for the higher ones. The results shown here, except for Alloys 70 and 80, are to be considered equilibrium hardnesses. Proof that this is the case for Alloy 90 is given by comparing the hardness on heating with that for isothermal ageing, Fig. 5. Within the accuracy of the measurements they are the same. Some evidence was found which suggested that the hardness would follow the same curve on cooling if this were done slowly enough. It has been convenient to show a break in these curves, although the data do not establish that the change is so abrupt. That the decrease above the break is real, and associated with a change in the solid solution which occurs rapidly, is shown by the hardnesses obtained by re-ageing just above the break for 1 h samples quenched from high temperatures. Except for possibly Alloy 80, all those so aged gave increases which placed the points very near the curves obtained from step heating.

It was shown in Fig. 8 that step heating was relatively unsatisfactory for resistance; so heating at 2° C per day was tried. The results for most of the alloys have been tabulated already in Table II. Figure 16 shows part of the data graphically. It was established that the region where the resistance is increasing towards the un-aged value is essentially an equilibrium process. The fact that on cooling at a much faster rate the same shape curve is obtained to some point past the break is partial evidence. Here, the resistance values do not break sharply, and no accurate value can be obtained for complete recovery. It is supposed that the presence of impurities was partly responsible for this in Alloy 41 and also for many of the rest. Moreover, for those alloys containing around 30% Cr the effect was considerably worse and only ranges for recovery could be established. This is no doubt partly connected with the loss of ferromagnetism simultaneous with re-solution. Figure 17 gives a better overall picture of the entire process. This was a purer alloy and the break as shown agrees very well with those for the hardness of the purer series.

Magnetic measurements were made with step heating analogous to the hardness and the results are

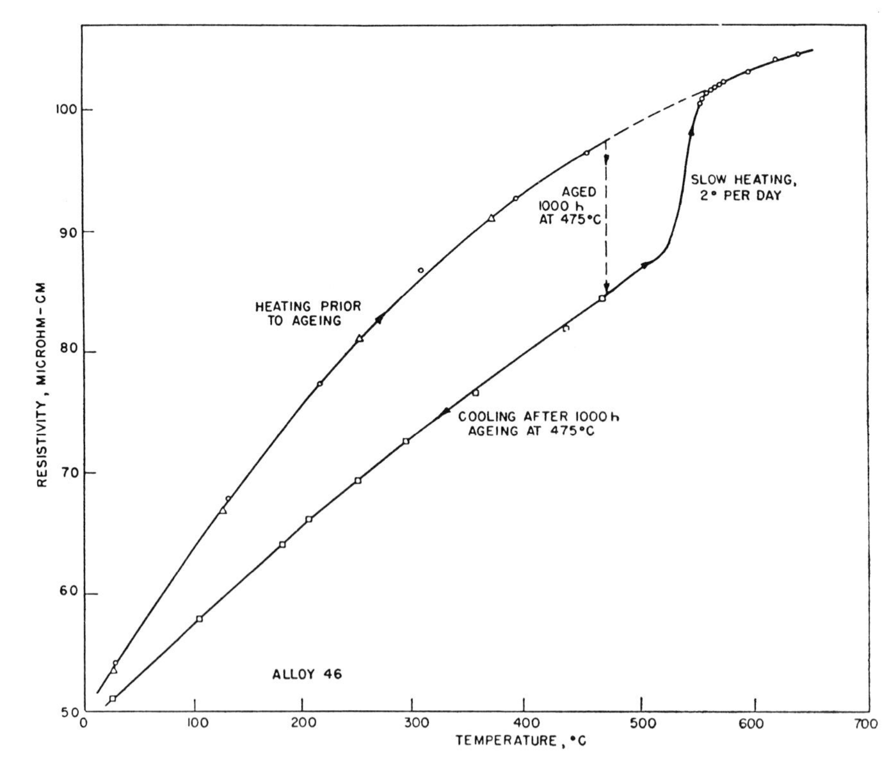

Fig. 17—Temperature/resistivity curves for a 46% Cr–Fe alloy before and after ageing

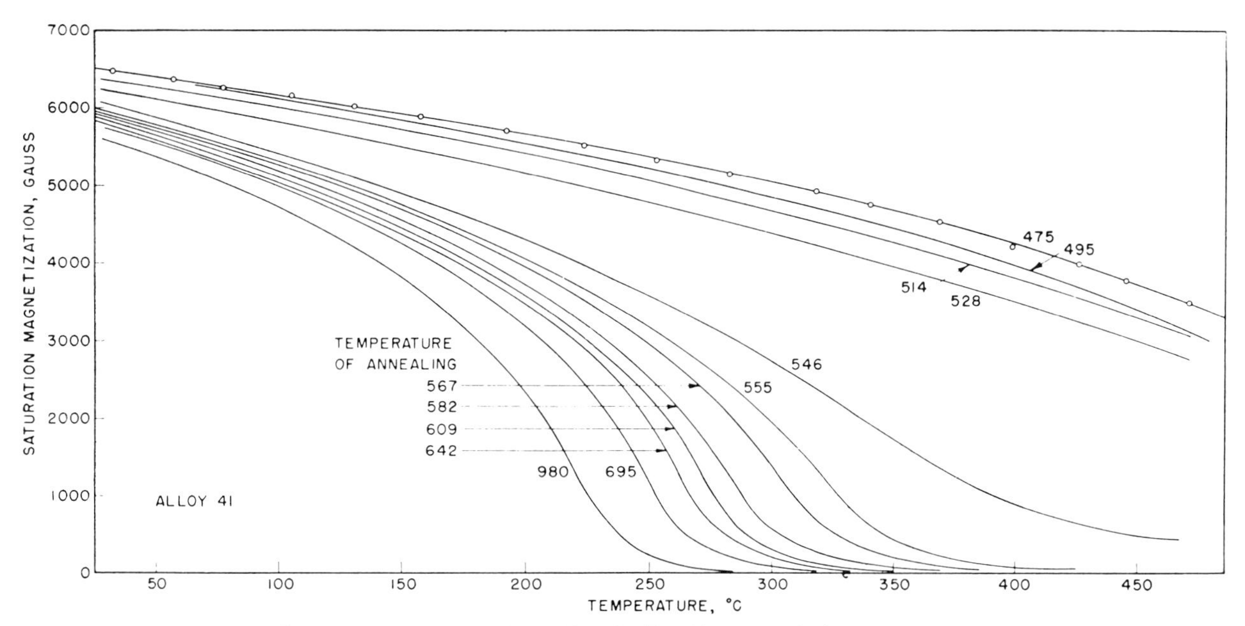

Fig. 18—Magnetization/temperature curves for a 53% Cr–Fe alloy annealed at successively higher temperatures

plotted in Figs. 18 and 19. The behaviour almost exactly parallels that for hardness. The tail-off above the break of the Curie temperature versus the temperature of holding is more pronounced than for the hardness; and it was found that a higher quenching rate from temperatures above 700° C could reduce the Curie temperature by at least 25° more.

Alloy 80 was heated at successively higher temperatures and the force required to separate the sample from a given magnet was measured and is plotted in Fig. 20. It is apparent that these data do not accurately give the temperature required for re-solution, although it is useful information in the absence of more extensive data on this alloy.

Sigma Phase Formation

It is well known that sigma will form in Fe–Cr alloys in the compositional and temperature range of interest in this study. It was thus necessary to make certain that none of the changes which were found could have been effects of σ-formation. As it turned out, the only alloys which showed sigma in any case were the resistivity samples which, for the first run, had been heavily cold reduced. In this case, no sigma could be found after ageing up to 510° C, but most of the samples contained appreciable sigma on reaching 550° C. The only graphical data reported here where sigma was formed are those for Alloy 82 in Fig. 8. This sample had a few per cent. of sigma on reaching 550° C. However, it could not be determined for this or any other case exactly what effect sigma had on resistivity and so the effect must have been slight. Undoubtedly, the reason that sigma was not formed in subsequent runs is that the material had been recrystallized and the holding times were somewhat shorter.

The question naturally arises as to whether sigma will form in very long times at the lower temperatures of ageing. If it will not, the mixture of the Fe-rich matrix and the Cr-rich precipitate is more stable and a eutectoid temperature must exist below which sigma is unstable. Unequivocal proof of the existence

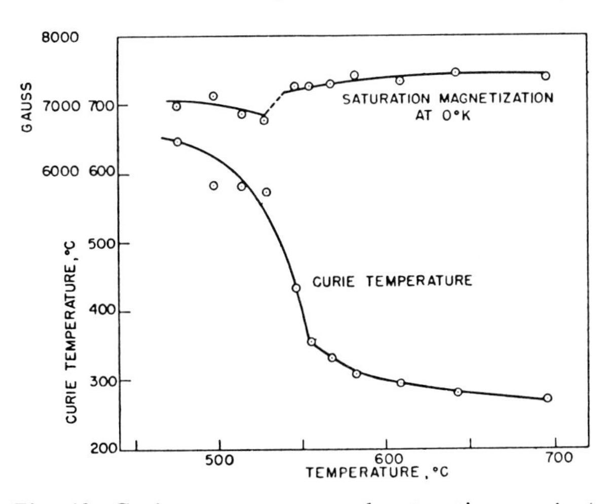

Fig. 19—Curie temperature and saturation magnetization of a 53% Cr–Fe alloy annealed at successively higher temperatures

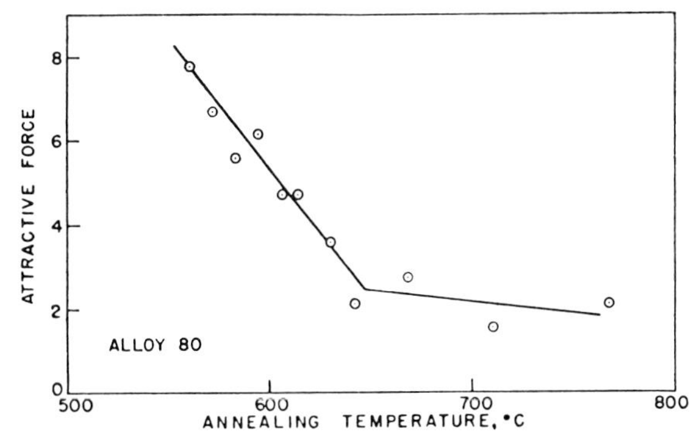

Fig. 20—Magnetic behaviour of a 80% Cr–Fe alloy with successively higher-temperature anneals

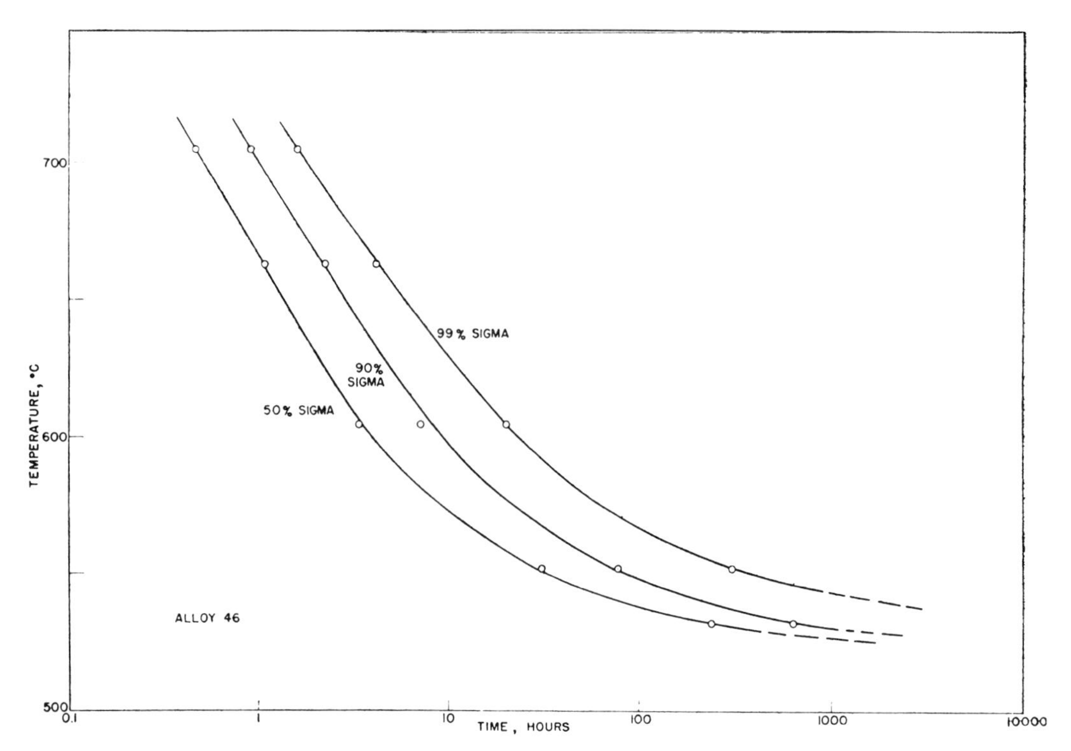

Fig. 21—Kinetics of sigma formation in filings of a 46% Cr–Fe alloy

of such a eutectoid would consist of evidence that sigma decomposes into the mixture of body-centred-cubic phases. Even if this eutectoid reaction were proved to exist by the above method, it would still remain to determine its temperature. An alternate method is to determine the kinetics curves for the formation of sigma at successively lower temperatures, a eutectoid temperature being demonstrated by the existence of increasing reaction times without bounds. For such a determination, filings of Alloy 46 were sealed in Pyrex tubing under vacuum and aged for successive time intervals. The formation of sigma could be followed reasonably well by measuring the attractive force between the filings and a permanent magnet. The results are plotted in Fig. 21. It is immediately evident that the reaction times become abnormally long at lower temperatures just as predicted. The best estimate for the eutectoid temperature is 520° C. It might be noted that the relative times for reaction to go from 90% to 99% becomes progressively longer with falling temperatures. This is explained by the existence of small regions falling outside the narrowing single-phase field as a result of normal freezing behaviour. A sample containing 99% σ and 1% ferrite was held at 490° C for 190 h with no apparent change.

In trying to fit together these data concerning the shape and position of the miscibility gap, the eutectoid temperature, and the sigma field according to Cook and Jones,[16] there was considerable uncertainty about the higher-Cr regions. It was planned to establish certain phase boundaries by reacting Alloy *K*590 (50% Cr) to sigma and the phase in equilibrium with it and then analysing this phase. It turned out that the rates of formation of sigma were very appreciably less than for Alloy 46 and equilibrium could not be approached in several hundreds of hours. This experiment produced only two pieces of useful information. Sigma was found after ageing at 530° C and there was a suggestion of a change of rate between 583° and 591° C. This possibility of a change in rate is the most direct evidence for a higher eutectoid temperature.

The work of Fisher *et al.*[1] has shown the particles

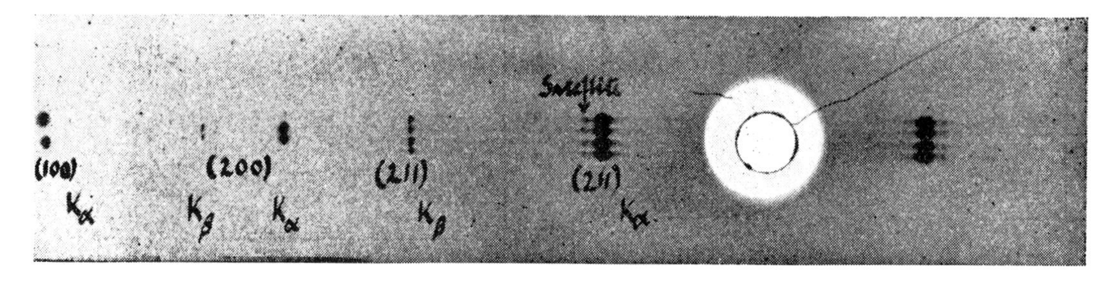

Fig. 22—Rotation pattern of single crystal of alloy 79 (30% Cr) aged 1·75 h at 475° C. Cr radiation, 0·17 min/degree exposure

formed during precipitation to be too small to be seen by a light microscope. They were able to clearly show the particles by an electron microscope only by special techniques. The present work required considerable microstructural work for control and σ detection, but no effects of ageing were apparent using electrolytic polishing. Some standard electron microscopy was done in this work, but no particles were clearly seen.

X-ray Diffraction

Several reports have been made of the change in nature of the diffraction patterns from aged Fe–Cr alloys. However, this work has shown that these effects must have been due to improper techniques since it has been impossible to find any effect whatsoever using great care. The one genuine effect which has been reported was due to the precipitation of a nitride[8] and does not appear to be directly associated with ageing. In this work use was made of Debye–Scherrer and single crystal techniques including Laue, rotating, and oscillating crystals. The most intensive series was made using a single crystal rotating or oscillating about a [110] axis through the positions for characteristic reflections.

Three more or less representative patterns are shown in Figs. 22 and 24. Of these, Fig. 22 represents a rotation pattern after ageing 1·75 h at 475° C using an exposure of 0·17 min/degree (the splitting resulted from the axis being slightly off from the [110] direction). Figure 23 represents the pattern after the sample had been given the following treatment: 1000 h at 475° C, 190 h at 510° C, 48 h at 500° C, 100 h at 450° C, and 150 h at 425° C. Although detail has been lost in reproduction, it is noticed that the only extra feature is the diffuse tail on the high-angle side of the spots. The spots themselves appear unchanged. This diffuse effect apparently is due to surface relaxation occurring during the ageing

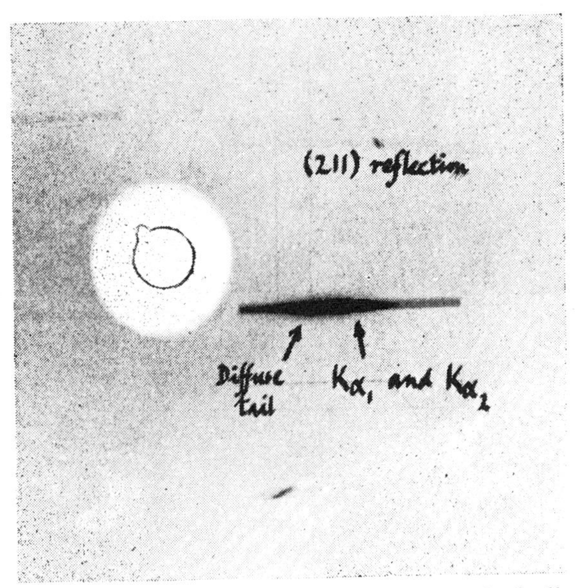

Fig. 23—**Oscillation pattern of single crystal of alloy 79 (30% Cr) after prolonged ageing showing diffuse effect. Cr-radiation (211) reflection, 8 min/degree exposure**

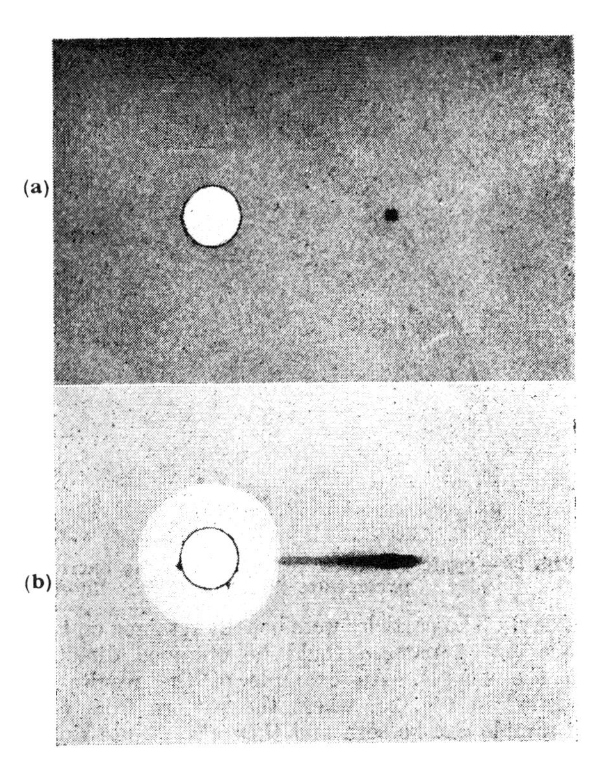

Fig. 24—**Oscillation patterns of single crystal of alloy 79 (30% Cr) after prolonged ageing with repolish (Cr-radiation, (211) reflection)**
(*a*) 0·27 min/degree (*b*) 16 min/degree

treatment because if the surface is again electropolished, this effect disappears as shown in Fig. 24. The thickness of this layer in which the iron-rich matrix has relaxed to a more or less strain-free state is estimated to be 1000–5000 Å thick.

Figure 24 gives two patterns of the sample after electropolishing and represents two extremes of exposure times. The degree of resolution is clearly indicated in Fig. 24*a*. Except for the surface effect discussed above, all patterns obtained were indistinguishable under visual examination when due consideration was given to the exposure times. Broadening and other diffuse effects must be very slight.

It might be noted on these patterns that there is a weak line at lower angles than the characteristic spots. This is a satellite line and has been found on patterns of other material using chromium radiation and definitely is not associated with this ageing.

By obtaining rotation patterns for the single crystal, it was possible to make accurate measurements of the lattice parameter. The average deviation was less than 0·01%, and so it is concluded that no change as great as 0·0003 Å took place. This corresponds roughly to the uncertainty of room temperature. The measured value of 2·8675 Å checks very well with the values of Preston[17] for a 30% alloy.

A simplified technique based upon the work of Fischer *et al.* for the extraction of the precipitate particles was used.

Here, a small region of the sample was exposed to a mixture of picric and hydrochloric acids in alcohol

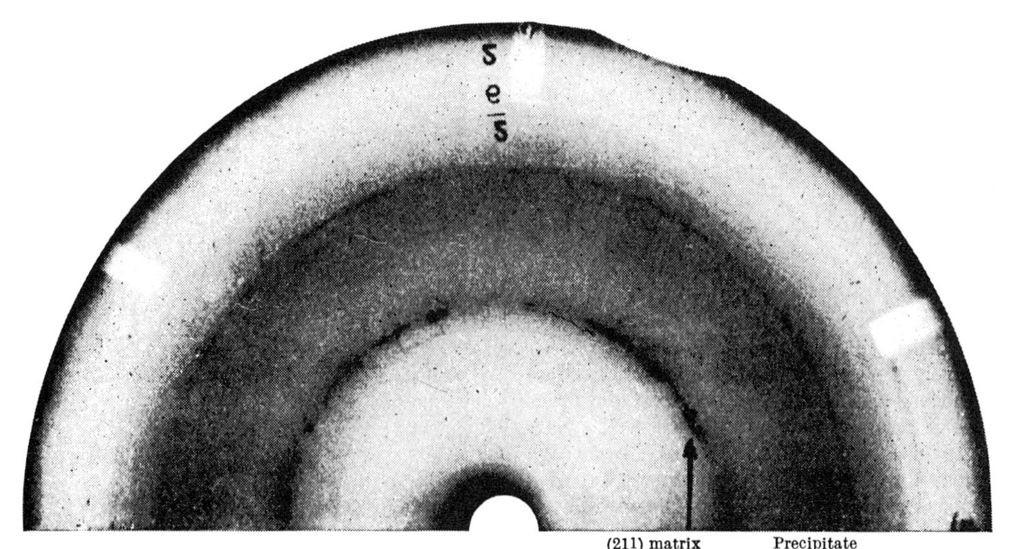

Fig. 25—Typical powder pattern from the extracted precipitate. The matrix gave the sharp lines, the precipitate gave the diffuse lines, of which the (211) is most prominent

for a day. The particles were usually retained on the surface and a pattern could be obtained directly from the sample. An example of the results is illustrated in Fig. 25, where the pattern from the bulk sample can be seen and then also diffuse rings just outside which are from the freed precipitate. The effect is particularly apparent on the innermost rings corresponding to the (211) reflection. The results for Alloy 90 in the hot-rolled condition are given in the following table:

Temperature of Ageing, °C	Time, h	Line Half Width of (211) Reflection,°	Particle Size, Size, Å
475	1000	none found	...
495	800	5	85
515	600	3	140
530	slow heating	2	200

The particle size has been estimated by using Scherrer's equation[18] and the results are probably accurate to better than a factor of two. The same technique was tried on the hot-rolled alloys from the Electro Metallurgical Company, but no results were obtained because the particles washed off the surface. Unaged samples naturally gave no residue. Accurate values of the lattice parameter were not obtained, although it appeared to be 0·0007 Å greater than the matrix. This corresponds at face value to a Cr content of 80%, according to Preston's data.[17]

DISCUSSION OF RESULTS

In the preceding sections the experimental results associated with the ageing of alloys of iron and chromium have been presented. But for a few exceptions, these results are in agreement with previous investigators. It was found that pure alloy did age, but no explanation for this difference from results of Becket[3] and Lena and Hawkes[8] was found. Also no X-ray diffraction effects were found in contrast to results of previous investigators.[4, 7, 8] This work has also added a number of new details concerning this ageing reaction.

This ageing, or 'embrittlement,' of Fe–Cr alloys is found to be the result of the precipitation of small, coherent particles rich in chromium as a result of the existence of a miscibility gap in the system. This idea is completely different from the proposal that the ageing resulted from the precipitation of some minor phase[12, 19] (oxide, carbide, nitride, phosphide, or sulphide), or that it was ordering.[6, 10, 20, 21] Also, it was not possible to confirm Lena and Hawkes' idea[8] that nitrogen was necessary in high-purity materials.

It is considered incorrect and unnecessary to term this precipitate a transition structure connected with the formation of sigma as proposed in the most common theory of ageing.[1, 2, 8, 22, 23] On the basis of

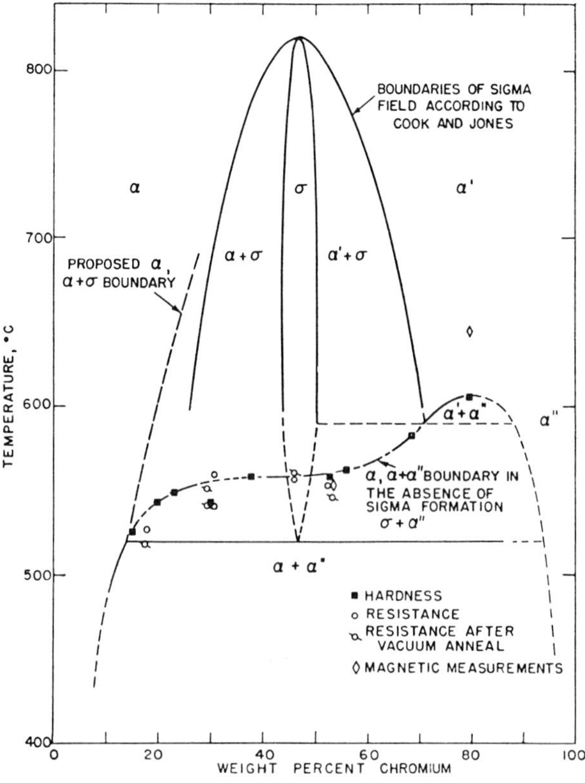

Fig. 26—Partial phase diagram of the Fe–Cr system

crystal structure, thermodynamics, and composition, it seems completely unreasonable to associate sigma with this Cr-rich phase.

Without exception, these results on ageing can be explained on the basis of a miscibility gap in the Fe–Cr system, as shown in Fig. 26. The sigma fields, according to Cook and Jones,[16] and the previously-mentioned eutectoid reactions, have been included. The miscibility gap, in the absence of sigma, has been drawn in according to the results of high-purity alloys, principally the hardness results in Figs. 14 and 15, although satisfactory agreement between hardness, resistivity, and Curie temperature has been shown. The presence of 1% Si and Mn together lowers the re-solution temperature about 10° C and lowers the bottom eutectoid temperature somewhat more since sigma is occasionally found in commercial alloys at as low as 480° C.[1, 24] No significant difference was found between alloys with low (0·002%) and high (0·38%) nitrogen contents.

In order that the lower eutectoid temperature, the miscibility gap, and the $\alpha/\alpha+\sigma$ boundaries agree in the left part of the diagram, it is necessary to re-position one line. It was believed that the gap and the eutectoid temperature were the more accurate, and hence a new position of the $\alpha/\alpha+\sigma$ boundary has been suggested. In reviewing the work of Cook and Jones,[16] it is not too surprising that their boundary could be off the required amount since the establishment of such a boundary by their method is very difficult.

Difficulties are also encountered in the high-Cr region, and the unsuccessful attempts to get additional data have been presented. It is obvious that the $\alpha+\sigma/\alpha$ boundary on the chromium side must be moved markedly to the right if it is to intersect the gap to give a single eutectoid temperature of 520° C. Reference to Cook and Jones' work suggests that this boundary may have been more accurately determined, so that an alternate course has been taken. It is supposed that a second eutectoid temperature exists at 590° C, as shown. Besides this indirect evidence, the kinetics of σ-formation in the 56% alloy also gives some hint of its existence. For the sake of completeness, it is regrettable that this high-chromium side could not be established with greater certainty.

The complete miscibility gap, in the absence of sigma, is seen to have a concave upward portion which has not been previously found for a miscibility gap. The results are rather conclusive in showing that this dip is real and is not associated with actual σ-formation. If it is supposed that this dip is associated with the potential to form sigma, then it must be concluded that the forces responsible for sigma can also modify the energy of the solid solution prior to any change in crystal structure.

This incompatibility between iron and chromium is somewhat surprising in that the size difference is so slight and normally both are considered to have the same valence in the metallic state. Assuming this, one then supposes that chemical energy must be responsible. In this case, however, it seems reasonable for magnetic effects to be fairly important. Indirect evidence includes the fairly small energy release on ageing,[10] the marked changes in Curie temperature on precipitation and clustering, the ferromagnetic state of iron and the anti-ferromagnetic state of chromium at room temperature. In the absence of some definite model it is not possible to know if magnetic effects could be responsible for the odd shape of the miscibility gap.

The temperature dependence of the hardness and Curie temperature and, to a much lesser degree, the resistivity which have been found above the miscibility gap (Figs. 14–17 and 19) are considered to result from increasing clustering with decreasing temperature. The high initial rates of precipitation are essentially in agreement with the idea that the nuclei are present at the start, a condition consistent with marked clustering. The notable case where short-range ordering has been found above a miscibility gap comes from the work of Flinn *et al.*[25] on the nickel–gold system where the large size difference was shown to be responsible. The present system could not be expected to behave similarly. It might be noted that it is essentially this clustering (accompanied possibly by minor precipitation) which caused the formerly unexplained changes in Curie temperature in this system as found by Adcock.[24]

Examination of the figures showing the changes in hardness, electrical resistance, and magnetic properties during isothermal ageing suggest that the process is rather simple and straightforward. These curves can be conveniently divided into three more or less distinct parts; first, a very rapid, decreasing rate, second, a more or less constant rate, and finally, a decreasing rate. Primary graphs show these stages fairly well, but the rate curve, Fig. 13, shows them best. The magnitude of the change of rate should be noted.

It has been pointed out that clustering is considered to exist in the alloys even prior to quenching, so that nucleation need not be considered as rate controlling, even for very short times. For this reason, it is not too surprising that the ageing process should start out at a high or maximum rate. However, it is somewhat surprising that the rate decreases by a factor of ten within the first hour (at 475° C) prior to a realization of more than 10% reaction. Two possible explanations of this have been considered. It might be supposed that some degree of clustering, distinct from precipitation, occurs in very short times in a fashion similar to that found above the miscibility gap. It seems more reasonable, however, that the excess vacancies which were quenched-in by rapid cooling increase the diffusion rate and thus give a high initial rate. It would thus be supposed that the excess number of vacancies reaches approximately equilibrium concentration in 1 h at 475° C. This picture is in qualitative agreement with the effects of quenching temperature, as shown in Fig. 4, in that a higher quenching temperature gave a higher initial rate. It is realized, however, that higher quenching temperatures would also change the degree of clustering so that a detailed analysis is not easy. On the basis that this first stage is due to excess vacancies, the apparent activation energy for the completion of this step of 55,000 cal-mol would seem too high. This should not be taken too seriously

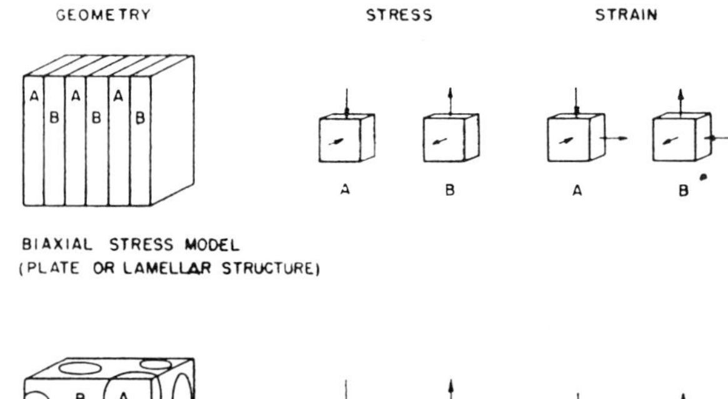

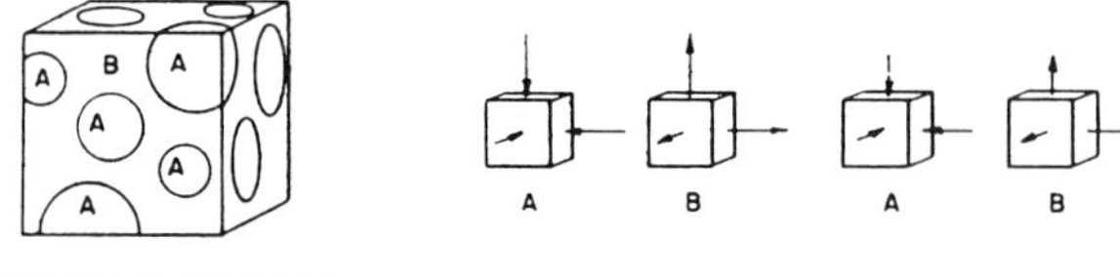

Fig. 27—Stress and strain models for biaxial and triaxial cases

until the process can be examined in greater detail.

Qualitatively, the later stages of ageing apparently occur by simple depletion of the solid solution. It is supposed that the change from the second stage to the third results from the interfering of depleted regions of adjacent particles, but this has by no means been proven. The marked parallel between hardness, resistivity, and Curie temperature is noteworthy in that this type of behaviour is uncommon in ageing reactions. This is due in part to the slow growth of the precipitate after precipitation, and the fact that it never loses coherency.

While overageing is very common in precipitation systems, it is not difficult to explain its absence in this system. The particle growth required for overageing can be expected to occur very slowly on the basis of considerations of driving force and diffusion. When the particles have reached a size of 100 Å, the driving force, interfacial energy, will be no more than 8% of the initial driving force. Since the distances for diffusion are now about twice that during precipitation, the rate of particle growth would not exceed 2% of the rate of precipitation. Such a slow process would not be detected by the experimental methods used.

Some particle growth was found in the particle size measurements, but this occurred when the temperature was increased, so that diffusion was much faster. It might be noted that the hardness decrease resulting from increasing temperatures is indicative of decreasing amounts of precipitate and lessened strains rather than particle growth. It was not possible to show any dependence of hardness on particle size within the scope of this work. Probably, the most significant reason why these particles do not lose coherency is that for such small particles the necessary dislocations would represent more energy than they could release.

The application of the diffusion data of Buffington *et al.*[27] along with the methods of Zener,[28] predicts particle sizes consistent with those observed.

It has become customary, as a result of theoretical work, to associate the initial shape of a particle with the degree of strain, as noted by Hardy and Heal.[29] On this basis, it is not surprising that particles in the Fe–Cr system start out spherical since the strain is rather small.[1] However, the problem does not seem to be well elucidated for there are other systems similar to this system yet the earliest detectable stages consist of coherent plates of about 20–40 Å thickness. Two examples are the Cu–Ni–Fe and the Cu–Ni–Co systems which have been extensively investigated.[30, 31, 32, 33] In the authors' opinion, these systems are rather well understood from a structural standpoint and the state of coherency strains as found by X-ray diffraction are represented in the top of Fig. 27 for a single set of (100) planes. Since the two phases are coherent on cube planes, one observes experimentally three sets of orientations of a double tetragonal structure, the 'c' axis being parallel to the original cube poles. This duplex structure has c/a ratios just less than, and just greater than, unity.

On the basis of the absence of X-ray diffraction effects for the aged Fe–Cr alloys, it must be assumed that both the matrix and precipitate are essentially isotropically strained to a common lattice parameter, that of the undecomposed alloys. The strain state which is believed to exist is illustrated at the bottom of Fig. 27. It is not possible to predict what the uniformity of strains should be for, say, a random collection of coherent spheres, since the only calculations have been based on a single particle. The lack of X-ray diffraction effects is offered as evidence that the proposed isotropic strain state is very nearly correct. This is in agreement with the theoretical work of Eshelby[37] for a collection of strain centres.

A simple calculation suggests that the strain energies are too small in this system to be very important in determining the structure or in causing a loss of coherency. If Preston's data[17] are used for the lattice parameter and an equi-atomic alloy were completely aged at 475° C, the following values are obtained. If the structure were plate-like, as in Cu–Ni–Fe, the strain energy at room temperature would correspond to 2·2 cal/mol, while an isotropic strain state would correspond to 5·6 cal/mol, where both results are based on the bulk elastic constants of iron. While the plate structure has significantly lower energy, in neither case is the magnitude considered important. If one estimates the lattice parameters at 475° C from available data,[34, 35] the strains can be expected to decrease by about a half, so that the above energy values would decrease by a factor of four, even if the modulus had not decreased. It seems likely that the importance of unequal expansion rates in such calculations has been overlooked in the past.

It seems unreasonable to explain the hardening during precipitation of these materials in terms of the usual dislocation model (see Hart[36] for a review of this subject), since it seems unreasonable to expect the particles to be able to withstand great shearing forces in that they are continuous with the matrix and the slip planes are undistorted as judged from the X-ray diffraction results. As an alternate explanation, it is proposed that the particles interact with the hydrostatic stress components of the edge dislocations since both the matrix and particles are considered to be hydrostatically stressed. An illus-

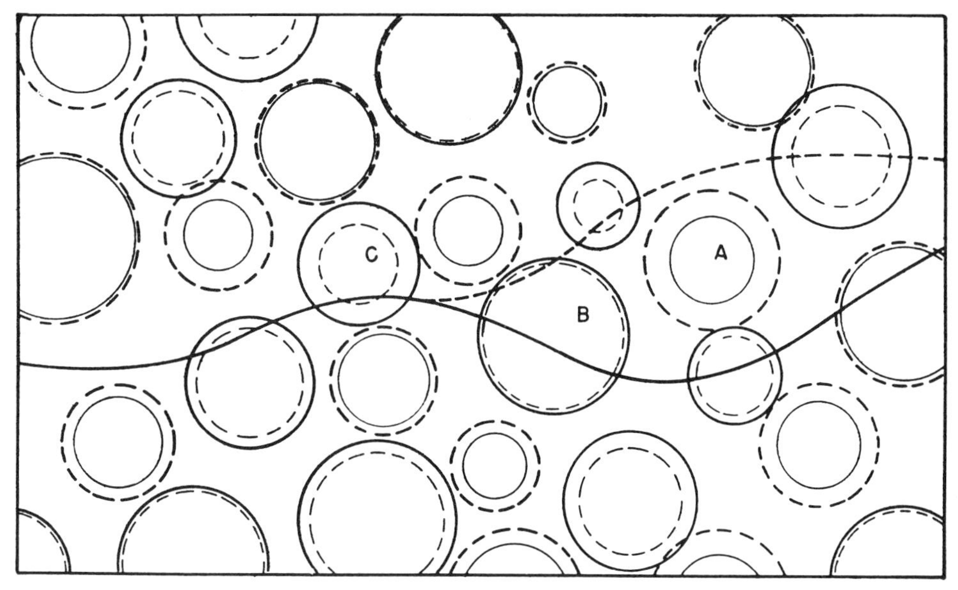

Fig. 28—Model of interaction of edge dislocation and hydrostatically stressed particles

tration of how such a model would cause hardening is given in Fig. 28. Here, the particles of chromium are considered to be spherical and in compression, while the iron-rich matrix is in tension and the extra plane of atoms of the dislocation is considered to extend beneath the plane of cut. Particle '*C*' lies almost wholly in the tension region of the dislocation and thus attracts it strongly. Particle '*B*' lies roughly centred on the slip plane and offers little interaction, while '*A*' repulses the dislocation strongly since it lies below the plane. At some earlier time it is supposed that the dislocation was in the dotted position until a stress was applied which forced the dislocation to roughly the centre of particle '*A*', where it then snapped into the position shown here and dissipated the supplied energy as lattice vibrations. One must then supply a greater amount of energy to slip aged material.

A detailed mathematical analysis of this model has shown it to be capable of predicting roughly the observed hardening and the apparent lack of dependence on particle size.

There are too few data on the resistivity of the Fe–Cr system, such that all the effects which have been found can be explained in detail. It is not known why the resistivity decrease at elevated temperatures is so much greater than at room temperature. The very slight rise in resistivity at 0° C during ageing at 475° C may have been due to particle size, but for longer times and higher temperatures the resistivity seemed independent of particle size. The time-dependent change of resistivity for aged samples in the ageing range is considered to result simply from diffusion as dictated by the phase diagram. The fact that it starts off proportional to the square root of time and appears to end as an exponential decay gives some credence to this idea. The apparent activation energy for this process of 36,000 cal/mol would seem too small, but the process should be examined in greater detail before this is concluded.

SUMMARY AND CONCLUSIONS

The principal conclusions which have been reached as a result of this work are:

(1) The so-called "475° C embrittlement of ferritic stainless steels" is actually an ageing or precipitation hardening phenomenon. This precipitation is a result of the existence of a miscibility gap in the Fe–Cr system at these temperatures. The precipitate may be considered to be body-centred-cubic chromium which contains small amounts of iron.

(2) This miscibility gap is connected by a eutectoid reaction at 520° C to the region where σ-phase is stable, and also, there seems to be a second eutectoid temperature at about 590° C. For the most part, the miscibility gap was studied in the absence of σ-phase, and would appear to have two inflexion points and to be skewed toward the chromium side.

(3) Chemical, or magnetic energies, or both, are considered to be responsible for this miscibility gap to the exclusion of either strain energy or valency effects.

(4) The presence of short-range clustering above the miscibility gap is demonstrated by the change in hardness and Curie temperature with temperature. This clustering is also considered to be partly responsible for rapid initial rates of precipitation in that the nuclei can be considered to be present at the start.

(5) The very high initial rate of ageing is considered to result from the presence of excess vacancies and is dependent on quenching temperature.

(6) There were no apparent differences in rate of precipitation between pure alloys and those containing the normal impurities. In particular, changing the nitrogen content from 0·0015% to 0·2% in an alloy produced no noticeable effect.

(7) The magnitude of this ageing decreased in high-chromium and low-chromium alloys. The type of behaviour was very similar for alloys between 20 and 60% chromium, but the kinetics appear to

change somewhat with higher chromium contents.

(8) No difference was apparent in the ageing behaviour of cold-rolled, hot-rolled, recrystallized, or large-grain material.

(9) On the basis of X-ray diffraction work, the particles appear to be larger the higher the temperature of precipitation. Particles formed at lower temperatures will grow at higher temperatures, but isothermal growth may be too slow to detect. The range of particle sizes which are expected are 300 Å and down.

(10) The hardness is dependent upon amount and composition of the precipitate, but it was not possible to show any dependence upon particle size from these data.

(11) The particles do not lose coherency at any time, and this is considered to be a result of the small strain energy and the small particle sizes. Even with very slow heating rates, the particles apparently go back into solution without losing coherency. On the basis of the absence of X-ray diffraction effects, it is concluded that both the matrix and precipitate are under essentially isotropic stresses and are strained to a common lattice parameter, that of the unaged material.

(12) It seems unreasonable that such particles could withstand substantial shears, and so it is suggested that hardening may arise from the interaction of the particles and matrix with the hydrostatic stresses around the dislocation.

(13) The presence of the normal amounts of silicon and manganese cause the resolution of the precipitate about 10–15° C below the temperature required for high-purity alloys of the same chromium content. Also, the presence of 5% Si will prevent the ageing of a 25% Cr alloy at 480° C.

(14) On the basis of this work, it would seem that the only practical method of eliminating this ageing commercially is by the investigation of useful ternary alloy additions which would lower the top of this miscibility gap.

Acknowledgments

This work was made possible through the financial assistance of the Allegheny Ludlum Fellowship. The Allegheny Ludlum Corporation also supplied extensive materials and services which are greatly appreciated.

Considerable assistance was received from the staffs of the Department of Metallurgical Engineering, Carnegie Institute of Technology, and the General Electric Research Laboratory. In particular, the assistance of the following people is acknowledged: A. J. Lena, E. J. Dulis, J. E. Goldman, and the late A. H. Geisler.

References

1. R. M. Fisher, E. J. Dulis, and K. R. Carroll: *Trans. Amer. Inst. Min. Met. Eng.*, 1953, vol. 197, pp. 690–695; Discussion: *Ibid.*, 1954, vol. 200, p. 663.
2. A. Bandel and W. Tofaute: *Arch. Eisenhüttenwesen*, 1941–1942, vol. 15, pp. 307–320.
3. F. M. Becket: *Trans. Amer. Inst. Min. Met. Eng.*, 1938, vol. 131, pp. 15–36.
4. J. J. Heger: *Metal Progress*, 1951, vol. 60, No. 2, pp. 55–61.
5. J. Hochmann: *Rév. Met.*, 1951, vol. 48, pp. 734–758.
6. Y. Imai and K. Kumada: *Science Reports of the Tohoku Imperial University, Sendai*, 1953, ser. *A*, vol. 5, pp. 220–226; pp. 520–32.
7. V. N. Krivobok: *Trans. Amer. Soc. Metals*, 1935, vol. 23, pp. 1–60.
8. A. J. Lena and M. F. Hawkes: *Trans. Amer. Inst. Min. Met. Eng.*, 1954, vol. 200, pp. 607–615.
9. C. E. MacQuigg: *Ibid.*, 1923, vol. 69, pp. 831–847.
10. H. Masumoto, H. Saito, and M. Sugihara: *Science Reports of the Tohoku Imperial University, Sendai*, 1953, ser. *A*, vol. 5, pp. 203–207.
11. H. D. Newell: *Metal Progress*, 1946, vol. 49, No. 5, pp. 977–1006.
12. G. Riedrich and F. Loib: *Arch. Eisenhüttenwesen*, 1941–1942, vol. 15, pp. 175–182.
13. M. Tagaya, S. Nenno, and Z. Nishiyama: *Nippon Kinkozu Gakkai-Si*, 1951, vol. *B*-15, pp. 235–236.
14. C. H. Samans: Discussion to H. S. Link and P. W. Marshall, *Trans. Amer. Soc. Metals*, 1952, vol. 44, pp. 561–564.
15. A. J. Lena: Personal Communication.
16. A. J. Cook and F. W. Jones: *J. Iron Steel Inst.*, 1943, vol. 148, pp. 217–226.
17. C. D. Preston: Appendix I to F. Adcock: *J. Iron Steel Inst.*, 1931, vol. 124, pp. 139–141.
18. C. S. Barrett: "Structure of Metals," 1943, McGraw-Hill, New York.
19. C. A. Zapffe: *Trans. Amer. Inst. Min. Met. Eng.*, 1951, vol. 191, pp. 247–248.
20. S. Takeda and N. Nagai: Lectures at Japan Inst. Metals, Tokyo and Nagoya, 1949.
21. P. Bastien and G. Pomey: *Compt. Rend.*, 1954, vol. 239, p. 1636.
22. C. A. Scharschu: *Metal Progress*, 1931, vol. 20, No. 7, pp. 59–63.
23. W. Dannohl: Discussion to Bandel and Tofaute and Riedrich and Loib, *Arch. Eisenhüttenwesen*, 1941–1942, vol. 319.
24. H. S. Link and P. W. Marshall: *Trans. Amer. Soc. Metals*, 1952, vol. 44, p. 549.
25. P. A. Flinn, B. L. Averbach, and M. Cohen: *Acta Met.*, 1953, vol. 1, No. 6, pp. 664–673.
26. F. Adcock: *J. Iron Steel Inst.*, 1931, vol. 124, pp. 99–146.
27. F. S. Buffington, L. D. Bakalar, and M. Cohen: Discussion to Birchenall and Mehl, *Trans. Amer. Inst. Min. Met. Eng.*, 1950, vol. 188, pp. 1374–1375.
28. C. Zener: *J. Appl. Phys.*, 1949, vol. 20, p. 950.
29. H. K. Hardy and T. J. Heal: "Progress in Metal Physics," pp. 143–278, Interscience, 1953, New York.
30. V. Daniel and H. Lipson: *Proc. Roy. Soc.*, ser. *A*, 1943, vol. 181, pp. 368–378.
31. A. H. Geisler: *Trans. Amer. Soc. Metals*, 1951, vol. 43, pp. 70–101.
32. A. H. Geisler and J. B. Newkirk: *Trans. Amer. Inst. Min. Met. Eng.*, 1949, vol. 180, pp. 101–120.
33. M. E. Hargreaves: *Acta Crystallographica*, 1949, vol. 2, p. 259; 1951, vol. 4, pp. 301–309.
34. A. B. Kinzel and R. Franks: Alloys of Iron and Chromium, 1940, Vol. II, McGraw-Hill, New York.
35. American Society for Metals, " Metals Handbook," 1948, Cleveland, Ohio.
36. E. W. Hart, " Theories of Dispersion Hardening," Relation of Properties to Microstructure, *Amer. Soc. Metals*, 1954, Cleveland, Ohio.
37. J. D. Eshelby: *Acta Met.*, 1955, vol. 3, No. 5, pp. 487–490.

Embrittlement of Ferritic Stainless Steels

T. J. NICHOL, A. DATTA, AND G. AGGEN

The mechanical properties and microstructures of commercial 11 to 29 pct Cr ferritic steels were examined as functions of aging times to 1000 h at 371, 482, and 593°C. Of the properties evaluated, changes in impact transition temperatures were the best measure of embrittlement. Embrittlement at 482°C occurs most rapidly in the 29 pct Cr alloy and somewhat more slowly in the stabilized 26 pct Cr alloy. The stabilized 18 pct Cr alloy embrittles much more slowly while little, if any, embrittlement was detected in a stabilized 11 pct Cr alloy. Embrittlement at 482°C was characterized by a rapid change in properties followed by a plateau region and then further property changes. The early property change is attributed to precipitation of interstitial compounds and the later change to classic 475°C embrittlement. The onset of 475°C embrittlement in the two highest Cr alloys was accompanied by clustering of Cr atoms along {100} planes indicative of spinodal decomposition. Concurrent with clustering there was also a change from turbulent slip to a more planar slip along {110} planes. Some embrittlement was observed after longer exposures at 371°C which was attributed to a combination of 475°C embrittlement and the precipitation of interstitial compounds. Two of the alloys also embrittled at 593°C, accompanied by optically observable precipitates. The precipitate in the stabilized 18 pct Cr alloy was identified as Laves (Fe_2Ti) phase. One of the precipitates in the 29 pct Cr alloy was identified as sigma phase.

In recent years, there has been an increase in the use and availability of commercial ferritic stainless steels. Ferritic stainless steels have always appeared attractive because of their relatively low cost, their good corrosion properties and their high resistance to stress-corrosion cracking. Usage and availability have been limited because of the poor toughness demonstrated by conventional ferritic stainless steels such as Types 430 and 446. Recent advances in stainless steel technology have made it possible to produce ferritic stainless steels with improved toughness. This has been accomplished in two ways: first, by utilizing the AOD (argon-oxygen decarburization) process to produce relatively low carbon (typically 0.02 pct) ferritic steels stabilized with either titanium or columbium (Nb) or a mixture of both, and second, by utilizing VIM (vacuum induction melting) processes to produce very low carbon (typically below 0.01 pct) ferritic steels.

The increased use of ferritic stainless steels has rekindled interest in the much studied phenomenon known as 475°C embrittlement. Embrittlement is noted at low temperatures (usually room temperature) after ferritic stainless steels have been exposed to temperatures in the range 371 to 510°C. The embrittlement phenomenon can occur during the manufacture of ferritic stainless steels and thus make production more difficult. Embrittlement is most likely to occur during the handling of primary and intermediate product forms such as ingots, slabs and hot rolled coils where rapid cooling is not practically feasible. Embrittlement can also occur during fabrication of finished products, particularly plate products where welding is necessary.

The 475°C embrittlement phenomenon has been recognized in ferritic stainless steels for many decades and was known to be enhanced by increasing chromium contents. The cause of embrittlement remained a puzzle for many years since it occurred in the absence of microstructural changes that could be detected by optical microscopy or by X-ray analysis. In 1953 Fisher *et al*,[1] using electron microscopy, showed that the embrittlement of a 27 pct Cr ferritic steel incurred after 10,000 h at 482°C was accompanied by the precipitation of a very fine Cr-rich body centered cubic phase with a lattice parameter close to that of the matrix. The phase, termed alpha prime (α') was about 200Å in diam and contained up to 80 pct chromium. To explain the presence of the phase, which was not predicted from existing Fe-Cr binary phase diagrams, Williams and Paxton[2] proposed a modified diagram having a solid state solubility gap below about 516°C. This diagram was further modified by Williams[3] and appears to be generally accepted now as a correct description of the Fe-Cr binary phase diagram.

The rate of 475°C embrittlement increases with increasing Cr content and appears to decrease with increasing purity. Although the phase diagram predicts that 475°C embrittlement could occur in steels with as low as 10 pct Cr, effects are rarely noted unless the chromium content exceeds 13 pct. Plumtree and Gullberg[4] attributed embrittlement in a 25 pct Cr alloy to the combined effect of the precipitation of chromium nitrides and carbonitrides and of α' with the former occurring more quickly and the latter becoming significant only after long aging periods (~ 500 h). Molybdenum has been reported to have no effect or to increase the rate of 475°C embrittlement.[5-7] The effect may be negligible at 14 pct Cr[5] but becomes significant at 18 pct Cr[5] and pronounced at 28 pct Cr.[6] Both titanium[5,7] and niobium[7] have

T. J. NICHOL is Manager, Technical Marketing, Tubular Products Division, Allegheny Ludlum Steel Corp., Wallingford, CT 06492.
A. DATTA, formerly with Allegheny Ludlum Steel Corporation, is now with Allied Chemicals Corporation, Morristown, NJ 07960.
G. AGGEN is Manager, Stainless and Alloy Metallurgy, Research Center, Allegheny Ludlum Steel Corp., Brackenridge, PA 15014.
Manuscript submitted June 19, 1979.

been reported to accelerate the embrittlement phenomenon.

The precipitation of α' in high purity Fe-Cr alloys can occur by either spinodal decomposition or by nucleation and growth depending upon composition and aging temperature. Spinodal decomposition is favored by high Cr contents and low aging temperatures. Lagneborg[8] proposed that an Fe-30 pct Cr alloy decomposed into a Cr-rich and Cr-depleted phase inside the spinodal at 475°C and outside at 550°C. The form of the precipitate was spherical at 475°C and disk shaped at 550°C, which he postulated was a consequence of the energetically more favorable disk shape being preferred at the higher temperature where nucleation is needed. DeNys and Gielen,[9] using Mössbauer spectroscopy, analyzed the decomposition of Fe-Cr binary alloys ranging from 20 to 50 pct Cr. They showed clearly that the 20 pct Cr alloy decomposed at 470°C by nucleation and growth while the 30, 40, and 50 pct Cr alloys decomposed spinodally.

Aging near 475°C has been shown to promote deformation by twinning in high Cr ferritic alloys.[10-14] Marcinkowski *et al*[10] using a 48 pct Cr alloy, showed that twinning became an increasingly important mode of deformation compared with that of slip, as the aging time at 500°C increased. Blackburn and Nutting[11] also observed twins in deformed samples of an Fe-21 pct Cr alloy after aging at 475°C. In a study of the mechanical properties of a low interstitial 29 pct Cr-4 pct Mo-2 pct Ni alloy, Nichol[12] observed that aging at 482°C promoted twinning although the propensity to twin also was observed to increase in annealed samples with increasing annealing temperature and, thus, grain size. Lagneborg[13] postulated that twinning is an important factor in the failure of embrittled Fe-Cr alloys. He observed cracks beginning in twin boundaries and twin-twin and the twin-grain boundary intersections. Kondyr *et al*[14] showed that increasing the aging time at 480°C led to a substantial increase in deformation by twinning which they correlated with a second period of embrittlement. In their experiments, deformation twins were absent in quenched samples and those aged for a short time at 480°C. The susceptibility to twinning increased considerably with chromium content for the same aging time.[14] A similar accelerating effect on twinning was observed for titanium in solid solution.[14]

The intent of this study is to compare the effect of isothermal heat treatments at 371, 482, and 593°C on the structure and mechanical properties of four commercial ferritic stainless steels. Three steels with chromium contents of 11 pct (Type 409), 18 pct (Type 439), and 26 pct (Type 26-1S) are titanium stabilized. The fourth steel is a low interstitial alloy containing 29 pct Cr, 4 pct Mo and 2 pct Ni. The rate of embrittlement was measured by changes in hardness, tensile and Charpy V-notch impact properties. Structures were examined by both optical metallography and transmission electron microscopy.

MATERIALS AND EXPERIMENTAL PROCEDURE

Material was selected from commercial heats of Types 409 (11 pct Cr), 439 (18 pct Cr), 26-1S (26 pct Cr-1 pct Mo), and 29-4-2 (29 pct Cr-4 pct Mo-2 pct Ni). The chemistries of the heats are shown in Table I. The thickness of the materials and the initial heat treatment are listed in Table II. The lighter gage material was used for tensile testing while the heavier gage was used for impact testing.

Material was heat treated at 371, 482, and 593°C for times up to 1000 h. After each heat treatment, the material was water quenched. Figure 1 demonstrates the location of the alloys and temperatures studied with respect to the diagram of Williams.[3] Tensile testing and Charpy V-notch impact testing were conducted in accordance with ASTM A-370 and E-8 specifications.

Samples examined by optical metallography were etched electrolytically in nitric acid. Thin foils for transmission electron microscopy were prepared by mechanically polishing the as-received samples followed by electropolishing to 0.25 mm thick coupons. Three mm diameter discs were punched out from the

Table I. Alloy Chemistries, Wt Pct

Designation	Type	Cr	Mo	C	N	Ti	Ni
11 Cr	409	11.42	NA*	0.007	0.015	0.26	0.16
18 Cr	439	18.20	NA	0.044	0.010	0.78	0.26
26 Cr-1Mo	AL26-1S	26.0	0.97	0.030	0.019	0.38	0.14
26 Cr-4Mo-2Ni	AL29-4-2	29.6	3.90	0.006	0.011	–	2.20

*Not analyzed.

Table II. Annealing Treatments

Designation	Thickness, mm	Initial Heat Treatment	Grain Size*
11 Cr	1.52	788°C, 5 min, W.Q.	6-8
18 Cr	1.52	871°C, 5 min, W.Q.	6-8
	6.86	871°C, 5 min, W.Q.	4-7
26 Cr-1Mo	1.83	871°C, 5 min, W.Q.	5-7
	3.18	871°C, 5 min, W.Q.	4-6
29 Cr-4Mo-2Ni	1.65	1121°C, 5 min, W.Q.	1-2
	4.32	1121°C, 5 min, W.Q.	>1-2

*ASTM grain size number.

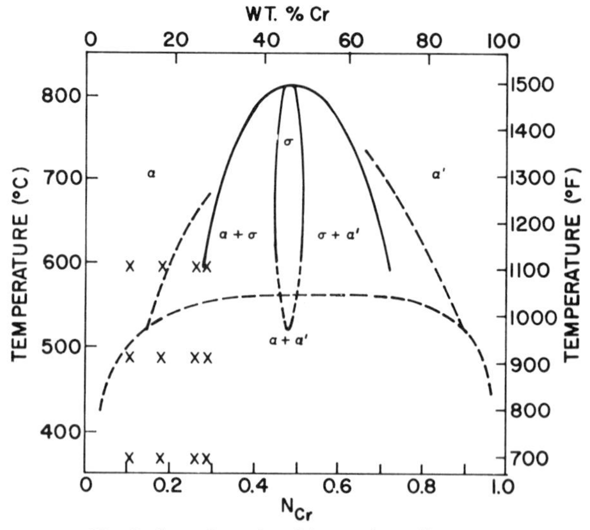

Fig. 1—Iron-chromium binary phase diagram.

prethinned stock and further electropolished in a twin-jet electropolisher at room temperature using 500 ml of glacial acetic acid, 90 g of chromium trioxide, and 20 ml of distilled water as the electrolyte. To examine the deformed structure samples were also prepared from the gage lengths of the tensile specimens which had been pulled to fracture. The thin foils were examined in a Siemens 102 Electron Microscope at 120 kV.

X-ray diffraction was performed on extracts from selected samples. Extraction was performed electrolytically in a 10 pct aqueous solution of ferric chloride by a procedure similar to that reported by Koh[15] except that leads were simply clipped to the electrodes. Extraction was performed at about 2 V and a current density of about 0.026 A/cm^2 for about 24 h. The extract was filtered through 0.8 μ cellulose acetate paper. Diffraction was performed on a General Electric XRD-5 Diffractometer using Cr radiation and a V filter. The X-ray tube operated at 45 kV and 15 mA. The pattern was obtained at a scan rate of 0.4 deg per min using a 3 deg beam slit and 0.3 deg receiving slit.

Table III. Annealed and Quenched Room Temperature Tensile Properties*

Designation	Hardness, R_B	0.2 Pct Y.S. MPa	0.2 Pct Y.S. Ksi	U.T.S. MPa	U.T.S. Ksi	Elong., Pct, in 5.1 cm (2 in.)
11 Cr	66	238.5	(34.6)	414.3	(60.1)	31.5
18 Cr	81	320.6	(46.5)	504.0	(73.1)	32.8
26 Cr-1Mo	79	307.5	(44.6)	497.7	(72.2)	32.3
29 Cr-4Mo-2Ni	92	541.9	(78.6)	639.8	(92.8)	19.8

*Longitudinal properties, water quenched from the anneal.

Table IV. FATT* for Annealed and Water Quenched Material

Designation	Gage, mm	Hardness, R_B	G.S. No.	FATT* C	FATT* F
18 Cr	6.86	83	4-7	38	100
26 Cr-1 Mo	3.18	85	4-6	21	70
29 Cr-4Mo-2Ni	4.32	97	>1-2	−67	−90

*Fracture appearance transition temperature.

RESULTS

A. Mechanical Properties

A.1. Annealed and Water Quenched Material. Table III shows the room temperature tensile properties of the materials studied in the annealed and water quenched condition. The 11 Cr alloy has the lowest yield strength of 238.5 MPa while the 29 Cr-4 Mo-2 Ni alloy has the highest of 541.9 MPa, *i.e.* more than double the 11 Cr level. The 11 Cr, 18 Cr and 26 Cr-1 Mo alloys have comparable tensile elongations in this condition while the 29 Cr-4 Mo-2 Ni alloy has approximately one-third less elongation, *i.e.*, 20 pct *vs* 30 pct.

Table IV describes the fracture appearance transition temperatures (FATT) for the 18 Cr, 26 Cr-1 Mo and 29 Cr-4 Mo-2 Ni alloys in the annealed condition. The 26 Cr-1 Mo alloy demonstrated a lower FATT than the 18 Cr alloy. The 26 Cr-1 Mo alloy, however, was tested at a lighter gage than the 18 Cr alloy and would probably have a higher transition temperature at a comparable gage. In spite of the large grain size, the 29 Cr-4 Mo-2 Ni alloy had the lowest FATT, *i.e.*, had the best toughness, of the materials studied. The −67°C FATT is lower than those for the other materials and is primarily attributed to the low level of carbon and nitrogen and absence of Ti (C,N) inclusions. The impact properties of the 11 Cr alloy were not measured.

A.2. Effect of 482°C Heat Treatments. Room temperature tensile tests were conducted on all four materials after heat treating at 482°C for times to 1000 h. The effect of heat treatment at 482°C on the 0.2 pct offset yield strength is shown in Fig. 2. The tensile properties of the 11 Cr alloy were not affected by heat treatments at this temperature. The 18 Cr and 26 Cr-1 Mo alloys show a two-stage effect, *i.e.*, a small increase in strength after a short aging time followed by a plateau and then a relatively large increment after longer aging times. The 29 Cr-4 Mo-2 Ni alloy also shows a small initial increase followed by a plateau. All samples of the 29 Cr-4 Mo-2 Ni alloy demonstrated serrated yielding and crackling noises during

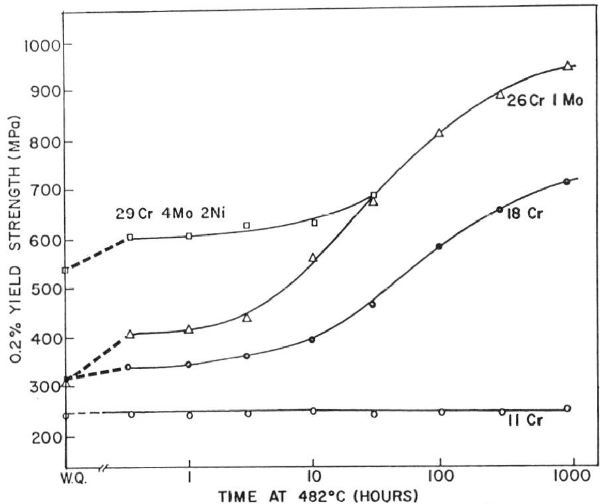

Fig. 2—Yield strength *vs* aging time at 482°C.

pulling which is indicative of deformation by twinning. Samples aged beyond 30 h broke before yielding.

Figure 3 illustrates the effect of aging time at 482°C on the tensile elongation. No significant effect is observed for the 11 pct alloy. The elongation of the 18 Cr alloy gradually decreases until after 1000 h it is approximately one-half the initial as-quenched value. The elongation of the 26 Cr-1 Mo alloy also gradually decreases to 19 pct at 100 h and then falls steeply to 4.5 pct at 300 h. The 29 Cr-4 Mo-2 Ni alloy shows the same drastic effect but at a shorter time; *i.e.*, from 17 pct after 10 h to 2.5 pct after 30 h. Samples aged for longer periods broke before yielding and exhibited no detectable elongation.

Figure 4 graphically summarizes the effect of exposures to 482°C on the FATT of the materials. The 11 Cr alloy was not studied. The two-stage behavior shown by the 0.2 pct offset yield strength (Fig. 2) is also evident when the FATT is used as a criterion. The initial increase in FATT varies from 35°C for the 18 Cr alloy to 80°C for the 26 Cr-1 Mo alloy. The 29 Cr-4 Mo-2 Ni alloy demonstrates a 40°C increase. All the

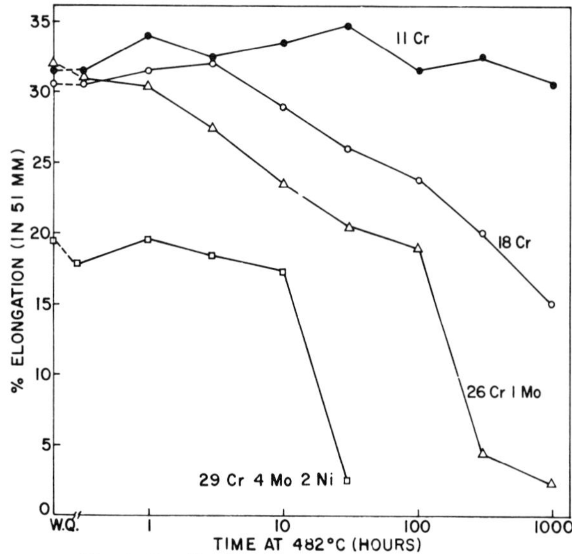

Fig. 3—Tensile elongation *vs* aging time at 482°C.

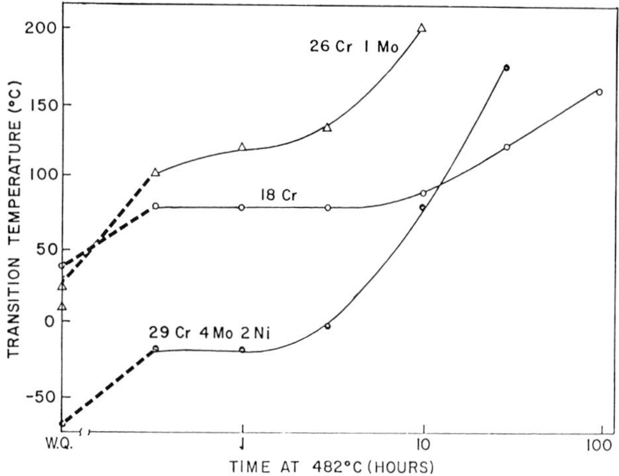

Fig. 4—Fracture appearance transition temperature (FATT) *vs* aging time at 482°C.

alloys show a plateau region with the 26 Cr-1 Mo and 29 Cr-4 Mo-2 Ni alloys showing rapid increases in FATT at times greater than three h at 482°C. The 18 Cr alloy shows a less rapid increase in FATT and the effect is not noted until after times greater than 10 h at temperature.

Table V lists the hardness values obtained from material heat treated at 482°C for times up to 1000 h. Like the tensile results, the 11 Cr alloy shows no significant change while the 18 Cr alloy starts to increase significantly after approximately 10 h. The 26 Cr-1 Mo alloy shows an initial increase followed by a plateau region and then a rapid rise after approximately 10 h. The 29 Cr-4 Mo-2 Ni alloy behaves in a similar fashion.

A.3. Effects of 371 and 593°C Heat Treatments. Figures 5 and 6 compare the effects of 371 and 593°C to that of 482°C on the 0.2 pct yield strength and FATT for the 18 Cr alloy. The yield strength increases after long times at 371°C, *i.e.*, greater than 100 h, however, the net increase after 1000 h at temperature is much less than the increase after 482°C, *i.e.* 76 MPa *vs* 407 MPa. Similarly the FATT increases after longer times at 371°C but never reaches the magnitude of the effect observed after 482°C exposures. Exposures at 593°C resulted in no effect or a slight decrease in yield strength, however, the FATT did increase after 300 and 1000 h at 593°C.

The effects of 371 and 593°C exposures on the yield strength of the 26 Cr-1 Mo alloy are compared to that of 482°C exposures in Fig. 7. Similar to the 18 Cr alloy, 371°C does result in a small increase in yield strength after long times while 593°C exposures did not significantly affect the yield strength. The effect of 371 and 593°C exposures on the FATT of the 26 Cr-1

Table V. Room Temperature Hardness *vs* 482°C Exposures

Time At 482°C, h	Hardness, R_B			
	11 Cr	18 Cr	26 Cr-1 Mo	29 Cr-4 Mo-2 Ni
0	66.0	81	79	91.5
0.3	68.5	83	87	100
1	65.5	82	87	100
3	67.5	84	86	98
10	67.0	86	92	100
30	66.5	92	100	27*
100	68.5	96	27*	26*
300	69.5	23*	30*	27*
1000	69.0	26*	30*	35*

*R_C.

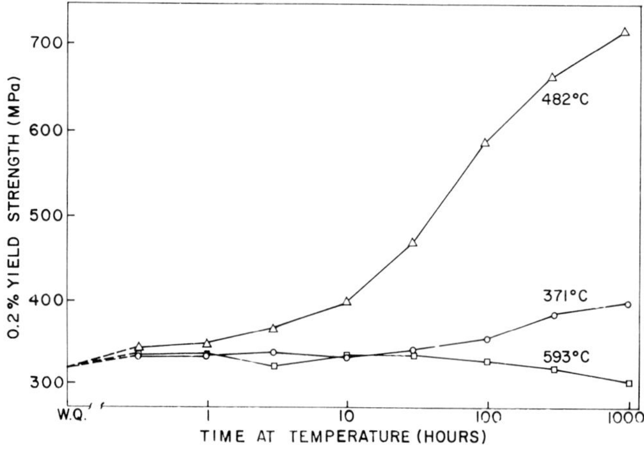

Fig. 5—Yield strength of 18 pct Cr alloy *vs* aging time.

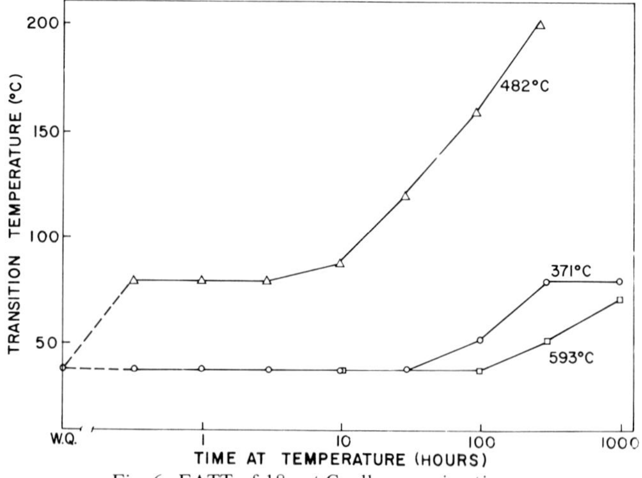

Fig. 6—FATT of 18 pct Cr alloy *vs* aging time.

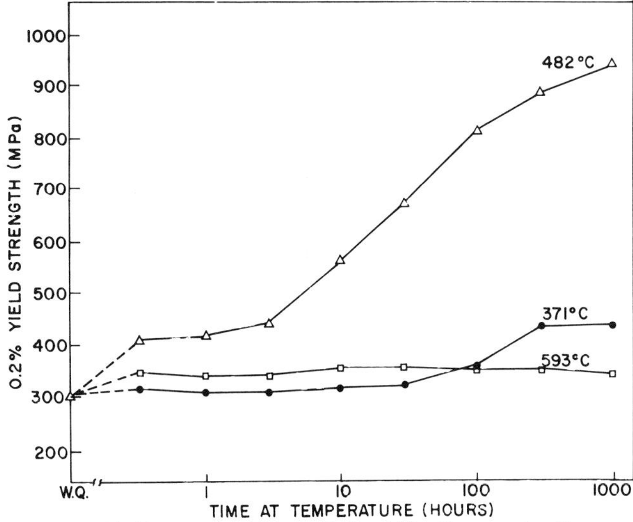

Fig. 7—Yield strength of 26 pct Cr-1 pct Mo alloy *vs* aging time.

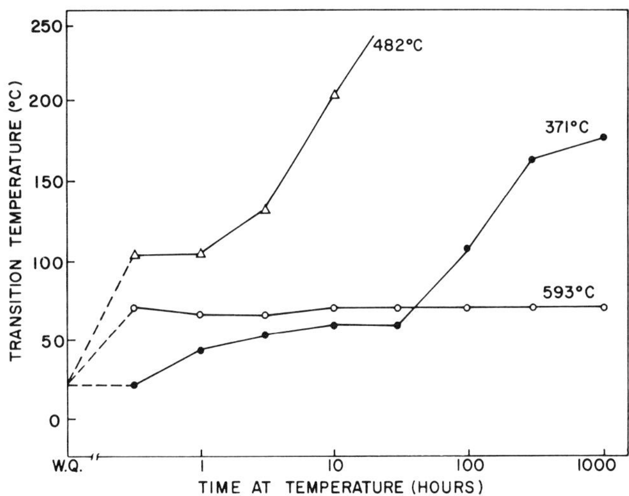

Fig. 8—FATT of 26 pct Cr-1 pct Mo alloy *vs* aging time.

Mo alloy is shown in Fig. 8. The 371°C results show the plateau behavior typical of the 482°C effects. There is a significant increase in FATT at times greater than 30 h at 371°C, *i.e.*, a 150°C increase in FATT after 1000 h at 371°C. The 593°C exposures resulted in a 27°C increase in transition temperature for all times studied.

The effects of 371 and 593°C exposures on the tensile elongation of the 29 Cr-4 Mo-2 Ni alloy are compared in Fig. 9 to those observed for 482°C exposures. Times up to 1000 h at 371°C had no significant effect, however, there is a precipitous drop in tensile elongation after 30 h at 593°C. A similar precipitous reduction in tensile elongation was noted after 10 h at 482°C. Figure 10 illustrates that the 482 and 593°C exposures are equally deleterious to the toughness of the 29 Cr-4 Mo-2Ni alloy. One hundred h exposures at either 482 or 593°C increase the FATT by approximately 275°C. Exposure at 371°C does result in an increase in FATT, however, the exposure time must be greater than 30 h and after 1000 h the increase in FATT is approximately 66°C.

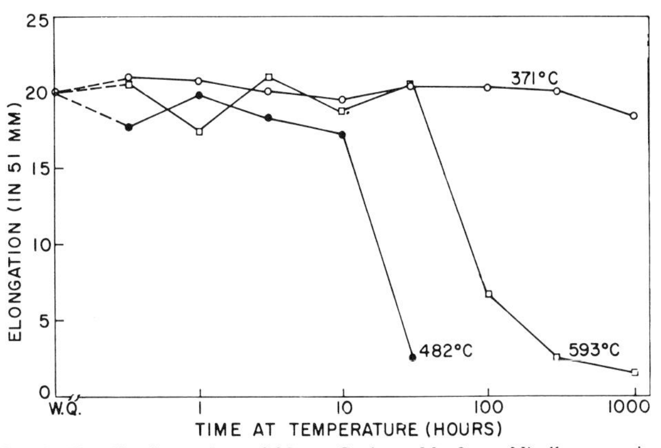

Fig. 9—Tensile elongation of 29 pct Cr-4 pct Mo-2 pct Ni alloy *vs* aging time.

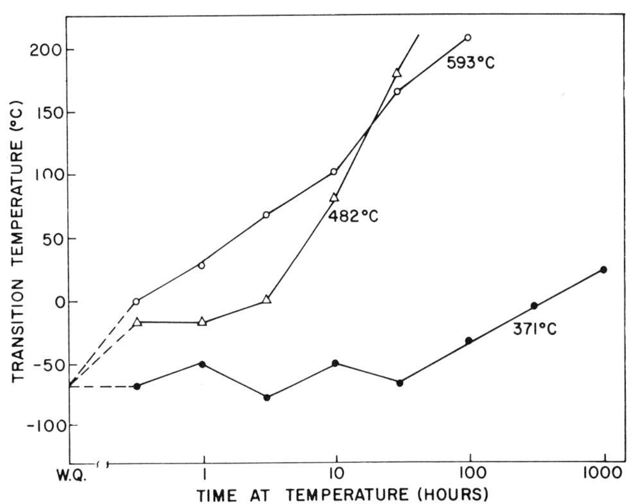

Fig. 10—FATT of 29 pct Cr-4 pct Mo-2 pct Ni alloy *vs* aging time.

B. Microstructure

B.1. Annealed and Water Quenched Material. Figure 11 demonstrates representative microstructures from annealed and quenched material. The 11 Cr, 18 Cr, and 26 Cr-1 Mo alloys are significantly different from the 29 Cr-4 Mo-2 Ni alloy. They have smaller grain sizes (ASTM Number 4 to 7 *vs* ASTM Number 1 to 2 for 29-4-2) and contain optically resolvable titanium nitrides, carbides and carbonitrides. The 29 Cr-4 Mo-2 Ni alloy, as listed in Table I, is a low interstitial ferritic stainless steel containing no deliberate additions of stabilizing elements such as titanium and, hence, does not contain compounds of these elements. The higher purity of 29 Cr-4 Mo-2 Ni combined with the higher annealing temperature results in the much coarser grain size.

B.2. Effects of 482°C Heat Treatment. No changes in structure after 482°C exposures are revealed by optical metallography. Figure 12 illustrates the microstructures of the 29 Cr-4 Mo-2 Ni after 100 h and the 18 Cr and 26 Cr-1 Mo alloys after 1000 h at 482°C. These materials are severely embrittled in this condition, but show no apparent changes in optical microstructures.

The microstructural changes resulting from 482°C exposures can only be detected through high resolution electron microscopy. Figure 13 illustrates the microstructure of as-quenched 29 Cr-4 Mo-2 Ni. There appears to be a faint strain contrast under a

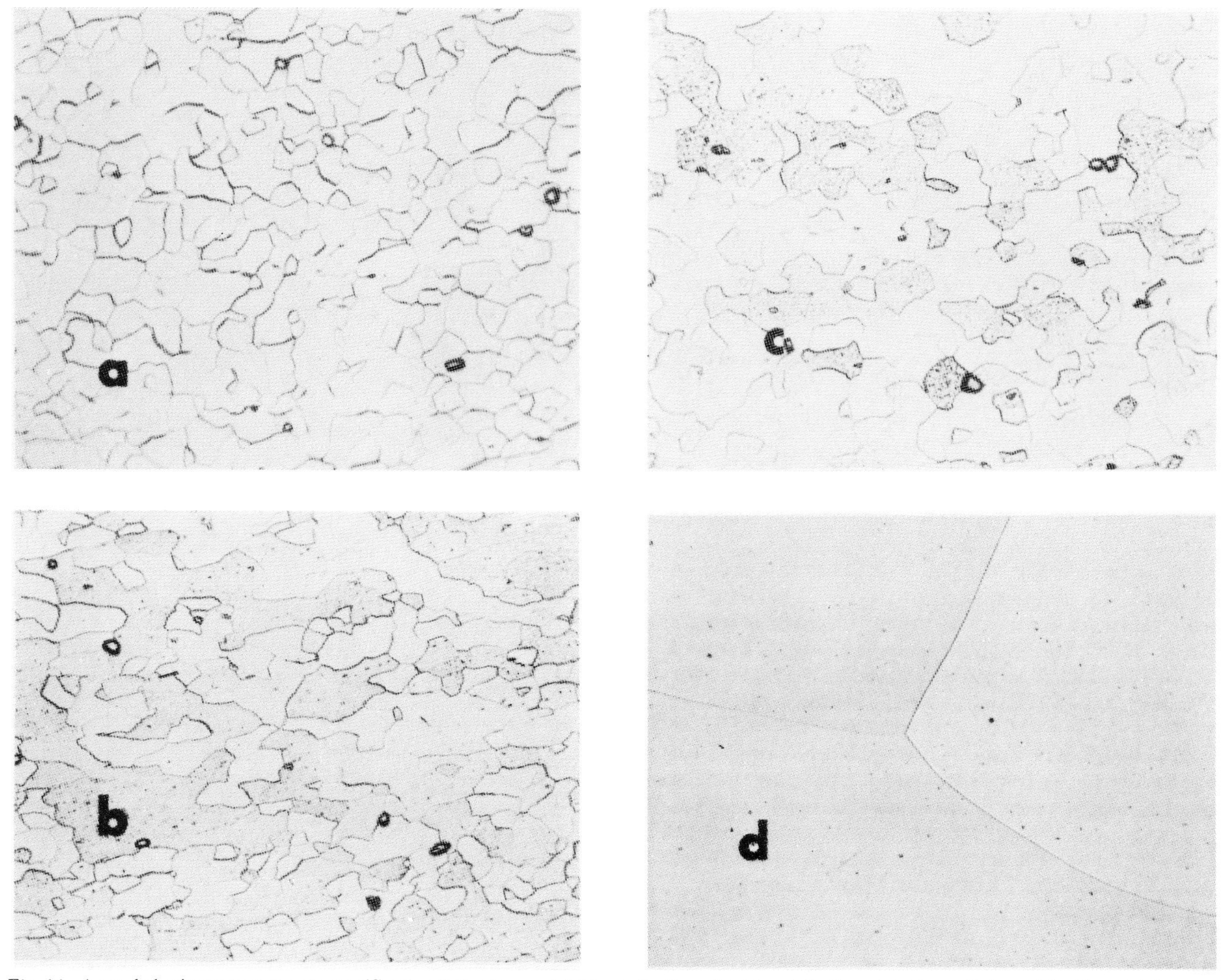

Fig. 11—Annealed microstructures at magnification 237.5 times: (*a*) 11 pct Cr alloy, (*b*) 18 pct Cr alloy, (*c*) 26 pct Cr-1 pct Mo alloy, (*d*) 29 pct Cr-4 pct Mo-2 pct Ni alloy.

strong two-beam condition indicating the very early stage of the formation of chromium rich clusters. When aged at 482°C, the as-quenched structure decomposes into chromium and iron rich bcc regions on account of the miscibility gap in the iron-chromium equilibrium system. The net matrix strain contrast originating from the chromium rich clusters is clearly imaged by the strain contrast as shown in Fig. 14. Both are from the same sample but with different foil orientations. Both the micrographs show two types of general precipitation, *viz* very fine wavy contrast striations along the traces of {100} matrix planes and large disc shaped precipitates which are also along {100} matrix planes. The fine wavelike clustering resembles the microstructure resulting from a periodic solute clustering in many other alloy systems which undergo spinodal decomposition.[16-19] However, there are no satellites flanking the matrix spots as shown in the adjoining diffraction pattern. Satellites in the diffraction pattern stem from the periodicity in lattice parameter and structure factor resulting from clustering of chromium atoms. This periodicity may not exist in iron-chromium systems owing to an extremely low mismatch parameter $\eta = 0.04$[20] and almost identical electron structure factors for iron and chromium atoms. The larger disc shaped precipitates do not give rise to any additional precipitate spots and presumably are α' as reported in the literature.[11] For a shorter aging time (30 h), the disc shaped precipitates are absent.

The kinetics of decomposition of as-quenched 26 Cr-1 Mo alloy is slower than that of 29 Cr-4 Mo-2 Ni. Figure 15 illustrates the microstructure of 26 Cr-1 Mo aged at 482°C for 100 h. There is no indication of clustering as evidenced from the absence of strain contrast striations in Fig. 15(*a*). The large idiomorphic particles in Fig. 15(*b*) are titanium carbonitrides but $Cr_{23}C_6$ particles were also observed elsewhere in the structure. These particles are present in the as-quenched state. Figure 16 illustrates the microstructure of 26 Cr-1 Mo aged at 482°C for 1000 h. The structure exhibits net matrix contrast striations along {100} matrix planes and is similar to 29 Cr-4 Mo-2 Ni aged for 100 h at the same temperature. $Cr_2(CN)$ particles are also present in the microstructure. In contrast to 29 Cr-4 Mo-2 Ni aged at 482°C (Fig. 14), aged 26 Cr-1 Mo does not show any disc shaped precipitates.

The structure of 11 Cr aged at 482°C for 1000 h is illustrated in Fig. 17. There is no clear evidence

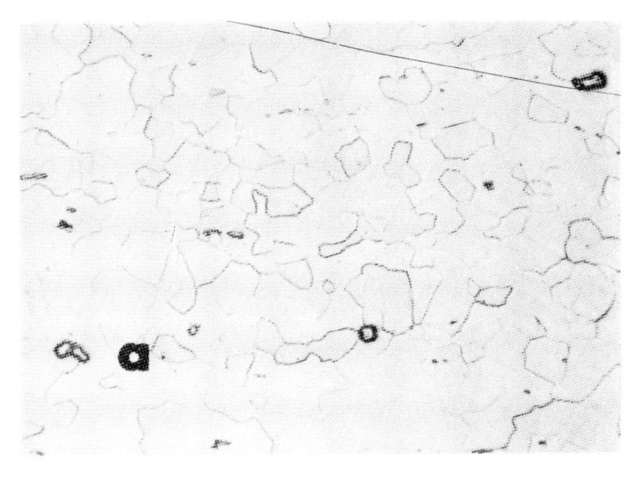

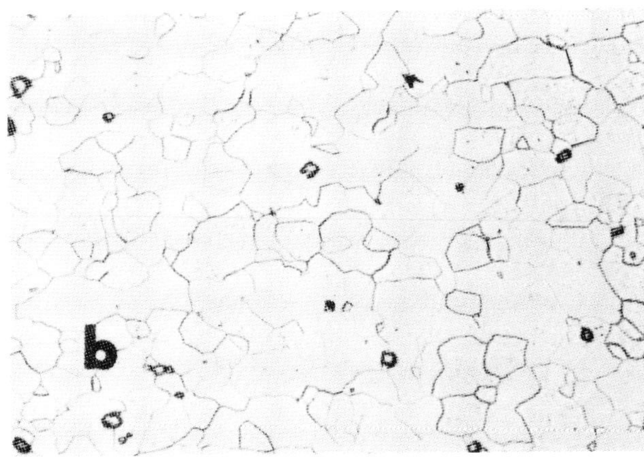

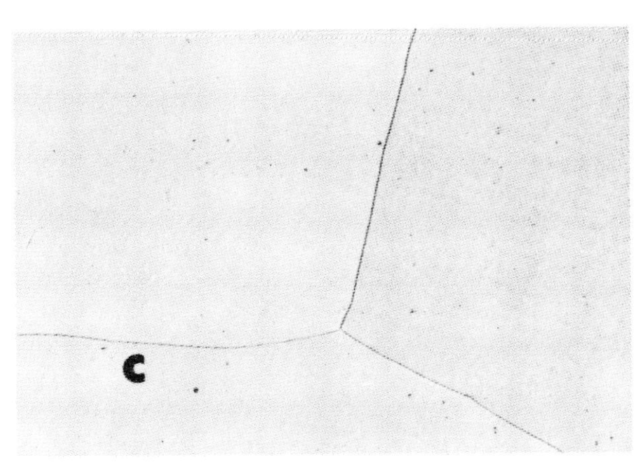

Fig. 12–482°C aged microstructures at magnification 235 times: (*a*) 18 pct Cr alloy aged for 1000 h, (*b*) 26 pct Cr-1 pct Mo alloy aged for 1000 h, (*c*) 29 pct Cr-4 pct Mo-2 pct Ni alloy aged for 100 h.

of the formation of fine chromium rich clusters as observed in 29 Cr-4 Mo-2 Ni (Fig. 14) and 26 Cr-1 Mo (Fig. 16). However, the structure does not appear to be homogeneous. The diffraction contrast reveals darker spherical regions. These probably are the incipient α' particles imaged by structure factor contrast.

Preliminary examinations of the as-deformed microstructure were made on gage sections from fractured tensile coupons. These show that there is a transition in the deformation mode when as-quenched 29 Cr-4 Mo-2 Ni is aged at 482°C for times which

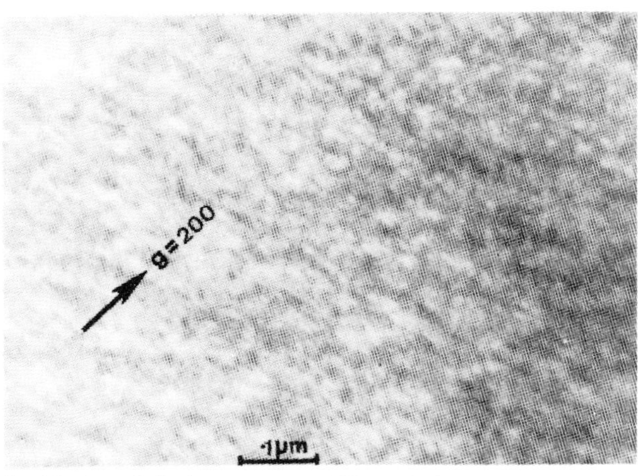

Fig. 13–Annealed microstructure of 29 pct Cr-4 pct Mo-2 pct Ni alloy foil: {001}, $g = \langle 200 \rangle$.

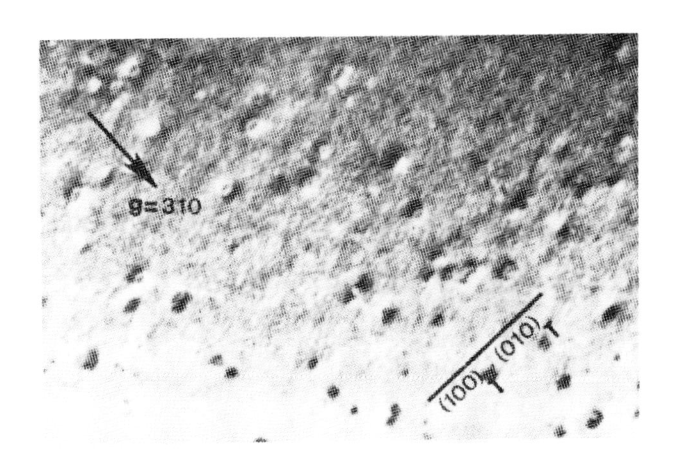

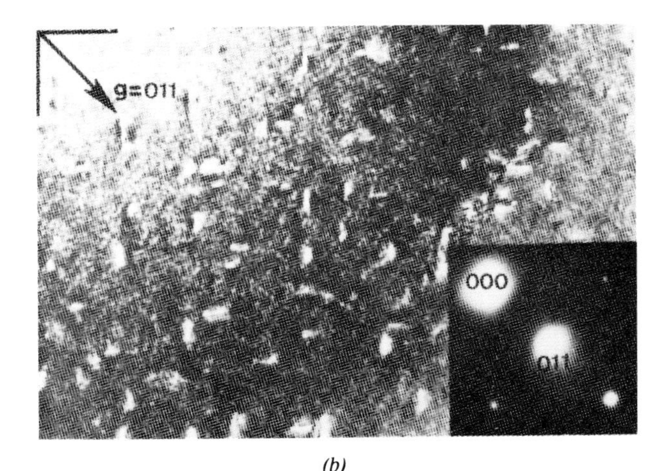

(*b*)

Fig. 14–Microstructure of 29 pct Cr-4 pct Mo-2 pct Ni alloy aged 100 h at 482°C under two different imaging conditions: (*a*) Foil: {130}, $g = \langle 310 \rangle$, (*b*) Foil: {100}, $g = \langle 110 \rangle$.

correspond to the observed embrittlement. Figure 18(*a*) illustrates the deformed structure of as-quenched 29 Cr-4 Mo-2 Ni. It is evident that slip is fairly homogeneous and turbulent. The well developed cell structure is indicative of profuse cross-slip. Figure 18(*b*) shows the deformed structure of the alloy when aged at 482°C for 30 h. The structure exhibits planar slip bands along the traces of {110} planes. The contrast

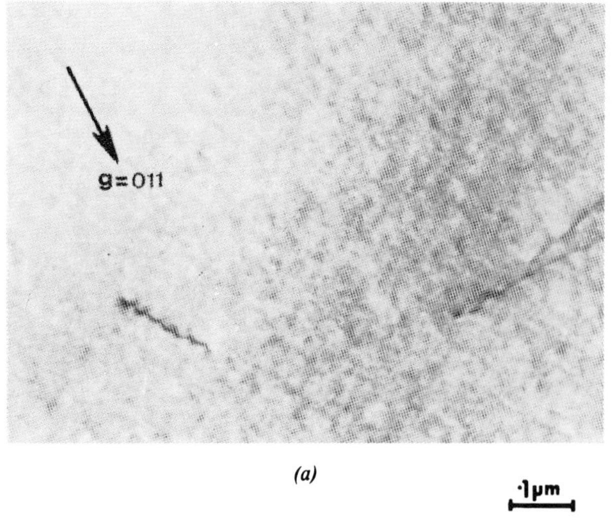

(a)

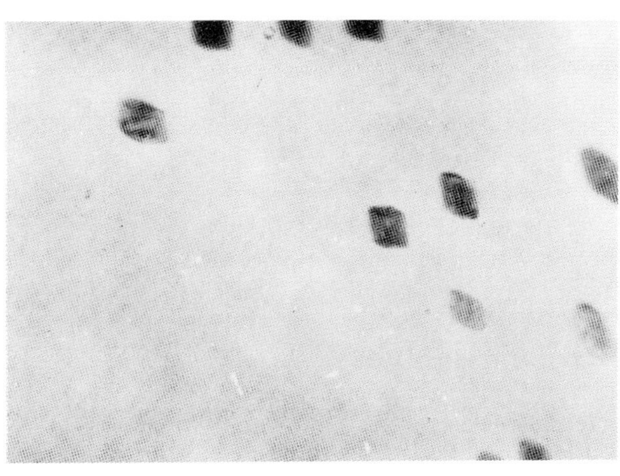

(b)

Fig. 15—Microstructure of 26 pct Cr-1 pct Mo alloy aged 100 h at 482°C: (a) Foil: {100}, g = <011>, (b) Titanium Carbonitride particles.

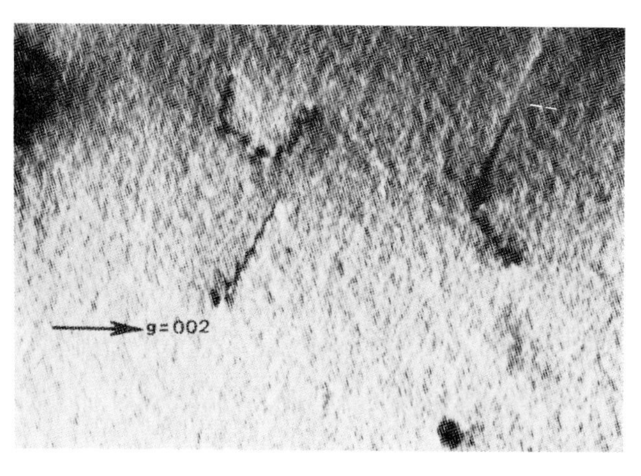

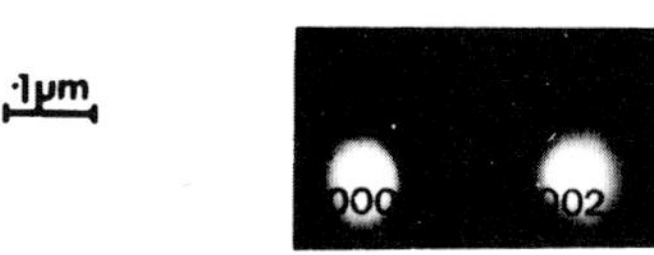

Fig. 16—Microstructure of 26 pct Cr-1 pct Mo alloy aged 1000 h at 482°C: Foil: {110}, g = <002>.

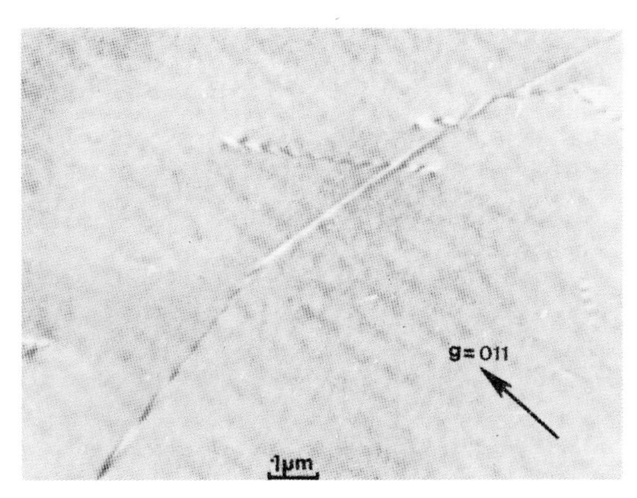

Fig. 17—Microstructure of 11 pct Cr alloy aged 1000 h at 482°C. Foil: {111}, g = <110>.

in the background results from the chromium rich regions or clusters. Concurrent with the planar slip mode the microstructure also exhibits fine deformation twinning as illustrated in Fig. 18(*c*). Embrittled 26 Cr-1 Mo also exhibits similar planar slip as illustrated in Fig. 19.

B.3. Effects of 371 and 593°C Heat Treatments. During exposures to 371°C, none of the alloys showed any change detectable in either optical or electron micrographs. The 11 Cr and 26 Cr-1 Mo alloys were similarly unaffected by the 593°C heat treatments as shown by Fig. 20(*a*). However, the 18 Cr and 29 Cr-4 Mo-2 Ni alloys showed distinct microstructural changes after prolonged times at 593°C as shown in Fig. 20(*b*) and (*c*). The 18 Cr alloy (Fig. 20(*b*)) shows a change from the as-quenched structure (Fig. 11) in that a precipitate is evident at the grain boundaries and within the matrix. X-ray diffraction of extracts showed that Laves Fe_2Ti is present (Table VII). The 29 Cr-4 Mo-2 Ni structure (Fig. 20(*c*)) is significantly different from either the as-quenched (Fig. 11) or the 482°C aged structure (Fig. 14). There is copious precipitation in the grain interior and at the grain boundaries. These large precipitates are illustrated in Fig. 21(*a*) and were electrolytically extracted for identification by X-ray diffraction. At least two precipitated phases were found to be present, one of which has been identified as sigma phase. Figure 21(*b*) illustrates the microstructure of 26 Cr-1 Mo aged at 593°C for 1000 h. The particles are $Cr_{23}C_6$ inherited from the as-quenched state and are imaged by their own precipitate spot.

DISCUSSION

A. Effect of 482°C Heat Treatments

One objective of this study was to determine the relative embrittlement rates at 475°C for four ferritic stainless steels. Exposures at 482°C were actually used for this purpose. In this study changes in impact transition temperatures and room temperature hardness, 0.2 pct offset yield strengths and tensile elongation were used as measures of embrittlement.

Changes in room temperature hardnesses or impact toughnesses have been most commonly used as

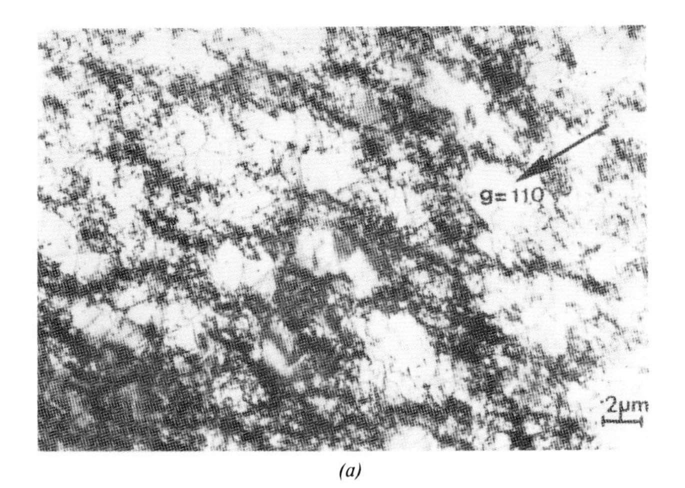

(a)

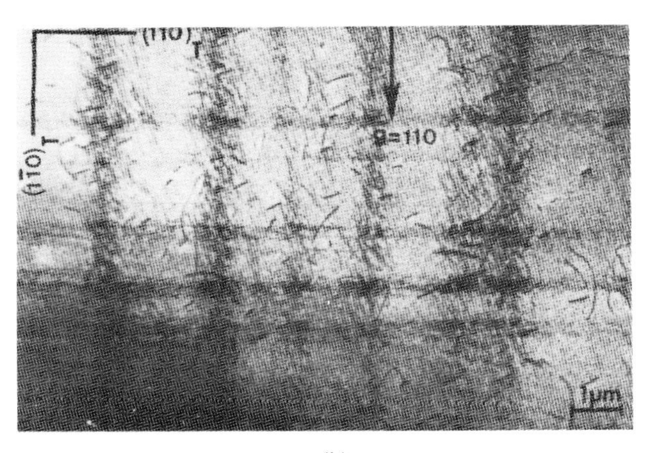

(b)

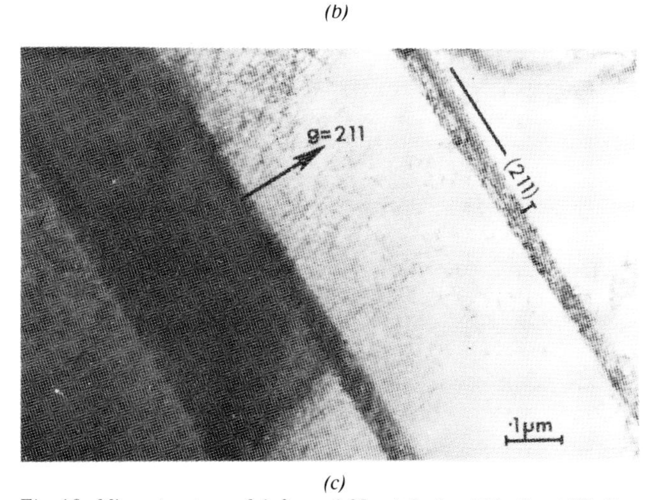

(c)

Fig. 18—Microstructure of deformed 29 pct Cr-4 pct Mo-2 pct Ni alloy: (*a*) As-quenched plus deformed, (*b*) Aged 30 h at 482°C plus deformed, (*c*) Aged 30 h at 482°C plus deformed.

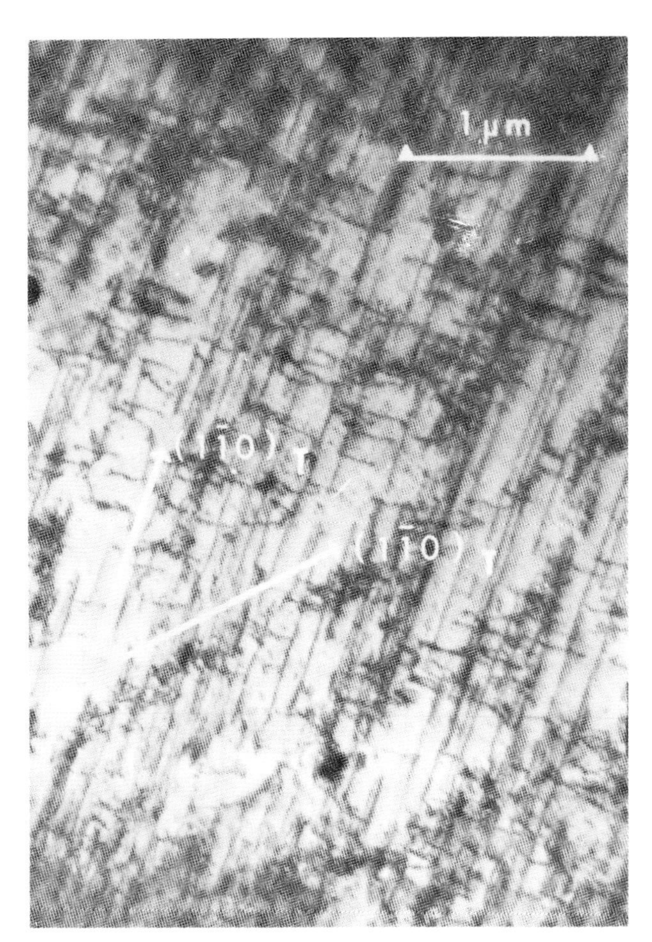

Fig. 19—Microstructure of 26 pct Cr-1 pct Mo alloy aged 1000 h at 482°C and deformed.

measurements of 475°C embrittlement but neither is very satisfactory. In the case of hardness measurements substantial embrittlement can occur with little or no hardness change and hardnesses can change considerably without severe deterioration in toughness. Impact testing at a single temperature can be ambiguous for ferritic stainless steels because they generally show sharp transitions with decreasing temperature from ductile shear failures with high energy absorption to brittle cleavage failures with very little energy absorption. If the initial annealed transition temperature is well below room temperature, say −50°C, the transition temperature could increase almost 75°C and not be detected by a room temperature impact test. On the other hand an alloy with an annealed FATT only slightly below room temperature may appear severely embrittled even though its FATT has risen only a few degrees Centigrade.

Change in offset yield strength alone is a poor measure of embrittlement. Although increases are usually associated with embrittling or hardening reactions, and *vice versa*, yield strengths can decrease simultaneously with severe embrittlement as measured by impact properties. Nichol[12] shows that this, indeed, does occur in the 29 Cr-4 Mo-2 Ni alloy upon exposures at 760°C. Elongation change is a somewhat better measure of embrittlement but is not too reliable unless values become very low.

With these limitations in mind we have used all four of these property measurements to define apparent embrittlement times. The time for embrittlement is arbitrarily defined for each property as follows:

i. Hardness—Time for hardness to increase 10 points on the Rockwell B scale. This corresponds to an ultimate tensile strength increase of 138 to 172 MPa;

ii. Yield Strength—Time for yield strength to increase approximately 50 pct from the initial (quenched) value;

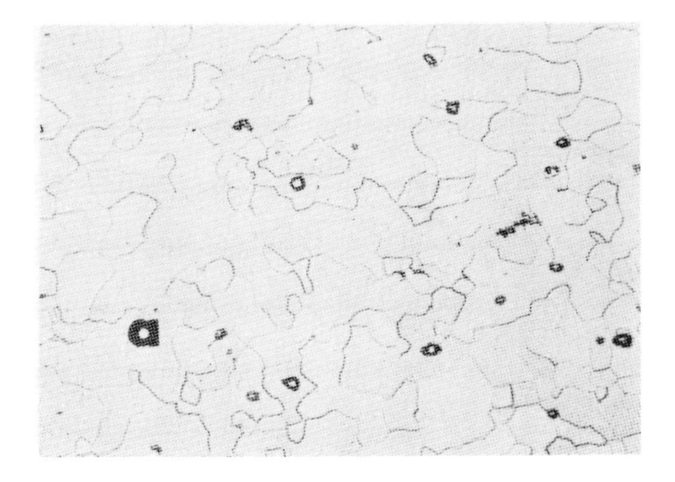

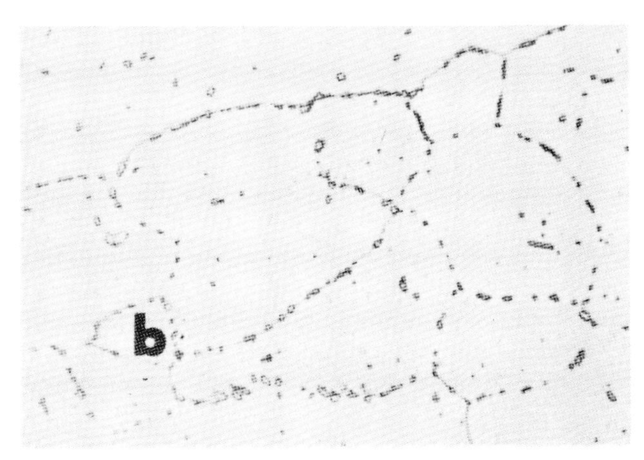

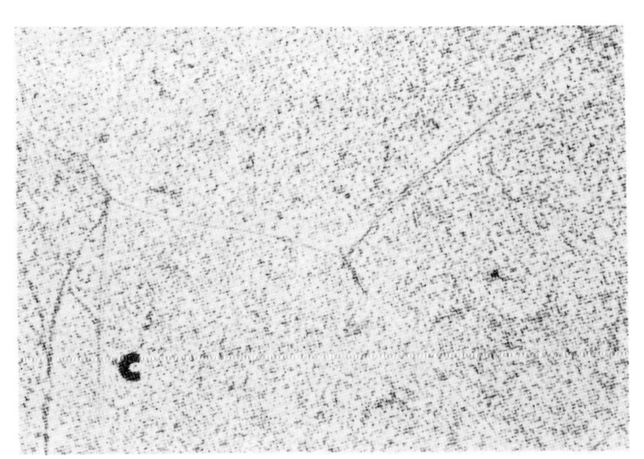

Fig. 20—Optical microstructures after aging 1000 h at 593°C: (*a*) 26 pct Cr-1 pct Mo alloy at magnification 235 times, (*b*) 18 pct Cr alloy at magnification 1410 times, (*c*) 29 pct Cr-4 pct Mo-2 pct Ni alloy at magnification 235 times.

iii. Tensile Elongation—Time for elongation to decrease to 50 pct of the initial value; and

iv. Fracture appearance transition temperature (FATT)—Time for the FATT to increase 100°C from the initial value.

Using these criteria the values in Table VI illustrate that the impact property changes revealed embrittlement at shorter times than most other properties. Independent of measurement technique the 26 Cr-1 Mo and the 29 Cr-4 Mo-2 Ni alloys embrittled at approximately the same rate which was faster than

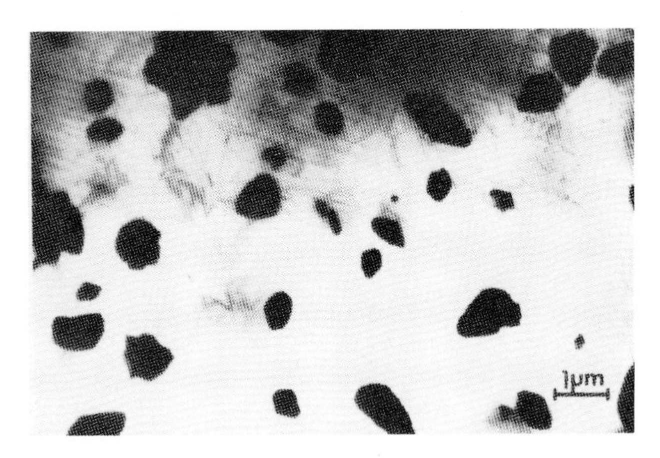

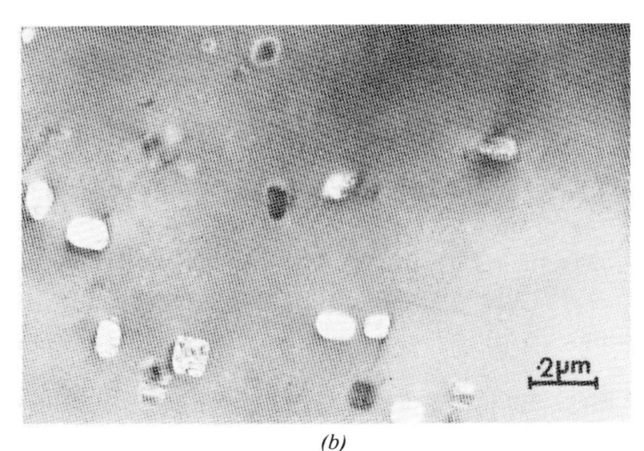

(*b*)

Fig. 21—TEM microstructures after aging 1000 h at 593°C: (*a*) 29 pct Cr-4 pct Mo-2 pct Ni alloy, (*b*) 26 pct Cr-1 pct Mo alloy.

Table VI. Embrittlement Times, h at 482°C

Type	Hardness*	0.2 Pct Y.S.†	Pct Elongation‡	FATT¶
11 Cr	>1000	>1000	>1000	–
18 Cr	100	100	1000	100
26 Cr-1Mo	30	10	200	10
29 Cr-4Mo-2Ni	30	–	30	10

*Increase of 10 R_B.
†Increase of 100 pct from quenched value.
‡Decrease to 50 pct of quenched value.
¶Increase of 150°C from quenched value.

the embrittlement of the 18 Cr alloy. No evidence of embrittlement was noted for the 11 Cr alloy for the times and properties evaluated. Of these four property measurements, changes in FATT appear to afford the most sensitive and unambiguous measure of embrittlement. Furthermore, the FATT is of practical significance when processing[21] or fabricating these alloys. If the FATT is in the region of room temperature or higher preheating is advisable to avoid cracking during cold rolling or fabrication.

Exposures to 482°C affect the mechanical properties of the three higher Cr alloys in the same characteristic fashion. From the initial water quenched condition to the shortest exposure time evaluated (20 min), there is a significant increase in yield strength (Fig. 2) and FATT (Fig. 4). This initial change is followed by an apparent plateau region of about 3 to 10 h where no further changes in properties occur. The plateau is followed by a third stage where rapid property changes occur and continue out to beyond 1000 h. The same sequence is observed from hardness changes (Table V) and, less obviously, from elongation changes (Fig. 3).

The origin of the initial, discontinuous changes in properties during the short aging time (20 min) at 482°C and, in two instances, at 593°C is not clear. The effect is observed most readily, perhaps, in FATT changes as illustrated by Figs. 6, 8, and 10. This initial change in FATT at 482°C is about 35 to 80°C. About the same increase is observed when air cooling these alloys from their annealing temperatures in place of water quenching.[12,21] The presence of the 3 to 10 h plateau before properties continue to change indicates that 475°C embrittlement is not responsible.

A reasonable alternative is that the initial property change is caused by the precipitation of small amounts of interstitial compounds, probably containing chromium. The stabilized 18 Cr alloy is not affected by a short exposure at 593°C, as shown in Fig. 6, whereas the stabilized 26 Cr-1 Mo alloy (Fig. 8) and the unstabilized 29 Cr-4 Mo-2 Ni alloy (Fig. 10) show large effects. The 18 Cr alloy not only contains less Cr but more Ti than the 26 Cr-1 Mo alloy and is, therefore, not expected to precipitate chromium-interstitial compounds as readily. Present specifications call for a minimum Ti/C ratio of 12 for the 18 Cr alloy and a minimum Ti/(C+N) ratio of 10 for the 26 Cr alloy to resist sensitization due to $Cr_{23}C_6$ precipitation in the as-welded condition. On these bases, the 18 Cr alloy is well within specifications whereas the 26 Cr alloy is not. In addition, as-welded corrosion tests on the latter material do, indeed, show that it is marginally or inadequately stabilized. Precipitation of interstitial compounds in the temperature range of 482 to 593°C is well known to cause embrittlement of stabilized and unstabilized ferritic alloys.[4,5] We were unable, however, to find any direct evidence of this from our own microstructural observations.

It is evident from the previous section that 475°C embrittlement occurs without any detectable change in the optical microstructure. A number of studies in the past[1,4,5,7,8] using transmission electron microscopy have attempted to characterize the finer microstructure attendant to 475°C embrittlement. The generally accepted explanation for this embrittlement is that it is caused by the precipitation of small spherical particles, ~200Å in diam, of a chromium rich (80 pct Cr) bcc phase α'. To identify the embrittling phase, most of the past studies used prolonged heat treatment times at 475°C necessary to form spherical α' particles which were subsequently extracted and identified by electron diffraction. Importantly, however, exposures as short as 10 to 100 h at 475°C can lead to a significant loss in ductility long before the formation of spherical α' particles. None of the previous studies has examined the microstructure corresponding to this onset of the embrittlement phenomenon. The present study, therefore, examines the characteristic microstructure at the onset of embrittlement and the accompanying change in the deformation mode.

During isothermal aging, the onset of embrittlement is evidenced by a sudden loss in ductility and rise in FATT and yield strength as shown in Figs. 2 to 4. These mechanical properties were used as guidelines for microstructural observation. For example, Fig. 3 shows that the 29 Cr-4 Mo-2 Ni alloy is severely embrittled at room temperature after ~100 h of aging at 482°C since it exhibits nil tensile elongation. There is also a corresponding increase in FATT indicating the inherent brittleness of the matrix. The onset of brittleness occurs without any detectable change in the optical microstructure (Fig. 12). High resolution electron microscopy, however, reveals structural differences. The embrittled sample, aged for 100 h at 482°C, is illustrated in Fig. 14(*a*) and (*b*) in two different imaging conditions. Both the micrographs show that the embrittled sample has decomposed into very fine chromium rich clusters as revealed by the wavelike net matrix contrast striations. The net displacement vector is along <100> indicative of clustering of chromium atoms on {100} planes. The structure is very similar to spinodally decomposed alloys[16-19] in many other systems reported in the literature. In addition to chromium rich clusters the microstructure also reveals platelets of α' along {100} matrix planes which may have formed *in situ* from the chromium rich clusters.

The embrittled 26 Cr-1 Mo alloy also exhibits similar microstructural behavior but longer exposures, *viz* ~1000 h, are required to cause severe embrittlement. The microstructure of the embrittled sample exhibits wavelike net matrix contrast striations along {100} traces. However, unlike the 29 Cr-4 Mo-2 Ni alloy larger platelets of α' on {100} matrix planes are absent. Hence, the platelets of α' are not responsible for the embrittlement phenomenon. A shorter exposure of 100 h at 482°C does not embrittle the 26 Cr-1 Mo alloy so severely as evidenced from the plot of tensile elongation *vs* aging time (Fig. 3). The corresponding microstructure is homogeneous and shows no sign of clustering. Neither alloy when aged at 371°C decomposes into chromium rich and poor clusters even with the longest aging time employed in the present study. Neither do they show as much embrittlement (Figs. 8 and 10).

Unlike the more highly alloyed materials the 11 Cr alloy does not apparently embrittle nor decompose into fine chromium rich clusters on {100} planes characteristic of spinodal decomposition. Although the 11 Cr alloy appears to be within the chemical spinodal at 482°C, a recent study[20] using Mössbauer spectroscopy shows that only alloys with chromium contents greater than 24 at. pct undergo spinodal decomposition. In more dilute alloys, chromium rich α' forms by nucleation and growth. After a long exposure (1000 h) at 482°C there is some evidence of clustering in the 11 Cr alloy imaged probably through structure factor contrast.

From the above discussion it is evident that significant embrittlement at 482°C is concurrent with microstructures consisting of wavelike clusters of chromium atoms. These wavelike clusters most probably derive from spinodal decomposition along the elastically soft <100> directions. The rate of decomposition is faster in the 29 Cr-4 Mo-2 Ni alloy because of a greater degree of supersaturation. Addition of nickel also enhances the decomposition kinetics[20] and hence the rate of embrittlement. At 371°C the kinetics of decomposition is too sluggish to allow detection of any clustering and embrittlement for the aging times employed in the present study.

Preliminary examination of the deformation mode in unbrittled and embrittled samples resulting from the exposure to 482°C has revealed an important difference in the deformation behavior. The embrittled samples, containing wavelike contrast striations due to the formation of chromium rich clusters, exhibited planar bands along the traces of {110} matrix planes. The unembrittled samples, without chromium rich clusters on {100} planes, exhibit more turbulent slip and cell formation. The planar slip mode in the embrittled samples leads to local stress concentration resulting from pile-ups at the barriers to dislocations, thereby enhancing susceptibility to brittle failure. Local stress concentration centers will, in turn, enhance deformation twinning which also accompany the formation of chromium rich clusters in the embrittled samples. A previous study[10] of the iron-chromium system has shown that twinning is preferred to slip during the later stages of aging and this induces embrittlement. However, the present study shows that twinning is not necessarily the cause for embrittlement since the 26 Cr-1 Mo alloy embrittles without any evidence of twinning. The planar slip mode, is, thus, a more fundamental deformation characteristic of the embrittlement phenomenon.

It is not clear in the present study, why such a transition from a turbulent to planar slip mode occurs concurrent with the formation of chromium rich clusters. Ordering in chromium rich clusters may induce planar slip as reported in other systems.[22] However, in agreement with previous studies,[10] there is no evidence of order (superlattice spots) in diffraction patterns of embrittled alloys, although the intensity of superlattice spots is expected to be extremely low owing to the close similarity of structure factors of chromium and iron atoms. Because of the ordered nature of the equilibrium phase[23] there exists a strong possiblity of ordering in chromium rich clusters or α'. Simultaneous ordering and clustering or spinodal ordering has been reported in other systems.[19-24] More detailed work is necessary to resolve the question of ordering in α'.

B. Effects of 371 and 593°C Heat Treatments

Exposures at 371°C to 1000 h result in relatively modest increases in yield strength and noticeable increases in FATT for the three higher Cr alloys (Figs. 5 to 8 and 10). The increase in yield strength of the 29 Cr-4 Mo-2 Ni alloy, not shown in the figures, was about the same as that of the 18 Cr alloy. The stabilized 26 Cr-1 Mo showed the greatest increase in FATT of about 150°C followed by the 29 Cr-4 Mo-2 Ni alloy and then the 18 Cr alloy. Embrittlement is attributed to both α' and interstitial compounds. In the case of the 26 Cr-1 Mo alloy the precipitation of intermetallic compounds is believed to be a major embrittlement factor at 371°C whereas it is of little significance, or absent, in the other two alloys. No microstructural changes, however, were observed after the 1000 h exposures.

Exposures of 1000 h at 593°C resulted in little change in yield strength of any of the higher chromium alloys. The 26 Cr-1 Mo shows a slight increase while the other two show a very small decrease. The FATT of the 18 Cr alloy shows no change after 100 h and then starts to rise fairly steeply. The 1000 h exposure is accompanied by a precipitate resolvable by optical microscopy (Fig. 20). The precipitate forms preferentially at grain boundaries but is also located within the matrix. Several attempts at extraction yielded only weak X-ray patterns. Work on a related alloy, however, made it possible to identify the phase as Fe_2Ti Laves ($MgZn_2$ type). One set of diffractions peaks matched that of FeO, (ASTM No. 6-0615) probably a contaminant arising during extraction. The X-ray diffraction data are given in Table VII. Of the five remaining weak peaks, two were identified as TiN (ASTM No. 6-0642). The appearance of this phase is believed responsible for the increase in FATT at the longer times.

The 26 Cr-1 Mo shows no change in FATT between 20 min and 1000 h nor does it exhibit any detectable change on microstructure after the 1000 h exposure from the as-quenched condition. It is mildly surprising that the more highly alloyed 26 Cr-1 Mo alloy appears more stable at 593°C than the 18 Cr alloy. We attribute this effect to the relatively high Ti content of the 18 Cr alloy. Other work at our laboratory has shown Ti to be a strong promoter of intermetallic phases in related ferritic alloys.

The 29 Cr-4 Mo-2 Ni alloy shows a continuous rise

Table VII. X-Ray Diffraction of Phases Present After 1000 h at 593°C

18 Cr-Ti Alloy			29 Cr-4 Mo-2 Ni Alloy		
d, Å	Rel. Int.	Identity*	*d*, Å	Rel. Int.	Identity†
2.497	60	FeO(111)	3.347	5	?
2.389	30	$Fe_2Ti(110)$	2.557	5	$\sigma(221)$
2.307	50	$Fe_2Ti(111)$	2.380	10	$\sigma(311)$
2.232	20	?	2.289	45	$\sigma(002)$
2.194	45	$Fe_2Ti(103)$	2.171	100	?
2.164	100	FeO(200)	2.138	90	$\sigma(410)$
2.120	25	TiN(200)	2.108	73	?
2.033	80	αFe, $Fe_2Ti(112)$	2.077	22	$\sigma(330)$
2.002	70	$Fe_2Ti(201)$	2.035	68	αFe, $\sigma(202)$
1.976	15	?	1.995	68	?
1.947	20	$Fe_2Ti(004)$	1.976	23	$\sigma(212)$
1.827	10	$Fe_2Ti(202)$	1.937	45	$\sigma(411)$
1.757	10	$Fe_2Ti(113)$	1.913	10	?
1.530	35	FeO(220), $Fe_2Ti(211)$	1.891	40	$\sigma(331)$
1.501	10	TiN(220)	1.872	4	?
1.339	15	$Fe_2Ti(123)$	1.847	4	$\sigma(222)$
1.305	45	$Fe_2Ti(115)$	1.769	4	$\sigma(312)$
1.300	25	FeO(311)	1.438	8	αFe
1.298	15	$Fe_2Ti(302)$	1.364	5	?
1.250	15	?	1.331	3	$\sigma(522)$
1.242	20	FeO(222), $Fe_2Ti(205)$	1.284	22	?
1.193	15	$Fe_2Ti(220)$	One or more unresolved peaks		
			1.263	17	$\sigma(532)$
			Several unresolved peaks		
			1.2107	21	$\sigma(720)$
			1.2033	8	$\sigma(551, 711)$
			1.1906	4	$\sigma(622)$
			One or two small, unresolved peaks		
			1.1746	29	αFe, $\sigma(721)$

*FeO-ASTM 6-0615 TiN-ASTM 6-0642, Fe_2Ti-Hexagonal, c_o = 7.797Å, a_o = 4.765Å, c/a = 1.636.

†σ = Tetragonal, c_o = 4.5795Å, a_o = 8.814Å, c/a = 0.5196.

in FATT as exposure time at 593°C is increased from 0.3 to 100 h. The embrittlement is associated with pronounced structural changes as illustrated in Fig. 20(*c*). The 1000 h exposure at 593°C produces a semicontinous to continuous precipitate at grain boundaries and a finer, dense precipitate within the grains. Extraction and X-ray diffraction of this material reveal that at least two phases are present, one of which was identified as sigma phase (Table VII). Examination under a scanning electron microscope showed that the extracted powder contained fine cubic to spherical particles 0.5 to 1.0 μ in size and coarse irregular particles about 10 times as large. Both, however, analyzed about 29 pct Mo, 24 pct Cr, 1.2 pct Ni, and the balance essentially Fe. The presence of the sigma and/or the unidentified phase is doubtlessly responsible for the rise in transition temperatures.

SUMMARY AND CONCLUSIONS

The effects of exposures to 1000 h at 371, 482, and 593°C on the mechanical properties of four stainless steels were measured. The properties evaluated were room temperature hardnesses and tensile properties and Charpy V-notch impact transition temperatures. Microstructures were examined by optical microscopy and, in selected cases, by transmission electron microscopy. Selected samples were deformed before and after 482°C exposures to determine the effects of 475°C embrittlement on microscopic deformation behavior. Attempts were made to identify precipitated phases by *in situ* electron diffraction of extractions. This work has led to the following conclusions:

1. The most sensitive and unambiguous property measurement, of those evaluated, for measuring embrittlement is the change in fracture appearance transition temperature. Changes in room temperature yield strength, elongation, hardness or impact strength are not very satisfactory.

2. Embrittlement at 482°C occurs most rapidly and severely in the high purity 29 Cr-4 Mo-2 Ni alloy. Embrittlement occurs slightly less rapidly in a Ti-stabilized 26 Cr alloy and much more slowly in a Ti-stabilized 18 Cr alloy. Little, if any, embrittlement takes place in a Ti-stabilized 11 Cr alloy upon exposures to 1000 h at 482°C.

3. Embrittlement at 482°C is characterized by an initial property change followed by a plateau region and then further property changes. The initial property change is attributed to the precipitation of interstitial compounds although this was not supported by microstructural observations. The later property changes are attributed to 475°C embrittlement (α' precipitation).

4. Early stages of embrittlement at 482°C of the 26 and 29 pct Cr alloys are accompanied by wavelike contrast striations indicative of chromium atom clustering on {100} planes. Upon embrittlement the tensile deformation mode changes from turbulent slip to a more planar slip along {110} matrix planes.

5. Some embrittlement, particularly at longer aging times, was observed at 371°C for the alloys containing 18 to 29 pct chromium. This is attributed to combination of α' embrittlement and interstitial compound precipitation. This conclusion, however, could not be supported by microstructural observation.

6. The 29 Cr-4 Mo-2 Ni and the stabilized 18 Cr alloys embrittle upon exposures at 593°C. In both cases embrittlement was accompanied by little change in yield strengths and the precipitation of optically observable phases. The precipitate in the 18 Cr alloy was identified as Laves Fe_2Ti phase. At least two phases precipitated in the 29 Cr-4 Mo-2 Ni alloy, one of which was identified as sigma phase.

ACKNOWLEDGMENTS

The authors appreciate the contributions of L. D. Bachman, D. E. Deemer, M. Gatial, J. F. Kiesel and R. K. Teorsky in helping to conduct the experimental work and preparing the manuscript.

REFERENCES

1. R. M. Fisher, E. J. Dulis, and K. G. Carroll: *Trans. AIME*, 1953, vol. 197, pp. 690-95.
2. R. O. Williams and H. W. Paxton: *J. Iron Steel Inst.*, 1957, vol. 185, pp. 358-74.
3. R. O. Williams: *Trans. TMS-AIME*, 1958, vol. 212, pp. 497-502.
4. A. Plumtree and R. Gullberg: *Met. Trans. A*, 1976, vol. 7A, pp. 1451-57.
5. P. J. Grobner: *Met. Trans. A*, 1973, vol. 4A, pp. 251-60.
6. H. Brandis, H. Kiesheyer, and G. Lennartz: *Arch. Eisenhuettenwes.*, 1975, vol. 46, pp. 799-804.
7. M. Courtnall and F. B. Pickering: *Met. Sci.*, 1976, vol. 10, pp. 273-76.
8. R. Lagneborg: *Trans. ASM*, 1967, vol. 60, pp. 67-78.
9. T. DeNys and P. M. Gielen: *Met. Trans.*, 1971, vol. 2, pp. 1423-28.
10. M. J. Marcinkowski, R. M. Fisher, and A. Szirmae: *Trans. TMS-AIME*, 1964, vol. 230, pp. 676-89.
11. M. J. Blackburn and J. Nutting: *J. Iron Steel Inst.*, 1964, vol. 202, pp. 610-13.
12. T. J. Nichol: *Met. Trans. A*, 1977, vol. 8A, pp. 229-37.
13. R. Lagneborg: *Acta Polytech. Scand.*, 1967, vol. 62, pp. 1-40.
14. A. I. Kondyr, A. N. Tkach, V. I. Astashkin, and M. F. Zamora: *Fiz. Khim. Mekh. Mater.*, 1974, vol. 10, pp. 24-28.
15. P. K. Koh: *Trans. AIME*, 1953, vol. 197, pp. 339-43.
16. E. P. Butler and G. Thomas: *Acta Met.*, 1970, vol. 18, pp. 347-65.
17. R. J. Livak and G. Thomas: *Acta Met.*, 1971, vol. 19, pp. 497-505.
18. P. E. J. Flewitt: *Acta Met.*, 1974, vol. 22, pp. 47-63.
19. A. Datta and W. A. Soffa: *Acta Met.*, 1976, vol. 24, pp. 987-1001.
20. H. D. Solomon and L. M. Levinson: *Acta Met.*, 1978, vol. 26, pp. 429-42.
21. G. Aggen: *Stainless Steel '77*, R. Q. Barr, ed., pp. 79-87, Climax Molybdenum Co., 1979.
22. R. K. Ham and L. M. Brown: *Strengthening Methods in Crystals*, A. Kelly and R. B. Nicholson, eds., pp. 9-135, Appl. Science, 1971.
23. E. O. Hall and S. H. Algie: *Met. Rev.*, 1966, vol. 11, pp. 61-88.
24. S. M. Allen and J. W. Cahn: *Acta Met.*, 1975, vol. 23, pp. 1017-26.

Physical and Welding Metallurgy of Chromium Stainless Steels

A review of published and unpublished information on martensitic and ferritic stainless steels with emphasis on their physical and welding metallurgy, 885° F. (475° C.) Brittleness, sigma-phase embrittlement, High-Temperature Embrittlement, notch-sensitivity, and the effects of various alloying elements.

by Helmut Thielsch

SUMMARY

DURING the past ten years, many new developments have greatly increased the understanding of the physical and welding metallurgy of the chromium stainless steels. With the development of new and improved alloys, many additional applications have been found. Moreover, because most of these chromium stainless steels are more economical than the austenitic chromium-nickel stainless steels, it is likely that future research will increase their usefulness still further. The present nickel shortage adds additional emphasis to the importance of the chromium stainless steels which are generally not alloyed with nickel.

The published and unpublished information, which has been made available by the leading laboratories, research investigators, and welding engineers, is reviewed, analyzed, and interpreted. Particular emphasis has been placed upon European developments which were the result of nickel shortages and which, therefore, caused the necessity to substitute, whenever possible, chromium stainless steels for austenitic chromium-nickel stainless steels.

The information has been divided in three major sections: (I) Physical Metallurgy, (II) Effects of Alloying Elements, and (III) the Welding of Chromium Stainless Steels.

Physical Metallurgy

Depending upon structure, these steels are generally divided into three classes: (1) the martensitic grades, (2) the ferritic grades, and (3) the ferritic-austenitic grades.

The martensitic stainless steels usually contain between 10 and 14% chromium. They are primarily used in applications in which a high mechanical strength is required in favor of ductility and toughness. These stainless steels also represent the most economical types (see Table 1).

The ferritic stainless steels usually contain between 14 and 30% chromium. Because of their good corrosion and oxidation resistance, these steels are particularly used in many refinery and chemical process applications, primarily at intermediate and elevated temperatures. Since the fully ferritic stainless steels have a coefficient of expansion which is similar to glass, these steels are also used in considerable quantities in indus-

Helmut Thielsch is Technical Assistant, Welding Research Council, New York, N. Y.

This report was prepared under the auspices of the Literature Advisory Committee of the Welding Research Council. G. E. Doan, *Chairman;* T. S. Fuller, L. E. Grinter, C. E. Jackson, E. M. MacCutcheon, W. Spraragen, David Swan, R. David Thomas, Jr., J. L. Walmsley and Helmut Thielsch, *Secretary.*

trial applications where direct metal to glass seals or joints are to be made, as, for example, in television tubes.[213]

The ferritic-austenitic grades usually contain between 24 and 30% chromium in addition to the presence of one or more austenitizers, such as nickel or manganese, which are responsible for retaining up to 50% austenite at room temperature. Although these steels are not as popular as the martensitic or ferritic types, they are produced in limited quantities, particularly in Europe. These ferritic-austenitic alloys are used in certain chemical process applications where ductility, toughness, and weldability are of primary importance.

In addition to the formation of austenite and martensite, the chromium stainless steels may become susceptible to one or more of the phenomena which produce drastic changes in the physical properties of the normally ferritic alloys. Distinction should be made between four major types: (1) 885° F. (475° C.) Brittleness, (2) sigma-phase precipitation, (3) High-Temperature Embrittlement of ferrite, and (4) the notch-sensitivity of the ferritic alloys. The nature of these phenomena are discussed and possible explanations are reviewed or suggested.

The 885° F. (475° C.) Brittleness occurs when ferritic chromium stainless steels are heated between 750 and 1000° F. (400 and 540° C.). Sigma-phase embrittlement, which seems to be related to 885° F. (475° C.) Brittleness, occurs when chromium-iron alloys containing between 15 to 20% and 70% chromium are exposed between 950° F. (510° C.) and 1300 to 1500° F. (705 to 815° C.). The High-Temperature Embrittlement occurs when alloys with fully ferritic structures are exposed to temperatures above 2100° F. (1150° C.). This embrittlement is also observed in ferritic stainless castings and weld deposits. A suitable heat treatment between 1350 and 1450° F. (730 and 790° C.) may remove the effects of the embrittlement. Notch-sensitivity at room temperature is of concern in ferritic alloys containing over 16 to 18% chromium. At room temperature, these higher chromium alloys tend to exhibit brittle behavior. This may be overcome by slightly elevated working and fabricating temperatures.

Effects of Alloying Elements

Aluminum additions are made to stainless steels to serve one or several functions: (1) to serve as a ferrite former, (2) to improve scaling resistance, (3) to cause grain refinement, and (4) to develop desirable electrical properties.

Carbon, at a given chromium content, is the principal element which affects the amount of martensite which may or may not be present. Steels, which are hardenable because of martensite formation, require special precautions in welding operations.

Chromium is of primary importance and is used to produce either a martensitic or ferritic structure in the steel. Moreover, chromium improves corrosion and scaling resistance considerably.

Nickel, as well as *manganese*, is useful to cause the retention of austenite in steels containing over 24% chromium. As such, nickel improves toughness and weldability.

Nitrogen additions are rarely made to chromium stainless steels. Their primary function is to produce a fine grain size.

Titanium effectively produces grain refinement and, in small quantities, improves the properties of weld deposits.

Welding of Chromium Stainless Steels

Although arc welding is preferred in the welding of chromium stainless steels, satisfactory weld deposits may be obtained with most of the other welding procedures.

The low-carbon martensitic stainless steels (for example Type 410) should receive a preheat treatment between 600 to 800° F. (315 to 425° C.) and a postheat treatment between 1300 and 1400° F. (705 and 760° C.). The higher carbon martensitic grades (over 0.25% carbon) are not readily weldable and will require special precautions which depend on composition, the welding process used, and other factors.

Although in ferritic stainless steels preheat treatments are not always necessary, a preheat treatment between 300 and 400° F. (150 and 205° C.) is advisable, particularly in steels containing over 25% chromium.

Although, ordinarily, satisfactory results may be obtained by welding the chromium stainless steels with electrodes having compositions identical to those of the parent metal, austenitic chromium-nickel electrodes are generally preferred.

INTRODUCTION

Stainless steels are primarily selected because they offer good resistance to the attack of many chemical

Table 1—Comparative Costs of Stainless Grades (Type 410 = 100)[156]

Type	*Sheets*	*Cold-rolled strip*	*Bar and wire*
301	114	113	124
302	114	122	124
302-B	120	133	128
303	. .	. .	135
304	120	130	130
305	127	139	135
308	137	144	148
309	158	189	180
310	180	233	244
316	161	204	200
317	194	244	248
321	138	165	148
347	152	180	167
403	109	124	113
405	105	119	107
406	115	141	120
410	100	100	100
414	102	102	100
416	102	124	102
420	123	161	124
430	108	102	102
430-F	. .	. .	104
431	109	104	102
440-A, B, C	123	161	124
446	152	222	141

corrodents and to the action of oxidizing and other detrimental gases, because of their appearance, or because they offer in addition to these properties strength, hardness, creep resistance, ductility, or toughness. However, not all of these features can be found in every stainless alloy. Thus, the selection of a particular grade depends upon the manufacturing and fabricating processes to which the steel will be subjected. Selection also depends upon the service requirements of the finished stainless part. For example, in a few corrosive applications, the presence of nickel in a stainless steel may actually be detrimental.

Economic considerations may also enter into the selection of a particular grade. The comparative costs of the various A.I.S.I. stainless steels are listed in Table 1. The economical advantages of the chromium stainless steels (400 series) over the austenitic chromium-nickel grades (300 series) are quite apparent.

This review is not only concerned with the particular grades covered by the standards of the American Iron and Steel Institute (A.I.S.I., Table 2) and the Alloy Casting Institute (A.C.I., Table 3), but also contains references to many other alloys. Thus, the more important recent European developments are included.

Table 2—Chromium Stainless Steels: A.I.S.I. Standard Type Designations

	Chemical composition, %							
Designation	*C*	*Mn, max.*	*Si, max.*	*P, max.*	*S, max.*	*Cr*	*Ni*	*Other elements*
403	0.15 max.	1.00	0.50	0.040	0.030	11.50–13.00	...	Turbine quality
405	0.08 max.	1.00	1.00	0.040	0.030	11.50–13.50	...	Al 0.10–0.30
406	0.15 max.	1.00	1.00	0.040	0.030	12.00–14.00	...	Al 3.50–4.50
410	0.15 max.	1.00	1.00	0.040	0.030	11.50–13.50	...	...
414	0.15 max.	1.00	1.00	0.040	0.030	11.50–13.50	1.25–2.50	...
416	0.15 max.	1.25	1.00	...	...	12.00–14.00	...	P, S, Se min. 0.07, Zr, Mo max. 0.60
420	Over 0.15	1.00	1.00	0.040	0.030	12.00–14.00	...	...
430	0.12 max.	1.00	1.00	0.040	0.030	14.00–18.00	...	...
430-F	0.12 max.	1.25	1.00	...	...	14.00–18.00	...	P, S, Se min. 0.07, Zr, Mo max. 0.60
431	0.20 max.	1.00	1.00	0.040	0.030	15.00–17.00	1.25–2.50	...
440-A	0.60–0.75	1.00	1.00	0.040	0.030	16.00–18.00	...	Mo 0.75 max.
440-B	0.75–0.95	1.00	1.00	0.040	0.030	16.00–18.00	...	Mo 0.75 max.
440-C	0.95–1.20	1.00	1.00	0.040	0.030	16.00–18.00	...	Mo 0.75 max.
446	0.35 max.	1.50	1.00	0.040	0.030	23.00–27.00	...	N_2 0.25 max.

Table 3—Chromium Stainless Steels: Alloy Casting Institute Type Designations

	Chemical composition, %							
Designation	*C*	*Mn, max.*	*Si, max.*	*P, max.*	*S, max.*	*Cr*	*Ni*	*Other elements*
CA-15	0.15 max.	1.00	1.50	0.04	0.04	11.5–14	1 max.	Mo 0.5 max.*
CA-40	0.20–0.40	1.00	1.50	0.04	0.04	11.5–14	1 max.	Mo 0.5 max.*
CB-30	0.30 max.	1.00	1.00	0.04	0.04	18–22	2 max.	....
CC-50	0.50 max.	1.00	1.00	0.04	0.04	26–30	4 max.	....
HC	0.50 max.	1.00	2.00	0.04	0.04	26–30	4 max.	Mo 0.5 max.*
HD	0.50 max.	1.00	2.00	0.04	0.04	26–30	4–7	....

* Molybdenum not intentionally added.

I. Physical Metallurgy

PHASE RELATIONS

The primary phase relations of the chromium stainless steels are principally centered upon the formation of austenite and/or ferrite. Also, in alloys containing over 15 to 20% chromium, the possible appearance of the sigma phase between 1000 and 1500° F. (540 and 815° C.) has to be considered. Moreover, in high-chromium alloys, which contain a relatively large proportion of austenite formers, some austenite may be retained at room temperature.

Although the phase relations generally pertain only to the end points of reactions, the suppression of structural changes, intermediate formation products, and transformation products may also be of significance.

THE GAMMA LOOP

Upon quenching or air cooling from temperatures above 1600° F. (870° C.), the common chromium stainless steels, which do not contain other alloying elements, may exhibit a structure containing martensite or ferrite or both. In the chromium grades, this primarily depends upon the percentages of chromium and carbon which are present in the steel. The amount of each may be estimated from Bain's[52] "Gamma Loop" which, for each particular analysis, shows how much austenite may form at elevated temperatures.

Effects of Alloying Elements

Because most of the common martensitic and ferritic

stainless grades contain no alloying element other than chromium, the effects of chromium upon the extent of the Gamma Loop (Fig. 1) are of greatest interest. In the absence of carbon and other austenite-forming elements, ferrite begins to appear as the chromium content is increased above about 11.5%. Moreover, the austenite plus ferrite region is very narrow.[4] With increasing carbon content, the fully austenitic region widens somewhat and is extended at 0.1% carbon to about 12.5% chromium. The two-phase region is widened considerably more to about 18% chromium. A higher carbon content may extend the limits of both the single- and two-phase regions still further, but has more effect on the two-phase region.

Effects similar to those produced by carbon are produced by other austenite formers such as nitrogen, nickel, and manganese. On the other hand, the ferrite formers aluminum, silicon, molybdenum, etc., tend to reduce the extent of the Gamma Loop. The relative effectiveness of various elements was discussed previously.[3]

Martensite Transformation

In steels with up to 14% chromium, martensite transformation from the austenite occurs at temperatures below 800 to 900° F. (425 to 480° C.). Thus, even on air cooling from temperatures at which austenite has formed (i.e., within the Gamma Loop), most of the austenite transforms into martensite. This, of course, is due to the extreme sluggishness of the austenite to ferrite transformation which occurs when the alloy is held at temperatures between 900 and 1600° F. (480 and 870° C.).

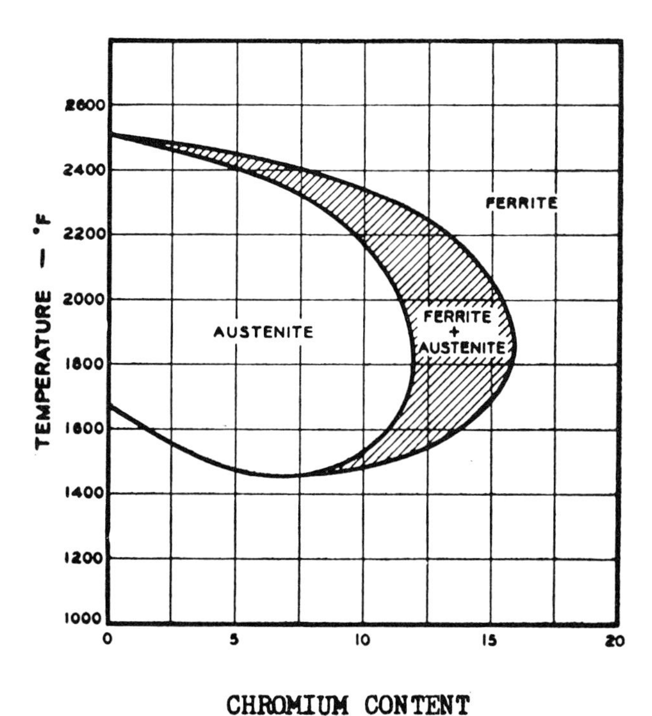

Fig. 1 The "Gamma Loop" in stainless irons containing 0.05% carbon (after Bain)[52]

Retained Austenite

In alloys, in which the presence of austenite formers extends the gamma region to compositions containing more than 16% chromium, increasing proportions of austenite may be retained at room temperature. This is caused by the increased sluggishness of transformation which is due to the higher alloy content and in particular to the tendency of chromium to stabilize austenite at room temperature.[2]

It should be remembered, however, that the ability of a particular alloy to form austenite at a chromium content exceeding 16% depends upon the amount and type of the austenitizers present. Thus, nitrogen, nickel, and carbon are most effective in retaining austenite. In fact, if sufficient quantities of these elements extend the two-phase region of the Gamma Loop to above 23% chromium, all the austenite present at elevated temperatures may be retained on quenching. Manganese, which also extends the Gamma Loop, does not seem to stabilize the austenite as efficiently as nitrogen or nickel. Thus, alloys, containing over 23% chromium and forming some austenite because of the presence of manganese, experience some transformation of part of the austenite into martensite.

If the austenite, which has been retained at room temperature, is exposed to temperatures between 750 and 1470° F. (400 and 800° C.), all or most of the austenite tends to break down into ferrite which contains carbides. Only in a few of the highly alloyed chromium stainless steels may the austenite be sufficiently stable to resist this breakdown. However, even in these alloys, some of the retained austenite tends to transform into ferrite and carbides.

Although the retention of austenite is beneficial because it improves certain room-temperature properties, alloys which form austenite at elevated temperatures show hot-working characteristics which are considerably lower than those exhibited by the fully ferritic steels.

FERRITIC STAINLESS STEELS

The ferritic stainless steels, which because of com position exhibit an essentially ferritic structure at any temperature, represent the most important series of the chromium stainless steels.[122]

Generally, the ferritic stainless steels are separated into two groups. The alloys of medium chromium content usually range between 14% and 22 to 23% chromium, whereas the high-chromium stainless steels contain between 22 and 30% or more chromium. Moreover, in the steels of medium chromium content, carbon, because of its tendency to extend the austenite region to a higher chromium content, is usually kept below 0.10 or 0.12%. In the high-chromium stainless steels, a somewhat higher carbon content may be tolerated and is present in commercial grades because of the cost which would be involved in producing very low-carbon ferritic stainless steels.

Source: *The Welding Journal*, 30, May 1951, 209-s–250-s

MARTENSITIC STAINLESS STEELS

The stainless steels which principally form martensite are also known as the hardenable stainless steels. In the absence of the major alloying elements which form austenite, these hardenable stainless grades usually include the alloys containing between 10 and 14% chromium.* When there is a need for highest values of yield and tensile strengths, the use of alloys which form as much martensite as possible may be quite advantageous. Thus, these steels are used the most in applications requiring the combination of hardenability with good corrosion resistance. However, these grades are also used at elevated temperatures as heat-resisting materials.

As the amount of martensite also depends upon the amount of austenite formed at elevated temperatures, these steels should be quenched from temperatures at which the Gamma Loop has its greatest extent. This temperature lies at 1800° F. (980° C.) for steels containing about 12% chromium and increases to 2200° F. (1205° C.) if 22% chromium and a sufficient quantity of austenite formers are present.

To avoid the formation of martensite, these steels have to be furnace cooled very slowly and held preferably above 1200° F. (650° C.). Thus, cooling rates of 50° F. (28° C.) per hour are usually recommended from 1600° F. (870° C.) down to 1200° F. (650° C.). The maximum permissible cooling rates will vary somewhat with the alloy content.

If, because of more rapid cooling rates, the steel has become martensitic, the martensite may be softened by soaking the steel thoroughly for about 2 hr. between 1300 and 1400° F. (705 and 760° C.). Heating at a lower temperature is not sufficient to cause complete softening and, in fact, may have a detrimental effect on the physical properties. Austenitizing temperature[167] and carbon content may cause some variations in the temperature limits of the "softening" range of the martensitic chromium stainless steels.

Although tempering below 1300° F. (705° C.) causes softening, the actual effects are similar to those produced by aging processes. Thus, a steel which contained:

	%
C	0.18
Mn	0.30
Si	0.42
Cr	12.4
Ni	0.17

exhibited, after extended drawing at temperatures between 840 and 1100° F. (450 and 595° C.), a parameter of over 0.001 Angstrom Units less than a specimen which had been annealed at 1470° F. (800° C.).[42] Although Palatnik and Barkov[42] felt that this was caused by the precipitation of finely dispersed particles from the supersaturated (martensitic) (acicular) ferritic solid solution, it seems more likely that a pre-precipitation process is actually responsible. This would consist of the formation of segregations of solute atoms in the ferrite matrix and straining the matrix in these regions by the existence of a coherent state. Similar explanations have been given in recent years by various authors to explain strain aging and temper brittleness in mild steels, and the 885° F. (475° C.) and High-Temperature Embrittlements which occur in the ferritic stainless steels.

In these martensitic stainless steels, this "segregation" or "coherent state" hypothesis is well supported by the fact that a tempering treatment between 850 and 1100° F. (455 and 595° C.) causes considerable embrittlement (Table 4) and reduces corrosion resistance, Fig. 2.[157] It should be noted that, whereas toughness is at a minimum at tempering temperatures of about 900° F. (480° C.), corrosion resistance is at its lowest at a temperature of 1100° F. (595° C.). The precipitates which form in this temperature range are unlikely to produce severe (if any) detrimental changes. On the other hand, the formation of atomic segregations with accompanying lattice straining is known to cause embrittlement and to reduce corrosion resistance considerably.

If steels have been embrittled by exposure between 850 and 1100° F. (455 and 595° C.), their toughness and corrosion resistance may again be improved by annealing between 1300 and 1500° F. (705 and 815° C.).[122]

When these steels have been heated between 1600 and 2100° F. (870 and 1150° C.), the amount of martensite that forms on quenching generally can be determined quite readily with the use of hardness readings. Thus, a fully martensitic structure will have a hardness of over 400 to 500 Brinell. Annealing reduces the hardness again to about 200 Brinell.

Effects of Carbon Content

In the martensitic grades, the air-hardening tendency depends on the carbon and chromium content of the steel. For commercial martensitic stainless steels, the effects of the carbon and chromium content on the free ferrite at a temperature of 1800° F. (980° C.) are shown

Table 4—Effects of Tempering Temperature on the Charpy V-Notch Impact Toughness of a Martensitic, 16% Chromium–2% Nickel Stainless Steel (Type 431)*

Tempering temperature, ° F.	*Toughness, ft.-lb.*
400	81
500	77
600	65
700	57
800	47
900	11
1000	20
1100	78
1200	82

* Minimum toughness values at 900° F. (480° C.) have also been reported for Type 410 chromium stainless steels.[179]

* The usual upper limit for the wrought, low-carbon, chromium stainless steels which may be welded is 14% chromium. A higher carbon content will raise the chromium limit. Thus, a chromium stainless steel with 1% carbon and 18% chromium (A.I.S.I. Type 440C) would be hardenable and may form a fully martensitic structure on quenching from 1800° F. (980° C.). Such a high-carbon stainless steel exhibits poor welding characteristics.

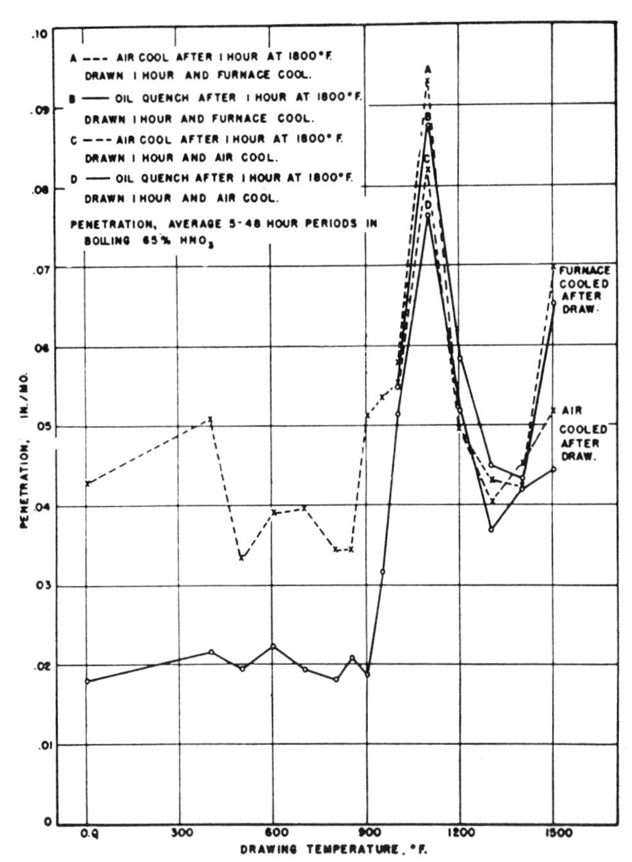

Fig. 2 Effect of drawing temperature on the corrosion rates in 65% boiling nitric acid of a 12% chromium stainless casting[157]

in Fig. 3.[189] Since, at a constant chromium content, increases in carbon content decrease the quantity of ferrite, which remains at austenizing temperatures, the air-hardening tendency increases with the carbon content. That is, the amount and hardness of the martensite becomes greater with increases in carbon content.

In the 12% chromium stainless grades (Type 410) containing no additional alloying elements, two major types are differentiated on the basis of their carbon content. The low-carbon type (sometimes referred to as "modified") contains less than 0.08% carbon and is only partially hardenable. On the other hand, the ordinary grade, containing generally 0.08 to 0.15% carbon, usually is primarily martensitic and hardenable.

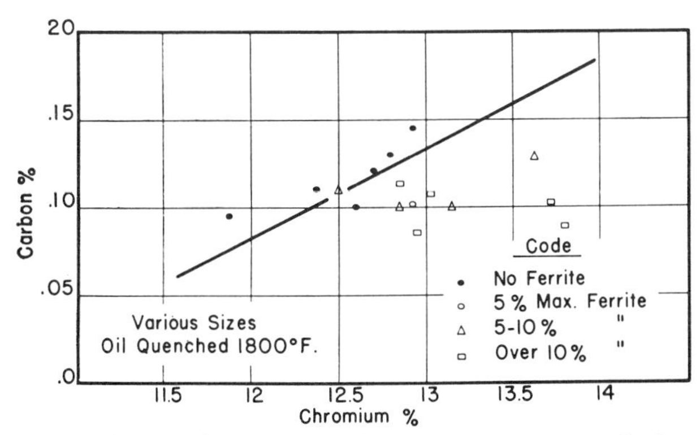

Fig. 3 Effects of carbon and chromium content on the free ferrite in a Type 416 stainless steel[189]

Effects on Corrosion Resistance

Corrosion resistance is greatest in the martensitic (fully hardened) condition. Moreover, it increases with chromium content. Tempering between 900 and 1100° F. (480 and 595° C.) decreases corrosion resistance considerably.[109, 122, 129, 157, 171, 173] The probable reason for this and the reduction in toughness may well be related to changes in the distribution of the carbon atoms which occur during the decomposition of martensite. Thus, this phenomenon may possibly be explained by the earlier mentioned segregation hypothesis. Corrosion resistance is again improved at higher temperatures at which chromium carbides would be precipitated. On the other hand, the older hypothesis may be valid instead. This suggests that actual precipitation of chromium carbides during martensite decomposition at temperatures between 900 and 1100° F. (480 and 595° C.) may locally reduce the chromium content around each precipitated carbide. Corrosion resistance may then be restored in some measure by heat treatments at higher temperatures which permit chromium diffusion to the impoverished area.[199]

STRUCTURAL PHENOMENA AFFECTING PHYSICAL PROPERTIES

In addition to the formation of austenite and martensite, the chromium stainless steels may become susceptible to one or more phenomena which produce drastic changes in the physical properties of the normally ferritic alloy. Distinctions should be made between four major types: (1) 885° F. (475° C.) Brittleness, (2) sigma-phase precipitation, (3) High-Temperature Embrittlement of ferrite, and (4) the notch-sensitivity of ferritic alloys.

Apart from composition, the principal factors affecting these phenomena are the temperature to which the steels have been heated either during the heat treatment or during service and the time of exposure to such temperature.

"885"° F. (475° C.) BRITTLENESS

This type of embrittlement occurs in chromium-iron alloys which contain between 15 and 70% chromium and are subjected to prolonged heating between 750 and 1000° F. (400 and 540° C.). The embrittlement derives its name from the fact that its effects are most severe on exposure to temperatures around 885° F. (475° C.).

Hardness readings are generally used as criteria to determine if 885° F. (475° C.) Brittleness has occurred. However, as this phenomenon seems to take place preferentially in the grain boundaries, hardness readings may not be sufficiently sensitive to reveal the initial stages of this embrittlement. Thus, impact* or

* In the ferritic stainless steels containing over 16 to 18% chromium, care should be taken that this 885° F. (475° C.) Brittleness is not confused with the notch-sensitive impact behavior of the higher chromium alloys.

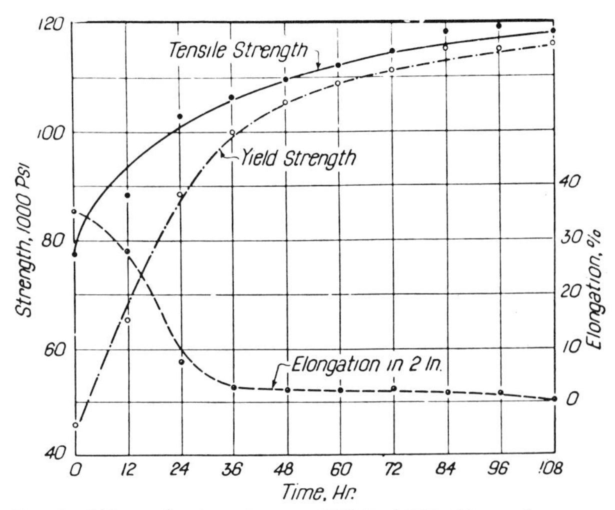

Fig. 4 Effect of aging time at 885° F. (475° C.) on the room-temperature tensile properties of 27% chromium-iron alloy[10]

bend tests may be more suitable since they reveal this embrittlement before an increase in hardness may be detected. Ductility in the transverse direction is also seriously altered during the early stages of the embrittlement.

Effects of Exposure Time and Cooling Rates

The increase in tensile and yield strengths after exposure at 885° F. (475° C.) is accompanied by a reduction in impact toughness and ductility. This is shown in Fig. 4 for a 27% chromium steel,[10,135] in which the prolonged exposure time at 885° F. (475° C.) increases the room-temperature strength and decreases the elongation considerably. The tensile strength during this time increases by 50% and the yield strength increases over 150% above that of the material in annealed form.

Ordinarily, hardness tests, tensile tests, and bend tests on specimens which are not notched require many hours or days at 885° F. (475° C.) before noticeable embrittlement occurs. Notched specimens, on the other hand, may reveal this embrittlement in much shorter periods. This was shown by Zapffe and Worden[194,202] on a 26% chromium stainless steel which showed embrittlement in a notch-bend test within the first half-hour of exposure at 885° F. (475° C.).

Once the high-chromium stainless steels have been embrittled by exposure at 885° F. (475° C.), the effects are evident at room temperature as well as at temperatures up to about 1000° F. (540° C.). This is apparent from Fig. 5 which shows the results of short-time tensile tests at temperatures up to 1400° F. (760° C.) on a 27% chromium steel.[136] Although reduction of area and elongation improve at testing temperatures above

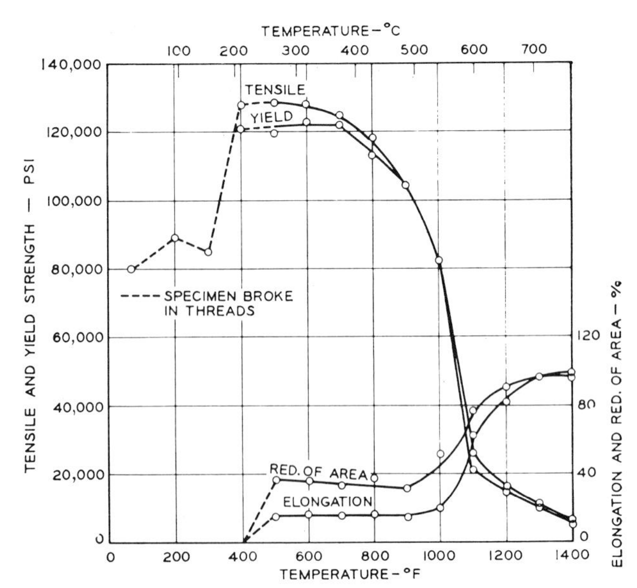

Fig. 5 Effect of testing temperature on the short-time tensile properties at elevated temperatures of a 27% chromium stainless steel which had been embrittled by aging for 500 hr. at 885° F. (415° C.).[136] *(Specimens were heated 1 hr. at testing temperature before pulling)*

500° F. (260° C.) the steel does not regain its characteristic ductility until testing temperatures above 1000° F. (540° C.) are reached. The effects of the 885° F. (475° C.) Embrittlement disappear at and above 1000° F. (540° C.). In the lower chromium ferritic stainless steels with, for an example, a chromium content of 17%, the effects of the 885° F. (475° C.) Embrittlement are likely to be less severe than they were shown in Fig. 5 for a 27% chromium alloy. Nevertheless, after long holding periods at the embrittling temperatures, a 17% chromium-iron alloy can be expected to be completely embrittled, as subsequently shown by impact data in Table 6.

In the common chromium-iron alloys, ordinary cooling rates through the embrittling range generally do not adversely affect the physical properties appreciably. However, in heavy sections, in which cooling rates are extremely slow, detrimental effects may occur. Moreover, some alloying elements also reduce the exposure periods necessary to cause embrittlement. For titanium-bearing alloys, this is evident from Fig. 6, which

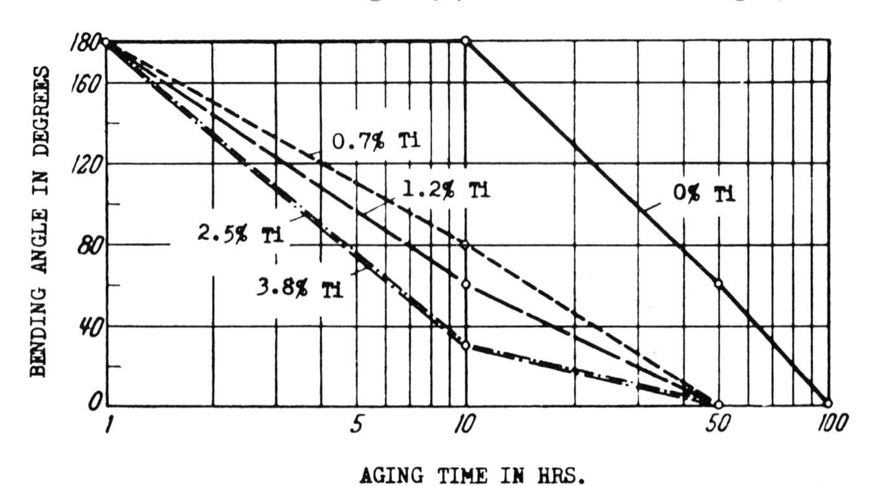

Fig. 6. Effects of aging time and titanium content at 885° F. (475° C.) on the room-temperature bending properties of 24% chromium steels[41]

shows how increases in the titanium content up to 3.8% increase the rates of embrittlement of 24% chromium alloys held at 885° F. (475° C.) for various periods.

Although air cooling of thin sections from elevated temperatures prevents 885° F. (475° C.) Embrittlement, heavier sections of highly susceptible compositions may have to be quenched from temperatures above 1100° F. (595° C.).

Severely cold-worked alloys seem to be more susceptible to 885° F. (475° C.) Embrittlement than annealed or tempered grades.[181]

Effects of Alloying Elements

Small amounts of other elements may modify the degree of 885° F. (475° C.) Brittleness and/or the range in which it occurs. Thus, Riedrich and Loib[41] reported that titanium above 0.9% and columbium above 2.4% tended to increase the severity of embrittlement noticeably. However, the elements did not appear to lower the range of chromium content in which precipitation occurred. Silicon, molybdenum, carbon, and aluminum seem to produce similar effects. Manganese, on the other hand, in amounts above 3%, seems to raise by several per cent the amount of chromium necessary before embrittlement can be observed. Nickel in small amounts is believed to increase[23] and nitrogen is believed to decrease[210] the degree of 885° F. (475° C.) Brittleness. It should be remembered, however, that the balance between these elements is extremely sensitive. Thus, it is quite difficult to point to any one element as being responsible for causing failure in the field under service applications at temperatures between 750 and 1000° F. (400 and 540° C.).

Effects on Corrosion Resistance

Corrosion resistance in acid solutions is also reduced[10,141] by the 885° F. (475° C.) Embrittlement. In fact, the effects of 885° F. (475° C.) Embrittlement seem of considerably greater severity than if sigma-phase precipitation had occurred instead. This is shown in Table 5 for a 27% chromium iron after it was immersed in boiling 65% nitric acid. However, after "sigmatizing" periods of 6000 hr. or longer, the specimen should be expected to exhibit higher corrosion rates than after the service exposure of 2900 hr. reported in Table 5.

Distinction from Sigma-Phase Precipitation

The fact that 885° F. (475° C.) Brittleness differs in some ways from sigma-phase precipitation was indicated by hardness determinations by Houdremont[5] whose results are shown in Fig. 7. Similar data are reported by other investigators.[23] The alloys containing 17.9 and 28.4% chromium did not reveal any hardening

Table 5—Effects of 885° F. (475° C.) and Sigma-Phase Embrittlement on Corrosion Resistance of a 27% Chromium-Iron Immersed in 65% Boiling Nitric Acid[10, 141]

Condition	*Penetration per month, in.*
Annealed 1 hr. at 1600° F. (870° C.), water-quenched	0.000744
Embrittled 500 hr. at 885° F. (475° C.)	0.002580
Embrittled 6000 hr. at 885° F. (475° C.)	0.009090
Sigma precipitated in 2900-hr. service	0.002080

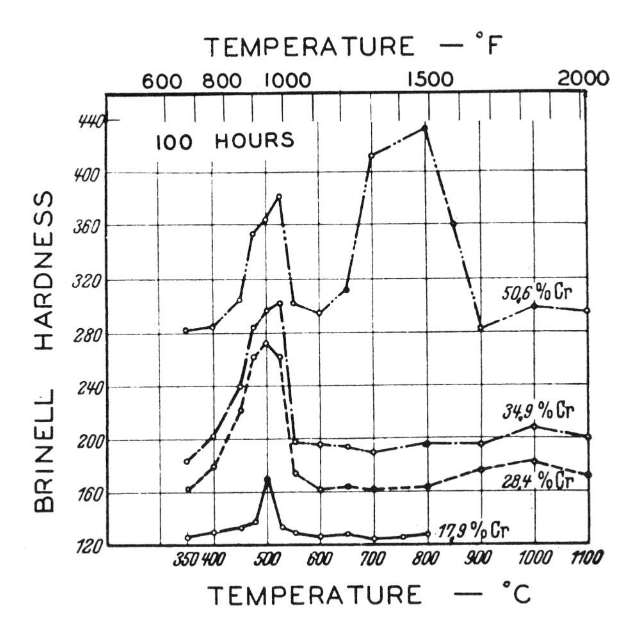

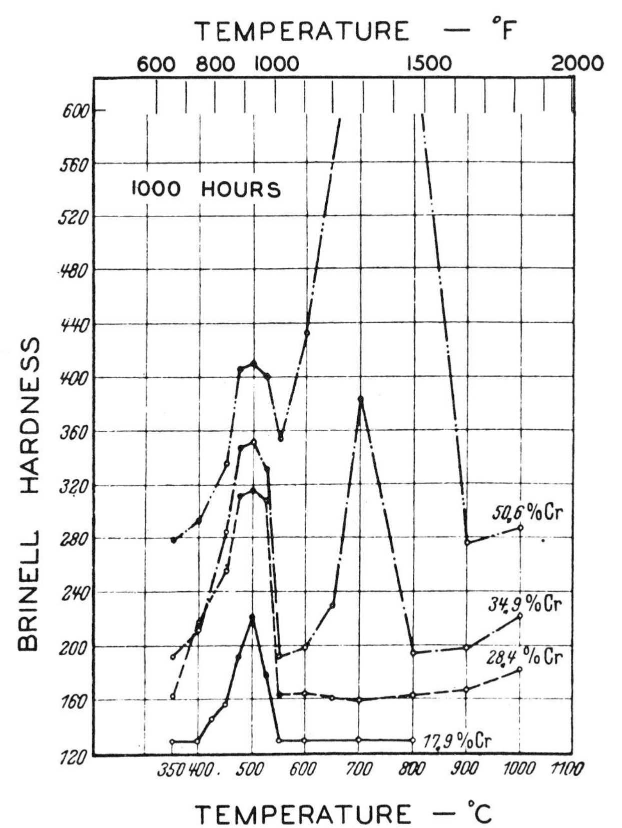

Fig. 7 Effects of temperature and aging time on the hardness increase caused by 885° F. (475° C.) Brittleness and sigma-phase precipitation in several chromium-iron alloys[5]

because of sigma-phase precipitations, although still longer periods should cause some precipitation in the 28.4% chromium steel. On the other hand, the precipitation caused by 885° F. (475° C.) Brittleness considerably increased hardness in these alloys. The effects of temperature are also apparent in Fig. 7.

Causes of 885° F. (475° C.) Brittleness

Although definite information as to the nature of this embrittlement is still lacking, it has been associated by most investigators with a grain boundary phenomenon of some kind.[4, 5, 10, 23, 41, 124, 140] Actually it is likely that the embrittlement occurs within the grains as well since fractures of embrittled alloys usually are transgranular instead of intergranular. However, the rate at which the phenomenon forms, which causes the embrittlement, may be much higher along the grain boundaries. This may explain why certain etching solutions usually reveal a grain boundary widening. This disappears again when the alloy is heated to temperatures at which the 885° F. (475° C.) Embrittlement is removed.[10, 142] However, why this brittleness should widen the grain boundaries on electrolytic etching in 10% oxalic acid,[41] for example, is not quite clear.[73]

As it occurs even in highly purified chromium-iron alloys,[10, 23, 72, 74] 885° F. (475° C.) Brittleness cannot be associated with precipitation of carbide or nitride particles. The suggestions that the embrittlement may be caused by the precipitation of chromium phosphides[23] or chromium monoxides[177, 202, 208] have received little, if any, support by other investigators.

A possible answer may have been given by Bandel and Tofaute[23] and by others,[135–138, 142] who compared the 885° F. (475° C.) Brittleness to the room-temperature aging of Duraluminum alloys (see also Newell[10]). If the relations are similar, 885° F. (475° C.) Brittleness might well be due to stress conditions caused by the formation of atomic complexes or disturbances short of a precipitation of the FeCr phase. The atomic aggregates are likely to be of higher chromium content than the ferrite matrix. They do not form a stable crystalline structure, but might be thought of as an intermediate transition stage between the solution of chromium in ferrite and the final sigma phase.

If 885° F. (475° C.) Embrittlement is caused by possible clusters of chromium atoms at the grain boundaries and, to a lesser extent, within the grains, this brittleness might then be compared to the High-Temperature Embrittlement discussed subsequently. However, although the 885° F. (475° C.) Embrittlement is particularly severe in the grain boundaries, the High-Temperature Embrittlement occurs within the grain. Moreover, because of temperature differences and the greater diffusibility of carbon in iron, the High-Temperature Embrittlement occurs at much faster rates.

Impact and hardness data reported by Wilder and Light[85] from tests on a 17% chromium and 0.09% carbon specimen exposed for long periods at various elevated temperatures also indicate that there is some relation between 885° F. (475° C.) Brittleness and sigma-phase embrittlement. Their results are repeated in Table 6. The hardness increase at 900° F. (480° C.) is in good agreement with the severe reduction in impact toughness. The reduction in impact toughness which occurred after the alloy was held 10,000 hr. at 1050° F. (565° C.) indicates sigma precipitation which must be extremely fine because of the lack of an increase in hardness values. Moreover, it also seems significant that, on extremely long holding periods, the composition range in which sigma precipitates approaches that of 885° F. (475° C.) Brittleness. The change in hardness does suggest a change in the distribution of the chromium atoms. Thus, it might well be that the aging mechanism is similar to strain aging which may occur in many mild steels. For example, as is true in strain aging, long exposure of the chromium stainless steels at 885° F. (475° C.) does not result in overaging, that is a return to ductile behavior and a reduction in the hardness.

In recent years a number of investigators have suggested that strain aging in mild steels is caused by atomic disturbances which are primarily due to nitrogen or carbon.

Removal of Effects of 885° F. (475° C.) Embrittlement

Alloys, in which 885° F. (475° C.) Embrittlement has occurred, may be returned to their normal ductile state by heating them for short periods at temperatures above 1100° F. (595° C.).[23] For compositions not susceptible to sigma-phase precipitation in relatively short periods of exposure, temperatures up to 1470° F. (800° C.) are often recommended.[22] However, in alloys with over 25% chromium, in which sigma precipitation may occur more rapidly, temperatures nearer 1100° F. (595° C.) should be employed. Thus, Newell[10] and Heger[142] reported that for one steel, containing 27% chromium, 1 hr. at 1100° F. (595° C.) or 5 hr. at 1050° F. (565° C.) was sufficient to remove the effects of the

Table 6—Effects of Temperature and Time Upon Impact Toughness and Hardness of 17% Chromium Steel[85]

	Before exposure	*Exposed 400 hr.* 900° F. (480° C.)*	*Exposed 1000 hr. 1050° F. (565° C.)*	*Exposed 1000 hr. 1200° F. (650° C.)*	*Exposed 10,000 hr. 900° F. (480° C.)*	*Exposed 10,000 hr. 1050° F. (565° C.)*	*Exposed 10,000 hr. 1200° F. (650° C.)*
Charpy keyhole impact (ft.-lb.)	46	0.3	32	34	1	3	4
Brinnell Hardness values (converted)	175	197	175	167	269	163	153

* From data by H. D. Newell[133] on a very similar alloy which contained 0.07% C and 17% Cr.

embrittlement. In these same alloys, extended times of several thousands or more hours at temperatures above 1000° F. (540° C.) resulted in the formation of the sigma phase.[142] This is apparent from Fig. 8, in which are shown the effects of time and temperature upon the softening of a 27% chromium steel which had been embrittled by holding for 500 hr. at 885° F. (475° C.). It is noteworthy that after 1000 hr. at 1200° F. (650° C.) the elongation begins to drop again because of the initiation of sigma precipitation.

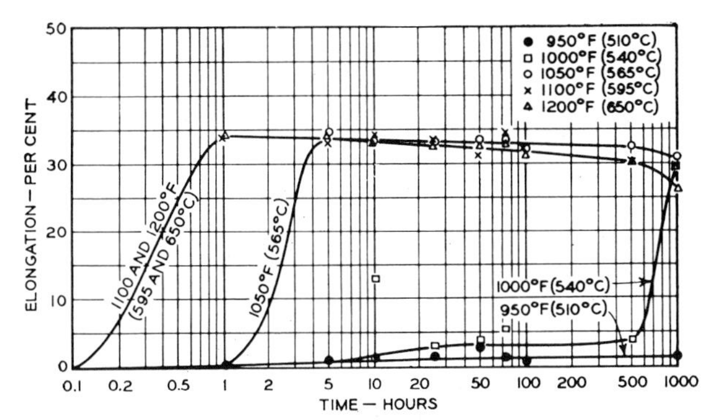

Fig. 8 Effects of annealing time and temperature on the recovery (elongation) of a 27% chromium steel which had been embrittled by aging for 500 hr. at 885° F. (475° C.)[142]

Effects of Welding

Ordinarily, it is doubtful that welding operations may cause 885° F. (475° C.) Embrittlement. Only in welding operations of heavy sections of highly susceptible alloys may the deposition of multiple-bead deposits cause some brittleness. In those alloys, a postannealing treatment above 1100° F. (595° C.) may be advisable, unless brittleness can be prevented by the control of the interpass temperatures.

In very heavy sections, preheating prior to welding should be practiced with care on chromium stainless steels with over 15 to 16% chromium, as preheating would reduce cooling rates through the 885° F. (475° C.) Embrittlement range considerably.

Effects in Service Application

Because of the 885° F. (475° C.) Embrittlement, ferritic chromium-iron alloys containing over 16% chromium should not be used between 700 and 1050° F. (370 and 565° C.) when ductility and toughness are desired. This, for example, is true in refinery applications where failures have occurred in 17% chromium stainless steels (Type 430) exposed near 885° F. (475° C.).[124, 132]

SIGMA-PHASE EMBRITTLEMENT

The sigma phase, described in a previous review[3] in austenitic chromium-nickel stainless steels, also appears as a hard and highly brittle, nonmagnetic phase in certain chromium stainless alloys. Although in these steels the sigma phase is generally labeled FeCr, other elements may go into solution with this phase. It is generally agreed that the sigma phase exhibits a tetragonal unit cell.[175, 182, 192, 205]

The phase relations of the sigma phase in highly purified chromium-iron alloys are shown in Fig. 9.[29] Pure sigma (FeCr) may be obtained by means of appropriate heat treatments in alloys containing between 42 and 48% chromium. A duplex structure of both ferrite and sigma may occur when alloys containing as little as 17% and as much as 70% chromium are exposed to temperatures between 1000° F. (540° C.) and 1500° F. (815° C.). The actual temperature limits are difficult to ascertain because they are influenced by alloying elements and by the fact that the transformation rates are extremely slow in the lower chromium alloys—particularly at temperatures below 1100° F. (595° C.). For example, Heger[60, 138] showed that in a 27% Cr-Fe alloy a very fine precipitate had formed after aging for 1000 hr. at 950° F. (510°C.).

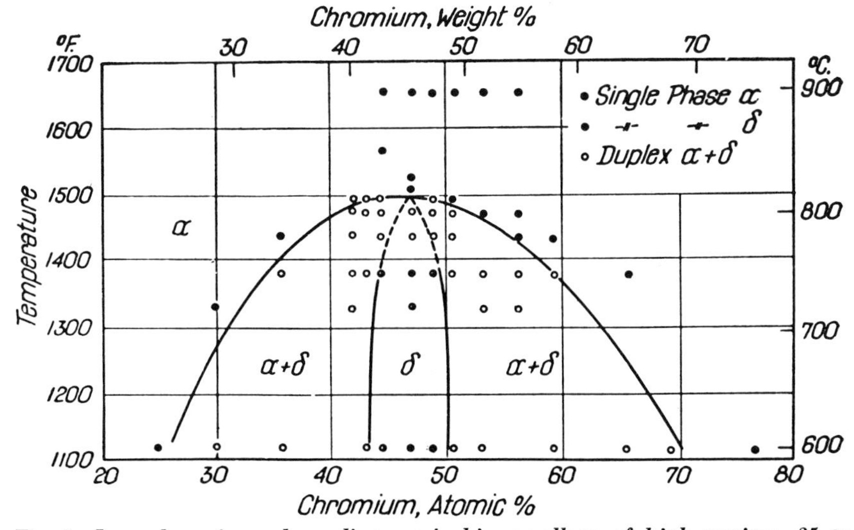

Fig. 9 Iron-chromium phase diagram in binary alloys of high purity; 25 to 76% chromium. This diagram shows pure sigma phase from 44 to 50% chromium and mixed alpha and sigma phase from about 25 to 44% of chromium with phase boundaries for temperature interval of 1100 to 1500° F. (595 to 815° C.)[29]

Rates of Sigma Formation

In most chromium-iron alloys, the rates of sigma formation are sufficiently slow that castings as well as weld deposits are unlikely to exhibit the sigma phase. This is apparent from Fig. 10 in which are shown the time required for sigma to form in alloys which contain:[159]

	%
C	0.04– 0.08*
Mn	0.31– 0.72
Si	0.40– 0.80
Cr	15.37–33.03
Ni	0.04– 0.08
Mo	0.00– 0.09

* A higher carbon content would reduce the rates of sigma formation.

Specimens, which were air-cooled from temperatures above the sigma-phase range, should also be unlikely to contain sigma. These considerations do not necessarily hold true in cold-worked chromium stainless steels and in some of the austenitic chromium-nickel stainless steels.[2, 3]

Effects of Alloying Elements

Most of the ferrite formers such as molybdenum, silicon,[102, 103, 106, 159] etc., shift the sigma-precipitation range to a lower chromium content. In addition, many of these elements tend to increase or decrease the temperatures required for the precipitation of sigma.

Similar effects are also produced by the austenite stabilizers nickel[3, 69, 113, 144, 146] and manganese.[114, 144, 159, 212] This seems to be caused primarily by the ability of the sigma phase to absorb up to 10% of nickel[113] and as much as 35% of manganese.[114] Nickel also raises the temperature range for sigma formation from 1500 to 1700° F. (815 to 925° C.).[193]

Nitrogen seems to be ineffective in the chromium stainless steels. Similarly, carbon in excess of its solubility limit in ferrite depletes the ferrite of chromium by forming chromium-rich carbides. This reduces the rate of sigma formation and causes a shift of the ferrite-ferrite plus sigma boundary to higher values of total chromium.[159]

Effect of Cold Work

Cold work also enhances the precipitation of the sigma phase considerably.[3, 10, 60, 106, 160, 162]

Houdremont's results, reported graphically in Fig. 7, show the extreme sluggishness of the sigma transformation in chromium-iron alloys which do not contain any other alloying additions. After 100 hr., sigma had begun to form in the 50.6% chromium alloy, but had not started in the 34.9% chromium alloy. After 1000 hr., sigma formation was probably complete in the 50.6% chromium alloy and was just starting in the 34.9% chromium alloy. Sigma did not form in either the 28.4 or the 17.9% chromium alloy. Every investigator of these plain chromium-iron alloys has stated that cold deformation is a prerequisite for the formation of sigma in any reasonable time.[199] For example, peening the welds of chromium-iron alloys susceptible to sigma formation will start this formation, which otherwise might never begin.[199]

Effects on Mechanical Properties

In commercial alloys, the presence of sigma precipitation generally is highly undesirable. Although catastrophic failures in service resulting from sigma-phase embrittlement have occurred only in a few instances, serious losses in toughness and ductility may be responsible for cracking during certain fabricating operations. As an extreme example, Oliver[193] mentions that low-carbon, 27% chromium stainless steels can break like plate glass when dropped a few feet on a floor after being exposed for 500 hr. at 1350° F. (730° C.).

Ordinarily, sigma-phase embrittlement may seriously reduce creep strength and stress-to-rupture properties at elevated temperatures. However, the effects of the sigma phase depend not only upon the quantity but also upon the size and distribution of the particles and upon the temperature at which they are formed. Thus, a fine, well-dispersed sigma precipitate has been found to be beneficial in certain valve steels which are used at temperatures up to 1400° F. (760° C.).

Effects in Castings and Weld Deposits

In the primarily ferritic stainless compositions, sigma seems to form more rapidly in wrought alloys than in castings[162] or weld deposits.[163] According to Gilman[162] the rate of sigma formation seems to depend upon the rate of cooling during initial solidification.

Solution of Sigma

Unlike the austenitic chromium-nickel steels, where complete solution of sigma is accomplished at temperatures above 1900 to 2000° F. (1040 to 1095° C.),[3] in the chromium stainless steels, sigma solution may be accomplished with relatively short holding periods, for example, 1 hr. at or above 1600° F. (870° C.). However, the presence of some elements such as nickel, molybdenum, and/or manganese reduces solution rates and may necessitate longer holding periods or higher temperatures.

For steels in which sigma has formed, the most suitable annealing temperatures will vary for each par-

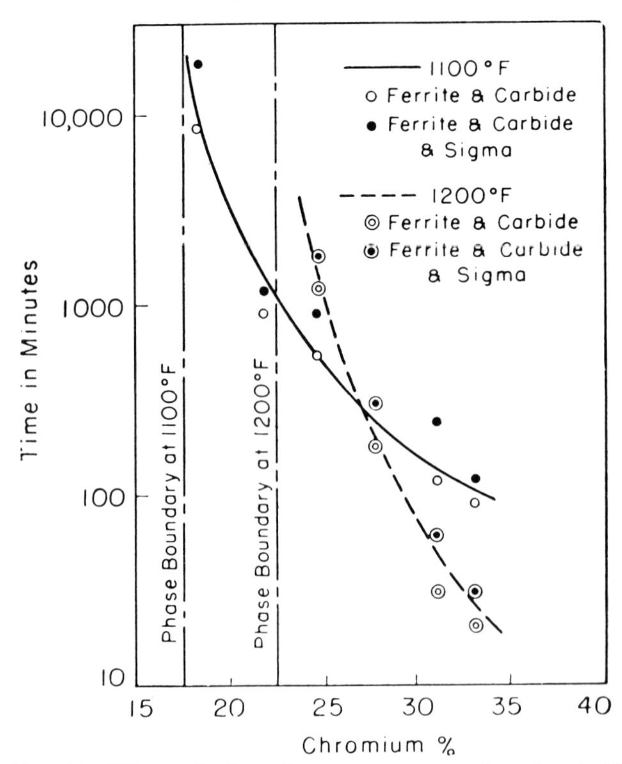

Fig. 10 Effect of chromium content on the threshold times of sigma formation at 1100 and 1200° F. (595 and 650° C.)[159]

ticular alloy. With 27% chromium steels, best results seem to be obtained by first annealing the alloy for at least 1 hr. at temperatures between 1400 and 1600° F. (760 and 870° C.) and then water quenching it.[139]

The effects of various annealing temperatures between 1200° and 1600° F. (650 and 870° C.) upon the mechanical properties of a 27% chromium steel are shown in Fig. 11.[139] Although these test results show a surprisingly high ductility for the air-cooled specimens, they may be misleading as far as mill experiences are concerned. Thus, as Heger and Cordovi[139] pointed out that, for the cold draw of tubes, an air cool from temperatures between 1400 and 1600° F. (760 and 870° C.) will invariably give unsatisfactory results, and only by employing a drastic water quench can sufficient ductility be obtained. Moreover, the inherent notch-sensitivity, which is due to surface conditions, is beneficially affected by and shows an even greater need for rapid cooling from the annealing temperatures. On the other hand, Bungardt[174] reported that experiences in Germany indicate that air cooling of 27% chromium stainless steels as wrought products and as seamless tubes is satisfactory. This latter contention also seems to be supported by other investigators.[199]

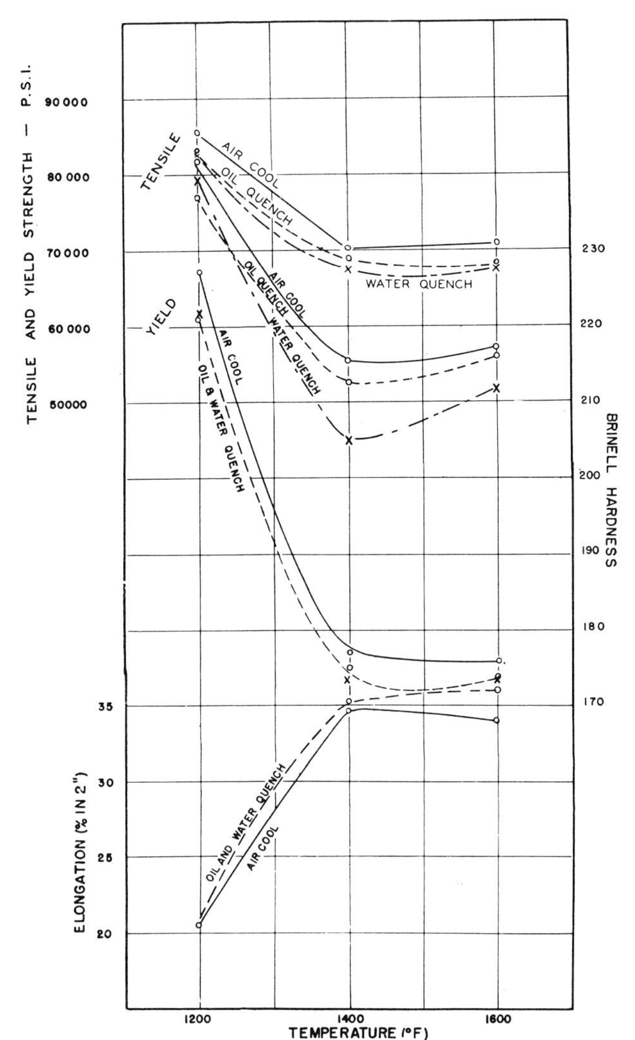

Fig. 11. The effects of annealing temperature and cooling rate on the mechanical properties of a 27% chromium steel[139]

HIGH-TEMPERATURE EMBRITTLEMENT

Chromium stainless steels which are or become fully ferritic above 2100° F. (1150° C.) are susceptible to another so-called High-Temperature Embrittlement.[61] Since the effects of this embrittlement may be removed by suitable heat treatments at intermediate temperatures, it is not ordinarily observed in wrought steels. However, castings or weld deposits, in the as-cast or as-welded state, may show a severe embrittlement, which may be removed by subsequent heat treatments. Wrought ferritic grades may also show this embrittlement when the steel has been exposed to and air-cooled or water-quenched from temperatures above 2100° F. (1150° C.).

In the ferritic stainless castings, the brittleness is most severe in thin sections where the cooling rates are most rapid. Some improvements may be noted when the lowest possible pouring temperatures are used since the degree of brittleness increases with pouring temperature.[54, 77] On the other hand, casting temperatures must be sufficiently high to allow the molten metal to flow into all parts of the mold.[77] In weld deposits, the embrittlement occurs in the ferritic weld and in that part of the adjacent heat-affected zone which has been heated to temperatures above 2100° F. (1150° C.).[61]

Effects of Structure

The High-Temperature Embrittlement seems to be accompanied by severe grain growth, although a large grain size alone will not account for the embrittlement. As this severe grain growth cannot occur unless the austenite and the carbides have been dissolved, its association with the High-Temperature Embrittlement is apparent. Thus, carbide stabilizers, which raise the temperature at which these particles go into solution, seem to increase proportionally the minimum temperature at which embrittlement can be noted.[9, 88]

Solution of carbides, which promotes consequent grain growth at the lower temperatures around 1830 to 2010° F. (1000 to 1100° C.), is enhanced by the length of holding. Moreover, as the chromium content increases, the solution temperatures of carbon seem to be lowered somewhat. Thus, an alloy containing 24.8% chromium and 0.20% carbon was severely embrittled after holding for 100 hr. at 1830° F. (1000° C.).[88] This is shown in Fig. 12. The beneficial effects of carbide stabilizers and of the presence of some austenite are also apparent in this figure.

Thus, the temperature limit above which the High-Temperature Embrittlement occurs primarily depends upon the structure. Embrittlement may occur in a ferritic matrix as soon as the carbides have gone into solution. This, however, is a function of time. Consequently, whereas ordinary fabricating operations (welding, casting, etc.) and standard heat treatments are not likely to cause embrittlement, in ferritic steels at temperatures below 2100° F. (1150° C.), long exposures at temperatures as low as 1800° F. (980° C.) when accompanied by carbide dissolution (no austenite present) are likely to cause this embrittlement.

Effects on Mechanical Properties

In Fig. 13 are shown the effects of the High-Temperature Embrittlement on the mechanical properties of a partially ferritic stainless steel containing:

	%
C.	0.067
Mn	0.46
Si	0.37
Ni	0.33
Cr	12.91

The drop in toughness and the simultaneous increase in tensile strength and hardness between 1600 and 2200° F. (870 and 1205° C.) are related to the formation of austenite at these temperatures which, on quenching, transforms to martensite. These effects are most pronounced at 1800° F. (980° C.), where the largest amount of austenite is formed. Above 1800° F. (980° C.), austenite tends to transform into (delta) ferrite and, at 2200° F. (1205° C.), all of the austenite has transformed into (delta) ferrite. The carbides, which had initially gone into solution in austenite, remain in solution in (delta) ferrite. Since above 2200° F. (1205° C.) this alloy had become fully ferritic, it became subject to severe grain growth and, simultaneously, to embrittlement.

Hardness determinations also exhibit a considerable increase in hardness values. This occurs, however, at somewhat higher temperatures than the embrittlement.

Since, as in the martensite-ferritic grades, hardness increases may be caused either by the formation of martensite or the High-Temperature Embrittlement, the hardness values alone cannot be used as criteria to identify the other mechanical properties of these stainless steels.

Removal of Embrittlement

Although it does not reduce the excessive grain size,

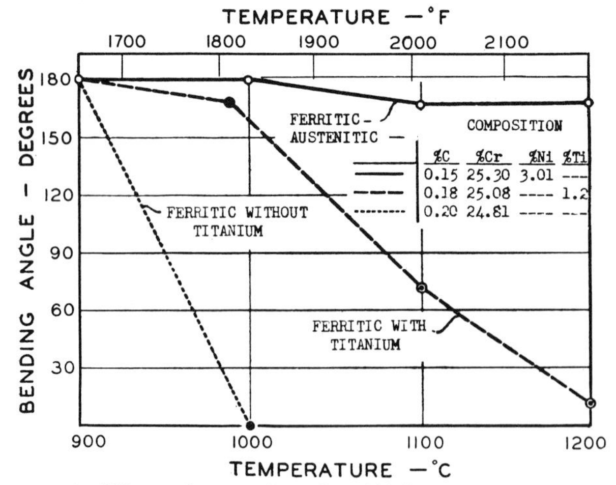

Fig. 12 Effect of annealing for 100 hr. at temperatures between 1650 and 2190° F. (900 and 1200° C.) on the bending properties of 2-mm. thick sheet specimens of different composition[88]

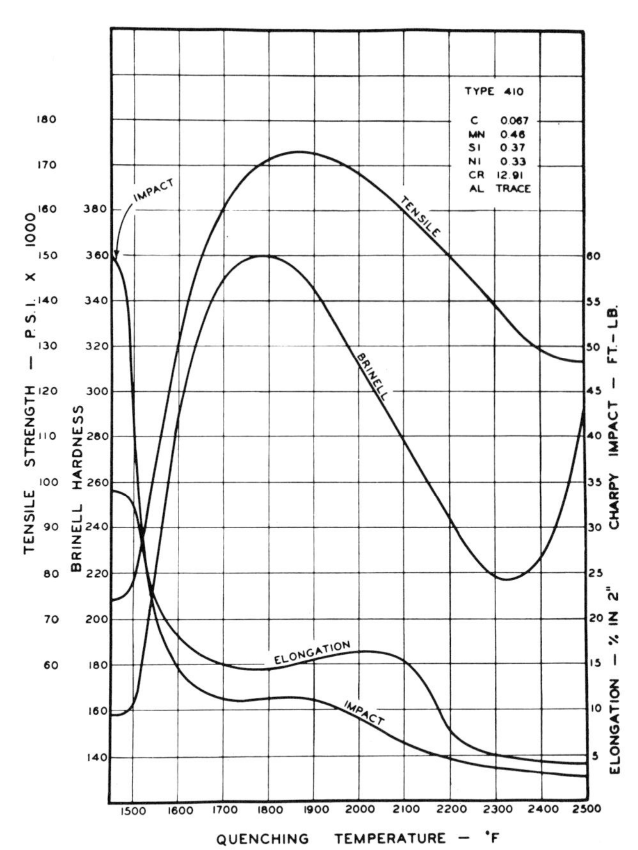

Fig. 13 Effect of quenching temperature on the mechanical properties of a modified (low-carbon) 13% chromium stainless steel

annealing embrittled alloys between 1350 and 1450° F. (730 and 790° C.)* does remove most of the detrimental effects of the High-Temperature Embrittlement.[61] This is evidenced by the precipitation of the carbides which had gone into solution in the ferrite matrix. The annealing treatment is particularly important in castings and weld deposits of the ferritic stainless steels which have been embrittled by exposure to temperatures above 2200° F. (1205° C.). Although welds are somewhat improved by the deposition of small multiple deposits which produce some annealing effect, a full post-annealing treatment should be employed wherever a restoration of the ductile properties of the steel is desired.

It is now apparent that the degree of embrittlement in castings and weld deposits depends on two factors, (1) the amount of time required to cool the steel through the high-temperature (embrittling) range and (2) the amount of time in the intermediate-temperature range which allows the reprecipitation of carbides. Thus, castings and weld deposits should be allowed to cool as rapidly as possible through the high-temperature range and to cool slowly* through the intermediate-temperature range. The first factor supports the reason why the casting temperatures should be kept as low as possible, and the second, why the severity of embrittlement decreases as the section size of the casting is in-

* Although lower and higher temperatures would also remove the effects of this embrittlement, this temperature range is selected for its beneficial effects on the other metallurgical characteristics of the steel.

creased. However, it should be remembered that rapid cooling through the high-temperature embrittling range is not sufficient to give the alloy its characteristic ductile properties. Thus, thin stainless castings, although they were cooled rapidly, are still extremely brittle.

Causes of Embrittlement

Since the High-Temperature Embrittlement is accompanied by a severe grain growth which was caused by the complete disappearance of the austenite phase as well as the solution of carbides and other possible particles, a relation should seem apparent. During cooling, most of the carbon does not reprecipitate as carbides. Thus, at room temperature, the ferrite phase is supersaturated with carbon. Apparently, the carbon is not uniformly distributed in the ferrite lattice, but is grouped in atomic clusters as has been suggested for certain age-hardening alloys.[79, 80] This hypothesis is supported by the increase found in hardness values after the carbides have gone into solution.†

Annealing between 1350 and 1450° F. (730 and 790° C.) causes the reprecipitation of carbides, and, thus, removes the carbon clusters and, consequently, causes the disappearance of the embrittlement and the increase in hardness values. Therefore, when toughness is a desirable feature, weld deposits on ferritic stainless steels should preferably be annealed in this temperature range.[100,] ‡

NOTCH-SENSITIVITY

Unlike the austenitic stainless steels, the ferritic stainless grades may become highly notch sensitive.[4, 10, 74] In this respect, they may well be compared to mild and low-alloy steels.

At room temperature, this notch effect is reflected by tensile as well as by impact properties. It does not occur in impact tests on specimens without notches. This was brought out by Newell[10] and Heger[143] who showed the effects of a full V-notch* and no notch on the impact properties of a 27% chromium iron. Their test results at various temperatures are shown in Fig. 14. It is apparent that, in this composition, the transition temperature from ductile-to-brittle behavior in notched specimens occurs above 300° F. (150° C.). The effects of notches upon strength and elongation of standard 0.505-in. diameter tensile specimens made from 27% chromium steels are shown in Fig. 15.[143] Thus, although notches raise the tensile and yield strengths, they reduce ductility considerably. The notch has the effect of restraining contraction so that the entire area

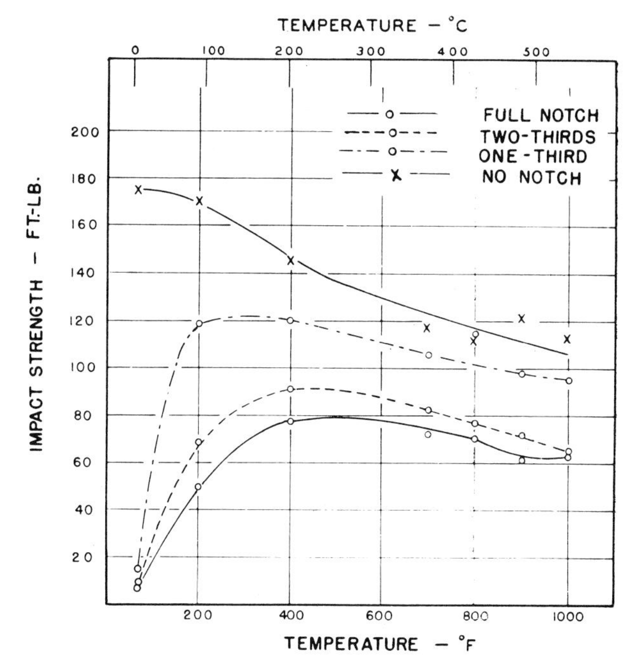

Fig. 14 Effects of testing temperature and notch depth on the V-notch impact properties of a 27% chromium-iron alloy[143]

of the specimen is effective in carrying the maximum load, whereas, without the notch, the specimen undergoes local contraction. In both cases, the tensile strength is figured as the unit stress based on the *original* area.[199]

It should also be remembered that the actual transition-temperature ranges may be somewhat raised by (1) increases in the speed of testing, (2) increases in the sharpness of the notch, (3) increases in the width of the test specimen, and (4) the presence of residual strains in the test specimen.[83, 108] In addition, prior heat treatments may also affect the transition temperatures.

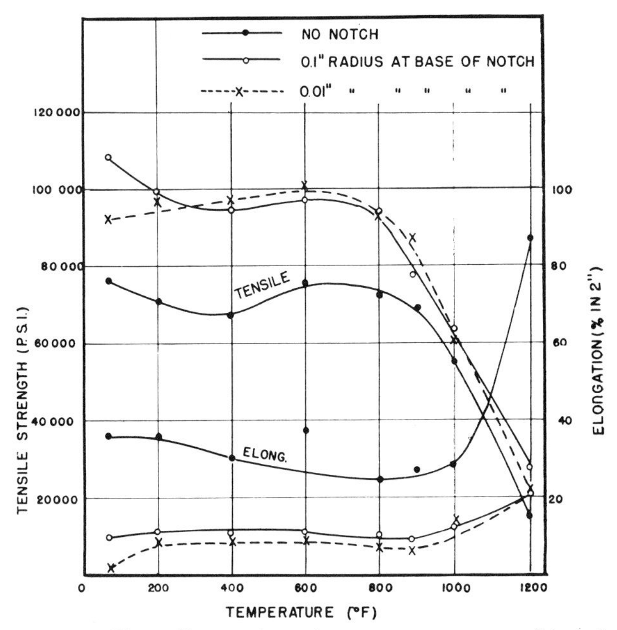

Fig. 15 The effects of testing temperature and notch radius on the short-time tensile properties of a 27% chromium-iron alloy[143]

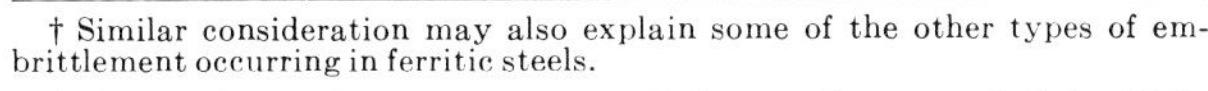

† Similar consideration may also explain some of the other types of embrittlement occurring in ferritic steels.

‡ Effects of annealing temperature and time on the removal of the High-Temperature Embrittlement of a 17% chromium stainless steel were recently shown by Kiefer.[190]

* Full V-notch conforming with A.S.T.M. Designation E23-41T; 1/3 and 2/3 depth notches were produced by machining the notched side of the specimens to approximately 1/3 or 2/3 of the usual dimension. The specimens with no notch had the notch fully removed by machining the 0.315 in. thickness; this leaves the area of the specimens at the base of the notch constant in all tests.[10]

Effects of Chromium Content

In the commercial ferritic chromium stainless steels, the transition from ductile-to-brittle behavior at room temperature occurs in alloys containing between 16 and 18% chromium,[74] Fig. 16.

As the chromium content is decreased, the transition range is shifted to lower temperatures. Results by Lincoln[74] are shown in Fig. 17. They demonstrate quite clearly the effects of chromium in alloys on impact specimens with a modified Charpy notch.

Commercial steels with a chromium content of more than 18% have their transition range above room temperature. Thus, the 27% chromium steels exhibit transition temperatures between 150 and 300° F. (65 and 150° C.), Fig. 14.

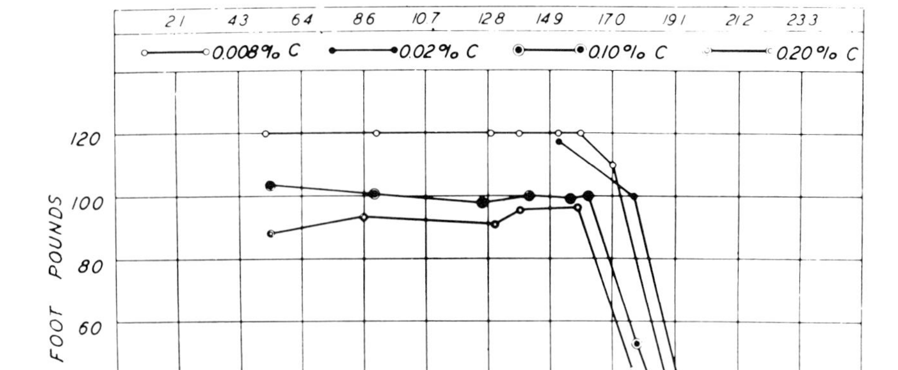

Fig. 16 Effects of variation in chromium and carbon content on the notch-impact toughness of commercial chromium stainless steels[74]

Whereas the above considerations are typical of commercial grades, vacuum-melted alloys of high purity exhibit highly superior properties.

Thus, Binder and Spendelow[152] showed that, in vacuum-melted, chromium-iron alloys, the ductile-to-brittle transition of Izod impact specimens at room temperature occurred only in alloys containing 35 to 40% chromium, Fig. 18. In fact, the impact toughness increased with chromium content up to 25 to 30%. The decrease in toughness of alloys with over 35% chromium may also suggest sigma formation.[199]

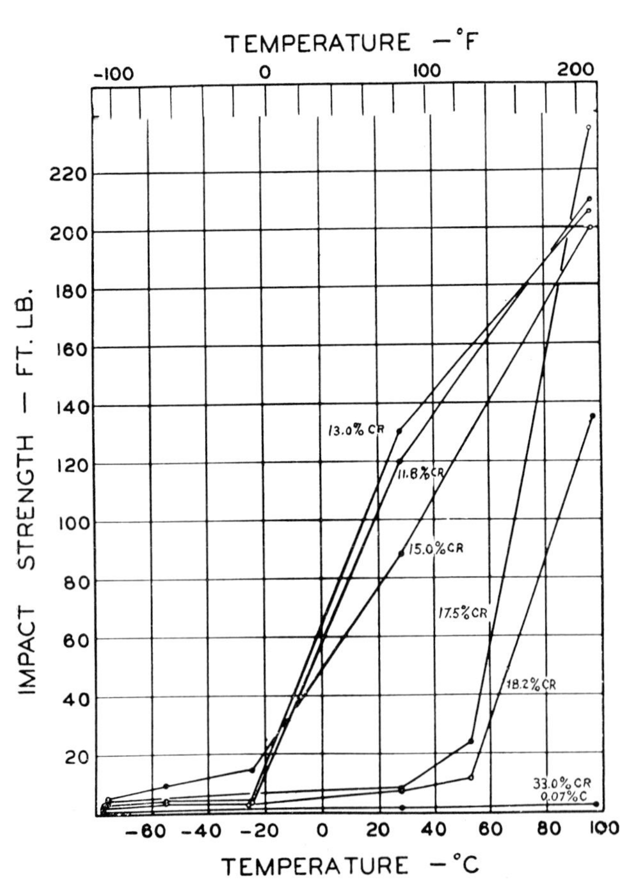

Fig. 17 Notch-impact toughness of various chromium-iron alloys containing less than 0.01% C[74]

Effects of the Carbon and Nitrogen Content

Until recently, it had generally been believed that this severe notch-sensitivity occurred independent of carbon and nitrogen content.[4, 74] However, recently Hochmann[97, 176] and Binder and Spendelow[152] have shown conclusively that, when the total carbon and nitrogen content are reduced to below 0.01%, the impact strength of 25% chromium-iron alloys will be of the order of 80 to 100 ft.-lb. The effects of the total carbon and nitrogen content upon the ductile-to-brittle

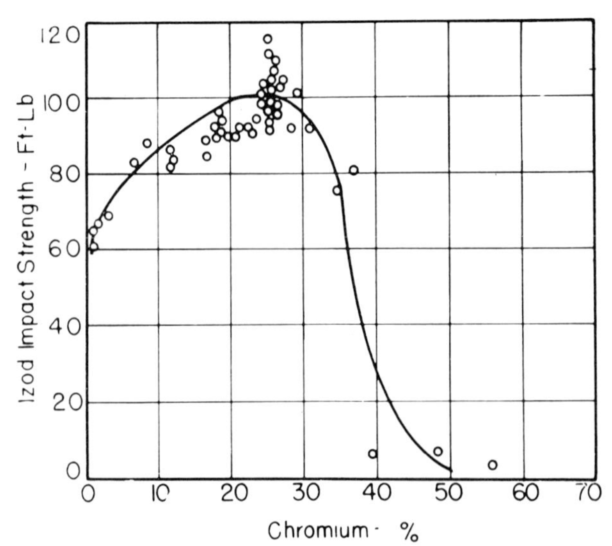

Fig. 18 Effect of chromium content on the impact strength of high purity, vacuum-melted, chromium-iron alloys[152]

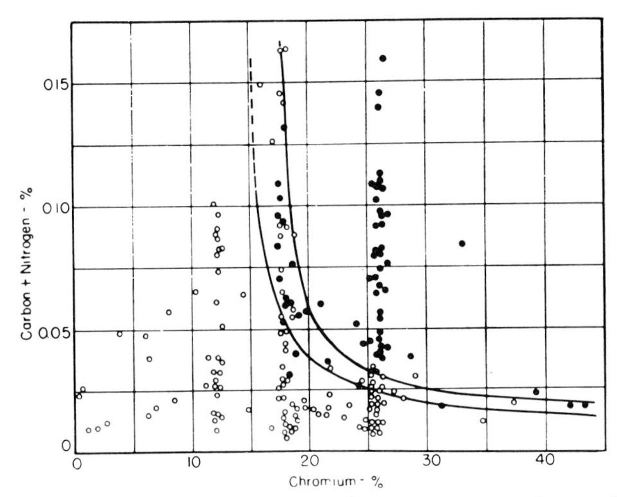

Fig. 19 Influence of carbon and nitrogen on toughness of chromium-iron alloys. Open circles — high-impact strength alloys; solid circles—low-impact strength alloys[152]

transition of Izod impact specimens at room temperature are shown in Fig. 19 for alloys of different chromium content. That the effects of carbon and nitrogen are additive is apparent from Fig. 20.

Unfortunately, at the present time, it is only possible to produce these highly pure alloys by melting in a vacuum or under an inert atmosphere and by using selected raw materials.[152]

It is quite likely that only the uncombined amounts of carbon and nitrogen are effective in exhibiting this strong influence upon transition temperatures. If compounds, such as carbides, nitrides, etc., are formed by these elements, only the remaining free or dissolved quantities of carbon and nitrogen should be held responsible for affecting transition temperatures. On the other hand, it also might seem possible that postulates similar to those which are suggested as one of the explanations of temper brittleness might apply here. If the carbides or nitrides, which precipitate in the grain boundaries, are of a critical size and distribution, brittleness, as shown by the notched-bar test, might be exhibited by the higher chromium stainless steels as well as by other ferritic steels.

If the nitrogen content exceeds 0.08 to 0.10%, the properties of the chromium stainless steels are again improved. Thus, in commercial air-melting practice, where total carbon and nitrogen levels below 0.01% cannot be obtained, the impact strength and ductility of steels containing upward of 20% chromium can be improved by the addition of some 0.10 to 0.25% nitrogen.[63, 152] It may well be that the beneficial effects of the higher nitrogen content are caused by the precipitation of nitride particles. For alloys containing 17 to 19% chromium, the beneficial effects of nitrogen are apparent from Fig. 21. The low-carbon-nitrogen specimens (Zone I) exhibited Izod impact values between 80 and 100 ft.-lb.; the specimens, which by their carbon and nitrogen content belonged in Zone II, generally were extremely notch sensitive and exhibited impact values below 10 ft.-lb.; and the steels of high-nitrogen content (Zone III) showed intermediate impact values between 30 and 50 ft.-lb.[152]

It seems likely that, as the nitrogen (and possibly carbon) content is increased above 0.10%, the tendency of these elements to form compounds increases also. Or, if sufficient quantities of these particles form, the notch-embrittling action of the remaining uncombined quantities of carbon and nitrogen will be reduced.

Although nothing has been said about the effects of phosphorus and oxygen, it is likely that the presence of more than the usual quantities of these elements would also be effective in enhancing the action of carbon and nitrogen.

Fractographic Analysis

Zapffe,[30, 98, 194] in fractographic studies, related the inferior notch-impact resistance to a block-like crystal architecture which offers a minimum of resistance to the propagation of fracture. However, as this crystal pattern does not lower the impact resistance of the unnotched specimens (Fig. 14), Zapffe's explanation does not seem sufficient. Thus, this "block-like crystal architecture" would not appear in fractures of unnotched impact specimens tested at room temperature. If, however, the appearance of the fractures is correlated with triaxial stress conditions, a relationship may be developed between the block-like micrographic fracture pattern (and not the inherent crystal architecture) and the low impact values of the notched impact specimens.

Effects of Grain Size

Grain size has a considerable effect on the tempera-

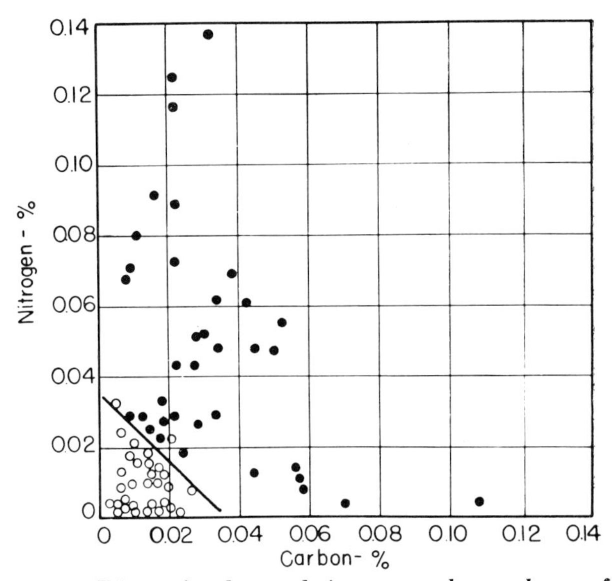

Fig. 20 Effects of carbon and nitrogen on the toughness of 24 to 26% chromium-iron alloys. Open circles—high-impact-strength alloys; solid circles—low-impact strength alloys[152]

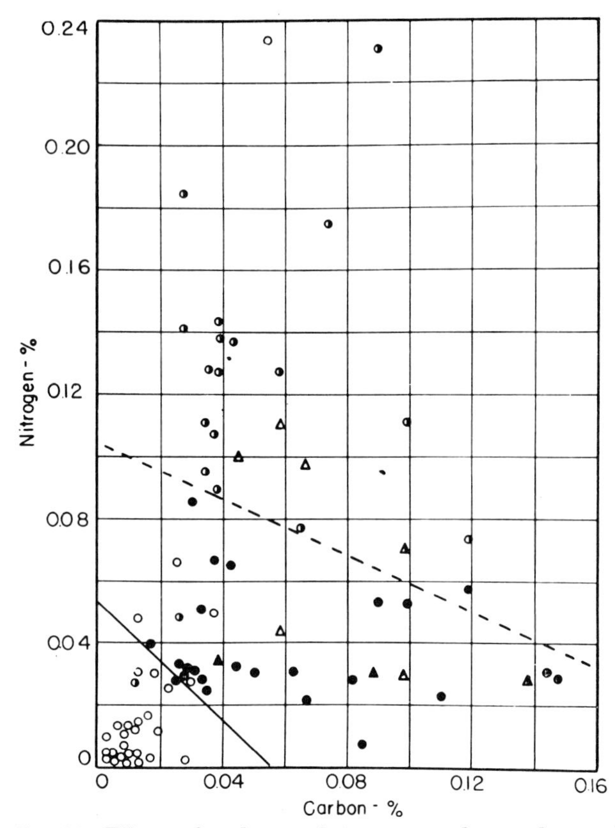

Fig. 21 Effects of carbon and nitrogen on the toughness of 17 to 19% chromium-iron alloys. Open circles—high impact-strength alloys; solid circles—low impact-strength alloys; semisolid circles—intermediate impact-strength alloys; triangles represent commercial arc-melted steels[152] (Zones I, II and III repersent low, medium and high nitrogen content, respectively)

ture range in which transition from high to low notch toughness occurs.[87, 134] This is demonstrated in Fig. 22 which gives data collected by Scherer[134] on an alloy containing 0.10% C, 18% Cr, and 1.8% Mo. The large grain size was obtained by heating the steel at 2100° F. (1150° C.).

This is also important in welding where, because of excessive grain growth, the heat-affected zone may have become highly notch sensitive. If the adjacent parent metal is fine grained, its transition zone may lie below room temperature, and, therefore, the parent metal ordinarily would not be considered to be notch sensitive. Of course, these considerations primarily depend upon the composition of the alloy.

Effects on Applications of High-Chromium Stainless Steels

At temperatures which lie above the transition range in which the change from ductile-to-brittle behavior occurs, the high-chromium stainless steels exhibit their natural ductility and toughness. Thus, the high-chromium stainless steels find their major application at elevated temperatures.[112]

In room-temperature applications, the high-chromium stainless steels should be used only when notch-sensitivity is not a factor in the use of the material.

Effects on Design

Designs and parts made from these notch-sensitive stainless steels should be simple and eliminate as much as possible notches and other surface irregularities. Thus, key-ways, re-entrant angles, and sharp radii should be avoided whenever possible.[10] Scratches and other surface defects can also reduce toughness considerably.

Effects on Forming

Forming operations should preferably be performed at temperatures above the transition curve. Thus, depending upon composition, temperatures between 400 and 600° F. (205 and 315° C.) are most suitable,[53, 129, 134, 143] and drastic forming and working operations may be applied satisfactorily. Lower temperatures may be satisfactory for forming operations on relatively thin sheets. Thus, on 17% chromium stainless steel sheets, temperatures above 140° F. (60° C.) have proved to be satisfactory.[174] More highly alloyed steels would again require higher temperatures.[174]

Many of the lighter forming operations, such as stamping, are performed at ordinary atmospheric temperatures or on steels which have been heated in boiling water. For example, spinning and drawing operations of round and rectangular television tube envelopes, automobile grill parts, and automotive trim parts are now being fabricated from stainless steels which have not received previous heat treatments.

GRAIN SIZE

Although some grain growth may occur on heating the alloys to temperatures above 1700 to 1800° F. (925 to 980° C.), the rates do not become appreciable until temperatures are reached at which a fully ferritic structure is obtained. Moreover, the inhibiting compound particles (usually carbides) must be dissolved in the ferritic matrix.

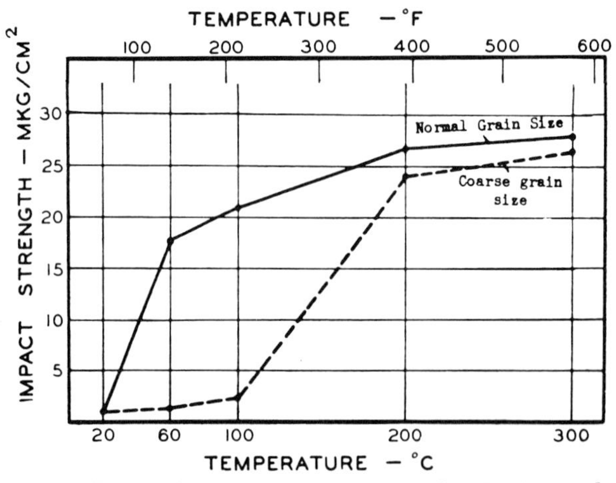

Fig. 22 Effects of testing temperature and grain size on the notch-impact toughness of an 18% chromium stainless steel[134]

The martensitic grades containing 11.5 to 12.5% chromium and over 0.10% carbon (Type 410) are, therefore, unlikely to experience severe grain growth. Although (delta) ferrite will form from the austenite as these alloys are heated to temperatures above 1900° F. (1040° C.), some austenite will remain even as high as 2500° F. (1370° C.). However, Type 410 alloys of lower carbon and/or higher chromium content may become fully (delta) ferritic at lower temperatures. Thus, a Type 410 (modified) grade containing 12.2% chromium and 0.06% carbon became fully ferritic at 2400° F. (1315° C.). Raising the chromium content to 12.7% (same carbon) would make the grade fully ferritic at 2300° F. (1260° C.).

The addition of ferrite formers such as aluminum, molybdenum, etc., or a higher chromium content will reduce the temperature at which the matrix becomes fully ferritic further and, if sufficient quantities of the ferritizer are present, the alloy may never become austenitic. In these alloys, the solution of carbides and other well-dispersed compound particles will determine the temperature at which appreciable grain growth occurs. Depending on the composition, the chromium carbides will go into solution between 2100 and 2300° F. (1150 and 1260° C.).

The time element, although it has little effect on the solution temperature, is an important factor in the determination of the subsequent degree of grain growth. Thus, induction-heating tests,[61] in which specimens were above 2100° F. (1150° C.) for only a few seconds, showed that the carbides had been completely dissolved and that some grain growth had occurred. Longer exposure would increase the grain size severely, as carbide solution is a prerequisite to this grain growth. The severity of the High-Temperature Embrittlement depends upon the carbide solution and not on the grain growth.

Alloying elements may effect the solution temperature of the carbides in two ways. First, they may form carbides themselves which dissolve at higher temperatures than do the chromium carbides. This is accomplished by columbium, titanium, and tantalum,[7–9, 17, 18, 62] which, therefore, are effective additions since they are of a particular benefit in grades which are to be welded. Other elements, although they do not form carbides themselves, may raise the temperature at which the chromium carbides go into solution. This is believed true in the case of molybdenum. Aluminum, between 0.15 and 0.30%, does not seem to have noticeable effects on the solution temperatures. However, in larger amounts (above 1%), aluminum even seems to decrease these temperatures.

EFFECTS OF ELEVATED TEMPERATURES

Effects of Hot Working

The temperatures at which rolling or forging operations are performed seem to exert a considerable influence on the mechanical properties, particularly the impact toughness.

For example, Schaufus[34] reported the results of tests on a free-machining stainless steel (0.07% C, 0.31% S, 0.76% Ni, 16.25% Cr) that had been rolled from 3-in. billets to 1 in. between 2210° F. (1210° C.) and 1700° F. (925° C.). The specimens exhibited Izod impact values varying from 6 to 12 ft.-lb. However, an initial temperature of 1925° F. (1050° C.) and a finishing temperature of 1000° F. (540° C.) resulted in an Izod impact toughness between 41 and 46 ft.-lb.

Figure 23 gives data by Schiffler and Hirsch[94] who showed the effects of finishing temperature upon impact toughness of a 17% chromium stainless steel (0.08% C, >0.5% Ti). These results indicate that the finishing temperatures should lie below 1650° F. (900° C.) in order to obtain a satisfactory toughness.

STRESS-CORROSION CRACKING

Generally, the martensitic and ferritic stainless steels are considerably less susceptible to stress-corrosion cracking than the austenitic chromium-nickel stainless steels.[3] For example, studies by Scheil, Zmeskal, Waber, and Stockhausen[166] showed that the martensitic

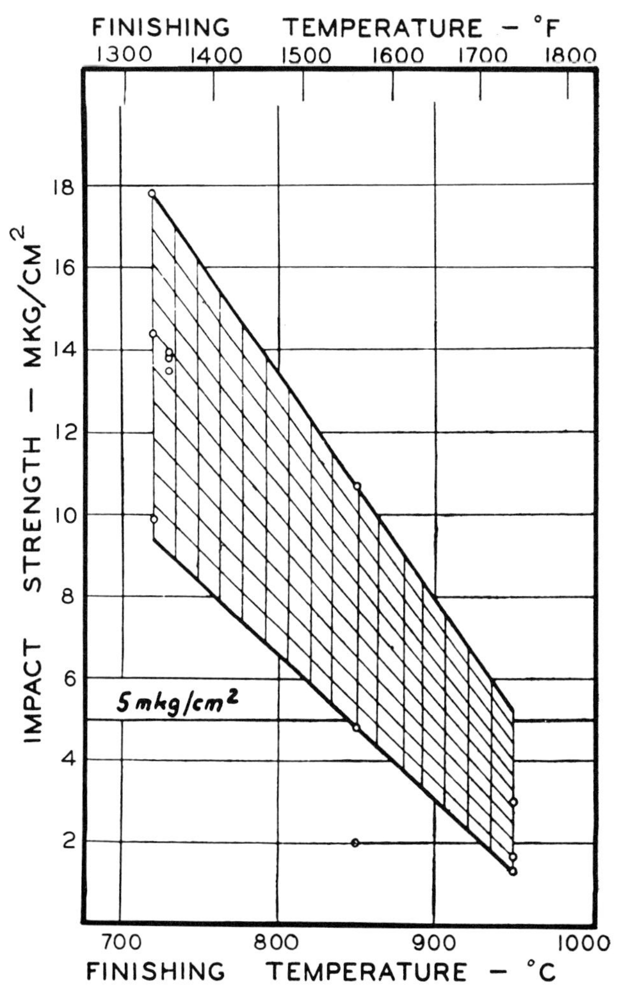

Fig. 23 Effect of the final rolling temperature on the notch-impact toughness of a 17% chromium stainless steel[94]

12% and the ferritic 18% chromium stainless steels were not susceptible to stress-corrosion cracking in those boiling chloride solutions which attacked the austenitic chromium-nickel stainless grades. On the other hand, steam has been found to be responsible for stress-corrosion cracking of martensitic stainless steels. Thus, Nathorst[169] has described such failures in steam turbine blades made of steel containing 0.14% carbon and 14.1% chromium. The cracks were intercrystalline and tended to follow the boundaries between the ferrite grains and the tempered martensite. The main direction of the cracks was at right angles to the ferrite streaks. These results may well be related to the embrittling action on steel of hydrogen which is derived from the steam atmosphere.[164, 168]

Similarly, Uhlig[27] reported that, in martensitic stainless steels, stress-corrosion cracking may occur under any condition that favors the discharge of hydrogen ions on the surface. He suggested that hydrogen, upon entering the alloy lattice, induces stresses of a magnitude sufficient to initiate a crack which, when aided by existing stresses, propagates itself throughout the cross section of the specimen.

Uhlig[27] reported further that the martensitic stainless specimens, after a brief cathodic polarization in NaCl or H_2SO_4, were exceedingly brittle. However, on standing in air these specimens regained their normal ductility. This behavior is typical of hydrogen embrittlement.[27]

II. Effects of Alloying Elements

ALUMINUM

Aluminum additions are made to stainless steels to serve one or several functions: (1) to serve as a ferrite former, (2) to improve scaling resistance, (3) to cause grain refinement, or (4) to develop desirable electrical properties.

Effect as a Ferrite Former

Of the various elements which stabilize ferrite, aluminum is the most powerful. Thus, most authors[10, 65] find aluminum about 10 to 15 times as effective as chromium in its tendency to form ferrite.

From 0.10 to 0.30% of aluminum are added commercially* to low-carbon (0.08% C max.) stainless steels containing between 11.5 and 13.5% chromium.[28, 124, 132] The aluminum addition will make the normally martensitic grades highly ferritic.

The small aluminum additions to the 11.5 to 13.5% chromium stainless steels do not reduce the toughness[4] noticeably. Moreover, as these steels are not notch sensitive at room temperature, they may be useful in many applications in which the ferritic stainless steels of higher chromium content are unsuitable. Although some authors[124] feel that these steels are not susceptible to 885° F. (475° C.) Embrittlement, it seems that on long exposures at 885° F. (475° C.) some embrittlement may occur nevertheless. The susceptibility to this embrittlement is more appreciable for the alloys of higher chromium content: e.g., near 13.5% chromium. For example, many cases have been encountered[181] where 11.5 to 13.5% chromium stainless steels embrittled in petroleum-refinery applications where exposure was to temperatures of about 825° F. (440° C.) [885° F. (475° C.) Embrittlement]. In every case, the material had either a chromium content somewhat above 13% or it contained aluminum (that is, it was Type 405). Such evidence is generally based on very extended exposures on the order of 10,000 hr. or more.

These 11.5 to 13.5% chromium steels are not susceptible to sigma-phase embrittlement which occurs at higher temperatures. They consequently have found use in oil-field equipment operated at elevated temperatures.

Because all ferritic stainless steels seem susceptible to the High-Temperature Embrittlement, the aluminum-bearing steels should not be heated to temperatures exceeding 2200° F. (1205° C.). Castings or welded sections should be annealed between 1350 and 1450° F. (730 and 790° C.) to remove the detrimental effects of the embrittlement.

Schiffler[64] suggested further additions of silicon and titanium to produce alloys containing 11–14.5% Cr, 0.15–0.5% Al, 0.6–1.2% Si, and 0.1–0.5% Ti. The total amount of aluminum and silicon was not to exceed 0.9 to 1.5%. The titanium addition served to produce a fine grain size in this fully ferritic alloy. The addition of about 0.50% molybdenum to 12% chromium-0.27% aluminum alloys (Type 405) does not seem to be beneficial to the mechanical properties.[183]

Effects Upon Scaling Resistance

Aluminum additions of one or more per cent are primarily made to alloys used at temperatures above 1800° F. (980° C.).[75, 118, 125, 207] Because the presence of aluminum considerably improves the scaling resistance at elevated temperatures, aluminum-bearing stainless

* A.I.S.I. Type 405.

steels are particularly suitable for service in atmospheres supporting oxidation or contaminated with sulphur.[93, 155] In such atmospheres and at temperatures between 1470 and 2280° F. (800 and 1250° C.), the aluminum-bearing chromium stainless steels are considerably superior to the silicon-bearing alloys of similar chromium content.[174] Specifications for standard German steels are given in Table 7.

The presence of aluminum also reduces susceptibility to carburization. The improved resistance to oxidation and carburization is utilized by the use of 20–2 (Cr-Al) alloys in acetic anhydride applications.

Although stainless steels containing several per cent of aluminum exhibit superior scaling resistance, the physical properties of these alloys are generally affected detrimentally. For example, these alloys tend to be extremely brittle.[35] Creep strength at elevated temperatures is also reduced considerably.[125] The primary reason for the low mechanical properties is due to the strong affinity of aluminum for oxygen and nitrogen, if they are present. Thus, the high-aluminum-bearing steels usually contain numerous alumina inclusions which weaken the alloy seriously and allow pitting in corrosive environments.[125] Hot-working or fabricating operations are similarly affected.

However, when these high-aluminum-bearing chromium stainless steels are properly made, they may exhibit good mechanical properties. According to Foley,[199] the method of adding aluminum in the large amounts required must be one which prevents the excessive formation of alumina. In the production of a steel used for exhaust values (0.40% carbon, 11% chromium, 1.25% nickel, and 1.5% aluminum), the best method is to melt the aluminum under a cryolite protective covering and then pour molten steel through the cryolite.[199] When these alloys are made this way, recovery is high and the resulting bars are free from excessive alumina inclusions. If the steel is rolled improperly, it can become very brittle.[199] For example, if the steel bars were heated only to 2000 or 2050° F. (1095 to 1120° C.), not higher, they would come out of the finishing pass in the rolling mill at around 1600° F. (870° C.) or below. This treatment produced soft bars which could be cold sheared to valve stem lengths without annealing, but which were unsatisfactory. When the stem lengths were heated for upsetting, excessive grain growth occurred at some point along the temperature gradient. At some time during subsequent handling, the stem would break just under the head of the valve and would show grains which were 1/4 in. and larger—sometimes the full section of the stem would be composed by only two grains. Thus, this method of rolling, which saved the expense of an anneal, had to be abandoned and a method utilizing higher rolling temperatures was used. With higher rolling temperatures, bars came off the rolls hard and had to be annealed before they could be sheared. However, no grain growth occurred as a result of the subsequent heating for upsetting.[199]

One bad feature of these steels is the difficulty encountered in making use of their scrap.[199] Unless practically all of the aluminum is oxidized out when the scrap is remelted, the resulting steel is likely to pull and be hot short.[199] Properly treated scrap can be used in the production of high-frequency, induction-furnace alloys which contain aluminum and for castings in which aluminum can be tolerated. Obviously, the use of this scrap is definitely limited.[199]

Effects on Electrical Properties

The addition of several per cent of aluminum to stainless steels provides certain desirable electrical properties.[204, 206] The relatively poor creep resistance, mechanical properties, and corrosion resistance of such grades are of no consequence in the electrical applications in which these alloys serve commendably.

Effects of Grain Size

Aluminum in amounts up to 0.05% is extremely effective in producing fine-grained castings and weld deposits. In larger amounts, aluminum additions become less effective in refining grain structure and, in fact, begin to cause grain coarsening. Thus, castings which contain over 1% aluminum usually exhibit an extremely coarse grain size.

Effects on Structure

A chemical analysis of the aluminum present in a particular alloy may not give a true indication of the actual

Table 7—Compositions and Welding Recommendations of Standard German Aluminum-Bearing, Wrought Stainless Steels[184]

Steel No.*	Composition, % C	Si	Cr	Ni	Al	Welding recommendations: Electrode classification	
4713†	<0.12	0.6–0.9	6.3– 6.8	...	0.6–0.9	4716, 4828	Postheat treatment not necessary
4724†	<0.12	1.0–1.3	12.5–13.5	...	0.8–1.1	4723, 4828	Postheat treatment not necessary
4742†	<0.12	0.8–1.1	17.5–18.5	...	0.8–1.1	4772	Postheat treatment at 1380° F. (750° C.) and air cool
						4828	Postheat treatment not necessary
4762†	<0.12	1.3–1.6	23.0–25.0	...	1.3–1.6	4772	Postheat treatment at 1380° F. (750° C.) and air cool
						4842	Postheat treatment not necessary

* Classification of Verein Deutscher Eisenhüttenleute.[185]
† Recommended for electric-arc, gas, spot and (roll) seam welding. Not recommended for electrodes required for atomic-arc welding.

location of the ferrite-austenite equilibrium region. In the melt, aluminum exhibits a strong affinity for nitrogen and tends to crystallize out as aluminum nitrides. Moreover, such crystals may form or may grow when the solid alloy is heated in nitrogen or even in ordinary atmospheres at relatively high temperatures for long periods of time; as, for example, 1000 hr. at 2200° F. (1205° C.).[101, 102] These nitrides are stable even at welding temperatures.

In addition to nitrides some aluminum will be retained in the form of oxides. In fact, almost all of the chromium stainless steels contain about 0.015 to 0.025% aluminum in the "combined" form, that is, as oxides. Of course, only the "free" aluminum, which is in solution, is effective in stabilizing ferrite.

Effects of Welding

Since aluminum is readily oxidized, stainless electrodes used in the electric-arc and oxyacetylene processes generally do not contain aluminum additions. A considerable quantity of the aluminum, if present, would be lost as slag. However, the shielded-arc processes, which protect the weld metal by the use of atomic hydrogen or inert gases, would prevent the loss of aluminum.

The low-aluminum Type 405 alloys (0.10 to 0.30% Al) are ordinarily welded with 18–8 (Cr-Ni) type electrodes.

The heat-resisting, high-aluminum types (0.6 to 2.2% Al) are generally welded with silicon-bearing ferritic or austenitic electrodes. Recent German welding recommendations for standard wrought steels are summarized in Tables 7 and 8. The absence of aluminum in the electrodes is compensated for by the larger quantity of chromium. Arc welding is preferred to gas and resistance welding since it minimizes the exposure times at welding temperatures.[184]

BISMUTH

As in the austenitic chromium-nickel stainless steels,[2] bismuth additions also increase the machinability of the chromium stainless grades.[31] In welding, it can also be assumed that the bismuth, present in the electrode wire, would tend to volatilize quite readily.

CARBON

In the chromium stainless steels, for a given chromium content, carbon is the principal element which affects the extent of the gamma loop.

With the exception of the cutlery steels, the carbon content in stainless steels generally does not exceed 0.20% in the high-chromium grades. In the lower chromium alloys, the carbon content generally is kept below 0.15%. Some chromium stainless grades even contain less than 0.08% carbon. The primary reason is that, by restricting the formation of austenite, hardenability will be prevented.

In the absence of other carbide formers (titanium, columbium, etc.), carbon tends to combine with chromium to form the chromium carbides Cr_7C_3 and Cr_4C (now taken to be $Cr_{23}C_6$). Actually, because iron is generally dissolved in the carbides, the formulas are generally expressed as $(CrFe)_7C_3$ and $(CrFe)_4C$, respectively.[24]

At elevated temperatures up to 1000° F. (540° C.), carbon somewhat improves the short-time tensile properties and the creep resistance. This does not hold true above 1000° F. (540° C.) where the carbides tend to coalesce.

Carburization of ordinary low-carbon stainless steels (15 to 30% chromium) is of primary importance when good wearing properties of the surface are desired, and the ductility and toughness, characteristic of the low-carbon stainless core, are to be retained. Greatest surface hardness is obtained by quenching carburized steels from about 1525° F. (830° C.). Higher temperatures are undesirable because the resulting retention of austenite would reduce the hardenability and wearing characteristics of the carbide case.[117] This process, however, does not seem to be used industrially. Carburization also decreases the corrosion resistance.[174]

Effects in Martensitic Stainless Steels

In the martensitic stainless steels* which contain about 12% chromium, the effects of carbon and of tempering treatments at 600, 1200, and 1500° F. (315, 650, and 815° C.), respectively, are shown in Fig. 24.[157] Whereas 1200° F. (650° C.) is sufficient to allow the softening of the martensite of steels containing less than 0.1% carbon, higher temperatures would be necessary to allow the softening of the alloys of higher carbon content. Complete softening is accomplished only at temperatures between 1350 and 1450° F. (730 and 790° C.). This range varies somewhat with composition. Thus, whereas the alloys of lower carbon content may also be softened at 1500° F. (815° C.), Fig. 24, the alloys of higher carbon content require the more limited

* True in wrought steels as well as in castings and weld deposits.

Table 8—Electrodes Recommended for Welding Aluminum-Bearing, Chromium Stainless Steels[184]

Electrode classification	*Composition, %*				
	C	*Si*	*Cr*	*Ni*	*Other*
4716	<0.12	0.6–0.8	8.5–9.5	...	...
4723	<0.12	1.3–1.8	14.0–15.0	...	...
4774	<0.12	1.3–1.6	28.0–30.0	...	...
4828	0.10–0.20	1.8–2.3	19.0–20.0	9.0–10.0	...
4842	0.10–0.20	0.9–1.2	23.0–25.0	19.0–20.0	2.0–2.3 Mn

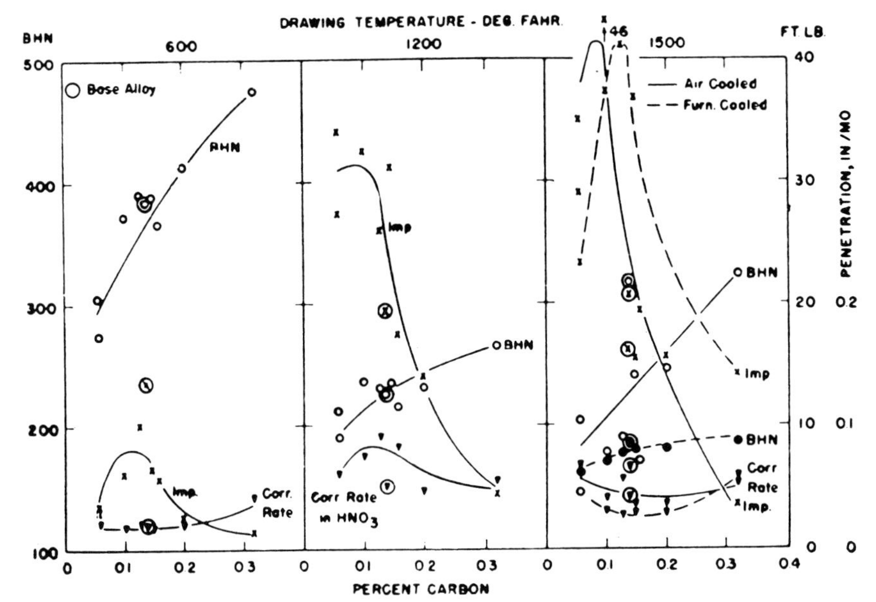

Fig. 24 Effect of carbon content on mechanical properties and resistance to corrosion in boiling 65% nitric acid of martensitic 12% chromium stainless castings[157]

range of 1350 to 1450° F. (730 to 790° C.). These considerations are particularly important when toughness is of primary concern.

The cutlery steels, which generally contain over 30% carbon, are not readily weldable.[178] They exhibit high-air-hardening characteristics and are susceptible to cracking. Preheating between 600 and 800° F. (315 and 425° C.) is required in most cases (see also pages 244-s to 246-s).

CHROMIUM

The commercial importance of chromium is primarily due to the beneficial effects of this element upon corrosion and scaling resistance.[4, 98, 158]

The effects of chromium upon the mechanical properties of various chromium-iron alloys are shown in Fig. 25.[125] The alloys were annealed in order to introduce favorable over-all physical characteristics. Thus, martensite in the hardenable chromium alloys was softened to ferrite. Moreover, in the ferritic grades, the 885° F. (475° C.) Embrittlement was prevented. Sigma-phase precipitation did not occur in the 25% chromium alloy, because, in this grade, sigma generally forms only on prolonged heating. The notch-impact toughness of the higher chromium alloys at room temperature is extremely low. This was discussed earlier.

Whereas at elevated temperatures chromium acts as a ferrite former, at room temperature chromium seems to stabilize austenite. This is due to the fact that as the chromium content is increased transformation becomes more sluggish. Thus, quenched, high-chromium steels (over 25% Cr), which, because of the presence of austenite formers such as nickel, nitrogen, and/or manganese, formed austenite above 1800° F. (980° C.), tend to retain some of the austenite at room temperature.

The coefficient of expansion of the chromium stainless steels is about 10% less than that of the carbon steels, and 50 to 70% less than that of the austenitic chromium-nickel stainless steels. Moreover, the coefficient of expansion decreases as the chromium content is increased.[110]

COLUMBIUM

Columbium, as does titanium, exhibits a strong affinity for carbon and tends to form carbides. Moreover, columbium acts as a potent ferrite stabilizer. Both of these effects tend to restrict austenite formation. Because of these characteristics, columbium, in amounts about 4 to 8 times the carbon content, has been suggested as an addition to ferritic stainless steels[130] and to welding electrodes.[119] However, commercial use of columbium additions to the chromium stainless steels is not made.

Effects on Physical Properties

The addition of columbium to martensitic stainless steels seems particularly undesirable. Results by Schoefer[157] on the effects of 0 to 1.28% columbium on

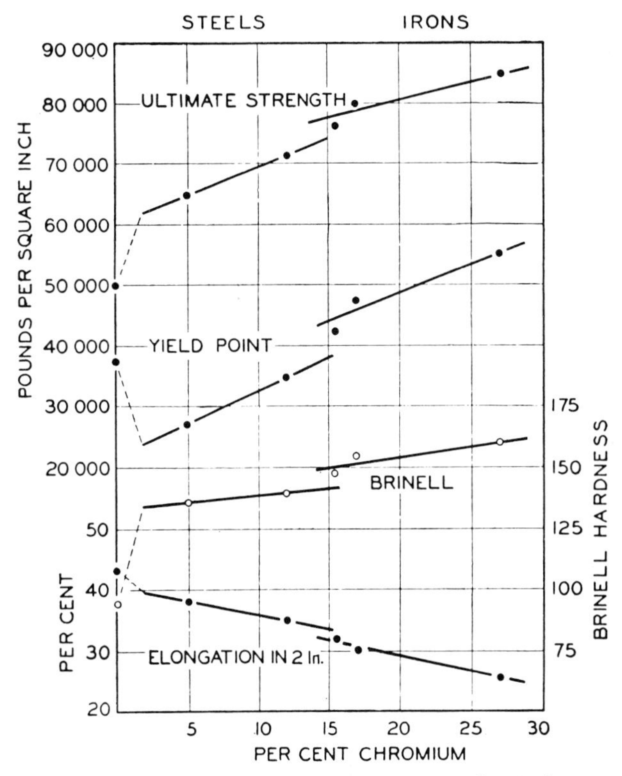

Fig. 25 Effect of chromium content on mechanical properties at ordinary temperature of annealed iron-chromium alloys containing approximately 0.10% carbon[125]

the hardness and toughness of 12% chromium steels are shown in Fig. 26. The addition of 1.28% columbium made the alloy almost fully ferritic. Tensile strength decreased from about 200,000 psi. for the alloy which contained no columbium to 40,000 psi. for a columbium content of 1.28%.

In chromium stainless steels which are used at elevated temperatures, the presence of columbium is also undesirable because it is believed to promote fatigue failures at elevated temperatures considerably.

Effects on Welding

In weld deposits made from ferritic stainless electrodes which contain over 20% chromium, as much as 1.25% columbium does not improve the microstructure by the elimination of dendritic solidification of the weld metal. Titanium, on the other hand, in quantities of as little as 0.35% is beneficial and completely eliminates dendritic microstructures.[145]

Columbium carbides go into solution at temperatures above those at which chromium carbides dissolve. Thus, columbium-bearing, ferritic chromium stainless steels, are less susceptible to the High-Temperature Embrittlement and may be arc welded without subsequent postheat treatments.[78] (See discussion on titanium.)

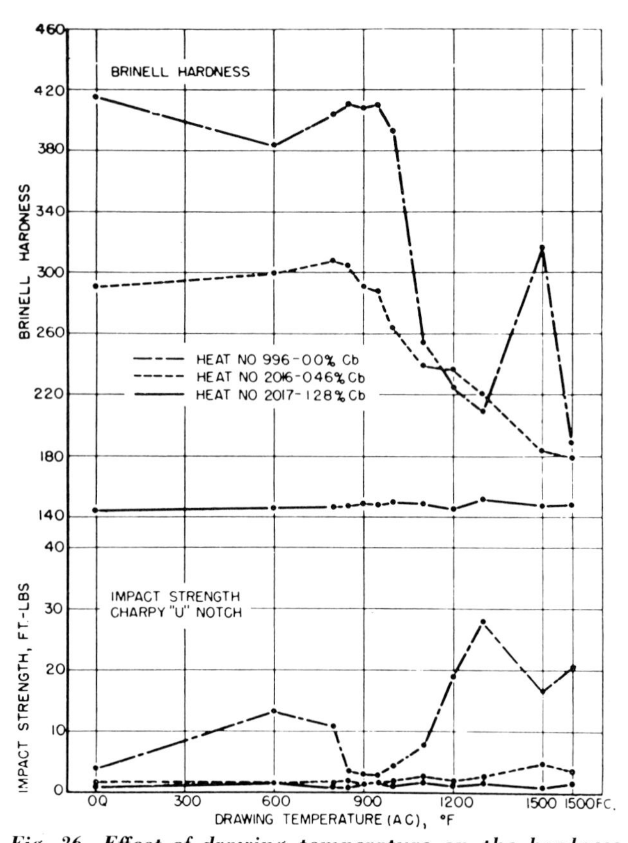

Fig. 26 Effect of drawing temperature on the hardness and impact toughness of martensitic 12% chromium stainless castings with varying columbium content[157]

Effects on Corrosion Resistance

Since ferritic chromium stainless steels may become susceptible to intergranular corrosion in certain corrosive environments, the addition of columbium, which acts as a carbide stabilizer by forming columbium carbides, may be beneficial.

Stabilization is believed[174] to be particularly important in the welding of ferritic stainless steels in which the heat-affected zone becomes highly susceptible to intergranular corrosion.

HYDROGEN

A great many investigations have been made in recent years on the effects of hydrogen in mild and alloy steel weldments.[197] However, very little has been published on the role played by hydrogen in martensitic and ferritic stainless steels. In part, this neglect may be ascribed to the fact that the martensitic and ferritic stainless steels are commonly welded with austenitic chromium-nickel electrodes which are generally not susceptible to hydrogen embrittlement. Similarly, dry martensitic or ferritic chromium stainless arc-welding electrodes are likely to produce weld deposits free from hydrogen embrittlement.

Circular groove tests recently made by Steinberger, De Simone, and Stoop[195] showed that hydrogen may cause embrittlement in Type 410 martensitic stainless weld deposits on modified S.A.E. 4320 material. Atomic-hydrogen welds were found to be susceptible to cold cracking when no postheat treatment was used or when cooling, arrested at 350 or 700° F. (177 or 370° C.), was immediately followed in each case by a heat treatment of 1 hr. However, arresting of cooling at 600° F. (315° C.) and postheating for 1 hr. at 600° F. (315° C.) prevented cracking in the circular groove test.

The observation that a 1-hr. postheat treatment was beneficial at 600° F. (315° C.) and not at 700° F. (370° C.) may well be related to the presence of residual austenite in the weld metal (see discussion on page 245-s). In these martensitic stainless steels, the austenite-to-martensite transformation is relatively sluggish. Thus, after 1 hr. (or possibly longer), the Type 410 weld deposit probably contained some retained austenite. Since hydrogen is more soluble in austenite than in ferrite, it concentrates in higher proportions in the austenite. When the small austenite patches transform at room temperature to martensite, the hydrogen is released locally, and, because the hydrogen cannot diffuse away at sufficiently rapid rates, the local internal pressures, caused when small hydrogen concentrations exist, are generally sufficient to cause cracking in the weld metal.[196] At 600° F. (315° C.), the diffusion rate of hydrogen is estimated to be about 500 times as fast as at room temperature.[196] Thus, since after 1 hr. at 600° F. (315° C.) less (if any) austenite was retained than was after 1 hr. at 700° F. (370° C.), more hydrogen could diffuse from the weld deposit. Consequently when the balance of the austenite (if any was still present) transformed at room temperature, the resulting hydrogen concentrations were not sufficiently severe to cause cracking. Cooling the weldment to 350° F. (177° C.) and postheating for 1 hr. apparently did not allow sufficient hydrogen to leave this Type 410 weld deposit in 1 hr. Consequently cold cracking could occur. Although these results were obtained in weld deposits produced by atomic-hydrogen welding, martensitic deposits made from wet coated arc-welding electrodes are also likely to exhibit hydrogen embrittlement.

The use of argon-arc or submerged-arc welding procedures prevented or minimized hydrogen absorption by the weld metal. Similar results were obtained with lime-coated (low-hydrogen) electrodes. The weldments made by these latter processes were not susceptible to cold cracking and, thus, did not require arrested cooling cycles and postheat treatments.

MANGANESE

Manganese, like nickel, acts as an austenite former. At ordinary temperatures, manganese is somewhat beneficial to the toughness of martensitic and ferritic stainless steels.[47] It increases creep strength slightly at moderately elevated temperatures. However, above 950° F. (510° C.), manganese has little effect on creep resistance.

As in the austenitic stainless steels, manganese also improves the hot-working characteristics of the chromium stainless steels.

The oxidation resistance is not improved by manganese.[125] In fact, the presence of several per cent of manganese seems to increase oxidation rates.[120]

The corrosion resistance also seems to be reduced by manganese additions.[5]

Effects on Structure

Manganese expands the gamma region at elevated temperatures. Thus, although an alloy containing 0.10% carbon, 23 to 24% chromium, and 3% manganese is fully ferritic, 5 to 7% manganese makes the alloy partially austenitic above 1700° F. (925° C.). In this respect, the effects of manganese are similar to nickel. However, the manganese-bearing austenite is not as stable as the nickel-bearing austenite. Thus, in the high-chromium stainless steels containing 23 to 24% chromium and up to 15% manganese, all or at least part of the austenite transforms into martensite.[46] Nickel, under equal conditions, would inhibit this austenite-to-martensite transformation.

Nitrogen additions to manganese-bearing chromium stainless steels also would help in the stabilization of the austenite.[5]

Effects on Welding

Electrodes containing about 1% manganese in the core wire tend to produce a somewhat higher manganese content in the weld deposit. This is due to the presence of some manganese compounds in the electrode coating. However, in electrodes containing from 3 to 7% manganese, about $^1/_{10}$ less manganese will be present in the deposit—in spite of the small gain from the coating.[38]

The presence of small amounts of manganese (1 to 2%) affects the weldability beneficially,[38, 91] particularly in martensitic grades.[91] Thus, manganese, as a ferro-alloy in the electrode, improves striking the arc, melting of the electrode, formation of the arc and welding pool, formation of the weld bead, and multibead welding.[38] (Small amounts of silicon may accomplish the same results more effectively.)

A higher manganese content (4 to 5%) in electrodes, which do not contain other alloying elements such as nickel and nitrogen, produces unsatisfactory results.[38] Depending upon the chromium content, the weld deposits will exhibit a ferritic-martensitic structure. They are extremely crack sensitive at the grain boundaries because of the high coefficient of expansion of the manganese-bearing austenite which, on cooling, transforms into martensite.[38, 71] The ductility of the higher manganese-bearing chromium stainless electrodes is considerably improved as the nickel content is raised to above 5.5%,[203] i.e., until the composition is such, that the weld deposit remains austenitic down to atmospheric temperatures.

Moreover, in the almost fully ferritic chromium-manganese alloys, the heat effects produced by welding alone are sufficient to cause considerable 885° F. (475° C.) Embrittlement.[38]

MOLYBDENUM

Molybdenum, in addition to stabilizing ferrite, is also effective in improving to a considerable extent the corrosion resistance of the martensitic and ferritic chromium stainless steels.[59, 81, 92, 104, 126, 134, 153, 171, 174, 198] In this respect, molybdenum acts in the same way as it does in austenitic stainless steels.[2]

As a ferrite former, molybdenum is about four times as effective as chromium. Thus, 2% molybdenum suppresses, almost completely, the formation of austenite in alloys containing over 11% chromium.

Martensitic Grades

The molybdenum-bearing martensitic stainless steels are primarily used in turbine blades, spindles, valves, etc., in contact with nitric and organic acids.

In partially martensitic grades containing 12 to 15% chromium, it is claimed[4] that the presence of about 0.5% molybdenum refines the grain size. In these steels (13% chromium), molybdenum also improves somewhat the creep strength at elevated temperatures[174] and the room-temperature impact toughness of alloys tempered between 900 and 1600° F. (480 and 870° C.).[157]

Ferritic Grades

The molybdenum-bearing ferritic stainless steels are considerably used in the textile and chemical industries where they exhibit a higher corrosion resistance than many other alloys in exposures to fatty and many other organic acids. Moreover, a high-chromium alloy (<0.15% C, 24–26% Cr, 2.3–2.6% Mo, 1.5–2.0% Ti) is recommended for applications in solutions containing chlorine, for example, calcium hypo-chloride, perchlorox, etc.[187]

In wrought ferritic stainless steels, intergranular corrosion is generally minimized by the addition of a carbide-stabilizing element such as titanium.

Since ferritic stainless castings alloyed with molybdenum show a large grain size and considerable brittleness,[40] they should receive a final heat treatment between 1300 and 1470° F. (705 and 800° C.).[188]

About 1% molybdenum seems to give the optimum beneficial resistance to creep at 1100 and 1200° F. (595 and 650° C.).[40] However, many commercial alloys contain somewhat higher percentages of molybdenum since their superior corrosion resistance to many environments is more important than their elevated temperature strength.

Effects of Welding

The martensitic molybdenum-bearing stainless steels are generally not recommended for welding operations. If welding is necessary, suitable preheat and postheat treatments should be made (see pages 244-s to 246-s).

Although, in ferritic grades, molybdenum seems to improve the weldability of the ferritic stainless steels,[59] the effects of the High-Temperature Brittleness are reduced only slightly.[78] Thus, preheat and postheat treatments are generally required.[40]

If postheat treatments are not possible, molybdenum-bearing grades, containing over 18% chromium, should also be alloyed with small amounts of columbium or titanium to minimize the detrimental effects of the High-Temperature Embrittlement.[134] A ferritic 17% chromium steel containing about 1.8% molybdenum, which is stabilized with titanium (or columbium or tantalum), should be welded either with low-carbon (<0.07%) or with columbium-stabilized Type 18–10–2 (Cr-Ni-Mo) electrodes.[174]

NICKEL

Nickel additions to chromium stainless steels are primarily made to widen the gamma region at elevated temperatures. Without phase changes, nickel additions would produce few beneficial effects other than a slight improvement of notch toughness.

Effects on Structure

Because nickel widens the gamma region, martensite may form in alloys containing more than 16% chromium. Nickel is also more effective than manganese in stabilizing austenite since it allows some residual austenite to be retained at room temperature. This is primarily due to the increased sluggishness of transformation caused by the nickel addition. Cold work would reduce the stability of the residual austenite and enhance its transformation into martensite.

The amounts of martensite and retained austenite at room temperature in chromium-nickel alloys primarily depend upon the nickel content. Thus, 16–2 (Cr-Ni) alloys may be fully martensitic on cooling,[45] although ordinarily a fair amount of ferrite is present in the structure.[165]

The martensitic chromium-nickel steels are highly unsuitable for cold-forming and welding operations unless they have been annealed between 1300° and 1450° F. (705 and 790° C.). Moreover, as nickel retards transformation rates considerably,[45, 122] these alloys require prolonged heat treatments.

A 24% chromium steel, which contains 3% nickel and is partially austenitic at elevated temperatures, retains all of the austenite at room temperature. Thus, the alloys which contain at least 24% chromium and 3% nickel are known also as ferritic-austenitic stainless steels.[100] A lower nickel and/or chromium content would reduce austenite stability and, thus, permit an increased martensite formation.

For a 30% chromium alloy the effects of nickel are shown in Fig. 27.[161] Whereas, at a chromium content of 12%, the maximum austenitizing temperatures were about 1800° F. (980° C.), 30% chromium-iron alloys exhibit austenitizing temperatures of about 2300° F. (1260° C.). Although a 30–1 (Cr-Ni) alloy, quenched from 2300° F. (1260° C.), would be expected to contain almost 15% austenite, at least 2 to 3% nickel should be present to make the alloy partially austenitic over a greater temperature range.

Because nickel enhances sigma formation, it should not be added in excessive amounts to the high-chromium steels. The effects of nickel and manganese addi-

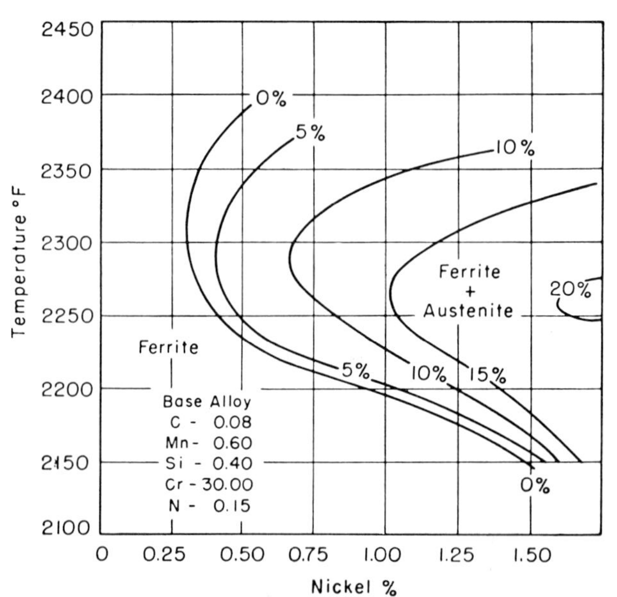

Fig. 27 Effects of nickel and temperature on the structure of a 30% chromium-iron alloy[161]

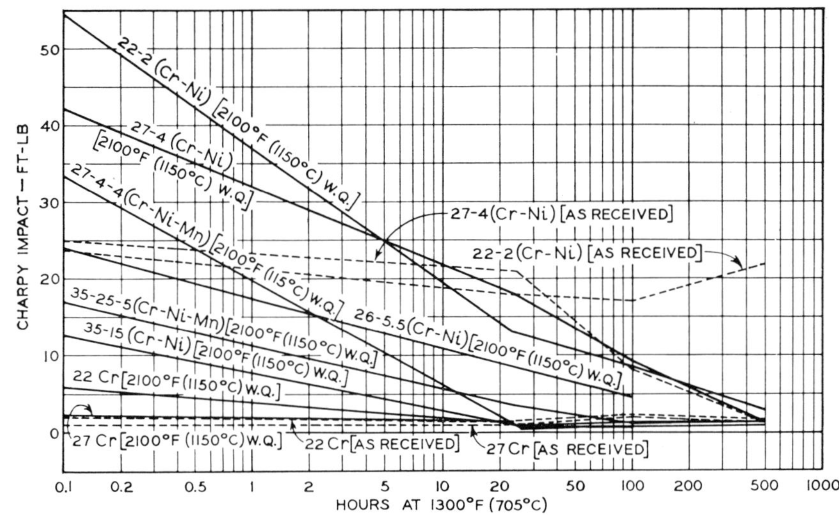

Fig. 28 Effect of aging time at 1300° F. (705° C.) on the impact toughness of various wrought stainless steels[147]

tions on sigma embrittlement of wrought steels are brought out[147] in Fig. 28. Thus, the 26–5.5 (Cr-Ni) steel shows lower toughness because of sigma formation than does the 27–4 (Cr-Ni) steel. The 27–4–4 (Cr-Ni-Mn) steel alloyed with manganese is still further embrittled down to values of 1 ft.-lb. after an exposure of 25 hr. at 1300° F. (705° C.). The reason why the 22 and 27% chromium steels show such low values of toughness even in the as-quenched state is due to the characteristic notch-sensitivity of the fully ferritic structures. Of course, these alloys may also form sigma after long exposures at precipitation temperatures. Carbide precipitation may also be partly responsible for the embrittlement.

Similar results were reported for weld deposits by Norén[163] who found that, in 26–4 (Cr-Ni) type compositions, sigma forms quite readily if the nickel content is increased to about 7%. The rates of sigma formation are further accelerated if about 1.5% molybdenum, 3–4% manganese, or 2% cobalt are added. In such compositions, strong sigma formation could be observed after 2 to 6 hr. at 1380° F. (750° C.). Ordinarily, 26–4 (Cr-Ni) weld metal does not exhibit any strong tendency toward sigma formation.

These results show that, in the alloys susceptible to sigma formation, the beneficial effect of a ferrite-austenite structure may be considerably reduced by exposure to temperatures between 1000 and 1500° F. (540 and 815° C.).

Effects on Physical Properties

In the absence of a phase change, nickel additions contribute little to the physical properties of either martensitic or fully ferritic stainless steels.

The effects of nickel additions to *martensitic stainless steels* are shown in Figs. 29 and 30, respectively.[157] In these alloys, which contain about 12% chromium, nickel is somewhat beneficial on the impact toughness of alloys tempered at temperatures up to about 1000° F. (540° C..) At 1200° F. (650° C.), a nickel content of 2% produces slightly lower impact values. Higher impact values are produced by alloys with a 4% nickel content, Fig. 29. After reheating at 1500° F. (815° C.), the specimens with the higher nickel content exhibited lower impact values and markedly higher hardness values. The increase in hardness may be related to the formation of austenite, which on cooling, transforms into martensite. The tensile strength of alloys, quenched from 1700 to 2000° F. (925 to 1095° C.) would not be noticeably affected. However, because nickel lowers the (initial) transformation temperature and reduces the rate of the austenite-to-ferrite transformation of alloys cooled to temperatures between 1350 and 1450° F. (730 and 790° C.), specimens which were air cooled to room temperature from above 1500° to 1700° F. (815° to 925° C.) would show greater hardenability for the higher nickel alloys.

In fully *ferritic steels,* nickel would have little effect upon the room-temperature and high-temperature strength. The notch toughness is only slightly improved, probably because the transition temperatures from ductile to brittle notch behavior were lowered (rendering carbon and nitrogen somewhat inactive).

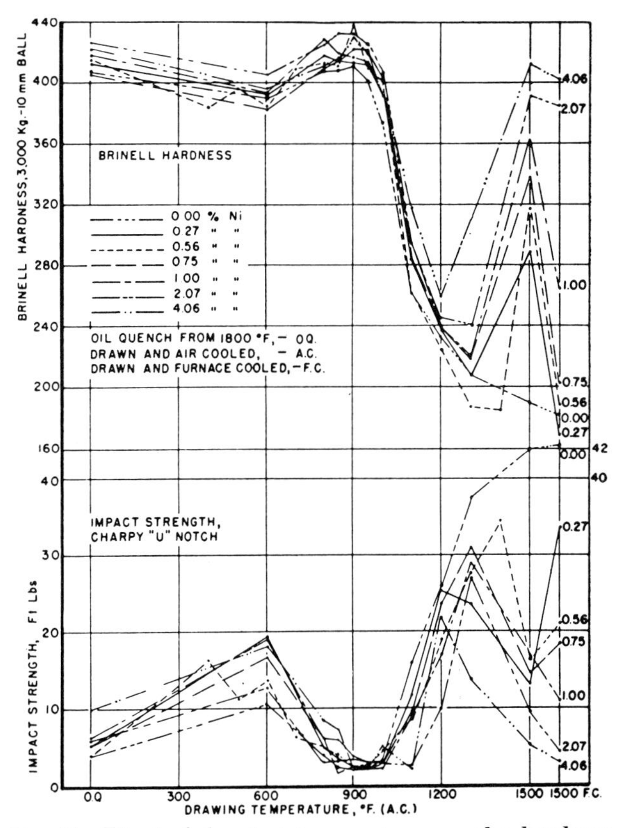

Fig. 29 Effect of drawing temperature on the hardness and impact toughness of martensitic 12% chromium stainless castings with varying nickel content[157]

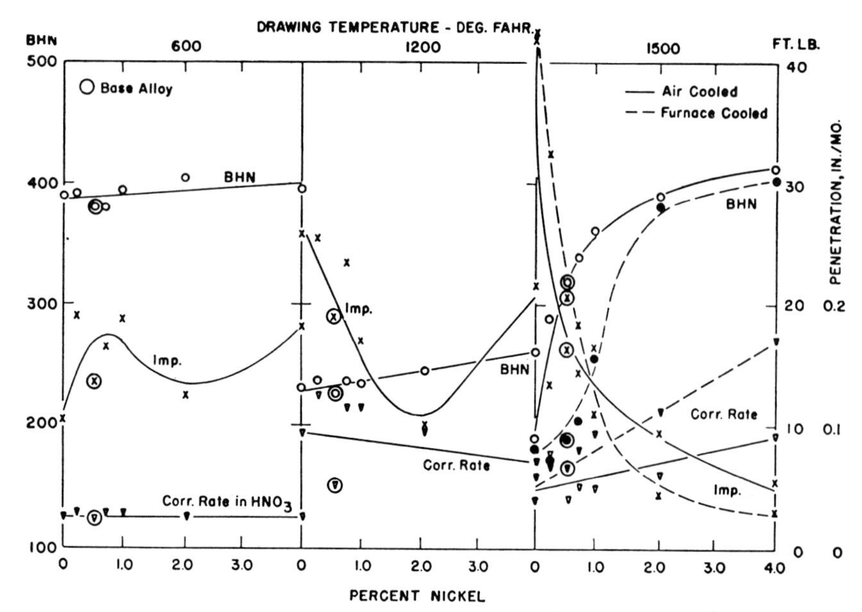

Fig. 30 Effect of nickel content on the mechanical properties and resistance to corrosion in boiling 65% nitric acid of martensitic 12% chromium stainless castings[157]

However, the notch-impact toughness is considerably improved when the addition of nickel (and other austenite formers) to the high-chromium steels causes the retention of austenite at room temperature.[5, 33, 34, 146, 147] This is also apparent from Fig. 28 for the as-quenched Charpy impact specimens which have not been exposed at 1300° F. (705° C.). Although a wrought 27% chromium steel exhibited an average notch-impact toughness of about 2 ft.-lb., the addition of 4% nickel increased the average toughness to about 42 ft.-lb.

Nickel additions do not improve high-temperature corrosion[125] or scaling[65, 125, 158] resistance unless a fully austenitic structure is obtained. However, in allowing martensite to form in alloys of higher chromium content, hardenability may be combined with the beneficial effects of improved corrosion and scaling resistance which is due to the higher chromium content.

Effects on Welding

Even small amounts of nickel (above 0.3 to 0.5%) considerably improve the properties of weld deposits.[36, 37] Larger quantities of nickel are also beneficial, particularly when a ferritic-austenitic weld deposit is produced.

Weld deposits showing ferritic-martensitic structures should be annealed for several hours between 1350 and 1450° F. (730 and 790° C.). If austenitic 18–8 (Cr-Ni) wires are used, preheating at 750° F. (400° C.) is advisable in addition to the postannealing treatment.[45]

Good results are obtained with electrode compositions producing weld deposits of sufficiently high chromium and nickel content in which all the austenite, forming at elevated temperatures, is retained at room temperatures. For example, electrodes containing 25 or 26% chromium and 4% nickel are satisfactory.[65, 100] Further improvement is obtained by the addition of some nitrogen. Thus, Kienberger[38] reported that electrodes, containing a minimum of 23% chromium, gave highly satisfactory weld deposits when alloyed with 3% nickel and 0.25% nitrogen. Although often advisable, postheat treatments on these deposits may not be necessary unless the heat-affected zone of the parent metal has been detrimentally affected by the welding operation.

Eberle,[146] however, showed that 27–4 (Cr-Ni) weld metal displayed in the as-welded condition erratic and low impact resistance which was inconsistent with the ferritic-austenitic microstructure of the deposit and with the characteristics of the corresponding wrought material (Fig. 28). He ascribed the unsatisfactory Charpy impact resistance of 2.5 ft.-lb. (Specimen 10, Table 9) to weld defects and not to the material itself. The more highly alloyed 26–5.5 (Cr-Ni) and the 27–4–4 (Cr-Ni-Mn) alloys exhibited impact values of 8.3 and 10.5 ft.-lb., respectively, in the as-welded condition.[146, 148] It might well be that the reason for the low notch-impact values is that the notch-impact transition temperatures of the welds are reduced to room temperature only by the ferritic-austenitic structure, whereas in wrought materials of similar composition, the notch-impact transition temperatures are reduced to considerably lower temperatures. This would explain why Charpy impact tests performed at 300° F. (150° C.) resulted in values above 20 ft.-lb. for these same three steels.

Table 9—Charpy Impact Properties of Various Weld Deposits* on 28% Chromium-Base Metal[148]

Weld	*Type electrode*	*As-welded tested at room temp., ft.-lb.*	*As-welded tested at 300° F. (150° C.), ft.-lb.*	*Held 100 hr. at 1300° F. and tested at 300° F. (150° C.), ft.-lb.*
1	28 Cr [High N]	1.5	16.5	16.5
2	28 Cr [Ti†]	1.7	21.3	10.2
3	28 Cr [Ti† + Ni‡]	2.7	30.3	7.2
4	35–15 (Cr-Ni)	8.25	. . .	. . .
5	25–20 (Cr-Ni)	34.8	36.3	29.0
6	23 Cr	1.5	22.0	24.0
7	23–2 (Cr-Ni)	3.6	22.6	18.7
8	26–5.5 (Cr-Ni)	8.3	21.0	8.3
9	27–4–4 (Cr-Ni-Mn)	10.5	22.0	1.0
10	27–4 (Cr-Ni)	2.5	24.0	34.0
11	25–12–Cb (Cr-Ni-Cb)	27.0	34.3	2.3
12	25–12 (Cr-Ni)	24.0	24.0	21.3
13	15–35 (Cr-Ni)	47.0	46.0	35.0
14	29–15 (Cr-Ni)	23.3	22.0	5.0
15	29–9 (Cr-Ni)	13.7	20.3	1.0
16	25–12 (Cr-Ni)	17.7	23.2	11.3

* Compositions given in Table 10.
† Added as FeTi in weld groove between beads.
‡ Added in weld groove between beads.

Table 10—Chemical Composition of Electrodes and Weld Metal[148]

Weld	Electrode, %						Deposited weld metal, %						
	C	*Mn*	*Si*	*Cr*	*Ni*	*N*	*C*	*Mn*	*Si*	*Cr*	*Ni*	*N*	*Others*
1	...	...	...	27.59	...	0.177	0.13	0.86	0.04	26.6	0.20	0.185	...
2	...	...	...	27.62	...	0.171	0.14	0.82	0.09	26.0	0.20	0.173	1.0 Ti*
3	...	...	...	27.62	...	0.171	0.15	1.23	0.09	25.9	1.27†	0.171	0.07 Ti*
4	0.13	1.54	1.30	34.41	15.69	...	0.14	1.53	0.09	31.4	9.34	1.36	...
5	0.07	...	...	26.40	21.00	...	0.17	1.41	0.05	26.0	16.9	0.117	...
6	0.13	0.90	0.25	22.95	...	0.15	0.15	1.45	0.03	23.1	0.20	0.189	...
7	0.14	0.86	0.26	22.95	2.14	0.15	0.14	1.18	0.03	22.9	1.83	0.163	...
8	0.23	0.96	1.04	26.34	5.64	...	0.31	1.30	0.09	25.7	5.0	0.049	...
9	0.13	4.40	0.62	27.09	4.07	0.163	0.12	3.76	0.28	26.9	3.76	0.172	...
10	0.10	0.64	0.57	26.80	4.07	0.158	0.12	1.03	0.26	26.3	3.92	0.199	...
11	0.08	1.34	0.67	24.82	14.32	1.15 Cb	0.14	1.31	1.02	24.4	12.5	...	0.74 Cb
12	0.08	1.09	0.33	24.32	15.02	...	0.18	1.49	...	24.6	11.3	...	...
13	0.08	0.11	1.76	20.26	34.80	...	0.10	0.95	...	21.0	28.6	...	...
14	0.11	1.19	0.60	28.70	15.00	...	0.11	1.12	0.24	27.7	12.1	0.106	...
15	0.18	1.66	0.33	29.05	13.47	...	0.13	1.26	0.44	28.0	9.57	0.13	...
16	...	...	...	25.34	14.61	...	0.33	2.03	...	27.3‡	10.7	...	...

* Added as ferrotitanium in weld groove between beads.
† Added in weld groove between beads.
‡ Cr added through electrode coating.

In the high-chromium steels which also contain 3 to 6% nickel, sigma forms more slowly in the weld deposit than it does in the corresponding wrought grades.[146] This is due to the nonhomogeneous nature of the weld deposit which requires longer periods for precipitation and produces a finer distribution of sigma particles which, as such, affect the mechanical properties less detrimentally. Thus, Eberle[146] reported that after only 100 hr. exposure of 27–4 (Cr-Ni) weld metal at 1300° F. (705° C.) could the initial stages of sigma-phase formation be observed.

NICKEL AND MOLYBDENUM

Nickel and molybdenum additions are sometimes made to chromium stainless steels containing between 25 and 30% chromium. The amount of nickel varies between 3 and 5% and the amount of molybdenum between 1 and 1.5%.

Although Kiefer[39] stated that the 27–4–Mo (Cr-Ni-Mo) type alloy is ferritic, it is likely that austenite will be retained down to room temperature.[82, 99, 104] Such a steel is produced in considerable quantity in Sweden and is always found to contain, at room temperature, a structure of ferrite and austenite.[165]

In these high-chromium steels, nickel is primarily added to improve toughness, whereas molybdenum serves to improve corrosion resistance.[4] Thus, these steels are used in the sulphite (paper) industry in fittings, pipe lines, pump shafts, etc.[104]

However, these particular compositions are highly susceptible to sigma-phase and 885° F. (475° C.) Embrittlement[133, 154] which occurs more rapidly in these steels than in the regular ferritic 28% chromium steels.[39] Thus, Foley[199] stated that, if the 27–4–1 (Cr-Ni-Mo) alloys were hot worked well above the range in which sigma could form, plates could be cold sheared and punched without difficulty. If hot-working operations were performed in the temperature range in which sigma forms, the steel would become quite brittle and would crack if the plates were accidentally dropped a distance ten feet from the crane. Satisfactory plates are produced when the billets are heated hot enough to finish at a temperature above 1700° F. (925° C.).[199] Billets for wire drawing must be rolled and finished hot. If intermediate annealing treatments are necessary during cold-drawing operations, it is essential to heat the billets above the sigma-formation temperature range in order to prevent enbrittlement.[199]

Highest toughness values are obtained on water-quenched alloys which were heated for short periods (15 min.) at 1900 to 2100° F. (1040 to 1150° C.). Such treatments result in Izod impact values from 60 to 110 ft.-lb. Lower temperatures and/or air cooling reduce impact toughness considerably. Thus, rapid cooling from 1700° F. (925° C.) of a steel containing 0.08% C, 27.5% Cr, 4.5% Ni, and 1.5% Mo resulted in an Izod impact toughness of only 8 ft.-lb.[99]* A double heat treatment which consisted of holding the steel for 48 hr. at 1350° F. (730° C.), furnace cooling, reheating to 1700° F. (925° C.), and rapid cooling gave a toughness of 16 ft.-lb.[99]

NITROGEN

Willful additions of nitrogen are rarely made in the commercial melting practice of martensitic or ferritic stainless steels. The primary purpose for the addition of nitrogen to these steels is to produce a grain refinement and/or a widening of the gamma region. One of the major exceptions is 28% chromium stainless-steel welding wire which is often specified to show a nitrogen content of 0.10 to 0.20%. This addition is made in the steel mill.[201] Wrought grades containing 23 to 27% chromium may have to meet similar specifications.[210]

The limited beneficial effects of nitrogen on the chromium-bearing stainless steels are due to its solubility, which increases with chromium content. Thus, the solubility of nitrogen amounts to approximately 0.17 and 0.38% at a chromium content of 15 and 25%,

* Original information from Bulletin *Working Data for Carpenter Stainless Steels*, Carpenter Steel Co., Reading, Pa. (1946).

respectively.[50, 57, 66, 105] However, excessive, undissolved quantities of nitrogen are as detrimental to the stainless steels as they are to the carbon steels in which nitrogen absorption is negligible. Thus, excessive amounts of nitrogen in stainless steels tend to produce unsound gassy ingots[11] and porous weld deposits.[38] Moreover, the undissolved quantities of nitrogen somewhat increase the hardness and strength of stainless alloys, although ductility and impact toughness[63] are reduced severely. The detrimental effects of an excessive nitrogen content explain why, in commercial melting practice, where it would be difficult to control the nitrogen content (as an alloying element) accurately, nitrogen additions are rarely made.

As a safe generalization, it is usually recommended that nitrogen should not amount to more than one-hundredth of the chromium content.[63]

Effects on Structure

In its tendency to form austenite, nitrogen is almost as effective as carbon. Thus, nitrogen widens the extent of the gamma loop considerably, particularly the two-phase, austenite-ferrite region.[5, 48, 57, 111] Consequently, alloys, in which nitrogen amounts to approximately one-hundredth the chromium content, will be composed of a duplex structure at elevated temperatures above 1800 to 2000° F. (980° to 1095° C.). Thus, 0.3% nitrogen widens the two-phase region to about 30% chromium.[57]

Because the sluggishness of the austenite-to-martensite transformation increases with chromium content, some austenite may be retained in quenched specimens. Thus, alloys with 23 or more per cent of chromium, which are quenched from above 2000° F. (1095° C.), exhibit a duplex structure, containing only austenite and ferrite when they contain a sufficient quantity of carbon and nitrogen.

On tempering the compositions having retained austenite, little change occurs up to 750° F. (400° C.). However, at temperatures between 750 and 1470° F. (400 and 800° C.) the austenite breaks down into ferrite which contains carbides and/or nitrides.[11, 67, 149]

Decomposition of residual austenite between 750 and 1470° F. (400 and 800° C.) can only be prevented by adding nickel and manganese. This, for example, was accomplished by adding 1.5% nickel and 3% manganese to alloys containing 25% chromium and 0.20 to 0.25% nitrogen. Such a steel showed a 50% austenitic structure which was not susceptible to sigma-phase precipitation.[67] These and other studies indicate that nitrogen reduces the susceptibility to sigma formation by shifting the precipitation range to a higher chromium content.[57]

Effects on Grain Size

A sometimes useful benefit derived from nitrogen additions to chromium stainless steels is the considerable grain refinement which may be obtained.[4, 11, 15, 57, 72, 112, 149, 171] Actually these beneficial effects are due to the formation of austenite[170] or certain nitrides of such elements as aluminum or titanium which in the form of fine nitrides particles would tend to inhibit grain growth. In fully ferritic compositions, uncombined nitrogen is not likely to have any beneficial effect on the grain size.

In castings and weld deposits, the fine grain size is primarily caused by minute nitride particles which act as crystallization centers (seed crystals) in the molten metal[38] and prevent columnar crystallization.[15] In castings, however, nitrogen loses its effectiveness to cause grain refinement if the pouring temperature of the molten metal is increased.[77] This is probably due to the dissolution of the seed crystals.

The particles are formed because of the strong affinity of nitrogen for various elements particularly aluminum, titanium, chromium, vanadium, etc., which may be present as alloying or residual elements in these stainless steels.

Generally, the nitrogen is introduced in the form of a cyanide,[25, 44] nitride particles as AlN[43] or other nitrided aluminum compounds,[43] or in the form of nitrogen-bearing ferrochromium.[4, 6, 15, 26, 57]

Nitrogen also inhibits grain growth in cast and wrought chromium stainless steels exposed to elevated temperatures.[11] Here, the presence of a duplex structure is even more effective than the nitride particles. Of course, severe grain growth will occur at temperatures at which the austenite has transformed into (delta) ferrite and the nitride particles have gone into solution. The critical temperatures are a function of the total composition.

Effects on Corrosion Resistance

Nitrogen does not seem to effect the general corrosion rates of the chromium stainless steels detrimentally.[4, 48, 84] In fact, nitrogen seems to be beneficial in improving resistance to pit corrosion.[48, 84, 149] Quenching these nitrogen-bearing stainless steels from 2100° F. (1150° C.) produces better results than are obtained on quenching from either 1560 or 2460° F. (850 or 1350° C.).

Effects on Physical Properties

In addition to improving the quality and soundness of castings[6] and weld deposits, nitrogen, when it is not present in excessive undissolved quantities, is also beneficial to the mechanical and fabricating properties of cast and wrought grades.[10, 11, 15, 25, 207, 210] Moreover, the fine-grained, nitrogen-bearing stainless steels exhibit a superior machinability—both from the point of view of cutting speed and surface finish.[11]

Although nitrogen, as does nickel, improves notch toughness,* the actual mechanism is not yet apparent.

[74, 172] * Improvement in notch toughness as used here means that, in addition to higher impact values, the transition range from ductile-to-brittle behavior is shifted to lower temperatures.

Table 11—Effects of Nitrogen Content and Heat Treatments Upon Notch Toughness[74]

Alloy composition, % Cr	N_2	C	Treatment	Modified Charpy, ft.-lb.
17.84	0.207	0.01	Normalized from 1650° F. (900° C.)	3–4
			Annealed at 1400° F. (760° C.), air-cooled	15.0
			Very carefully annealed, slowly cooled	72.0
18.26	0.145	0.006	Normalized from 1650° F. (900° C.)	120.0
			Quenched in oil and drawn at 1400° F. (760° C.)	113.0
			Annealed at 1400° F. (760° C.), air-cooled	83.0

Thus, at each composition, a definite amount of nitrogen seems to produce the most beneficial results.

Results of tests conducted by Lincoln[74] on 18% chromium stainless steel alloys made partially martensitic by nitrogen additions are repeated in Table 11. It is apparent that, in this composition, best results are obtained with 0.145% nitrogen. The normalized specimens primarily contained ferrite and a sorbite-type structure.

Further careful studies are needed to clear up these conflicting results and to explain the action of particular amounts of nitrogen in improving notch toughness.

In alloys containing over 23% chromium and a duplex structure of ferrite and some austenite, best results are obtained by a water quench from 2010 to 2190° F. (1100 to 1200° C.).[11, 149] This treatment suppresses transformation of the austenite to martensite.

In certain applications, it may be particularly desirable to use steels in which some austenite has been retained, since the presence of austenite causes a very considerable improvement in the notch toughness. Results by Colbeck and Garner[11] on 25% chromium stainless alloys are shown in Fig. 31. Small additional amounts of nickel improved notch toughness still further.[11] After heat treatments between 750 and 1400° F. (400 and 1040° C.) which caused the decomposition of the retained austenite, the specimens became quite brittle again.[11]

In addition to the presence of residual austenite, notch-impact toughness is also improved by the fine grain size produced by the presence of nitrogen.

In the intermediate temperature range above 300° F. (150° C.), partially martensitic alloys, containing 12 to 16% chromium, up to 0.2% carbon, and between 0.05 to 0.2% nitrogen, are found to be of high strength, ductility, and toughness.[16]

The heat resistance is not affected adversely by nitrogen.[11, 16, 70] In fact, Sissener[35] found a 25% chromium–0.3% nitrogen steel very suitable for use as hearth material at 2010 to 2190° F. (1100 to 1200° C.) in CO atmospheres.

In chromium steels containing 25% chromium, a nitrogen content of about 0.25% improves hot-forging[70] characteristics. Cold drawing[70, 76] also can be performed with much greater ease because of the refined grain size of the nitrogen-bearing alloys. However, excessive amounts of nitrogen (over 1/100 the chromium content) would be undesirable and introduce considerable hardness and stiffness.[76]

Effects of Welding

In welding these stainless steels, nitrogen improves weldability considerably when amounting to about 1% of the chromium content.[63, 70, 84] In this respect, the effects of nitrogen are similar to those produced by small additions of nickel.

In stainless grades containing 17 to 20% chromium, the presence of nitrogen (about 0.15%) reduces the susceptibility to grain growth in the heat-affected zone.[57] However, toughness is not improved unless subsequent annealing treatments are employed between 1200 and 1400° F. (650 and 760° C.).[67] This is primarily due to formation of martensite, which should be changed to ferrite if high toughness is desired.

In weld deposits made by arc welding with 25% chromium electrodes which also contained 0.24% nitrogen, annealing improved the elongation from 7 to 10%.[63] In this composition, because slight amounts of martensite may be assumed to be present in the ferritic-austenitic matrix, toughness should also be improved by annealing.

In steels with a chromium content near 25%, weldability is improved still further if small amounts of nickel are added to the nitrogen-bearing electrodes.[38, 84] This is primarily due to the complete retention of austenite in the weld deposits.

Although nitrogen is ordinarily added directly to either the core wire[51] or the electrode coating in the form of nitrogen-bearing alloys or compounds, nitrogen

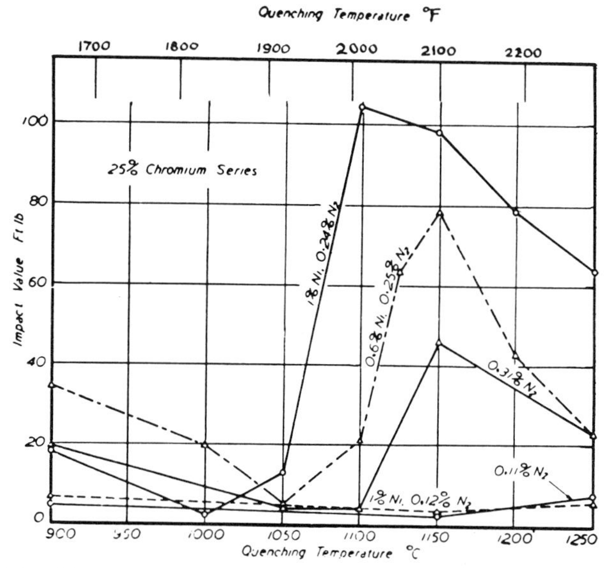

Fig. 31 Effect of quenching temperature on the impact toughness of 25% chromium stainless steels[11]

may also be absorbed from the atmosphere. By reducing the electrode coating, the protective action, produced by the sheathing effect of the gas and the slag, is decreased.[51] This results in increased absorption of the atmospheric nitrogen by the weld metal.

Thus Portevin[12, 19] effectively controlled the nitrogen content of arc-welded deposits by varying the thickness of the electrode coating. Whereas a bare electrode (28% chromium) gave 0.52% nitrogen in the weld deposits, a heavy coating reduced the nitrogen content to 0.17%. Oxyacetylene welding, because of the sheathing effect of the oxyacetylene gas, gave a maximum of 0.12% nitrogen. However, in arc-welding operations, it is best practice to control the nitrogen content of the weld deposit by a suitable adjustment of the composition of the core wire and coating of the welding electrodes.

PHOSPHORUS

Although it has been claimed[49] that between 0.1 and 2.5% phosphorus gives chromium stainless steels containing 13 to 40% chromium and 0.1 to 2.5% carbon very good working properties and a high resistance to wear, it is doubtful that this is true of phosphorus additions above 0.25%, which are not feasible commercially. Because phosphorus tends to produce hot shortness, the higher phosphorus steels are likely to show poor mechanical properties and poor hot-working characteristics. In martensitic grades containing 13% chromium, phosphorus improves creep strength.[174]

Phosphorus severely decreases corrosion resistance in chromium stainless steels which have been heated at temperatures above 1650° F. (900° C.). The low corrosion rates are believed to be due to the formation of a "reactive layer" which formed along the grain boundaries above 1650° F. (900° C.).[95] For example, after welds of a steel containing:

	%
C	0.10
Mn	0.51
Si	0.34
P	0.19
Cr	16.35
Ni	0.25

had been heated for more than 1 hr. at 2010° F. (1100° C.), whole grains fell out after exposure to 60% HNO_3. Annealing between 1290 and 1470° F. (700 and 800° C.) again restored the corrosion resistance.[95]

Phosphorus-bearing chromium stainless steels are not produced commercially.

SILICON

In chromium stainless steels, silicon is primarily useful in improving scaling resistance at elevated temperatures. The advantages are particularly noticeable in steels exposed to sulphur-bearing atmospheres.[153, 186]

Thus, silicon-bearing, stainless wrought and cast alloys are extensively used as furnace parts at temperatures above 1100° F. (595° C.) in which high mechanical properties are generally not required. Moreover, in castings, silicon additions are beneficial because their presence improves fluidity.[77]

Effects on Structure

Silicon, as a ferrite former, is about four times as effective as chromium in its tendency to restrict the gamma region. Thus, in ferritic chromium stainless steels, up to 3% silicon is used, at times, as a substitute for chromium.[38]

In the ferritic-austenitic stainless steels (containing over 23% chromium), the substitution of silicon for part of the chromium is not practical for alloys used at room temperature, because silicon does not increase the sluggishness of the austenite-to-martensite transformation as well as chromium does.

Effects in Castings

In martensitic 12% chromium stainless castings, Schoefer[157] showed that silicon somewhat reduces the impact toughness of castings which are tempered at temperatures up to about 1200° F. (650° C.), Fig. 32. In wrought martensitic grades, silicon may not be as detrimental.

In ferritic-austenitic castings containing about 27% chromium and 5% nickel, silicon may have detrimental effects. In these alloys, silicon may cause a considerable reduction in scaling resistance[165] which may be related to the fact that a high-silicon content produces a tendency to excessively large grain formation.

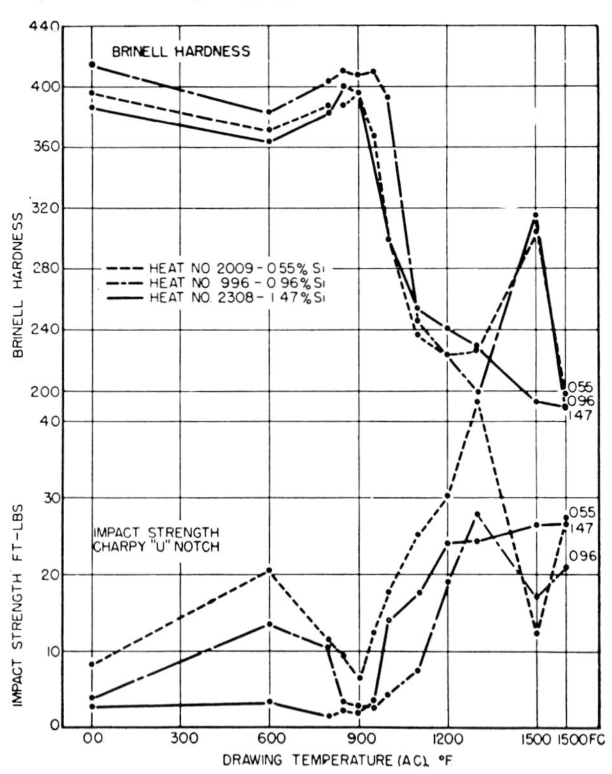

Fig. 32 Effect of drawing temperature on the hardness and toughness of martensitic 12% chromium stainless castings with varying silicon content[157]

Effects at Elevated Temperatures

Although wrought, silicon-bearing, chromium stainless steels are generally not as resistant to scaling at elevated temperatures as aluminum-bearing stainless steels,[174] such alloys are nevertheless used considerably. Castings for elevated temperature service generally contain silicon and not aluminum. The reason for this is the fact that aluminum is readily oxidized, and thus, the production of aluminum-bearing castings is difficult.

Effects of Welding

Although silicon improves the weldability slightly, its beneficial[78, 89] effects on the High-Temperature Embrittlement of ferritic stainless alloys are too small to obviate postheat treatments of grades used at room temperature. However, since the silicon-bearing grades are generally used above 1100° F. (595° C.), these service temperatures will remove the effects of the High-Temperature Embrittlement. Sigma-phase embrittlement may also occur in alloys which contain sufficient quantities of chromium, silicon, and other elements at suitable temperatures (see pages 218-s to 220-s).

Recent German specifications for heat-resisting, wrought and cast, silicon-bearing alloys as well as their welding characteristics are summarized in Tables 12 and 13.

SULPHUR

Sulphur, as in the austenitic chromium-nickel stainless steels,[3] improves the free-machining characteristics of the martensitic and ferritic chromium stainless steels. These beneficial effects of sulphur are primarily due to the formation of sulphide particles. They are responsible for the considerable commercial use of the sulphur- (and/or selenium-) bearing stainless steels where good machinability is of primary importance. On the other hand, toughness and ductility (bendability[194]) are considerably lowered by the presence of the sulphide particles which, in certain applications, may offset the advantages of good machinability.

The martensitic grades containing 11 to 14% chromium and 0.05% sulphur should first be heated to a temperature, below the gamma loop, of about 1450° F. (790° C.) to obtain large carbide particles. (The sizes and shapes of the sulphide particles are fixed by the composition of the alloy, the casting and solidification conditions, and by the nature and amount of reduction during rolling and other fabricating operations. Annealing at 1450° F. (790° C.) is not likely to affect the sulphide inclusions.) After air cooling from 1450° F. (790° C.) the alloy should be reheated to temperatures at which a small amount of austenite forms, which, after rapid quenching, transforms into martensite.[20] As the carbides should not go into solution completely, temperatures near 1580° F. (860° C.) seem most suitable. According to Malcolm,[20] this slightly hardened condition gives superior machinability. However, the advantage of forming even a small amount of martensite seems highly questionable. In tool steels, the presence of residual martensite is accepted to be detrimental to machinability.

Table 12—Composition and Welding Recommendation of German Silicon-Bearing, Wrought and Cast, Chromium Stainless Steels[184, 186]

*Steel No.**	*Type*	Composition, % *C*	*Si*	*Cr*	*Ni*	Welding recommendations: *Electrode classification*	
4712	Wrought	<0.12	2.2–2.5	5.8–6.2	...	4723, 4828	Postheat treatment not necessary
4722	Wrought	<0.12	2.0–2.3	12.5–13.5	...	4723, 4828	Postheat treatment not necessary
4741	Wrought	<0.12	1.9–2.2	17.5–18.5	...	4772	Postheat treatment at 1380° F. (750° C.) and aircool
						4828	Postheat treatment not necessary
4821	Wrought	0.15–0.25	1.0–1.3	24.0–26.0	3.5–4.5	4821, 4842	Postheat treatment not necessary
4710	Cast	0.25–0.35	2.0–2.5	5.5–6.5	...	4723	Postheat treatment necessary if fabricating operations follow welding
4740	Cast	0.40–0.60	1.3–1.8	16.5–17.5	...	4772	
4745	Cast	0.50–0.70	1.3–1.8	21.0–23.0		4772	Weld while red hot between 1300 and 1475° F. (700 and 800° C.) and cool slowly in furnace
4776	Cast	0.50–0.70	1.3–1.8	28.0–30.0			
4777	Cast	1.20–1.40	1.3–1.8	28.0–30.0			
4823	Cast	0.30–0.50	1.0–1.5	26.0–28.0	3.5–4.5	4772, 4821	Small parts may be welded while cold, otherwise welding while red hot 1300–1475° F. (700–800° C.) is recommended

* Classification of Verein Deutscher Eisenhüttenleute.[185]

Table 13—Electrodes Recommended for Welding Silicon-Bearing, Chromium Stainless Steels[184, 186]

Electrode classification	Composition, % *C*	*Si*	*Cr*	*Ni*	*Other*
4723	<0.12	1.3–1.8	14.0–15.0	...	...
4772	<0.12	1.3–1.6	28.0–30.0	...	...
4821	0.15–0.25	0.3–0.5	24.0–26.0	3.5–4.5	...
4828	0.10–0.20	1.8–2.3	19.0–20.0	9.0–10.0	...
4842	0.10–0.20	0.9–1.2	23.0–25.0	19.0–20.0	2.0–2.3 Mn

Similar heat-treating cycles about 50° F. (30° C.) higher are recommended by Malcolm[21] for ferritic stainless steels.

The addition of nitrogen improves the machinability still further. This is due to the effects of nitrogen which widens the gamma loop and refines the grain structure. Thus, Schaufus[32, 33, 34] found that alloys with 14.5 to 18% chromium, 0.2–0.5% sulphur, 0.5 to 2% nickel, and 0.06 to 0.3% nitrogen contained between 30 and 60% of ferrite.

During welding, the ferritic sulphur-bearing stainless steels tend to develop porosity, small checks, or leaks. Thus, welding is not recommended,[55, 178] unless tightness is not a factor.

TITANIUM

Titanium tends to restrict the extent of the gamma loop considerably.[13] This is primarily caused by the affinity of titanium for carbon and nitrogen, which neutralizes the austenitizing tendency of these elements by causing the formation of titanium carbides and titanium nitrides. In addition, the free titanium acts as a strong ferrite stabilizer.

Effects on Grain Size

Titanium is also highly effective in producing a very substantial grain refinement in cast irons[13] and in weld deposits.[145, 150] As such, titanium cyano-nitrides have been suggested as a very suitable addition.[44]

The ability of titanium to cause grain refinement is probably due to the formation of titanium oxides and titanium nitrides which, like aluminum oxides or nitrides, act as seed crystals. This may also explain the fact that columbium is relatively ineffective in causing grain refinement, because, in these steels, columbium does not readily form columbium oxides or columbium nitrides.

In castings[13] and in weld deposits,[145, 150] the presence of a small amount of titanium (0.20 to 0.45%) is highly effective in suppressing the formation of the characteristic columnar dendrites. However, larger quantities of titanium are detrimental because they cause a decrease in the strength of cast alloys and decrease the fluidity of the molten metal.[171] Cast alloys, therefore, rarely contain willful additions of titanium.

Effects on Physical Properties

Corrosion resistance and fabricating properties are somewhat improved by titanium additions.[95] Thus, Becket and Franks[115] produced seamless tubing of very high ductility from ferritic stainless steels containing up to 30% chromium, <0.3% carbon, and titanium amounting to two to four times the carbon content.

In steels which contain over 0.10 to 0.15% carbon and over 0.50% titanium, streaks of TiC particles may be present. They may give trouble in hot-forming operations.

Effects on Welding

Titanium, because of its strong affinity for carbon, somewhat reduces the susceptibility to the High-Temperature Embrittlement. Thus, when present in amounts at least six times the carbon content, titanium is sometimes recommended as an addition to welding electrodes. Care, however, has to be taken for it seems that, when the electrode contains over 0.50% titanium, difficulties may be expected. Thus, preliminary tests by Schaeffler,[150] made with titanium-bearing 28% chromium electrodes, indicated the poor operating characteristics of these high-titanium electrodes, since they produced weld deposits which contained entrapped slag and porosities. Thus, it may be necessary to limit the titanium additions to about 0.20 to 0.40% titanium and the carbon content in the electrode to about 0.07 to 0.10% carbon.

Similar tests were performed by Tikhodeev and Fedotov[68] who developed a coating which contained 35% fluorspar, 25% marble, 20% of 23% ferrotitanium, 20% of 49% ferroaluminum,* and a water-glass binder. The weld deposits from these coated, high-chromium electrodes contained about 35% Cr, 0.23% Ti, and 0.12% Al and exhibited a finely grained structure. The tensile strength of the as-welded deposit amounted to about 82,500 psi., which approached that of the parent metal. The ductility (because of the High-Temperature Embrittlement) remained low, but was found to be improved by heating and quenching from 1380 to 1470° F. (750 to 800° C.).[68]

Another titanium-bearing coating that is claimed[58] to produce satisfactory welds from high-chromium steel electrodes contained 37% marble, 10% dolomite, 32% fluorspar, 4.5% caustic soda, 5.5% of 75% ferromanganese, 2% of 75% ferrosilicon, 5.5% of 20% ferrotitanium, 3.5% starch, and 24% water-glass.

According to Tofaute[78, 123] and others,[107] the presence of titanium in wrought, fully ferritic steels, containing more than 17% chromium and up to 0.10% carbon, reduces the effects of the High-Temperature Embrittlement sufficiently in electric-arc welding with Cr-Ni electrodes to obviate postheat treatments. As is shown in Table 14, this is not true[78, 123] in gas welding of thin sheets where, because of longer exposure at high temperatures, embrittlement is sufficiently severe to require postheat treatments.

Nevertheless, under ordinary welding conditions, considerable difficulties seem to exist when welding wrought, titanium-bearing, ferritic stainless steels. Thus, Nathorst[165] reported that many welded constructions resulted in leaking pipes, storage tanks, and other containers. Nathorst[165] suggested that in addition to the High-Temperature Embrittlement some kind of intercrystalline corrosion seemed to be respon-

* Aluminum powder has also been suggested.[121]

Table 14—Mechanical Properties of Welded Chromium Steel Plates 0.27 In. (6 mm.) Thick[78]

Composition of base plate	*Welding technique*	*Yield strength, psi.*	*Tensile strength, psi.*	*Elongation % 11.3 $\sqrt{f}$*	*Bending angle, degrees*
0.04% C, 0.44% Si, 0.36% Mn	Gas welded with electrode of same composition	44,000	57,000	4.0	20
18.2% Cr	Electric-arc welded with electrode of same composition	55,000	65,000	6.4	22, 22, 41
	Gas welded with 18–8 (Cr-Ni) electrode	43,000	53,000	9.1	49, 40, 40
	Electric-arc welded with 18–8 (Cr-Ni) electrode	57,000	68,000	10.6	90, 99, 99
0.06% C, 0.44% Si, 0.41% Mn	Electric-arc welded with electrode of same composition	57,000	73,000	5.9	15, 33
16.8% Cr, 0.88% Ti	Gas welded with 18–8 (Cr-Ni) electrode	51,000	71,000	6.0	33, 42, 66
	Electric-arc welded with 18–8 (Cr-Ni) electrode	56,000	78,000	15.2	132, 180, 180

sible for the inferior properties of the welded, titanium-bearing, chromium stainless steels.

Such welds may be made reliable by a short heat treatment between 1350 and 1450° F. (730 and 790° C.). For example 5 min. at 1385° F. (750° C.) proved to be sufficient.[165] In practice such a treatment may be accomplished locally. However, if the welds can be heat treated at these temperatures, the titanium (or columbium) addition is not necessary[165] as it was primarily intended to dispense with the heat treatment.

These difficulties with titanium-bearing ferritic stainless steels when welded with the 18–8 type steel electrodes do not seem to have been encountered by a German metallurgist,[174] who showed considerable enthusiasm for such steels. The beneficial characteristics of titanium-bearing stainless seem quite important, because in welding large stainless structures, postheat treatments between 1350 and 1450° F. (730 and 790° C.) are often not possible. However, in order to minimize overheating of the heat-affected zone, the interpass temperature should be controlled. This may be accomplished by the use of small austenitic electrodes. Moreover, fast rates of electrode travel in addition to intermittent cooling of the weld are highly advisable. Such careful welding procedures may well explain the differences of opinion in regard to the welding characteristics of the titanium-bearing ferritic stainless steels. Moreover, for best results the titanium-bearing, 17% chromium stainless steels should be welded either with columbium-bearing 18–8 (Cr-Ni) electrodes (Type 347) or with low-carbon (<0.07%), 18–8 (Cr-Ni) electrodes.[174]

TUNGSTEN

Tungsten is about twice as effective as chromium in stabilizing ferrite.[65] However, tungsten induces in these steels a susceptibility to temper brittleness.

The addition of 2.5 to 3.5% tungsten to martensitic compositions (10–12% Cr) considerably reduces the softening rates of martensite when the steel is heated at about 930° F. (500° C.).[4] These alloys, because of increased sluggishness, retain their (martensitic) strength at elevated temperatures more effectively. Because of these characteristics, tungsten-bearing martensitic stainless steels have been suggested for oil-cracking service.[4]

Tungsten additions (2.5–3.15%) have also been suggested[128] to a martensitic grade containing 0.12–0.15% C, 0.5–2.0% Ni, 12–14% Cr, and fractional amounts of manganese and silicon. After heating the steel for 8 hr. at approximately 2010° F. (1100° C.), quenching, reheating 4 hr. at 1200° F. (650° C.), and air cooling, the alloy is useful as a turbine bolting material.

Welding of martensitic tungsten-bearing stainless steels requires particular precautions to prevent crack formation during cooling.[91] Thus, the steel should be heated to 600 to 800° F. (315 to 425° C.), prior to welding, and annealed between 1200 and 1400° F. (650 and 760° C.) after welding has been completed. The weld should not be allowed to cool between these treatments.

Tungsten has also been suggested[116] as addition to ferritic stainless welding electrodes. Thus, weld deposits from these electrodes are believed[116] to exhibit a fine grain structure and a high elastic limit.

III. Welding Chromium Stainless Steels

GENERAL CONSIDERATIONS

The weldability of the chromium stainless steels is influenced by various factors. Of primary concern are the properties required of the weld deposit. Next, comes the ability or inability to produce welds which can be improved by use of certain heat treatments. Finally, variables introduced by welding technique, design, shrinkage, etc., must be considered.

Properties of Weld Deposits

In general, the weld deposits made from stainless electrodes will exhibit properties which are similar to castings of identical composition. However, because of the greater cooling rates and the resulting finer grain size, weld deposits generally are more ductile than castings of corresponding compositions. Electrodes, having compositions which produce martensitic structures, would give deposits exhibiting high hardness, tensile, and yield strengths, yet very low impact toughness and ductility. The properties of ferritic deposits would depend upon composition and the various embrittling behaviors discussed earlier.

Use of Austenitic Chromium-Nickel Electrodes

Austenitic electrodes are often used to avoid the various problems which arise in using martensitic and ferritic materials and to obviate the necessity of post-heat-treating assembled structures.

Their use, however, does not guarantee satisfactory properties. Thus, the detrimental effects of welding upon the heat-affected zone of a ferritic parent metal, susceptible to the High-Temperature Embrittlement, were already pointed out. In addition, the more highly alloyed austenitic weld deposit shows a number of properties different from those of the chromium stainless steels. In some cases, these differences are highly undesirable and may even lead to service failure of welded sections.

The coefficient of expansion of the austenitic chromium-nickel steels is about 50 to 70% greater than the coefficient of the chromium stainless grades. This considerable difference may cause high stresses during repeated heating and cooling cycles and ultimately lead to failure in the welded joint. However, in actual practice, failures between austenitic and ferritic stainless steels are very rarely caused by the difference in coefficients of expansion. The relatively low yield strength of the austenitic weld bead minimizes the chance that a sufficient concentration of stress will develop in areas adjacent to the ferrite or the brittle martensite zone of the parent metal. Thus, the soft austenite acts as a cushion or spring to distribute the stress uniformly. Only in cases where a great many heating cycles (thermal fatigue) are experienced by the dissimilar metal joint have failures been caused or accentuated by the difference in the coefficients of expansion.

When shrinkage stresses are of significance, it may be advisable, after removal of the slag, to stretch each separate pass by careful hammering or shot peening. This procedure, however, is rarely applied in commercial practice.

The proper annealing heat treatment of the chromium-nickel weld metal and the chromium parent metal differ somewhat. Thus, the chromium-nickel deposit requires annealing temperatures [above 1850° F. (1010° C.)] which are higher than the temperatures recommended for the chromium stainless steel[14] [1350 to 1450° F. (730 to 790° C.)].

Dilution of the austenitic stainless alloy in the fusion zone may be responsible for the formation of an undesirable martensitic layer between the ferritic parent metal and the austenitic weld deposit. In fact, in extreme cases the whole weld deposit may become partially martensitic.

Although the martensite layer may appear in welds made from 18–8 and 25–12 (Cr-Ni) electrodes, this is rarely the case in deposits from 25–20 (Cr-Ni) electrodes. Moreover, high ductility and toughness are generally not of primary consequence when chromium stainless steels are used. Thus, the formation of a martensitic fusion zone may not be critical.

The corrosive environment may also require some attention.[112] Thus, electrolytic couples may cause greater attack on the less noble chromium steel. However, this may only be serious in a very few applications. For example, this electrolytic type of attack presents a serious problem in many oil-field and paper-processing applications.

These considerations will determine the use of one of four types of electrodes in welding the chromium stainless steels. They are (1) martensitic chromium electrodes, (2) ferritic chromium electrodes, (3) ferritic-austenitic electrodes containing small amounts of austenite-forming elements, and (4) austenitic chromium-nickel electrodes. Their characteristic effects on welding the martensitic and ferritic chromium stainless steels vary considerably.

Effects of Welding Process

In welding chromium stainless steels, arc welding is generally preferred.[14, 127] This is due to the fact that arc welding allows (1) a closer control of carbon content, (2) a more highly localized application of an extremely high welding temperature which gives a narrower heat-affected zone, and (3) the fact that electrode coatings prevent contamination from the atmosphere.[14] For example, oxide inclusions in weld deposits may produce serious difficulties.

With the proper precautions, satisfactory weld deposits may be produced with some of the other welding processes.

Oxyacetylene welding may be the least suitable process because of possible carbon pickup which would tend to make the weld martensitic. If oxyacetylene welding is necessary, austenitic welding rods should be used. Moreover, oxyacetylene welding is limited to thin sheet materials (below $^3/_{16}$ to $^1/_4$ in.) and is not suitable for all types of position welding. Welding of flat sections seems to produce best results.

Atomic-hydrogen welding produces sound weld metal. However, the major disadvantage of atomic-hydrogen welding is the wide heat-affected zone which is developed.[90, 131]

Submerged-melt welding is also useful in joining chromium stainless steels by automatic means. Sound weld deposits are obtained by using either austenitic or ferritic types of wire. Perhaps the greatest difficulty is predicting the chromium recovery in the weld metal. Such losses of chromium and of other alloying elements are primarily dependent upon the type of flux used. Arc length also has some effect.

WELDING MARTENSITIC STAINLESS GRADES

In welding the chromium stainless steels containing up to 14% chromium, primary differentiation has to be made in their carbon content. The amount of carbon determines whether the alloy may be designated as fully or partially martensitic. In the absence of other alloying elements, the steels containing over 0.08% carbon generally represent the fully martensitic grades. If the carbon content is below 0.08%, the steel is partially ferritic.

Fully Martensitic Grades

These stainless steels exhibit a high hardening tendency which increases with carbon content. Their welding characteristics are quite poor because the severe hardening, which occurs in the heat-affected zone of the parent metal, results in large residual stresses. These may easily result in cracking of the already brittle martensite.[90]

Cracking may be minimized by preheating these steels prior to welding at 600 to 800° F. (315 to 425° C.). These temperatures should be retained during welding. Moreover, a postheat treatment between 1300 and 1400° F. (705 and 760° C.) should follow directly after welding without cooling the section. The heat-treated sections should be cooled in air. Under these conditions, very ductile welds may be obtained.[90] Although it generally is preferred practice to have the final heat-treating temperatures above 1300° F. (705° C.), good results may be obtained above 1200° F. (650° C.). Such a procedure has been used in the welding of heavy pump barrels made of 11 to 13% chromium stainless steels (Type 410) and welded with 12% chromium stainless electrodes.[209] Since the whole welding operation required four days of two-shift welding by two welders, the pump barrels were stored at night in a furnace in order to maintain the preheat temperatures.[209]

Partially Martensitic Grades

The presence of soft ferrite in the otherwise martensitic structure decreases the over-all hardenability of the steel. This reduces cracking[87, 90] when welding thin sheets below about $^3/_{16}$ in.; however, in heavier sheets and in plates, cooling rates and interpass temperatures should be controlled. Nevertheless, the partially martensitic grades are much more readily adapted to welding procedures than the fully martensitic chromium steels.

Thus, although advisable, preheating may not be necessary. Postannealing treatments, as in the fully martensitic grades, are highly important if ductility and toughness are of consequence.

Use of Fully Martensitic Electrodes

Electrodes which produce martensitic stainless weld deposits are primarily used in applications where the hardening characteristics of the material are desired, while, at the same time, the corrosion and heat-resisting properties of the steel are retained.[96, 127] On the other hand, the requirements for ductility must be relatively low. Thus, typical application are valves, gears, etc.

Preheat and postheat treatments are necessary when ductility and toughness are desired and when it is important to prevent cracking, which may be due to hydrogen or other causes.

In the martensitic stainless steels containing up to 0.15% carbon, a preheat temperature at about 300° F. (150° C.), in many cases, seems to produce best results. If the weldment is to be ductile and tough, that is, if it is to be primarily ferritic, a postheat treatment between 1300 and 1450° F. (705 and 790° C.) should immediately follow the welding operation. However, if the high strength of the martensite is more important, postheat treatments are not necessary when no, or only negligible, quantities of hydrogen are present in the weld metal. When hydrogen has been absorbed (see page 231-s), a postheat treatment should immediately follow welding at temperatures and of a sufficient duration to insure that almost all of the austenite transforms into martensite. For Type 410 stainless weld metal, 1 hr. at 600° F. (315° C.) seems to be sufficient.[195]

Detailed experiments were reported by Norén[56, 191] on two martensitic electrodes containing:

	C	*Mn*	*Si*	*P*	*S*	*Cr*
No. 1	0.10	0.25	0.80	0.030	0.030	11.5
No. 2	0.25	0.30	0.80	0.030	0.030	13.0

The weld deposit from electrode No. 1 showed about 20% ferrite and 80% martensite, whereas electrode No. 2 gave a fully martensitic deposit. At temperatures above 930° F. (500° C.), softening will be initiated. However, as shown in Fig. 33, full softening is not accomplished until temperatures above 1290° F. (700° C.) are reached.

The effects of tempering at 1290° F. (700° C.) and full hardening [quenching from 1740 to 1830° F. (950 to 1000° C.)] are reported in Table 15.

Each application will require its own welding procedures and processes. For example, the edges of propeller blades[180] made from martensitic stainless steels (Type 410) are most efficiently joined by an initial deposit (Type 410 weld metal) made by heliarc welding which is followed by electric-arc welding with coated electrodes. Whereas the initial heliarc "tie-in" bead always appeared sound, subsequent beads made by

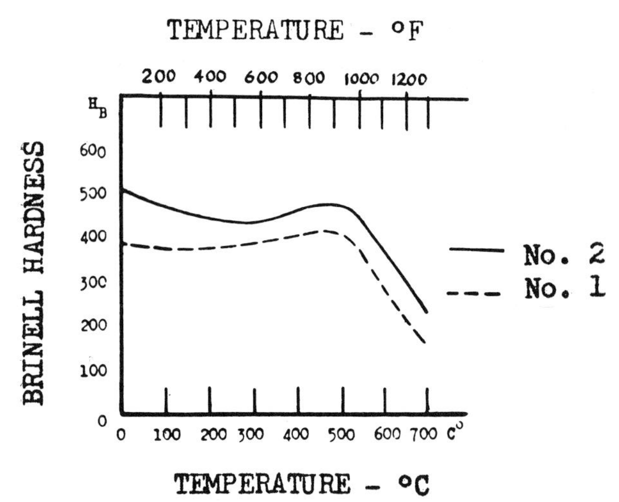

Fig. 33 Effect of tempering temperature on the Brinell Hardness of martensitic stainless weld metals[56]

heliarc welding showed considerable porosity. However, subsequent deposits by electric-arc welding produced highly satisfactory weldments.[180]

Retained Austenite

The understanding of the physical metallurgy of the fully and partially martensitic chromium stainless steels is somewhat complicated by the fact that some austenite, which may be retained at intermediate temperatures, may transform, upon subsequent tempering, into martensite. If this occurs, the beneficial effects of the tempering treatment may be considerably reduced.

In these steels, it is a preferable practice to use a preheat temperature of about 300° F. (150° C.) instead of over 600° F. (315° C.), as ordinarily recommended. At the higher temperature, martensite formation will not occur in the heat-affected zone, the width of which increases with the preheating temperature. Thus, at the higher temperatures, as much as 50% retained austenite has been found after welding.[163] On cooling of the welded section from the tempering temperature (the postheat treatment), some of the retained austenite will transform into martensite. This will make the heat-affected zone relatively brittle.[163] When hydrogen is present, hydrogen embrittlement can readily occur.

On tempering between 1200 and 1300° F. (650 and 705° C.), which is often used commercially, a large amount of the retained austenite will begin to transform into ferrite and carbide. This transformation starts in the grain boundaries. The stable carbide in this temperature range is the high-chromium carbide $(CrFe)_4$ C. Consequently, the ferrite formed in the grain boundaries will be low in chromium. At the same time, the primary martensite will also soften into ferrite and carbide.[163]

This results in a considerable susceptibility to intergranular corrosion which may be minimized by a tempering heat treatment between 1300 and 1450° F. (705 and 790° C.).[163] In this temperature range, the lower chromium carbide $(CrFe)_7C_3$ is the stable carbide. Since, at these temperatures, the diffusion rate is also higher for chromium as well as other possible alloying elements, the susceptibility to intergranular corrosion will be reduced and the general corrosion resistance will be higher because of the higher chromium content in the ferrite.[163]

A relatively quick method used to determine whether the weld has transformed to martensite is to check it with a magnet.[178] (This test will not work if an austenitic chromium-nickel electrode has been used to make the weld.) If the weld is magnetic, the transformation has taken place. If it is nonmagnetic, transformation to martensite has not taken place and the weld should be allowed to cool until it becomes magnetic. The tempering (postheat) treatment should be made as soon as the weld has cooled enough to become magnetic.

Use of Partially Martensitic Electrodes

Because the crack sensitivity is reduced with decreases in the carbon content, electrodes containing less than 0.08% carbon are preferred to weld the fully and partially martensitic grades, particularly as no preheating is required when using these low-carbon, chromium stainless electrodes.

Postheat treatments between 1300 and 1450° F. (705 and 760° C.) should still be employed, particularly when ductility and toughness are of primary concern.[4, 87, 89]

Use of Austenitic Chromium-Nickel Electrodes

Welding with these electrodes is necessary if a heat treatment subsequent to welding is not possible. Small diameter electrodes should be used in order to minimize heating in the parent metal and minimize dilution in the fusion zone. Moreover, time should be allowed for cooling of the welds between passes. Although some martensite does still form in the heat-affected zone, the

Table 15—Effects of Tempering and Hardening of Martensitic Stainless Steel Weld Deposits[56]

	Tensile strength, psi.		*Elongation % (L = 5d)*		*Reduction in area, %*	
Electrode	*Tempered 1 hr. at 1290° F. (700° C.)*	*Water quench from,* ° F.*	*Tempered 1 hr. at 1290° F. (700° C.)*	*Water quench from,* ° F.*	*Tempered 1 hr. at 1290° F. (700° C.)*	*Water quench from,* ° F.*
No. 1	122,000	188,000	13	3.6	47	10
No. 2	166,000	229,000	8.7	1.2	22.5	3

* Deposit No. 1 quenched from 1740° F. (950° C.) and deposit No. 2 quenched from 1830° F. (1000° C.).

weld itself will be tough and ductile. If proper care has been taken, the heat-affected zone of the parent metal may also exhibit sufficient toughness and ductility to give satisfactory service in most applications. Thus, McClow[86] described welding the low-carbon, 12 to 14% chromium stainless lining of a fractionating tower with small columbium-bearing 18-8 (Cr-Ni) electrodes $^1/_8$ to $^5/_{32}$ in. in diameter. In this application a postannealing heat treatment was not necessary. Generally, however, heat treatments between 1300 and 1450° F. (650 and 760° C.) are preferred as long as these temperatures do not produce detrimental effects in the austenitic weld metal.

For best results it is preferred practice to weld the 12–14% chromium stainless steel lining with a 25–20 or 25–12 (Cr-Ni) electrode. This would minimize the effects of dilution which might be detrimental if 18–8–Cb (Cr-Ni-Cb) electrodes had been used instead. Thus, the formation of a possible austenite-martensite fusion zone between the chromium stainless steels and the austenitic weld metal is less detrimental in 25–20 (Cr-Ni) weld deposits than it is in 18–8 (Cr-Ni) weld deposits.

WELDING FERRITIC GRADES

As the ferritic stainless steels are not subject to air hardening, they are far less susceptible to crack formation in the welded section. However, because these steels are susceptible to several types of embrittlement, their welding characteristics should be understood.

Of greatest concern are the effects of notch-sensitivity and of the High-Temperature Embrittlement.

As was shown earlier, notch-sensitivity exists at atmospheric temperatures in commercial alloys containing over 16 to 18% chromium. Only the addition of austenite-forming elements and the use of suitable heat treatments may change the ductile-to-brittle transition range enough to allow the use of the higher chromium alloys in room-temperature applications requiring no notch-sensitivity. However as, in the majority of applications, the ferritic stainless steels are used at elevated temperatures, well above the ductile-to-brittle transition temperature range, notch-sensitivity does not give concern.

As the chromium stainless steels, which are or become fully ferritic at temperatures above 2100° F. (1150° C.), are susceptible to the High-Temperature Embrittlement, postannealing at intermediate temperatures is usually necessary. This heat treatment may be dispensed with only when alloys containing carbide-stabilizing elements are welded using the precautions discussed earlier.

Sigma-phase precipitation and 885° F. (475° C.) Brittleness should concern the welding engineer only when unusual temperature conditions exist. Thus, 885° F. (475° C.) Brittleness may occur in alloys with over 15% chromium exposed near 885° F. (475° C.) for several hours. Ordinary cooling rates, therefore, may not be detrimental. However, to be on the safe side, preheating of the ferritic chromium stainless steels prior to welding, if practiced, should be employed with care and should not exceed 300 to 400° F. (150 to 205° C.). Similarly, in multibead welding the interpass temperature should not exceed 300 to 400° F. (150 to 205° C.).

Sigma-phase precipitation is not found in welded chromium stainless steels, unless the alloys (over 22% chromium) are exposed for long periods (100 hr. or so) between 1000 and 1500° F. (540 and 815° C.).

Detailed welding experiments on 28% chromium base metal were reported by Fowler.[148] The results obtained from 16 different welding electrodes are summarized in Tables 9 and 10. The Charpy impact tests at 300° F. (150° C.) have the advantage that they do not show the notch-sensitivity which is brought out when testing is done at room temperature. The impact tests in the specimens which were heated for 100 hr. at 1300° F. (705° C.) reflect the effects of sigma precipitation.

Use of Ferritic Welding Electrodes

Ordinarily, satisfactory results are obtained by welding the ferritic stainless steels with electrodes having compositions identical to those of the parent metal. Postannealing treatments are essential if ductility is of importance, unless service at elevated temperatures produces similar tempering effects. Nevertheless, 25–20 or 25–12 (Cr-Ni) electrodes are generally preferred for the welding of ferritic stainless steels which contain over 23% chromium.[200]

Because brittleness in ferritic weld deposits increases with higher chromium content, preheat treatments at 300 to 400° F. (150 to 205° C.) are always advisable in steels containing over 25% chromium and may be necessary.

Use of Ferritic Welding Electrodes

Weld deposits which contain a small amount of austenite within the ferrite matrix are generally superior to fully ferritic deposits made from the chromium stainless-steel electrodes discussed in the previous section. Although these ferritic-austenitic weld deposits do not exhibit the superior ductility and toughness found in austenitic weld deposits made from chromium-nickel electrodes, the ferritic-austenitic weld deposits nevertheless give highly satisfactory service in most applications.

The major advantage of welds made from ferritic-austenitic electrodes is that their coefficient of expansion is of a magnitude which is similar to that of the ferritic stainless parent metal. This is of particular importance in service applications in which the welded equipment experiences constant or even occasional cooling cycles (thermal fatigue). In welded sections made with austenitic chromium-nickel electrodes, in which the coefficient of expansion of the austenitic weld deposit exceeds the coefficient of the ferritic parent metal by 50 to 70%, serious residual stress and warpage may result.

Kienberger,[38] after conducting a detailed study of many alloyed chromium stainless electrodes, recommended the use of the following type composition:

	%
C	0.10 max.
Si	0.8–1.0
Mn	3
Cr	22–24
Ni	1.5
N_2	0.22 max.

The combination of the three alloying additions, manganese, nickel, and nitrogen, seemed to produce the most satisfactory weld deposits, from the standpoint of both weldability and physical properties.

Table 16 gives compositions of a typical electrode and of a weld deposit made from the same electrode.

The mechanical properties of electrode and weld deposit are listed in Table 17.

As the ferritic stainless steels are generally used in equipment employed at elevated temperatures, the room-temperature properties listed in Table 17 should be reinterpreted to give higher values of ductility and toughness.

Use of Austenitic Chromium-Nickel Welding Electrodes

These electrodes are used in many applications because the austenitic weld metal is not susceptible to the High-Temperature Embrittlement. However, possible problems of distortion should be analyzed carefully.

Although the austenitic weld deposit itself may not require postannealing heat treatments, the heat-affected zone of the ferritic parent metal may have experienced changes in its physical properties which would make postheat treatments necessary. In those cases, the effects of heat treatments upon the inherent characteristics of the austenitic weld metal as well as on the ferritic parent metal must be considered. Thus, in many of these "dissimilar" metal joints a suitable heat treatment for the chromium stainless steel may detrimentally affect the austenitic stainless weld metal. For these reasons, it is usually necessary to use heat treatments above 1600° F. (875° C.).[98]

In welding the high-chromium stainless steels containing over 25% chromium, good mechanical properties of the welded joint may be obtained by depositing three root beads from 27–4 (Cr-Ni) electrodes and by completing the weld with 25–20 (Cr-Ni) electrodes, Fig. 34.[151] Although some dilution occurred in the 25–20 (Cr-Ni) weld metal which caused the formation of some ferrite, the over-all weld deposit exhibited good mechanical properties. They are reported in Table 18.

In corrosive applications, in which it would be detrimental to have austenitic chromium-nickel weld metal adjacent to a ferritic stainless steel base which is exposed to a corrosive solution, it may be advisable to deposit 30% chromium filler weld metal on that side of the weldment which is exposed to the corrosive solution. The balance of the weld may be made with 25–20 or 25–12 (Cr-Ni) electrodes.[200]

Table 16—Compositions of an Arc-Welding Electrode and a Weld Deposit Containing a Ferritic-Austenitic Structure[38]

	C	*Si*	*Mn*	*Cr*	*Ni*	N_2	*Ti**
Electrode	0.1	0.68	3.62	24.0	1.0	0.268	...
Deposit	0.1	0.14	2.70	23.0	1.1	0.270	0.016

* From electrode coating.

Table 17—Mechanical Properties of a Ferritic-Austenitic Welding Electrode and Weld Deposit.[38]

	Electrode	*Deposit*
Yield strength, psi.	...	78,800
Tensile strength, psi.	147,900	102,800
Elongation ($L = 5d$), %	30.0	25.6
Reduction of area, %	62.0	33.0
Impact, kg./cm.²	...	2.0
Brinell Hardness	...	227

Welding of High-Chromium Ferritic Stainless Castings

Ferritic stainless castings containing 26 to 30% chromium may require special considerations when satisfactory weldments are to be made. This is particularly true in alloys which contain over 0.30% carbon. Such alloys should preferably be welded while red hot at temperatures between 1300 and 1470° F. (705 and 800° C.) and should be furnace cooled slowly.[188] Moreover, if nickel is present (about 3.5 to 4.5%), these alloys, when used in corrosive environments, should receive an additional postheat treatment above 1300° F. (705° C.) in order to minimize susceptibility to intergranular corrosion[188] and to the High-Temperature Embrittlement.

In welding the high-chromium cast steels, electrodes recommended in German specifications[188] are listed in Table 19.

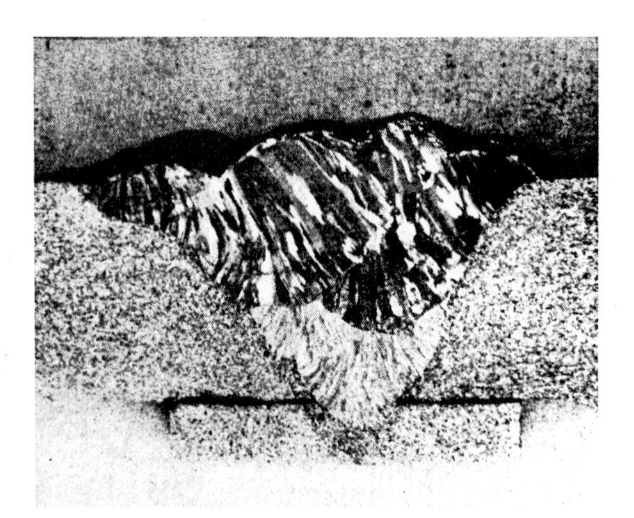

Fig. 34 Section through experimental weld on a 27% chromium stainless steel simulating weld construction of actual joint. Magnification 2½ ×[151]

Table 18—Summary of Test Results[151]

1. *Guided Jig Bend Tests:*
   Angle of Bend, Deg.: 180, 180, 180, 180, 180, 180.

2. *Transverse Weld Tensile Tests:*

	Yield point, psi.	*Tensile strength, psi.*	*Elongation in 2 in., %*	*Reduction of area, %*	*Location of fracture*
	55,200	86,100	16.0	44.7	Outside of weld
	75,200	93,700	10.0	27.6	Outside of weld
	49,200	89,400	13.5	33.0	Outside of weld
Av.	59,870	89,730	13.2	35.1	

3. *Charpy Impact Tests:*

		Notched specimens, ft.-lb.	*Unnotched specimens, ft.-lb.*
(*a*) As-welded—tested at room temp.		12.0	220, 56
		16.0	187, 86
		17.5	
	Av.	15.2	
(*b*) After 500 hr. at 1300° F. (705° C.) tested at room temp.		10.0	76.5
		12.5	330
		14.0	
	Av.	12.2	
(*c*) After 500 hr. at 1300° F. (705° C.) tested at 300° F.		15.5	
		18.5	
		23.0	
	Av.	19.0	
(*d*) After 500 hr. at 885° F. (475° C.) tested at room temp.		8.0	
		8.5	
		13.0	
	Av.	10.0	

Table 19—Composition of German Electrodes Used for Arc Welding High-Chromium Stainless Cast Steels[188]

Electrode type	Composition, %					
	C	*Si*	*Cr*	*Ni*	*Mo*	*Cb*
4551	<0.07	1.3–1.6	18.5–19.5	8.5–9.5	...	7–12 × C
4772	<0.12	1.3–1.6	28.0–30.0	...	...	
4821	0.15–0.25	0.3–0.5	24.0–26.0	3.5–4.5	...	

ACKNOWLEDGMENTS

Major contributions were received from: K. Bungardt, Deutsche Edelstahlwerke, A.-G.; F. B. Foley, the International Nickel Co.; H. D. Newell, The Babcock and Wilcox Co.; T. Norén, Elektriska Svetsnings-aktiebolaget; R. D. Thomas, Jr., H. C. Campbell, and A. L. Schaeffler, Arcos Corp. In addition useful comments and information were also supplied by: M. Baeyertz, Armour Research Foundation; D. C. Buck, United States Steel Co.; C. L. Clark, The Timken Roller Bearing Co.; G. F. Comstock, Titanium Alloy Mfg. Div., National Lead Co.; E. J. Dulis, United States Steel Corp.; C. T. Evans, Jr., Elliott Co.; J. J. Heger, United States Steel Co.; F. H. Keating, Imperial Chemical Industries, Ltd.; J. Kerr, C. F. Braun and Co.; A. Leroy, Institut de Soudure; G. E. Linnert, Armco Steel Corp.; K. E. Luger, K. E. Luger Co.; H. Nathorst, Sandvikens Jernverks Aktiebolag; J. M. Parks, Armour Research Foundation; S. J. Rosenberg, U. S. Department of Commerce; G. V. Smith, United States Steel Corp.; W. J. Burling Smith, The Mond Nickel Co., Ltd.; A. W. Steinberger, Curtiss-Wright Corp.; C. Vollers, Shell Laboratorium (Netherlands); C. A. Zapffe, Consulting Metallurgist.

Bibliography

1. Spraragen, W., and Claussen, G. E., THE WELDING JOURNAL, **18,** Research Suppl., 65-s to 107-s (1939).
2. Thielsch, H., *Ibid.*, **29,** Research Suppl., 361-s to 404-s (1950).
3. Thielsch, H., *Ibid.*, **29,** Research Suppl., 577-s to 621-s (1950).
4. Kinzel, A. B., and Franks, R., *The Alloys of Iron and Chromium, Vol. II. High Chromium Alloys*, McGraw-Hill Book Co., New York (1940).
5. Houdremont, E., *Handbuch der Sonderstahlkunde*, Springer Verlag, Berlin (1943).
6. Smith, E. K., *Metal Progress*, **37,** 49–54 (1940).
7. Zeyen, K. L., and Lohmann, W., *Schweissen der Eisenwerkstoffe*, 2nd ed., Verlag Stahleisen (1948).
8. Tofaute, W., *Z. Ver. deut. Ing.*, **81,** 1117–1122 (1937).
9. Hougardy, H., and Riedrich, G., *Tech. Mitt. Krupp, Forschungsber.*, **30,** 547–560 (1937); also in *Metallwirtschaft*, **10,** 1329–1342 (1937).
10. Newell, H. D., *Metal Progress*, **49,** 977–1004 (1946).
11. Colbeck, E. W., and Garner, R. P., *J. Iron Steel Inst.* (*London*), **139,** 99–146 (1939); also in *Heat Treating Forging*, **25,** 286–292, 346–347, 360 (1939); also *Engineering*, **147,** 726–729 (1939).
12. Portevin, A., *J. Iron Steel Inst.* (*London*), **139,** 137–140 (1939).
13. Poboril, F., *Ibid.* (*London*), **139,** 142–144 (1939).
14. Mueller, R. A., Carlson, I. H., and Seabloom, E. R., THE WELDING JOURNAL, **23,** Research Suppl., 12-s to 22-s (1944).
15. Samarin, A. M., Korolev, M. L., and Paisov, I. V., *Metallurg*, **13** (11), 80–83 (1938). (Brutcher Translation No. 1766.)
16. Franks, R., U. S. 2,140,905 (Dec. 20, 1938).
17. Hougardy, H., *Maschinenbau Betrieb*, 411–412 (Aug. 1938).
18. Hougardy, H., *Elektroschweissung*, **9,** 5–9 (1938).
19. Portevin, A., and Séférian, D., *Rev. mét.*, Memoires, **34,** 225–237 (1937).
20. Malcolm, V. T., U. S. 2,262,690 (Nov. 11, 1941).
21. Malcolm, V. T., U. S. 2,218,973 (Oct. 22, 1940).
22. Rafalovich, T. N., *Teoriya i Prakt. Met.*, **11** (12), 59–63 (1939).
23. Bandel, B., and Tofaute, W., *Arch. Eisenhüttenw.*, **15,** 307–320 (1941–42). (Brutcher Translation No. 1893.)
24. Westgren, A., Phragmén, G., and Negresco, T., *J. Iron Steel Inst.* (*London*), **117,** 383–400 (1928).
25. Field, A. L., U. S. 2,378,397 (June 19, 1945).
26. Loik, P. A., *Liteĭnoe Delo*, **12,** 28 (1941).
27. Uhlig, H. H., *Metal Progress*, **57,** 486–487 (1950).
28. Newell, H. D., *Ibid.*, **51,** 617–626 (1947).
29. Cook, A. J., and Jones, F. W., *J. Iron Steel Inst.* (*London*), **148,** 217–226 (1943).
30. Zapffe, C. A., *Rev. mét.*, **44,** 91–96 (1947).
31. Peoples, R. S., Pray, H. A. H., U. S. 2,273,731 (Feb. 17, 1942).
32. Schaufus, H. S., U. S. 2,384,567 (Sept. 11, 1945).
33. Schaufus, H. S., U. S. 2,384,565 (Sept. 11, 1945).
34. Schaufus, H. S., U. S. 2,384,566 (Sept. 11, 1945).

35. Sissener, J., *Foundry Trade J.*, **78**, 341–345 (1946).
36. Arness, W. B., U. S. 2,310,341 (Feb. 9, 1943).
37. Arness, W. B., U. S. 2,306,421 (Dec. 29, 1942).
38. Kienberger, H. P., Dissertation, Tech. Hochschule, Braunschweig, Germany (1944).
39. Kiefer, G. C., *Metal Progress*, **40**, 59–62 (1941).
40. Allegheny Ludlum Steel Corp. and Babcock & Wilcox Tube Co., *Metals and Alloys*, **18**, 55–62 (1943).
41. Riedrich, G., and Loib, E., *Arch. Eisenhuttenw.*, **15**, 175–182 (1941–42). (Brutcher Translation No. 1249.)
42. Palatnik, L. S., and Barkov, V. N., *Metallurg*, **15**, (11–12), 68–70 (1940). (Brutcher Translation No. 1306.)
43. Kirichenko, I. D., Russ. 58,509 (Dec. 31, 1940).
44. Comstock, G. F., *Metal Progress*, **33**, 269–274 (1938).
45. Watkins, S. P., *Ibid.*, **44**, 99–103 (1943).
46. Schmidt, M., and Legat, H., *Arch. Eisenhüttenw.*, **10**, 297–306 (1936–37). (Brutcher Translation No. 494.)
47. Legat, H., *Metallwirtschaft*, **17**, 509–513 (1938). (Brutcher Translation No. 654.)
48. Schottky, H., *Z. Metallkunde*, **39**, 120–122 (1948).
49. Rubensdorffer, F., Ger. 666,627 (Nov. 18, 1938).
50. Brick, R. M., and Creevy, J. A., *Metals Technol.*, **7**, Tech. Publ. No. 1165 (April 1940).
51. Mirt, O., Dissertation, Tech. Hochschule, Graz, Austria (1938).
52. Bain, E. C., *Trans. Am. Soc. Steel Treat*, **9**, 9–32 (1926).
53. Spencer, L. F., *Steel Processing*, **34**, 127–133, 153, 156–157 (1948).
54. Jones, A. C., *The Book of Stainless Steels*, Am. Soc. Metals, Cleveland, pp. 276–277 (1935).
55. Anonymous, *Machine Design*, **17**, 175–178 (Feb. 1945).
56. Norén, T., *Svetsaren*, **12**, 103–109 (1947).
57. Krainer, H., and Mirt, O., *Arch. Eisenhüttenw.*, **15**, 467–472 (1941–42). (Brutcher Translation No. 1321.)
58. Shashkov, A. N., *Avtogennoe Delo*, (3), 10–11 (1940).
59. Maurer, F., *Metal Progress*, **31**, 535–536 (1937).
60. Heger, J. J., *Ibid.*, **49**, 976–B (1946).
61. Thielsch, H., THE WELDING JOURNAL, **29**, Research Suppl., 122-s to 132-s (1950).
62. Hougardy, H., and Schierhold, P., *Z. wirtschaft. Fertigung*, **43**, Nr. 3, 24–31 (1939).
63. Franks, R., *Trans. Am. Soc. Metals*, **23**, 968–987 (1935).
64. Schiffler, H. J., Ger. 731, 161 (Dec. 3, 1942).
65. Thielemann, R. H., *Proc. Am. Soc. Test. Mat.*, **40**, 788–804 (1940).
66. Krainer, H., and Nowak-Leoville, M., *Arch. Eisenhüttenw.*, **15**, 507–513 (1941–42). (Brutcher Translation No. 1853.)
67. Riedrich, G., *Ibid.*, **15**, 514–515 (1941–42).
68. Tikhodeev, G. M., and Fedotov, L. E., *Metallurg*, (2) 31–37 (1940).
69. Foley, F. B., *Bull. Research Inst. Temple Univ.*, 7–32 (Nov. 14, 1947).
70. Wright, E. C., *Trans. Am. Soc. Metals*, **23**, 987–988 (1935).
71. Rapatz, F., and Hummitzsch, W., *Arch. Eisenhüttenw.*, **8**, 555–556 (1934–35). (Brutcher Translation No. 2202.)
72. Becket, F. M., *Trans. Am. Inst. Min. Met. Engrs.*, **131**, 15–36 (1938).
73. Dannöhl, W., *et al.*, *Arch. Eisenhüttenw.*, **15**, 319 (1941–42). (Brutcher Translation No. 1884.)
74. Lincoln, R. A., Dissertation, Carnegie Institute of Technology (1935).
75. Kantzow, G. A.v., and Nordström, B. G. O., U. S. 2,288,660 (July 7, 1942).
76. Newell, H. D., *Trans. Am. Soc. Metals*, **23**, 988–989 (1935).
77. Korschan, H. L., *Tech. Mitt. Krupp.*, *Tech. Ber.*, **9**, 1–15 (1941).
78. Tofaute, W., *Ibid.*, **6**, 17–24 (1938).
79. Wassermann, G., *Metallwirtschaft*, **23**, 387–391 (1944).
80. Mitsche, R., *Berg- und Hüttenm. Monatsh.*, **93**, 163–165 (1948); also in *Arch. Metallkunde*, **3**, 299–300 (1949).
81. Electro Metallurgical Co., Brit. 528,729 (May 5, 1940).
82. Persson, O., *Svetsaren*, **10**, 47–64 (1945).
83. Vanderbeck, R. W., and Gensamer, M., THE WELDING JOURNAL, **29**, Research Suppl., 37-s to 48-s (1950).
84. Rapatz, F., *Stahl u. Eisen*, **61**, 1073–1078 (1941). (Brutcher Translation No. 1255.)
85. Wilder, A. B., and Light, J. O., *Trans. Am. Soc. Metals*, **41**, 141–166 (1949).
86. McClow, W. W., *Chem. and Met. Eng.*, **50**, 134–135 (Nov. 1943).
87. Henry, O. H., Claussen, G. E., and Linnert, G. E., *Welding Metallurgy*, 2nd ed., AMERICAN WELDING SOCIETY, New York (1949).
88. Riedrich, G., *Stahl u. Eisen*, **61**, 852–860 (1941). (Brutcher Translation No. 1129.)
89. Henrion, E., *Rev. Soudure (Belgium)*, **2**, 65–71, 111–118, 166–169 (1946); **3**, 14–17 (1947).
90. Bulletin, *Welding Avesta Stainless Steels*, Avesta Jernverks Aktiebolag, Avesta, Sweden.
91. Shashkov, A. N., *Avtogennoe Delo*, **12** (6), 20–23 (1941).
92. Tofaute, W., *Tech. Mitt. Krupp*, *Tech. Ber.*, **7**, 31–35 (1940).
93. Guillet, L., *Genie Civil*, **117**, 25–29 (1941).
94. Schiffler, H. J., and Hirsch, W., *Tech. Zentr. prakt. Metallbearbeit.*, **49**, 373–376 (1939).
95. Levin, I. A., and Novitskaya, M. A., *Trudy Konferentsiĭ Korziĭ Metal.*, **2**, 260–267 (1943).
96. Bulletin, *General Information on ESAB's New OK Hard Facing Elektrodes*, Elektriska Svetsningsaktiebolaget, Göteborg, Sweden.
97. Hochmann, J., *Compt. rend.*, **226**, 2150–2151 (1948).
98. Zapffe, C. A., *Stainless Steels*, Am. Soc. Metals, Cleveland, Ohio (1949).
99. Archer, R. S., Briggs, J. Z., and Loeb, C. M., Jr., *Molybdenum*, Climax Molybdenum Co., New York (1948).
100. Norén, T., *Svetsen*, **7**, 44–60 (1948).
101. Houdremont, E., and Bandel, G., *Arch. Eisenhüttenw.*, **11**, 131–138 (1937–38).
102. Bandel, G., *Ibid.*, **11**, 139–144 (1937–38).
103. Andersen, A. G. H., and Jette, E. R., *Trans. Am. Soc. Metals*, **24**, 375–419 (1936).
104. Bulletin, *Uddeholm Corrosion, Acid and Heating Resisting Steels*, Uddeholm Aktiebolag, Uddeholm, Sweden (1947).
105. Kootz, T., *Arch. Eisenhüttenw.*, **15**, 77–82 (1941–42). (Brutcher Translation No. 1316.)
106. Jette, E. R., and Foote, F., *Metals & Alloys*, **7**, 207–210 (1936).
107. Ericson, C., *Svensk Papperstidn.*, **46**, 23–30, 46–50 (1943).
108. Maurer, E., and Mailander, R., *Stahl u. Eisen*, **45**, 409–423 (1925).
109. Jackson, R., and Sarjant, R. J., *J. Inst. Petroleum*, **34**, 445–485 (1948).
110. Jones, J. A., and Heselwood, W. C., *J. Iron Steel Inst. (London)*, **137**, 361–382 (1938).
111. Tofaute, W., and Schottky, H., *Arch. Eisenhüttenw.*, **14**, 71–76 (1940–41); also in *Tech. Mitt. Krupp.*, *Forschungsber.*, **3**, 103–110 (1940).
112. Sands, G. A., *Chem. and Met. Eng.*, **49**, 81–91 (April 1942).
113. Schafmeister, P., and Ergang, R., *Arch. Eisenhüttenw.*, **12**, 459–464 (1938–39). (Brutcher Translation No. 915.)
114. Schafmeister, P., and Ergang, R., *Ibid.*, **12**, 507–510 (1938–39). (Brutcher Translation No. 1054.)
115. Becket, F. M., Franks, R., U. S. 2,139,538 (Dec. 6, 1938).
116. Electro Metallurgical Co., Ger. 681,719 (Sept. 7, 1939).
117. Widawski, E., *Arch. Eisenhüttenw.*, **11**, 195–198 (1937–38).
118. Erakhtin, V., *Stal*, **8** (7), 62–66 (1938).
119. Franks, R., Can. 385,091 (Nov. 14, 1939).
120. Kluke, R., *Arch. Eisenhüttenw.*, **11**, 615–618 (1937–38).
121. Erastov, V. I., and Khmel'nitskaya, R. B., *Khim. Mashinostroenic*, **9** (8–9), 27–31 (1940).
122. Bulletin, *Fabrication of U. S. S. Stainless and Heat Resisting Steels*, United States Steel Corp., 1948 ed.,
123. Tofaute, W., *Tech. Mitt. Krupp*, *Tech. Ber.*, **8**, 76–82 (1940).
124. Scheil, M. A., *Metal Progress*, **52**, 91–102 (1947).
125. Technical Bulletin No. 6-E, *Properties of Carbon and Alloy Seamless Steel Tubing for High-Temperature and High-Pressure Service*, The Babcock and Wilcox Tube Co. (1948).
126. Rocha, H. J., *Tech. Mitt. Krupp*, *Forschungsber.*, **3**, 191–198 (1940).
127. Pospíšil, R., *Z. Schweisstechnik*, **39**, 159–166 (1949).
128. Wheeler, A. W., U. S. 2,349,319 (May 23, 1944).
129. Heger, J. J., *Steel*, **123**, 71–75, 90, 92, 94 (Oct. 25, 1948).
130. Franks, R., U. S. 2,183,715 (Dec. 19, 1939).
131. Bungardt, K., *Tech. Zentr. prakt. Metallbearbeit*, **49**, 706, 708, 753, 754, 756 (1939).
132. Jackel, W. J., *Corrosion*, **1**, 83–93 (1945).
133. Newell, H. D., Private Communication, The Babcock & Wilcox Tube Co., Beaver Falls, Pa.
134. Scherer, R., *Chem. Fabrik*, **13**, 373–379 (1940).
135. Heger, J. J., "The Effect of Time on Physical Properties and Hardness of 27% Chromium-Iron Alloy on Aging at 885° F.," Report No. RR-3-4 Sept. 8, 1944, The Babcock and Wilcox Tube Co., Beaver Falls, Pa.
136. Olzak, Z. E., "Short-Time Elevated Temperature Properties of Annealed Versus Embrittled 27% Chromium-Iron Alloy," Report No. RR-3-2, June 5, 1944, and Supplement No. 1, June 24, 1944, The Babcock and Wilcox Tube Co., Beaver Falls, Pa.
137. Heger, J. J., "An Investigation of the Cause of 885° F. Brittleness in 27% Chromium-Iron Alloy," Report No. RR-3-1, March 1, 1944, The Babcock and Wilcox Tube Co., Beaver Falls, Pa.
138. Heger, J. J., "An Investigation of the Cause of 885° F. Brittleness in 27% Chromium-Iron Alloy," Report No. RR-3-1, Supplement No. 1, Sept. 19, 1945, The Babcock and Wilcox Tube Co., Beaver Falls, Pa.
139. Heger, J. J., and Cordovi, M. A., "Effect of Various Thermal Treatments on the Properties of 27% Chromium-Iron Alloys," Report No. RR-3-8, Sept. 26, 1945, The Babcock and Wilcox Tube Co., Beaver Falls, Pa.
140. Olzak, Z. E., "Effect of Long-Time Heating on Hardness and Structure of 27% Chromium-Iron Alloy as Determined by Gradient Bar Heating Test," Report No. RR-3-3, Aug. 4, 1944, The Babcock and Wilcox Tube Co., Beaver Falls, Pa.
141. Manfre, J. A., "Corrosion Resistance of Annealed Versus Embrittled 27% Chromium-Iron Alloy in Boiling 65% Nitric Acid," Report No. RR-3-5, Sept. 23, 1944, The Babcock and Wilcox Tube Co., Beaver Falls, Pa.
142. Heger, J. J., "The Effect of Time and Temperature in Removing 885° F Embrittlement in 27% Chromium-Iron Alloy," Report No. RR-3-7, April 5, 1945, The Babcock and Wilcox Tube Co., Beaver Falls, Pa.
143. Heger, J. J., "The Effect of Notching on the Tensile and Impact Strength of Annealed 27% Chromium-Iron Alloy at Room and Elevated Temperatures," Report No. RR-3-6, March 28, 1945, The Babcock and Wilcox Tube Co., Beaver Falls, Pa.
144. Olzak, Z. E., "The Effect of Alloy Composition on Sigma Phase Precipitation in 27% Chromium-Iron Catalyst Tubes, First Step Dehydrogenation, Plains Plant," Phillips Petroleum Co., Borger, Tex., Report No. RR-4-2, Aug. 25, 1944, The Babcock and Wilcox Tube Co., Beaver Falls, Pa.
145. Schaeffler, A. L., and Thomas, R. D., Jr., "Metallurgical Examination of 28% Chrome Weld Metal with Additions of Titanium and Columbium," Experiments No. 512C and 613, Jan. 29, 1945, Arcos Corp., Philadelphia, Pa.
146. Eberle, F., "Suitability of High Chromium and High Chromium-Nickel Alloys as Electrode Material for Welding 27% Chromium-Iron Subsequently to Be Exposed to Service Temperatures of About 1300° F.," Report No. 2786, Oct. 29, 1942, EMD-1128, The Babcock and Wilcox Tube Co., Beaver Falls, Pa.
147. Eberle, F., "Suitability of High Chromium-Nickel Alloys as Electrode Material for Welding 27% Cr Iron Subsequently to Be Exposed to Service Temperature of About 1300° F.," Report No. 2779, Oct. 15, 1942, The Babcock and Wilcox Tube Co., Beaver Falls, Pa.
148. Fowler, F. A., "Tests Exploring the Different Methods of Welding 28% Chromium-Base Metal in Order to Obtain the Highest Possible Toughness in Weld Metal," Report No. 2771, Sept. 17, 1942, and Report No. 2771, Supplement 1, Oct. 26, 1942, The Babcock and Wilcox Tube Co., Beaver Falls, Pa.
149. Eberle, F., "The Properties of 22 Cr-High Nitrogen Alloys Containing Small Additions of Nickel and Their Limitations for Service at Elevated Temperatures," Report No. 2805, Mar. 10, 1943, The Babcock and Wilcox Tube Co., Beaver Falls, Pa.
150. Schaeffler, A. L., "Metallurgical Investigation No. 2 of Titanium Bearing Chromend 28," Experiment No. 512, Apr. 8, 1946, Arcos Corp., Philadelphia, Pa.
151. Eberle, F., "Mechanical and Microstructural Properties of Weldments of 27% Cr-Iron Made According to Actual Fabrication Procedures as Developed for Phillips Petroleum Co. Installations—EMO 1170," Report No. 2801, Feb. 4, 1943, The Babcock and Wilcox Tube Co., Beaver Falls, Pa.
152. Binder, W. O., and Spendelow, H. R., Jr., "The Influence of Chromium on the Mechanical Properties of Plain Chromium Steels," 1950 Preprint, Am. Soc. Metals.
153. Bulletin, *Rostfritt stål*, Fagersta Bruks A. B., Fagersta, Sweden, 1944.
154. Bergsman, E. B., and Ericsson, C., *Värmländska Bergsmannaföreningens Annaler*, 31–74 (1942).
155. Hummitzsch, W., *Schweisstechnik*, **2**, 119–129 (1948).
156. DuMond, T. C., *Materials and Methods*, **31**, 83–98 (May 1950).
157. Schoefer, E. A., *Alloy Casting Bulletin No. 12* (October 1948).
158. Colegate, G. T., *Metal Treatment*, **17**, 93–101, 109 (1950).
159. Shortsleeve, F. J., and Nicholson, M. E. "Transformations in Ferritic Chromium Steels between 1100 and 1500° F. (595 and 815° C.)," 1950 Preprint, Am. Soc. Metals.
160. Heger, J. J., *Symposium on the Nature, Occurrence, and Effects of Sigma Phase* (A.S.T.M.), 75–78 (1951).
161. Post, C. B., and Eberly, W. S., "Formation of Austenite in High-Chromium Stainless Steels," 1950 Preprint, Am. Soc. Metals.
162. Gilman, J. J., "Hardening of High-Chromium Steels by Sigma Phase Formation," 1950 Preprint, Am. Soc. Metals.
163. Norén, T., Private communication, Elektriska Svetsningsaktiebolaget, Göteborg, Sweden.
164. Zapffe, C. A., *Trans. Am. Soc. Metals*, **40**, 315–352 (1948).
165. Nathorst, H., Private communication, Sandvikens Jernverks Aktiebolag, Sandviken, Sweden.
166. Scheil, M. A., Zmeskal, O., Waber, J., and Stockhausen, F., THE WELDING JOURNAL, **22**, Research Suppl., 493-s to 504-s (1943).
167. Tholander, E., *Jernkontorets Ann.*, **132**, 367–419 (1948).
168. Zapffe, C. A., and Phebus, R. L., "Embrittlement of Stainless Steel

by Steam in Heat Treating Atmospheres," 1950 Preprint, Am. Soc. Metals.
169. Nathorst, H., *Jernkontorets Ann.*, **134,** 97–119 (1950); Welding Research Council, *Bull. No. 6* (Oct. 1950).
170. Bungardt, K., *Stahl u. Eisen*, **70,** 582–596, 607 (1950).
171. Roesch, K., *Ibid.*, **70,** 596–606 (1950).
172. Baerlecken, E., *Ibid.*, **70,** 607 (1950).
173. Dannöhl, W., *Ibid.*, **70,** 606–607 (1950).
174. Bungardt, K., Private communication, Deutsche Edelstahlwerke, Krefeld, Germany.
175. Smith, G. V., *Iron Age*, **166,** 63–68 (Nov. 30, 1950); 127–132 (Dec. 7, 1950).
176. Hochmann, J., *Bull. cercle études métaux*, **5,** 221–281 (1949).
177. Zapffe, C. A., Private communication, Consulting Metallurgist, Baltimore, Md.
178. *Procedure Handbook of Arc Welding Design and Practice*, 9th ed., The Lincoln Electric Co., Cleveland, Ohio (1950).
179. *Timken Technical Bull. No. 33*, Timken Roller Bearing Co., Canton, Ohio.
180. Steinberger, A. W., Private communication, Curtiss-Wright Corp., Caldwell, N. J.
181. Luger, K. E., Private communication, K. E. Luger Co., Houston, Tex.
182. Duwez, P., and Baen, S. R., *Symposium on the Nature, Occurrence, and Effects of Sigma Phase* (A.S.T.M.), 48–54 (1951).
183. Résumé of High Temperature Investigations Conducted During 1943–44, The Timken Roller Bearing Co., pp. 73–77.
184. Verein Deutscher Eisenhüttenleute, "Hitzebeständige Walz- und Schmiedestähle," Stahl-Eisen-Werkstoffblatt, No. 470-49 (Apr. 1949).
185. Verein Deutscher Eisenhüttenleute, Güteschlüssel für Stahl und Eisen.
186. Verein Deutscher Eisenhüttenleute, "Hitzebeständiger Stahlguss," Stahl-Eisen-Werkstoffblatt, No. 471-49 (Apr. 1949).
187. Verein Deutscher Eisenhüttenleute, "Nichtrostende Walz- und Schmiedestähle," Stahl-Eisen-Werkstoffblatt, No. 400-49 (Apr. 1949).
188. Verein Deutscher Eisenhüttenleute, "Nichtrostender Stahlguss," Stahl-Eisen-Werkstoffblatt, No. 410–49, (Apr. 1949).
189. Clark, C. L., Private communication, The Timken Roller Bearing Co., Canton, Ohio.
190. Kiefer, G. C., *Engineering Experiment Station News*, The Ohio State University, **22,** 21–24, 44–51 (June 1950).
191. Norén, T., *Svetsaren*, **13,** 34–45, 49–61 (1948).
192. Menezes, L., Roros, J. K., and Read, T. A., *Symposium on the Nature, Occurrence, and Effects of Sigma Phase* (A.S.T.M.), 71–74 (1951).
193. Oliver, D. A., *Metal Progress*, **55,** 665–667 (1949).
194. Zapffe, C. A., and Worden, C. O., The Welding Journal, **30,** Research Suppl., 47-s to 54-s (1951).
195. Steinberger, A. W., De Simone, B. J., and Stoop, J., *Ibid.*, **29,** 752–764 (1950)
196. Sims, C. E., *Ibid.*, **30,** 52–53 (1951).
197. "Hydrogen in Weldments," critical review to be published in The Welding Journal.
198. Juretzek, H., *Giesserei*, **29,** 217–226, 243–249 (1942).
199. Foley, F. B., Private communication, The International Nickel Co., Bayonne, N. J.
200. Seabloom, E. R., *Welding Engr.*, **31,** 44–49 (Oct. 1946).
201. Campbell, H. C., Private communication, Arcos Corp., Philadelphia, Pa.
202. Zapffe, C. A., Worden, C. O., and Phebus, R. L., *Stahl u. Eisen*, **71,** 109–119 (1951).
203. Fuke, V., *J. Japan Welding Soc.*, **18,** 65–69 (1949).
204. Signora, M., and Baldi, F., *Met. Italiana*, **42,** 255–260 (1950).
205. Dickens, G. J., Douglas, A. M. B., and Taylor, W. H., *Nature*, **167,** 192 (1951).
206. Thomas, H., *Z. Metallkunde*, **41,** 185–190 (1950).
207. Bungardt, K., *Stahl u. Eisen*, **71,** 273–283 (1951).
208. Zapffe, C. A., *J. of Metals*, **3,** 247–248 (1951).
209. Burt, F. M., *Welding Engr.*, **35,** 28–30, 34 (March 1950).
210. Hoyt, S. L., *Metal and Alloys Data Book*, Reinhold Publ. Corp., New York, pp. 186–187 (1943).
211. Tofaute, W., and Bandel, G., *Inst. hierro y acero*, **3,** 177–186 (1950)
212. Shortsleeve, F. J., and Nicholson, M. E., *Symposium on the Nature, Occurrence, and Effects of Sigma Phase* (A.S.T.M.), 79–80 (1951).
213. Rose A. S., *Metal Progress*, **57,** 761–764 (1950).

Improving the Engineering Properties of Ferritic Stainless Steels

R. H. Kaltenhauser

FERRITIC STAINLESS STEELS have generally taken a backseat to the more versatile austenitic stainless steels. With increasing costs of nickel and alloying elements, the ferritic steels are receiving increased interest. During the several decades when the austenitics held prime importance, steel producers were continuing research and development work on the ferritics to improve their engineering properties. Thus, to consider today's ferritic stainless steels as replacements for austenitic stainless steels, or for other materials, users should reacquaint themselves with the properties of these materials.

There are a number of service conditions of such severe corrosivity that austenitic stainless steels must be used in preference to the ferritics. However, there are many more applications for stainless steeels which have less severe service conditions and in which the unique properties of the austenitics are not required. Additional applications exist where ferritics have property advantages over the austenitics, including improved machinability, higher thermal conductivity, lower thermal expansion, and immunity to chloride stress-corrosion cracking. There are an increasing number of applications where the higher cost of austenitic stainless steels, compared with that of the ferritics, makes the austenitics economically unsuitable. Many present applications for austenitic stainless steels used ferritics during early development. Thus, when considering the suitability of ferritics for some applications, the technical facets may already be known, and a reappraisal can be based on the current price difference between two grades.

Mr. Kaltenhauser is Head, Stainless Steel Technology Section, Research Center, Allegheny Ludlum Industries Inc., Brackenridge, Pa.

Metallurgical Considerations

Thielsch summarized the major metallurgical objections to the use of ferritic stainless steels as engineering materials (1). These are 885 F (475 C) embrittlement, sigma phase, high temperature embrittlement, and notch sensitivity. Each of these lowers the ductility and toughness of the ferritic stainless steels at room temperature. These conditions are manifest to the fabricator and user in various restrictions on operations such as hot-forming, brazing, welding, and heat treating in order to minimize high-temperature exposure. The sequence of fabrication operations, and often the design, consider the relatively low toughness and ductility of the material. Cold-forming operations must also be commensurate with possibly reduced ductility and toughness. Final solutions to these problems have not yet been obtained, but fabrication and service can work around them.

The technical literature contains many discussions of these metallurgical problems. Suggestions are given to minimize the occurrence of the problem and to remove the detrimental effects by appropriate heat treatment. For example, the first three metallurgical conditions can be re-

Source: *Metals Engineering Quarterly,* 11(2), May 1971, 41-47, © 1971 American Society for Metals

moved by heat treating at approximately 1400 F (760 C).

Two additional metallurgical considerations which have received less attention are the formation of intergranular martensite in ferritic weld metal and heat-affected zones and the susceptibility to intergranular corrosion of ferritic weld metal and heat-affected zones. It is surprising that these have received such scant attention in the literature. Proper control and correction in ferritic stainless steels will significantly improve their engineering properties and increase the range of applications.

Commercial Ferritic Stainless Steels

The present commercial ferritic stainless steels are shown in Table 1. The basic grade is Type 430, which has a completely ferritic structure when in the annealed condition. Since the principal alloying element is chromium, the iron-chromium phase diagram describes the metallurgical structures present, Fig. 1. It indicates that at 16% chromium the structure would be completely ferritic at all temperatures. However, the addition of carbon and residual elements to the levels found in commercial steels increases the austenite loop and significantly broadens the ferrite-plus-austenite region (2).

During repeated hot-working and annealing operations used in producing stainless steels, most of the austenite is gradually decomposed into ferrite and carbides. To avoid production and fabrication problems associated with the formation of hard martensite, a completely ferritic steel is desirable. The analysis ranges shown in Table 1 result in only small amounts of austenite being formed during production of the steels. However, during the thermal cycle of a welding operation, austenite will be formed interdendritically in the weld metal and at the grain boundaries in the weld heat-affected zone. With such a distribution, even a small amount of martensite can significantly decrease ductility and toughness.

The tendency to form martensite in weld metal can be determined by making the calculation shown in Table 2. The relative austenitizing and ferritizing strengths of the various alloy elements (after Thielemann, et al.) (3) have been used to calculate a ferrite factor. The high and low range of these factors for each of the ferritic steels is shown. Metallographic observations have been made of the welded structures for the presence or absence of martensite to determine the ferrite factor below which austenite will form. These calculations have only considered the austenitizing ability of the various alloying elements, and have not taken into consideration second phases which may be formed, such as titanium carbides or aluminum nitrides. Corrections could be made for these phases based upon the stoichiometric equivalents of the elements making up the second phase. Such calculations are not significantly better than the simplified calculation.

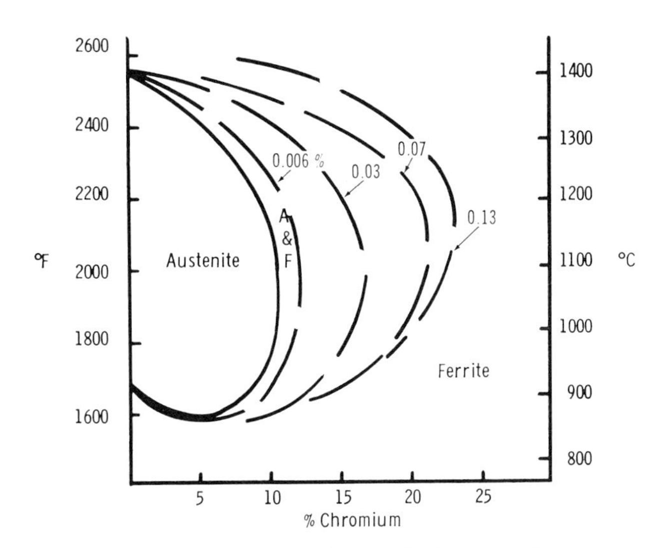

Fig. 1. Effect of percent carbon and nitrogen (given equal weight) on austenite and ferrite regions.

The ferrite factor below which martensite is present is not the same for the various alloys. Thus, observations must be made on each alloy and cannot be transferred from one system to another. This information illustrates the relatively wide range of the metallurgical balance that exists in these commercial alloys and the relative position of the composition which is completely ferritic in a welded structure.

11.5% Cr Ferritic Stainless Steel

A low-chromium stainless steel has a ferritic structure which avoids production and fabrication problems. Calculations of its ferrite stability predict either the presence or absence of some martensite. For example, when SAE Type 51409, or muffler steel, is used in applications requiring welding and forming, the presence of martensite would decrease the formability of the welded article. The higher ductility of a martensite-free weld is illustrated by the percent elongation and smallest radius for a bend as shown in Table 3. Since martensite in the as-welded condition would be untempered, it probably would lower impact and fatigue properties of the weldment.

The weld metal and heat-affected zone structure of Type 51409 gas tungsten-arc welded strip material is shown in Fig. 2 and 3. The material welded in Fig. 2 contains a considerable amount of interdendritic martensite in the weld metal and in the heat-affected zone. The material in Fig. 3 has a higher ferrite factor and produced only a trace of martensite in the weld metal, with none in the weld-heat-affected zone. The ductility of these two types of weld

Table 1 Composition of Commercial Ferritic Stainless Steels

Type	C	Cr		Other
SAE 51409	0.08 max.	10.5	11.75	6 × C min Ti
AISI 405	0.08 max.	11.5	14.5	0.2 Al
AISI 429	0.12 max.	14.0	16.0	
AISI 430	0.12 max.	16.0	18.0	
AISI 434	0.12 max.	16.0	18.0	0.75/1.25 Mo
HWT	0.07 max.	17.75	18.50	12 × C min Ti

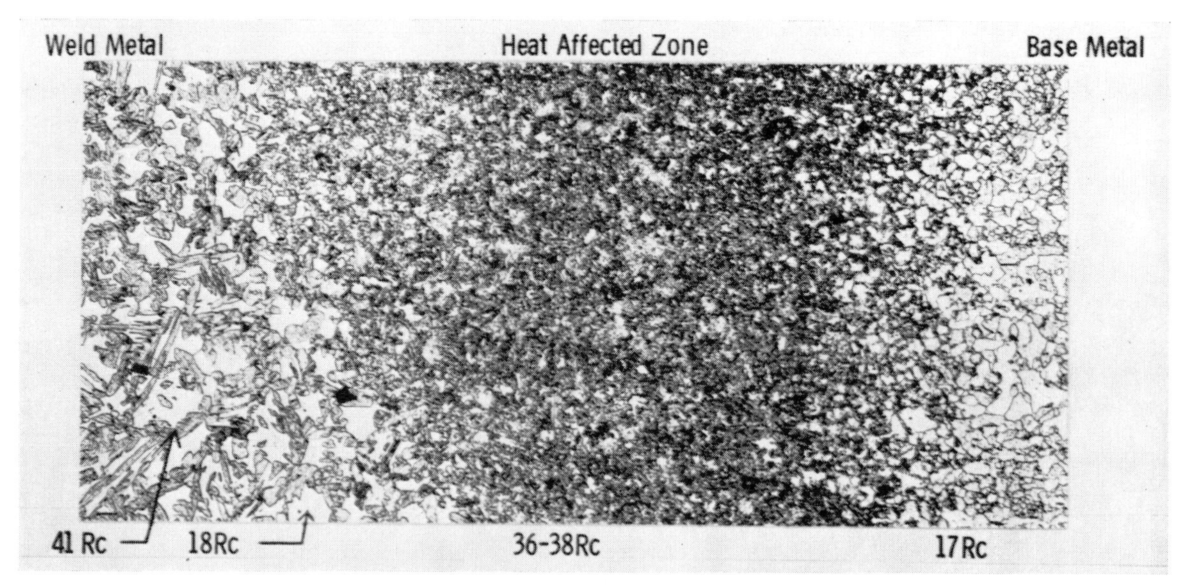

Fig. 2. Weld metal and heat-affected zone of gas tungsten-arc-welded SAE 51409 steel. Note amount of interdendritic martensite.

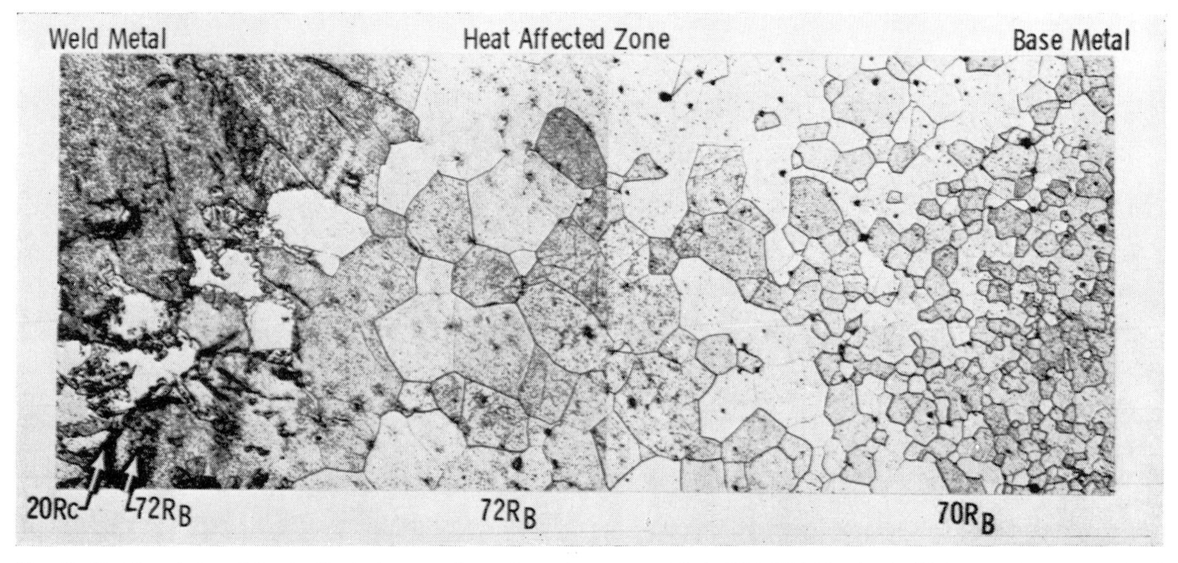

Fig. 3. Weld metal and heat-affected zone of gas tungsten-arc-welded SAE 51409 steel. Note greater amount of ferrite and small amount of martensite compared to Fig. 2.

Table 2. Calculation of Ferrite Factor for Ferritic Stainless Steels *

$F(\alpha) = -40(C + N) - 2Mn - 4Ni + Cr + 6Si + 8Ti + 4Mo + 2Al$

Type	Range of F(α)		Martensite in Weld when F(α)
405	3.5	19	
51409	6.5	20.5	<13.5
429	6.0	20.0	
430	8.0	22.0	<17
434	11.0	27.0	
HWT	12.5	30.5	<13.5

* Modified Thielemann's constants

structures would be similar to that shown in Table 3. Both weld structures are shown in the as-welded condition with no postweld heat treatment. For applications such as welded tubing or containers, which require the as-welded material to be formed, a weld structure similar to Fig. 3 might successfully make the part, while that represented in Fig. 2 might crack during forming.

Another application of this grade of stainless is the manufacture of tubing by high-frequency resistance welding. This procedure produces a weld structure similar to that of a flash or upset butt weld. The molten weld metal is extruded from the joint, and only a weld-heat-affected zone remains. The structure of the solidified weld metal is of no direct importance to service performance. However, the martensite in the weld-heat-affected zone may affect straightening or sizing operations, while martensite in the weld flash may influence the hot scarfing which removes the weld flash. Most of the welding applications of this grade have been for light-gage sheet and strip. One application requires material approaching ¼-in. thickness as a structural framework for shipping containers. This material is press-brake formed, and welded to itself and to carbon steel castings using the gas metal-arc welding process with 308 filler metal.

The SAE 51409 composition was first developed for use in automotive exhaust mufflers. The automotive manufacturers wished to increase the service life from that of galvanized and aluminized carbon steel. The corrosion rates of several materials are demonstrated in the Walker test, Fig. 6, which is a cyclic test in a synthetic muffler condensate. The performance of SAE 51409 is better than that of the Type 410 stainless. This improvement in corrosion performance is a result of titanium addition. (The influence of this addition is also manifest in improved oxidation resistance.) Even though commercial heats of muffler steel usually contain lower chromium than heats of Type 410 stainless, the titanium addition more than compensates for the lower chromium to give the improved corrosion and oxidation performance.

Low-chromium ferritic stainless steels also have two additional metallurgical advantages compared with the higher-chrome materials. At this level of chromium, the steel is not adversely affected by 885 F (474 C) embrittlement. Exposures up to 100 hr between 700 and 1000 F (371 and 538 C) resulted in no significant embrittlement or loss of corrosion resistance in both annealed and cold-worked SAE 51409 steel. In addition, the alloy is not susceptible to intergranular corrosion in its weld metal and weld-heat-affected zone. Welded samples have been exposed 30 days to corrosion environments, simulating different underground transformer case environments. These included synthetic seawater, brackish water, a neutral fertilizer mixture, swamp water containing, 0.05% sulfuric acid, mine water containing sulfuric acid with iron sulfate, and a solution of an alkali carbonate. Serious pitting was encountered in the synthetic seawater and the brackish water, especially at the liquid level. Some general attack was also noted in the sulfuric acid solution, but no pitting. Most significant of these results was the total absence of any localized attack in the weld metal or the weld-heat-affected zone. Type 430 stainless in the as-welded condition exposed to the same test conditions would be expected to show local intergranular attack in the weld areas.

A number of mildly corrosive applications are successfully using welded SAE 51409 with no evidence of preferential attack in the weld or heat-affected zones. These include culverts in acid mine drainage areas, auto mufflers, underground transformer cases, dry-fertilizer storage tanks, coal washing and handling equipment, furnace and space-heater chambers, and heat-exchanger welded tubing.

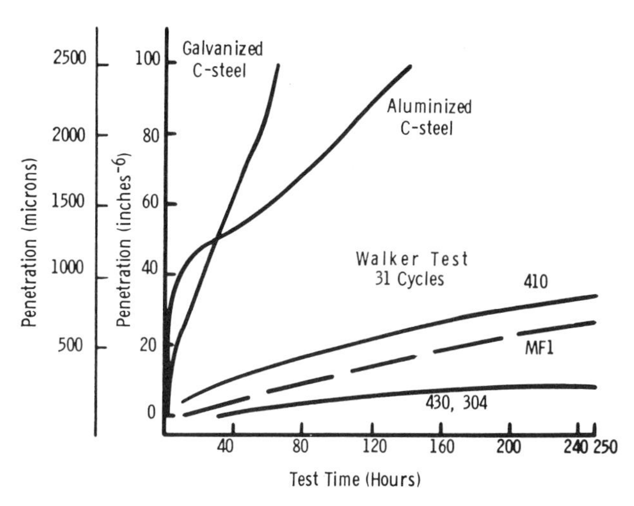

Fig. 4. Results of synthetic auto-muffler condensate corrosion test.

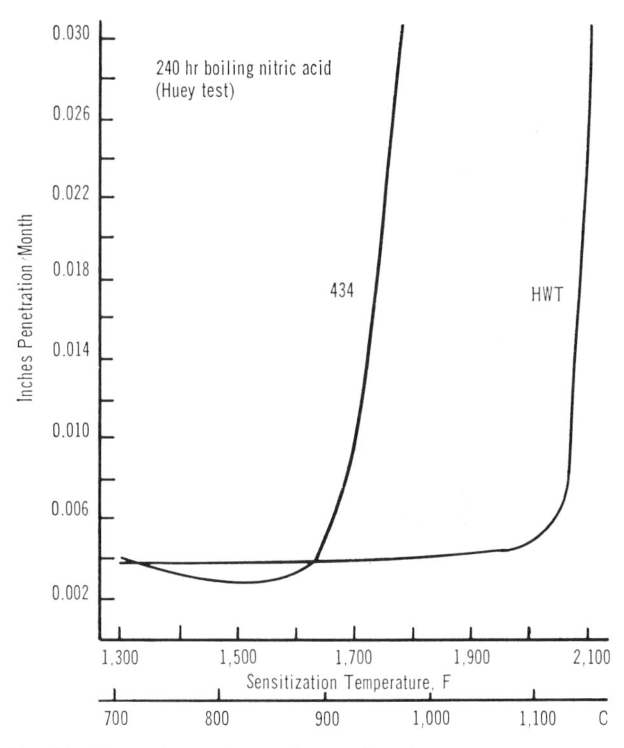

Fig. 5. Corrosion rates and sensitization temperatures in ferritic stainless steels.

17% Chromium Ferritic Stainless Steels

Unfortunately, the popular Type 430 stainless is susceptible to intergranular corrosion in the weld metal and weld-heat-affected zone. This susceptibility is not apparent

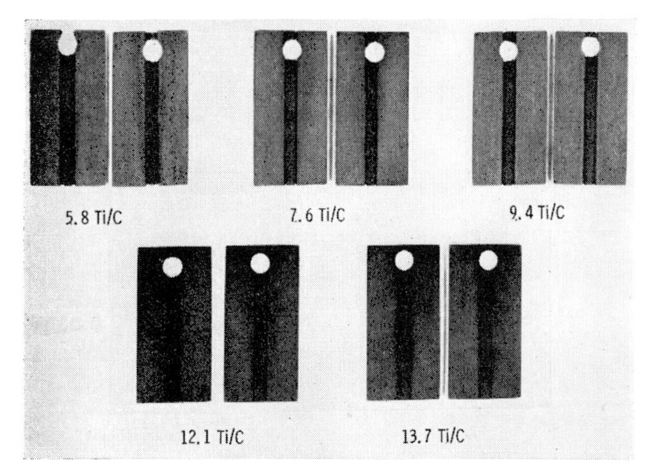

Fig. 6. Effect of Ti/C ratio on intergranular attack in 240-hr, boiling nitric acid tests.

under relatively mild corrosion conditions. For example, welded Type 430 stainless steel kitchen sinks have been used extensively without showing any evidence of this susceptibility. However, the range of corrosion environments to which welded, ferritic, stainless steels are susceptible is apparently much broader and includes less-severe environments than those to which welded, austenitic, stainless steels are susceptible. The mechanisms of this sensitization and various accelerated corrosion test data have been presented by Lula and other investigators (4). The susceptibility to intergranular corrosion in accelerated laboratory tests extended down to carbon and nitrogen levels lower than those commercially obtainable in 1954. More recently, Bond (5, 6) summarized work in this area and reported that nitrogen levels greater than 0.022% and carbon levels greater than 0.012% were sufficient to sensitize the materials after heat treatments at temperatures higher than 1700 F (927 C). Both investigators determined that titanium can stabilize material against this intergranular attack if the corrosion media is not too severe (highly oxidizing).

This type of intergranular attack has often been erroneously related to the appearance of martensite in the weld and heat-affected zone of ferritic steel, such as those illustrated in Fig. 2 and 3. As previously described in Table 2, the tendency for a particular heat of 430 to form martensite can be determined by calculating its ferrite factor. Modern techniques, such as the electron microscope and X-ray diffraction studies of extracted constituents, were utilized by Bond to establish the mechanism of this susceptibility, and confirmed that attack occurs when martensite is not present.

Stabilization Against Sensitization

As pointed out by various researchers, stabilization through additions of titanium does not completely eliminate susceptibility to this type of attack, but rather increases the sensitization temperature. This can be illustrated by the data for accelerated testing shown in Fig. 5. When annealed material is heat-treated at the temperatures indicated, the corrosion rate increases when the sensitization temperature is reached. Type 434 reaches its sensitization temperature at about 1700 F (927 C). The stabilized HWT modification (shown in Table 1) has a much higher sensitization temperature, about 2050 F (1221 C, than the unstabilized Type 434. In practice this improvement would have the effect of narrowing the susceptible area of a weld-heat-affected zone to include only that area of the base metal which had exceeded the necessary sensitization temperature. Although this is a significant improvement, it is not sufficient in all corrosion environments.

A decrease in the degree of sensitization in accelerated tests can also be observed with increasing ratios of titanium to carbon, Fig. 6. Weld metal and weld-heat-affected-zone intergranular attack in boiling nitric acid was suppressed when the Ti/C ratio exceeded 9.4. A shortcoming of such severe accelerated tests is that they may produce a relatively high rate of general corrosion which can then mask the occurrence of local attack. To further investigate this, corrosion tests were conducted in less severe media.

Longer testing is required to firmly determine the influence of titanium stabilization upon the susceptibility of these steels to intergranular attack in various other corrosive media. Accordingly, intergranular corrosion tests were conducted for 72 and 500 hr in boiling, acidified, copper-sulfate solution. The 72-hr tests were in the Krupp Solution, while the 500-hr tests had copper shot added as specified in ASTM A262, Practice E. Tests were run on welded material and on material which had been sensitized by annealing at 2100 F (1149 C). The titanium-to-carbon ratio was varied from approximately 10.5 to 13 in these heats. Unstressed welded samples were exposed for 72 and 500 hr in the boiling copper-sulfate test, and were then bent around a 2T radius in the weld-heat-affected zone. Welded Type 430 stainless fractured during bending after 72-hr exposure in this test. At the end of both exposure periods, the heat of stabilized ferritic stainless steel with the Ti/C ratio of 10.5 exhibited intergranular attack in the weld-heat-affected zone and cracked during bending. The remaining heats, with ratios of 11.3 and higher, showed no intergranular attack. To more closely simulate service conditions, a boiling hot-water test was developed. Welded samples were formed into stressed U-bends and immersed in a boiling solution of water containing 0.1% NaCl, 0.1% Na_2SO_4, 0.05% $MgSO_4$, and Cu shot. Welded Type 430 stainless developed intergranular attack within 3 days in this test. The welded heat of HWT stainless with the 10.5 Ti/C ratio developed intergranular attack after 4 months. The remaining three heats were continuously tested for 24 months, but did not develop any intergranular attack. Samples of these three heats, which had been sensitized at 2100 F (1140 C), also passed the 24-month exposure. The sensitized samples from the 10.5 Ti/C ratio heat-cracked in 1 to 8 weeks.

Table 3. Longitudinally Welded Properties of SAE 51409 Strip

Structure of Weld	0.2% YS, ksi	UTS, ksi	% Elong.	Min Bending R/T for 135°
Continuous interdendritic martensite	39	71	26	2.5
No martensite	36	57	34	0.4

Service data obtained on hot water tanks have substantiated the validity of these simulated laboratory tests. Gas-fired hot water tanks made from the stabilized ferritic stainless steel have been in service in various parts of the United States for 11 years. Some tanks have been recalled for thorough examination, but no evidence of intergranular attack has been found. Additional tanks remain in service to accumulate longer performance information.

Pitting and Crevice Corrosion

Service applications, such as hot water tanks or plumbing, anticipate the successful performance of the material for a long period of time, such as 30 or 40 years. Even though the resistance of welded material to intergranular attack can be demonstrated, such long-time service also requires the fabricated items to resist local pitting and crevice corrosion. The design and fabrication of stainless steel items can significantly influence the development of conditions suitable for crevice corrosion and pitting. For example, the design of the gas-fired hot water tanks previously discussed used a concave-downward bottom dome. The crevices thus formed at the bottom of the tank can collect solid matter which could intensify the stagnant conditions which usually exist in a crevice. This area is also subject to the most severe thermal gradients during the heating phase of its operation. Improvements in design to reduce the susceptibility for these types of corrosion would include elimination of the crevice by reversing the domed bottom of the tank and use of an electric immersion heater instead of external gas firing. In spite of these controversial design considerations, approximately 100 such tanks have successfully performed during 11 years of consumer service in various areas of this country representing a variety of water conditions. Of approximately 40 tanks examined after 6 months to 3 years of service, serious pitting and crevice corrosion were found in only four tanks. Duplicate tanks are still in service. The waters in these areas also were among those in the United States having the highest contents of chlorides and solids. Based upon these service experiences, the performance of the stabilized ferritic stainless steel HWT for consumer hot water tanks is rated as highly satisfactory.

Pitting Resistance of Ti-Stabilized Ferritic Stainless Steels

One method of comparing the pitting resistance of several alloys is to measure the relative breakthrough or pitting potential of the materials with the potentiostat. Several materials were tested in a variety of chloride solutions with concentrations varying up to 10,000 ppm. These levels were selected to represent not only potable waters but also industrial waste waters, such as in mine drainage. Drinking water in some parts of the United States may contain up to 500 ppm chlorides. Mine drainage water in various parts of western Pennsylvania may contain high quantities of sulfates and chlorides at pH levels as low as 2.5.

A range of breakthrough potentials was obtained on each of the alloys tested. This is the result of natural variations in surface condition of the steels and polarization effects during the electrochemical reaction. The minimum breakthrough potential of the titanium-stabilized, 11 Cr material (SAE 51409) was significantly higher than the titanium-free, 410 stainless which contains 12 Cr. Similarly, the

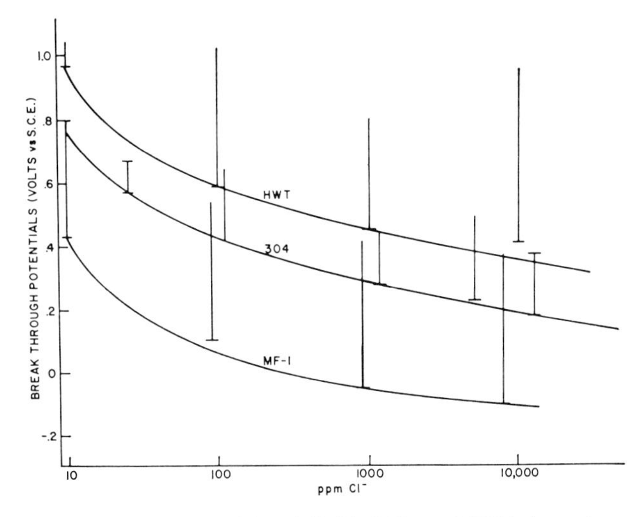

Fig. 7. Pitting potentials of HWT, 304, and MF-1 in various chloride compounds.

pitting potential of the titanium-stabilized, HWT alloy was significantly higher than that of the titanium-free, Type 430 stainless. The chrome-nickel Type 304 stainless is also shown for comparison in Fig. 7. The higher pitting potential of the 18-Cr, titanium-stabilized, ferritic, stainless steel compared with that of the 18Cr-8Ni, Type 304 stainless was unexpected. However, repeated testing and observations of service conditions have confirmed these results.

Welded tubing of these three alloys has been evaluated as domestic water piping in a number of tests and service installations. As an example of the correlation between the potentiostatic pitting potential data and water tubing service, a 6-month examination of a water system revealed that the HWT alloy contained no rusting or pitting on the interior surface of the tubing. Type 304 stainless in the same system contained scattered rust spots, while the lower-chromium titanium-stabilized alloy (SAE 51409) had additional rusting and pitting spots dispersed over its interior surface. Thus, the service experience correlated directly with the pitting potential data.

This particular water system was constructed to simulate conditions of installation and use in a typical home. Several weeks elapsed between soldering of the first connections and the introduction of water into the completed system. Since most soldering fluxes for stainless steels produce pitting, the residue should be promptly washed from the soldered joint. Unfortunately, this necessary cleaning step cannot be promptly and effectively done in building construction. Various laboratory simulations and operating installations of soldered stainless steel plumbing systems have been made and examined after different service conditions and times. Commercial soldering fluxes for stainless steels initiated pitting or rusting on the interior surfaces in most of the test conditions.

A recently developed soldering flux for stainless steels does not cause pitting when in contact with the stainless after soldering. This flux has a phosphoric-acid base which is active enough to clean the stainless at the soldering temperature, but upon cooling to ambient temperature the flux residue is not corrosive. Stainless steel water tubing, soldered with this phosphoric acid-based flux and exposed to the same variety of test conditions, did not contain any rusting or pitting.

Improved Ferritic Stainless Steels

In summary, several significantly improved ferritic stainless steels, which complement Type 430, are now in use. Type 434 stainless with its 1% Mo, offers improved corrosion resistance for increasingly severe service, such as automotive trim in areas using de-icing salts and slag on winter highways. The 11% Cr ferritic steel (SAE 51409) provides suitable corrosion resistance for less-severe applications in which nonstainless materials are unsuitable. In addition, this steel is relatively insensitive to 885 F embrittlement and intergranular sensitization from welding. By proper balancing of their compositions, the ferritic stainless steels can be kept completely ferritic in the weld metal and weld-heat-affected zones. The absence of intergranular martensite significantly increases their formability and corrosion resistance. Addition of the proper level of Ti in the 18-Cr ferritic stainless (HWT) stabilizes the welded alloy against intergranular corrosion attack in many common service environments. Stainless steels can be soldered with a recently developed phosphoric acid-based flux whose residue does not cause pitting of the stainless.

These improvements offer fabricators new lower-cost methods of providing suitable stainless quality for an increasing variety of maintenance-free, long-life applications. Further improvements in the fabrication and corrosion performance of ferritic stainless will appear in the near future. New melting facilities are coming into use to provide new lower levels of C and N in all stainless steels. The vacuum-induction, electron-beam, melting process will provide a ferritic stainless with low enough carbon to avoid intergranular sensitization during welding ($<0.01\%$). Vacuum-refining melting processes will produce stainless steels with significantly lower levels of C and N than in present AISI stainless steels. With only a very minor addition of a stabilizing element, ferritic steels melted by this process will achieve stabilization against intergranular corrosion attack. The lower levels of C and N will also provide improved toughness and ductility. By reducing the level of Ti required for stabilization, the toughness and cleanliness of the ferritic stainless steels will be improved.

With these improved properties now available, the ferritic stainless steels should be carefully evaluated as candidate materials for an ever-increasing number of corrosion resistant applications.

REFERENCES

1. H. Thielsch, Physical and Welding Metallurgy of Chromium Stainless Steels, Weld. J., May 1951.
2. E. Baeiecken, et al., Investigations Concerning the Transformation Behavior, the Notched Impact Toughness and the Susceptibility to Intercrystalline Corrosion of Iron-Chromium Alloys with Chromium Contents to 30%, Stahl und Eisen, 81, No. 12 (1961), 768.
3. R. Thielemann, Some Effects of Composition and Heat Treatment on the High Temperature Rupture Properties of Ferrous Alloys, ASTM Proceedings, Vol. 40, 1940.
4. R. A. Lula, A. J. Lena and G. C. Kiefer, Intergranular Corrosion of Ferritic Stainless Steels, ASM Trans. 46 (1954), 197.
5. A. P. Bond, Mechanisms of Intergranular Corrosion in Ferritic Stainless Steels, Trans. AIME, 245 (October 1969), 2127.
6. A. P. Bond and E. A. Lizlovs, Intergranular Corrosion of Ferritic Stainless Steels, J. Electrochem. Soc., 116 (September 1969), 1305.

Mechanism of High Temperature Embrittlement and Loss of Corrosion Resistance in AISI Type 446 Stainless Steel★

*J. J. DEMO**

Abstract

The ferritic alloys, particularly AISI Type 446 steel with its high chromium content, have desirable properties of corrosion resistance, low raw material cost, and resistance to stress corrosion cracking (SCC); yet they are not widely used in construction work because of the damaging effects of high temperature exposures (such as welding) on their corrosion resistance and ductility. This work describes the causes for their loss of corrosion resistance and ductility, and defines changes in composition and heat treatment that would improve their material properties.

The binary iron-chromium body centered cubic ferritic stainless steels have been known for more than 30 years; however, except for the AISI Type 430 steel, their use has been limited. AISI Type 446 steel contains 23 to 27% Cr, the highest Cr content of the commercially available ferritic alloys. This alloy has several properties which make it commerciallv attractive. When annealed, it displays excellent corrosion resistance to a variety of corrosive solutions, does not stress crack in chloride containing environments,[1] and has a relatively low material cost. However, its corrosion resistance and ductility are severely weakened by high temperature exposures such as welding. Therefore, it has been used as a construction material far less than have the austenitic grades of stainless steel which contain chromium and nickel. Figure 1 shows the results of welding on the ductility and corrosion resistance of AISI Type 446 steel.

This paper deals with the reasons for the loss of corrosion resistance and ductility that follow exposure of ferritic stainless steels to high temperatures. An attempt was made to determine which of several previously proposed theories, or parts of theories, might be applied to these problems, or might be combined with recent data into a new theory to explain the high temperature problems of ferritic stainless steels. After the factors responsible for property loss are defined, it may be possible to alter such variables as composition or heat treatment to produce better materials.

Commercial and experimental heats of chromium alloys containing 26% iron (Fe-26% Cr) were chosen for this study because of their greater corrosion resistance than the 18% Cr Type 430 steel.

Surveys by Thielsch[2] and the ASTM[3] have summarized the literature on this high temperature problem, as well as other problems of embrittlement in ferritic stainless steels, such as 475 C (887 F) and sigma phase embrittlement. In this paper, only the effects of high temperature exposure on both the corrosion resistance and ductility are discussed.

Previous investigators have not studied concurrently the relationships of high temperature exposure to loss of ductility and to loss of corrosion resistance. Two theories have been advanced to explain high temperature embrittlement: (1) the segregation or coherent state theory postulates that embrittlement results from a clustering or segregating of carbon atoms in the ferrite matrix.[4] This clustering occurs when rapid cooling prevents precipitation of carbon as carbides from solid solution. Annealing at 730 to 790 C (1346 to 1454 F) causes the carbon to precipitate as carbides, thus removing the coherent carbon clusters which cause embrittlement, and (2) according to the martensite mechanism,[5] regions relatively high in carbon transform first to austenite at high temperatures, then, during rapid cooling, to brittle martensite. Annealing at 649 to 788 C (1200 to 1450 F) tempers the martensite and causes chromium carbide to precipitate.

Several experimenters have postulated reasons for loss in corrosion resistance following exposure of high chromium ferritic stainless steels to high temperatures. Houdremont and Tofaute[6] postulated that a carbon rich austenite forms at the sensitizing temperature. Upon cooling, easy to dissolve iron carbides precipitate at the grain boundaries where the supersaturated austenite is in contact with ferrite. Annealing at approximately 760 C (1400 F) converts the iron carbides to chromium carbides

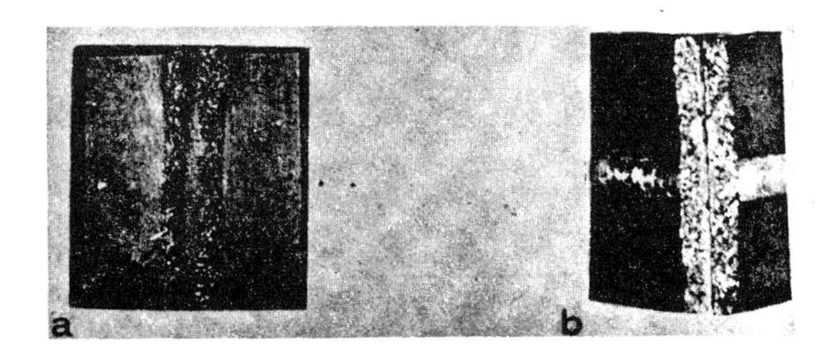

FIGURE 1 — Effect of welding on the ductility and corrosion resistance of AISI Type 446 steel (27.7% Cr, 950 ppm C, 770 ppm N). (a) = corrosion resistance following exposure to boiling 50% H_2SO_4-$Fe_2(SO_4)_3$ for 24 hours, and (b) = ductility.

★Presented at CORROSION/71, March 22-26, Chicago, Ill.
*Engineering Materials Laboratory, Engineering Research Division, E. I. du Pont de Nemours & Co., Inc., Wilmington, Del.

TABLE 1 – Chemical Compositions of Alloys

	Commercial Alloys		
	AISI Type 446 SA2	AISI Type 304	Typical Experimental Alloy E72
Cr	27.2	18.4	26.0
Ni	0.35	9.1	0.062
C	0.095	0.057	0.014
N	0.077	0.035	0.004
Mn	0.59	1.46	1.37
Si	0.28	0.43	0.92
P	0.016	0.025	0.015
S	0.013	0.013	0.009

which resist chemical attack and therefore make the metal immune to intergranular corrosion.

Hochmann[7] also postulates that austenite formation at the sensitizing temperature is necessary to explain intergranular corrosion. However, he believes that the grain boundary austenite phase may be so impoverished in chromium that it readily dissolves.

In contrast to these theories, the recent and perhaps most comprehensive study of this subject by Lula, Lena, and Kiefer[8] rejects any mechanism of intergranular corrosion in ferritic steels that requires the formation of austenite. They pointed out that no operation to prevent austenite formation (i.e., increasing the chromium content, adding Ti, Si, or V, or reducing the carbon content) makes steel immune to intergranular attack after high temperature exposure. They hypothesize that precipitation of either a carbide or nitride phase occurs in the grain boundary during the cooling of the supersaturated ferrite matrix. The precipitate, by straining the grain boundary matrix, makes these areas susceptible to intergranular attack. Annealing at 650 to 815 C (1202 to 1499 F) relieves the stresses caused by the precipitated phases, thus restoring corrosion resistance.

The effects of heat treatment on the corrosion resistance of austenitic and ferritic type stainless steels can also be compared. The heat treatments which cause sensitization (loss in corrosion resistance characterized by severe intergranular attack with eventual grain dropping) in commercial ferritic chromium grades and in austenitic chromium-nickel steels, differ. The ferritic steels sensitize during rapid cooling from temperatures above 950 C (1742 F). The austenitic stainless steels sensitize following a slow cool or a hold in the 400 to 800 C (752 to 1472 F) range. An austenitic stainless steel that is rapidly cooled from temperatures above 950 C is not sensitized.

TABLE 2 – Time in Seconds Required for Samples to Cool to Selected Temperatures from 1100 C

	Cooling Methods(1)			
Temperature	Water Quench	Air Cooled	Furnace(2) Cooled	Programmed Furnace Cool(3)
800 C (1472 F)	0.17	14	1560	7200
500 C (932 F)	0.46	66	6400	14,400

(1)Sample size: 1 x 1 x 0.1 in.
(2)Furnace power turned off at 1100 C (2012 F).
(3)Programmed cooling 20 C/8 min (i.e., 2.5 C/min).

TABLE 3 – Effect of Thermal Treatments on the Corrosion Resistance and Ductility of AISI Type 446 and 304 Steels

Sample Designation		Corrosion Resistance,(1) mpy (Exposure hrs)	% Elongation (1 in)
AISI Type 446 Steel			
1	As-received	30 (120)	25
2	30 min, 1100 C (2012 F), W.Q.	780 (24)	2
3	30 min, 1100 C (2012 F), A.C.	800 (24)	27
4	30 min, 1100 C, Slow Cool(2) to 850 C (1562 F), W.Q.	25 (120)	33
5	30 min, 1100 C, W.Q. + 30 min, 850 C, W.Q.	42 (120)	27
AISI Type 304 Steel			
6	As-received	28 (120)	78
7	30 min, 1100 C (2012 F), W.Q.	29 (120)	84
8	30 min, 1100 C (2012 F), A.C.	38 (120)	85
9	120 min, 677 C (1250 F), A.C.	265 (48)	84

(1)Boiling 50% H_2SO_4-$Fe_2(SO_4)_3$.
(2)Cooling rate 20 C/8 min.

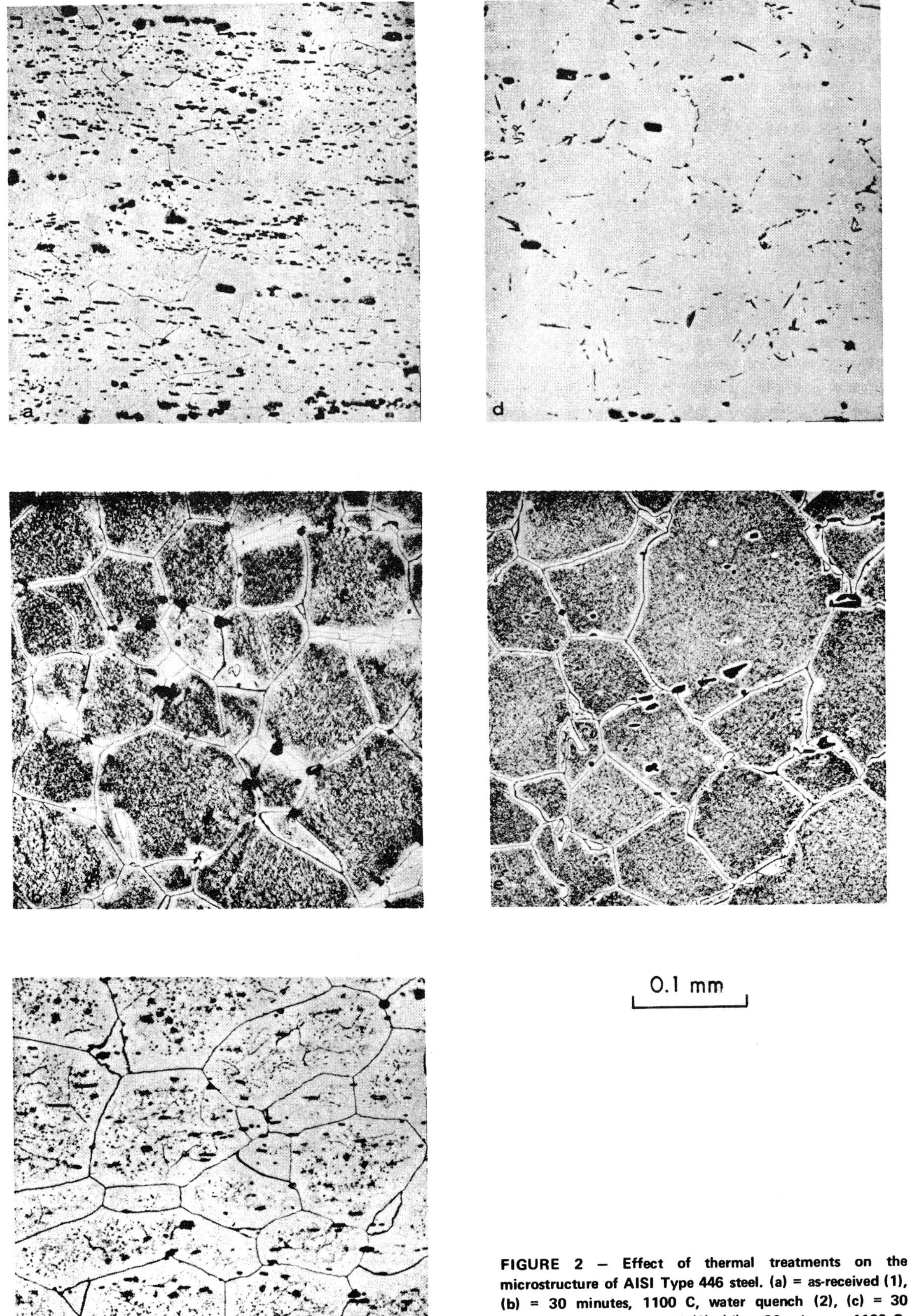

FIGURE 2 — Effect of thermal treatments on the microstructure of AISI Type 446 steel. (a) = as-received (1), (b) = 30 minutes, 1100 C, water quench (2), (c) = 30 minutes, 1100 C, air cool (3), (d) = 30 minutes, 1100 C, water quench plus slow cool to 850 C, then water quench (4), and (e) = 30 minutes, 1100 C, water quench plus 30 minutes, 850 C, water quench (5).

The factors causing sensitization in austenitic stainless steels have been studied by many workers, especially by Bain *et al*[9] in 1933. They hypothesized that sensitization occurred in austenitic stainless steels when carbon was precipitated from solid solution in the form of chromium rich carbides during the cooling through the 400 to 800 C range. These carbides draw heavily upon the chromium of the adjacent matrix causing intergranular corrosion. More recent work has given support that chromium depletion in the neighborhood of the carbides is the primary cause of intercrystalline attack in austenitic stainless steels.[10]

The effects of heat treatment on the mechanical properties of ferritic and austenitic stainless steels differ also. High temperature exposure severely embrittles commercial ferritic steels; but affects the ductility of austenitic steels only slightly.[9]

Experimental Procedures

Samples

Typical compositions of the two commercial alloys, AISI Type 446 ferritic stainless steel and AISI Type 304 austenitic stainless steel, and of the experimental Fe-26% Cr alloys, are shown in Table 1.

Samples of the commercial AISI Type 446 steel were obtained by slicing 100 mil thick sections from a 2 x 2 inch hot rolled, annealed, wrought bar stock. (Usually, the anneal consists of heating to a temperature of 788 to 871 C (1450 to 1600 F) long enough to allow thorough soaking followed by a water quench.) From these slices, corrosion coupons 1 x 1 inch were cut in such a way that the 1 x 1 inch surface was parallel to the rolling direction. Samples of commercial AISI Type 304 steel were cut from 0.1 inch thick sheet stock, cold rolled, and annealed at the mill.

Samples of the experimental Fe-26% Cr alloys, made from electrolytic chromium and iron, were produced as 600 gm buttons in a vacuum arc melting furnace. Controlled additions of nitrogen and carbon, in the form of Cr_2N and spectrographic Grade C, respectively, were made to the experimental heats. At 1260 C (2300 F), the experimental buttons were hot rolled to 100 mil thickness. The resulting sheets were annealed for 30 minutes at 850 C (1562 F) and then water quenched.

Heat Treatments

Samples 1 x 1 inch were heat treated at selected temperatures in flowing helium in a tube furnace. Depending on the experiment, the samples were water quenched, air cooled, or cooled at two rates in the furnace. Cooling rates for the various cooling modes were measured. The times required to reach 800 C and 500 C (932 F), respectively, from 1100 C (2012 F) for the various cooling modes are recorded in Table 2.

Corrosion Tests

Untreated and heat treated samples were wet belt ground to an 80 grit finish to remove surface irregularities and scale. The cleaned, weighed, and measured samples were supported in glass cradles and were exposed to 600 cc of a boiling solution of 50% H_2SO_4 containing 25 gm $Fe_2(SO_4)_3 \cdot XH_2O$. This test is described in ASTM Bulletin No. 229 for austenitic stainless steels.[11] Recent work by Streicher[12] has shown that it is also suitable for assessing susceptibility to intergranular corrosion in ferritic alloys such as AISI Type 430 and Type 446.

Ductility Tests

After selected heat treatments, samples were machined into flat, tensile specimens of 1 inch gage length. The tensile samples were tested in an Instron testing machine at a cross-head speed of 0.05 in/min. The ductility of welded samples such as shown in Figure 1 was evaluated by bending it through a 180 degree angle along a line transverse to the weld axis according to the guided bend test procedure, and the apparatus described in the ASME Pressure Vessel Code.[13]

Precipitate Identification

The precipitates were analyzed and identified by transmission electron diffraction of thinned sections, X-ray diffraction of residues extracted by a bromine-methanol technique,[14] and chemical analysis of the residue.

TABLE 4 – Effect of Slow Cooling[(1)] to Various Temperatures on Corrosion Resistance of Commercial AISI Type 446 Steel (All Samples Initially Heated 30 Minutes at 1100 C (2012 F)

Sample Designation	3	2	10	11	12	13	14
	Air Cooled 1100 C (2012 F)	Water Quenched 1100 C (2012 F)	Furnace[(2)] Cooled 1100 C (2012 F)	Furnace Cooled To and Water Quenched From 1000 C (1832 F)	900 C (1652 F)	800 C (1472 F)	700 C (1292 F)
Corrosion[(3)] Rate, mpy	800	780	202	767	27	20	18
Exposure, Hours	(24)	(24)	(72)	(120)	(120)	(120)	(120)

(1) Cooling rate 20 C/8 minutes.
(2) Furnace power turned off at 1100 C.
(3) Boiling 50% H_2SO_4-$Fe_2(SO_4)_3$, 120 hours exposure unless otherwise noted.

Experimental Results (AISI Type 446 Steel)

Effect of Thermal Treatment

Several 1 x 3 x 0.1 inch pieces of AISI Type 446 commercial steel were given a high temperature treatment at 1100 C and subjected to various cooling treatments. The pieces were first tensile tested, then the ends of the samples were corrosion tested as outlined above. The results of these tests are shown in Table 3.

The as-received sample 1 of Type 446 steel displayed good corrosion resistance and ductility. (Good corrosion resistance throughout this paper means freedom from intergranular attack.) Following treatment at 1100 C, samples 2 and 3 suffered severe intergranular attack regardless of whether they had been water quenched or air cooled. However, they recovered their corrosion resistance when they were either very slowly cooled in the furnace from 1100 to 850 C (sample 4); or were quenched from 1100 C, reheated to 850 C, and water quenched (sample 5). Only sample 2, which was exposed to 1100 C and water quenched, suffered a loss of ductility. The reasons for this will be discussed later. All other samples, regardless of their corrosion resistance, had ductility equivalent to that of sample 1.

In contrast, sample 7 of Type 304 steel, which was water quenched from 1100 C, exhibited a corrosion resistance and room temperature ductility that were essentially like those of the as-received sample 6. When sample 9 was heated at 677 C (1051 F), the sensitizing temperature, its corrosion rate as expected increased, but with no reduction in ductility.

Microstructures

For austenitic chromium-nickel stainless steels, there is generally a high correlation between susceptibility to intergranular attack and the presence of intergranular precipitates. To determine if a comparable correlation exists for the AISI Type 446 steel described above, the microstructures of the heat treated samples were examined (Figure 2).

The as-received sample 1 displayed a structure of relatively fine grain size with clean grain boundaries, but with large precipitates in the grain faces. Sample 2, heated to 1100 C and water quenched, exhibited a series of dramatic changes: a fine continuous intergranular precipitate, a very fine precipitate in the grain faces exposed by sectioning, and precipitate-free zones as well as patches of unetched materials at the grain boundaries. Analogously, sample 3, which was air cooled from 1100 C showed an intergranular precipitate, but no discrete patches of an unetched phase, and significantly less fine precipitate in the grain faces than did sample 2. Sample 4, which was exposed to 1100 C and then slowly cooled to 850 C had no fine precipitate in the grain faces, but a discontinuous thick intergranular precipitate was observable. Sample 5 was heated to 1100 C, quenched, reheated at 850 C, and quenched. Its structure was little different from that of sample 2. In sample 5, a continuous intergranular precipitate and a fine precipitate in the grain faces were apparent. The outlines of the previous patches of an *unetched* phase remained, but with observable precipitates in them.

These data show that there is not a complete one to one correlation between the loss of corrosion resistance and

TABLE 5 – Effect of Hold Time at 850 C (1562 F) on Corrosion Resistance of AISI Type 446 Steel Sensitized[1] at 1100 C

Hold Time, min, at 850 C, W.Q.	Corrosion Resistance,[2] mpy	ASTM Grain Size
Water Quenched from 1100 C	789	<1
5	65	1
15	58	2-4
30	42	3-6
60	59	5-7
120	48	3-5
180	53	2-4
300	54	<1-2

[1] All samples were exposed for 30 minutes at 1100 C and water quenched.
[2] Boiling 50% H_2SO_4-Fe_2SO_4, 120 hours exposure unless otherwise noted.

TABLE 6 – Effect of Multiple Thermal Treatments on the Corrosion Resistance of Commercial AISI Type 446 Stainless Steel

Sample Designation	Heat Treatment	Corrosion Resistance,[1] mpy (Exposure hr)
2	30 min, 110 C (2012 F), W.Q.	780 (24)
5	30 min, 1100 C (2012 F), W.Q. + 30 min, 850 C (1562 F), W.Q.	42 (120)
15	30 min, 1100 C, W.Q. + 30 min, 850 C, A.C.	111 (120)
16	30 min, 1100 C, W.Q. + 30 min, 850 C, W.Q. + 10 min, 800 C (1472 F), W.Q.	30 (120)
17	30 min, 1100 C, W.Q. + 30 min, 850 C, W.Q. + 10 min, 600 C (1112 F), W.Q.	147 (24)

[1] Boiling 50% H_2SO_4 + $Fe_2(SO_4)_3$.

the presence of a continuous intergranular precipitate. The corrosion resistance of sample 5, for example, which was reheated to 850 C was recovered although a continuous intergranular precipitate was still present. Further, a loss in ductility cannot be correlated to either corrosion resistance or to the presence of a precipitate. Both the air cooled sample 3 and the 850 C annealed sample 5 had continuous intergranular precipitates and good ductility, but sample 3 showed poor corrosion resistance.

Effect of Slow Cooling from 1100 C

The excellent corrosion resistance of sample 4 (Table 3) which was exposed to 1100 C and then slowly cooled to 850 C at 20 C/8 min (i.e., 2.5 C/min), is not inconsistent then with the presence of an agglomerated intergranular precipitate. To determine the effect of temperature and cooling rate on corrosion resistance recovery, additional 446 type samples were heated at 1100 C, then slowly cooled at 20 C/8 min to various temperatures from which they were quenched. Other samples were cooled to room temperature in a turned off furnace to determine the effect of a cooling rate intermediate between an air cool and the controlled furnace cool. The results of the corrosion test appear in Table 4.

Samples slowly cooled from 1100 C to the range 700 to 900 C (1292 to 1652 F) and then quenched have good corrosion resistance (samples 12, 13, 14). Although the corrosion resistance of the furnace cooled sample 10 was poor, it was still considerably better than that of the air cooled (sample 3) and water quenched samples (sample 2) whose temperatures were more rapidly reduced. This observation indicates that a time-dependent healing process must occur at an intermediate temperature range which is below 1000 C (1832 F).

Effect of Holding Time at 850 C

The corrosion resistance of samples sensitized at 1100 C can be recovered, not only by slow cooling from this temperature, but by reheating to 850 C, a temperature within the normal annealing range for ferritic stainless steels. To measure the speed of corrosion resistance recovery, 1 x 1 inch samples of AISI Type 446 steel were sensitized for 30 minutes at 1100 C, water quenched, reheated for different periods at 850 C, and again water quenched. The samples were corrosion tested, and companion samples were mounted for metallographic examination. The corrosion results are shown in Table 5. The process responsible for recovery of corrosion resistance is rapid, as indicated by the sample held for only 5 minutes that recovered its corrosion resistance. No significant increases in corrosion resistance were noted for the longer hold times up to 5 hours. At all hold times, the microstructures were similar, showing a fine precipitate in the grain faces, and a continuous intergranular precipitate.

The effect of holding time at 850 C on grain size was interesting (Table 5). A 60 minute holding at 850 C produced the smallest grain size, a refinement which is unusual in a material that does not undergo transformation. A possible reason for this apparent grain refinement is discussed later.

Identity of Precipitates

To clarify the way in which precipitates affect corrosion resistance, those that appear after thermal treatments which sensitize, and those treatments which recover corrosion resistance, have been identified. In this study, bromine-methanol extraction techniques[14] with X-ray diffraction, transmission electron diffraction of thinned sections, and chemical analysis of residues were used. Regardless of the thermal treatment used and the samples' corrosion performance, the precipitates were found to be a mixture of $M_{23}C_6$ and β-Cr_2N. Fe_3C has not been observed using either electron diffraction of thinned sections or X-ray diffraction of extracted residues. Thus, it would appear that the mechanism of intergranular corrosion[6] is not dependent on formation of Fe_3C.

Effect of Multiple Heat Treatments on Corrosion Resistance

Since our data showed that samples sensitized at 1100 C could be desensitized when reheated to 850 C and water quenched, the question arises, could corrosion resistance be recovered if the samples were cooled more slowly from 850 C? Further, would samples quenched from 850 C retain their corrosion resistance if given subsequent lower temperature heat treatments? To answer these questions, samples of the Type 446 commercial alloy were sensitized at 1100 C, water quenched, reheated at 850 C, and then slowly air cooled. Other samples, given the same heat treatment and reheat at 850 C, were heated for short periods at 600 and 800 C (1112 and 1472 F). The results of this work are shown in Table 6.

TABLE 7 – Effects of Thermal Treatment on the Ductility and Corrosion Resistance of Fe-26% Cr Experimental Alloys[(1)]

Sample Designation	Heat Treatment	Corrosion Resistance,[(2)] mpy (Exposure hr)	% Elongation (1 in)
18	30 min, 1100 C, W.Q.	25 (120)	30
19	30 min, 1100 C, A.C.	549 (48)	32
20	30 min, 1100 C, Slow Cool[(3)] to 850 C, W.Q.	25 (120)	30

(1) See Table 1 for a typical composition.
(2) Boiling 50% H_2SO_4-$Fe_2(SO_4)_3$.
(3) Cooling rate, 20 C/8 min.

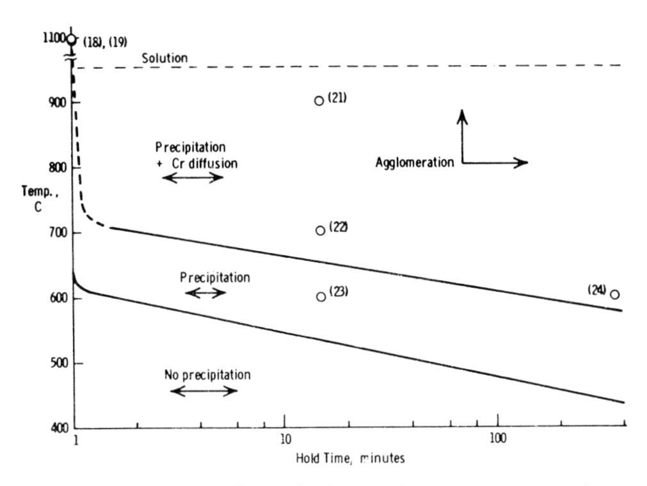

FIGURE 3 — Effect of time and temperature on the corrosion resistance of an experimental Fe-26% Cr alloy (C + N = 180 ppm).

The sample that was air cooled following the 850 C annealing treatment (sample 15) displayed a reduced corrosion resistance as compared with the water quenched sample (sample 5). The desensitized sample 5 was then given a third heat treatment (sample 16) at 800 C, which had no appreciable effect on its corrosion resistance. However, when the third heat treatment on sample 5 was carried out at 600 C (sample 17), a severe loss in corrosion resistance resulted.

What happens at 850 C to improve corrosion resistance even though a continuous intergranular precipitate appears in the microstructure? What happens at 600 C to cause sensitization? These two key questions suggested by the data presented thus far will be discussed next.

Consideration of Results

Mechanism of Loss in Corrosion Resistance

Background. By analogy to austenitic stainless steels, the foregoing data seem to indicate that ferritic Type 446 steel is sensitized by the precipitation of chromium rich carbides and nitrides in the temperature region of 600 C. If this is true, it must be explained why samples rapidly cooled from 1100 C sensitize, and why sensitized samples annealed at 850 C show good corrosion resistance despite the presence of a continuous intergranular precipitate.

A study of the extent of damaging precipitate produced around 600 C is not possible with the commercial alloys because even rapid quenching is not sufficient to maintain the carbon and nitrogen in solid solution following the high temperature exposure. A sample was sought, which, after a high temperature treatment, would be corrosion resistant but supersaturated in carbon and nitrogen. By study of carbon and nitrogen supersaturated samples, the range of heat treatment which causes intergranular precipitation and loss of corrosion resistance could be established in much the same way that the time-temperature, chromium-carbide precipitation envelope was defined for the chromium-nickel austenitic stainless steels.[9] It was speculated that such an alloy would be prepared if the carbon and nitrogen contents in the 26% Cr-Fe alloys were lowered below the typical values of 1000 ppm.

When Fe-26% Cr alloys of the interstitial levels listed in Table 1 were heat treated and quenched from 1100 C, they showed excellent corrosion resistance (Table 7, Sample 18). However, their corrosion resistance was poor when they were air cooled (Table 7, Sample 19), a fact which is consistent with the appearance of intergranular precipitation at intermediate temperature levels.

Definition of Sensitization Region. To define the temperature range of sensitization or the precipitation envelope, a series of tests were run using the experimental alloys with reduced carbon and nitrogen levels (Table 1). Pieces 1 x 1 inch were heated at 1100 C for 30 minutes and then water quenched. These alloys had good corrosion resistance, but were supersaturated in dissolved carbon and nitrogen. They were subjected to a second heat treatment consisting of different times at temperatures from 400 to 1000 C (752 to 1832 F). The samples were water quenched and corrosion tested. It was observed the time-temperature relationship caused some samples to lose corrosion resistance. More than 70 samples were tested, and from the resultant data, a time-temperature band was found where exposure of samples caused a loss of corrosion resistance (Figure 3).

Similar experiments were repeated with samples of the experimental alloy which were originally heated for 30 minutes at 1100 C and air cooled. These samples had poor corrosion resistance. As above, these samples were also subjected to a second heat treatment consisting of different times at temperatures from 400 to 1000 C. From the corrosion data, it was possible to draw a line connecting temperatures and time, where exposure of samples in the region above this line restored corrosion resistance. This line coincided with the high temperature line described above, which was generated by testing samples originally water quenched from 1100 C. Those results are presented in the form of a time-temperature corrosion resistance diagram in Figure 3.

It is believed that the line marking the low temperature side of the time-temperature band reflects the minimum time and temperature required for precipitation of chromium carbides and nitrides to occur, with formation of chromium depleted zones. The line marking the high temperature side of this band reflects the minimum time and temperature relationship necessary for sufficient diffusion of chromium to occur to heal the areas depleted from precipitation. These processes of sensitizing and self healing in the ferritic stainless steels are analogous to the process discussed in detail by Strawström and Hillert[15] on the depleted zone theory of sensitization in austenitic stainless steels. An important point is that re-solution of the precipitate is not necessary for recovery of corrosion resistance. As long as the chromium depleted areas are recovered by chromium diffusion, corrosion resistance will be restored, even though a precipitate may be observed. These points are validated by the microstructural characterization of selected samples described below.

Relationship of Microstructure to Corrosion Resistance. The time and temperature relationship of the sensitization envelope is defined in Figure 3. Also in this figure, several heat treated specimens are shown, whose microstructures are presented in Figure 4. These results are for the experimental alloys of lower carbon and nitrogen levels. All samples, unless noted, were treated for 30 minutes at 1100 C and water quenched before subsequent heat treatment.

When sample 18 of this experimental alloy was heated at 1100 C and quenched (see temperature position 1100 C in Figure 3, and micrograph 18, Figure 4), it showed good corrosion resistance, which is consistent with its clean grain boundary structure. After sample 19 was air cooled from 1100 C, it showed poor corrosion resistance, which is consistent with the appearance of a continuous intergranular precipitate. Sample 21, reheated at 900 C (1652 F), had good corrosion resistance and a discontinuous intergranular precipitate. Comparison of sample 21 with sample 18 provides good evidence that intergranular precipitation does occur at 900 C, followed by rapid diffusion of chromium, and then the observed agglomeration of the chromium rich carbides and nitrides. Sample 20, reheated at 700 C (1292 F), exhibited good corrosion resistance even though it had a continuous intergranular precipitate. At this temperature, the rapid precipitation of chromium rich carbides and nitrides was compensated for by almost simultaneous chromium diffusion which healed the depleted areas. A discontinuous intergranular precipitate was not observed in this microstructure because the hold time at 700 C was not long enough to permit agglomeration.

Sample 23 reheated at 600 C for a short time, exhibited a fine precipitate on slip bands in the grain faces, and a precipitate free zone along the grain boundaries. More importantly, a fine continuous intergranular precipitate was observed. The jagged edges of the etched grain boundaries indicated that precipitation had occurred within the plane of the boundary and transverse to it. This is a particularly damaging condition because depletion not only occurs on opposite sides of the precipitate formed within the plane of the grain boundary, but also within the volume perpendicular to the plane of the boundaries.

These observations are consistent with the findings of Lewis and Hattersley[16] on the precipitation of $M_{23}C_6$ in austenitic steels, and with Raymond's[17] more recent data on the effects of the mode of chromium carbide precipitation on the sensitization of Incoloy Alloy 825.[(1)] When ferritic stainless steels are heated at 600 C, the carbon and nitrogen in solid solution rapidly precipitated as chromium rich carbides and nitrides on high energy surfaces such as found in the grain faces and grain boundaries. Since diffusion of chromium is much slower at 600 C, the areas depleted in chromium during the rapid precipitation were not healed, hence the high corrosion rate.

Sample 24, held for 6 hours at 600 C, displayed good corrosion resistance. No fine precipitate appeared in the grain faces as it did in sample 23 and the intergranular precipitate was located almost wholly within the grain boundary plane. The longer hold time permitted enough chromium to diffuse after precipitation so that the depleted areas were healed, and the sample recovered its corrosion resistance.

[(1)]Trademark for International Nickel Co., Inc., Huntington Alloy Products Division.

These data cast doubt on the theory that intergranular corrosion is caused by precipitates which strain the grain boundaries. If a short term heat treatment at 600 C causes loss of corrosion resistance through precipitation induced stress, it would not be expected that a longer heat treatment at this temperature would cause corrosion resistance recovery.

Mechanism of Sensitization in Ferritic Stainless Steels. Based on the above data, a mechanism for intergranular corrosion of ferritic stainless steels exposed to high temperatures is proposed. When ferritic alloys are heated above about 950 C, carbon and nitrogen are dissolved in solid solution. If their levels are low enough, as is the case with the experimental alloys, they can be kept in solid solution by a rapid quench. For these quenched alloys, the corrosion resistance is good in contrast to the resistance observed with the commercial alloys. However, when the experimental samples are air cooled, chromium rich carbides and nitrides precipitate from solid solution at the grain boundaries. Adjacent, then, to the precipitate is a zone which has a far lower chromium content than does the rest of the matrix, and which is, therefore, more susceptible to intergranular attack.

When the quenched samples supersaturated with carbon and nitrogen are reheated from 500 to 950 C, chromium rich carbides and nitrides precipitate. However, in the temperature range 700 to 950 C, almost as fast as precipitation occurs, the rapid diffusion of chromium in the ferrite heals the impoverished areas at the grain boundaries; therefore, corrosion resistance is good despite the observed presence of an intergranular precipitation.

In the temperature range 500 to 700 C, where precipitation of chromium rich carbides and nitrides also occurs, chromium diffusion rates are so reduced that healing of the chromium impoverished areas does not occur except during long hold periods.

For the commercial ferritic Type 446 stainless steels with much higher carbon and nitrogen contents, the same considerations can explain the observed effects of heat treatment on corrosion resistance and microstructure. Of particular note is the observation that even a rapidly quenched commercial alloy has poor corrosion resistance. The formation of chromium rich carbides and nitrides cannot be prevented, even with a rapid quench from high temperature in these alloys of high interstitial content. This behavior can be attributed to the high driving force for

TABLE 8 – Selected Diffusion Rates for Chromium and Carbon in Ferrite and Austenite

Temp, C	Ferrite Chromium[(1)]	Ferrite Carbon[(2)]	Austenite Carbon[(3)]
	Diffusion Rate, cm²/sec		
500 (932 F)	2.7×10^{-15}	3×10^{-8}	–
700 (1292 F)	1.7×10^{-12}	8×10^{-7}	4×10^{-9}
900 (1652 F)	1.3×10^{-10}	1×10^{-5}	8×10^{-8}

[(1)]Ref. 16.
[(2)]Ref. 17.
[(3)]Ref. 18.

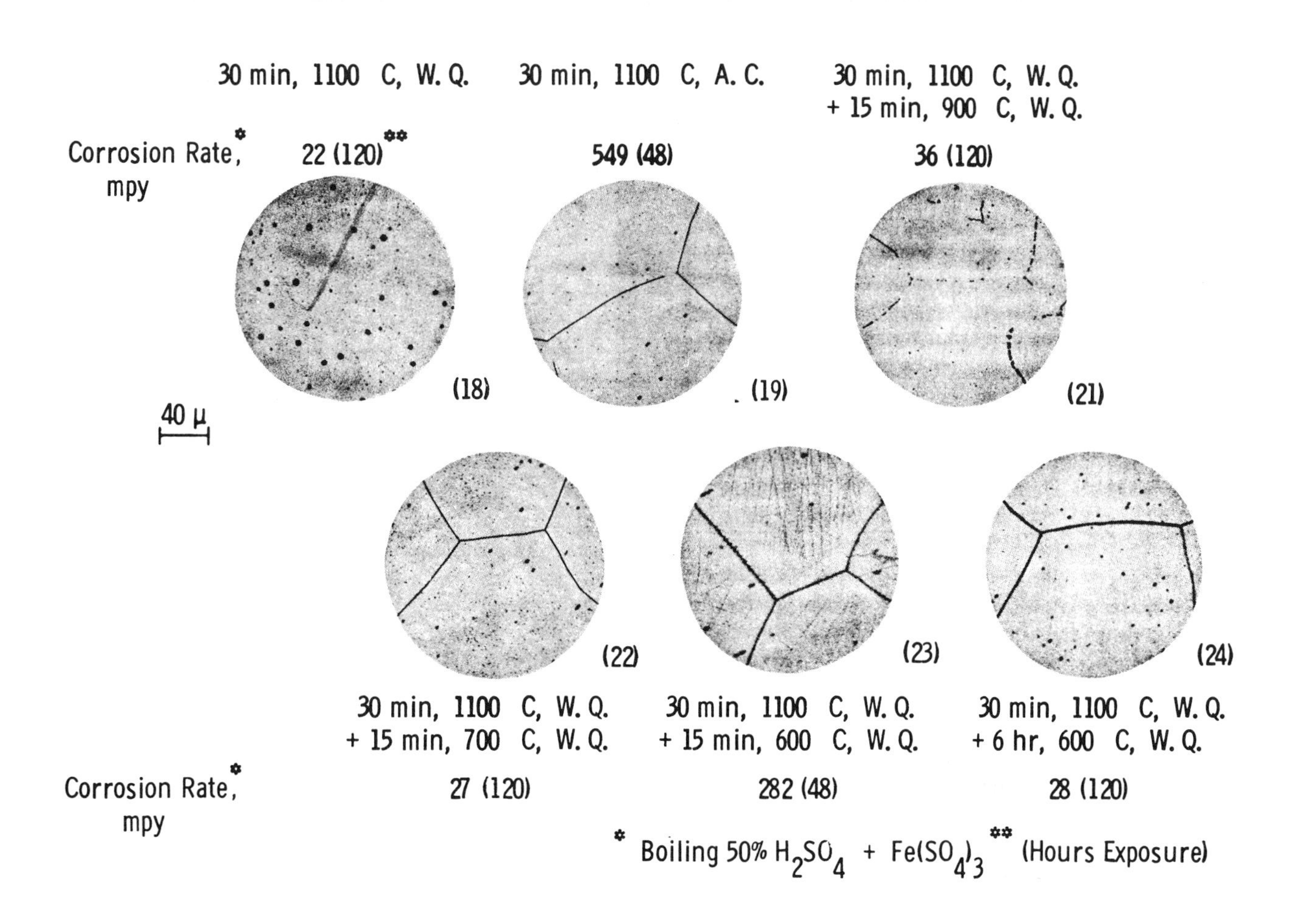

FIGURE 4 – Effect of thermal treatment on microstructure and corrosion resistance of experimental Fe-26% Cr alloys (C + N = 180 ppm).

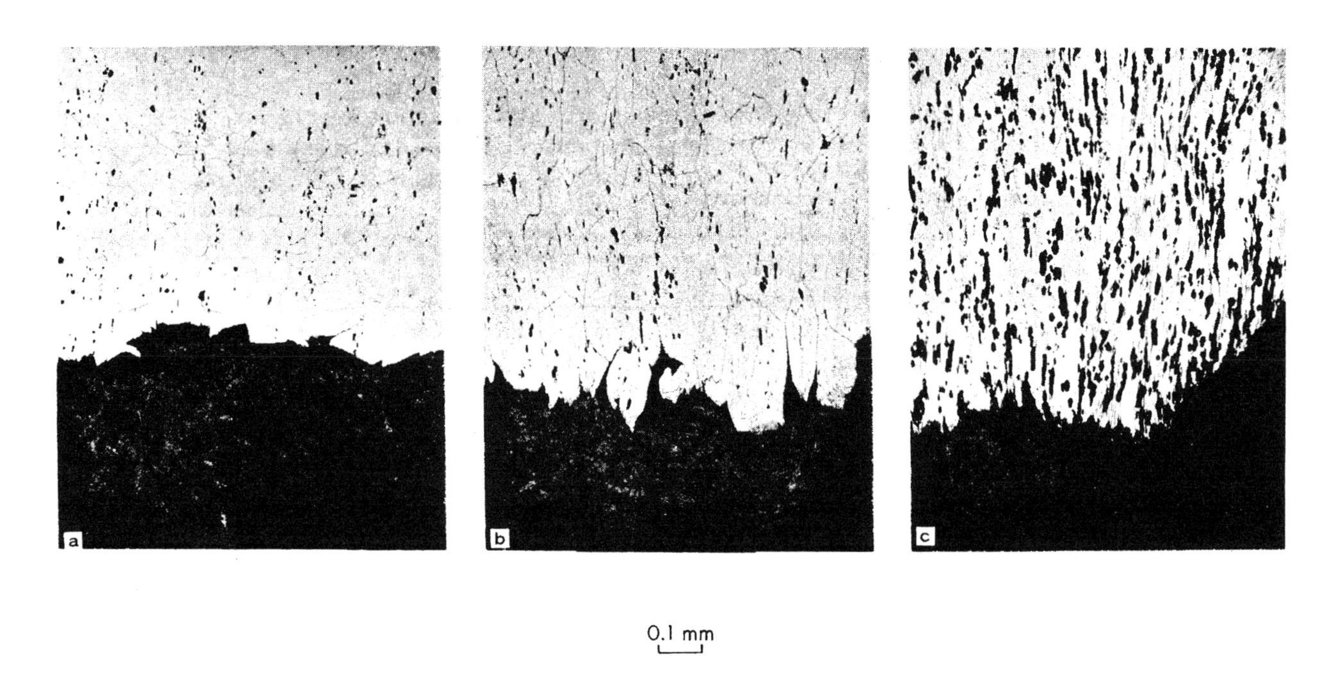

FIGURE 5 – Fracture edges of heat treated AISI Type 446 stainless steel after tensile testing (951 ppm C, 770 ppm N). (a) = 30 minutes, 1100 C, water quench (2), (b) = 30 minutes, 1100 C, air cool (3), (c) = 30 minutes, 1100 C, slow cooled (20 C in 18 minutes) to 850 C, water quench (4).

TABLE 9 – Calculated Times for Appreciable Diffusion of Carbon and Chromium Over Distances of 15μ and 3μ Respectively

Temp, C	Ferrite Chromium	Ferrite Carbon	Austenite Carbon
	Time for Appreciable Diffusion,[1] min		
500 (932 F)	600,000	1	–
700 (1292 F)	900	0.05	10
900 (1652 F)	11	0.004	0.5

[1]For appreciable diffusion, $x/\sqrt{Dt} \cong 1$ (Ref. 19).

precipitation in the intermediate temperature range caused by a combination of supersaturation; rapid diffusion of chromium, carbon, and nitrogen; and rapid rates of precipitation.

Chromium depleted areas adjacent to the chromium rich precipitates of carbon and nitrogen so reduce the corrosion resistance of these areas that rapid intergranular corrosion is observed following exposure to corrosive environments.

Likewise, the corrosion resistance of the sensitized commercial alloy can be restored by heating in the standard annealing range of 700 to 950 C. As described before, rapid diffusion of chromium *smooths out* the depleted areas adjacent to the grain boundaries. Therefore, consistent with the microstructure shown in Figure 2, and the corrosion data for commercial alloys in Table 3, it is noted that corrosion resistance is restored even though a grain boundary precipitate is observed.

Diffusion Rate Effects. To illustrate the role that diffusion rates play in this mechanism, the diffusion rates for chromium in a ferritic alloy of composition Fe-25% Cr, and the diffusion rates for carbon in both alpha iron containing 0.22% C, and in an austenitic matrix are tabulated in Table 8.[18-20] A magnitude drop of three to five orders in the diffusion rates of carbon and chromium in ferrite from 500 to 900 C may be noted. Further, a three orders of magnitude slower diffusion rate for carbon in a face centered cubic matrix as compared to its diffusion in ferrite, is noted. Using these diffusion rates, and the approximation that $x/\sqrt{Dt} \cong 1$ for appreciable compositional changes to occur by diffusion,[21] it is possible to calculate the time required for diffusion to occur over areas that will be sensitized. In the equation, $x/\sqrt{Dt} \cong 1$, x is the diffusion distance, D is the diffusion rate, and t is the time for diffusion over the distance x.

Since the extent of the chromium impoverished areas cannot be determined with the electron microprobe, it may be assumed that they are less than 3μ wide. Using this distance, the approximation equation, and the diffusion rates, it is calculated that appreciable diffusion of chromium occurs within minutes at the high temperatures, but requires hours at the low temperatures. The times calculated are shown in Table 9.

While only approximate, this confirms the theory that a substantial change in chromium concentration over small areas does occur quickly at higher temperatures and that very little change occurs in finite times at the lower temperatures. These same calculations may be applied to the diffusion of carbon. It may be assumed that the uniform unetched bands at the grain boundaries in samples 2 and 5 of Figure 2 represent depleted areas from which the carbon has diffused to form chromium carbides during the rapid quench from temperature. The width is estimated to be 15μ. Using this number, the diffusion rates for carbon, and the approximation formula, the time necessary for large compositional changes in carbon to occur as a function of temperature can be calculated. These data also are shown in Table 9. These approximations tell us that the diffusion of carbon in ferrite over microscopic distances occurs within fractions of seconds at the high temperatures, and within minutes at the lower temperatures. In comparison, significant carbon diffusion in austenite takes place within only minutes at high temperatures. Thus, sensitization takes fractions of seconds in ferritic stainless steel, and recovery only minutes, as compared to austenitic stainless steels where sensitization takes minutes and recovery takes hours.

This approximate analysis of the periods required for diffusion to occur over reasonable distances to bring about substantial compositional change is consistent with our observations. The relatively short times required for diffusion of chromium, carbon, and nitrogen in ferrite help clarify why the high interstitial commercial ferritic stainless steels cannot be quenched fast enough to prevent diffusion and precipitation of chromium rich carbides and nitrides. It also helps explain why austenitic stainless steels can be quenched without precipitation of these phases.

Comparison of Fe-26% Cr Stainless Steel to Austenitic Stainless Steel. When Fe-26% Cr stainless steel is exposed to high temperatures, and is then held or slow cooled through a temperature range of 500 to 700 C, it suffers severe intergranular attack because of the precipitation of chromium rich carbides and nitrides.

If the interstitial levels are low, samples quenched following high temperature exposure will have good corrosion resistance because the carbon and nitrogen are retained in solid solution. This situation is, of course, analogous to the observations made on austenitic stainless steels. For the commercial ferritic alloys with high carbon and nitrogen levels, on the other hand, water quenching is not fast enough to prevent sensitization by precipitation of chromium rich carbides and nitrides.

Whatever the interstitial level, the corrosion resistance of sensitized ferritic stainless steels may be recovered by heating in a temperature range from 700 to 950 C where in minutes, rapid chromium diffusion heals the depleted areas. Healing sensitized austenitic stainless steels by this heat treatment is not practical because hours are required for sufficient chromium diffusion to heal the depleted areas. As described by Nielsen, Mahla[14] and Bain *et al,*[9] corrosion resistance of sensitized austenitic stainless steels is recovered by heating above the solutioning temperature to dissolve the carbides and nitrides, then rapidly cooling, which maintains them in solid solution. As described above, this treatment cannot be used for commercial ferritic stainless steels because their higher diffusion rates and higher degrees of supersaturation allow precipitation to occur even during a rapid quench.

Of the theories for sensitization mentioned earlier, these data mainly support Hochmann's idea that intergranular corrosion occurs when chromium depleted areas at the grain boundaries are attacked. However, attack does not occur because a low chromium austenite phase is present as he suggests. In fact, as our experiments with the low interstitial experimental alloys show, intergranular corrosion problems are manifested in this alloy even though austenite cannot be formed. Rather, the precipitation of chromium rich carbides and nitrides cause immediately adjacent areas to be depleted in chromium. These low chromium areas are rapidly dissolved and intergranular corrosion is observed. The data reported here for Fe-26% Cr alloys, combined with recently reported work by Bond[22] on Fe-17% Cr alloys, strongly support the mutual idea that intergranular corrosion in ferritic stainless steels occurs in chromium depleted areas which result when chromium rich carbides and nitrides precipitate at the grain boundaries.

Mechanism for Loss in Ductility

Background. As noted earlier (Table 3), loss in ductility is observed only when the commercial alloys are exposed to high temperatures and quenched. Commercial samples more slowly cooled from high temperatures (air cooled or slow cooled), or subsequently annealed at 850 C, show excellent ductility. In addition, the experimental alloys (Table 7) with much lower carbon and nitrogen content do not lose ductility when quenched. From these observations, it appears that the loss of ductility in the commercial alloys must be associated both with the high interstitial content and cooling rate following exposure to high temperature. To examine the microstructural differences, brittle and ductile samples of the commercial alloy were examined metallographically.

Fracture Edges. To examine differences in the mode of failure, cross sections of fractured commercial samples which had been heat treated in different ways were polished and very lightly etched (Figure 5). The water quenched sample 2 (brittle) shows predominantly intragranular cleavage with no evidence of localized deformation. The air cooled sample 3 (ductile) shows intergranular cleavage with some localized deformation and elongation of the grains. The slow cooled sample 4 (ductile) shows a classical fibrous shear structure with considerable localized plastic deformation and elongation.

Since both the air cooled and water quenched samples had intergranular precipitates (Figure 2), and since the fracture in the water quenched samples was predominantly intragranular, intergranular precipitation does not grossly affect ductility in ferritic stainless steel. This observation is

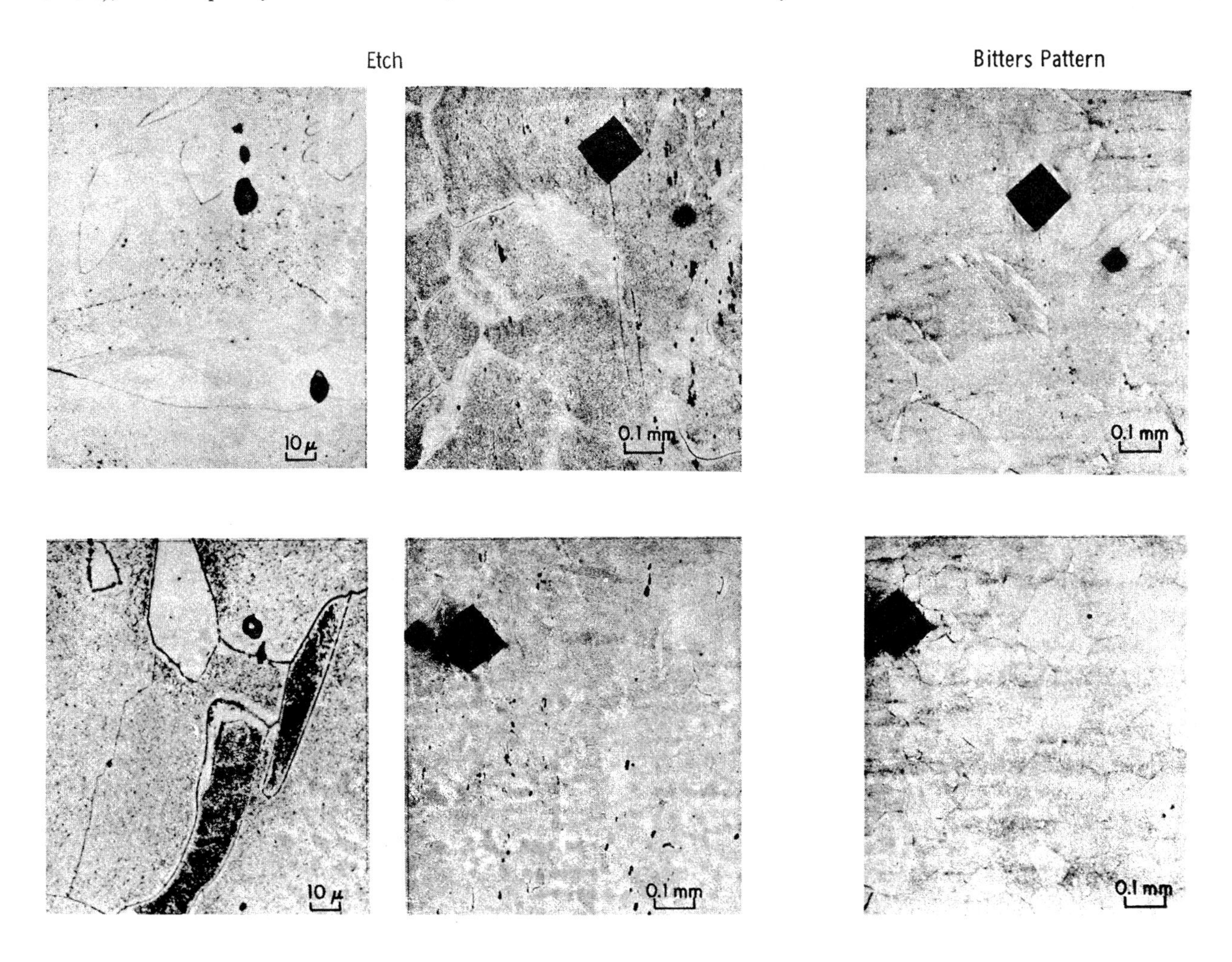

FIGURE 6 — Comparison of etched microstructure and bitters pattern replica of Type AISI 446 steel given two heat treatments: top, 30 minutes, 1100 C, water quench (2), and bottom, 30 minutes, 1100 C, water quench plus 30 minutes, 850 C, water quench (5).

TABLE 10 — Effect of Thermal Treatment on Microhardness of Selected Areas in AISI Type 446 Steel (VHN-10 g)

	Matrix	Second Phase
30 min, 1100 C (2012 F), W.Q.	111 ±10	133 ±11
30 min, 1100 C, W.Q. + 30 min, 850 C (1562 F), W.Q.	101 ±3	104 ±5

consistent with similar observations made of austenitic stainless steels. An austenitic alloy sensitized with a grain boundary precipitate of chromium rich carbides has excellent ductility (Table 3).

Microstructure. Examination of the samples shown in Figure 2 which have been heavily etched shows that the water quenched (sample 2), the water quenched annealed (sample 5), and the air cooled samples (sample 3) all have intergranular precipitates. The water quenched and the water quenched annealed samples, in addition to the dense dot-like precipitates in the grain faces, have discrete patches of essentially unetched materials adjacent to the grain boundaries.

The second phase material was investigated further and it was shown that the patches in the water quenched samples were nonmagnetic. This was done by first etching the polished sample, locating an area with a micro-hardness indentor, then repolishing the sample to a metallographic finish. Then colloidal magnetic iron oxide[23] was applied to the polished surface to make a Bitter's pattern replica of the magnetic and nonmagnetic areas on the surface. The etched structure and the magnetic powder pattern of the corresponding areas are shown in Figure 6 for a water quenched sample (sample 2), and a water quenched annealed sample (sample 5). Equivalent areas may be located in relationship to the microhardness indents. The unetched patches in the water quenched samples are nonmagnetic and are presumed to be austenite. Similar areas in the water quenched annealed samples are magnetic.

In addition, under high magnification, it was found that a lamellar precipitate had appeared in the second phase areas of sample 5. These observations suggest that the nonmagnetic areas present after water quenching quickly revert at 850 C to magnetic ferrite and carbides. The presence of austenite in a ferritic material with 26% Cr is perhaps unexpected; however, as shown by Nehrenberg and Lillys[24] and Baerlecken *et al,*[25] patches of austenite can form in high chromium ferritic alloys as a result of localized concentrations of carbon and nitrogen. Since poor ductility occurs only in samples containing this nonmagnetic second phase, it could be concluded that the presence of the duplex structure is responsible for the brittleness. As suggested by Mogford[26] for duplex structures, perhaps there is a large stress concentration at the ferrite-austenite interface which could cause differential yielding which would in turn lead to rapid propagation of microcracks. The possibility of differential yielding under an applied load is suggested by microhardness measurements.

A 10 gm load was used to measure the Vicker's

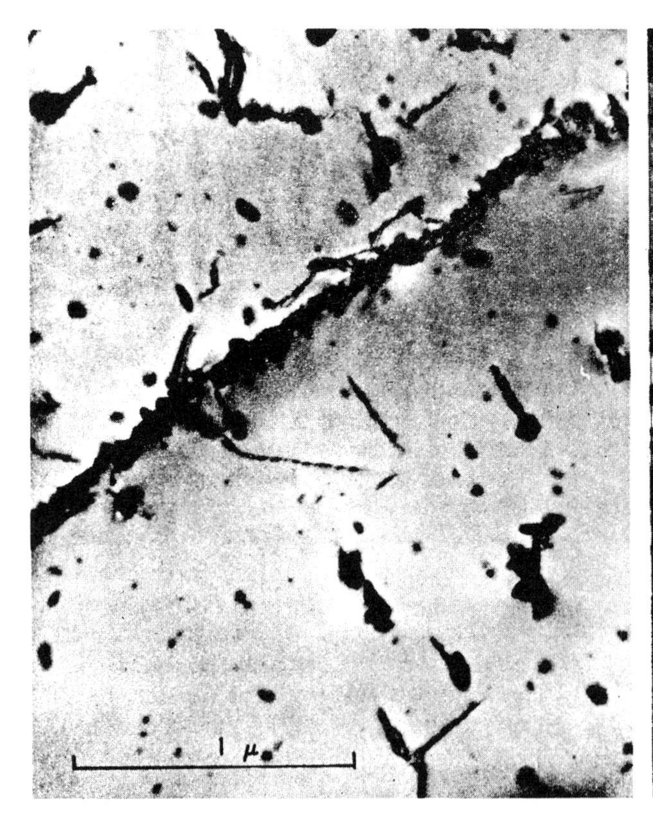

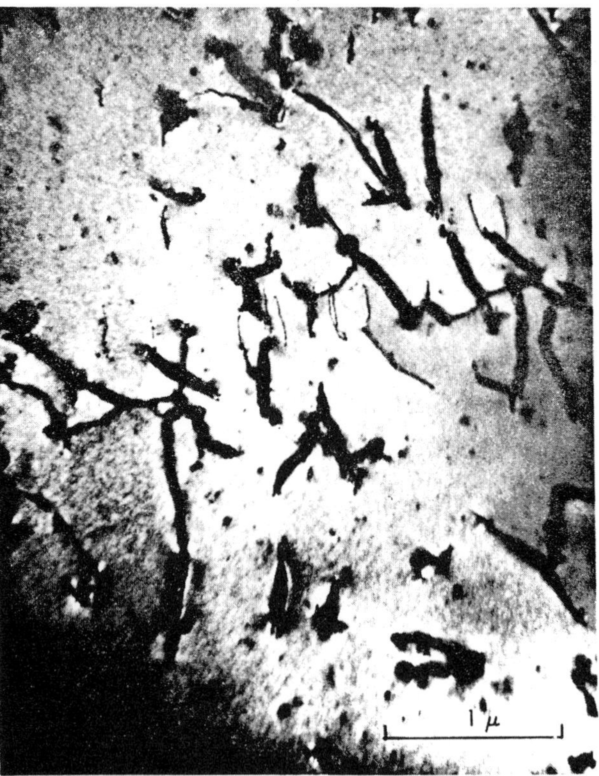

FIGURE 7 — Transmission electron micrograph of AISI Type 446 steel heated 30 minutes at 1100 C and water quenched (2).

microhardness values of the matrix and the patches of second phase material. The results are shown in Table 10. The measurements show that the patches of nonmagnetic second phase material in the water quenched samples are harder than the matrix, while in the water quenched annealed samples, the microhardness values are the same within experimental error.

Electron Microscopy. Limited electron microscopy thin film transmission work on the commercial alloys suggests another possibility for explaining the brittleness of the water quenched samples. Thin film transmission techniques were applied successfully to commercial alloys which had been water quenched and air cooled from 1100 C. These transmission electronmicrographs are shown in Figures 7 and 8, respectively.

For the water quenched sample, precipitates are noted in the grain boundaries. More importantly, however, precipitates are located on nearly all dislocations. Such a locking of the dislocations would prevent their movement during deformation, and thus cause the poor ductility observed. On the other hand, as Figure 8 shows, the dislocations in the air cooled sample are generally free of precipitates and are, therefore, presumably free to move during deformation. As noted, this sample has good ductility.

Why, then, are there precipitates on the dislocations in the water quenched samples and not on the air cooled samples even though both have grain boundary precipitates and poor corrosion resistance? Further, why does precipitation in austenitic stainless steels in the sensitizing range occur predominantly at grain boundaries and not at carbide precipitation nuclei such as dislocations in the grain faces? For some reason, as noted by Bain *et al*,[9] the relief of carbon supersaturation in austenitic stainless steels occurs almost wholly by rejection of Cr carbides at the grain boundaries. This phenomenon in austenitic stainless steel is analogous to what is observed in the air cooled commercial AISI Type 446 ferritic steel. Looking at the optical micrographs in Figure 2, the general absence of a fine dot-like precipitate in the grain faces of the air cooled

FIGURE 8 – Transmission electron micrograph of AISI Type 446 steel heated 30 minutes at 1100 C and air cooled (3).

sample (sample 3), may be noted along with a continuous precipitate at the grain boundaries.

Two possibilities can be proposed to explain these observations. In the commercial ferritic alloys, the high interstitial contents and high supersaturation serve as a driving force during quenching for rapid precipitation on all high energy surfaces, whether these be grain boundaries or dislocations. During a slow cool, the longer time available to relieve supersaturation perhaps allows diffusion of carbon and nitrogen to the grain boundary areas where the supersaturation is relieved by precipitation on the preferred high energy surfaces of the grain boundaries. An alternate possibility might be that dislocation nucleation during the short quench period occurs simultaneously with rejection of interstitial carbides and nitrides to relieve supersaturation. With the slower air cool which reduces thermal stresses, perhaps relief of supersaturation by precipitation on a high energy surface, such as a grain boundary, occurs before dislocations are nucleated by the effects of thermal stresses.

It would seen reasonable, on the basis of these observations and the importance of dislocation motion to deformation, to associate the poor ductility of a quenched AISI Type 446 stainless steel to precipitate locking of dislocations, a phenomenon analogous to that of precipitation hardening. However, the role of the duplex structure, observed in the quenched sample, in affecting ductility must also be considered. Further thin film electron transmission microscopy work and experiments with alloys containing low interstitial contents, but increasing nickel content (to give a duplex structure in the absence of precipitated phases), may help to define the relative importance of the observed duplex structure and the precipitate locked dislocations on the poor ductility of rapidly cooled commercial ferritic stainless steels.

Grain Size Reduction. The apparent reduction in grain size (Table 5) when the commercial sample is annealed may also be explained by the presence of a duplex structure. The small ferritic islands that form when the nonmagnetic areas transformed during the 850 C anneal, grow at a faster rate than the larger ferritic grain present originally. This transformation and growth process would cause the observed grain size refinement followed at longer annealing time by grain growth.

Conclusions

Ferritic alloys of iron containing 26% Cr become sensitized following high temperature exposures such as welding because the carbon and nitrogen dissolved in solid solution precipitate at intermediate temperatures, leaving chromium depleted areas in the grain boundaries. This mechanism of intergranular corrosion is strikingly similar to the sensitization mechanism for chromium-nickel austenitic stainless steels. However, because diffusion rates are higher and interstitial solubilities are lower in the ferritic materials than in the austenitic, the effects of thermal treatment differ, and the correlation between good corrosion resistance and the absence of intergranular precipitates must be modified. The high interstitial levels of commercial ferritic alloys make it impossible to prevent damaging precipitation by any cooling method including water quench. At lower interstitial levels, however, Fe-26% Cr alloys quenched from high temperatures have excellent corrosion resistance. The effect of thermal treatment and cooling rate on the corrosion properties of low interstitial level alloys is analogous to that observed for the austenitic stainless steels.

This work shows that $M_{23}C_6$ and Cr_2N precipitate in the temperature range from approximately 500 C to the solutioning temperature of 950 C. Sensitization or loss in corrosion resistance is caused by holding or cooling through 500 to 700 C. At temperatures above 700 C, almost as fast as damaging precipitation occurs, rapid chromium diffusion heals the depleted areas adjacent to the grain boundary precipitates. Therefore, sensitized samples which are annealed above 700 C show good corrosion resistance despite the presence of a continuous intergranular precipitate. In fact, at high temperatures, and for long hold times, the continuous precipitate is agglomerated. At temperatures below 700 C, where precipitation also occurs, and diffusion of chromium is slow, corrosion resistance is poor because the chromium depleted zones do not heal unless very long hold times are used.

The loss of ductility caused by heating commercial AISI Type 446 steels to high temperatures and quenching may be associated with one or both of two mechanisms. First, the small amount of austenite formed at high temperature and retained during quenching may cause large localized stress concentrations which, under an applied tensile stress, could cause formation and propagation of small localized cracks. Second, it would be expected that observed precipitates on the dislocations in the quenched steel would immobilize them, thus severely reducing the capacity of the steel for deformation.

The sensitization and loss in ductility which follow high temperature exposure of AISI Type 446 steel can be minimized in several ways. It can be recovered by a short time anneal in the recommended commercial annealing range of 788 to 872 C, followed by a rapid cool. Alternately, an alloy with both good corrosion resistance and ductility is produced when a commercial alloy is slowly cooled to approximately 800 C and then water quenched. Finally, a reduction in carbon and nitrogen considerably below the present levels in commercial ferritic alloys would eliminate the problem of brittleness and intergranular corrosion induced by high temperature exposure.

The effects of interstitial contents on the properties of ferritic stainless steels have received considerable attention over the past two years. An alloy E-Brite(2) 26-1 made to low interstitial levels by electron beam technology is reported[27] to have good as-welded ductility and intergranular corrosion resistance. A recent publication by R. J. Hodges[28] describes the effect of cooling rate and alloy composition on intergranular corrosion in high purity ferritic stainless steels.

(2) A product of Airco Vacuum Metals, Berkeley, California.

Acknowledgment

The author expresses his appreciation to R. L. Colicchio for performing the many critical heat treatments and corrosion tests; and to J. L. Youngblood for performing the thin-film transmission electron metallography.

References

1. A. P. Bond and H. J. Dundas. *Corrosion,* **24,** 344 (1968).
2. H. Thielsch. *Welding J.,* **30,** 209s (1951).
3. L. Rajkay. *Proc. ASTM,* **67,** 158-169 (1967).
4. H. Thielsch. *Welding J.,* **34,** 22s (1955).
5. T. A. Pruger. *Steel Horizons,* **13,** 10 (1951).
6. E. Houdremont and W. Tofaute. *Stahl und Eisen,* **72,** 539 (1952).
7. J. Hochmann. *Rev. Met.,* **48,** 734 (1951).
8. R. A. Lula, A. J. Lena, and G. C. Kiefer. *Trans. ASM,* **46,** 197-223, Discussion, 223-230 (1954).
9. E. C. Bain, R. H. Aborn, and J. J. B. Rutherford. *Trans. Am. Soc. Steel Treating,* **21,** 481-509 (1933).
10. V. Cihal and I. Kasova. *Corrosion Science,* **10,** 875 (1970).
11. M. A. Streicher. *ASTM Bulletin,* **229,** 77-86 (1958).
12. M. A. Streicher. E. I. du Pont de Nemours & Co., Inc., Wilmington, Del. Private Communication on the Applicability of the Ferric Sulfate-Sulfuric Acid Test to Ferritic Alloys.
13. ASME Pressure Vessel Code, Sec. 9, p. 59 (1965).
14. E. M. Mahla and N. A. Nielsen. *Trans. ASM,* **43,** 290-314, Discussion, 314-322 (1951).
15. C. Strawström and M. Hillert. *J. Iron and Steel Inst.,* **207,** 77 (1969).
16. M. H. Lewis and B. Hattersley. *Acta Met.,* **13,** 1159 (1965).
17. E. L. Raymond. *Corrosion,* **24,** 180 (1968).
18. H. W. Paxton and T. Kunitake. *Trans. Met. Soc. AIME,* **218,** 1003-1009 (1960).
19. R. P. Smith. *Trans. Met. Soc. AIME,* **224,** 105-111 (1962).
20. C. Wells, W. Batz, and F. Mehl. *J. Metals, Trans. AIME,* **188,** 553-560 (1950).
21. L. S. Darken and R. W. Gurry. *Physical Chemistry of Metals,* McGraw-Hill Book Co., New York, p. 444 (1953).
22. A. P. Bond. *Trans. Met. Soc. AIME,* **245,** 2127-2134 (1969).
23. F. Bitter. *Phys. Rev.,* Ser. 2, 1903-1905 (1931).
24. A. E. Nehrenberg and P. Lillys. *Trans. ASM,* **46,** 1176-1203, Discussion, 1203-1213 (1954).
25. E. Baerlecken, W. A. Fischer, and K. Lorenz. *Stahl und Eisen,* **81,** 768 (1961).
26. I. L. Mogford. *Met. Rev.,* **12,** 49 (1967).
27. Data Sheet, E-Brite 26-1, Airco Vacuum Metals, Berkeley, Calif. (1969) October.
28. R. J. Hodges. *Corrosion,* **27,** 119 (1971).

Mechanisms of Intergranular Corrosion in Ferritic Stainless Steels

A. Paul Bond

Two series of 17 pct Cr iron-base alloys with small, controlled amounts of carbon and nitrogen were vacuum-melted in an effort to determine the mechanisms of intergranular corrosion in ferritic stainless steels. An alloy containing 0.0095 pct N and 0.002 pct C was very resistant to intergranular corrosion, even after sensitizing heat treatments at 1700° to 2100°F. However, alloys containing more than 0.022 pct Ni and more than 0.012 pct C were quite susceptible to intergranular corrosion after sensitizing heat treatments at temperatures higher than 1700°F. This corrosion was observed after the usual exposure tests and after potentiostatic polarization tests. Electronmicroscopic examination of the alloys susceptible to intergranular corrosion revealed a small grain boundary precipitate; this precipitate was absent in the alloys not susceptible to such corrosion. The electronmicrographs indicate that intergranular corrosion of ferritic stainless steels is caused by the depletion of chromium in areas adjacent to precipitates of chromium carbide or chromium nitride. It also seems likely that the precipitates themselves are attacked at highly oxidizing potentials. Confirmation of the proposed mechanisms was obtained in tests on air-melted ferritic stainless steels containing titanium. The titanium additions greatly reduced susceptibility to intergranular corrosion at moderately oxidizing potentials but had no beneficial effect at highly oxidizing potentials.

A major obstacle to the use of ferritic stainless steels has been their susceptibility to intergranular corrosion after welding or improper heat treatment. It appears that sensitization of ferritic stainless steel occurs under a wider range of conditions than for austenitic steels. In addition, a greater number of environments lead to damaging intergranular corrosion of sensitized ferritic stainless steels than to sensitized austenitic steels.

The chromium depletion theory of intergranular corrosion is widely accepted for austenitic stainless steels[1,2] although there are some objections.[3] On the other hand, several alternative mechanisms proposed for ferritic stainless steels include precipitation of easily corroded iron carbides at grain boundaries,[4] grain boundary precipitates that strain the metal lattice,[5] and the formation of austenite at the grain boundaries.[6] The application of the chromium depletion theory to ferritic stainless steels has been discussed extensively by Bäumel.[7]

The present investigation was undertaken to determine which of the proposed mechanisms can be substantiated with experimental data obtained on ferritic stainless steels. High-purity 17 pct Cr alloys containing small controlled additions of carbon or nitrogen were therefore prepared, and then examined electrochemically and metallographically.

A. PAUL BOND is Research Group Leader, Climax Molybdenum Co. of Michigan, Ann Arbor, Mich.

This manuscript is based on a talk presented at the symposium on New Developments in Stainless Steel, sponsored by the IMD Corrosion Resistant Metals Committee, Detroit, Mich., October 14-15, 1968.

EXPERIMENTAL PROCEDURES

Materials. Two series of experimental alloys were prepared from electrolytic iron and low-carbon ferrochromium using the split-heat technique. In this technique, the base composition is melted, and part of the melt is poured off to produce an ingot. To the balance of the melt, the required addition is made and the next ingot cast. This process is repeated until a series of the desired compositions is cast. By this procedure the impurity levels are essentially constant within each series. All the alloys in the carbon-containing series were melted and cast in vacuum. The base composition in the nitrogen series was melted and cast in vacuum; subsequent ingots in the series were melted with additions of high-nitrogen ferrochromium, and cast under argon at a pressure of 0.5 atmosphere. Two additional alloys were produced starting with normal purity materials. They were induction-melted while protected by an argon blanket and cast in air. Table I gives the composition of the alloys.

The 2-in.-diam ingots produced were hot-forged and hot-rolled to a thickness of 0.3 in. and then cold-rolled to 0.15 in. All specimens were annealed at 1450°F for 1 hr. The indicated sensitizing heat treatments were performed on annealed material. All heat treatments were followed by a water quench.

Specimen Preparation. For the 65 pct nitric acid test, 1 by 2 by 0.14-in. specimens were wet-surface ground to remove surface irregularities and polished through 3/0 dry metallographic paper. For the modified Strauss test, $\frac{3}{8}$ by 3 by 0.14-in. specimens were similarly prepared. Immediately prior to testing, the

Table I. Compositions of the Alloys

Alloy	Composition, pct			
	Cr	Mo	C	N
270A	16.76		0.0021	0.0095
270B	16.74		0.0025	0.022
270C	16.87		0.0031	0.032
270D	16.71		0.0044	0.057
271A	16.81		0.012	0.0089
271B	16.76		0.018	0.0089
271C	16.69		0.027	0.0085
271D	16.81		0.061	0.0071
4073*	18.45	1.97	0.034	0.045
4075†	18.5	2.0	0.03	0.03

*Also contains 0.47 pct Ti and 0.13 pct Si.
†Also contains 0.11 pct Ni.

specimens were washed in acetone and then in a solution of detergent in hot water, rinsed in distilled water, dipped in methanol, dried, and weighed.

Nitric Acid Tests. Tests in 65 pct nitric acid were performed according to ASTM Designation A262-64. The corrosion cells consisted of 1-liter Erlenmeyer flasks fitted with cold-finger condensers. The test specimens were immersed in 700 ml of boiling 65 pct nitric acid solution for 48 hr, washed in distilled water, rinsed in methanol, dried, and weighed. They were then exposed in fresh solution for another 48-hr period. This procedure was repeated for a total of five 48-hr periods.

Modified Strauss Tests. Modified Strauss tests[8,9] were performed in 1-liter Erlenmeyer flasks fitted with cold-finger condensers. Each specimen was suspended in a vented glass cradle made from a 30-ml beaker. A total of 50 g of heavy copper turnings was packed around the sample in the cradle. Care was taken to ensure good electrical contact between turnings and test piece. The baskets containing the copper and the specimen were immersed in 700 ml of solution containing 16 pct sulfuric acid and 6 pct copper sulfate by weight. The test was conducted in the boiling solution for 24 hr. After the corrosion exposure the specimens were cleaned ultrasonically, first in distilled water and then in methanol.

Table II. Results Obtained from the Modified Strauss Test on the 17 pct Cr Alloys Containing Carbon and Nitrogen

Alloy	Carbon, pct	Nitrogen, pct	Temperature, °F, of Final Heat-Treatment*	R/R_0†	Bend Test‡
270A	0.0021	0.0095	1450	1.004	P
			1700	1.04	C
			1900	0.944	P
				1.035	
			2100	1.046	C
				1.135	
270B	0.0025	0.022	1450	1.006	P
			1700	1.372	F
			1900	1.623	F
				4.48	
			2100	3.7	F
270C	0.0031	0.032	1450	1.005	P
			1700	1.17	F
				1.198	
			1900	1.292	F
				1.26	
			2100	4.88	F
270D	0.0044	0.057	1450	1.006	
			1700	1.097	F
			1900	1.133	
			2100	1.087	F
				1.102	
271A	0.012	0.0089	1450	1.002	
			1700	1.83	F
			1900	3.83	F
				Disintegrated	
			2100	Disintegrated	
271B	0.018	0.0089	1450	0.999	
			1700	1.55	F
			1900	2.04	F
				2.03	
			2100	Disintegrated	
271C	0.027	0.0085	1450	1.002	
			1700	1.336	F
			1900	1.392	F
				1.445	
			2100	3.35	F
271D	0.061	0.0071	1450	1.002	P
			1700	1.312	F
			1900	1.099	F
				1.151	
			2100	1.76	F

*For 1 hr.

†Ratio of specimen resistance after corrosion test to specimen resistance before test.

‡After the resistance measurements were completed, the specimens were bent through 180 deg around a radius equal to the specimen thickness and examined at X20. P = no cracks visible; C = small cracks visible; and F = specimen fractured.

To evaluate the behavior of the test specimens, the electrical resistance of the specimens was measured before and after testing by passing a constant current of about 1 amp through the specimen and through a standard specimen of similar material and dimensions in series with it. A potentiometer was used to measure the IR drop across the test specimen and across the standard specimen. Since measurements on the standard and test specimens were completed in less than 1 min, the effects of variations in current or temperature were largely eliminated by normalizing the test specimen resistance against the standard specimens.

Some resistance measurements were supplemented by a bend test. After exposure the specimen was bent through 180 deg around a radius equal to the specimen thickness and examined for cracks at a magnification of X20. If the specimen fractured or if cracks were visible, a similar specimen, given the identical heat treatment but not exposed to the corrosive environment, was bent and examined in the same manner. Each time a specimen was reported as cracking or fracturing in the bend test, the control specimen was found to sustain the bend test without cracking.

Polarization Measurements. The polarization measurements were made with the aid of a potentiostat. Standard procedures were followed. The electrolyte was 1N sulfuric acid saturated with argon. The temperature was 24 ± 1°C.

RESULTS

The results of the modified Strauss test on the two series of high-purity alloys containing carbon and nitrogen are shown in Table II. None of the alloys was subject to intergranular corrosion when annealed at 1450°F, which is in accord with the standard practice of annealing ferritic stainless steels at 1300° to 1500°F.[10]

The resistance change cannot be readily converted into a direct measure of the extent of intergranular attack. The specimen geometry and grain size are very important in determining the resistance change associated with a given amount of intergranular corrosion. Correlation of the resistance changes with the bend tests, metallographic examination, and the results of the boiling nitric acid tests to be subsequently reported aids in the interpretation of the resistance change. On this basis, a resistance change of less than 1 pct indicates that no intergranular corrosion has occurred; a change of 1 to 5 pct corresponds to superficial grain boundary attack; and an increase of more than 10 pct shows that severe intergranular corrosion has occurred.

Alloy 270A, containing no deliberate additions of carbon or nitrogen, showed practically no intergranular attack after sensitizing treatments at 1700° or 1900°F for 1 hr. One of the specimens sensitized at 2100°F was only slightly attacked, but the other suffered definite intergranular corrosion.

The smallest deliberate additions of nitrogen (Alloy 270B) or of carbon (Alloy 271A) were sufficient to cause definite intergranular corrosion in the modified Strauss test after sensitizing treatments at 1700°, 1900°, or 2100°F. Larger additions of carbon or nitrogen did not greatly increase the extent of intergranular attack on sensitized specimens. In fact, the alloy containing the most nitrogen (270D) was less severely attacked than the two alloys (270B and 270C) with lower nitrogen contents. In addition, the alloy containing the largest carbon content (271D) was less severely attacked than the alloys (271A, 271B, and 271C) containing less carbon.

Metallographic examination of these alloys after the sensitizing heat treatment showed that a large amount of martensite was present in the alloys with the highest nitrogen or carbon contents, Fig. 1. Thus, it appears that carbon or nitrogen contents great enough to promote the formation of large amounts of austenite at the sensitizing temperature lead to less severe intergranular attack than do slightly smaller amounts.

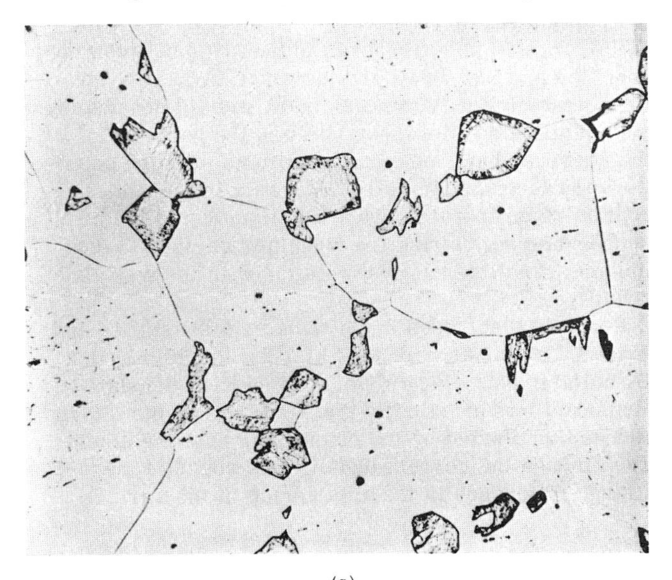

(*a*)

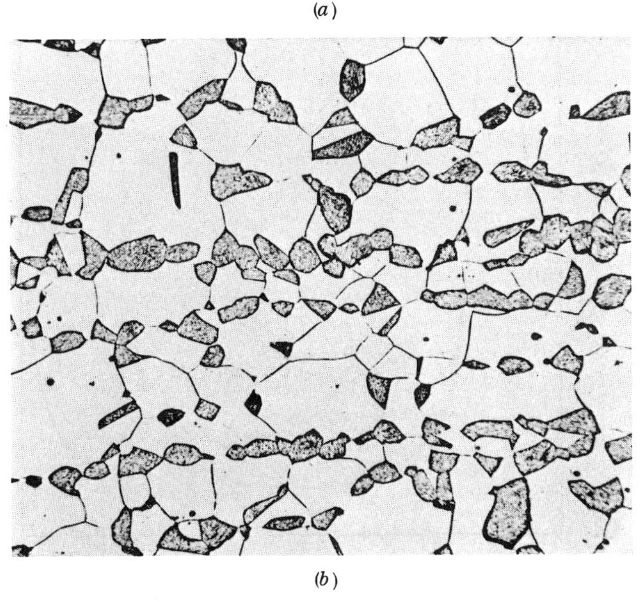

(*b*)

Fig. 1—17 pct Cr alloys, water quenched from 2100°F. Magnification 177 times. (*a*) Alloy 270D (0.0044 pct C and 0.057 pct N) (*b*) Alloy 271D (0.061 pct C and 0.071 pct N).

The standard boiling 65 pct nitric acid test was also performed on a representative set of alloys and heat treatments, Table III. In this test, a high corrosion rate and especially an increase in corrosion rate with successive exposure are indicative of intergranular corrosion. The results of this test were in complete accord with the results from the Strauss test, *i.e.*, the alloys annealed at 1450°F were not subject to intergranular attack. Alloy 270A (no added carbon or nitrogen) was only slightly subject to intergranular corrosion after heat treatment at 2100°F, but highly resistant after treatment at 1700° or 1900°F. The smallest carbon or nitrogen addition rendered the alloys susceptible to intergranular corrosion after sensitizing heat treatments. Finally, the greater carbon or nitrogen contents were somewhat less damaging than the smaller amounts.

In addition to the high-purity series, two alloys simulating commercial purity were examined. They were of similar composition except one contained 0.47 pct Ti. These alloys were included so that the effect of a stabilizing agent could be evaluated. Previous experience suggests that the 2 pct Mo in these two alloys does not have a decisive influence on their intergranular corrosion susceptibility in comparison to unmodified 17 pct Cr stainless steels.

The results of the corrosion tests, Tables IV and V, show that the titanium-free steel is susceptible to

Table III. Results of the Boiling 65 pct Nitric Acid Test on Some of the 17 pct Cr Alloys Containing Carbon and Nitrogen

			Corrosion Rate, mdd, of Specimens in Successive 48-Hr Periods after 1-Hr of Heat-Treatment at the Indicated Temperature			
Alloy	Carbon, pct	Nitrogen, pct	1450°F	1700°F	1900°F	2100°F
270A	0.0021	0.0095	137	127	106	164
			161	127	160	251
			172	189	144	319
			181	230	171	419
			182	254	159	479
270B	0.0025	0.022	133	226		103
			162	520		368
			173	924		848
			174	1390		1121
			180	1658		1365
270D	0.0044	0.057	157		194	95
			186		411	226
			186		596	321
			200		783	482
			200		901	660
271A	0.012	0.0089	116	366	96	
			136	392	572	
			132	852	2712	
			128	1179	3232	
			118	1329	2475	
271D	0.061	0.0071	130	195	107	
			154	420	624	
			152	696	1827	
			162	861	2545	
			137	925	2766	

Table IV. Results of the Boiling Sulfuric Acid + Copper Sulfate Test on 18 pct Cr - 2 pct Mo Alloys with and without Titanium

Alloy	Temperature, °F, of Final 1-Hr Heat-Treatment	Test Medium	Duration of Test, hr	R/R_0*	Bend Test†
4075	1500	16 pct H_2SO_4-6 pct $CuSO_4$	24		P
(18 pct Cr-2 pct Mo-0 pct Ti)	1700	16 pct H_2SO_4-6 pct $CuSO_4$	24		F
4073	1500	16 pct H_2SO_4-6 pct $CuSO_4$	24		P
(18 pct Cr-2 pct Mo-0.47 pct Ti)	1700	16 pct H_2SO_4-6 pct $CuSO_4$	24		P
	1900	16 pct H_2SO_4-6 pct $CuSO_4$	24		P
	2100	16 pct H_2SO_4-6 pct $CuSO_4$	240	0.99	P
	2100	50 pct H_2SO_4-6 pct $CuSO_4$	24	0.996	P

*Ratio of specimen resistance after corrosion test to specimen resistance before test.

†After the resistance measuerments were completed, the specimens were bent through 180 deg around a radius equal to the specimen thickness and examined at X20. P = no cracks visible; F = specimen fractured.

intergranular corrosion after heat treatment at 1700°F or higher. On the other hand, the steel containing 0.47 pct Ti did not undergo intergranular attack in the Strauss test even after heat treatment at 2100°F. This was true even when the test was made more severe by increasing the exposure time tenfold or by increasing the sulfuric acid concentration from 16 to 50 pct. In contrast, this same steel heat-treated at 1700°F experienced severe intergranular attack in boiling 65 pct nitric acid. Thus, titanium can prevent intergranular corrosion when the oxidation potential of the corrodent is not too high, but is ineffective in boiling nitric acid. This is in accord with the observations of Lula *et al.*[5]

Controlled potential electrochemical measurements were also made on these alloys. The results of potentiodynamic scanning of electrodes of Alloy 271A are shown in Fig. 2. It can be seen that over much of the potential region investigated there is little difference in anodic dissolution rate between sensitized and annealed material. The peak corresponding in active dissolution (the potential range from −0.5 to −0.3 v) is essentially identical in the two conditions. The sensitized steel does exhibit a secondary peak in the passive region at about −0.05 v that is not shown by the annealed alloy. The remainder of the passive region (about −0.3 to +0.9 v) is substantially the same in both steels. The beginning of the transpassive region is also the same for the two heat treatments. Alloys containing larger amounts of carbon behaved in a similar manner except that as the carbon content increased, the effects of sensitization on the behavior in the passive potential region also increased. At the higher carbon contents, sensitization not only led to a secondary peak around −0.05 v but also give rise to currents that were larger throughout the passive region than were displayed by the corresponding annealed electrodes.

Potentiodynamic scans of the highest purity alloy (270A) showed no differences in behavior between the annealed and the sensitized material. Actually, the alloys containing only nitrogen additions did not display any significant differences between the behavior of the sensitized and annealed conditions in these potentiodynamic scanning tests. Thus only in the alloys containing deliberate carbon additions could the potentiodynamic polarization technique be used to distinguish sensitized alloys from those in the annealed condition.

Polished electrodes made from both sensitized and annealed material were held at 1.2 v for 90 min in a 1N sulfuric acid electrolyte. The current density measured was practically the same for all the alloys and was unaffected by the sensitizing heat treatment. Even though the current density was constant there was a large difference in the appearance of the various

Table V. Results of the Boiling 65 pct Nitric Acid Test on 18 pct Cr-2 pct Mo Alloys with and without Titanium

Alloy	Titanium, pct	Corrosion Rate, mdd, of Specimens in Successive 48-Hour Periods after 1-Hr Heat-Treatment at the Indicated Temperature		
		1500°F	1700°F	1900°F
4075	0.0	98	126	
		71	588	
		74	774	
		77	1080	
		84	824	
4073	0.47	72		2070
		87		8500
		108		
		139		
		164		

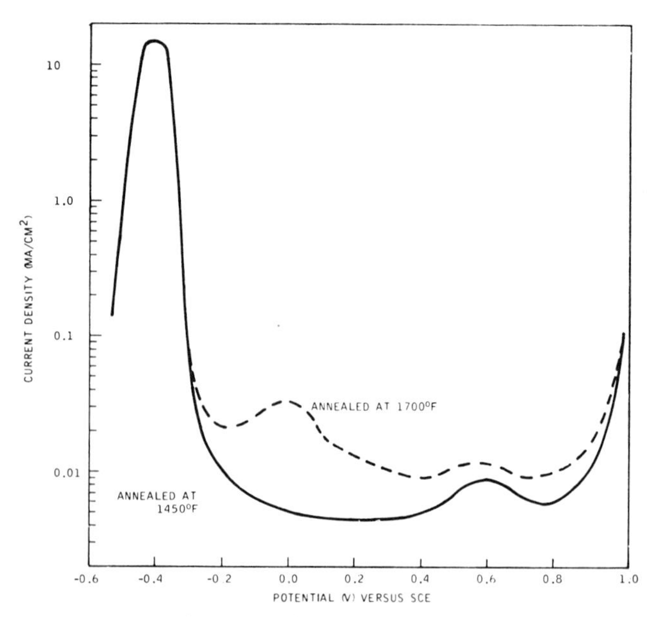

Fig. 2—Potentiodynamic polarization curves obtained at a scanning rate of 600 mv per hr in 1N sulfuric acid at 24°C on Alloy 271A (16.76 pct Cr, 0.012 pct C, and 0.0089 pct N) heat-treated at 1450° and at 1700° F.

electrodes after this treatment. All the alloys, except 270A, heat-treated at 1700°F exhibited definite preferential grain boundary attack. None of the alloys annealed at 1450°F showed any selective grain boundary corrosion. The appearance of representative test electrodes, Fig. 3, shows that even the smallest de-

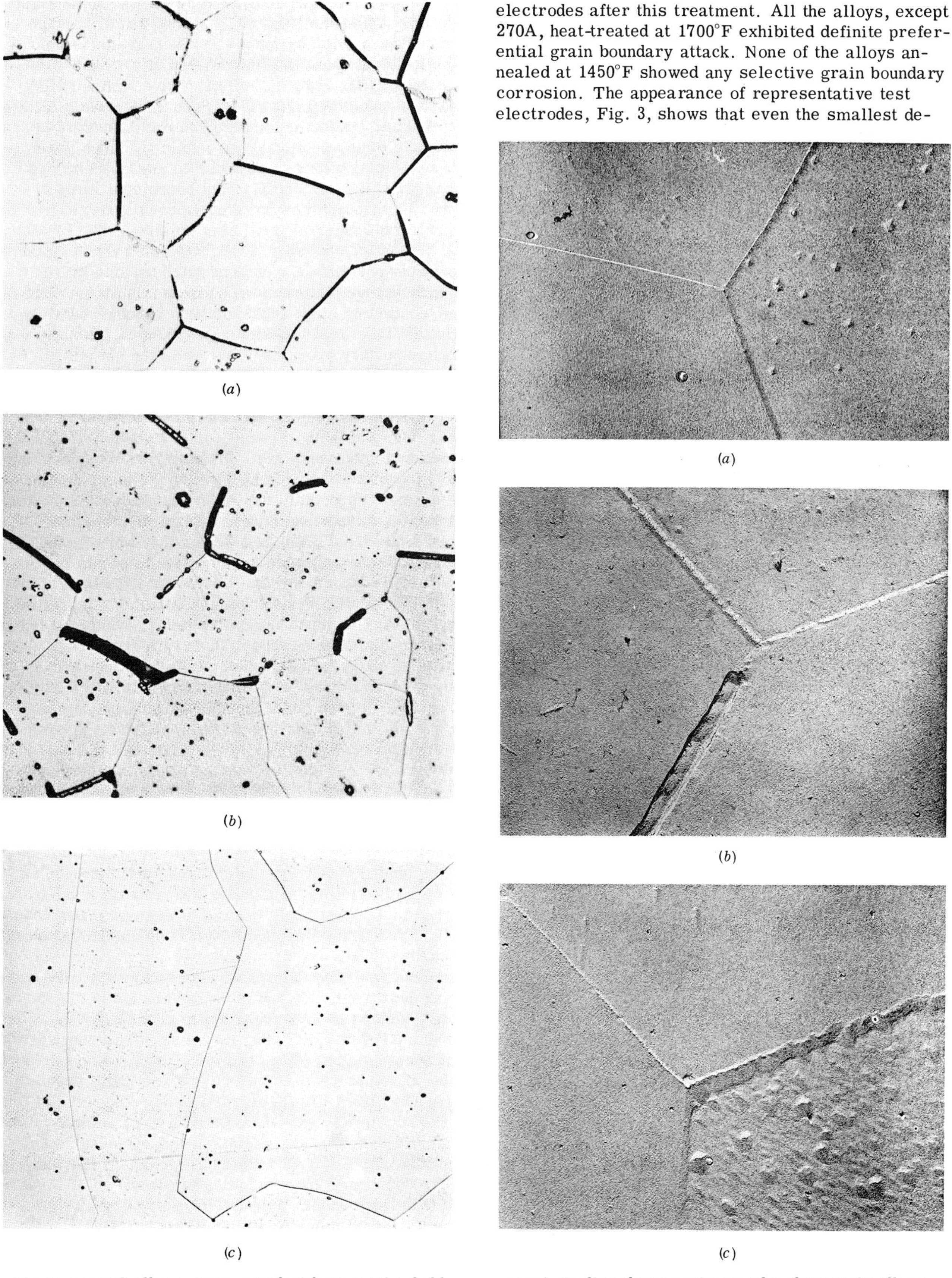

Fig. 3—17 pct C alloys, water quenched from 1700°F, held for 90 min at 1.2 v vs SCE in 1N sulfuric acid at 24°C. (*a*) Alloy 271A (0.0126 pct C and 0.0089 pct N) (*b*) Alloy 270B (0.0025 pct C and 0.022 pct N) (*c*) Alloy 270A (0.0021 pct C and 0.0095 pct N). Magnification 341 times.

Fig. 4—Replica electron micrographs of 17 pct Cr alloys water quenched from 1700°F. Magnification 9150 times. (*a*) Alloy 270A (0.0021 pct C and 0.0095 pct N) (*b*) Alloy 270B (0.0025 pct C and 0.022 pct N) (*c*) Alloy 271A (0.012 pct C and 0.0089 pct N).

liberate nitrogen addition led to definite intergranular corrosion although not all grain boundaries were attacked. Thus there is complete correlation between the compositions and heat treatments that show grain boundary attack at 1.2 v in 1N sulfuric acid at room temperature and those that show intergranular corrosion in the Strauss and Huey tests.

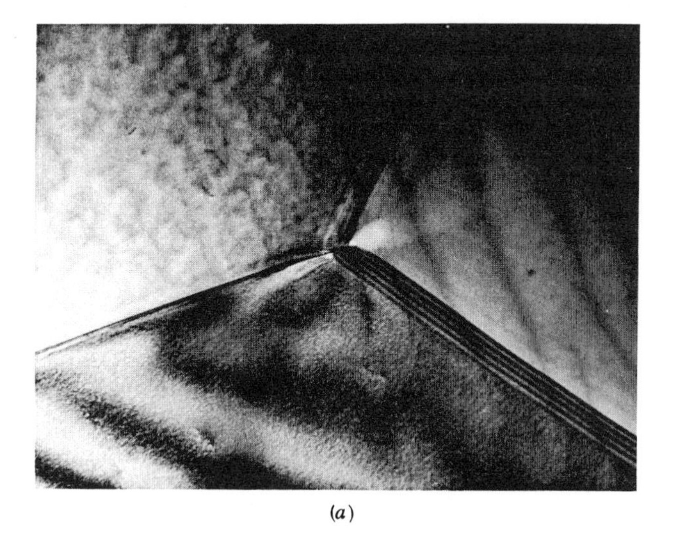

(a)

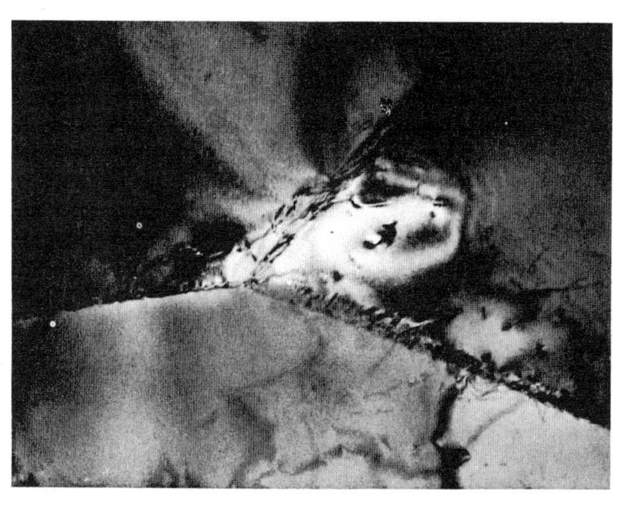

(b)

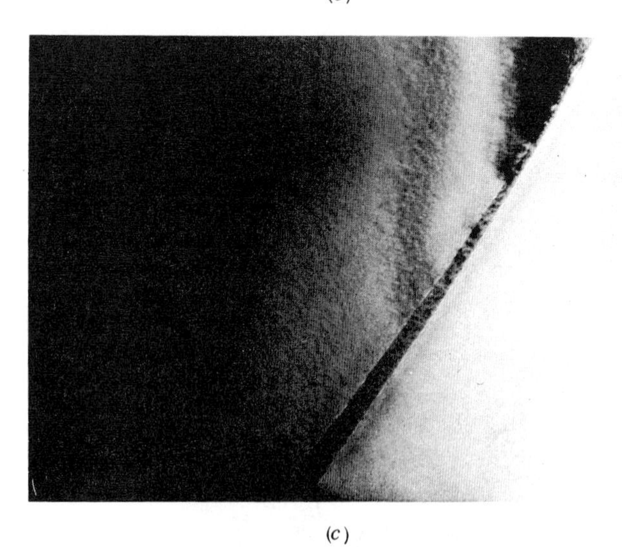

(c)

Fig. 5—Transmission electron micrographs of 17 pct Cr alloys water quenched from 1700°F. Magnification 22,800 times. (a) Alloy 270A (0.0021 pct C and 0.0095 pct N) (b) Alloy 270B (0.0025 pct C and 0.022 pct N) (c) Alloy 271A (0.012 pct C and 0.0089 pct N).

Alloys 270A, 270B, and 271A were examined after heat treatment at 1450°, 1700°, and 2100°F using both replica and transmission electron microscopy. No evidence of any continuous or semicontinuous grain boundary precipitates was found in any of the alloys after annealing at 1450°F. Only occasional large spherical precipitates were seen, usually not in grain boundaries. After a sensitizing treatment at 1700°F, the highest purity alloy (270A) remained free of grain boundary precipitates; evidence of very fine grain boundary precipitates could be seen in both the nitrogen-containing alloy (270B) and the carbon-containing alloy (271A). Heat treatment at 2100°F led to some grain boundary precipitation even in the highest purity alloy (270A) investigated. Examples of the microstructures after a 1700°F anneal are shown in Figs. 4 and 5. No evidence of grain boundary martensite was seen in any of these three alloys.

Several specimens were examined, after undergoing intergranular corrosion in the Strauss test, to determine what happened in the grain boundary precipitates during the corrosion process. Extraction replicas were made from individual grains that were dislodged on bending the corroded specimens. As shown in Fig. 6, considerable quantities of carbides remained at the grain boundaries of the carbon-containing alloys after corrosion. Electron diffraction patterns obtained from these precipitates matched the Cr_7C_3 pattern very well. The appearance of these precipitates indicates that they have undergone very little chemical attack in the corrosive environment. Similar examination of the alloys containing deliberate nitrogen additions showed a very fine grain boundary precipitate that did not yield a clear enough electron diffraction pattern to allow identification.

DISCUSSION

The tendency of ferritic stainless steels to become sensitized to intergranular corrosion in aggressive environments is closely related to the amounts of carbon and nitrogen in the alloy. Very small amounts of carbon (0.012 pct) or nitrogen (0.02 pct) were sufficient to cause intergranular corrosion after a sensitizing heat treatment. These same amounts of carbon or nitrogen were also sufficient to produce grain boundary precipitates after the same heat treatment that sensitized the alloys to intergranular corrosion. Thus, it seems reasonable to associate grain boundary precipitates with intergranular corrosion.

The results of the electron microscopy showed clearly that martensite is not necessary to cause sensitization to intergranular corrosion. In fact, the alloys containing large amounts of carbon or nitrogen, and thus large amounts of martensite, were less severely corroded than were the alloys containing somewhat lower amounts of carbon or nitrogen. This behavior must be the result of the partitioning of carbon and nitrogen that occurs when appreciable austenite is formed at the sensitizing temperature. The austenite will be enriched in carbon and nitrogen while the ferrite is depleted in these elements. On cooling, the austenite transforms to martensite, probably still retaining much of the carbon and nitrogen in solution. As a final result, it appears that a less continuous network of grain boundary precipitates forms under these conditions than when the carbon or nitrogen content is just sufficient to saturate the ferrite at the sensitizing temperature, but is insufficient to cause appreciable austenite to form.

The idea that the grain boundary precipitates cause intergranular corrosion by forming excellent sites for the cathodic half of the corrosion reaction is not valid under the conditions of the Strauss or potentiostatic tests. In the Strauss test, the stainless steel specimen is coupled in a large area of metallic copper surrounded by cupric ion. This forms a cathode that cannot be polarized and that maintains the potential at 0.06 v.[9] Thus, behavior of cathodic sites on the stainless steel could have no effect on the corrosion reaction under these conditions. The same reasoning applies to the potentiostatic tests in which the test electrode was maintained at the very oxidizing potential of 1.2 v. In these tests, too, an external electrode serves at the cathode and there is no way that possible cathodic sites on the test electrode could influence its dissolution behavior.

Another possible mechanism whereby grain boundary precipitates could cause intergranular corrosion involves the strain occasioned by the precipitated particles. The fact that the titanium-containing steel (4073) had grain boundary precipitates after heat treatment at 2100°F but did not undergo intergranular corrosion in the Strauss test indicates that more than the stress associated with grain boundary precipitates is required to cause intergranular corrosion in these alloys. The question of the effect of residual stresses on corrosion rate has been studied rather extensively for iron and for stainless steels.[11-13] A recent study has shown that the anodic polarization of a carbon steel is hardly influenced by microstructure.[14] Considerable indirect evidence shows that the stress field set up by the intergranular precipitates could not account for the observed degree of preferential grain boundary corrosion.

Another proposed cause of intergranular corrosion is preferential dissolution of the grain boundary precipitates themselves. However, work on austenitic steels has shown that chromium carbides at the grain boundaries are not attacked in the Strauss test,[3] and the present study shows this is also true of ferritic steels, Fig. 6. In general, it has been observed that except under highly oxidizing conditions carbide and nitride precipitates in iron-base alloys are cathodic to the base metal,[13] which is shown by the behavior of the titanium-containing alloy. When this alloy was heat-treated at 1900°F, an intergranular precipitate, presumably titanium carbonitride, was present and attacked by boiling 65 pct nitric acid. However, when the alloy in this condition was tested in the Strauss test, no evidence of intergranular corrosion was obtained. Thus, it seems clear that intergranular corrosion under conditions represented by the Strauss test is not due to preferential dissolution of grain boundary precipitates. Nevertheless, under highly oxidizing conditions, direct attack on the grain boundary precipitates may at least contribute to intergranular corrosion.

The intergranular corrosion that occurs in sensitized ferritic stainless steels under conditions typified by the Strauss test is best explained by the chromium depletion theory. The temperatures (1700°F and up) that cause sensitization to occur correspond to markedly increased solubility of carbon and nitrogen in ferrite.[15,16] On cooling, this ferrite becomes supersaturated with carbon or nitrogen and chromium-rich carbides or nitrides precipitate. This precipitation occurs at such a high temperature and at such a rapid rate that even in water-quenched specimens of moderate dimensions the precipitate is evident. However, it has been reported that quenching small-diameter wire into iced brine does prevent precipitation and sensitization.[5] The chromium-rich precipitates that do form are located preferentially at grain boundaries. The areas in the alloy immediately adjacent to these precipitates are depleted of chromium.

Iron-chromium alloys containing less than about 12 pct Cr are more difficult to passivate than those of higher chromium contents. The polarization measurements of Shiobara *et al.*[17] show this most clearly. Their results indicate that this difference in behavior with chromium content would be most noticeable in the potential range of –0.2 to 0.3 v vs SCE. The potential corresponding to the Strauss test is about in the middle of this range. These data strongly suggest that the hump in the anodic passivation current, found in this work for some sensitized alloys, Fig. 2, is due to the presence of regions of lower chromium content than in the bulk alloy.

The behavior of the alloy containing titanium can be explained on this basis. Titanium carbides and nitrides are formed in preference to the chromium-rich precipitates so that no chromium-depleted zone is formed. However, a grain boundary network of titanium carbides evidently formed after heat treatment at 2100°F. This network was severely attacked in the boiling 65 pct nitric acid test. This result is in agreement with results that have been obtained on titanium stabilized austenitic stainless steels.[18]

The mechanism of intergranular corrosion in ferritic stainless steels is very similar to that generally ac-

Fig. 6—Electron micrograph of an extraction replica of the corrosion-fracture surface of Alloy 271C sensitized at 2100°F after the 24-hr modified Strauss test. Magnification 4675 times.

cepted as applying to austenitic stainless steels. In both types of steel, depletion of chromium adjacent to the grain boundaries plays a major role, and this depletion can result from the precipitation of chromium-rich carbides or nitrides. The differences in the conditions that lead to this precipitation are due to differences in solubilities of carbon and nitrogen in ferrite or austenite. The solubility of carbon or nitrogen in ferrite is much smaller than it is in austenite at a given temperature. Thus austenite can retain the usual amounts of carbon and nitrogen in solution down to relatively low temperatures, so low that easily obtainable cooling rates can suppress precipitation altogether. On the other hand, the precipitation reactions in ferritic steels occur at higher temperatures and are so rapid that it is very difficult to attain a fast enough quench to prevent precipitation. Nevertheless, the net result is the same in austenitic and ferritic stainless steels. If chromium-rich precipitates are allowed to form, corresponding chromium-depleted regions are formed, and these depleted regions are subject to preferential corrosion in some environments.

ACKNOWLEDGMENT

The author expresses his appreciation to V. Biss for performing the electron microscopy and electron diffraction work during this investigation.

REFERENCES

[1] A. Bäumel *et al.*: *Corrosion Science*, 1964, vol. 4, p. 89.

[2] H. H. Uhlig: *Corrosion and Corrosion Control*, John Wiley and Sons, Inc., New York, 1963, p. 266.

[3] R. Stickler and A. Vinckier: *Corrosion Science*, 1963, vol. 3, p. 1.

[4] E. Houdremont and W. Tofaute: *Stahl ü. Eisen*, 1952, vol. 72, p. 539.

[5] R. A. Lula, A. J. Lena, and G. C. Kiefer: *Trans. ASM*, 1954, vol. 46, p. 197.

[6] J. Hochmann: *Rev. Met.*, 1951, vol. 48, p. 734.

[7] A. Bäumel: *Arch. Eisenhüttenw.*, 1963, vol. 34, p. 135.

[8] L. R. Scharfstein and C. M. Eisenbrown: *Advances in the Technology of Stainless Steel and Related Alloys*, Special Technical Publication 369, p. 235, American Society for Testing Materials, Philadelphia, 1965.

[9] M. A. Streicher: *J. Electrochem Soc.*, 1959, vol. 106, p. 161.

[10] *Metals Handbook*, eighth ed., American Society of Metals, Metals Park, Ohio, 1964, vol. 2, p. 244.

[11] Z. A. Foroulis and H. H. Uhlig: *J. Electrochem Soc.*, 1964, vol. 111, p. 522.

[12] T. P. Hoar and J. C. Scully: *J. Electrochem Soc.*, 1964, vol. 111, p. 349.

[13] H. H. Uhlig: *Corrosion*, 1963, vol. 19, p. 231t.

[14] M. E. Komp and H. E. Trout, Jr.: *Corrosion*, 1968, vol. 24, p. 11.

[15] E. Baerlecken, W. A. Fischer, and K. Lorenz: *Stahl ü Eisen*, 1961, vol. 81, p. 768.

[16] W. Tofaute, C. Kuttner, and A. Buttinghaus: *Arch. Eisenhüttenw*, 1936, vol. 9, p. 607.

[17] K. Shiobara, Y. Sawada, and S. Morioka: *Trans. Japanese Inst. Metals*, 1965, vol. 6, p. 58.

[18] *Intergranular Corrosion of Chromium-Nickel Stainless Steel*, Progress Report 1, Welding Research Council Bulletin 93, January 1964.

T. J. Nichol[1] *and J. A. Davis*[1]

Intergranular Corrosion Testing and Sensitization of Two High-Chromium Ferritic Stainless Steels

REFERENCE: Nichol, T. J. and Davis, J. A., **"Intergranular Corrosion Testing and Sensitization of Two High-Chromium Ferritic Stainless Steels,"** *Intergranular Corrosion of Stainless Alloys, ASTM STP 656,* R. F. Steigerwald, Ed., American Society for Testing and Materials, 1978, pp. 179–196.

ABSTRACT: The effect of several parameters (alloy chemistry, thermal history, and solution chemistry) on the resistance of two new ferritic stainless steels to intergranular corrosion was investigated. The susceptibility of Types 26-1S and 29-4 to intergranular corrosion depended on the alloy chemistry and the particular test used to evaluate susceptibility. For both alloys, the severity of the standard tests increased in the following order: copper sulfate ($CuSO_4$)-16 percent sulfuric acid (H_2SO_4), nitric acid (HNO_3)-hydrogen fluoride (HF), $CuSO_4$-50 percent H_2SO_4 and ferric sulfate ($Fe_2(SO_4)_3$)-50 percent H_2SO_4. Sensitization was produced in these alloys by either welding or high-temperature annealing followed by air cooling. The most reliable evaluation method was bending of the material combined with optical examination.

KEY WORDS: stainless steels, intergranular corrosion, ferritic stainless steels, sensitizing, evaluation, tests

The susceptibility of conventional ferritic stainless steels to intergranular corrosion after welding operations has been one of the major problems preventing extensive utilization of these steels. The sensitized regions of ferritic stainless steel weldments are normally the weld metal itself and a narrow region adjacent to the weld metal. The mechanism responsible for intergranular corrosion is generally accepted as being the same as that in austenitic stainless steels. The formation of chromium compounds, usually carbides or nitrides or both, at the grain boundaries results in the creation of chromium-depleted regions adjacent to the grain boundary particles [*1*].[2] Intergranular corrosion occurs because of the relative difference

[1] Technical marketing manager and research specialist, stainless and alloy steel metallurgy, respectively, Allegheny Ludlum Steel Corporation, Research Center, Brackenridge, Pa. 15014.

[2] The italic numbers in brackets refer to the list of references appended to this paper.

in corrosion resistance between the chromium-depleted regions and the matrix.

Lula, Lena, and Kiefer [2] suggested that intergranular corrosion could be prevented in ferritic stainless steels by lowering the carbon content or by stabilizing the carbon with titanium. Two newer ferritic stainless steels which represent these alternatives are Type 26-1S [3], a 26Cr-1Mo steel stabilized with titanium, and Type 29-4 [4], a 29Cr-4Mo steel melted to low levels of both carbon and nitrogen. Bond and Dundas [5] have determined the amount of titanium required in ferritic stainless steels to prevent intergranular corrosion in a boiling copper sulfate-sulfuric acid ($Cu\ SO_4$-$H_2\ SO_4$) solution. For 26Cr-1Mo alloys having combined carbon and nitrogen levels up to 0.04 percent, they claim the minimum titanium necessary to prevent intergranular corrosion is equal to 0.15 percent plus 3.7 times the carbon plus nitrogen content. Pollock, Sweet, and Collins [6] also investigated the intergranular corrosion resistance of Type 26-1S and concluded that a minimum titanium ratio of ten times the carbon plus nitrogen contents is preferred to provide resistance. ASTM specifications for Type 26-1S (ASTM Grade XM33) list a minimum titanium of seven times the carbon plus nitrogen content. Lula et al [2] demonstrated that ferritic stainless steels containing titanium were resistant to intergranular corrosion in the boiling $Cu\ SO_4$-$H_2\ SO_4$ solution but not to boiling 65 percent nitric acid (HNO_3). Baumel [7] subsequently showed that titanium carbides were attacked directly by boiling 65 percent HNO_3. Streicher [4] used a boiling ferric sulfate ($Fe_2(SO_4)_3$)-$H_2\ SO_4$ solution to determine the carbon and nitrogen contents necessary to prevent intergranular corrosion of welded 28Cr-4Mo alloys. The levels determined were 100 ppm for carbon and 200 ppm for nitrogen, with the additional requirement that the total carbon plus nitrogen must be less than 250 ppm.

Several studies [1, 2, 8–15] have utilized various heat treatments (other than welding) to produce sensitization in ferritic stainless steels. The critical parameters are temperature, time at temperature, and cooling rate from elevated temperatures. Three temperatures are important for sensitization: an elevated temperature (referred to as the sensitization temperature) which corresponds to the temperature where significant amounts of carbon and nitrogen are soluble in the ferrite matrix, an intermediate temperature where titanium and columbium carbides and nitrides form, and a lower temperature which corresponds to the temperature at which chromium carbides and nitrides form. Time at temperature controls both the dissolution process (at or above the sensitization temperature) and the degree of chromium depletion (at the precipitation temperature). Cooling rates from the sensitization temperature or higher control the balance between chromium depletion resulting from precipitation and chromium leveling resulting from chromium diffusion [12].

The difference in corrosion rate between chromium-depleted regions and the ferritic matrix has been attributed to simply the dependence of the passive corrosion rate on the chromium content of the ferrite.

Streicher [*14*] and Steigerwald [*15*] demonstrated the very large effect of chromium content on the general corrosion rate (in the passive state) in a boiling $Fe_2(SO_4)_3$-H_2SO_4 acid solution. Streicher's [*14*] work demonstrated that chromium-depleted regions are probably less than 25 percent chromium in high-chromium ferritic steels and greater than 12 percent chromium in 16 percent chromium steels. Streicher also emphasized that the effectiveness of any solution in delineating susceptibility to intergranular corrosion depends on sensitivity of that solution to chromium content. Bond [*1*] and Bond and Lizlovs [*9*] demonstrated that sensitized and nonsensitized ferritic stainless steels exhibit different anodic polarization characteristics in 1 *N* H_2SO_4. They found that the peak corresponding to active dissolution is unaffected by sensitization; however, sensitized steels exhibited a secondary peak in the passive range as well as higher current densities than nonsensitized steels. Anodic polarization only showed effects of sensitization resulting from carbon increases and not nitrogen.

These studies have shown that the susceptibility of ferritic stainless steels to intergranular corrosion depends on (*a*) the corrosive media in which the steels are exposed, (*b*) the chemistry (particularly carbon and nitrogen) of the steel, and (*c*) the thermal history of the steel.

The purpose of this study was to investigate the parameters (chemistry, thermal history, and solution chemistry) which determine the susceptibility of two new ferritic stainless steels to intergranular corrosion. The initial work utilized welded materials having variable compositions previously determined [*4, 5*] in one standard test to represent both resistant and susceptible material. The behavior of welded specimens in four standard laboratory intergranular corrosion tests was determined. In addition, the effects of sensitization temperature and cooling rate were investigated for the different compositions of material. Finally, anodic polarization studies in $CuSO_4$-50 percent H_2SO_4 were conducted.

Experimental Procedure

Materials and Preparation

Strip material 0.17 cm (65 mil) thick from several heats of Types 26-1S and 29-4 was obtained in the annealed condition 387 cm per minute per centimetre of thickness (60 in. per minute per inch of thichness) at 954°C (1750°F) for Type 26-1S and 1066°C (1950°F) for Type 29-4). The Type 26-1S material represented a range of compositions from Ti/C + N = 3.5 to Ti/C + N = 18.5, while the Type 29-4 encompassed compositions from 34 to 380 ppm of carbon and 27 to 265 ppm of nitrogen. The specific carbon, nitrogen, and titanium levels are tabulated in the Results section of this paper. Welded specimens were prepared using conventional tungsten inert gas procedures. Helium was used as the shielding and backup gas. Heat treatments were conducted in a furnace containing an argon atmosphere. All specimens were machined (both edges and surface) to a standard 5.1 by 2.5 by 0.15 cm (2 by 1 by 0.060 in.) size having a 50 rms surface finish. All specimens were degreased in hot alkaline cleaner rinsed in distilled water, dipped in methanol, dried, and weighed prior to testing.

Corrosion Tests

Cupric Sulfate Test—This test was conducted in accordance with ASTM Recommended Practices for Detecting Susceptibility to Intergranular Attack in Stainless Steels (A 262, Practice E). Specimens were exposed to a boiling aqueous solution of 6 percent by weight $CuSO_4$ plus 16 percent by weight H_2SO_4 for a 24-h period. The specimens were in galvanic contact with copper shot.

Nitric Acid-Hydrofluoric Acid (HF) Test—This test, except for exposure time, was conducted in accordance with ASTM A 262, Practice D. Duplicate specimens were tested in a 10 percent HNO_3-3 percent HF solution at 70°C (158°F) for two 4-h periods (a fresh solution was used for each period).

Modified Cupric Sulfate Test—This test, except for solution chemistry [*16*], was conducted in accordance with ASTM A 262, Practice B. Specimens were exposed for one period of 120-h duration in a boiling solution consisting of 71.4 g of reagent grade $CuSo_4 \cdot 5H_2O$ and 600 ml of 50 percent H_2SO_4. A clean, solid piece of copper was immersed in the solution (not in contact with the specimens) throughout the test period. The first use of this test for Type 26-1S was described by Pollock et al [*6*].

Ferric Sulfate Test—This test was conducted in accordance with ASTM A 262, Practice B. Specimens were exposed for one period of 120-h duration in a boiling solution consisting of reagent grade $Fe_2(SO_4)_3$ and 50 percent H_2SO_4.

Polarization Tests—Anodic polarization measurements were conducted in a glass cell modified for sheet specimens. Specimens were clamped against a circular Teflon[3] gasket, exposing 1 cm^2 of surface area to the corrosive media. A platinized platinum electrode was used for a counter electrode, and a saturated calomel electrode was used as a reference electrode with potentials recorded through a salt bridge. Anodic polarization curves were run potentiodynamically at a scan rate of 32 V/h. The rapid scan rate was used to avoid extensive dissolution in the active region of the anodic polarization curve. Anodic polarization curves were generated in boiling 2 *N* H_2SO_4 and in the four intergranular test solutions. Activa-

[3] Trade name of the du Pont de Nemours Company.

tion areas were determined by measuring the area under the primary active anodic peak or the secondary active anodic peak.

Evaluation Procedures

All specimens were weighed after testing and a corrosion rate determined. Specimens were also examined at ×30 magnification for evidence of grain dropping. In addition, most specimens were bent through 180 deg around a radius equal to the specimen thickness and examined for cracks or fissures.

Welded specimens were rated as passed (that is, resistant to intergranular corrosion) if no grain dropping was observed at ×30 and no cracks or fissures were observed after bending. A failure thus constituted either grain dropping or fissures on bending. Corrosion rates were not used for evaluation of welded specimens.

Heat-treated specimens were evaluated in a fashion similar to welded specimens; however, correlations between corrosion rate and evidence of intergranular corrosion were made.

Results

Table 1 lists the results for welded Type 26-1S material in the four different evaluation tests. All materials subjected to the $Fe_2(SO_4)_3$-50 percent H_2SO_4 test failed. The results for the three other tests delineate three areas of behavior. Materials having a Ti/C + N ratio lower than 6 failed all three tests. Materials having a Ti/C + N ratio between 6 and 9 demonstrated different responses, depending on the particular test that, is, the same material may pass one test and fail the other two. Materials having a Ti/C + N ratio greater than 9 passed the three tests. Figure 1 shows the appearance of the bent materials having Ti/C + N ratios of 5.4, 6.2, 7.4, and 9.1 after testing in the three different tests. The variable response of the materials having Ti/C + N ratios of 6.2 and 7.4 is evident.

Table 2 lists the results for welded Type 29-4 material in the four evaluation tests. Although the results from test to test are more consistent than the welded Type 26-1S results (Table 1), there is still disagreement between the tests for two heats. In this case, the two heats passed the $CuSO_4$-16 percent H_2SO_4 test and failed the three other tests. The Type 29-4 materials can be also delineated into the three areas, that is, materials that pass all the tests, materials that fail all the tests, and materials whose resistance is dependent on the test.

Figure 2 shows the bent specimens from two heats of Type 29-4 in the four evaluation tests. All the tests except the $CuSO_4$-16 percent H_2SO_4 test resulted in very definitive grain dropping in the heat-affected zone of the material having 100 ppm of carbon and 150 ppm of nitrogen. The

TABLE 1—*Results for welded Type 26-1S.*

Composition, percent by weight							
C	N	Ti	Ti/C + N	Test 1 $CuSO_4$ 16% H_2SO_4	Test 2 HNO_3 HF	Test 3 $CuSO_4$ 50% H_2SO_4	Test 4 $Fe_2(SO_4)_3$ 50% H_2SO_4
0.017	0.031	0.17	3.5	F	F	F	F
0.017	0.031	0.26	5.4	F	F	F	F
0.028	0.037	0.40	6.2	P	F	F	NT
0.024	0.030	0.40	7.4	P	P	F	NT
0.030	0.019	0.38	7.8	P	P	F	F
0.041	0.034	0.62	8.3	P	P	F	NT
0.019	0.030	0.42	8.6	P	P	F	F
0.025	0.019	0.40	9.1	P	P	P	NT
0.024	0.028	0.52	10.0	P	P	P	NT
0.073	0.017	1.06	11.8	P	P	P	NT
0.020	0.024	0.57	13.0	P	P	P	F
0.040	0.018	0.77	13.3	P	P	P	NT
0.012	0.017	0.44	15.2	P	P	P	NT
0.054	0.010	1.06	16.6	P	P	P	F
0.020	0.020	0.74	18.5	P	P	P	F

NOTE— P—no grain dropping or fissures on bending.
F—grain dropping or fissures on bending.
NT—not tested.

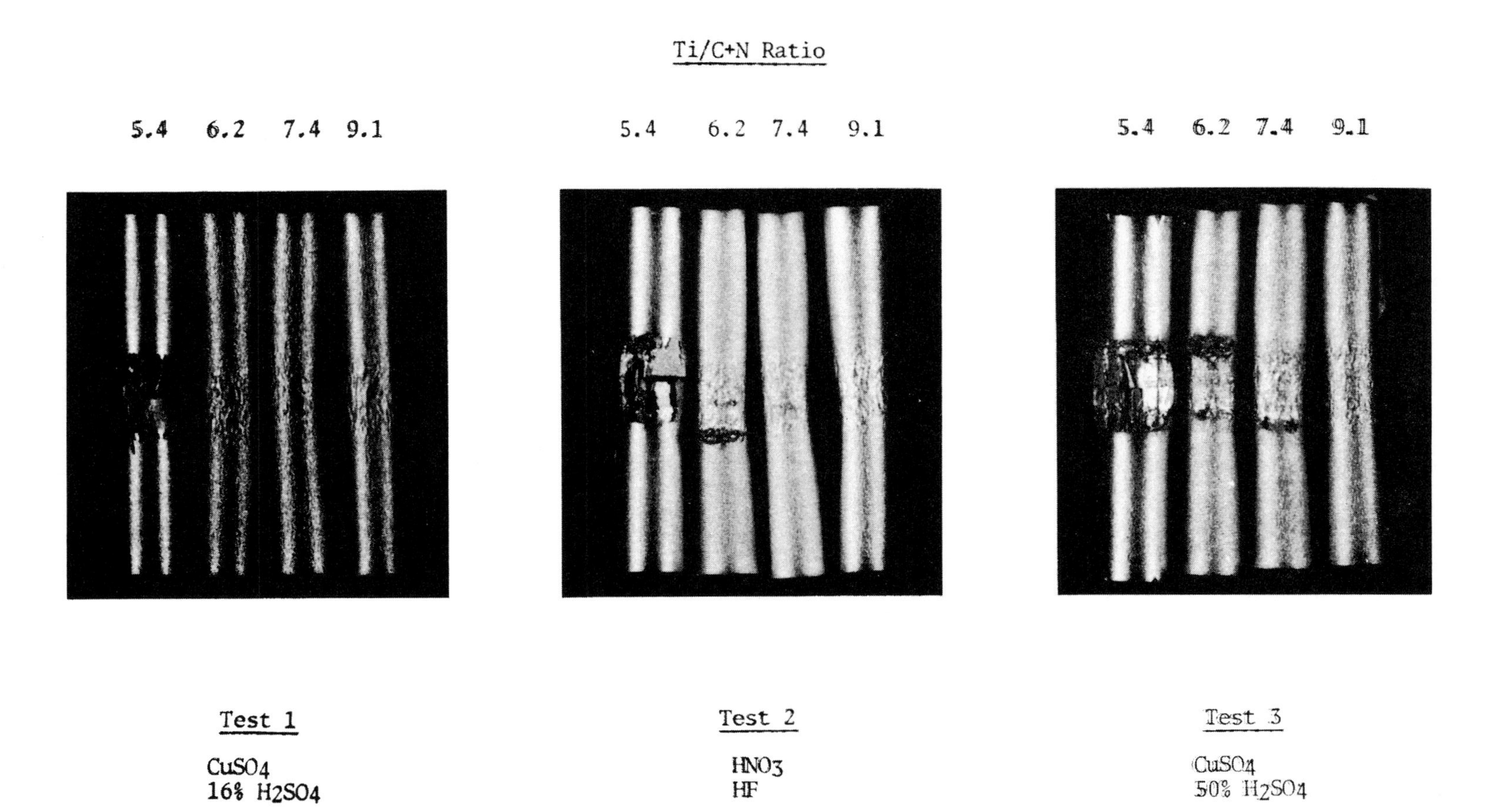

FIG. 1—*Effect of Ti/C + N ratio and evaluation test on the susceptibility of welded Type 26-1S specimens to intergranular corrosion.*

TABLE 2—*Results for welded Type 29-4.*

Composition, ppm			Test 1 $CuSO_4$	Test 2 HNO_3	Test 3 $CuSO_4$	Test 4 $Fe_2(SO_4)_3$
C	N	C+N	16% H_2SO_4	HF	50% H_2SO_4	50% H_2SO_4
380	150	530	F	F	F	F
180	160	340	F	F	F	F
110	38	148	NT	NT	F	F
100	150	250	P	F	F	F
45	265	310	P	F	F	F
42	250	292	P	P	P	P
46	190	236	P	P	P	P
44	118	162	P	P	P	P
40	150	190	P	P	P	P
16	140	156	P	P	P	P
34	27	61	P	P	P	P

NOTE—P = no grain dropping or fissures on bending.
F = grain dropping or fissures on bending.
NT = not tested.

$CuSO_4$-16 percent H_2SO_4 test did not result in grain dropping, and no cracks or fissures were evident after bending. The appearance of the weld metal is a result of the bending. This is similar to an orange-peel effect that is commonly observed after deformation of material having a large grain size. In Fig. 2, the material having 42 ppm carbon and 250 ppm nitrogen shows this orange-peel effect; however, no cracks or fissures were present.

The effect of heat-treatment temperature and cooling rate were determined for 26-1S having a Ti/C+N ratio of 7.8. Material was heat treated for 5 min at temperatures ranging from 871 to 1260°C (1600 to 2300°F) and then quenched into water, cooled in ambient air, or cooled in a bed of vermiculite. The corrosion rates were determined in the $CuSO_4$-50 percent H_2SO_4 test and are shown in Fig. 3. Higher corrosion rates were observed as the heat-treatment temperature exceeded 1038°C (1900°F). Air cooling resulted in the highest corrosion rates. Both water quenching and cooling in vermiculite reduced the corrosion rates for material heat treated above 1038°C (1900°F). In addition, water quenching raised the apparent sensitization temperature. Figure 4 illustrates the effect of Ti/C+N ratio on the corrosion rate in $CuSO_4$-50 percent H_2SO_4 after elevated temperature exposure followed by air cooling. The apparent sensitization temperature is increased as the Ti/C+N ratio is increased.

Anodic polarization curves are shown in Fig. 5 for 26-1S exposed to boiling $CuSo_4$-50 percent H_2SO_4. The curves shown in Fig. 5 are for material having Ti/C+N ratios of 7.8 and 11.8, and heat treated at 1149°C

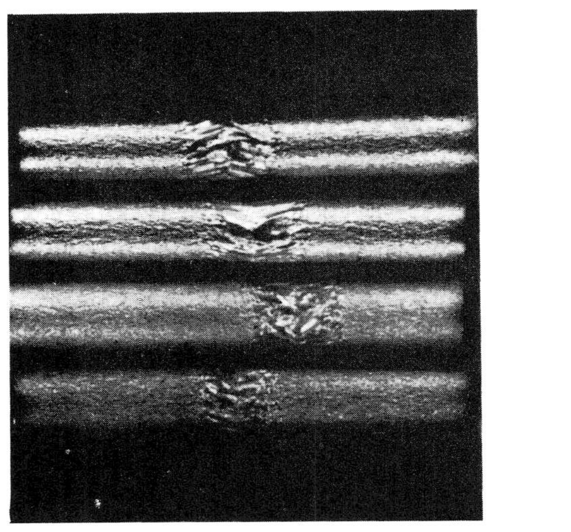

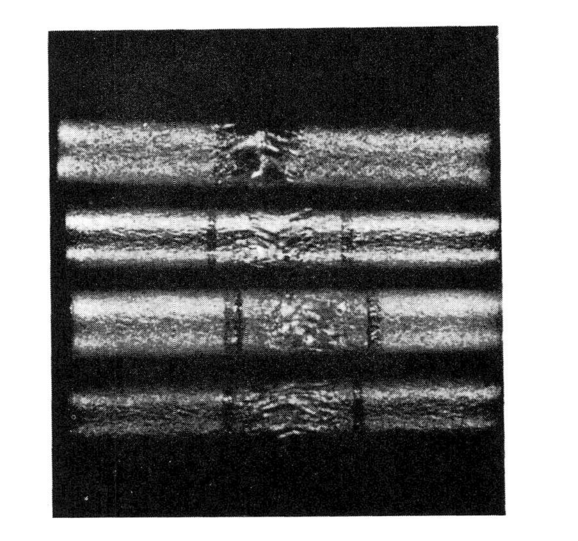

FIG. 2—*Effect of carbon, nitrogen, and evaluation test on the susceptibility of welded Type 29-4 material to intergranular corrosion* $(1.0 \text{ mil/year} = 2.5 \times 10^{-2} \text{ mm/A})$.

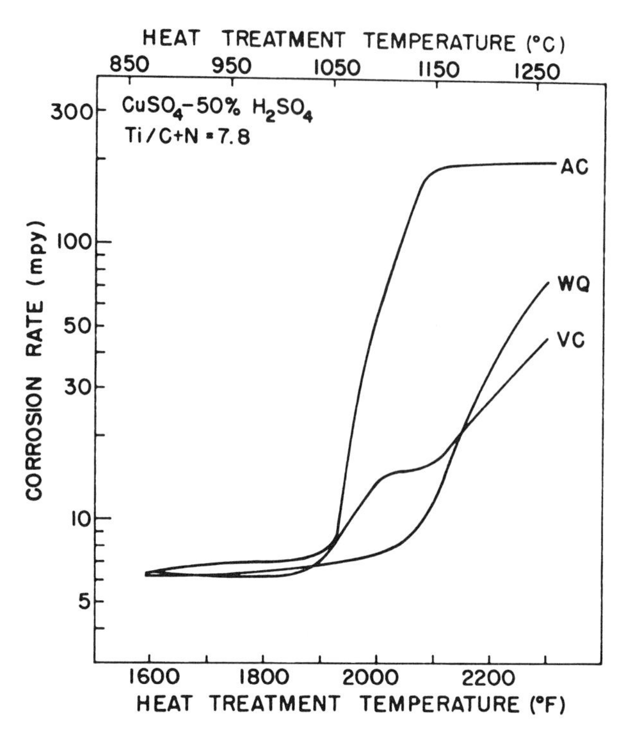

FIG. 3—*Effect of cooling rate and heat-treatment temperature on the corrosion rate of Type 26-1S. AC = air cool, WQ = water quench, VC = vermiculite cool (1.0 mil/year = 2.5×10^{-2} mm/A).*

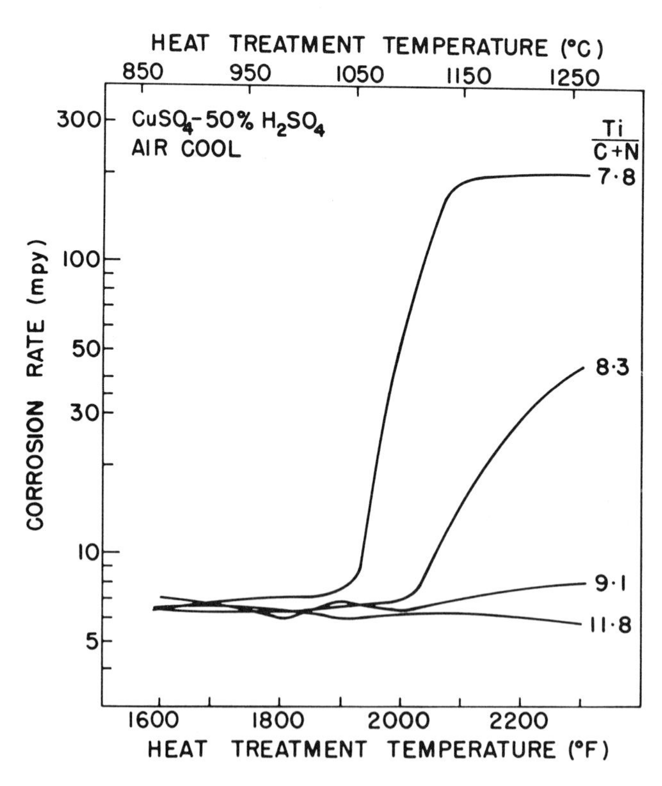

FIG. 4—*Effect of temperature and Ti/C + N ratio on the corrosion rate of Type 26-1S.*

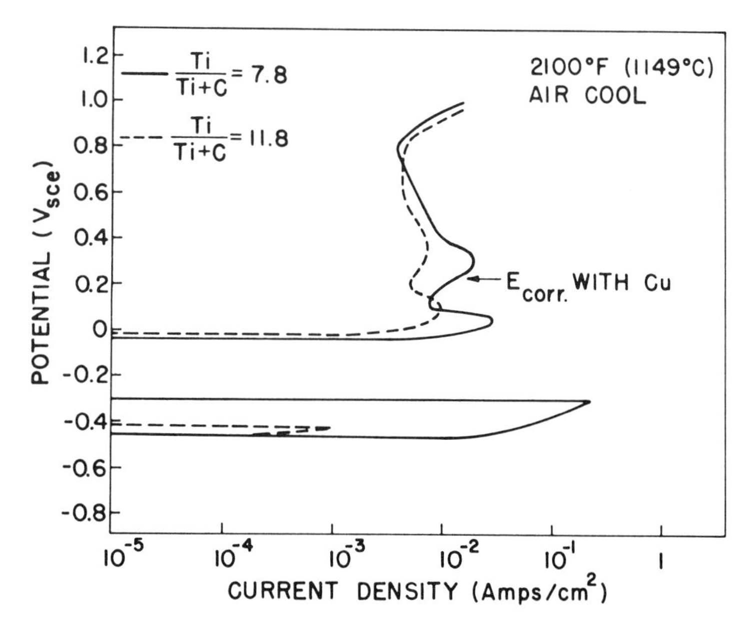

FIG. 5—*Effect of Ti/C + N on the anodic polarization of Type 26-1S in $CuSO_4$-50 percent H_2SO_4.*

(2100°F), followed by cooling in air. The critical current density, primary activation area, and secondary activation area were determined for the four Ti/C + N ratios shown in Fig. 4 for the 1149°C (2100°F) heat treatment, and the primary activation area as a function of Ti/C + N is shown in Fig. 6. A semilog relationship was observed between primary activation area and Ti/C + N.

The results in Table 3 illustrate the effect of elevated temperature (5 min at temperature) and cooling rate (air cool or water quench) on the corrosion rate of several Type 29-4 heats in the $Fe_2(SO_4)_3$-50 percent H_2SO_4 test. Water quenching again raises the apparent sensitization temperature (for the 180-ppm carbon material) and decreases the corrosion rate at all temperatures. The susceptibility of the 29-4 heats to high corrosion rates is more sensitive to the carbon content than the nitrogen content. This effect is shown in Fig. 7 for material heat treated at 1121°C (2050°F), air cooled, and tested in either the $Fe_2(SO_4)_3$-50 percent H_2SO_4 solution or the $CuSO_4$-50 percent H_2SO_4 solution. The materials having variable carbons contained approximately 150 ppm of nitrogen, while the materials having variable nitrogen contained approximately 40-ppm carbon. There is good correlation between the corrosion rates determined in $CuSO_4$-50 percent H_2SO_4 test and the $Fe_2(SO_4)_3$-50 percent H_2SO_4 test.

The effect of heat treating in the low-temperature sensitizing range was determined at one temperature 593°C (1100°F) for three heats of Type 29-4. Material was initially heat treated at 1121°C (2050°F) and then water quenched. The effect of time at 593°C (1100°F) on the corrosion rate in the $Fe_2(SO_4)_3$-50 percent H_2SO_4 test is shown in Fig. 8. The high-carbon material (110 ppm) demonstrated an increased corrosion rate after 30 min, followed by a decrease in corrosion rate for longer times. The high corrosion rate at 30 min was much less than the corrosion rate observed when this material was air cooled from 1121°C (2050°F) (Table 3). The high-nitrogen material (250 ppm) demonstrated a relatively constant corrosion rate up to 100 min and then increased slightly after 300 min. The corrosion rate of the low-carbon (34 ppm) and nitrogen (27 ppm) material was not affected by time at 593°C (1100°F).

The effect of the different tests on the corrosion rates of several 29-4 heats is shown in Table 4. All materials were heat treated at 1121°C (2050°F) and air cooled. For material that is resistant to intergranular corrosion, the corrosion rates for the HNO_3-HF test, the $CuSO_4$-50 percent H_2SO_4 test, and the $Fe_2(SO_4)_3$-50 percent H_2SO_4 test are similar, while the $CuSO_4$-16 percent H_2SO_4 test results in corrosion rates that are an order of magnitude lower. For material that is susceptible to intergranular corrosion, the corrosion rates increase in the order: $CuSO_4$-16 percent H_2SO_4 test, HNO_3-HF test, $Fe_2(SO_4)_3$-50 percent H_2SO_4 test, and $CuSO_4$-50 percent H_2SO_4 test. The corrosion rate in the $CuSO_4$-50 percent H_2SO_4 test is one to three orders of magnitude higher than the corrosion rate in the $CuSO_4$-16 percent H_2SO_4 test.

Discussion

The requirement for reliable laboratory intergranular corrosion tests and evaluation criteria designed specifically for ferritic stainless steels has developed with the introduciton of several new ferritic stainless steels. Reliable laboratory tests are required in the development stage to delineate the required compositional limits for the alloys and in the commercial product stage to establish product acceptance criteria. The data in Tables 1 and 2, for Types 26-1S and 29-4, clearly show that the present laboratory intergranular corrosion tests do not yield comparable results when the same materials are tested in the as-welded condition. If the number of materials that failed a given test is used as a measure of test severity, then the tests increase in severity from Test 1 to 2 to 3 to 4.

Both Types 26-1S and 29-4 have compositonal ranges which represent materials that either pass all tests or fail all tests. The intermediate composition ranges (characterized by varied results from test to test) are the composition of concern. In order to determine the suitability of materials within these compositional ranges for use in the as-welded condition, the laboratory tests will have to be supplemented by and correlated with actual service enviroments. The results indicate that, for Type 26-1S, there is a minimum ratio of $Ti/C+N$ required to provide intergranular corrosion resistance, and the absolute minimum level depends on the evaluation test

TABLE 3—*Effect of heat treatment on the corrosion rate (mil/year)*[c] *of Type 29-4 in* $Fe_2(SO_4)_3$*-50%* H_2SO_4.

Composition, ppm			Water Quench			Air Cool		
C	N	C + N	1010°C (1850°F)	1121°C 2050°F)	1232°C (2250°F)	1010°C 1850°F)	1121°C 2050°F)	1232°C 2250°F)
40	150	190	4.0	4.7	5.8	4.5	5.2	6.0
100	150	250	3.8	5.7	3.9	8.2	47.3[b]	64.4[b]
180	160	340	4.0	4.4[b]	21.2[b]	5.3	127.1[b]	D[a]
110	38	148	3.9	7.6[b]	5.5[b]	4.6	51.0[b]	D[a]
34	27	61	4.0	4.5	4.4	3.9	5.0	6.0
40	150	190	4.0	4.7	5.8	4.5	5.2	6.0
42	250	292	3.8	4.4	5.9	4.5	5.1	7.5[b]
45	265	310	3.0	7.7[b]	3.7	5.0	7.6[b]	16.2[b]

[a] Dissolved.
[b] Grain dropping or fissures on bending.
[c] – 1.0 mil/year = 2.54×10^{-2} mm/A.

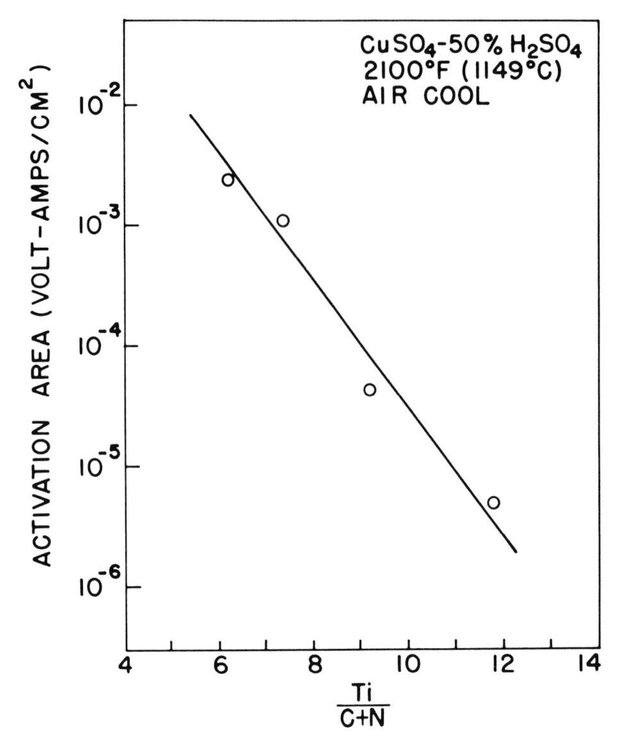

FIG. 6—*Relationship between primary activation area and Ti/C + N ratio in Type 26-1S.*

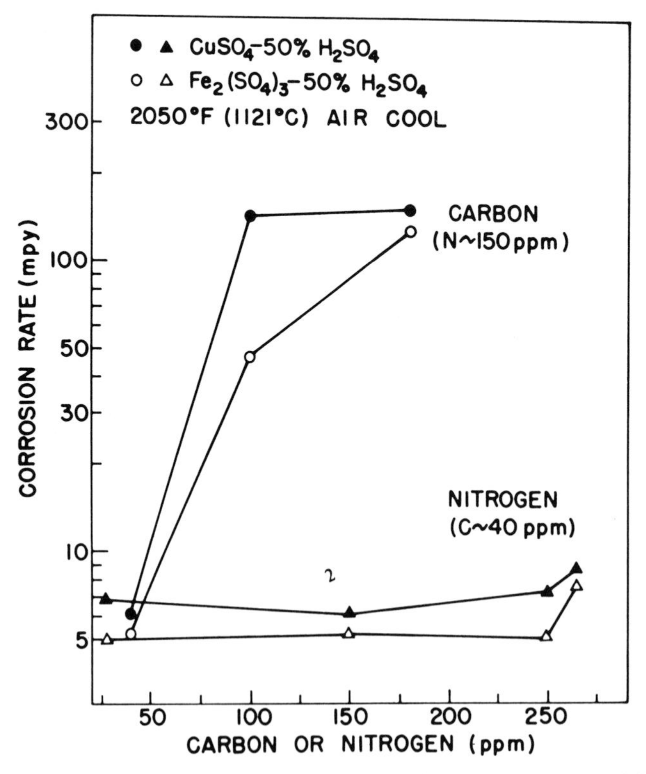

FIG. 7—*Effect of carbon and nitrogen content on the corrosion rate of heat-treated Type 29-4 (1.0 mil/year = 2.5 × 10^{-2} mm/A).*

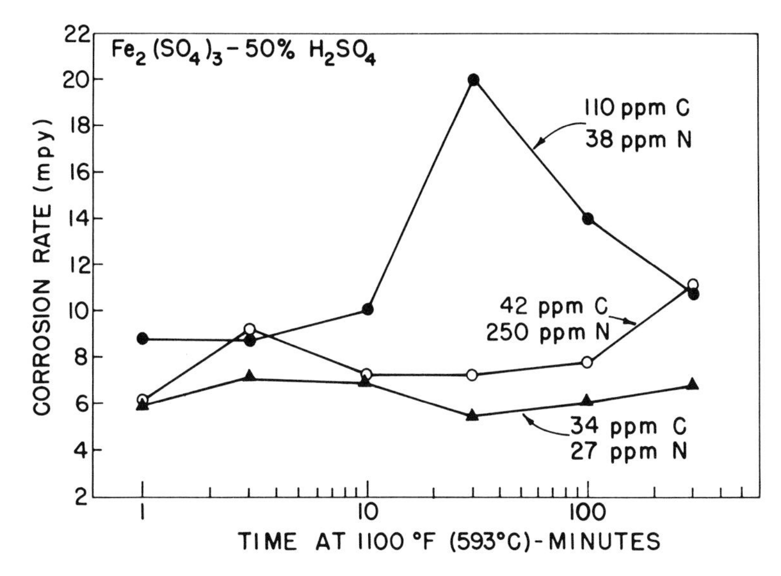

FIG. 8—*Effect of time at 593°C (1100°F) on the corrosion rate of heat-treated (1121°C (2050°F), water quenched) Type 29-4 (1.0 mil/year = 2.5 × 10^{-2} mm/A).*

utilized. For Type 29-4, the carbon and nitrogen levels must be maintained below certain levels, and the tolerance for nitrogen is higher than for carbon [*4*]. Again, the particular chemistry requirement depends on the evaluation test used.

The results obtained with heat-treated specimens illustrate the dependence of a sensitization temperature on chemical composition and thermal history. Figures 3 and 4 show that with Type 26-1S the observed sensitization temperature increases when the Ti/C + N ratio is increased and also when water quenching is employed rather than air cooling or cooling in

TABLE 4—*Corrosion rates (mil/year)*[c] *of heat-treated*[a] *Type 29-4.*

Composition, ppm			Test 1 $CuSO_4$	Test 2 HNO_3	Test 3 $CuSO_4$	Test 4 $Fe_2(SO_4)_3$
C	N	C + N	16% H_2SO_4	HF	50% H_2SO_4	50% H_2SO_4
40	150	190	0.3	3.8	6.1	5.2
100	150	250	0.4 [b]	7.9 [b]	142.1	47.3 [b]
180	160	340	5.3 [b]	31.8 [b]	151.3	127.1 [b]
110	38	148	4.9 [b]	30.8 [b]	61.6 [b]	51.0 [b]
34	27	61	0.2	4.2	6.9	5.0
40	150	190	0.3	3.8	6.1	5.2
42	250	292	0.3	3.2	7.3	5.1
45	265	310	0.4	4.7 [b]	8.7	7.6 [b]

[a] 1121°C (2050°F), 5 min, air cool.
[b] Grain dropping or fissures on bending.
[c] 1.0 mil/year = 2.5 × 10^{-2} mm/A.

vermiculite. The occurrence of an upper sensitization temperature is presumably related to the chemical and thermal conditions which produce a sensitized microstructure. This is the temperature at which significant amounts of carbon and nitrogen are in solution and thus available to form chromium carbides or nitrides on cooling to room temperature. The effect of Ti/C + N ratio is thus a result of both increased stability (that is, slower dissolution kinetics) and increased precipitation kinetics of titanium carbide and titanium nitride as the Ti/C + N ratio increased. The increase in apparent sensitization temperature for water quenching (Fig. 3) is a result of a reduction in the amount of precipitation with a faster cooling rate. The results in Table 3 illustrate that water quenching (contrasted with air cooling) also increased the observed sensitization temperature in Type 29-4. The highest carbon material sensitized to a greater degree than the lower carbon steels, even at 1232°C (2250°F), when water quenched.

The existence of a lower sensitization temperature and the dependence on chemistry was shown for Type 29-4 in Fig. 8. This represents the temperature range where carbon or nitrogen, or both, which has been retained in solution by water quenching from an elevated temperature, precipitates as chromium compounds. The decrease in corrosion rate as the time at temperature increases is a result of chromium rediffusion after precipitation is complete. This removes the chromium gradients which cause intergranular corrosion. A similar effect was noted in Fig. 3 where cooling in vermiculite (as compared to water or air cooling) reduced the corrosion rate. The maximum corrosion rate observed at the lower sensitization temperature is much less than the corrosion rates observed after high-temperature exposure of the same material. This is probably a result of the use of only one temperature, 593°C (1100°F), which may not be the temperature which causes the maximum intergranular corrosion rate. This temperature would depend on the chemical composition, particularly, chromium, molybdenum, carbon, and nitrogen.

The different response of the same material in the four corrosion tests (Tables 1 and 2) could be a result of several phenomena. Some of the tests may selectively attack components of the microstructure that other tests do not. The work of Lula, Lena, and Kiefer [2] and Baumel [7] has shown that nitric acid may attack chromium-depleted regions as well as titanium carbides and nitrides; whereas, a $CuSO_4$-16 percent H_2SO_4 solution apparently only attacks regions depleted in chromium. The susceptibility to intergranular attack in $Fe_2(SO_4)_3$-50 percent H_2SO_4 of high Ti/C + N ratio heats of 26-1S is attributed to the attack of titanium carbides or nitrides by this highly oxidizing solution. The resistance to intergranular corrosion in the other solutions of the same material indicates the absence of a significant amount of chromium-depleted regions.

A second cause for variability between tests is the difference in their ability to detect susceptibility to intergranular attack associated with chromium depletion. Streicher [14] has shown that the sensitivity of the test to chromium gradients depends on the sensitivity of the corrosion rate in the media on chromium content. The data in Table 4 demonstrate that the $CuSO_4$-16 percent H_2SO_4 test yields much lower corrosion rates than either the $CuSO_4$-50 percent H_2SO_4 test or the $Fe_2(SO_4)_3$-50 percent H_2SO_4 test. As the absolute corrosion rate decreases (for example, from Test 3 to Test 1), the relative sensitivity to chromium content decreases, and thus the ability to detect susceptibility to intergranular corrosion resulting from chromium depletion also is reduced.

A third observed phenomenon which relates to the relative sensitivity of the tests was shown in Fig. 5. Anodic polarization tests in the test media used for immersion testing revealed a difference in both the active dissolution behavior and the passive dissolution behavior. The activation area in the active region correlated with the corrosion rates in the immersion tests (Fig. 8). The significance of this observation is not understood, and further polarization studies are needed.

Conclusions

The purpose of this study was to investigate the parameters (alloy chemistry, thermal history, and solution chemistry) which determine the susceptibility of two new ferritic stainless steels to intergranular corrosion. The results are relevant to the development of evaluation procedures for these alloys and to the understanding of the causes of intergranular corrosion in ferritic stainless steels. Based on the work described in this study, it can be concluded that:

1. Depending on the alloy chemistry, particularly carbon, nitrogen, and titanium, sensitization can be produced in these alloys by either welding or high-temperature annealing, followed by air cooling. Sensitization can also occur with dual heat treatments; however, the combination of heat treatments required depends very much on chemistry of the material being studied.
2. The most reliable evaluation method, independent of the corrosive media, was bending of the material combined with optical examination for evidence of grain dropping. Corrosion rate in the absence of optical examination was not reliable.
3. For Types 26-1S and 29-4, the severity of the standard tests increased in the following order: $CuSO_4$-16 percent H_2SO_4, HNO_3-HF, $CuSO_4$-50 percent H_2SO_4, and $Fe_2(SO_4)_3$-50 percent H_2SO_4. The significance of the tests with regard to chemistry requirements of the alloys will have to be resolved by correlation with results obtained in actual service environments.
4. The results are consistent with models relating intergranular corrosion in ferritic stainless steels to the presence of regions depleted in chromium.

The attack of high Ti/C + N ratio heats of Type 26-1S in $Fe_2(SO_4)$ is attributed to the direct attack of titanium compounds.

Acknowledgments

The authors wish to acknowledge the contributions of the staff of Allegheny Ludlum Steel Corporation's Research Center to this work: L. Bachman and B. Clover for preparation of alloys and specimen preparation; P. Pavlik for coordination of corrosion testing; C. Canterna, J. Cook, and E. Vrotney for conducting corrosion tests; G. Aggen and M. Johnson for helpful comments and assistance in data analysis.

References

[*1*] Bond, A. P., *Transactions*, American Institute of Mining, Metallurgical and Petroleum Engineers, Vol. 245, 1969, p. 2127.
[*2*] Lula, R. A., Lena, A. J., and Kiefer, G. C., *Transactions*, American Society for Metals, Vol. 46, 1954, p. 197.
[*3*] Wright, R. N., *Welding Journal*, Vol. 50, 1971, p. 434S.
[*4*] Streicher, M. A., *Corrosion*, Vol. 30, No. 3, p. 77.
[*5*] Bond, A. P. and Dundas, H. J., "Stabilization of Ferritic Stainless Steels," paper presented at the Symposium on New Higher Chormium Ferritic Stainless Steels, held at ASTM Committee Week, Bal Harbour, Fla., 6 Dec. 1973.
[*6*] Pollock, W. I., Sweet, A. J., and Collins, J. A., "On Utilizing High-Chromium Ferritic Alloys in the Chemical Process Industry," paper presented at Symposium on New Higher Chromium Ferritic Stainless Steels, held at ASTM Committee Week, Bal Harbour, Fla., 6 Dec. 1973.
[*7*] Bäumel, A., *Stahl and Eisen*, Vol. 84, 1964, p. 798.
[*8*] Baerlecken, E., Fischer, W. A., and Lorenz, K., *Stahl and Eisen*, Vol. 81, 1961, pp. 768-778.
[*9*] Bond, A. P. and Lizlovs, E. A., *Journal of the Electrochemical Society*, Vol. 116, 1969, p. 1305.
[*10*] Hodges, R. J., *Corrosion*, Vol. 27, No. 3, 1971, p. 119.
[*11*] Hodges, R. J., *Corrosion*, Vol. 27, No. 4, 1971, p. 164.
[*12*] Demo, J. J., *Corrosion*, Vol. 27, No. 12, 1971, p. 531.
[*13*] Rarey, C. R. and Aronson, A. H., *Corrosion*, Vol. 28, No. 7, 1972, p. 255.
[*14*] Streicher, M. A., *Corrosion*, Vol. 29, No. 9, 1973, p. 337.
[*15*] Steigerwald, R. F., *Metallurgical Transactions*, Vol. 5, 1974, p. 2265.
[*16*] Streicher, M. A., *Corrosion*, Vol. 20, 1964, p. 57t.

Recent Studies into the Mechanism of Ridging in Ferritic Stainless Steels

HUNG-CHI CHAO

THE ferritic stainless steels, such as AISI Types 430 (17 pct Cr) and 434 (17 pct Cr, 1 pct Mo), are used extensively in many applications requiring forming and drawing operations, such as kitchen sinks and automobile trim and other decorative items. However, an undesirable surface condition, known as "ridging" or "roping," often develops during forming. This undesirable defect, which always occurs parallel to the sheet rolling direction, appears on the surface of a formed part as narrow, raised areas similar to corrugations. Ridging is detrimental to the appearance of decorative items, and expensive grinding and polishing operations are required to eliminate it.

In recent years, a number of mechanisms[1-6] have been proposed to explain ridging, all based on considerations of crystal plasticity and texture observation. Most recently, a plastic-buckling mechanism was proposed[6] which seems to indicate that ridging is not caused solely by anisotropic plastic flow. The present paper presents the results of recent studies to clarify certain conflicting points among various proposed mechanisms.

Experimental Results and Discussion. As observed by Chao,[1,2] by Ohashi,[4] and by Pouillard and Osdoit,[7] bands of cube-on-face texture groups (CF) such as {001}⟨011⟩, {117}⟨011⟩, {115}⟨011⟩, and {113}⟨011⟩ and bands of cube-on-corner texture groups (CC) such as {111}⟨011⟩, {111}⟨112⟩, and {554}⟨225⟩[2] exist in the rolling direction of ferritic stainless sheet. During straining along any direction in the plane of the sheet, the anisotropic plastic flow of these highly banded structures can be described by using the strain ratio (r value). The r values for textures common to ferritic material are well documented in the literature.[8] Fig. 1 shows r values calculated for both CF and CC textures as a function of direction of testing.[9] For all directions of stressing, the CC-textured grain offers two to three times more thinning resistance (r = 2 to 3) than the highest value for the CF-textured grain (r = 1). Ridging should be expected for all directions of stressing.

Commercial Type 434 stainless-steel sheet (with the composition shown in Table I and texture components shown in Fig. 2) was strained in tension. The tensile axis of the specimens was along directions 0, 30, 45, 60, and 90 deg from the rolling direction. The surface undulations of the strained samples are shown in Fig. 3. Notice that all ridges are along the rolling direction. The appearance of the rolling-direction ridges on the transversely strained sample could not be produced by the plastic buckling mechanism for ridging as proposed by Wright.[6] A tensile stress applied in the transverse direction can hardly produce compressive stress to cause buckling of the (001)[110] texture to give ridges along the rolling direction.

In a previous study,[2] relatively little ridging was found in a sample (No. 436H) with a texture almost

Table I. Chemical Composition of AISI Type 434 Stainless-Steel and Carbon-Steel Specimens Used in This Study

	Wt Pct		
	Type 434 Stainless	SK Steel	Rimmed Steel
C	0.088	0.040	0.030
Mn	0.36	0.30	0.37
P	0.019	0.007	0.004
S	0.007	0.015	0.007
Si	0.58	0.015	0.006
Cu	0.045	*	*
Ni	0.16	*	*
Cr	17.10	*	*
Mo	1.04	*	*
N	0.030	0.009	0.002
Al soluble	*	0.031	0.004
Al total	*	0.037	0.008

*Not determined.

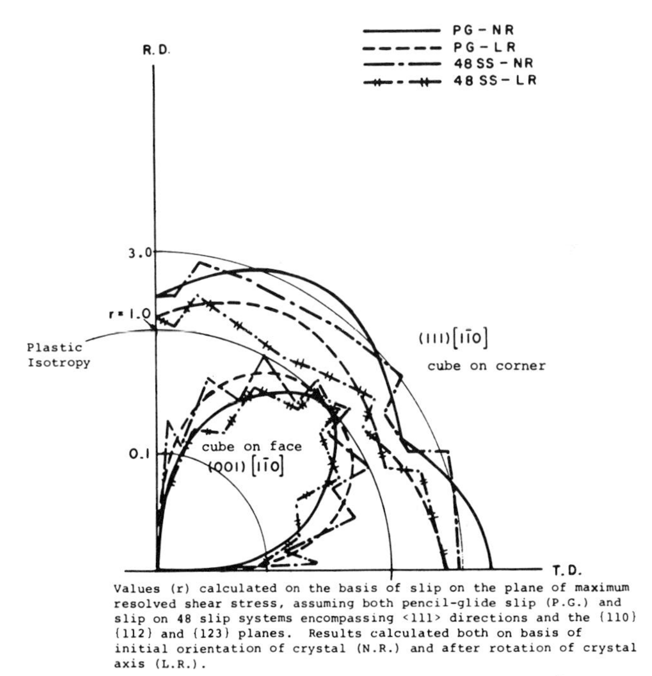

Values (r) calculated on the basis of slip on the plane of maximum resolved shear stress, assuming both pencil-glide slip (P.G.) and slip on 48 slip systems encompassing <111> directions and the {110} {112} and {123} planes. Results calculated both on basis of initial orientation of crystal (N.R.) and after rotation of crystal axis (L.R.).

Fig. 1—Polar plot of calculated r-values for (001)[1̄10] and (111)[1̄10] textures. From Ref. 9.

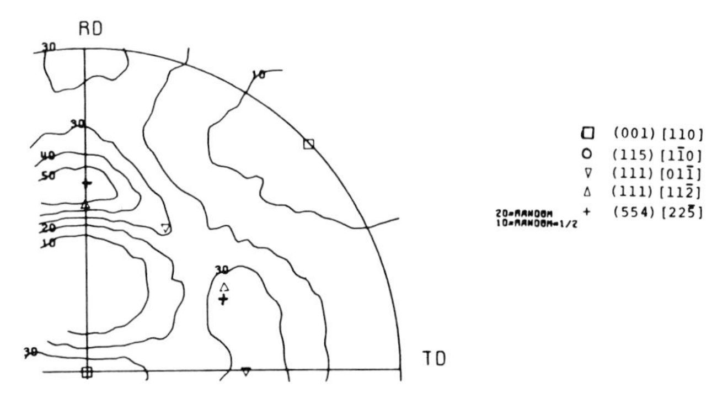

Fig. 2—Pole figure of specimen shown in Fig. 3.

HUNG-CHI CHAO is Associate Research Consultant, United States Steel Corp. Research Laboratory, Monroeville, Pa. 15146.

Manuscript submitted June 16, 1971.

Source: *Metallurgical Transactions*, 4, April 1973, 1183-1186, © 1973 American Society for Metals

Table II. Correlation of Textures With Degree of Ridging in Ferritic Stainless Steels

Sample No.[a]		Composition, Pct						Other Significant Alloying Elements	CC[e] Component	CC[e] Fiber-Background Texture	Other Texture Components[f]	Degree of Ridging Determined by the Room-Temperature Tensile Test
		C	Cr	Ni	N	Mn	Si					
U-8X1938	S[b]	0.064	17.39	–	0.037	–	–	0.48 Cb	X2.8	X1.7	CE	1[d]
	M[c]								X2	X1.2	CE	
U-9X0766	S	0.064	17.5	0.25	0.037	0.43	0.26	0.50 Cb	X1.5	X1.2	CE	1
	M								X1.5	X1	CE	
U-9X0566	S	0.078	16.4	0.25	0.043	0.35	0.22		X1	X0.5	CE	2
	M								Not developed	X0.5	CF and its nearby textures	
U-9X0722	S	0.072	17.2	0.25	0.054	0.44	0.34		X1.5	X1	CE	3
	M								Not developed	X0.7	CF and its nearby textures	
C-430X	S	0.060	15.21	0.20	–	0.67	0.66	0.41Cb, 0.01Ti, 0.02Mo	X4	X3	CE	0
	M								X2.1	X1.6	CE	
C-CUCA	S	0.086	16.0	0.35	0.063	0.47	–	0.39Cb, 0.036Ti, 0.028Al	X1.7	X1.3	CE and some CF and its nearby textures	3
	M								X1.4	X0.8	CE and some CF and its nearby textures	

a. U group and C group are from two commercial producers.
b. S denotes texture of surface portion of sample.
c. M denotes texture of midthickness portion.
d. Degree of ridging is classified from 0 to 4 where 0 indicates no or very slight ridging and 4 indicates most severe ridging.
e. CC denotes cube-on-corner textures.
f. CE denotes cube-on-edge textures and CF denotes cube-on-face textures.

Fig. 3—Ridging of Type 434 specimens pulled along the rolling direction, along the transverse direction, and along directions 30, 45, and 60 deg away from the rolling direction. Specimens were obtained from the same sample sheet.

free of (100)[110] texture. Additional experimental data from various commercial AISI 430-type stainless steel, Table II, offers further evidence to support this observation. Of the six ferritic stainless steels tested, those with CF and nearby texture components ridged whereas those with a well-developed CC fiber texture and no CF components showed little or no ridging. This same effect occurs in carbon steel. Fig. 4 compares the ridging behavior of low-carbon rimmed steel with that of low-carbon special killed (SK) steel, the compositions of which are shown in Table I. The rimmed steel, which has an extremely low r value, and a mixture of CC and CF texture, Fig. 5(a), ridges whereas the SK steel, which has a highly CC textured structure, Fig. 5(b), does not ridge.

The fact that ridging always appears with the banded CF texture in the CC texture matrix indicates that the mechanism proposed by Takechi and coworkers[5] cannot be generalized, although it might be true for a particular case. The samples they used had a strong (100)[011] texture component along with $\{111\}\langle 011\rangle$ and $\{211\}\langle 011\rangle$. Ridging must be shown to occur in a sample without a (100)[011] component before their proposed mechanism can be accepted as general.

Conclusion. New experimental evidence was obtained to verify current theories of the mechanism of ridging. The observation of ridging in transverse direction tension tests rules out the "buckling" theory.[6] The fact that ridging correlates with the relative amount of CF and CC texture in carbon steels as well as stainless steels supports the proposed "mixed-texture bands" of different component theory[2,3] but not the theory of "mixed texture rotations" of the same component.[5] Ridging should always appear in the rolling direction, irrespective of direction of strain because ridging is

Fig. 4—Surface appearance of rimmed steel (left) and SK-steel (right) sheet after 15 pct straining.

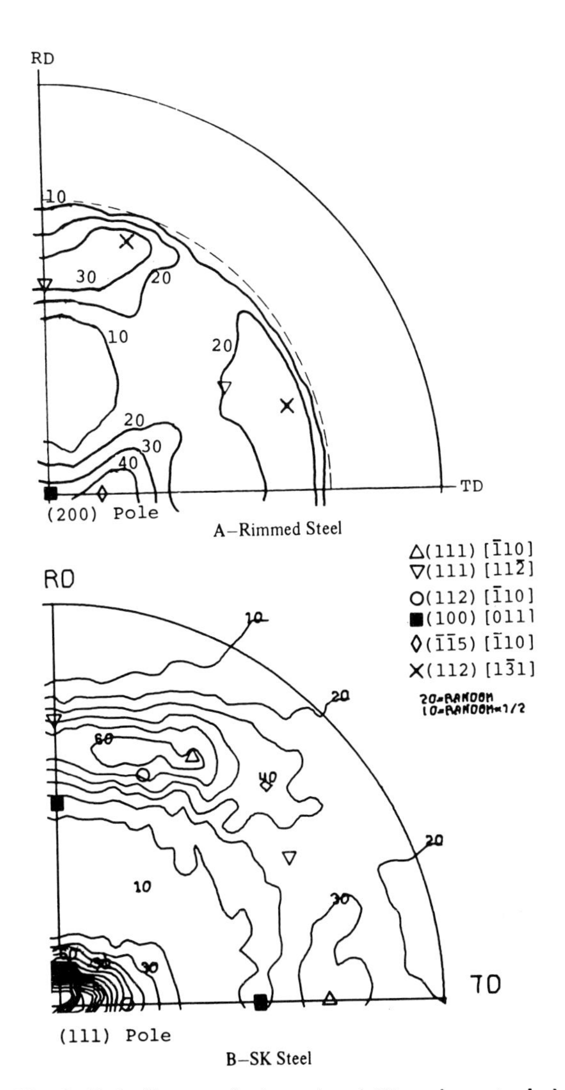

Fig. 5—Pole figures of rimmed and SK carbon-steel sheet.

due to the anisotropic flow of the mixture of the banded structures of two or more highly anisotropic orientations coexisting along the rolling direction.

The author is grateful to Dr. Waldo Rall and Mr. P. A. Stoll for supplying the X-ray data on commercial stainless steels listed in Table II, to Dr. P. R. Mould for offering the carbon steel samples, and to Drs. K. G. Brickner and J. D. Defilippi for their helpful suggestions after reviewing the manuscript. The critical review of Professor W. F. Hosford is also deeply appreciated.

1. H. C. Chao: *The Mechanism of Ridging in Ferritic Stainless Steels,* presented at the 96th Annual Meeting of the AIME, Los Angeles, California, February 23 1967.
2. H. C. Chao: *ASM Trans. Quart.,* 1967, vol. 60, pp. 37-50.
3. H. C. Chao: *Trans. Amer. Soc. Met.,* 1967, vol. 60, p. 549.
4. N. Ohashi: *J. Jap. Inst. Metals,* 1967, vol. 31, pp. 519-25.
5. H. Takechi, H. Kato, T. Sunami, and T. Nakayama: *Trans. Jap. Inst. Metals,* 1967, vol. 8, pp. 233-39.
6. R. N. Wright: *Met. Trans.,* 1972, vol. 3, pp. 83-91.
7. E. Pouillard and B. Osdoit: *Rev. Met.,* 1969, vol. 66, pp. 763-69 (BISRA Translation No. 8192, May 1970).
8. D. J. Blickwede: *Trans. Amer. Soc. Met.,* 1968, vol. 61, pp. 653-79.
9. R. W. Vieth and R. L. Whiteley: *Anisotropy and Tensile Test Properties and Their Relationship to Sheet Metal Forming,* in *Proceedings,* Int. Deep Drawing Research Group (IDDRG), London, June 3, 1964.

Development of Pitting Resistant Fe-Cr-Mo Alloys*

MICHAEL A. STREICHER*

Abstract

Because of the resistance of iron-chromium stainless steels to chloride stress corrosion, this alloy system was used as a base for developing superior resistance to various forms of corrosion by means of alloying with molybdenum, nickel, and the six metals of the platinum group. The effects of these alloying elements were evaluated by accelerated laboratory tests for pitting, intergranular, general, and stress corrosion. The optimum ductility and resistance to pitting, intergranular, and stress corrosion were found for an alloy of Fe-28% Cr-4% Mo with carbon not exceeding 0.010% and nitrogen below 0.020% (C+N $<$0.025%). This alloy resists pitting and crevice corrosion in 10% $FeCl_3 \cdot 6H_2O$ at 50 C (122 F) with six crevices on the specimen surfaces, and it resists all intergranular attack on a welded specimen in the boiling ferric sulfate-50% H_2SO_4 test. Addition of 2% Ni to this alloy extends its general corrosion resistance in oxidizing and organic acids to boiling 10% H_2SO_4 and 1% HCl, in which it is also self-repassivating. The nickel addition makes the alloy subject to stress corrosion cracking (SCC) in the boiling 45% $MgCl_2$ test, but not in the NaCl wick test, which more nearly simulates plant exposures and cracks 18Cr-8Ni stainless steel. Additions of small amounts of any of the six platinum metals, *e.g.,* 0.020% Ru, also make the 28Cr-4Mo alloy passive in boiling 10% H_2SO_4. But only ruthenium, iridium, and osmium do this without impairing pitting resistance in halide solutions. For self-repassivation, 0.50% Ru is required, and this amount makes the alloy subject to stress corrosion in the boiling 45% $MgCl_2$ test, but not in the wick test.

Austenitic Fe-Cr-Ni stainless steels are subject to stress corrosion cracking (SCC) in chloride environments, such as those encountered in heat exchangers using natural waters for cooling. This has led to interest in the use of ferritic (Fe-Cr) stainless steels for such applications. For example, AISI Type 430 and 446 stainless steels (Table 1) are much more resistant to SCC than the austenitic alloys, but welding reduces their ductility and their resistance to stress corrosion and intergranular attack.[1,2] There is also a need for greater resistance than now available in ferritic alloys to pitting and crevice corrosion on surfaces exposed to cooling waters and to general corrosion in certain mineral and organic acids on surfaces exposed to process solutions.

This investigation was undertaken in 1966 to develop ductile, weldable, ferritic stainless steels having superior resistance to chloride pitting, intergranular attack, and general corrosion in acids without loss in resistance to chloride SCC.

Ideally, experimental alloys should be evaluated in plant tests. However, for such tests, relatively large amounts of material are required along with testing times of 6 to 18 months in the case of cooling water environments. For these reasons, it was necessary to develop and apply rapid and, therefore, very severe laboratory tests to evaluate experimental alloys. The choice of these tests is critical, because it determines the results of the investigation, its duration, and its costs.

The need for rapid results eliminates the use of laboratory tests which simulate plant conditions. Instead, the tests must be based on some knowledge of the mechanism of the various forms of corrosion. By the selection of very severe testing conditions and criteria for evaluation, the gap between laboratory test results and service performance can be narrowed but not closed. Ultimate proof of utility of promising compositions can be provided only by tests in plant environments and service in operating equipment.

Because superior pitting resistance was the primary aim, the experimental procedure consisted of preparing 400g laboratory alloys in sheet form for testing in chloride solutions. When compositions were found which met the severe criteria for resistance to pitting and crevice corrosion at elevated temperatures, they were then tested to establish the limits for optimum ductility, resistance to intergranular attack, SCC, and general corrosion.

Preparation of Alloys

Alloying Materials

High purity materials, low in carbon and nitrogen, were weighed out to melt 400g ingots. Iron, chromium, molybdenum, nickel, copper, aluminum, niobium, titanium, zirconium, tantalum, manganese, and silicon were added as pure elements in the form of flakes, sheet, rod, powder, and lumps. Deliberate additions of nitrogen were made in the form of Cr_2N or high nitrogen ferrochrome and of carbon as high carbon ferrochrome. Recovery of alloying elements was within 0.1g, both in the 400g ingot and in rolled sheet.

*Submitted for publication August, 1973.

*E. I. du Pont de Nemours & Company, Inc., Engineering Materials Laboratory, Experimental Station, Wilmington, DE 19898

Because analytical results on alloys were within only ±0.2% for Cr and ±0.1% for Mo, the concentrations of alloying elements weighed out were used for plotting data rather than the results of analyses. All carbon and nitrogen analyses are based on chips made from sheet material. These elements are present as impurities in all commercial and laboratory alloys. Tests were made in this investigation to determine the maximum concentrations which can be present without impairing the corrosion resistance and ductility of the alloys. Titanium, niobium, and aluminum were determined by X-ray fluorescence.

Melting

The alloying elements were melted in a vacuum induction furnace made by Vacuum Industries, Inc. For support in the furnace, the crucibles, 2-3/8 inches in diameter, 4 inches high, made of high purity, recrystallized alumina, were placed in a 400 ml, Vycor 1100 beaker with Fiberfax high temperature insulation around the sides and fine granular zirconia at the bottom. This assembly was dried by heating in a vacuum oven at 140 C (284 F). The weighed alloying elements were then placed in the crucible, and the assembly was placed in the center of the induction coil of the vacuum furnace. The furnace was sealed and evacuated to 10^{-3} to 10^{-5} Torr. The power was turned on and increased gradually to minimize thermal shock. When melting was about to begin, the chamber was filled with gettered argon to a pressure of 5 inches of Hg in order to minimize vaporization of alloying elements.

Available power (25 to 29 kW) limited the maximum temperature to about 2800 F (1538 C). This temperature is only 100 to 200 F (38 to 93 C) above the Fe-Cr and Fe-Mo liquidus lines. To assure dissolution of all components, the maximum temperature was held 15 to 20 minutes. At the completion of the melting operation, the heat was cast into a copper mold through a fire-brick funnel. After cooling, the cylindrical ingot was removed and the hot top containing the shrinkage cavity was cut off. This left an ingot 1-1/4 inches in diameter and about 2-3/4 inches long.

Processing

The ingot was soaked for 3 hours at 2200 F (1204 C) in an electric furnace (air atmosphere) and then forged to a rectangular cross section, 2-1/2 x 1/2 inch. The forged ingot was heated to 2150 F (1177 C) and rolled to a thickness of 100 mils in light passes with four reheating periods. The usable product of these operations was about 10 inches long and 1-1/2 inches wide. After the rolling operation, the sheet was placed in a furnace for final heat treatment of 1 hour at 2000 F (1093 C), to recrystallize and homogenize the structure, and then quenched in water. Heat treating scale was removed by sandblasting. Alloys containing elements (*e.g.*, Ti and Nb) intended to combine with carbon and nitrogen were given a final heat treatment of 2 hours at 1650 or 1750 F (899 or 954 C).

Various specimens for corrosion, mechanical, and analytical tests were cut on a mechanical saw. The cut specimens were ground to an 80 grit finish on a water cooled silicon carbide belt. To investigate the effect of welding on corrosion resistance and on mechanical properties, autogenous welds were made with an electric arc using a tungsten tip shielded with argon. The back of the specimen was also shielded with argon to reduce oxidation and to minimize pickup of nitrogen.

Pitting Resistance

Two oxidizing chloride solutions, potassium permanganate-sodium chloride and ferric chloride, were selected for simple immersion tests to determine resistance to pitting. The severity of attack was progressively increased by raising the test temperature. Exposure of a number of familiar, commercially available alloys to the two chloride solutions over a range of temperatures provided data on their pitting resistance and on the characteristics of the test solutions. It also provided a basis for comparing the resistance of new compositions with commonly available materials.

The two pitting tests may be viewed as simple methods for exposing alloys to conditions of constant corrosion potential over a range of temperature. It is shown below that the corrosion potentials happen to be essentially the same in the two solutions, +0.60 V vs SCE, but because of the differences in chloride concentration, pH, and the presence of other anions and cations, the two solutions differ in their action on alloys.

Permanganate-Chloride Test

Hydrated manganese oxides have frequently been found to be a constituent of water-side deposits on heat exchanger tubes.[3] In a plant along the Ohio River, a Carpenter 20Cb-3 tube with a manganese-rich deposit failed by pitting in 6 months. Spectrographic analysis of the deposits showed 1 to 5% Fe, 5 to 25% Mn, and 10 to 50% Zn as the main constituents. The combination of manganese and zinc suggests a mineral origin for the water deposit. There are several minerals which contain both zinc and manganese.[4] Manganese oxide deposits have also been attributed to the action of bacteria which concentrate the soluble manganese in the water into insoluble oxides.[5] Analysis of the river water gave seasonal variation of 0.1 to 1.3 ppm Fe, 0.1 to 2.1 ppm Mn, 0.03 to 0.28 ppm Zn, 10 to 160 ppm Cl, and pH = 6.4 to 7.3.

To prevent accumulation of organic slimes in heat exchangers, the river water is chlorinated for short periods several times per day.

The reaction mechanism responsible for this very rapid failure of a relatively pitting resistant stainless steel appeared to be similar to that proposed in a previous analysis of pitting attack associated with manganese rich deposits. Long and Rice[5] found some permanganate in these deposits along with hydrated MnO_2. They concluded that the permanganate is formed during chlorination by the oxidation of manganous ions to manganic and permanganate ions. They also found in a laboratory test that Type 316 stainless steel pits in less than 24 hours in a $KMnO_4$-NaCl solution at room temperature. Scharfstein[6] verified these findings on Carpenter 20Cb-3[(1)] steel. It seems that pitting is most likely to take place at hydrated manganese oxide deposits which form on tubes, especially

---

(1)Trademark, Carpenter Technology Corp., Reading, PA.

Source: *Corrosion*, 30(3), March 1974, 77-91

TABLE 1 – Pitting Tests on Commercial Alloys

Alloy	Composition						Pitting Corrosion					
							$KMnO_4$-NaCl(1)				$FeCl_3 \cdot 6H_2O$(2)	
	Stainless Steels Iron Base											
	Cr	Ni	Mo	Cu	Si	Cb	RT†	50 C	75 C	90 C	RT†	50 C
AISI 430	16.2	–	–	–	–	–	F	–	–	–	F	–
AISI 446	25.5	–	–	–	–	–	F	–	–	–	F	–
AISI 304	18.4	9.1	–	–	–	–	F	–	–	–	F	–
AISI 310	25	20	–	–	–	–	F	–	–	–	F	–
AISI 316	17.5	12.8	2.3	–	–	–	F	–	–	–	F	–
AISI 316L	18.4	12.9	3.0	–	–	–	F	–	–	–	F	–
Carpenter 20Cb-3	19.8	34.6	2.3	3.3	–	0.6	F	–	–	–	F	–
USS 18-18-2	18	18	–	–	2	–	F	–	–	–	F	–
SP-2*	18	10	2.5	–	2.5	–	R	R	F	–	R	F
	Copper Alloys											
	Cu	Zn	Ni	Mn	Sn							
Admiralty Brass	71	28	–	–	1		F_G	–	–	–	F_G	–
Cartridge Brass	70	30	–	–	–		F_G	–	–	–	F_G	–
Cupro-Nickel	90	–	10	–	–		F_G	–	–	–	F_G	–
Monel	30	–	67	1.1	–		F	–	–	–	F_G	–
	Nickel-Chromium Alloys											
	Ni	Cr	Mo	Fe	Cu	Ti						
Inconel 600	77	15	–	7	0.5	–	R	F	–	–	F	–
Inconel 625	59	22	9.0	5	–	0.4	R	R	R	R	R	F
Incoloy 800	32	21	–	47	0.75	0.4	R	F	–	–	F	–
Incoloy 825	42	21	3.0	30	1.8	1.0	R	R	F	–	F	–
	Nickel-Base Alloys											
	Cr	Mo	Fe	Cu	Co							
Hastelloy C	16	16	6	–	–		–	R	R	R	R	R
Hastelloy B	1	28	5	–	2.5		F_G	–	–	–	F_G	–
Hastelloy G	22	6.5	20	2.5	–		–	–	–	R	F	–
	Pure Metals											
Nickel							F_G	–	–	–	F_G	–
Molybdenum							–	–	–	F_G	–	–
Titanium							–	R	R	R	R	R
Tantalum							–	–	R	R	R	R
Niobium							–	–	R	R	R	R

(1) 2% $KMnO_4$-2% NaCl—no crevices (pH = 7.5).
(2) 10% $FeCl_3 \cdot 6H_2O$—with crevices (pH = 1.6).
NOTE: R = no pitting, F = failed by pitting, F_G = failed by general corrosion, * = with 0.2% N and carbon $<$0.030%,[9] and † RT = room temperature.

on the hottest portions. Then, when chlorine reacts with the deposit, this strong oxidizing agent converts solid manganese dioxide into soluble permanganate ions.[7,8] Chloride ions from the cooling water and from reduction of chlorine, together with the permanganate ions, are concentrated at the surface of the tubes by adsorption on the deposits. They react with the stainless steel to form pits. On the basis of these findings, a $KMnO_4$-NaCl test was selected for this investigation.

In preliminary tests, the effects on pitting of concentration of permanganate, chloride, and variations in pH were investigated. A Type 304 stainless steel, which pits readily in this solution at room temperature, and the SP-2 steel (Table 1), which is resistant up to 50 C, were used for these tests. At 75 C (167 F), it was found that the results on Type 304 steel are unchanged when the NaCl concentration is varied from 0.4% to 5.0% and the $KMnO_4$ concentration from 2.0% to 5.0%. Also, variations of pH of 2% $KMnO_4$ + 2% NaCl solution at 50 C from a pH of 2.0 to 12.0 were without appreciable effect on the results. Only below a pH of 2.0, obtained by addition of HCl, is there general corrosion of SP-2 at 50 C. Above a pH of 12, obtained by addition of NaOH, pitting is inhibited on Type 304 steel at 50 C.

On the basis of these tests, a 2% $KMnO_4$-2% NaCl

solution whose pH was adjusted to 7.5 with NaOH was selected for all tests. Large test tubes, 11-1/2 inches long by 1-1/2 inches in diameter, with 150 ml of solution, were immersed in a thermostatically controlled water bath. The test tubes were covered with a one hole, No. 8 rubber stopper containing a 4 inch (medicine dropper) glass tube for venting and condensation. The specimens, about 1 x 2 inches and 0.08 inch thick, were ground to an 80 grit finish. No crevices were applied to the specimens. Rubber bands, such as used for ferric chloride tests, are attacked by the permanganate solution and can, therefore, not be used.

Pitting attack in this solution is readily detected by the extensive formation of insoluble manganese oxides, which form a coating over the entire surface of the specimen. The alloy (M) dissolves at anodic sites which are created when there is localized removal of protective, hydrated oxide films by the peptizing action of chloride ions.[9]

$$3M - 6e^- \rightarrow 3M^{+2} \quad \text{(anodic reaction)}$$

There is an electrochemically equivalent reaction at cathodic sites resulting in the formation of an insoluble coating of manganese oxides (Figure 1).

$$2MnO_4^- + 4H_2O + 6e^- \rightarrow 2MnO_2 + 8OH^- \quad \text{(cathodic reaction)}$$

If a corroding specimen is left in the solution for an extended period, all the permanganate is eventually reduced to insoluble precipitate. The color of the solution changes from the characteristic permanganate-purple to colorless and the pH from 7.5 to 13.0.

When this reaction takes place in a heat exchanger, rather than in a laboratory test, there is regeneration of the manganese dioxide, which, if it deposits on the surface, can then serve again as a source for permanganate ions during a subsequent chlorination cycle.

The corrosion potential of resistant stainless steel in 2% $KMnO_4$-2% NaCl solution at 90 C (194 F) was found to be +0.61 V (SCE), which is also the redox potential measured on a platinized platinum electrode.

To reveal the extent of pitting attack under the manganese dioxide deposit, specimens were immersed in an inhibited oxalic-sulfuric acid solution:[10] 900 ml H_2O, 27.4 ml 96.5% H_2SO_4, 14.4 g oxalic acid, 0.2 g Alkanol WXN,[(2)] and 0.2 g diorthotolylthiourea.

The cleaned specimens clearly reveal pitting attack (Figure 1). All evidence of pitting, which was readily revealed by the appearance of some manganese oxide coating, led to a rating of *failed* (F). Resistant alloys had no deposit, only a light brown oxide film (Figure 1). Most nonresistant alloys pitted within 24 hours and were removed from solution. Specimens which were resistant during the first 24 hour period were left on test for at least 14 days and in some cases for periods up to 12 months. Alloys which resisted pitting at room temperature were then tested at successively higher temperatures up to 90 C to simulate thermal conditions in heat exchangers.

[(2)]Du Pont registered trademark, alkyl-aryl sodium sulfonate.

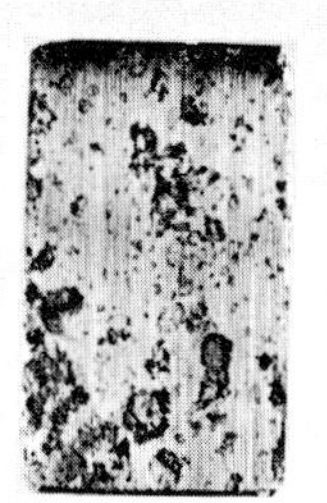

FIGURE 1 – Permanganate-chloride test specimens. Left: Pitted AISI 430 specimen with coating of manganese oxides. Exposed for 15 hours at 90 C. Center: Pitted AISI 430 specimen after removal of the coating in inhibited acid. Right: Typical resistant Fe-Cr-Mo alloy specimen after as much as 16 months at 90 C. Rinsed in water, no acid treatment.

Results

None of the commercially available austenitic stainless steels of Table 1 are resistant to pitting at room temperature, nor are Type 430 and 446 alloys. The only resistant iron-base alloy was the SP-2 composition, a modified Type 316L alloy, developed in a previous investigation.[9] It was known to be resistant in ferric chloride solution at room temperature and was found in this investigation to be resistant up to 50 C in the permanganate-chloride test.

Corrosion on metals and alloys which are not resistant to oxidizing environments is by shallow pitting which merges into general corrosion. These are copper-base alloys, Hastelloy B, nickel, and molybdenum. Only Inconel 625, Hastelloy C, Hastelloy G, titanium, tantalum, and niobium are resistant at 90 C.

The forms of corrosion observed on cupro-nickel and Carpenter 20Cb-3 tubes in heat exchangers using Ohio River water were similar to those found in the permanganate-chloride test. Because on the cupro-nickel tubes pitting is shallow and has the appearance of general corrosion, this alloy can be used for limited periods in this environment. On Carpenter 20Cb-3 alloy, pits are small in diameter and deep. These observations support the pitting mechanism based on the formation of permanganate ions by chlorine in the water.

Ferric Chloride Test

Pitting tests were also made in the widely used $FeCl_3$ solution for comparison with the $KMnO_4$-NaCl results and to provide information on the effect of crevice conditions. Two small Teflon[(3)] blocks were placed on the front and back surfaces and held in place by two elastic bands stretched at 90 degrees to each other (Figure 2). The result on a 1 x 2 inch, 0.08 inch thick specimen is two sharp crevices at the top and bottom where the elastic touches the specimen, two somewhat less sharp crevices where the elastic around the midsection touches the specimen, and two crevices under the Teflon blocks. Contraction of the elastics provides constant crevice conditions as the metal corrodes at the points of contact. The test specimens were

[(3)]Registered trademark, E. I. du Pont de Nemours & Co., Inc., Wilmington, Delaware.

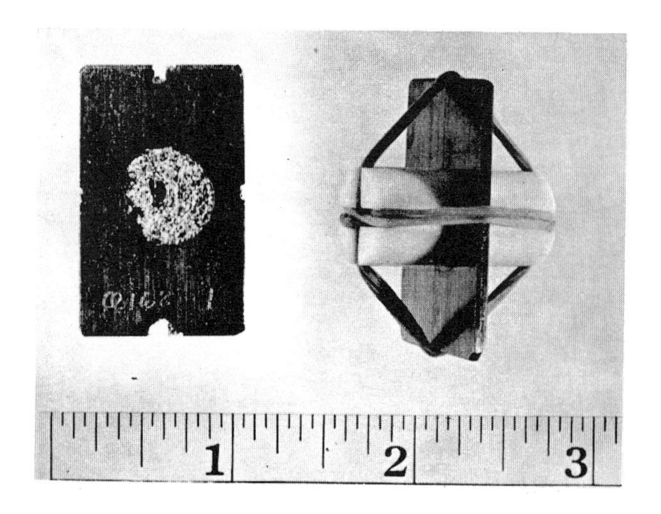

FIGURE 2 – Ferric chloride test specimens. Right: Specimen with six crevices ready for testing. Left: Specimen after testing with attack at all four crevices formed by elastics and under Teflon block.

ground to an 80 grit finish and immersed in a solution of 10% $FeCl_3 \cdot 6H_2O$.

The Teflon blocks (Figure 2) are 1/2 inch in diameter, 1/2 inch high with perpendicular grooves, 1/16 inch wide and 1/16 inch deep, in the top surface to hold the two rubber bands. Two loops of a No. 12 (1-3/4 inch) rubber band are used around the middle of the specimen and two loops of a No. 14 (2 inch) band from top to bottom.

In preliminary tests at room temperature, it was found that if an alloy with a crevice pits, it will eventually also pit without a crevice, but the exposure time required to reveal this may be as long as 4 months. With crevices on the specimens, there was pitting within 24 hours on almost all susceptible alloys. Therefore, all subsequent tests were made with crevices. Resistant alloys were exposed for weeks, and in some cases, as long as 12 months without suffering any pitting action. Only those specimens which were completely free of all pitting and crevice corrosion at all six crevices were rated as resistant (R in Table 1). All others were rated as *failed* (F).

Ferric chloride solutions severely corrode stainless steels because they contain a large concentration of chloride ions, which, together with the relatively low pH (1.6) of this solution, readily break down passive films. As a result, metal is exposed to the solution and will dissolve either to form a protective film by reaction with the solution or to form a pit (anodic reaction),

$$M - 2e^- \rightarrow M^{+2}$$

if there is an efficient, electrochemically equivalent, cathodic reaction which consumes the electrons liberated by this process. In $FeCl_3$ solutions, the conversion of ferric to ferrous ions

$$2Fe^{+3} + 2e^- \rightarrow 2Fe^{+2}$$

on the entire unpitted portion of the specimen provides a very efficient cathodic reaction. Surfaces with crevices are the most susceptible to pit formation because the normal process of continuous film repair at points of breakdown is prevented by limited access of bulk solution, which leads to a drop in pH, and by the resulting electrochemical currents, which produce even higher concentrations of chloride ions.

The redox potential (platinized platinum electrode) and the steady state corrosion potential of pitting resistant alloys were found to be in the range +0.60 to 0.68 V (SCE) in 10% $FeCl_3 \cdot 6H_2O$ solution at 50 C, almost the same potential as measured for the 2% $KMnO_4$-2% NaCl solution.

Ferric chloride tests were made in large test tubes, 11-1/2 inches long and 1-1/2 inches in diameter, with vented rubber stoppers as described in the section on the permanganate-chloride test. The test tubes were heated by immersion in a thermostated water bath.

Results

From Table 1, it is apparent that the commercial ferritic and austenitic stainless steels pit in the $FeCl_3$ solution at room temperature. Among the 18-8 steels, only the SP-2 alloy is resistant. At 50 C, only Hastelloy C,[4] titanium, tantalum, and niobium are resistant in the $FeCl_3$ test. Inconel 625[5] is resistant at room temperature. Copper-base alloys, Hastelloy B, nickel, and molybdenum, which are not resistant in oxidizing environments, suffer general corrosion. Both Hastelloy C and titanium are also resistant at 65 C (149 F) but fail by crevice corrosion at 75 C.

Comparison of the results on Inconel 625 and Hastelloy G in the two test solutions shows that the $FeCl_3$ test at 50 C is more severe than the $KMnO_4$-NaCl test at 90 C. Subsequent results on laboratory Fe-Cr-Mo alloys confirmed this finding.

The results on the SP-2 alloy provided the starting point for the present investigation. This alloy is markedly superior in both pitting tests to all the common austenitic and ferritic alloys, but it is subject to chloride SCC. To avoid this form of corrosion, attempts to improve pitting resistance were based on the iron-chromium system.

Iron-Chromium-Molybdenum Alloys

The first laboratory heat was based on the SP-2 composition,[9] and 2% each of molybdenum and silicon were added to Fe-20Cr. This alloy proved to be resistant in the $KMnO_4$-NaCl test at room temperature, *i.e.*, superior to all the stainless steels of Table 1. To improve its resistance, the chromium content was increased. Silicon was gradually replaced by molybdenum because the reproducibility of pitting resistance on these alloys was inferior to the Fe-Cr-Mo alloys. Eventually, all tests were made at 90 C in the $KMnO_4$-NaCl solution and 50 C in the $FeCl_3$ solution. The performance of numerous Fe-Cr-Mo alloys in these two tests is shown in Figure 3.

Without molybdenum, none of the alloys, even with as much as 40% Cr, are resistant in either of the pitting tests. The amount of molybdenum required to obtain pitting

(4) Hastelloy, registered trademark, Cabot Corp., Kokomo, IN.
(5) Inconel and Incoloy, registered trademark, The International Nickel Co., Huntington, WV.

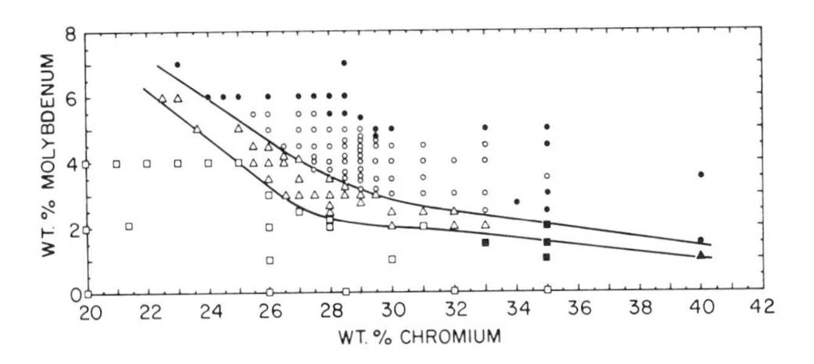

FIGURE 3 – The effect of chromium and molybdenum in Fe-Cr-Mo alloys on resistance to pitting. △▲: Resistant in permanganate-chloride test at 90 C. ○●: Resistant with crevices in ferric chloride test at 50 C, and in the permanganate-chloride test at 90 C. □■: Pitted in both tests. Note: The solid points represent compositions which failed by cracking during bending or by stress corrosion (Figure 7).

resistance decreases sharply with increasing chromium content between 22 and 28% Cr, and more gradually above this range. At a given concentration of chromium, the amount of molybdenum required to provide resistance to pitting in the ferric chloride test is greater than that required in the permanganate-chloride test. For optimum pitting resistance, alloys should have compositions falling in the area above the upper line in Figure 3. All the alloys shown in this figure were prepared from high purity materials with carbon and nitrogen contents below 100 ppm C, 200 ppm N, and C+N <250 ppm.

Recently introduced alloys with 26Cr-1Mo[11,12] and 28Cr-2Mo[13-15] are not resistant in either of the tests of Figure 3. The 26Cr-1Mo alloy is resistant only up to 50 C in the $KMnO_4$-NaCl test, and with crevices, is not resistant at room temperature in $FeCl_3$ solution. The properties of an alloy with 35Cr[16] without molybdenum are the same in the two pitting tests as those of the 26Cr-1Mo alloy. The 28Cr-2Mo alloy is resistant up to 50 C in the $KMnO_4$-NaCl test and at room temperature in the $FeCl_3$ test.

The Effect of Carbon and Nitrogen

Intergranular Corrosion

Iron-chromium alloys containing carbon and/or nitrogen in excess of certain small concentrations are subject not only to changes in mechanical properties, *e.g.*, an increase in the transition temperature,[17] but also to changes in corrosion resistance, such as susceptibility to pitting, intergranular attack, and to SCC depending on heat treatments.[2,18] Precipitation of chromium rich carbides and nitrides primarily at grain boundaries creates chromium depletion in adjacent zones. In a previous investigation,[2] it was shown that the boiling ferric sulfate-sulfuric acid test[19,20] provides a simple, rapid method for detecting susceptibility to intergranular attack in Fe-Cr alloys. This method was used in the present investigation to determine the maximum concentrations of carbon and nitrogen which can be tolerated in pitting resistant Fe-Cr-Mo alloys without causing susceptibility to intergranular attack. Tests were made on 1 x 2 x 0.1 inch specimens exposed for 120 hours to the ferric sulfate test.

FIGURE 4 – Welded specimens after exposure in ferric sulfate-sulfuric acid test. Left: No intergranular attack; 27.5 Cr, 5.5 Mo, 15 ppm C, 195 ppm N. Center: Severe intergranular attack on weld metal; 29.0 Cr, 4.7 Mo, 856 ppm C, 219 ppm N. Right: Severe intergranular attack in fusion zone and weld metal; 28.5 Cr, 4.2 Mo, 0.3 Al, 22 ppm C, 388 ppm N.

To facilitate fabrication of equipment, alloys should be weldable without requiring subsequent heat treatment to prevent intergranular and other forms of corrosion during plant service. Therefore, an autogenous weld was made on all test specimens providing weld metal, fusion zone, heat affected zone, and unaffected base metal for test. Details of the welding technique are described above under *Processing.* The specimens were ground to an 80 grit finish.

Because the weldment simulates materials and thermal cycles as encountered in plant equipment, such as heat exchanger tubes, evaluation of test specimens for evidence of susceptibility to intergranular attack was based on very strict criteria. Weight loss measurements could not be used because intergranular attack is sometimes localized in one small element of the weldment, *e.g.*, the fusion zone or the weld metal. After 120 hours in the ferric sulfate test, specimens were examined in a binocular microscope at 40X for undermined and dislodged grains. Specimens having more than three grains dislodged on either side of a 1 x 2 inch specimen were rated as *failed.* The weight loss resulting from a few undermined grains could not be readily detected. Figure 4 contains macrographs of a resistant alloy and two examples of severe intergranular attack at weldments. Only a very narrow zone of attack is required at the grain boundaries to produce undermining and dislodgement of grains.

Data on the effect of carbon and nitrogen on susceptibility to intergranular attack and other properties are given in Table 2. It is apparent that (1) both carbon and nitrogen, when present in excess of certain limits, can cause susceptibility to intergranular attack; (2) this limit is lower for carbon than for nitrogen; (3) the heat affected zone is most readily *sensitized, i.e.*, made subject to intergranular attack, and the base plate is the least sensitive; and (4) increasing the concentration of molybdenum increases the tolerance for nitrogen (and carbon), note 26Cr-1Mo, Alloys 487 and 530.

To avoid sensitization during welding the carbon content should not exceed 100 ppm and the nitrogen content should be less than 200 ppm, with C+N <250 ppm.

TABLE 2 — Effect of Carbon and Nitrogen on the Properties of Fe-Cr-Mo Alloys

Code	Alloy Composition				Ferric Sulfate-Sulfuric Acid Test(1)			Pitting Tests		Bend Test	Stress Corrosion(2)
	Cr (%)	Mo (%)	C (ppm)	N (ppm)	Heat Affected Zone	Weld	Base Plate	$KMnO_4$-NaCl (90 C)	$FeCl_3$ (50 C)	(U-bend)	($MgCl_2$)
628	31.5	3.0	7	235	F	R	R	R	R	F	—
620	38.0	2.5	20	438	F	R	R	R	F	F	—
599	33.0	3.0	109	68	F	F	R	R	F	P	R
589	32.0	2.5	22	215	F	F	R	R	F	F	—
586	28.5	4.5	66	230	F	R	R	R	R	P	F
549	27.5	5.5	15	195	R	R	R	R	R	P	R
531	28.5	4.5	334	25	F	F	R	R	F	F	—
526	28.5	4.2	10	5	R	R	R	R	R	P	R
530	26.0	1.0	15	90	F	F	R	F	F	P	R
487	26.0	1.0	26	204	F	F	F	F	F	P	R
497	28.0	3.5	29	209	F	R	R	—	F	P	—
492	27.0	5.0	21	283	—	—	—	—	R	P	F
464	28.5	4.0	22	239	F	R	R	R	F	F	—
461	28.5	4.0	189	89	F	F	F	R	F	F	—
460	28.5	4.0	171	70	F	F	F	R	F	P	F
452	28.5	3.0	33	267	F	F	R	R	F	—	—
450	27.5	3.0	14	204	R	F	F	R	F	—	—
409	29.0	4.7	856	219	F	F	R	R	F	F	—
408	29.0	4.7	48	327	F	F	F	R	F	F	—
510	29.5	4.0	5	170	R	R	R	R	R	P	R
526	28.5	4.2	10	5	R	R	R	R	R	P	R
537	28.5	4.5	23	133	R	R	R	R	R	P	R
494	27.0	6.0	10	305	R	R	R	R	R	F	—

(1) 120 hour ferric sulfate-sulfuric acid test.
(2) Stress corrosion test in boiling 45% magnesium chloride solution on welded U-bend specimen.
NOTE: R = resistant, F = failed, and P = passed, ductile when weldment is bent over 0.366 in diameter mandrel in bend test.

Titanium and niobium are sometimes added to stainless steels to combine with carbon and nitrogen and thereby prevent susceptibility to intergranular corrosion caused by chromium depletion. Addition of titanium and niobium to Fe-28Cr-4Mo alloys containing carbon and/or nitrogen in excess of the limits given above reduces their ductility and resistance to SCC in the $MgCl_2$ test, and therefore, cannot be used in the high molybdenum alloys to prevent intergranular attack.

Pitting Resistance

When the nitrogen and/or carbon content are high enough to cause susceptibility to intergranular attack on sensitized specimens, the pitting resistance in the 50 C $FeCl_3$ test is also impaired in most cases. However, in the 90 C $KMnO_4$-NaCl test, only sensitization caused by high nitrogen (645 ppm) results in pitting attack. When most of the high carbon (430 ppm) or high nitrogen (645 ppm) content is put into solid solution by annealing, there is no impairment of pitting resistance except on the high carbon alloy in the $FeCl_3$ test. (These results were obtained on two 28Cr-4Mo alloys not listed in Table 2.)

These and other data show that both carbon and nitrogen should be kept as low as possible and that for optimum properties carbon should not exceed the limits given above. A high (0.1% or 1000 ppm) carbon content in the Fe-30Cr-3Mo alloy described by Wilde[21] appears to be the reason for the susceptibility of this alloy to crevice corrosion both in $FeCl_3$ solution and in a sea water exposure.

Stress Corrosion Cracking

In a concurrent investigation,[18] it is shown that Fe-Cr alloys which are subject to intergranular attack because of their carbon content may also be susceptible to SCC in magnesium chloride and even in dilute sodium chloride solutions. It has also been shown by Bond and Dundas[22] that additions of less than 1% nickel or copper to Fe-Cr-Mo alloys make them susceptible to SCC in magnesium chloride. Thus, it is not the crystal structure, ferrite, or austenite, which makes an alloy susceptible to SCC, but the composition of these two phases. Both ferrite and austenite in stainless type alloys may be susceptible or resistant to SCC in a given environment, depending on the composition of these phases.

FIGURE 5 — U-bend specimen after testing 2400 hours in $MgCl_2$ (155 C) solution. Alloy: 28.5 Cr-4.0 Mo; 20 ppm C, 25 ppm N. No cracking even though there was extensive, shallow pitting. 4X

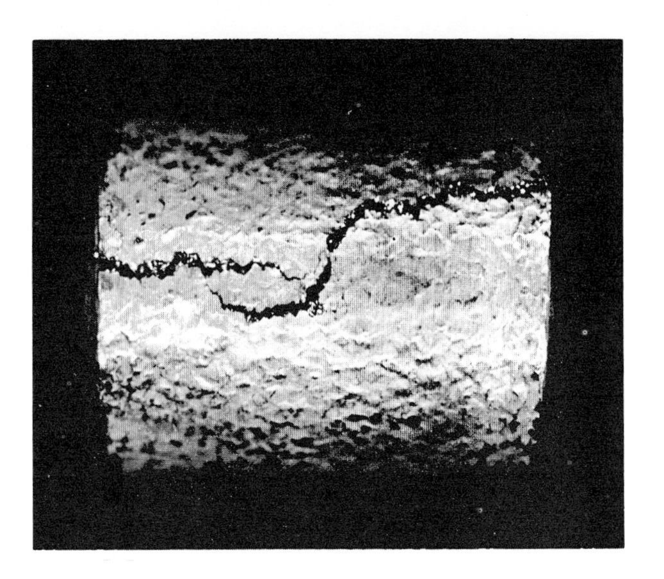

FIGURE 6 — U-bend specimen after testing 117 hours in $MgCl_2$ (155 C) solution. Alloy: 28.5 Cr-7.0 Mo, 9 ppm C, 153 ppm N. Stress corrosion cracking even though there was very little corrosion. 4X

Susceptibility to SCC was determined by testing U-bend specimens in boiling [155 C (311 F)] 45% $MgCl_2$ solution. Apparatus and procedures have been described previously.[23] All tests were made on specimens, 3 x 0.75 x 0.1 inch, with an autogenous, 3-inch long weld running lengthwise in the center of most specimens. Ground, 80 grit, blanks were bent over a 0.366 inch diameter mandrel. The arms were then pulled parallel in a vise and held in place by a Hastelloy C bolt, which was insulated from the specimen by Teflon washers. Forming the specimen into a U-bend involves plastic deformation of wrought sheet and the various elements of the weldment and provides a test of the ductility of the alloy. Some alloys cracked during bending and were, therefore, not tested in the magnesium chloride test. Those that could be bent were immersed in the test solution and examined periodically with a binocular microscope (40X) for evidence of cracking. Specimens which resisted cracking were tested for 100 days. Testing on those which cracked was terminated as soon as crack formation was clearly established.

Pitting in the 45% $MgCl_2$ solution at 155 C is severe on most alloys, *i.e.*, there was considerable corrosion involving generation of hydrogen (Figure 5). Even though ferritic stainless steels are subject to hydrogen embrittlement, this appears not to have been the cause of cracking on those alloys which failed in $MgCl_2$ (Figure 6). Bond and Dundas[22] have also concluded that hydrogen embrittlement and SCC are separate phenomena in ferritic steels. This does not exclude the possibility that hydrogen embrittlement may be a contributing factor in some cases of SCC.

Results on ductility and stress corrosion tests are shown in Table 2 and Figures 3 and 7. The same factors which reduce ductility also tend to make these alloys susceptible to SCC. These factors are (1) increased concentrations of alloying elements, chromium and molybdenum, beyond certain limits (Figures 3 and 7) and (2) high concentrations of carbon and nitrogen (Table 2).

As in the case of Fe-Cr alloys,[18] some of the Fe-Cr-Mo alloys which are subject to intergranular attack in the ferric sulfate test because of high carbon or nitrogen contents also fail by stress corrosion in the $MgCl_2$ test (alloys No. 460, 492, 586 in Table 2). However, the two 26Cr-1Mo alloys (No. 487 and 530 with nitrogen contents of 90 and 204 ppm) were subject to severe intergranular attack in the ferric sulfate test, but were immune to cracking in a 2400 hour test in $MgCl_2$. The low molybdenum content of these alloys reduced their tolerance for carbon and nitrogen as compared with 28Cr-4Mo alloys, and, as a result, made them subject to severe intergranular attack. In contrast, alloys with 4% Mo can tolerate up to 200 ppm N before they begin to become susceptible to intergranular attack in the tests of this investigation. Comparison of results on the two groups of high nitrogen alloys (487, 530, 492, 586) suggests the possibility that not only chromium nitrides and consequent chromium depletion, but also that nitrogen in solid solution contributes to cracking susceptibility. Because chromium nitride precipitation takes place more readily in alloys with low concentrations of molybdenum (1%), the amount which remains in solid solution in these alloys after sensitizing heat treatments is lower than in the 4% Mo alloys. Furthermore, nitrogen in solid solution may reinforce the action of high concentrations of molybdenum.

When the molybdenum content exceeds certain limiting concentrations, the Fe-Cr-Mo alloys become subject to stress corrosion in the $MgCl_2$ test. The upper, safe limit for molybdenum decreases with increasing chromium content (Figure 7). Several alloying elements, copper, nickel, and ruthenium, also affect the resistance of Fe-Cr-Mo alloys to SCC in the $MgCl_2$ test. Data on these are given below.

General Corrosion in Acids

In addition to chloride pitting and stress corrosion

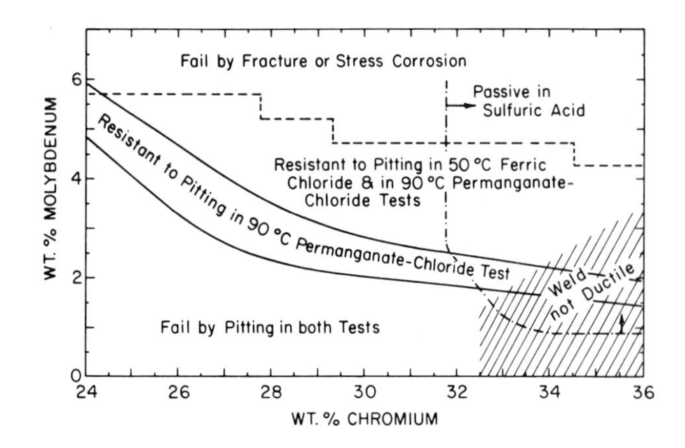

FIGURE 7 — Properties of Fe-Cr-Mo alloys. Pitting tests: 10% $FeCl_3 \cdot 6H_2O$ at 50 C (with crevices), 2% $KMnO_4$-2% NaCl at 90 C. Stress corrosion test: 45% boiling (155 C) $MgCl_2$ on welded U-bend specimens. Sulfuric acid: boiling 10% by weight. C <100 ppm, N <200 ppm, C+N <250 ppm.

resistance, stainless steels also should have the widest possible range of resistance to acid environments. To provide an indication of resistance to general corrosion by acids, tests were made in boiling mineral acids (oxidizing and reducing) and in several organic acids on some Fe-Cr-Mo alloys and, for comparison, on a number of commercial alloys.

The effect of chromium and molybdenum concentration on resistance to 10% boiling H_2SO_4 is shown in Figure 7. All compositions shown in this figure to the left and below the line at 32% Cr are active in this acid and corrode at a high rate, *e.g.*, 52,180 mils/yr in the case of an Fe-28Cr-4Mo alloy. There is profuse evolution of hydrogen. Alloys on the other side of this line are passive in this solution and have a relatively low rate of corrosion, 56 mils/yr on an Fe-33Cr-3Mo alloy. However, while exposed to the acid, these alloys can be activated by contact with an iron rod. They are not self-repassivating and remain active when the contact is broken. Only removing and rinsing the specimens in water restores the passive state when they are re-immersed.

For subsequent tests, a specific Fe-Cr-Mo alloy was selected from the zone of optimum combination of properties (Figure 7), *i.e.*, in the area containing the alloys which are resistant in both pitting tests and are immune to fracture on bending and to SCC. The limits for this alloy were 28.5 to 30.5% Cr and 3.5 to 4.2% Mo, with carbon and nitrogen as low as possible, but not to exceed 100 ppm C, 200 ppm N, and 250 ppm C+N at the surface of fabricated material. These ranges for chromium and molybdenum were selected to provide optimum pitting resistance, ductility, and resistance to stress corrosion. Amounts of silicon (0.5%) and manganese (0.8%) commonly present in commercial stainless steels were without deleterious effect on the properties of the 28Cr-4Mo alloy.

The 28Cr-4Mo alloy was used for acid corrosion tests (Table 3). Some of the data on commercial stainless steels were derived from a previous investigation,[24] as were the testing conditions. If a corrosion rate of 50 mils/yr is used to separate resistant from nonresistant alloys, it is apparent that, except in boiling 10% sulfuric acid, the 28Cr-4Mo alloy has excellent resistance in all acids tested. It is superior to commercial austenitic and ferritic stainless steels and is most like Carpenter 20Cb-3 alloy. Hastelloy C has relatively high rates in oxidizing acids. Titanium is not resistant in solutions containing sulfuric (or sulfamic) acid or in two of the organic acids.

The Effect of Nickel

To produce resistance to boiling 10% H_2SO_4, a series of Fe-Cr-Mo alloys was made with increasing amounts of

TABLE 3 — Comparison of General Corrosion of Alloys in Boiling Acids(1)

Alloy	General Corrosion(2) (mils per year): Nitric (65%)	50% Sulfuric with Ferric Sulfate	Sulfamic (10%)	Formic (45%)	Acetic (20%)	Oxalic (10%)	Sodium Bisulfate (10%)	Sulfuric Acid (10%)
AISI 430	20	312	144,000	84,700	3,000	6,400	91,200	252,000
AISI 446	8	36	150,000	9,700	0	7,000	64,800	270,000
AISI 304	8	23	1,300	1,715	300	570	2,760	16,420
AISI 316	11	25	75	520	2	96	170	855
Carpenter 20Cb-3	8	9	16	7	2	7	11	43
Hastelloy C	450	240	8	5	0	8	8	17
Titanium	1	140	285	873	0	950	250	6,290
Fe-28% Cr-4% Mo	2	6	0	1	0	13	9	52,180
Fe-33% Cr-3% Mo	—	—	—	—	—	—	—	56
Fe-35% Cr	6	6	0	3	0	4	—	74,000(3)

(1)Acid concentrations in wt%.
(2)The length of testing time varied with the corrosion rate, 10 minutes for the high rates and 10 days for the low rates.
(3)Performance of Fe-35% Cr alloys in sulfuric acid is variable. Some are passive, and some are active.
NOTE: See Table 1 for compositions.

nickel. These alloys were tested in H_2SO_4, in the pitting tests, and for stress corrosion resistance in the $MgCl_2$ test (Table 4). This table shows that the 28Cr-4Mo alloy becomes passive when 0.25 Ni is added. However, at this same concentration of nickel, the (unwelded) alloy loses its resistance to SCC in the $MgCl_2$ test. Larger additions of nickel reduce the passive rate in boiling 10% H_2SO_4 from 56 to about 10 mils/yr. The time to failure in the $MgCl_2$ test is progressively reduced from more than 100 hours to less than 3 hours. With 2% Ni or more, the alloy becomes self-repassivating. However, at 2.5% Ni and over, resistance to pitting in the ferric chloride test is lost. Otherwise, all pitting results in Table 4 are in accordance with Figure 7, *i.e.*, additions of less than 2.5% Ni did not impair pitting.

Data in Table 4 also show that lowering the molybdenum content of 28.5 Cr alloys (with nickel) makes the alloys active. Similarly, reducing the chromium content to 27% with 4 Mo and 0.4 Ni results in a loss in passivity. Increasing the molybdenum content to 6% in a 25 Cr alloy with 0.4 Ni does not restore passivity. Thus, the effect of nickel depends both on the chromium and on the molybdenum content of the alloy. An alloy (28.5Cr-2.0Ni) without Mo, but containing 2% Ni is active in boiling 10% H_2SO_4. Note that the time to failure in the $MgCl_2$ test is greatly increased when the chromium content is reduced to values which result in a loss of passivity in H_2SO_4.

Cobalt additions to 28Cr-4Mo alloys also produce passivity in H_2SO_4, but in place of 0.25 Ni, more than 1.5 Co are needed. These passive alloys also crack in the $MgCl_2$ test, and their resistance to pitting is impaired in both tests.

Because both nickel and copper may be present as residual elements in stainless steel, additional tests were made on welded specimens to determine the maximum concentrations of copper and nickel which can be present without loss in stress corrosion resistance in the $MgCl_2$ test (Table 5). These tests show that as much as 0.40 Cu or 0.15 Ni can be tolerated without impairing resistance to stress corrosion in the $MgCl_2$ test. (Note that in the unwelded alloy slightly more nickel, 0.20%, can be tolerated, Table 4.) Also, when both elements are present, 0.15 Ni plus 0.15

TABLE 4 – Effect of Nickel Additions to Fe-Cr-Mo Alloys

Alloy No.	Composition[1]			Boiling 10% Sulfuric Acid		Pitting Corrosion		Stress Corrosion[4]
	Cr	Mo	Nickel	State	Corrosion Rate (mils/year)	$KMnO_4$-NaCl[2]	$FeCl_3$[3]	(not welded)
181	28.0	4.0	0.10	active	63,000	R	R	resistant
239	28.0	4.0	0.20	active	59,700	R	R	resistant
217	28.0	4.0	0.25	passive	56	R	R	failed
183	28.0	4.0	0.30	passive	52	R	R	cracked after 119 hrs
191	28.0	4.0	0.40	passive	29	R	R	cracked after 261 hrs
241	28.0	4.0	0.5	passive	24	R	R	cracked after 16 hrs
245	28.5	4.0	1.5	passive	6	R	R	cracked in <16 hrs
681	28.5	4.2	1.8	passive	11	R	R	cracked
664	28.5	4.2	2.0	passive*	8	R	R	cracked in 3 hrs
658	28.5	4.2	2.5	passive*	10	R	F	–
649	28.5	4.2	3.0	passive*	9	R	F	–
193	28.5	3.0	0.4	passive	–	R	F	–
194	28.5	2.0	0.4	passive	–	F	F	–
195	28.5	1.0	0.4	active	–	F	F	–
680	28.0	2.5	0.25	passive	–	R	F	cracked in <17 hrs
641	28.0	2.0	0.25	active	–	F	F	–
184	28.0	4.0	0.4	passive	–	–	–	cracked after 6 hrs
233	27.0	4.0	0.4	active	–	R	F	cracked after 447 hrs
232	26.0	4.0	0.4	active	–	R	F	resistant
231	25.0	4.0	0.4	active	–	F	F	cracked after 447 hrs
213	25.0	6.0	0.4	active	–	R	R	–

(1) Wt%.

(2) 2% $KMnO_4$-2% NaCl at 90 C.

(3) 10% $FeCl_3 \cdot 6H_2O$ at 50 C with crevices.

(4) Magnesium chloride test (45%) on unwelded specimen. Resistant = no cracking after 2400 hours of testing.

NOTE: * = Self-repassivating, – = not tested, R = resistant, no pitting, and F = failed, pitted.

Source: *Corrosion*, 30(3), March 1974, 77-91

Cu can be tolerated, *i.e.*, the presence of nickel reduces the tolerance limit of copper.

The Wick Test

The $MgCl_2$ solution is a very severe testing medium which has been used extensively for research on SCC, primarily on austenitic stainless steels. It is not representative of the vast number of chloride environments to which stainless steels are exposed in service. To provide data on stress corrosion resistance in an environment which approximates more closely the conditions encountered in plant exposures, welded specimens were exposed in the wick test developed by Dana and DeLong.[25,26]

A 7 x 2 inch specimen, 0.062 inch thick, containing a 7-inch long weldment in the middle is ground to a 120 grit finish and then bent into a U-shape with a 1 inch radius. To simulate a heated surface exposed to a chloride environment, an electrical current is passed through it while it is in contact with glass wool immersed in an NaCl solution (Figure 8). Evaporation of the solution concentrates NaCl on the hot surface. Type 304 stainless steel cracks within three days when the specimen is heated to 100 C while exposed to a solution containing 1500 ppm Cl added as NaCl.[27] In contrast, Carpenter 20Cb-3 alloy, which is known to be more resistant to cracking than Type 304 in many plant environments, but which also cracks in the $MgCl_2$ test, resists cracking in the wick test.

All of the alloys of Table 5 which are listed as *failed* in $MgCl_2$ were exposed in the wick test for at least 60 days. None of them cracked in this test. The 2Ni alloy was tested for 90 days at 100 C, and then for an additional 50 days at 110 C (230 F) without cracking. These results suggest that residual amounts of copper and nickel in the 28Cr-4Mo alloy may not reduce its resistance to SCC in plant environments and that alloying additions of 2Ni also may not make the alloy subject to SCC in such environments. Tests in plant service are needed to verify these laboratory findings.

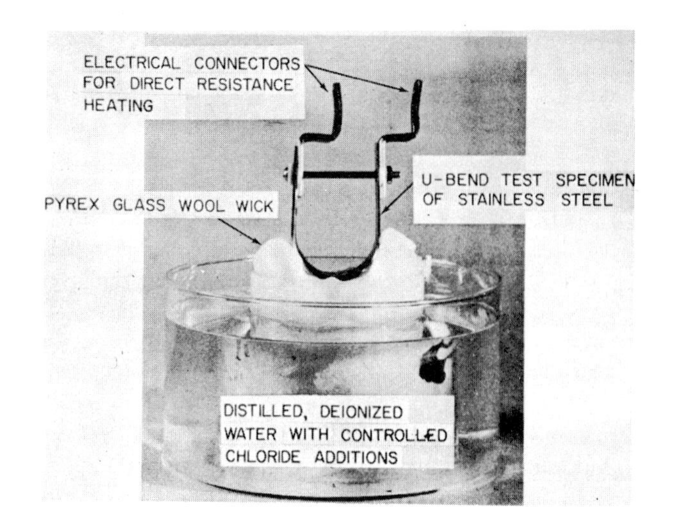

FIGURE 8 – Wick test specimen and apparatus (from Dana[26] and Warren[27]). The U-bend test specimen with a weld running lengthwise in the center is heated to 100 to 110 C. By capillary action chloride-bearing water rises to the surface of the specimen where its evaporation concentrates and crystallizes the chloride salt.

Addition of 2 Ni results in self-repassivation of the 28Cr-4Mo alloy in boiling 10% H_2SO_4 without impairing pitting resistance. Table 6 shows that this alloy also resists boiling 1% hydrochloric acid and is self-repassivating in this solution.

The Effect of Noble Metals

Platinum and palladium (0.1 to 1.0%) have been added

TABLE 5 – Effect of Nickel and Copper in Fe-28%Cr-4%Mo Alloy on Stress Corrosion in the Magnesium Chloride and the Wick Tests

Alloy	Stress Corrosion Cracking: Boiling 45% Magnesium Chloride Test (welded U-bend specimens)	Stress Corrosion Cracking: Wick Test (1500 ppm Cl as NaCl – 100 C) (welded specimens)
28.5% Cr-4.2% Mo		
Additions (wt%)		
0.15Ni	None (2136 hrs)	–
0.20Ni	Failed after 67 hrs	None (60 days)
2.00Ni	Failed in 16 hrs	None (90 days + 50 days at 110 C)
0.20Cu	None 2547 hrs)	–
0.30Cu	None (2544 hrs)	–
0.40Cu	None (2544 hrs)	–
0.60Cu	Failed after 432 hrs	None (90 days)
0.10Ni+0.10Cu	None (2547 hrs)	–
0.15Ni+0.10Cu	None (2544 hrs)	–
0.15Ni+0.15Cu	None (2544 hrs)	–
0.15Ni+0.20Cu	Failed after 432 hrs	None (60 days)
0.30Ni+0.30Cu	Failed in less than 48 hrs	None (60 days)
0.75Ru	Failed in less than 15 hrs	None (60 days)

TABLE 6 – Corrosion in Boiling Sulfuric and Hydrochloric Acids

Alloy	Corrosion Rate (mpy) 10% H_2SO_4	1% HCl
AISI 446	270,000	74,100
AISI 316	850	2,800
Hastelloy C-276[(1)]	10	11
Hastelloy G[(1)]	22	150
Carpenter 20Cb-3[(1)]	31-43	70
28Cr-4Mo-2Ni[(1)]	8-22	6

(1)Alloy is self-repassivating.

to 18Cr-8Ni and 26 Cr stainless steels by Tomashov and Chernova[28-30] to enhance their corrosion resistance in H_2SO_4. These noble metals, either dissolved in the stainless steel or coupled to it in the form of a sheet, polarize the stainless steel from the active into the passive range of electrode potentials. Similar results can be obtained by making the corroding specimen an anode in a circuit consisting of an external voltage and an inert cathode while they are immersed in the solution (anodic protection).

Several investigations have verified these effects of relatively large additions of platinum and palladium in 17 Cr[31,32] and 19Cr-13Ni[33] stainless steels.

In this investigation, initial attempts to make the 28Cr-4Mo alloy resistant to boiling 10% H_2SO_4 consisted of adding each of the six platinum group metals to this composition. It was found (Tables 7 and 8) that every one of these, when present in excess of a certain minimum concentration, passivates the 28Cr-4Mo alloy in boiling 10% H_2SO_4. The minimum concentration of the noble metal required varies with the element and is independent of atomic weights, *i.e.*, concentration in At%. The determining factors are probably the electrochemical properties of these elements, hydrogen overvoltage and exchange currents of cathodic reactions taking place at those sites in the surface where the noble metals are exposed to the solution. Additions of a noble metal in a concentration lower than that required to produce passivity actually increase the rate of corrosion of the 28Cr-4Mo alloy (52,180 mils/yr). This phenomenon was also observed by Biefer[31] on 17 Cr steel. Increasing the concentration *above* the minimum concentration required for passivity progressively decreased the passive corrosion rate.

The minimum concentrations of noble metals (0.005 to 0.020%) required to produce passivity in the 28Cr-4Mo

TABLE 7 – Effect on Corrosion of Platinum, Palladium, and Iridium Additions to Fe-28%Cr-4%Mo Alloy

Alloy	Composition[(1)] Noble Metal	Boiling 10% Sulfuric Acid State	Boiling 10% Sulfuric Acid Corrosion Rate (mils/year)	Pitting Corrosion $KMnO_4$-NaCl[(2)]	Pitting Corrosion $FeCl_3$[(3)]	Stress Corrosion[(4)] (not welded)
	Platinum					
290	0.005	active	58,000	R	R	–
318	0.006	passive	48	–	R	–
317	0.008	passive	39	–	R	–
295	0.010	passive	15	R	R	–
286	0.030	passive	6	R	R	resistant
276-A	0.05	passive	4	R	R	resistant
274	0.10	passive	2	R	R	–
265	0.20	passive	1	–	–	resistant
319	0.30	passive	2	R	R	–
	Palladium					
470	0.01	active	74,000	–	F	–
469	0.02	passive	4	–	F	–
259	0.05	passive	2	F	F	–
257	0.10	passive	2	F	F	–
251	0.20	passive	1	F	F	resistant
	Iridium					
329	0.01	active	–	–	–	–
573*	0.01	passive	112	R	R	–
473	0.02	passive	78	–	R	resistant (welded)
268	0.1	passive	13	R	R	resistant

(1)Wt% noble metal in Fe-28.5% Cr-4.0% Mo.
(2)2% $KMnO_4$-2% NaCl at 90 C.
(3)10% $FeCl_3 \cdot 6H_2O$ at 50 C with crevices.
(4)Magnesium chloride test in boiling 45% solution.
NOTE: R = Resists pitting, F = pitted, and * = alloy contains 4.2% Mo.

alloy when exposed to boiling 10% H_2SO_4 are considerably lower than the concentrations used in previous investigations (0.1 to 1.0%) in alloys with lower chromium contents. Data in Table 8 on ruthenium additions show that the chromium content primarily determines the minimum amount of noble metal required for passivation.

To obtain self-repassivation, more than 10 times the minimum concentration of ruthenium required for passivity is needed, *i.e.*, only the alloys with 0.5% or more ruthenium are self-repassivating in 10% acid. Of all the alloys containing noble metals which were tested in the $MgCl_2$ test, only those which were self-repassivating were subject to SCC in the $MgCl_2$ test, but not in the wick test (Table 5). This is in contrast to the effect of nickel described above. Nickel bearing alloys became subject to SCC at about the concentration (0.25%) of nickel required to make them passive in boiling 10% H_2SO_4, rather than at the 2% concentration of nickel required for self-repassivation. The concentration of ruthenium required for self-repassivation is a function of the concentration of H_2SO_4. In 1% acid, a 28.5 Cr-4.2 Mo-0.03 Ru alloy (not shown in Table 8) is self-repassivating, *i.e.*, it does not take 0.5 Ru as in the case of 10% H_2SO_4.

Passivation of 28Cr-4Mo alloy in H_2SO_4 by noble metal additions is useful only if pitting resistance is not impaired. From the data in Tables 7 and 8, it is apparent that palladium destroys the resistance in both pitting tests and that rhodium impairs resistance in the $FeCl_3$ test. The other four platinum metals are without effect on resistance in these two tests. Additional pitting tests were made in two other halide solutions, bromine-zinc bromide and sodium hypochlorite at room temperature (Table 9). The

TABLE 8 — Effect on Corrosion of Ruthenium, Rhodium, and Osmium Additions to Fe-28%Cr-4%Mo Alloy

	Composition[1]			Boiling 10% Sulfuric Acid		Pitting Corrosion		Stress Corrosion[4]
Alloy	Cr	Mo	Noble Metal	State	Corrosion Rate (mils/yr)	$KMnO_4$-NaCl[2]	$FeCl_3$[3]	
			Rhodium					
574	28.5	4.2	0.005	passive	14	R	F	–
327	28.5	4.0	0.01	passive	7	R	F	–
270	28.5	4.0	0.1	passive	1	R	F	resistant (not welded)
			Osmium					
326	28.5	4.0	0.01	active	57,800	–	–	–
575	28.5	4.2	0.015	active	76,600	R	R	–
471	28.5	4.0	0.02	passive	26	–	R	–
269-A	28.5	4.0	0.1	passive	65	R	R	resistant (welded)
			Ruthenium					
338	28.5	4.5	0.015	active	62,200	–	–	–
477-A	28.5	4.0	0.017	active	–	–	R	–
334	28.5	4.0	0.02	passive	60	R	R	resistant (welded)
474	28.5	4.0	0.20	passive	9	–	R	resistant (welded)
475	28.5	4.0	0.30	passive	2	R	R	resistant (welded)
683	28.5	4.2	0.5	passive*	3	–	R	cracked (17 hrs)
671	28.5	4.2	0.75	passive*	1	R	R	cracked
684	28.5	4.2	1.50	pasisve*	1	–	R	cracked
634	26.0	1.0	0.02	active	–	F	F	–
	26.0	1.0	0.20	passive	–	F	F	–
572	27.0	3.0	0.02	active	–	–	–	–
688	30.0	2.5	0.02	passive	38	–	–	–
476	28.5	4.0	0.01 0.10 Ni	passive	41	R	R	resistant (welded)

(1)Wt%.
(2)2% $KMnO_4$-2% NaCl at 90 C. R = no pitting, F = pitting.
(3)10% $FeCl_3 \cdot 6H_2O$ at 50 C with crevices. R = no pitting, F = crevice corrosion.
(4)Magnesium chloride test in boiling 45% solution. Resistant—no cracking after 2400 hrs.
NOTE: * = Self-repassivating.

bromine solution is more corrosive than the $FeCl_3$ test, *i.e.*, pitting occurs in both Hastelloy C and in the 28Cr-4Mo alloy with platinum. Iridium, osmium, and ruthenium do not impair pitting resistance in the bromine solution nor does ruthenium (the only element tested) in the hypochlorite solution. Because the addition required to produce passivity in 28Cr-4Mo alloy with ruthenium is less costly than with iridium or osmium, ruthenium is the preferred noble metal for this purpose.

Mutual reinforcement between ruthenium and nickel is illustrated by Alloy 476 in Table 8. The concentrations of both of these elements in this alloy are lower than the minimum required for inhibition when they are used alone, 0.02 Ru and 0.25 Ni, yet the alloy is passive in boiling 10% H_2SO_4.

Addition of 0.2% silver did not result in passivity. To produce passivity with gold, 0.2% were required, and at this concentration, resistance to pitting in the ferric chloride was lost, but not in the $KMnO_4$-NaCl test.

The influence of heat treatment on the microstructure and other properties of Fe-28Cr-4Mo and Fe-28Cr-4Mo-2Ni alloys, their resistance in sea water and alkaline environments, the effect of hydrogen, some mechanical properties and data on stress corrosion in various chloride solutions are described in subsequent publications.[18,34]

Conclusions

1. Tests on commercial alloys and a number of new Fe-Cr-Mo compositions show that pitting in 10% $FeCl_3 \cdot 6H_2O$ solution at 50 C on specimens containing crevices is more severe than on specimens immersed in a solution of 2% $KMnO_4$-2% NaCl at 90 C. The latter test was developed as an accelerated method for simulating the very corrosive conditions in natural cooling waters from which manganese deposits form.

2. Among nonferrous-base metals, only Hastelloy C (or C-276), titanium, tantalum, and columbium (Nb) resist pitting corrosion in the above tests. Commercially available austenitic and ferritic stainless steels, copper base, and nickel base alloys are all subject to attack in the two chloride pitting tests.

3. By varying the amount of chromium and molybdenum in iron, a range of concentrations was found which is resistant in both tests. An alloy with 28% Cr and 4% Mo is representative of this resistant range and was investigated in some detail.

4. For optimum resistance to intergranular corrosion, the carbon must be <100 ppm, nitrogen <200 ppm, and C+N <250 ppm. These limits were established by welding sheet specimens and determining their susceptibility to intergranular attack in the boiling ferric sulfate-50% sulfuric acid test.

It was found that the relatively high molybdenum content (4%) increases the tolerance for carbon and nitrogen, *i.e.*, the maximum tolerable limit for these elements is lower for alloys which contain appreciably lower amounts of molybdenum.

Carbon and nitrogen in excess of the above limits may also make the 28Cr-4Mo alloy subject to pitting attack in the ferric chloride and the permanganate chloride tests, reduce its weld-bend ductility, and make it subject to SCC in the 45% $MgCl_2$ test.

5. The 28Cr-4Mo alloy is resistant to general corrosion in oxidizing acids and in a variety of organic acids. Resistance to boiling 10% sulfuric acid is imparted by addition of 0.25% or more of nickel. When 2.0% Ni are added, the alloy also becomes self-repassivating, *i.e.*, when activated in boiling 10% H_2SO_4 or 1% HCl by contact with an iron rod, the alloy spontaneously repassivates itself after the contact is broken.

6. Nickel additions which make the 28Cr-4Mo alloy

TABLE 9 — Comparison of Pitting Resistance in Halide Media (R = resistant; F = pitted)

Alloy	Permanganate-Chloride at 90 C(1)	Ferric Chloride at 50 C(2)	Bromine-Bromide at Room Temp.(3)	Sodium Hypochlorite at Room Temp.(4)
AISI 316	F	F	F	F
Carpenter 20Cb-3	F	F	F	F
Hastelloy C	R	R	F	F
Titanium	R	R	R	R
Fe-35%Cr	F	F	F	F
Fe-28%Cr-4%Mo	R	R	R	R
Fe-28%Cr-4%Mo+Pd	F	F	F	—
Fe-28%Cr-4%Mo+Rh	R	F	F	—
Fe-28%Cr-4%Mo+Pt	R	R	F	—
Fe-28%Cr-4%Mo+Ir	R	R	R	—
Fe-28%Cr-4%Mo+Os	R	R	R	—
Fe-28%Cr-4%Mo+Ru	R	R	R	R

(1) 2% $KMnO_4$-2% NaCl.
(2) 10% $FeCl_3 \cdot 6H_2O$, with crevices.
(3) 54.5% Br_2 + 20.6% $ZnBr_2$.
(4) 0.1% NaClO with Teflon crevices.

resistant in boiling 10% H_2SO_4 also make it subject to SCC in boiling 45% $MgCl_2$. However, the alloy resists cracking in the NaCl wick test at 110 C, which simulates certain plant conditions. Type 304 stainless steel cracks within 1 to 3 hours in the 45% $MgCl_2$ test and within 3 days in the wick test.

7. Additions of any of the six platinum metals to 28Cr-4Mo alloy also make it passive in boiling 10% H_2SO_4. However, only ruthenium, iridium, and osmium do this without impairing resistance to pitting corrosion. Platinum, palladium, and rhodium reduce pitting resistance in a variety of halide solutions. While 0.02% Ru makes the alloy passive in boiling 10% H_2SO_4, 0.50% Ru is needed to produce self-repassivation. This amount also makes the alloy subject to SCC in the boiling 45% $MgCl_2$ test, but not in the wick test.

8. Two compositions which have optimum combinations of properties are Fe-28Cr-4Mo and Fe-28Cr-4Mo-2Ni. In both alloys, the carbon and nitrogen contents should be held as low as possible, and are not to exceed 100 ppm C and 200 ppm N (C+N 250 ppm). Both alloys are resistant in the two pitting tests: (1) 2% $KMnO_4$-2% NaCl at 90 C, and (2) 10% $FeCl_3 \cdot 6H_2O$ at 50 C. When welded (80 mil sheets), they are immune to intergranular attack in the 120 hour ferric sulfate-50% sulfuric acid test.

Both alloys are immune to SCC in the NaCl wick test. The 28Cr-4Mo alloy, but not the 2% Ni alloy, resists cracking in the 45% $MgCl_2$ test. To maintain this resistance, the nickel content must not exceed 0.15% and copper 0.40%. If both elements are present, these limits are 0.15% Ni + 0.15% Cu.

Additions of 2% Ni to the 28Cr-4Mo composition make the alloy resistant and self-repassivating in boiling 10% H_2SO_4 and 1% HCl. Both alloys resist a variety of oxidizing acids and organic acids.

Acknowledgment

The author wishes to acknowledge assistance with laboratory work by S. J. Kucharsey and A. J. Sweet for specimen preparation, microscopy, and the many corrosion tests, and by A. W. Vansant and R. F. Settine for the preparation of laboratory alloys. Plant pitting corrosion problems involving manganese oxide deposits were brought to the author's attention by W. Sansom. Data (Table 9) on corrosion tests in hypochlorite solution were supplied by B. M. Dusenbury and in bromine-zinc bromide solution by K. B. Keating.

References

1. W. G. Renshaw. Communication to the Welding Research Council (1966) June.
2. M. A. Streicher. The Effect of Carbon, Nitrogen, and Heat Treatments on the Corrosion of Fe-Cr Alloys, Corrosion, Vol. 29, p. 337 (1973).
3. N. A. Long. Recent Operating Experiences with Stainless Steel Condenser Tubes, Proceedings of the 1966 American Power Conference.
4. M. Fleischer. Glossary of Mineral Species, U.S. Geological Survey (1971).
5. N. A. Long and N. Rice. Private Communication, Technical Memorandum on Pitting of Stainless Steels (1964) November.
6. L. R. Scharfstein. Private Communication, Memorandum on Chlorination Effects (1965) March.
7. N. V. Sidgwick. Chemical Elements and Their Compounds, Oxford Press, Vol. 2, p. 1269 (1950).
8. J. J. Morgan. Principles and Applications of Water Chemistry, Edited by S. D. Faust and J. V. Hunter, John Wiley & Sons, Inc., New York, p. 600, Fig. 12 (1967).
9. M. A. Streicher. J. Electrochem. Soc., Vol. 103, p. 375 (1956).
10. M. A. Streicher. Corrosion, Vol. 28, p. 143 (1972).
11. R. J. Knoth, G. E. Lasko, and W. A. Matejka. Chem. Eng., Vol. 77, p. 170 (1970).
12. R. J. Hodges, C. D. Schwartz, and E. Gregory. Br. Corrosion J., Vol. 7, p. 69 (1972).
13. R. Oppenheim and G. Lennartz. Chem. Ind., Vol. 23, p. 705 (1971).
14. G. Lennartz and H. Kiesheyer. DEW Tech. Ber., Vol. 11, p. 230 (1972).
15. R. Oppenheim and H. Laddach. DEW Tech. Ber., Vol. 11, p. 71 (1971).
16. R. F. Steigerwald. Corrosion, Vol. 22, p. 107 (1966).
17. W. O. Binder and H. R. Spendelow. Trans. ASM, Vol. 43, p. 759 (1951).
18. M. A. Streicher. The Effect of Composition and Heat Treatment on Stress Corrosion Cracking of Ferritic Stainless Steels (In preparation).
19. M. A. Streicher. ASTM Bulletin No. 229, p. 77 (1958) April.
20. ASTM A262-70 in Part 3 of 1971 Annual Book of ASTM Standards, American Society for Testing and Materials, Philadelphia, Pa.
21. B. E. Wilde. Corrosion, Vol. 28, p. 283 (1972).
22. P. Bond and H. J. Dundas. Corrosion, Vol. 24, p. 344 (1968).
23. M. A. Streicher and A. J. Sweet. Corrosion, Vol. 25, p. 1 (1969).
24. M. A. Streicher. Corrosion, Vol. 14, p. 59t (1958).
25. A. W. Dana and W. B. Delong. Corrosion, Vol. 12, p. 309t (1956).
26. A. W. Dana. ASTM Bulletin No. 225, p. 196 (1957) October.
27. D. Warren. Proceedings of the Fifteenth Annual Purdue Industrial Waste Conference (1960) May.
28. N. D. Tomashov and G. P. Chernova. Dokl. Akad. Nauk SSSR, Vol. 89, p. 121 (1953).
29. N. D. Tomashov. Corrosion, Vol. 14, p. 229t (1958).
30. T. P. Hoar. Platinum Metals Review, Vol. 2, p. 117 (1958).
31. G. J. Biefer. Canadian Metallurgical Quarterly, Vol. 9, p. 537 (1970).
32. V. S. Agarwala and G. J. Biefer. Corrosion, Vol. 28, p. 64 (1972).
33. G. Bianchi, G. A. Camona, G. Fiori, and F. Mazza. Corrosion Science, Vol. 8, p. 751 (1968).
34. M. A. Streicher. Microstructures and Some Properties of Fe-28%Cr-4%Mo Alloys. Corrosion, Vol. 30, p. 115 (1974).

Immunity to chloride stress corrosion cracking makes 18Cr-2Mo, one of the new ferritics, a prime candidate for the collector panels used in solar water heating systems. These stainless steel panels are being tested at DSET Laboratories Inc., Phoenix, as a part of a field evaluation program sponsored by the American Iron and Steel Institute's Committee of Stainless Steel Producers.

The New Ferritic Stainless Steels

By Ralph M. Davison and Robert F. Steigerwald

MANY NEW Fe-Cr-Mo and Fe-Cr-Mo-Ni ferritic stainless steels have recently become commercially available. They offer an extraordinary range of corrosion resistance that peaks at a level not previously available in a stainless steel. This corrosion resistance, combined with a practical immunity to stress corrosion cracking, provides a new technically and economically powerful tool for dealing with corrosion problems.

As early as the 1950's, it was known that the Cr-Mo ferritic stainless steels offered excellent corrosion resistance. Their toughness and weldability, however, were inadequate owing to high interstitial (carbon and nitrogen) contents.[1] No method was available for economically producing the required low interstitial levels, such as $C + N < 0.05\%$.

The new ferritic stainless steels owe their existence to the development of new steelmaking technology: argon oxygen decarburization (AOD), vacuum induction melting (VIM), and vacuum oxygen decarburization (VOD). All three are capable of producing low interstitial content, ferritic stainless steels. The AOD process, the one most commonly associated with ferritic stainlesses, was introduced primarily because it provides a substantial increase in chromium yield.

There are very many combinations of chromium, molybdenum, and nickel that might be considered. A proliferation of grades, however, is constrained by produc-

Source: *Metal Progress*, 116(6), June 1979, 40-46,

tion and inventory of economics. The grades that have emerged as dominant — the ones discussed here — are naturally those that most effectively meet the corrosion resistance requirements of major market segments.

Ferritics Provide Four Levels of Corrosion Resistance

Four distinct levels of corrosion resistance are represented by the new ferritics.

The 18Cr-2Mo steel is a general purpose grade with corrosion resistance at least comparable and frequently superior to that of AISI type 304. The 26Cr-1Mo steels provide outstanding performance in a number of chemical process environments, such as caustic service. The newest of the new ferritics is 26Cr-3Mo-2.5Ni, a conventionally produced grade for seawater service. The vacuum melted 29Cr-4Mo steels provide extraordinary corrosion resistance, making them competitive with even titanium and nickel-base alloys. The compositions, properties, and corrosion resistance of these steels are characterized in the *Metal Progress Datasheet* in this issue.

General Purpose Ferritics: The common austenitic stainless steels are extremely useful materials for process equipment in many industries. However, their use is occasionally troubled by stress corrosion cracking (SCC). Although the actual number of failures may seem small relative to the attention given the problem, the sudden and possibly catastrophic nature of the failure makes SCC a serious concern. Among the standard ferritic grades, AISI type 430 has sometimes been considered as a solution to SCC. But type 430 has not often been used in process applications because, although immune to SCC when properly annealed, it has limited general and pitting corrosion resistance. What is required is a ferritic steel with corrosion resistance superior to that of type 304. The 18Cr-2Mo grade meets these requirements. In fact, its corrosion resistance sometimes approaches that of type 316.

The properties of the 18Cr-2Mo composition have been known for many years. Commercial production, however, hinged on the improvements in steelmaking technology mentioned above. Carbon is now held to below 0.025% and the steel is stabilized with titanium or columbium or, as typical of U.S. practice, with a combined Ti+Cb addition.

The steel is now included in ASTM A176, A240, and A268, and is covered by ASME Boiler Code Case 1825. Designated S44400 in the Unified Numbering System, it is known as 18-2, 444, and type 444. Leading U.S. producers are Jones & Laughlin Steel Corp.; Allegheny Ludlum Steel Corp., Div. Allegheny Ludlum Industries Inc.; Crucible Inc., Div. Colt Industries Inc.; and, to a lesser extent, Carpenter Technology Corp.; Universal-Cyclops Specialty Steel Div., Cyclops Corp.; and Al Tech Specialty Steel Corp.

The largest single application for 18Cr-2Mo has been in medium-duty catalytic converters for light trucks. Applications gaining in importance include solar panels (see lead photo), beer tankage, food processing equipment, chemical plant heat exchangers, and a variety of other processing equipment.

The steel has also achieved substantial market acceptance in Japan and Europe. The foreign products differ only in minor details of composition.

There is also a free machining version of 18Cr-2Mo. The resulfurized grade, designated S18200 or XM-34 in ASTM A582, has corrosion resistance superior to that of type 303 (a free machining version of type 304) and machinability at least equal to that of type 416 (a free machining version of type 410). This steel is offered commercially by CarTech as Project 70 182-FM and Universal-Cyclops as Uniloy 18-2FM.

Special Purpose Ferritics: The 26Cr-1Mo steels combine SCC resistance with chloride corrosion resistance significantly better than that of AISI type 316. There are two 26Cr-1Mo grades.

One is E-Brite 26-1, a proprietary Allegheny Ludlum alloy. Airco Vacuum Metals, Div. Airco Inc., developed the grade using electron beam refining to produce extremely low carbon and nitrogen levels. Allegheny Ludlum uses vacuum induction melting. Although initially produced without a stabilizer, all E-Brite 26-1 melted after November 1971 has a small addition of columbium. The stabilizer is not mentioned in ASTM or ASME specifications, but it is thought to be beneficial to the steel's properties. Allegheny Ludlum is proposing that the columbium requirement be added to applicable specifications. E-Brite is designated UNS S44625 and XM-27 in ASTM A176, A240, and A268. Low interstitial levels provide unusually good toughness levels for a highly alloyed Cr-Mo ferritic stainless steel.

The other 26Cr-1Mo grade is produced using the AOD process and a titanium stabilizer. Designated UNS S44626 and XM-33 in ASTM A176 and A268, the grade is sometimes called 26-1-Ti or 26-1S. Higher levels of carbon and nitrogen mmake it less tough than E-Brite 26-1. Consequently, 26-1-Ti is available only as light gage sheet or tubing where toughness is of limited importance. The AOD-produced grade is offered by Crucible and Allegheny Ludlum.

The 26Cr-1Mo steels have been successfully used in a variety of severe chemical environments. The most outstanding application has been the substitution of E-Brite 26-1 for Nickel 200 in caustic evaporators. The 26-1-Ti grade is also being evaluated for this application.

Table I—Crevice Corrosion of Stainless Steels in Oxygen-Saturated Chloride Solutions at 90 C (195 F)[1]

Solution	Weight Loss, mg/(dm^2 · day)[2] Type 434	Type 304	Type 316	18Cr-2Mo
200 ppm Cl−, 1 ppm Cu++	41.0	15.0	2.6	0.2
600 ppm Cl−, 1 ppm Cu++	241.0	11.8	3.5	0.8

1. From Ref. 3
2. Multiply weight loss values by 1.16 to obtain μg/(m^2 · s); by 0.12 to obtain oz/(ft^2 · yr).

Table II — Crevice Corrosion Behavior of Ferritic Stainless Steels in Low-Velocity Seawater at Ambient Temperature[1]

Nominal Composition, %			Probability of Initiation, %		Maximum Depth, mm[2]	
Cr	Mo	Ni	61 Days	272 Days	61 Days	272 Days
18	2	—	13	—	0.64[3]	—
26	1	—	0	—	0	—
26	1	—	3	—	0.43	—
28	2	—	0	0.8	0	0.14[4]
28	2	4	0	7.5	0	0.2
25	3.5	—	0	0	0	0
25	3.5	2	—	0[5]	—	0[5]
25	3.5	4	0	0	0	0
29	4	—	0	0	0	0
29	4	2	1.4	0	<0.02	0

1. From Ref. 5.
2. Multiply maximum depth values by 0.03937 to obtain inches.
3. Perforated.
4. Attack at fouling sites: 1.1 mm.
5. 186 days.

The fact that 26Cr-1Mo steels resist some seawater environments initially encouraged their use. However, they were found to fail rapidly in other seawater applications. Consequently, it is now thought that these grades lack an adequate safety margin to permit general use in seawater.

The 26Cr-1Mo grades are primarily U.S. developments, although the Japanese recently began production of similar grades for caustic service.

Seawater Ferritics: The dilemma of how to AOD-produce a tough stainless steel with an alloy content sufficient for seawater applications was resolved by adding nickel. As discussed later, the addition of 2 to 4% Ni sacrifices the steel's ability to resist SCC in boiling 42% $MgCl_2$, but does not lead to SCC in boiling acidified 25% NaCl or in the NaCl wick test.

Colt's Crucible Stainless Steel Trent Tube Divisions have recently introduced a 26Cr-3Mo-2.5Ni steel for seawater condenser service. The steel is called SC-1; the tubing made from it, SeaCure. Crucible is sponsoring the alloy for inclusion in ASTM specifications and is conducting extensive field qualification tests. Tests of this alloy in natural seawater by Climax Molybdenum Co., Div. Amax Inc., have shown excellent resistance to crevice corrosion and SCC. It is expected that the results of qualification testing will indicate that this grade represents a major competitor for titanium, Cu-Ni alloys, and the more highly alloyed austenitic stainless steels currently used for seawater condenser tubing.

A comparable grade also before ASTM is MONIT, a 25Cr-4Mo-4Ni steel produced by Granges Nyby AB, Sweden, and available in Europe and North America. A number of other grades with Cr-Mo-Ni combinations in this range are also available. They include a 28Cr-2Mo-4Ni steel from TEW (Thyssen Edelstahlwerke AG), West Germany, and several Japanese steels. These Cr-Mo-Ni steels are all produced by AOD or VOD and stabilized by an addition of titanium, columbium, or combinations of these and other elements. Conventional processing and relatively low-cost starting materials permit excellent economics compared with competing materials. Future applications are expected to be found in the pulp and paper industry and in pollution control equipment.

High Alloy Ferritics: The 29Cr-4Mo steel developed by Michael A. Streicher, Du Pont Co., can withstand extremely corrosive environments, particularly strong chloride solutions, that typically require nonferrous alloys. It is a low interstitial ferritic produced by vacuum induction melting without a stabilizer addition.[2] Adding 2% Ni produces a variation of the alloy with improved sulfuric acid resistance and toughness. Test results demonstrate that these steels have excellent resistance to corrosion by seawater and the severe environments encountered in paper mill bleach washers. Thanks to their low interstitial contents, these steels have good mechanical properties and they may be welded without risk of sensitization to intergranular

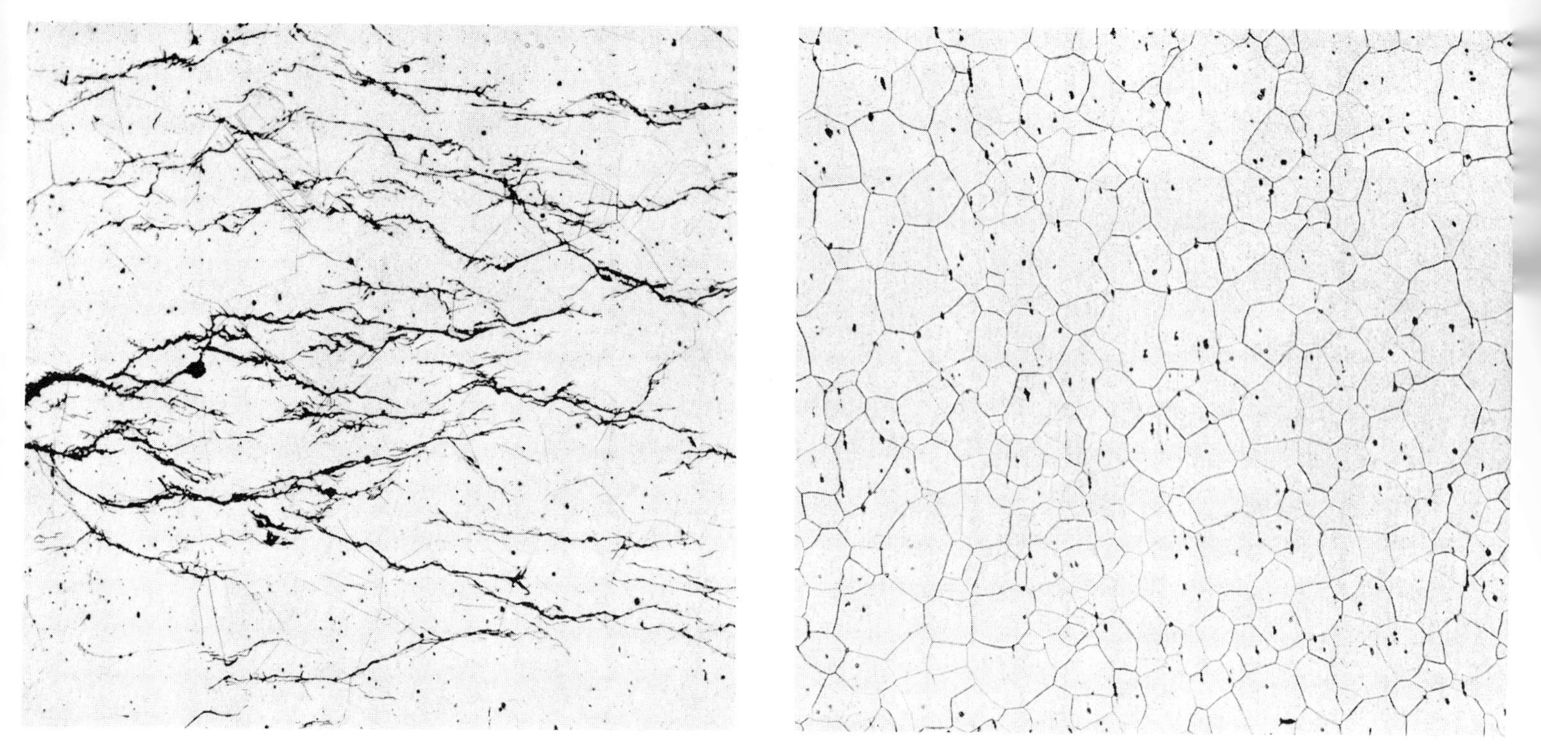

Fig. 1 — Numerous stress corrosion cracks are visible in the 100X photomicrograph (left) of an 18Cr-14Ni-1.43Mo austenitic stainless steel tested for 144 h in the boiling acidified 25% NaCl test. No new ferritic has exhibited SCC in this test even after very long exposures. The 18Cr-2Mo grade (right, 200X) is an example. It can withstand more than 2000 h of exposure without failing.

attack. Excellent shielding must, of course, be provided during welding.

Allegheny Ludlum is the primary licensee and producer of these steels. They're covered in ASTM specifications as UNS S44700 (29Cr-4Mo) and S44800 (29Cr-4Mo-2Ni).

Properties and Characteristics of the New Ferritics

The assumption of immunity to chloride stress corrosion cracking is a prime motive for developing new ferritic stainless steels. While this assumption may be valid from a practical point of view, R&D on ferritics has demonstrated that the nature of SCC is more complex than previously thought. Susceptibility to SCC may be varied by composition or thermomechanical history even for the so-called "immune" steels.[3] For example, relatively small amounts of copper or nickel can make ferritic susceptible to SCC in boiling 42% $MgCl_2$, particularly if the steel is in a welded or cold worked condition. However, the boiling 42% $MgCl_2$ test is overly severe for most applications.

More relevant is the NaCl wick test — results correlate well with SCC field experience.[4] In the test, a concentrated NaCl solution is conducted by a glass wool wick to the surface of a U-bend specimen. The specimen is heated by an electric current so that boiling of the solution occurs on its surface. The heat flow from the metal into the solution and the resultant concentration of the solution is an excellent simulation of the conditions shown by experience to be associated with SCC. Under these conditions, type 304 typically cracks in about three days and type 316 in about ten days. In no case has a new ferritic stainless steel shown cracking in wick tests run for 60 days, and even longer. Thirty days is the typical test time required to establish an adequate level of SCC resistance for most applications.

Another test, more convenient to run in the laboratory but well correlated with the wick test, is the boiling acidified 25% NaCl test (see Fig. 1). Here, a U-bend specimen is suspended in a boiling 25% NaCl solution acidified to pH 1.5. No new ferritic has exhibited SCC in this test even after very long exposures.

As noted, copper and nickel residuals can affect the SCC resistance of ferritic stainless steels. For 18Cr-2Mo, the 26Cr-1Mo grades, and 29Cr-4Mo, these elements must be at very low levels if it's necessary to pass the boiling 42% $MgCl_2$ test. However, these steels always pass the boiling acidified NaCl test. For 26Cr-3Mo-2.5Ni and 29Cr-4Mo-2Ni, resistance to SCC in boiling $MgCl_2$ has been intentionally sacrificed to provide the increased toughness and superior acid corrosion resistance associated with the nickel addition. These steels do pass the wick test and the boiling

Table III—Corrosion Resistance of 18Cr-2Mo Compared With Austenitic Stainless Steels[1]

Corrosive Medium	Temperature	Corrosion Rate, mils/yr[2]		
		Type 304	Type 316	18Cr-2Mo
20% acetic acid	Boiling	30	0.3	0.2
80% acetic acid	Boiling	—	—	0.2
20% citric acid	Boiling	0.4	0.5	0.3
30% formic acid	Boiling	81	29	34
45% formic acid	Boiling	—	—	212
20% lactic acid	Boiling	73	0.1	0.2
40% nitric acid	Boiling	2.4	2.2	2.3
1% oxalic acid	Boiling	—	—	20
3% oxalic acid	Boiling	110	57	57[3]
10% oxalic acid	Boiling	74	47	2340
50% phosphoric acid	Boiling	785	7.5	4.4
2% sulfuric acid	30 C (85 F)	1.3	0	400
25% sodium hydroxide	100 C (210 F)	1.1	1.9	7.6
35% sodium hydroxide	100 C (210 F)	2.2	1.6	20
60% sodium hydroxide	100 C (210 F)	3.0	2.7	24

1. From Ref. 3. Test conditions: no activation, 24 h exposure.
2. Multiply corrosion rate values by 0.0254 to obtain mm/yr; by 0.806 to obtain pm/s.
3. Active/passive behavior.

Table IV — Corrosion Resistance of 26Cr-1Mo Compared With Austenitic Stainless Steels[1]

Corrosive Medium	Temperature	Corrosion Rate, mils/yr[2]		
		Type 304L	Type 316L	26Cr-1Mo
Acetic acid, concentrated	Boiling	3.2	0.5	0.5
Acetic acid, concentrated+, 220 ppm Cl−	Boiling	270	180	20
25% sodium hydroxide	100 C (210 F)	2	1	0.1
30% fluosilic acid	Room	180	110	360
50% formic acid	Boiling	77	33	0.1
85% lactic acid	Boiling	22	0.1	0.1
85% lactic acid, 100 ppm NaCl	Boiling	32	0.4	0.1
75% nitric acid	Boiling	24	37	14
10% oxalic acid	Boiling	37	21	5
40% sulfuric acid	Room	340	71	300
50% zinc chloride	Boiling	16	0.9	0.5

1. From Ref. 6.
2. Multiply corrosion rate values by 0.0254 to obtain mm/yr; by 0.806 to obtain pm/s.

Source: *Metal Progress*, 116(6), June 1979, 40-46

acidified NaCl test and thus represent practical solutions to the problem of SCC in heat exchangers.

Pitting and Crevice Corrosion: Resistance to SCC must be accompanied by resistance to pitting and crevice attack by chlorides. The new ferritics are very resistant to chloride attack, generally in proportion to their alloy contents. It is believed that this resistance derives from the freedom to produce the higher chromium levels at which the molybdenum addition has its strongest effect. As a result, crevice corrosion resistance is achieved economically in terms of alloy additions.

The 18Cr-2Mo steel was designed for dealing with general purpose applications such as hot water tanks and fresh water heat exchangers. Table I shows that over the range of chloride levels that might be anticipated, the steel's crevice corrosion resistance can approach that of type 316.[3]

For the higher alloy steels, crevice corrosion must be evaluated in much more severe environments. One laboratory method proven to be rapid and reliable is the measurement of maximum temperature for resistance to crevice attack in 10% $FeCl_3 \cdot 6H_2O$ shown in the Datasheet. Correlation of these test results with practical environments is not always possible, but the relative resistance of the new steels is sharply defined.

Table II shows some results of a seawater test program being conducted by Climax Molybdenum.[5] Specimens with an artificial crevice are being exposed to low velocity natural seawater. Extensive biofouling provides additional severe crevice sites. The 26Cr-1Mo steels were found to be marginal in crevice corrosion resistance even for the short test. At longer times, the 28Cr-2Mo steel has shown slight attack at the artificial crevices and very severe attack at isolated biofouling sites. The 25Cr-3.5Mo and 29Cr-4Mo steels have remained unattacked in the longer test. The one incident noted (for 29Cr-4Mo-2Ni) is regarded as insignificant.

General Corrosion: The general corrosion resistance of the new ferritic stainless steels is compared with that of various standard grades and alloys in Tables III, IV, and V. The ferritics stack up well with some exceptions. For example, they have superior resistance to organic acids, but the nickelfree grades have very high corrosion rates in strong acids that exceed a critical concentration. This concentration is very low for 18Cr-2Mo. An even 29Cr-4Mo is severely attacked in boiling 10% H_2SO_4. However, the addition of 2% Ni to 29Cr-4Mo significantly slows the corrosion rate in this acid. Likewise, 26Cr-3Mo-2Ni shows excellent resistance to the lower sulfuric acid concentrations even under boiling conditions.

The 18Cr-2Mo steel is somewhat deficient in concentrated sodium hydroxide solutions. The resistance of 26Cr-1Mo, however, is excellent. This behavior has been confirmed by experience with E-Brite 26-1 in caustic evaporator applications.

Mechanical Properties: A useful approach to describing the properties of ferritic stainless steels is to compare them with those of the generally familiar austenitic type 304.

As a class, ferritics are characterized by higher yield strengths and lower ductility and work hardening rates. For example, the minimum annealed yield strength of 18Cr-2Mo is 40 000 psi (275 MPa) compared with 30 000 psi (205 MPa) for type 304. The elongation for annealed 18Cr-2Mo is typically 30%; for type 304, 50% is typical. Because of its lower work hardening rate, 18Cr-2Mo's tensile strength is about 70 000 psi (485 MPa); type 304's is 90 000 psi (620 MPa).

Increasing the alloy content of the Cr-Mo ferritics increases both the yield and tensile strengths, and slightly decreases the elongation. For example, 29Cr-4Mo's yield strength is typically 75 000 psi (515 MPa); its tensile strength, 90 000 psi (620 MPa). Most notable is the effect of nickel — a 2% addition results in a jump of about 10 000 psi (70 MPa) in yield and tensile strength for both the 26% and 29% Cr levels. Elongation is not increased by the nickel addition but the ductile-to-brittle transition temperature is significantly reduced.

Impact Resistance: Toughness limitations of ferritic stainless steels have always been a serious concern. The ferritic crystal structure is characterized by a ductile-to-brittle transition temperature. The effect of lowering carbon and nitrogen contents in the new ferritic stainlesses is to push this transition to higher alloy levels or to thicker sections. Toughness of the ferritics is a complex function of the chromium and molybdenum contents, the carbon and nitrogen contents, the stabilizing element and the amount used, the presence or absence of a significant nickel addition, the grain structure, and the thermal history.

All of the new grades are acceptably tough as both sheet an tubing.

The 18Cr-2Mo has also been successfully applied as thin plate, but special care must be taken in its production and fabrication. Because of their extremely low carbon and nitrogen levels, E-Brite 26-1, 29Cr-4Mo, and 29Cr-4Mo-2Ni may also be considered for thin plate applications. Special precautions during welding are required to avoid contamination which can decrease toughness.

The 26Cr-1Mo-Ti steel is limited to thin sheet because of the combination of high chromium content and the higher interstitial levels associated with AOD production. The 26Cr-3Mo-2.5Ni steel is also AOD produced, but the nickel addition makes it sufficiently tolerant of interstitials to permit production as both sheet and tubing. The production of this alloy as thin plate is being investigated.

Embrittlement: Ferritic stainless steels are known to be embrittled by long exposures to temperatures in the 650 to 1100 F (345 to 595 C) range. This phenomenon has been called "885 embrittlement" because the detrimental precipitation of the high chromium alpha prime phase proceeds most rapidly at 885 F (475 C). The problem is not eliminated in the new ferritics by the low interstitial levels.

Table V — Corrosion Resistance of 29Cr-4Mo and 29Cr-4Mo-2Ni in Boiling Acids [1]

	Corrosion Rate, mils/yr[2]					
Corrosive Medium	Type 316	Carpenter 20Cb3	Hastelloy [3]C	[4]Titanium	29Cr-4Mo	29Cr-4Mo-2Ni
65% nitric acid	11	8	450	1	2	—
50% sulfuric acid, ferric sulfate	25	9	240	140	6	—
45% formic acid	520	7	5	873	1	—
10% oxalic acid	96	7	8	950	13	—
10% sulfamic acid	75	16	8	285	0	0
10% sulfuric acid	855	43	17	6290	52 180	8

1. From Ref. 2
2. Multiply corrosion rate values by 0.0254 to obtain mm/yr; by 0.806 to obtain pm/s.
3. Trademark of Carpenter Technology Corp.
4. Trademark of Cabot Corp.

It may, in fact, be slightly worsened by the presence of stabilizers which appear to accelerate the embrittling reaction. Consequently, neither the new nor the standard ferritic stainless steels should be subjected to long exposures above 650 F (345 C) if room temperature toughness must be maintained.

The ferritic stainless steels also offer two physical properties that are of particular importance.

1. Their thermal conductivity is about 50% higher than that of the austenitics, or about half that of carbon steel.

2. Their thermal expansion coefficient is quite low, roughly equal to that of carbon steel and about 30% less than that of type 304.

Weldability: Early in the development of the new ferritics, weldability was recognized as a critical problem. Special attention was consequently given to this area. The result: the steels can all be welded, and they all retain their corrosion resistance in the as-welded condition. It is, however, essential that welding practices be designed to take into account the steels' special characteristics.

The most important factor in making good welds in the new ferritics is the maintenance of alloy purity. Because the high quality of these steels is based on low carbon and nitrogen contents, it is essential that these elements, along with oxygen and hydrogen, not be introduced into the metal. Careful shielding, front and back, with dry argon is necessary.

Another point is the fact that these steels are fully ferritic at all temperatures up to the melting point. Heat input should thus be maintained at as low a level as possible to limit grain growth. Preheating and postheating are usually unnecessary and frequently undesirable.

The ferritics are not more prone to welding defects than the austenitics, but their notch sensitivity makes any defects that are present much more detrimental. Special care must be taken to avoid undercutting or poor penetration. Thorough inspection of welds is highly desirable.

Although matching filler metals have been successfully used in a few cases, the most satisfactory results are obtained with low-carbon, austenitic stainless steel or nickel-base alloy fillers. The filler metal selected must be more corrosion resistant than the base metal in the intended service environment. Individual suppliers should be consulted for more specific recommendations on filler metal selection and welding practice.

Cleaning before and after welding is also important for assured weld quality and corrosion resistance. All hydrocarbon residues such as grease, oil, or solvents should be removed before welding to avoid contaminating the weld with carbon. Optimum corrosion resistance is obtained by mechanically or chemically removing the heat tint after welding.

References

1. "The Influence of Chromium on the Mechanical Properties of Plain Chromium Steels," by W.O. Binder and H.R. Spendlow Jr., *Trans. ASM*, 43, 1951, p. 759-772.
2. "Development of Pitting Resistant Fe-Cr-Mo Alloys," by M.A. Streicher, *Corrosion*, 30, 1974, p. 77-91.
3. "The New Fe-Cr-Mo Ferritic Stainless Steels," by R.F. Steigerwald, A.P. Bond, H.J. Dundas, and E.A. Lizlovs, *Corrosion*, 33, 1977, p. 279-295.
4. "Stress Corrosion Cracking Test," by A.W. Dana and W.B. DeJong, *Corrosion*, 12, 1956, p. 309t-310t.
5. "Stainless Steels for Seawater Service," by A.P. Bond, H.J. Dundas, S. Ekerot, and M. Semchysen, presented at Stainless Steel '77, a symposium sponsored by Climax Molybdenum Co. and Amax Nickel, Divisions of Amax Inc., London, 1977.
6. "E-Brite 26-1: Meeting the Challenge," Airco Vacuum Metals, Div. Airco Inc., Berkeley, Calif., 1975.

CREVICE CORROSION PERFORMANCE OF A FERRITIC STAINLESS STEEL DESIGNED FOR SALINE WATER CONDENSER AND HEAT EXCHANGER APPLICATIONS

C. W. Kovach, L. S. Redmerski
Colt Ind, Crucible Inc
P.O. Box 88, Pittsburgh, PA 15230
H. D. Kurtz, Colt Ind, Trent Tube Division
East Troy, Wisconsin 53120

SUMMARY

This paper describes the crevice corrosion performance of a new ferritic stainless steel designed to be economic and to have adequate resistance to saline waters so as to be suitable for electric utility condenser applications where brackish and sea-waters are used for cooling. The alloy is a 26% Cr stabilized steel containing 3% molybdenum and 2.5% nickel known as SC-1 (SEA-CURE*) stainless steel. In accelerated laboratory crevice corrosion tests the steel outperforms Type 316 stainless and is similar or better than AL-6X** stainless depending on the particular test employed. This same relative performance is maintained in long-term laboratory tests using environments designed to simulate high temperature-high chloride cooling waters encountered by condensers and heat exchangers. Long-term field tests in operating electric utility condensers have been very encouraging with no evidence of crevice corrosion initiation in test periods as long as twenty months. This new alloy therefore shows considerable promise for condenser and heat exchanger applications.

---

* Registered Trademark of Colt Industries, Trent Tube Division
** Registered Trademark of Allegheny Ludlum Industries, Inc.

Paper presented at Corrosion/80, Chicago, NACE. Reprinted with the permission of the authors and Crucible, Inc.

INTRODUCTION

Stainless steels offer considerable advantages as heat exchanger materials over nonferrous alloys because of their ability to resist corrosion from polluted water and erosion from high water velocities. However, the traditional stainless grades do not generally have adequate crevice corrosion resistance to highly saline water to make them useful for high performance condenser and heat exchanger applications. The corrosion resistance required for electric utility condensers has now been achieved with a new highly alloyed austenitic stainless, AL-6X, which has compiled an adequate service performance record.[1] However, the high molybdenum and nickel content and processing requirements make such alloys relatively expensive, and development work has therefore continued with ferritic alloys which meet both performance and economic objectives. Such an alloy has been developed in our laboratory and is now undergoing extensive field testing and initial commercial application with promising results. The alloy contains nominally 26% Cr-3% Mo-2.5% Ni-Ti/Cb stabilization and can be melted and processed with conventional steel mill equipment. It is identified by Crucible as SC-1 stainless (SEA-CURE™ Condenser Tubing) and has been assigned the UNS Number S44660. The purpose of this paper is to review this development primarily from the corrosion standpoint and to present field test results obtained from operating electric utility steam condensers. These field test results are evaluated in relation to various accelerated and long-term laboratory crevice corrosion tests in an effort to estimate performance over a broad range of water temperature-chloride conditions.

BACKGROUND

Efforts to develop stainless steels with improved chloride crevice corrosion resistance have generally utilized chromium and molybdenum additions in ferritic structured alloys that do not require high nickel contents so as to achieve alloy cost savings. M. A. Streicher showed that high purity ferritic alloys containing 28% Cr and 4% Mo perform very well in laboratory accelerated crevice corrosion tests,[2] while Pessall and Nurminen further defined similar Cr-Mo composition ranges which were predicted to resist hot seawater as would be encountered in desalination applications.[3] These alloys, however, require expensive vacuum melting procedures and processing is difficult because of embrittling phase formation. While a detailed field service performance data base has not yet been made available for these developmental alloys, initial test results are reported to be promising.[4]

Source: Paper presented at Corrosion/80, Chicago, NACE

Recent work with ferritic steels has centered upon higher interstitial-stabilized alloys. Stainless steels containing about 26% Cr - 3% Mo and Ni ranging from 0-4% were described in 1977 by both Bond[5] and Pinnow[6] who reported promising laboratory and seawater crevice corrosion test results. The significant feature of these steels is that titanium or columbium stabilization is used to allow conventional AOD melting and improve corrosion performance, and the Cr-Mo-Ni alloy balance is designed to minimize embrittling phase formation and maximize toughness. The need which now exists in the development of these steels is to devise methods for estimating service performance from the short-term laboratory tests and current-ongoing field test programs. This would be a relatively straight-forward task if ferritic stainless steels behaved similarly to austenitic steels in accelerated laboratory tests because the service performance record of the standard austenitic grades is reasonably well established. However, the ferritic steels as a group appear to perform better than austenitic steels in some accelerated crevice corrosion tests at an equivalent molybdenum content. This has not been demonstrated in service tests nor would it be anticipated from simple Cr-Mo content considerations.

In this work we have evaluated a number of steels using common accelerated chloride crevice corrosion tests. These data, and data from the literature are then evaluated in terms of alloy composition versus performance. We then conducted relatively long-term laboratory crevice corrosion tests under conditions intended to be similar to those encountered in major potential areas of application. In these tests artificial crevices were formed by using serrated plastic washers. Synthetic seawater salt was used to produce artificial seawater according to ASTM D-1141-52, and temperatures were 70°C and 95°C. Finally, the service performance of the new SC-1 ferritic stainless steel being tested in utility steam surface condensers is summarized and compared to the laboratory crevice corrosion data and field experience with existing commercial alloys.

ACCELERATED LABORATORY CREVICE CORROSION EVALUATIONS

A series of laboratory melted 26% chromium titanium stabilized ferritic stainless steels were used for crevice corrosion evaluation with accelerated laboratory tests. These steels contained variations in molybdenum and nickel content as shown in Table I. Processing of these steels included vacuum induction melting, forging and hot rolling to sheet; and then cold rolling with 50% reduction to a finished sheet thickness of 1.52 mm (0.060 in.). Final annealing was conducted at 1037°C (1900°F) for one hour with a water quench. This anneal produced a single phase ferrite microstructure in all of the alloys.

Crevice corrosion performance was evaluated according to the method of Brigham[7] using 10% $FeCl_3 \cdot 6H_2O$ and pH adjusted to 1 with HCl. Duplicate 120 grit ground specimens were exposed with rubber band-plastic plug crevices at 2.5°C temperature intervals for 24 hours. The test results evaluated in terms of weight loss are given in Table II. Visual and microscopic examination of the specimens showed that light etching at crevice sites began at weight losses of about 0.10 mg/cm^2 and definite crevice corrosion was apparent at weight losses of about 0.50 mg/cm^2. Using weight loss as a crevice corrosion criteria, weight loss is plotted as a function of test temperature and molybdenum in Figure 1. Once crevice corrosion has initiated weight loss increases linearly with test temperature for any given alloy. Extrapolating these curves to zero weight loss provides a measure of the crevice corrosion initiation temperature which ranges from less than 25°C at 1.0% molybdenum to about 46°C at 3.89% molybdenum.

Nickel variations over the range studied do not appear to effect the crevice corrosion initiation temperature since all weight loss plots extrapolate to the same initiation temperature as shown in Figure 2. Although the trend is not consistent, there is some indication that higher nickel content will increase the rate of crevice corrosion penetration. This could account for the slight detrimental nickel effect reported by Bond[8] for similar alloys. This detrimental effect of nickel, and also molbydenum, when present in large amounts is likely due to chi or sigma phase in the microstructure as has been observed in our work and reported by Streicher.[9]

The results of this study and data from other sources have been combined to develop a correlation of 10% ferric chloride crevice initiation temperature with molybdenum content for 26% chromium-titanium stabilized alloys which is shown in Figure 3. This correlation indicates a nearly linear relationship between crevice initiation temperature and molybdenum content ranging from 24°C at 1% molybdenum to 50°C at 4% molybdenum for 26% chromium-titanium stabilized alloys. This curve is described by the relationship: $CCT_f(^oC) = 14 + 9 \times \%\ Mo$ where CCT_f is the crevice corrosion initiation temperature for ferritic stainless steels in the 10% ferric chloride test. Steels with high nickel seem to have slightly inferior performance although the difference may not be great enough to be of consequence in commercial steels.

Since the field service performance of the standard and some newer austenitic stainless steels is well established it is now useful to compare accelerated laboratory crevice corrosion test results for the two groups of alloys. This has been done using the austenitic 18 to 20% chromium 10% ferric chloride crevice corrosion temperature (CCT) correlation with molybdenum developed by Brigham[7] where $CCT(^oC) = -(45 \pm 5) + 11 \times \%\ Mo$; and the correlation is shown as the lower solid curve in Figure 3. This comparison shows that as a group the 26% chromium ferritic steels outperform the chromium-nickel austenitic steels at a given molybdenum content in the 10% ferric chloride test. This relative behavior is also exhibited in some, but not all other accelerated crevice corrosion tests which have been used to compare the two groups of alloys. A compilation from the literature showing accelerated crevice corrosion comparative data for representative alloys is given in Table III. In this comparison the AL-6X steel also ranks below the ferritic alloys in the synthetic seawater potassium ferricyanide test but is ~~about equivalent in the potassium permanganate-sodium chloride~~ and polarization tests. The need then is to determine the predictive nature of these tests by evaluating the ferritic steels in long-term conditions more representative of field service, with comparison being made to the standard austenitic grades.

LONG-TERM LABORATORY CREVICE CORROSION EVALUATIONS

The crevice corrosion performance of SC-1 stainless has been evaluated in laboratory tests of six months duration in comparison to other austenitic stainless steels chosen to represent a range of crevice corrosion resistance in high chloride waters. All test specimens were taken from commercially produced mill annealed sheet having the compositions listed in Table IV. Crevice corrosion was evaluated using coupon immersion tests in distilled water - artificial sea salt (ASTM D-1141-52) solutions up to 95°C. Testing was conducted in 1,000 ml flasks equipped with reflux condensers. Specimens were prepared by 120 grit surface grinding and measured 2.54 x 5.08 x 0.024 cm with a center hole to accept the crevice assembly. Crevices were produced with serrated teflon plastic washers similar in design to that described by Anderson[10] except for only 12 crevice plateaus per washer or 24 per assembled coupon. Duplicate specimens were run for most test conditions providing a total of 48 crevice sites have a bold to crevice area ratio of 15:1.

The crevice coupons were evaluated in terms of percent sites affected and average and maximum depth of crevice penetration, the results are given in Table V. In all environments Type 316 ranks as the poorest performer both in terms of the number of affected sites and depth of penetration. Crevice corrosion developed at 70° and 95°C at seawater chloride concentrations as would be predicted from field experience and the accelerated laboratory tests.

The two new highly alloyed stainless steels' crevice corrosion behavior seems to depend on the severity of the test environment. In artificial seawater at 70°C, which appears to be the least aggressive, the ferritic SC-1 alloy has a lower frequency of affected crevice sites and only about two-thirds the depth of crevice penetration compared to the austenitic alloy. In the same environment at 95°C the two alloys are similar based on depth of crevice penetration. As a group, both of these alloys would appear to display similar performance in these environments and this performance is distinctly superior to that of Type 316. Although neither the ferritic SC-1 or the austenitic alloy are immune to crevice corrosion in these environments, the crevice corrosion rates are low enough to suggest that useful service performance can be anticipated.

FIELD CREVICE CORROSION PERFORMANCE

The field performance of SC-1 stainless, to this time, has been evaluated primarily in relation to its intended application as condenser tubing for electric utility condensers operating on saline waters. This test program has consisted initially of seawater coupon tests and then test installations in operating condensers. A summary of seawater coupon immersion test results is provided in Table VI where SC-1 stainless is compared to Type 316, 26-1S and Cru-6M stainless. In these tests no crevice corrosion has initiated on SC-1 or Cru-6M stainless after 14 months exposure. In the same tests Type 316 stainless developed a maximum 0.40 mm (0.016 inches) and 26-1S 0.06 mm (0.0024 inches) depth of crevice corrosion attack over the same time period.

Currently over twenty condenser tube test installations with SC-1 stainless have been made in the U.S.A. and in Europe and tubes are now being removed for examination as units come down for routine maintenance. A careful record of operating conditions is being maintained and the test tubes are being carefully examined to characterize all performance aspects. Field test location and conditions have varied widely as shown in the summary of all test data given in Table VII. This summary indicates that a 26% Cr-3% Mo ferritic stainless steel will resist waters with salinity in the range of seawater at water temperatures at least up to 40°C even under low velocity conditions where severe fouling can develop. An example of one of the more severe exposure conditions is shown in Figure 4 where the tube was subjected to stagnant conditions and severe deposits and marine fouling including barnacle growth. Barnacles are reported to represent one of the most severe types of natural crevices and yet no crevice corrosion initiated beneath the barnacles or other deposits, either in the base metal or in the welds of the test tubing. The condenser field test results therefore show, to the present time, no instance of pitting or crevice corrosion initiation with SC-1 stainless in utility condenser service on saline cooling water. Furthermore, the absence of corrosion initiation in test periods as long as 20 months strongly indicates that corrosion initiation will not take place regardless of exposure time under these service conditions. The results also indicate that SC-1 stainless is at least equivalent to AL-6X which has established an outstanding performance record in the same service.

In an effort to estimate the long-term crevice corrosion performance of SC-1 stainless in comparison to standard stainless grades and over a broader range of water chloride-temperature conditions, use has been made of Inco(11) spool test data and our own condenser field data for Type 304 and Type 316. These long-term data are plotted as a function of temperature and water chloride content in Figures 5 and 6 where regions in which crevice corrosion may be anticipated are defined. These curves for Type 304 and Type 316 are then shown in Figure 7 on a plot of the SC-1 field and long-term laboratory test data previously discussed. If the crevice corrosion temperature dependence of SC-1 is similar to that of the other steels, then an estimated crevice corrosion performance curve for SC-1 is established to the right of the field test data and below the long-term laboratory data points where some crevice corrosion occurred. While this curve will have to be substantiated by actual field experience it suggests that SC-1 stainless will resist crevice corrosion over a much broader range of water temperature-chloride conditions than does Type 316 stainless, and is essentially immune to crevice corrosion in seawater at ambient and near-ambient temperatures such as are encountered in steam condensers.

PROPERTY REQUIREMENTS FOR CONDENSER APPLICATIONS

In view of the good field crevice corrosion performance being demonstrated by SC-1 stainless it is worthwhile to briefly review other properties of this steel which can relate to condenser and heat exchanger applications. These properties for the most part derive from the ferritic structure which imparts some characteristics which are considerably different than those of the austenitic stainless steels (Table VII). The high yield strength and elastic modulus produces benefits in relation to the strength of rolled tube/tubesheets joints, vibration and fatigue resistance. The thermal properties produce heat transfer and thermal stress advantages compared to the austenitic stainless steels. And, the limited use of nickel and molybdenum in this alloy provides cost advantages over higher alloyed materials.

CONCLUSIONS

The crevice corrosion resistance of 26% chromium stabilized stainless steels increases nearly linearly with molybdenum content over the 1 to 4% Mo range as evaluated with the laboratory accelerated 10% ferric chloride test. Increasing nickel content in these steels over the 0-4% Ni range may slightly reduce the crevice corrosion temperature at high nickel levels, possibly because of the accelerated formation of second phases in the microstructure.

When evaluated with various accelerated laboratory crevice corrosion tests the performance of a 26% Cr-3% Mo ferritic stainless steels is consistently superior to that of Type 316, and depending on the particular test, either superior or equal to AL-6X which is a high-alloy 6% Mo austenitic stainless that has given excellent service performance in electrical utility condensers operating on saline waters. Long-term laboratory crevice corrosion tests indicate that these two alloys will have similar crevice corrosion resistance in various high temperature-high chloride waters.

A ferritic stainless steel containing 26% Cr-3% Mo-2.5% Ni, Crucible SC-1 (UNS S44660), has undergone extensive field testing as condenser tubing in electric utility condensers operating on high chloride waters. The absence of any crevice corrosion after service exposures as long as 20 months is consistent with laboratory data and suggests that this alloy can provide useful service in steam surface condensers.

Source: Paper presented at Corrosion/80, Chicago, NACE

REFERENCES

1. LaQue, F. L., "Qualification of Stainless Steel for OTEC Heat Exchanger Tubes," ANL/OTEC-001, Argonne National Laboratory, January 1979.

2. Streicher, M. A., "Development of Pitting Resistant Fe-Cr-Mo Alloys," Corrosion, Vol. 30, No. 3, pg 77 (1974).

3. Pessall, N. and Nurminen, J. I., "Development of Ferritic Stainless Steels for Use in Desalination Plants," Corrosion, Vol. 30, No. 11, pg 381 (1974).

4. Maurer, J. R., "Use of the New Stainless Alloys for OTEC Heat Exchanger," presented at the 6th OTEC Conference, Washington, D.C., 1979.

5. Bond, A. P., et al, "Stainless Steels for Seawater Service," Proceedings: Stainless '77, London, England, September, 1977.

6. Pinnow, K. E., "Progress in the Development of High Chromium Ferritic Stainless Steels Produced by AOD Refining," Ibid.

7. Brigham, R. J., "Temperature as a Crevice Corrosion Criterion," Corrosion, Vol. 30, No. 11, pg 396 (1974).

8. Bond, A. P., et al, "Corrosion Resistance and Mechanical Properties of Nickel-Bearing Ferritic Stainless Steels," Werkstoffe und Korrosion, Vol. 28, pg 536 (1977).

9. Streicher, M. A., "Microstructure and Some Properties of Fe-28% Cr-4% Mo Alloys," Corrosion, Vol. 30, No. 4, pg 115, (1974).

10. Anderson, D. B., "Statistical Aspects of Crevice Corrosion in Seawater," Galvanic and Pitting Corrosion - Field and Laboratory Studies, ASTM STP 576, American Society for Testing and Materials, pg 231 (1976).

11. Flint, G. N., "Resistance of Stainless Steels to Corrosion in Naturally Occurring Waters," Transactions of the 2nd Spanish Corrosion Congress.

TABLE I - Chemical Composition of Laboratory Ferritic Stainless Steels Used for Evaluation of the Effect of Molybdenum and Nickel Content on Crevice Corrosion in 10% Ferric Chloride

Heat No.	Var.	C	N	Mn	P	S	Si	Ni	Cr	Mo	Al	Ti
3D96	1 Mo	0.027	0.011	0.38	0.016	0.007	0.10	2.46	26.01	1.00	0.050	0.57
3D97	2 Mo	0.023	0.020	0.38	0.016	0.005	0.12	2.54	26.28	1.98	0.057	0.55
3D98	3 Mo	0.024	0.012	0.36	0.015	0.006	0.15	2.52	26.16	2.95	0.060	0.58
3E1	4 Mo	0.027	0.012	0.26	0.013	0.006	0.29	2.50	26.04	3.89	0.072	0.55
3E2	2 Ni	0.029	0.012	0.26	0.014	0.006	0.29	2.05	26.18	2.91	0.071	0.55
3D98	2.5 Ni	0.024	0.012	0.36	0.015	0.006	0.15	2.52	26.16	2.95	0.060	0.58
3E4	3 Ni	0.034	0.012	0.26	0.018	0.007	0.29	2.91	26.21	2.90	0.067	0.58
3E5	3.5 Ni	0.030	0.012	0.28	0.013	0.006	0.30	3.44	26.07	2.98	0.082	0.57

Table II - Ferric Chloride Crevice Corrosion Performance of Ferritic 26% Chromium Titanium Stabilized Stainless Steels with Nickel and Molybdenum Variations

Alloy Code	Composition (Wt. %)			Weight Loss at Indicated Test Temperature (°C)-(mg/cm²)										
	Cr	Ni	Mo	25.0	27.5	30.0	32.5	35.0	37.5	40.0	42.5	45.0	47.5	50.0
3D96	26.01	2.46	1.00	1.01	1.01	3.57	7.20	-	-	-	-	-	-	-
3D97	26.28	2.54	1.98	0.04	0.05	0.05	0.02	1.02	4.54	7.56	11.24	-	-	-
3D98	26.16	2.52	2.95	0.02	0.02	0.04	-	0.02	0.37	1.42	5.35	6.47	-	-
3E1	26.04	2.51	3.89	-	-	-	-	-	0.02	0.19	0.19	0.23	1.21	3.33
3E2	26.18	2.05	2.91	0.02	0.02	0.02	-	0.02	0.50	0.54	1.94	2.48	-	-
3E98	26.16	2.52	2.95	0.02	0.02	0.04	-	0.02	0.37	1.42	5.35	6.47	-	-
3E3	26.21	2.91	2.90	0.03	0.02	0.02	0.02	0.09	0.56	1.86	3.57	3.95	-	-
3E4	26.07	3.44	2.98	0.03	0.03	0.02	-	0.02	0.16	1.35	4.53	5.78	-	-

10% $FeCl_3 \cdot 6H_2O$ - pH 1 adjusted with HCl
Specimens Solution Annealed at 1037°C (1900°F) - WQ
Exposed 24 Hours with Plastic Plug - Rubber Band Crevices
Light Etching Evident at Weight Losses above ∿0.10 mg/cm²
Definite Crevice Attack at Weight Losses above ∿0.50 mg/cm²
Weight Loss Based on Total Specimen Area of ∿13 cm²

TABLE III - Summary Comparing Various Ferritic and Austenitic Stainless Steels in Accelerated Laboratory Tests Used to Evaluate Crevice Corrosion Performance

Alloy	Coupon Immersion Tests			Polarization Tests	
	10% Ferric Chloride Artificial Crevices CCT - °C	Syn. Seawater + 1% $K_3Fe(CN)_6$ Artificial Crevices CCT - °C	1% $KMnO_4$ - 2% NaCl No Crevices CPT - °C	Syn. Seawater Ec vs SCE mv at 90°C	Syn.Seawater with Crevice EC-EP vs SCE mv at 23°C
Type 316	<0	<23	<23	0.00	0.33
Type 317	<0	26	-	-	-
AL-6X	23	35	>90	0.18	0.03
Cru SC-1	40	45	>90	0.30	-
29Cr-4Mo	50	-	>90	0.33	0.05

Source: Paper presented at Corrosion/80, Chicago, NACE

TABLE IV - Grade Identification and Heat Analysis of Commercially Produced Stainless Sheet Used for Laboratory Crevice Corrosion Evaluations

Alloy	C	Mn	P	S	Si	Ni	Cr	Mo	Ti	N
Type 316	0.040	1.68	0.034	0.027	0.66	12.14	17.48	2.53	-	-
Cru 6M	0.016	1.72	0.012	0.004	0.36	24.81	20.43	6.18	-	-
Cru SC-1	0.010	0.20	0.025	0.007	0.20	2.25	25.34	3.10	0.40	0.024

TABLE V - Long Term Laboratory Crevice Corrosion Performance of Stainless Steels in Non-Aerated High Temperature-High Chloride Waters

Wrought Mill Annealed Sheet Specimens with Ground Surfaces.
Serrated Plastic Washer Crevices with 15 to 1 Bold:Crevice Area.
Exposure Time - 6 Months.

Alloy	Environment	pH	T(°C)	Crevice Sites Affected (%)	Average Crevice Depth - mm (in.)	Max Crevice Depth - mm (in.)
Type 316	Artificial Seawater[1]	7	70	25	0.065 (0.0026)	0.104 (0.0041)
Cru 6M	Artificial Seawater	7	70	8	0.065 (0.0026)	0.079 (0.0031)
Cru SC-1	Artificial Seawater	7	70	4	0.041 (0.0016)	0.041 (0.0016)
Type 316	Artificial Seawater	7	95	63	0.076 (0.0030)	0.132 (0.0052)
Cru 6M	Artificial Seawater	7	95	8	0.042 (0.0017)	0.058 (0.0023)
Cru SC-1	Artificial Seawater	7	95	21	0.050 (0.0020)	0.056 (0.0022)

(1) Prepared according to ASTM D-1141-52.

TABLE VI - Natural Seawater Crevice Corrosion Performance of SC-1 Stainless

Alloy	6 Months Exposure			14 Months Exposure		
	Percent of Sites Attacked	Depth of Attack-mm(in.)		Percent of Sites Attacked	Depth of Attack-mm(in.)	
		Range	Average		Range	Average
Type 316	50	0.24-0.91 (0.009-0.036)	0.56 (0.022)	8	0.12-0.40 (0.005-0.016)	0.25 (0.010)
26-1S	0	0	0	2	0.06 (0.0024)	0.06 (0.0024)
Cru-6M	0	0	0	0	0	0
Cru-SC-1	0	0	0	0	0	0

Exposure in Slow Flowing Seawater at Wrightsville Beach, N.C.
Water Velocity 0.6 m/sec (1.0 ft/sec) at 15-25°C (60-77°F)
Serrated Plastic Washer Crevices with 15 to 1 Bold to Crevice Area Ratio

TABLE VII - Crucible SC-1 (SEA-CURE) Condenser Tubing Field Test Performance

Site	Test Conditions			Duration Months	Test Results		
	Velocity m/sec (ft/sec)	Temp. ^oC (^oF)	Chloride 1,000 ppm		Surface Fouling	Pitting Corrosion	Crevice Corrosion
Chesapeake Bay	0-2.3 (0-7.5)	32 (90)	8-10	9	Severe	None	None
Mystic River	1.1-1.8 (3.6-5.9)	10-32 (50-90)	17	10	Moderate	None	None
Tampa Bay	1.4 (4.5)	25-29 (77-85)	13-18	11	Moderate	None	None
Tampa Bay	3.2 (10.8)	18-38 (64-100)	4-40	14	Light	None	None
Nueces Bay	2.0 (6.5)	10-40 (50-104)	16	9	Moderate	None	None
Chesapeake Bay	2.1 (7.0)	18-43 (65-110)	10	13	Light	None	None
Chesapeake Bay	0-2.3 (0-7.5)	32 (90)	8-10	20	Severe	None	None
Cedar Bayou	2.2 (7.1)	31-36 (87-97)	20	12	Light	None	None

Source: Paper presented at Corrosion/80, Chicago, NACE

TABLE VIII - Typical Physical and Mechanical Properties of Crucible SC-1 Stainless Steel

Physical Properties

Density	7.69×10^3 Kg/m^3 (0.278 lb/in^3)
Magnetic Permability	Ferromagnetic
Thermal Expansion (20-260°C, 68-500°F)	9.64×10^{-6} K^{-1} (5.38×10^{-6} in./in./°F)
Thermal Conductivity	16.4 W/mK (114 BTU in./ft^2 hr °F)
Specific Heat (20-100°C, 68-212°F)	0.50 KJ/Kg·K (0.12 BTU/lb °F)

Mechanical Properties

Yield Strength (0.2% Offset)	517 MPa (75,000 psi)
Tensile Strength	620 MPa (90,000 psi)
Elongation	32%
Hardness	95 RB
Modulus of Easticity in Tension	2.15×10^2 GPa (31.2×10^6 psi)

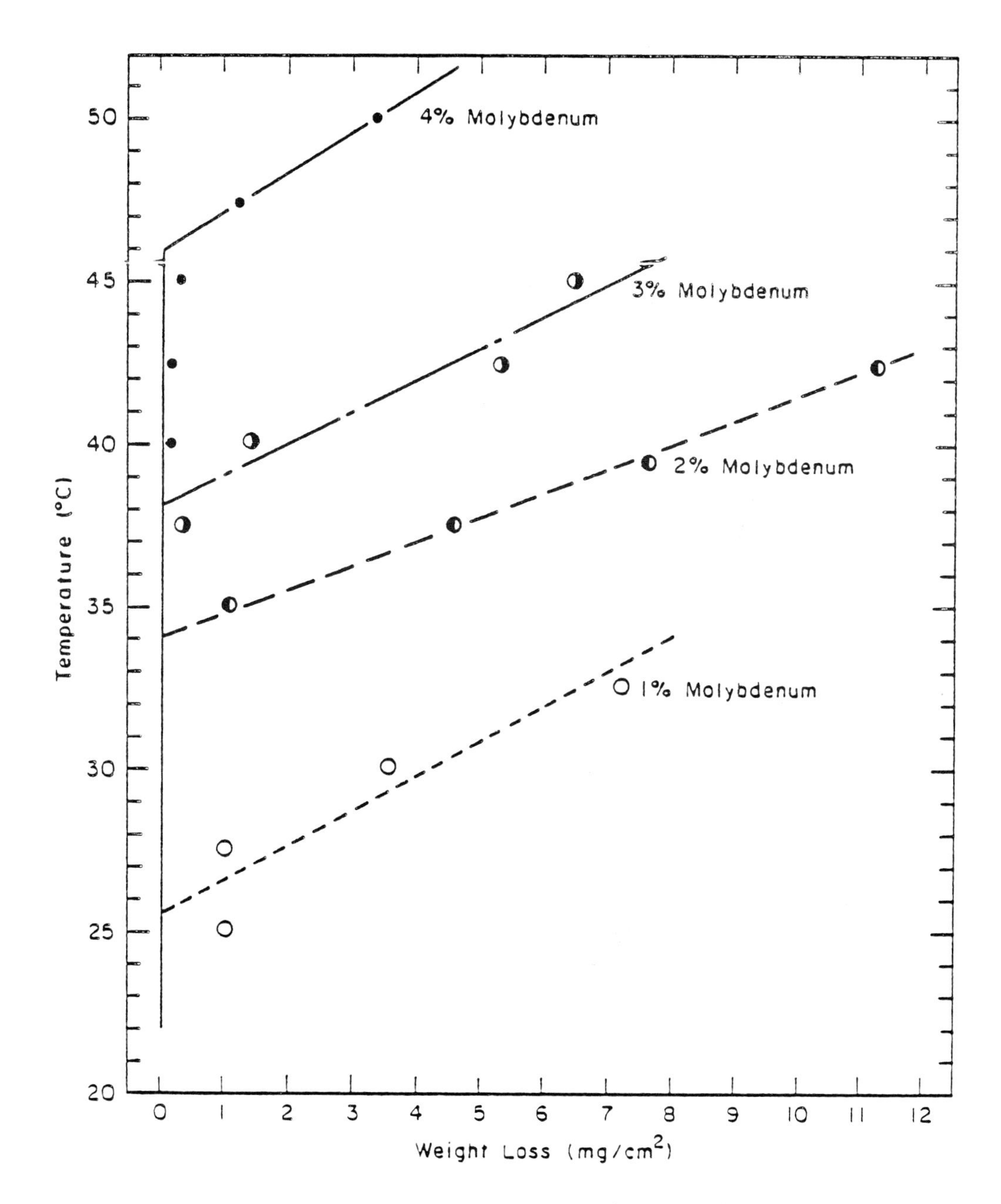

Figure 1. Effect of molybdenum on the ferric chloride crevice corrosion of ferritic 26 chromium-2.5 nickel-titanium stabilized stainless steels.

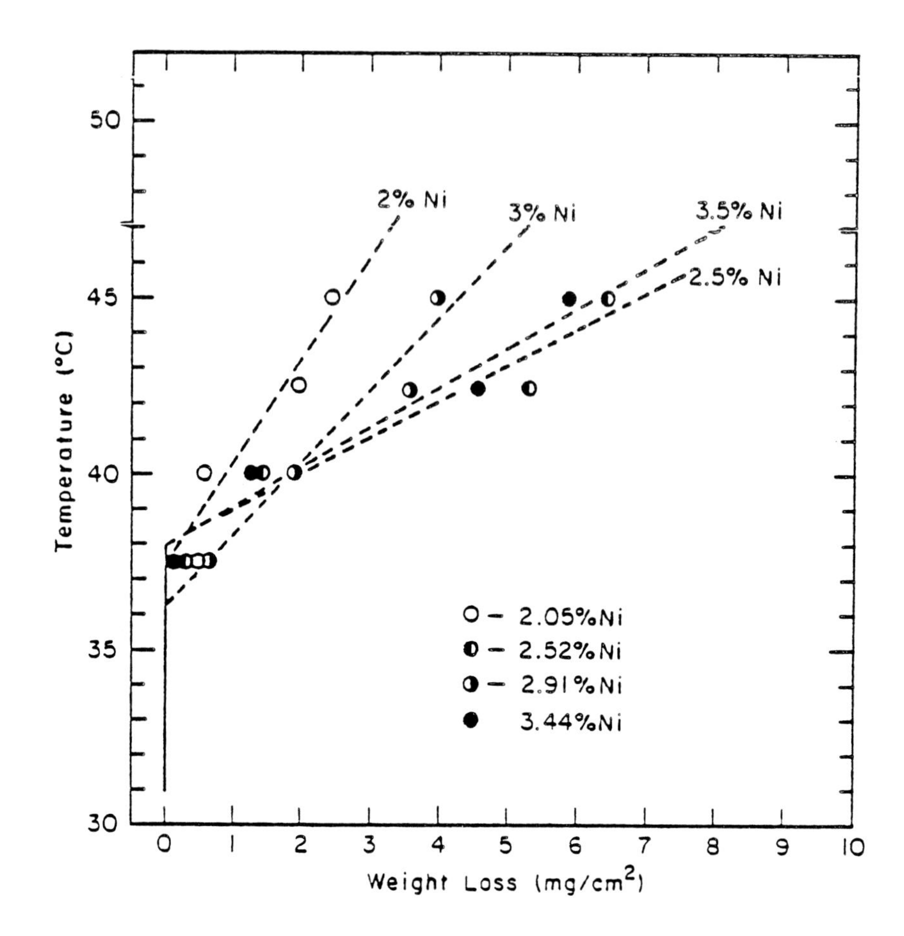

Figure 2. Effect of nickel on the ferric chloride crevice corrosion of ferritic 26 chromium-3 molybdenum-titanium stabilized stainless steels.

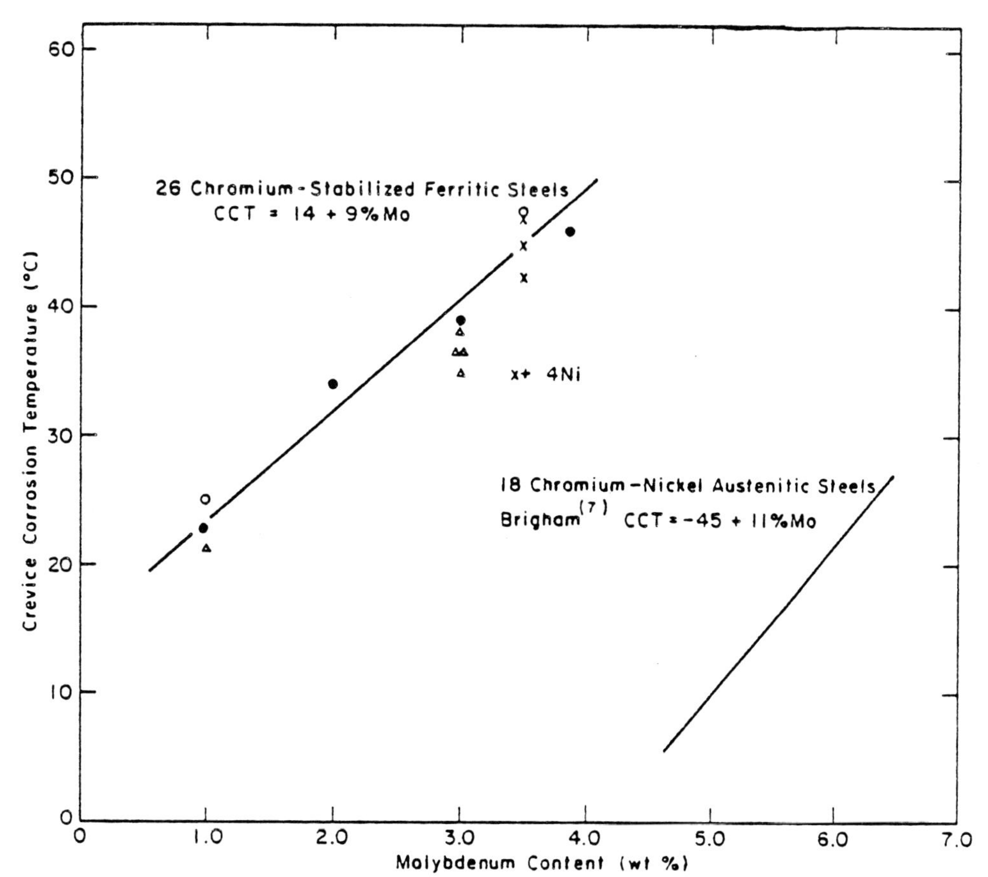

Figure 3. Effect of molybdenum content on the crevice corrosion temperature of ferritic and austenitic stainless steels in the 10% ferric chloride-24 hr crevice corrosion test.

- ● - 26Cr, 2-3Ni, Ti stabilized - this study
- o - 25Cr, 0Ni, Ti stabilized - Bond[4]
- + - 25Cr, 4Ni, Ti stabilized - Bond[4]
- x - 25Cr, 0-4Ni, Ti stabilized - Bond[4]
- Δ - 26Cr, 0-4Ni, Ti stabilized - Pinnow[5]

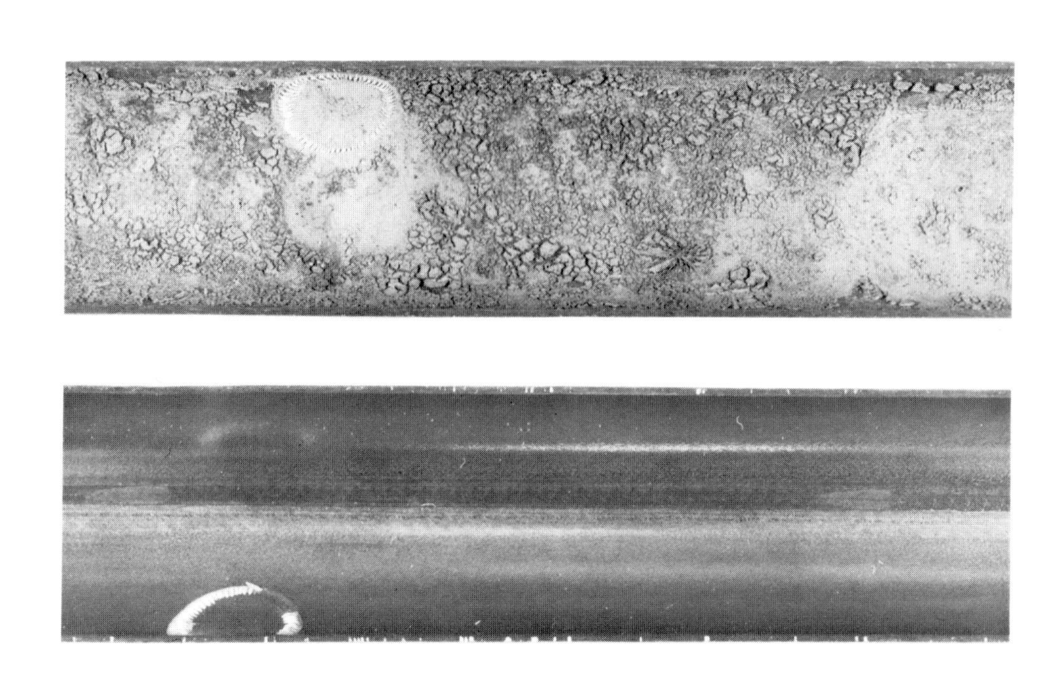

Figure 4. Crucible SC-1 (SEA-CURE™) condenser tubing exposed to brackish condenser cooling water for twenty months under flow and stagnant conditions showing the absence of crevice corrosion under deposits.

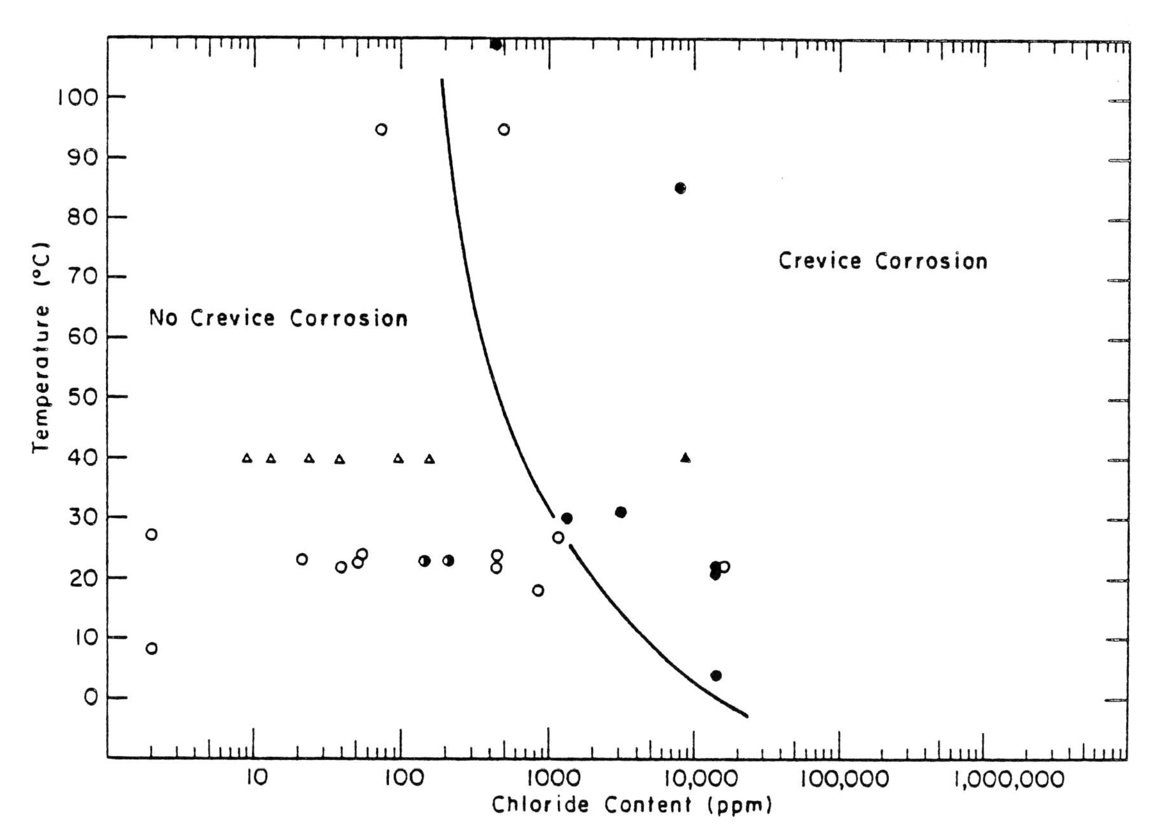

Figure 5. Field crevice corrosion performance of Type 304 stainless steel in high temperature-chloride waters.

▲ - Significant Crevice Corrosion - Condenser Tubing
△ - No Significant Crevice Corrosion - Condenser Tubing
● - Crevice Corrosion >0.04 mm (0.0015 in.)/yr-Inco Spool Test
◑ - Crevice Corrosion <0.04 mm (0.0015 in.)/yr-Inco Spool Test
○ - No Crevice Corrosion-Inco Spool Test

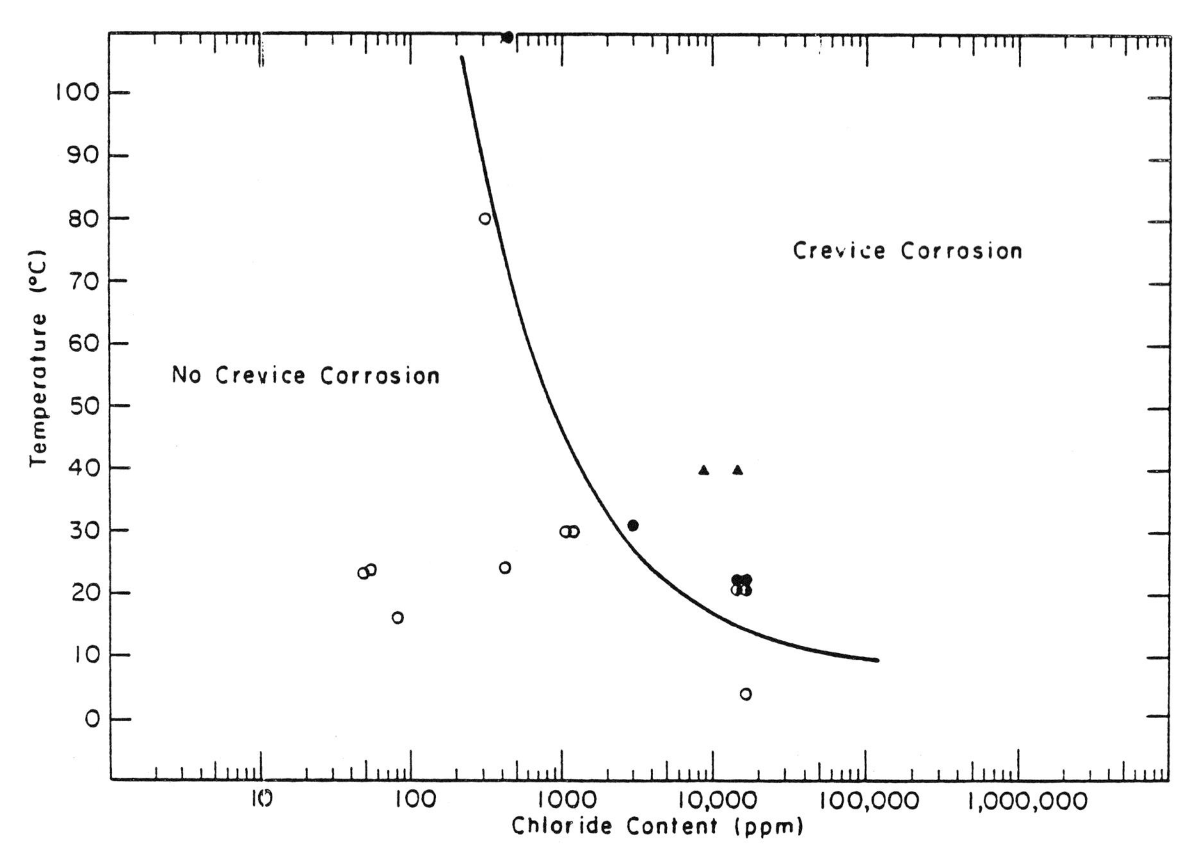

Figure 6. Field crevice corrosion performance of Type 316 stainless steel in high temperature-chloride waters.

- Same as Figure 5.

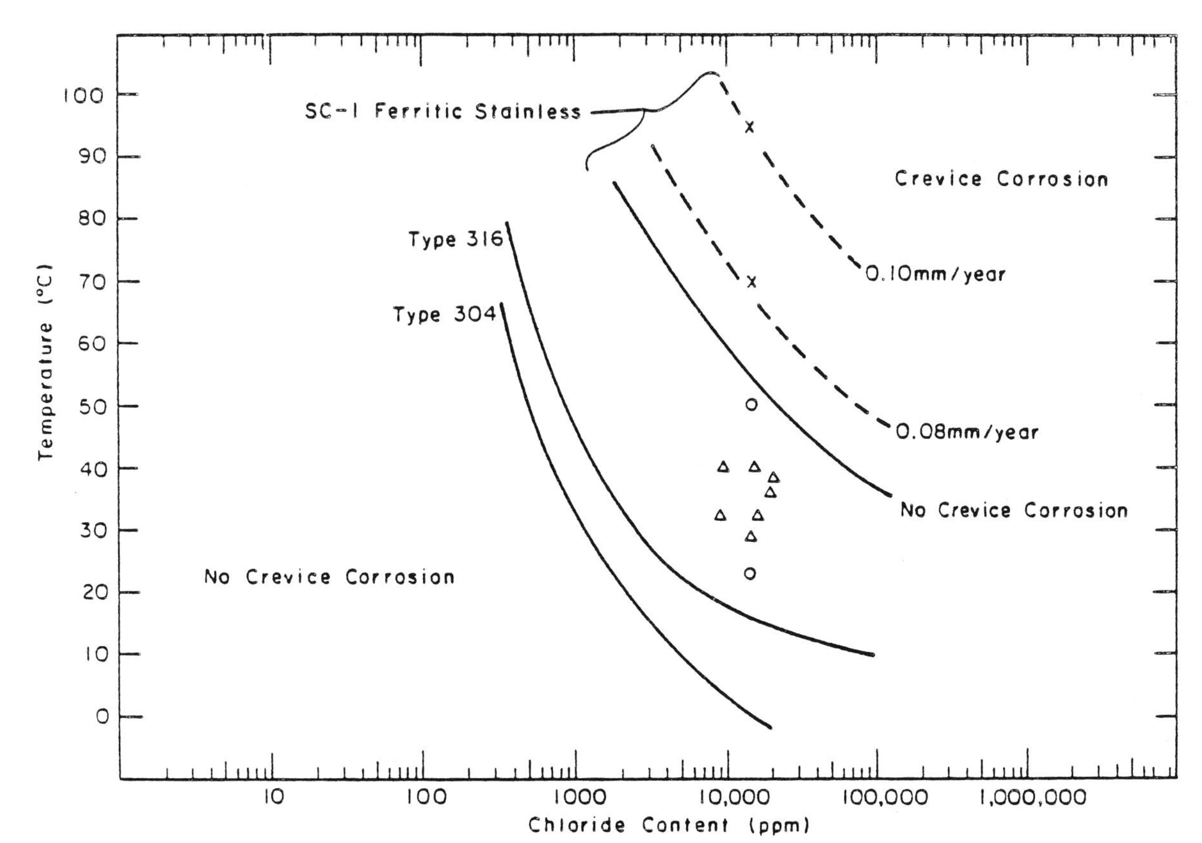

Figure 7. Field crevice corrosion performance of Crucible SC-1 ferritic stainless in high temperature-chloride waters.

Δ - Condenser Tubing
o - Seawater Crevice Tests
x - Laboratory Crevice Tests

Ferritic Stainless Steels

By Remus A. Lula

GENERALLY, FERRITIC STAINLESS STEELS provide about the same corrosion resistance as their austenitic counterparts at less cost — lower amounts of alloying elements are needed. Further, these grades have certain useful corrosion properties in their own right, such as resistance to chloride stress-corrosion cracking, corrosion in oxidizing aqueous media, oxidation at high temperature, and pitting and crevice corrosion in chloride media.

The newer ferritics, especially those with high chromium content, have become possible through vacuum and argon-oxygen decarburization, electron-beam melting, and large-volume vacuum induction melting. Compositions are listed below:

Type 409	0.05 C, 11 Cr, 0.5 Ti
Type 439	0.05 C, 18 Cr, 0.5 Ti
18SR (Armco)	0.05 C, 18 Cr, 0.4 Ti, 0.5 Ni, 1 Si, 2 Al
20-Mo (J&L)	0.02 C, 20 Cr, 0.5 Cb, 1.6 Mo
18-2	0.02 C, 18 Cr, 0.4 Ti, 2 Mo
26-1S	0.03 C, 26 Cr, 0.5 Ti, 1 Mo
E-Brite 26-1 (Airco)	0.002 C, 0.010 N, 26 Cr, 1 Mo
29Cr-4Mo (Du Pont)	0.004 C, 0.01 N, 29 Cr, 4 Mo
29Cr-4Mo-2Ni (Du Pont)	0.004 C, 0.01 N, 29 Cr, 4 Mo, 2 Ni

A discussion of present and potential applications can be simplified by grouping the alloys according to chromium level:

The 12% Cr steels, including type 409 and various proprietary modifications, are low in cost; formability and weldability are good. Usage thickness is limited to approximately 0.150 in. (3.8 mm) max if ductile-to-brittle transition temperature (DBTT) at room temperature or lower is needed. Atmospheric corrosion resistance is adequate for functional uses, but not for decorative applications.

Applications include automobile exhaust equipment, radiator tanks, catalytic reactors, containerization, culverts, dry fertilizer tanks, animal containment housings. Type 409 and its proprietary modifications can replace coated carbon steel and brass. Example: a radiator cap made of a low-residuals-modified type 409. Outstanding

ESCOA Fintube Corp., Pryor, Okla., uses MF-1 (AISI type 409) for fin strips of heat exchanger tubes.

Source: *Metal Progress*, 110(2), July 1976, 24-29, © 1976 American Society for Metals

Corrosion Resistance + Economy

formability permits replacement of brass in this instance.

The 18-20% Cr steels — such as type 439, 18Cr-2Mo, 18SR, and 20-Mo — resist chloride stress-corrosion cracking. Resistance to general and pitting corrosion is approximately equivalent to that of austenitic types 304 and 316; type 18SR has oxidation resistance equivalent to that of austenitic stainless steels. Limitations: sheet cannot exceed approximately 0.125 in. (3.2 mm) thickness if DBTT at room temperature or lower is needed. Embrittlement at 885 F (470 C) and low strength at high temperatures are also problems.

These grades are suitable for equipment exposed to aqueous chloride environments, heat transfer applications, condenser tubing for fresh water powerplants, food handling uses, and water tubing for domestic and industrial buildings. Type 439 can be used where corrosion resistance must be equivalent to that of type 304; in this respect, 18-2 and 20-Mo are closer to type 316. The ferritic steels add the benefit of stress corrosion resistance.

These steels can replace brass, bronze, cupronickels, and austenitic stainless steels. Example: gas-fired hot water tanks made out of welded type 439, selected because of its resistance to stress corrosion. Another example: flexible hose for connecting appliances to gas supply lines. It too is made of type 439, which demonstrates the excellent formability of this alloy in light gages.

26-1 steels are typified by the low-residual version, E-Brite 26-1, and the titanium-stabilized steel, 26-1S. They feature resistance to stress-corrosion cracking, oxidizing corrosion conditions, and chloride solutions. E-Brite 26-1 has good toughness in heavier sections. The 26-1S grade is limited to thicknesses of less than about 0.100 in. (2.5 mm) if DBTT at room temperature or lower is needed. It also embrittles at 885 F (470 C) and has low strength at elevated temperatures.

Applications: condenser tubing, heat exchangers, equipment for handling acids (organic acids in general) in the chemical and petrochemical industries, synthesis of urea, the pulp and paper industry.

29 Cr steels include 29Cr-4Mo and 29Cr-4Mo-2Ni. They resist stress-corrosion cracking, corrosion in acids, and chlorides better than do existing stainless steels. Another feature is excellent pitting and crevice corrosion resistance in seawater. The 29-4-2 type has good corrosion resistance in sulfuric acids, and both alloys are ductile in heavy sec-

Source: *Metal Progress*, 110(2), July 1976, 24-29

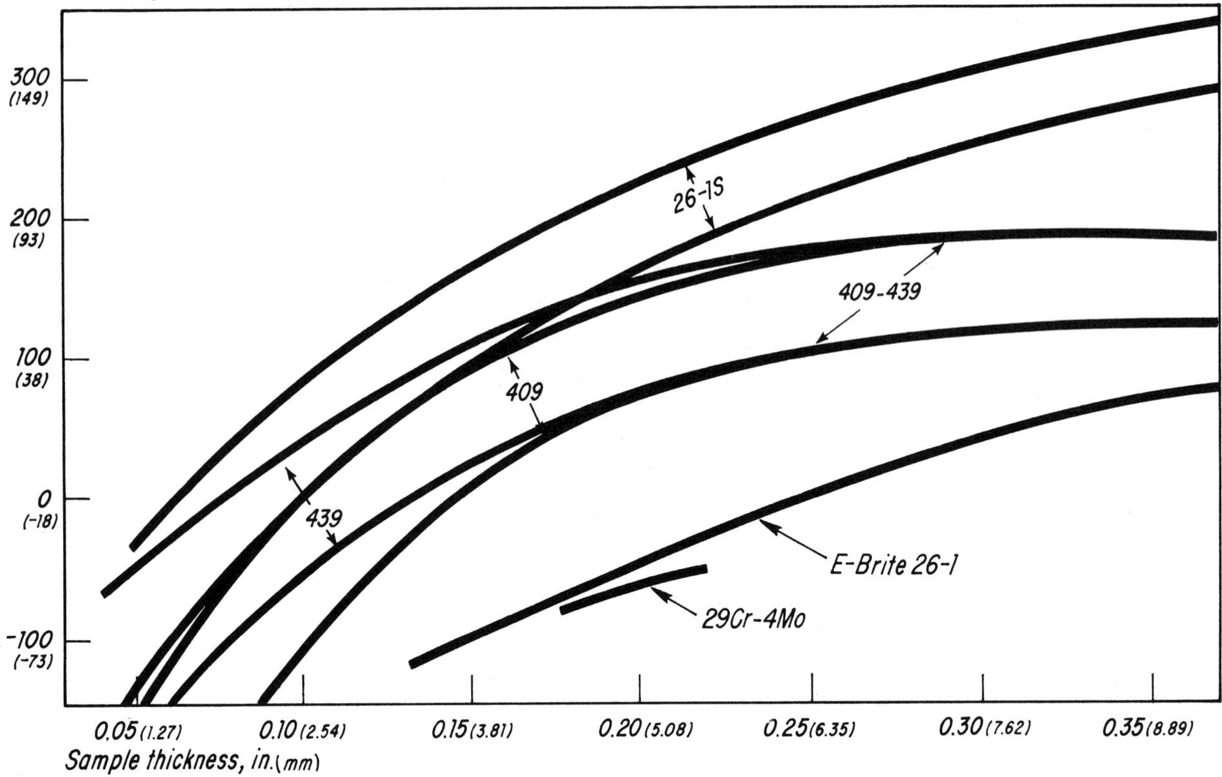

Fig. 1 — Ductile-to-brittle transition temperatures (DBTT) for ferritic stainless steels rise with section thickness. Bands for 26-1S, 409, and 439 indicate data scatter.

tions. Like others, they embrittle at 885 F (470 C) and have low strength at elevated temperatures.

They are used in equipment exposed to seawater, and in chemical industry equipment for highly corrosive acids or pitting media.

A Closer Look at the Alloys

Most of the new ferritics are either available commercially or in an advanced stage of development. Type 409 is an AISI standard grade, while types 439, 26-1S, and E-Brite 26-1 are being standardized by the ASTM.

Structures are completely ferritic at room and high temperature — via additions of titanium or columbium, or by melting to very low levels of carbon and nitrogen, or both. Such microstructures provide ductility and corrosion resistance in weldments.

Molybdenum improves pitting corrosion resistance, while silicon and aluminum add resistance to high-temperature oxidation. Type 18-2 is also available in a resulfurized version for machinability purposes; type 409 comes in several proprietary modifications, primarily developed for forming.

Ferritic steels containing above approximately 13% Cr precipitate alpha prime phase in the 650 to 1000 F (340 to 540 C) range — the maximum effect is at about 885 F (470 C). Precipitation rates vary with chromium content — the higher the chromium content, the faster the rate. Because precipitation hardening lowers room temperature ductility, it must be taken into account in both processing and usage of ferritic stainless steels, especially those with higher chromium content.

Sigma phase also forms above 1050 F (570 C) in types which contain chromium above approximately 20%. Chi phase has also been detected in molybdenum-containing types. The tendency to form sigma, chi, and alpha prime precipitates should always be considered when planning to use these alloys for long-time service at high temperature.

Transition Temperatures: Toughness or ductile-to-brittle transition temperature (DBTT), as measured by impact testing, is important in applying these steels. Impact strength is often low at ambient temperature, and the DBTT is above room temperature. Problems with the DBTT, however, are generally limited to heavier section sizes; in lighter gages, all the ferritic steels behave in a ductile manner at room temperature.

Figure 1 shows the DBTT, as determined with Charpy V-notch specimens. Note that it rises with thickness of the material for the titanium-bearing types (409, 439, and 26-1S) and the low-residual grades (E-Brite 26-1 and 29Cr-4Mo). The other grades are expected to follow a corresponding trend.

Transition temperatures of titanium-bearing steels are below room temperature up to approximately 0.1 to 0.2 in. (2.5 to 5.1 mm) thickness; consequently, these steels are ductile at ambient temperature only in light gages. This thickness dependence of the DBTT is caused primarily by the mechanical constraint associated with increasing thickness, which inhibits through-thickness yielding. Because heavier gages of titanium-bearing steels have transition temperatures above room temperature, they should be used selectively and with caution.

Titanium additions raise transition temperatures of ferritic steels, but are important for producing a ferritic structure, needed especially to prevent intergranular corrosion. Lowering carbon and nitrogen contents increases toughness of ferritic steels, even in titanium-bearing types.

The pronounced scatter band in the impact data of Fig. 1 is attributed to several metallurgical factors. The first is grain size — the smaller the size, the lower the transition temperature. Lacking an austenite-ferrite or martensite phase transformation, these steels can only be grain-refined by cold rolling followed by a recrystallization anneal. For this reason, thinner gage material, which receives more cold reduction, generally has smaller grains and is tougher.

Another factor is the cooling rate from the annealing temperature. Steels lower in chromium, such as type 409, are relatively insensitive to variations in cooling rate. In types containing 18, 26, and 29% Cr, fast cooling improves toughness because even short-time exposure to the alpha prime precipitation range caused by a slower cooling rate will cause some increase in the DBTT.

Low Residuals Important: Stainless steels with very low carbon and nitrogen contents have low ductile-to-brittle transition temperatures, and are thus more ductile, even in heavier sections. Examples are E-Brite 26-1 and 29Cr-4Mo, which are produced by electron-beam hearth refining process and by vacuum induction melting, respectively. Because their DBTT's are below room temperature even in heavier gages (Fig. 1), these low interstitial alloys are preferred in applications requiring ductile, heavy-gage materials.

Argon-oxygen decarburizing (AOD) melting and refining permit effective lowering of carbon and nitrogen to under 200 ppm each without the use of pure raw materials. These values are intermediate between those produced by electric arc melting and vacuum induction melting. Because they lower the DBTT, these levels of carbon and nitrogen prove useful in production of titanium-bearing steels.

Room temperature tensile properties of ferritic steels are shown in Table I. Variations in yield strength are mainly due to the solid-solution strengthening effect of chromium and molybdenum. These steels drop drastically in strength above 1000 F (540 C), indicating that they should be used with caution when load bearing at high temperature is important.

Weldability of ferritic steels is good; tensile properties, including elongation, are very similar to those of the base metal (Table I). The DBTT of weld metal shows only a small increase compared with that of the base metal in titanium-bearing steels, and is unchanged in the titanium-free low-carbon and nitrogen grades, providing proper shielding prevents absorption of interstitial elements.

The high-temperature oxidation resistance of ferritic stainless steels is equivalent to that of austenitic grades, while the thermal expansion is lower. Oxidation resistance can be enhanced by aluminum and silicon additions, as illustrated by 18SR. Experience shows that ferritic stainless steels perform very well in high-temperature applications where applied stresses are low.

Corrosion Resistance of Ferritics

Resistance to stress-corrosion cracking is the most obvious advantage of the ferritic stainless steels. In contrast with austenitic types, they resist chloride and caustic-stress corrosion cracking very well. Results of tests in boiling 42% $MgCl_2$ are shown below, indicating time to fracture:

Types 304, 305, 316, 321, and 347	<10 h
Type 310	<50
Type 439	>200 (no cracks)
26-1S	>200 (no cracks)
E-Brite 26-1	>1200 (no cracks)
29Cr-4Mo	200 (no cracks)

Nickel and copper residuals lower resistance of ferritic steels to stress corrosion, as determined in boiling 42%

Table I — Tensile Properties of Ferritic Stainless Steels

Base Alloy	Yield strength, psi (MPa)	Tensile strength, psi (MPa)	Elongation, %
Type 409	30 000-40 000 (210-280)	60 000-70 000 (410-480)	25-35
Type 439, 18-2	40 000-60 000 (280-410)	65 000-85 000 (450-590)	25-35
26-1S, E-Brite 26-1	50 000-55 000 (340-380)	75 000-80 000 (520-550)	25-35
29Cr-4Mo, 29Cr-4Mo-2Ni	70 000-80 000 (480-550)	75 000-85 000 (520-590)	25-35
Weld Metal			
Type 409	30 000-40 000 (210-280)	50 000-60 000 (340-410)	30-35
Type 439	45 000-55 000 (310-380)	70 000-75 000 (480-520)	25-35
18SR	61 000-64 000 (420-440)	84 000-88 000 (580-610)	19-28
E-Brite 26-1	55 000-60 000 (380-410)	65 000-70 000 (450-480)	10-25
26-1S	60 000-65 000 (410-450)	70 000-80 000 (480-550)	15-25
29Cr-4Mo	70 000-75 000 (480-520)	90 000-95 000 (620-650)	20-27

$MgCl_2$; also, the tolerable amount of these two elements is lower for the higher chromium steels. For instance, the Ni+Cu content of 26-1S has to be kept below approximately 0.30%, but can be slightly higher in type 439.

Boiling $MgCl_2$, however, is a very severe test; and it does not accurately represent conditions encountered in chemical plant exposure. The wick test is more realistic — the stressed sample is heated by an electric current while it is in contact with glass wool immersed in a NaCl solution. Even high-chromium ferritic steels with up to 2% Ni can survive the wick test.

Field experience with ferritic stainless steels confirms their resistance to cracking. A typical example is type 439, as used in domestic hot water tanks. No stress-corrosion failures have been reported in spite of the high chloride content in many areas. Similar results are claimed for type 439 heat exchanger tubing.

Intergranular Corrosion Resistance: Susceptibility of the ferritics to intergranular corrosion is due to chromium depletion, caused by precipitation of chromium carbides and nitrides at grain boundaries. Because of the lower solubility for carbon and nitrogen and higher diffusion rates in ferrite, the thermal cycle for sensitization is, however, different from that for austenitic steels. For this reason, the sensitized zone of welds in ferritic steels is in the weld and adjacent to the weld; in austenitic steels, it is located at some distance from the weld.

To eliminate intergranular corrosion, either reduce carbon to very low levels, or add titanium and columbium to tie up the carbon and nitrogen. In ferritic types, carbon and nitrogen have to be considerably lower than in austenitic steels. For example, total C+N of about 0.01% max is needed to prevent intergranular corrosion, with the exception of 29Cr-4Mo, which can tolerate about 0.025% max. This low level of residuals cannot be consistently attained by electric arc melting or argon-oxygen refining. Vacuum induction or electron beam melting have to be used. Low interstitial residuals are also essential for DBTT control, while titanium or columbium are needed to insure a ferritic structure. Thus, all the steels listed on p. 24 contain the features needed to resist intergranular corrosion.

The amount of titanium or columbium required is usually expressed as ratios to carbon or carbon plus nitrogen. These quantities should be at least stoichiometric, but preferably higher, to satisfy the various testing methods used for checking corrosion. Though steels with a Ti:C ratio of 6:1 resist nitric tests, type 439 requires a Ti:C ratio of 12:1 minimum for long-time exposure in boiling water.

Pitting and Crevice Corrosion: Pitting, an insidious localized type of corrosion occurring in halide media (particularly chlorides), can put installations out of operation in a relatively short time. Resistance to this type of corrosion is affected by such factors as chloride concentration, exposure time, temperature, and oxygen content. In general terms, resistance to pitting increases with chromium content. Molybdenum also plays an important role. It is considered equivalent to several percentages of chromium.

Relative pitting corrosion resistance can be determined in the laboratory by electrochemical techniques or by simple exposure in selected severe pitting solutions. Accelerated tests are more useful in determining comparative properties of various alloys (including those with known field performance) than they are in predicting field performance with any great accuracy. For instance, the minimum critical pitting potentials of types 439 and 304 in various chloride solutions are very close, indicating that they have similar pitting corrosion resistance, a fact confirmed by field experience.

The pitting potential of several ferritic stainlesses in 1 M NaCl at 25 C (70 F) was determined by investigators at Climax Molybdenum Co. Types 304 and 439 are very close (0.22 and 0.28 V [vs SCE]), as are 18-2 (0.43 V) and type 316 (0.48 V). No pitting could be produced in the higher-alloyed ferritic steels such as 26-1S, E-Brite 26-1, and 29Cr-4Mo.

A more severe test for pitting and crevice corrosion consists of exposing rubber-banded samples in a solution of 10% $FeCl_3 \cdot 6H_2O$ for 72 h at room temperature, and gradually at higher temperatures. Per cent weight loss from original is shown below:

	70 F	95 F
Type 409	14.5	—
Type 304	12.8	—
Type 439	10.9	—
18-2	4.67	—
Type 316	1.4	4.08
26-1S	0.030	13.0
E-Brite 26-1	0.001	5.05
29Cr-4Mo	0	0
Hastelloy C	0	0

A material free of pitting or crevice attack at room temperature in this test is considered to have good corrosion resistance in seawater. Any weight loss (expressed in percentage of sample weight) indicates that the material shows crevice corrosion, but the difference between the loss of various materials has only approximate comparative value.

A limited amount of seawater exposure confirms that 29Cr-4Mo is not attacked. Thus, this alloy belongs to a select group consisting of Hastelloy alloy C, Inconel 625, and titanium, which are considered to be immune to corrosion in seawater. Chloride-contaminated waters frequently encountered in industrial conditions can be handled by lower-alloyed materials, including ferritic steels discussed here (other than type 409).

General Corrosion Resistance: The atmospheric corrosion resistance of the ferritic steels discussed equals — and in many instances is superior to — that of type 304, again excepting type 409. Table II shows that the ferritic steels, starting with type 439, have good corrosion resistance to strongly oxidizing acids such as nitric acid. In organic acids, all the listed ferritic steels are superior to type 304. Type 439 and 18-2 are about equivalent to type 316, while the higher-chromium grades are superior to type 316.

The situation is different in reducing media; as illustrated by the 10% H_2SO_4 data, losses for ferritic steels are higher than for type 304 or type 316. A notable exception is 29Cr-4Mo-2Ni, which has been passivated by the 2% Ni addition.

Table II — General Corrosion Resistance of Ferritic Stainless Steels

	Rate, in./mo. (mm/yr)			
Alloy	**65% nitric acid**	**20% acetic acid**	**45% formic acid**	**10% sulfuric acid**
Type 439	0.00261 (0.795)	0.00027 (0.082)	0.03961 (12.07)	19.44236 (5926)
18-2	—	0.0002 (0.061)	0.0340 (10.36)	—
26-1S	0.00033 (0.101)	0.00001 (0.003)	0.00011 (0.034)	7.94675 (2422)
E-Brite 26-1	0.0005 (0.152)	—	—	—
29Cr-4Mo	0.00052 (0.158)	0.00001 (0.003)	0.00013 (0.0396)	Dissolved
29Cr-4Mo-2Ni	—	—	—	0.0004 (0.122)
Type 304	0.00067 (0.204)	0.0250 (7.62)	0.14292 (43.56)	1.36833 (417.1)
Type 316	0.00092 (0.280)	0.00017 (0.052)	0.04333 (13.21)	0.07125 (21.72)

Source: *Metal Progress*, 110(2), July 1976, 24-29

T. Nakazawa,[1] *S. Suzuki,*[1] *T. Sunami,*[1] *and Y. Sogō*[1]

Application of High-Purity Ferritic Stainless Steel Plates to Welded Structures

REFERENCE: Nakazawa, T., Suzuki, S., Sunami, T., and Sogō, Y., **"Application of High-Purity Ferritic Stainless Steel Plates to Welded Structures,"** *Toughness of Ferritic Stainless Steels, ASTM STP 706,* R. A. Lula, Ed., American Society for Testing and Materials, 1980, pp. 99-122.

ABSTRACT: This investigation deals with the practical application of high-purity 18Cr-2Mo-Nb ferritic stainless steel plates 6, 12, and 25 mm thick to welded structures.

In the V-notch Charpy test, the ductile-to-brittle transition temperature is about −30°C. The fracture toughness, K_c, determined by the deep-notch test is 350 kgf/mm$^2 \cdot \sqrt{mm}$ for 12-mm-thick plate at −100°C. In the ESSO test, which has a temperature gradient, the K_c-value obtained is more than 400 kgf/mm$^2 \cdot \sqrt{mm}$ at 0°C.

The toughness of the gas tungsten-arc welded joints using 316 L (C $<$ 0.02 percent) wire deteriorated to some extent, compared with that of the base metal. In the deep-notch test (notch position: fusion boundary, heat-affected zone), however, the K_c-value is more than 200 kgy/mm$^2 \cdot \sqrt{mm}$ at 0°C.

It is generally recognized that the safety of a weldable structural steel can be secured if its weld zone has resistance to brittle fracture initiation and the base metal has brittle fracture arresting properties. It can be concluded, therefore, that in thicknesses up to 12 mm this material can be used for welded structures at service temperature above 0°C because of its excellent fracture toughness.

Since the toughness of 18Cr-2Mo-Nb deteriorates with aging above 300°C or in cold-working in excess of 10 percent, and even further deteriorates in strain-aging, these effects must be taken into account in its fabrication.

KEY WORDS: ferritic stainless steels, deep-notch test, ESSO test with temperature gradient, fracture toughness, welded joint, cold-work, aging, precipitation treatment

In recent years, significant progress has been made on the development and industrial application of high-purity ferritic stainless steels. Strictly speaking, however, many problems remain concerning the practical use of these high-purity steels in thick plate and welded structures. The major prob-

[1]Assistant to manager, engineering metallurgist, manager, and research supervisor, respectively, Yawata Works, Nippon Steel Corp., Kitakyushu, Japan.

lem that has prevented the utilization of plates for welded structures has been the decrease of toughness by welding.

Improvements in toughness which result from increasing the purity of ferritic stainless steels were pointed out by Binder [*1*][2] and Baerlecken et al [*2*]. And the effects of stabilizing elements on toughness were investigated by Semchyshen et al [*3*] and Jarleborg [*4*].

Based on these results, the authors have made a systematic study [*5*] of the relationships between the carbon and nitrogen contents of this steel and the stabilizing elements such as titanium and niobium, because they relate to the toughness and the intergranular corrosion resistance of welded joints. As a result, it was found that the toughness of welded joints is improved by the addition of niobium and is at an optimum level when the Nb/(C + N) is around 10. It was also recognized that the steel becomes immune to intergranular corrosion if its niobium content is 0.02 percent or higher at $C + N \leq 140$ ppm.

Based on these results, we selected the following composition for plates applicable to welded structures: $\leq$ 140 ppm (C + N)−0.14Nb−19Cr−2Mo. This material has a corrosion resistance similar to that of Types 304 or 316 steels.

Successful production trials were carried out with this composition using the 54 431-kg (60 ton) vacuum oxygen decarburization (VOD) process. We succeeded in producing hot-rolled plates up to 12 mm thick. This paper discusses the applicability of the mill-produced plates to welded structures on the basis of the evaluation of the results of Charpy and wide-plate tests, and describes the effects of fabrication and service conditions on the steel properties.

Materials and Experimental Procedure

The materials used in this study were hot-rolled plates of 6, 12, and 25-mm thicknesses produced commercially. Chemical compositions are described in Table 1. Figure 1 illustrates the schematic flow diagram for the preparation of test specimens. The welded joint was prepared by gas tungsten-arc welding (GTAW) techniques using Type 316 L filler metal. Cold-working was induced by tensile strains of 5, 10, and 20 percent. Aging treatment was carried out at 300 ~ 550°C for 10 ~ 1000 h. As for strain-aging, specimens were strained to 10 percent in tension and then aged at 300 ~ 450°C for 300 h. After solid solution treatment (1250°C, 1 h; water-cooled), precipitation treatment was carried out at 400 ~ 1150°C for 2 h.

In order to study the fracture toughness of this steel, both Charpy and wide-plate tests were carried out. Since the brittle fracture characteristics of steels are evaluated normally in two different categories, that is, initiation of

[2]The italic numbers in brackets refer to the list of references appended to this paper.

TABLE 1—*Chemical composition (weight percent).*

C	Si	Mn	P	S	Cr	Mo	Nb	N
0.004	0.07	0.07	0.025	0.007	18.75	1.82	0.14	0.0085

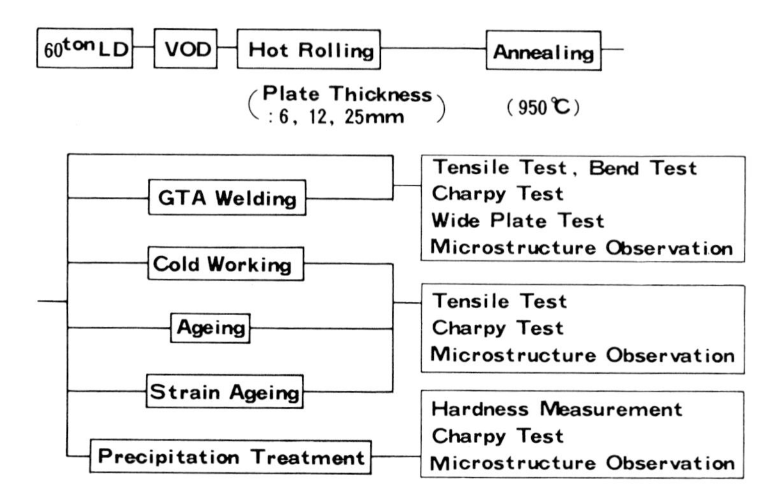

FIG. 1—*Flow diagram for the test procedure.*

brittle crack and propagation of brittle crack, the *V*- and *U*-notch Charpy tests and the deep-notch test [*6*] were conducted for the former and the press-notch Charpy test and the ESSO test with temperature gradient [*7*] for the latter.

Experimental Results and Discussion

Properties of Annealed Material

The microstructure of the steel annealed, shown in Fig. 2, consists of ferritic grains slightly extended in the rolling direction (*a*). Particles of about 0.1 μ in diameter were observed in the grains (*b*). These precipitated particles were identified as those of the *Z*-phrase (Cr_2-Nb_2-N_2) [*8*] and niobium carbonitride (carbon, nitrogen), indicating that both carbon and nitrogen were fixed by niobium.

The tension test results are given in Table 2. The proof stress and the tensile strength of the steel were slightly dependent upon the rolling direction, those in the rolling direction being smaller than those transverse to it. This test result may be attributed to the type of microstructure shown in Fig. 2*a*.

Figure 3 is a view of the bending tests, indicating that the steel is very ductile in this type of test.

Figure 4 provides the Charpy test results for specimens with a 2-mm V-notch (ASTM, E23, Type A), showing sharp ductile-to-brittle transition

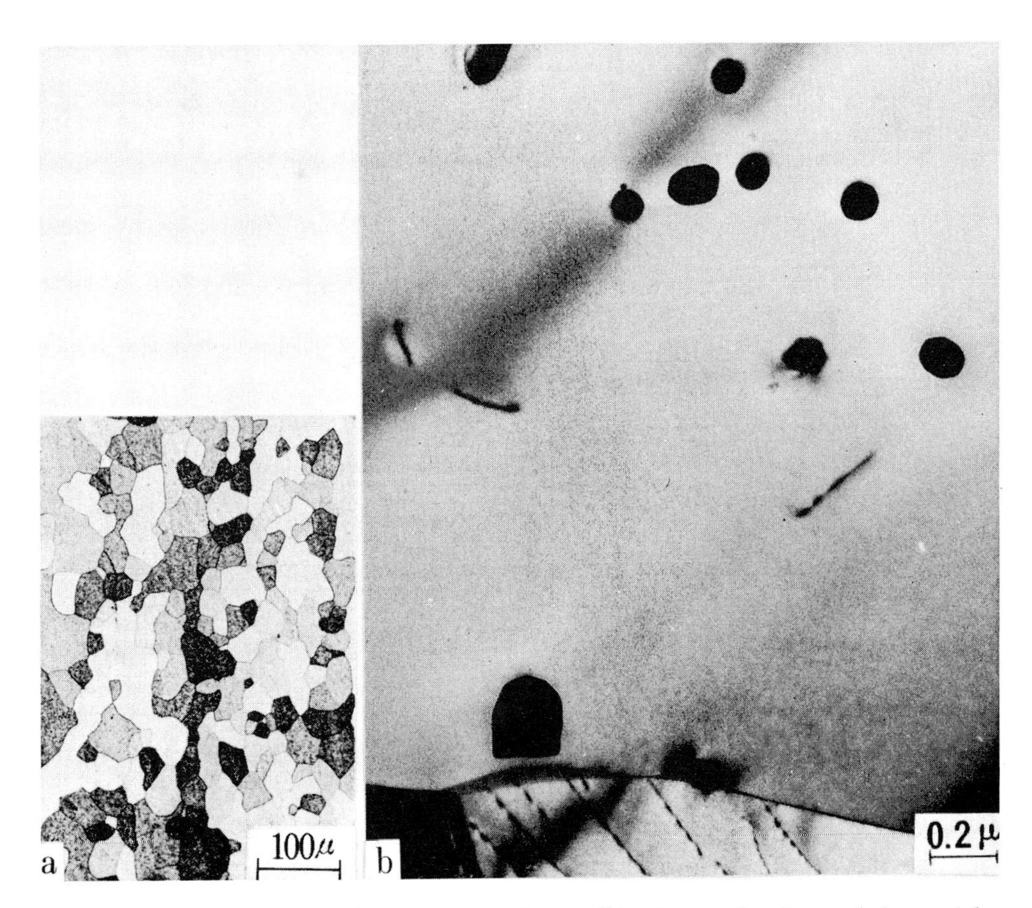

FIG. 2—*Optical* (a) *and transmission electron* (b) *micrographs of annealed material.*

TABLE 2—*Tensile properties of base metal.*

Plate Thickness mm	Orientation (1)	Tensile Tests (2)		
		0.2%Proof Stress kgf/mm²	Tensile Strength kgf/mm²	Elongation %
6	L	32	47	33
	T	36	52	27
12	L	33	47	40
	T	34	50	37
12	L	31	47	42
	T	32	51	41
12	L	32	48	40
	T	34	51	41

Notes: 1. L: Parallel to the rolling direction.
T: Transverse to the rolling direction.
2. Test Piece: JISZ2201. No.13B tension test specimen with 50mm gauge length.

FIG. 3—*View of the bend-tested specimens.*

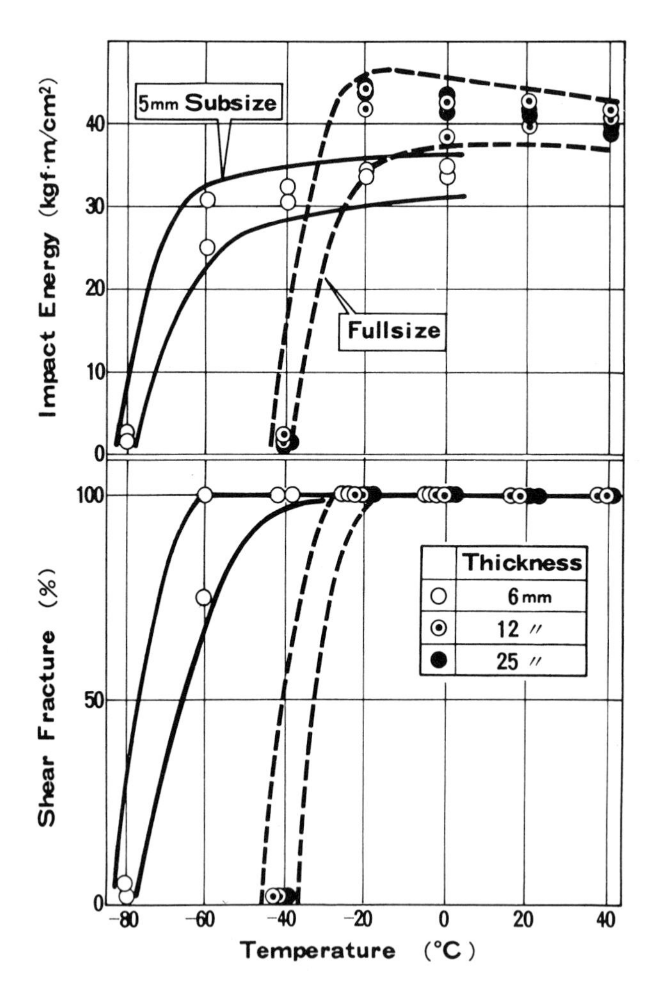

FIG. 4—*Transition curves for V-notch Charpy specimens of base metal.*

temperatures (DBTT's) for all the gages investigated. High impact absorbed energies—for example, 30 kgf·m/cm^2 at −60°C or higher for a plate thickness of 6 mm and 40 kgf·m/cm^2 above −20°C for plate thicknesses of 12 and 25 mm—were obtained. These high absorbed energies are probably due to the high purity of the steel [*1*]. The fractures observed within the brittle region were all of the cleavage type.

Test results for specimens with 2 mm deep and 5 mm deep U-notches are shown in Fig. 5. Both tests show a similar fracture transition temperature, which is about 20°C lower than that of V-notch tests.

Figure 6 shows the results of the deep-notch test [*6*]; the test specimen is also shown in Fig. 6. The fracture toughness value (K_c) is calculated from

$$K_c = \sqrt{\frac{2B}{\pi C} \tan \frac{\pi C}{2B}} \cdot \sigma \cdot \sqrt{\pi C}$$

where

B = half width of specimen,

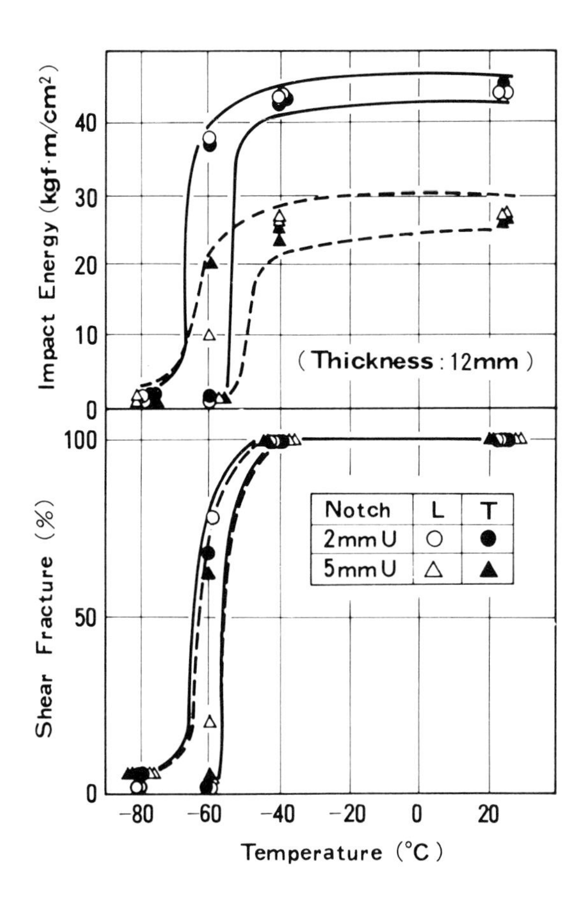

FIG. 5—*Transition curves for full-size U-notch Charpy specimens of base metal* (L: *parallel to the rolling direction;* T: *transverse to the rolling direction*).

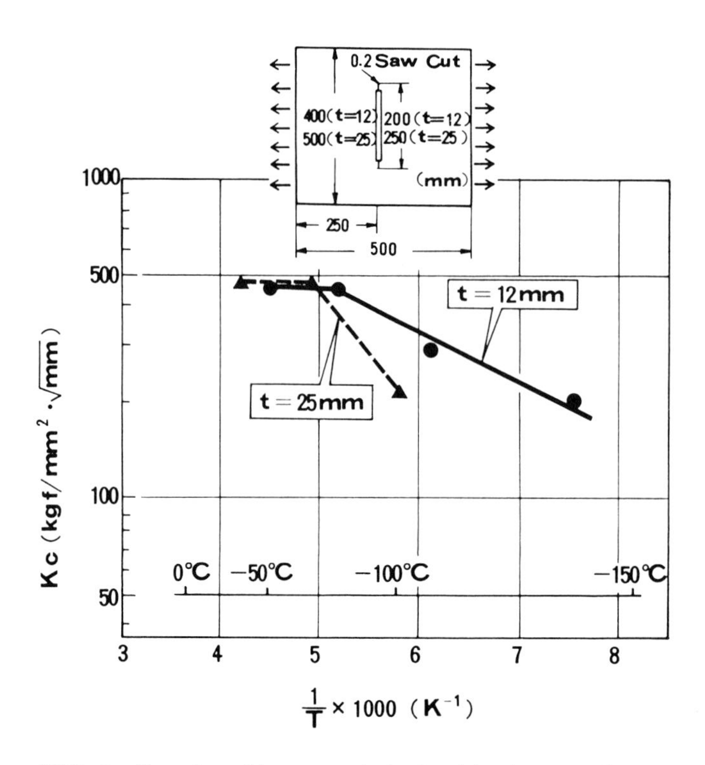

FIG. 6—K_c-*value of base metal obtained by deep-notch test.*

C = half length of notch, and
σ = mean stress.

Low-stress brittle fracture occurred only at temperatures below −80°C for the 12-mm-thick specimens and below −70°C for the 25-mm-thick specimens. The fracture toughness values (K_c values) obtained were 350 kgf/mm$^2 \cdot \sqrt{\text{mm}}$ for 12-mm-thick and 200 kgf/mm$^2 \cdot \sqrt{\text{mm}}$ for 25-mm-thick specimens, respectively, even at −100°C. The foregoing test results indicate that this steel in the annealed condition has a respectable resistance to brittle fracture initiation at 0°C.

Figure 7 shows the results of the press-notch Charpy test in the annealed condition. The DBTT in this test is slightly higher than in the V-notch Charpy test.

The results of the ESSO test with temperature gradient [7] are shown in Fig. 8. The test specimen is also shown in Fig. 8. The K_c-value is calculated from

$$K_c = \sqrt{\frac{2B}{\pi C} \tan \frac{\pi C}{2B}} \cdot \sigma \cdot \sqrt{\pi C}$$

where

B = width of specimen,

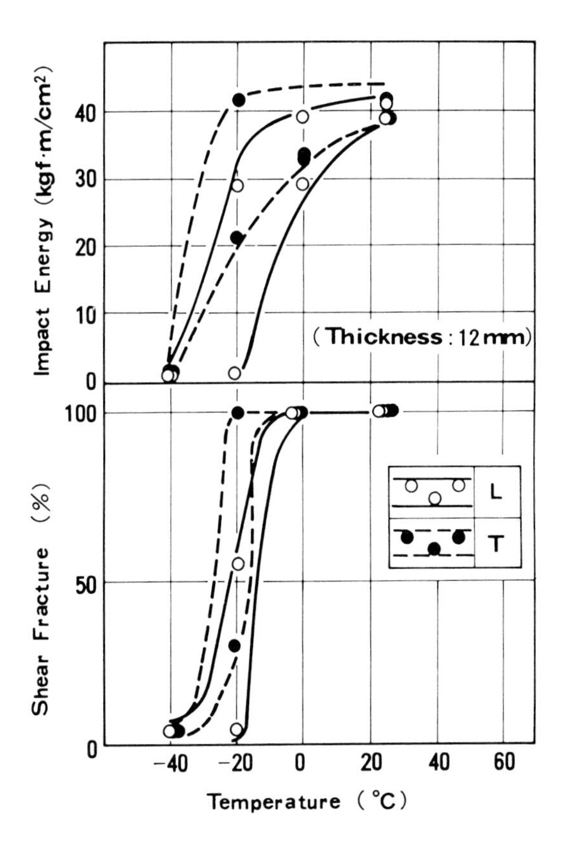

FIG. 7—*Transition curves for full-size press-notch Charpy specimens of base metal* (L: *parallel to the rolling direction;* T: *transverse to the rolling direction*).

C = length of arrested crack, and
σ = applied stress.

The K_c-value somewhat decreased with the increase of plate thickness. The K_c-values were on the same level as the K_c values of JIS G 3126·SLA 33B (ASTM, A537, Class A) steels used for liquefied propane gas storage tanks. Further, when compared with the level required by WES 136 of the Japan Welding Engineering Society for A (arrest) use, which is calculated on the assumption of the existence of a crack 100 mm long and a design stress of 10.5 kgf/mm^2, the K_c value of this steel at 0°C was found to be high. Figure 9 shows a fracture surface of the tested specimens. Brittle fracture propagated with little shear lip on the plates of 12 and 25-mm thicknesses, and arrested at 0°C and −2°C, respectively. From the aforementioned observations, it is supposed that this steel, in the annealed condition, has good crack arresting properties for large brittle cracks at 0°C or higher.

Properties of Welded Joints

The GTAW joints were made using Type 316 L (C < 0.02 percent) welding wire under procedures given in Table 3.

Figure 10 shows transmission electron micrographs (TEM's) of the heat-

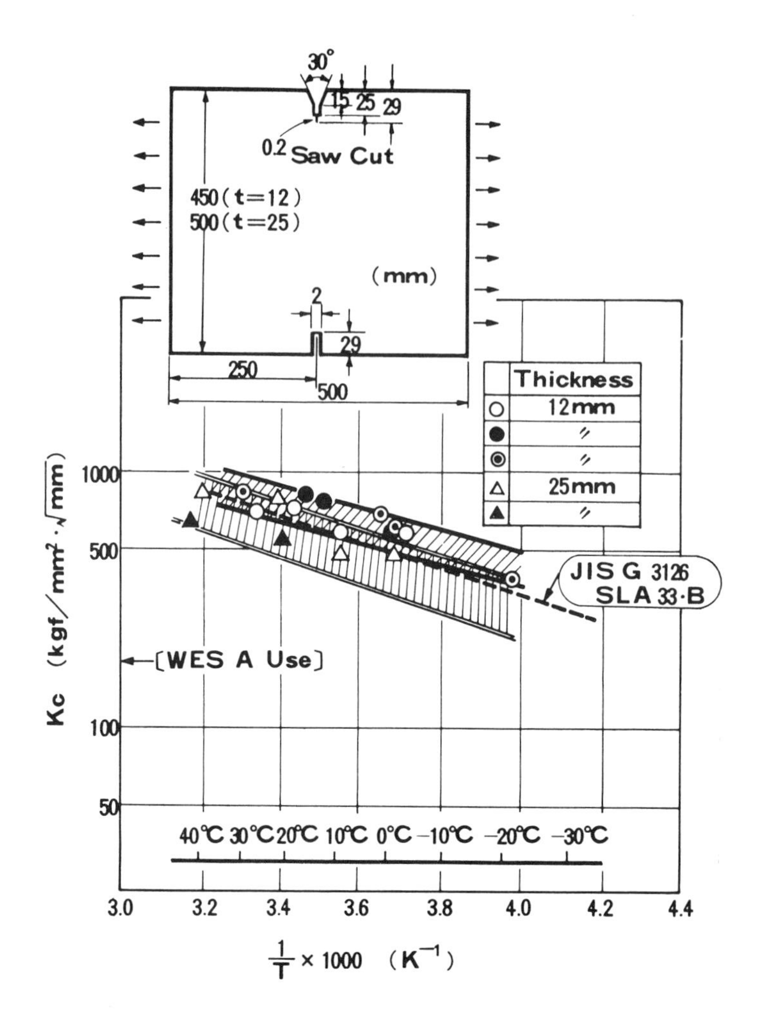

FIG. 8—K_c*-value of base metal obtained by ESSO test with temperature gradient.*

affected zone (HAZ). The particles of about 0.1 μ in diameter of the *Z*-phase and niobium carbonitride previously found in the annealed condition (Fig. 2) are not observed near the fusion boundary (*a*); instead, fine precipitates and an increase of dislocation density can be recognized. However, at a point 5 mm from the fusion boundary (*b*) the precipitate which appeared during annealing still remains and the dislocation density is significantly low. Hence, it can be considered that there occurred some microstructural changes such as an increase in the density of dislocations due to thermal stress and the taking in solution and reprecipitation of carbonitrides due to thermal cycle.

The results of the tension test, bending test, and Charpy test for the welded joints are given in Table 4. The tensile strength of welded joints was equal to that of the base metal. All the bend tests were free of cracks. The Charpy specimens were notched at the weld metal, fusion boundary, and HAZ. The DBTT of all specimens was lower than 0°C.

Figure 11 shows the results of the deep-notch test of the welded joints. As

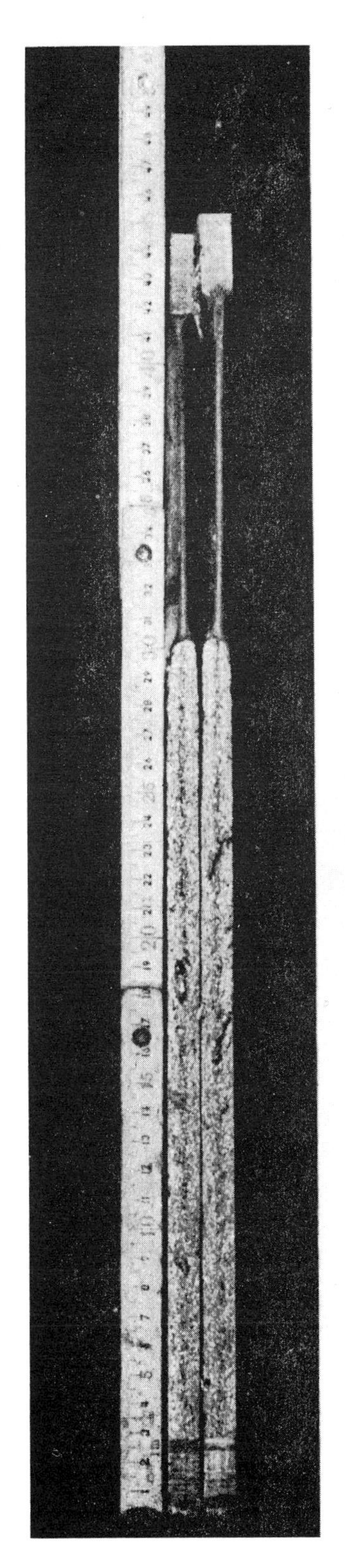

$t = 12mm$, $K_C = 685\ kg/mm^2 \cdot \sqrt{mm}$

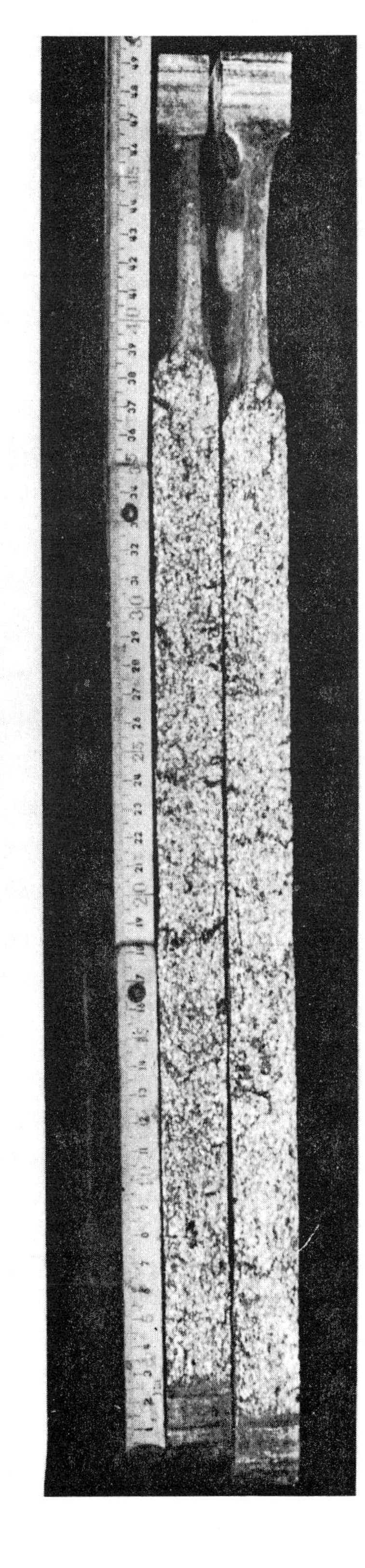

$t = 25mm$, $K_C = 479\ kg/mm^2 \cdot \sqrt{mm}$

FIG. 9—*Fracture appearance of the ESSO-tested specimens.*

TABLE 3—*Welding procedure.*

Plate No.	Thickness (mm)	Welding Condition					Test
		Geometry (mm)	Current (A)	Voltage (V)	Speed (mm/min)	Interpass Temp. (°C)	
1	6	60°, 6, 4	250	13	100	100 ~150	Tensile Bend
		30°, 6, 4	〃	〃	〃	〃	Charpy Deep Notch
2,4	12	45°, 12, 5	250	13	100	100 ~150	Tensile Bend
		30°, 12, 5	〃	〃	〃	〃	Charpy Deep Notch
3	12	45°, 12, 5	150	12	100	100 ~150	Tensile Bend
		30°, 12, 5	〃	〃	〃	〃	Charpy Deep Notch

Welding Wire : 316UL, 1.6φ
Position : Flat

shown in Fig. 11, specimens having the straight weld fusion boundary perpendicular to the plate surface were prepared, with the notch at the fusion boundary and in the HAZ 2 mm away from the fusion boundary. All the test results are plotted in Fig. 11. The K_c-values show a considerable amount of scatter. The K_c-value at the fusion boundary is somewhat higher than that in the HAZ. The K_c-value at 0°C, however, was over 200 $kgf/mm^2 \cdot \sqrt{mm}$. Compared with the K_c-value calculated on an assumption of the existence of a crack 100 mm long and a design stress of 10.5 kgf/mm^2, this value is still adequately high to withstand brittle fracture initiation. The K_c-value thus obtained is substantially lower, however, than that of the base metal shown in Fig. 6. The reduction in toughness is undoubtedly caused by the microstructural changes indicated in Fig. 10, namely, the change in the precipitation morphology of the carbonitrides and the increase in dislocation density.

As a general rule, the application of steels for welded structures is considered safe, on condition that the steel possesses resistance to brittle fracture initiation in welds and is able to arrest brittle fracture in base metal. As shown in Fig. 12, plates of this steel can be used safely for welded structures at 0°C or higher.

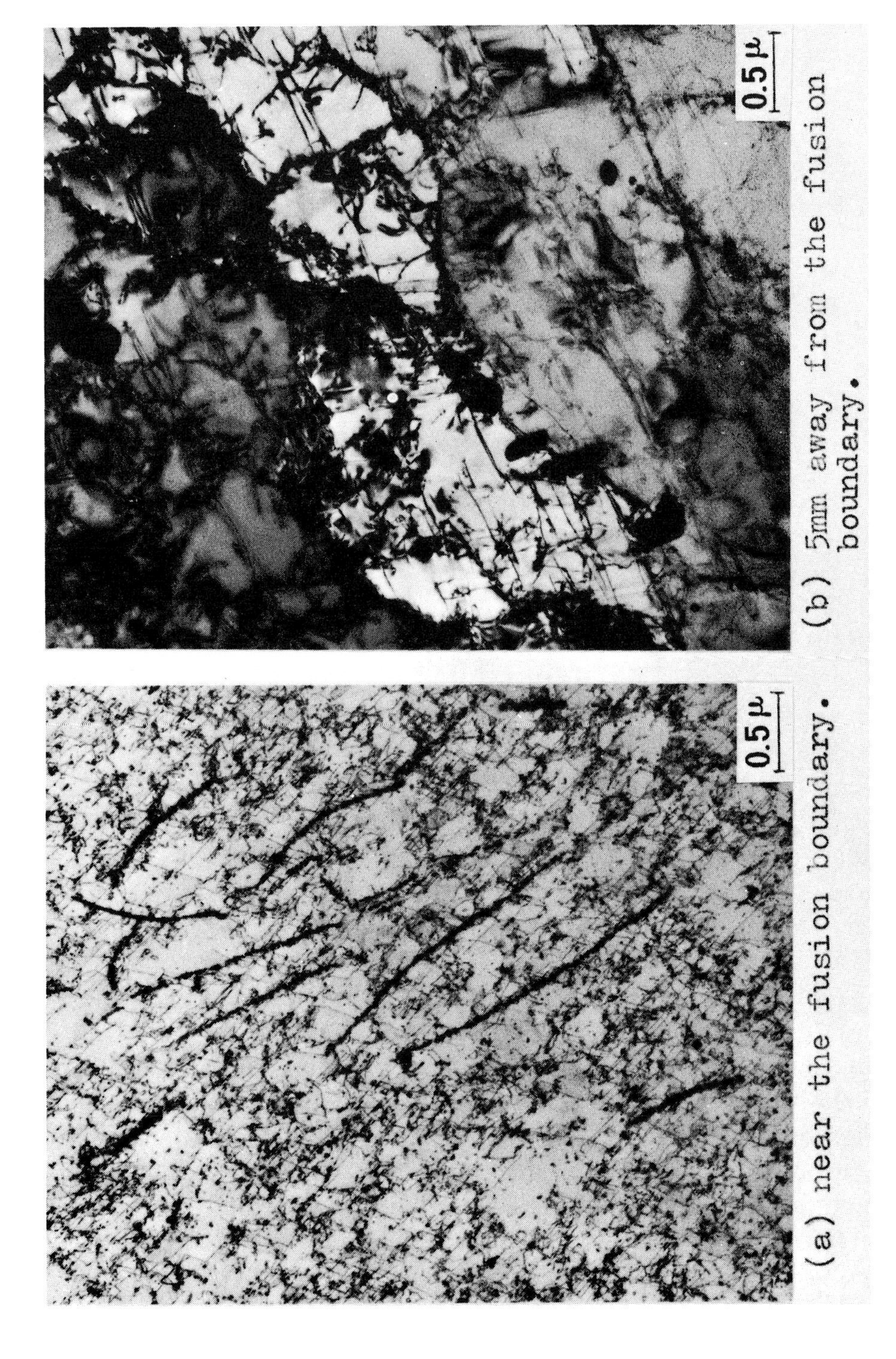

FIG. 10—*Transmission electron micrographs of the heat-affected zone in butt welded joint.*

TABLE 4—*Mechanical properties of welded joints.*

Plate No.	Thickness (mm)	Tensile Test			Bend Test	Charpy Test, Location of Notch (*)					
						D		A		B	
		T.S. (kgf/mm²)	El (%)	Break Position	(Roller Bend Bend Radius : 2t)	(°C) vTrE	(kgf·m/cm²) vE_0	(°C) vTrE	(kgf·m/cm²) vE_0	(°C) vTrE	(kgf·m/cm²) vE_0
1	6	49.7	20	Base Metal	Good	<-20	20	−60	28	−70	25
2	12	48.9	31	〃	〃	〃	20	−17	21	−20	25
3	12	50.4	30	〃	〃	〃	21	−10	22	−15	35
4	12	51.4	31	〃	〃	〃	19	−16	19	−20	23

(*) Location of Notch

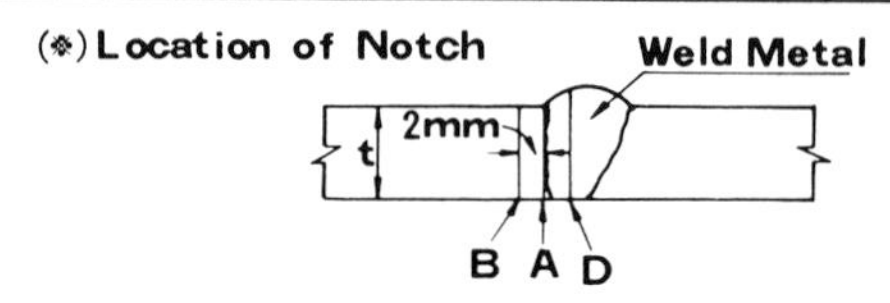

Effects of Cold-Working and Heat Treatment on Steel Properties

Effect of Cold-Working—The results of the tension test and the V-notch Charpy test on cold-worked specimens are summarized in Fig. 13. With the increase in the amount of tensile strain, the tensile strength and especially the proof stress increase while the ductility decreases.

The decrease of toughness is first observed at about a 10 percent strain. Close to about 20 percent strain, the DBTT rises to +20°C.

Effect of Aging—Shown in Fig. 14 is the behavior of change in tensile properties caused by aging. Practically no change was observed up to 400°C. The increase in strength and the decrease in ductility are shown when aging in the 450 to 500°C temperature range, with the maximum increase occurring at 475°C, where the tensile strength after aging for 1000 h attained 80 kgf/mm^2. Aging at 550°C, however, produced no changes in mechanical properties.

Figure 15 shows the effect of aging on the toughness of the steel. There is no change up to 350°C, even in 1000 h. At 400°C there is no change in toughness in 300 h, but in 1000 h the toughness shows some deterioration. The embrittlement is quite severe in the 450 to 500°C temperature range, with 475°C exposure being the most severe. In 1000 h at this temperature the DBTT is +180°C, an indication of typical 475°C embrittlement. Little change occurred at 550°C for 300 h, but slight deterioration was recognized at 1000 h.

Comparing the isostrength and the isotoughness curves in Figs. 14 and 15, it can be concluded that the effect of aging on tensile properties is quite similar to that on toughness.

The 475°C embrittlement of high-chromium steels has been widely studied [*9,10*] and is said to be caused by the precipitation of the chromium-rich α'-phase. A similar embrittlement was investigated by Grobner [*11*] on lower

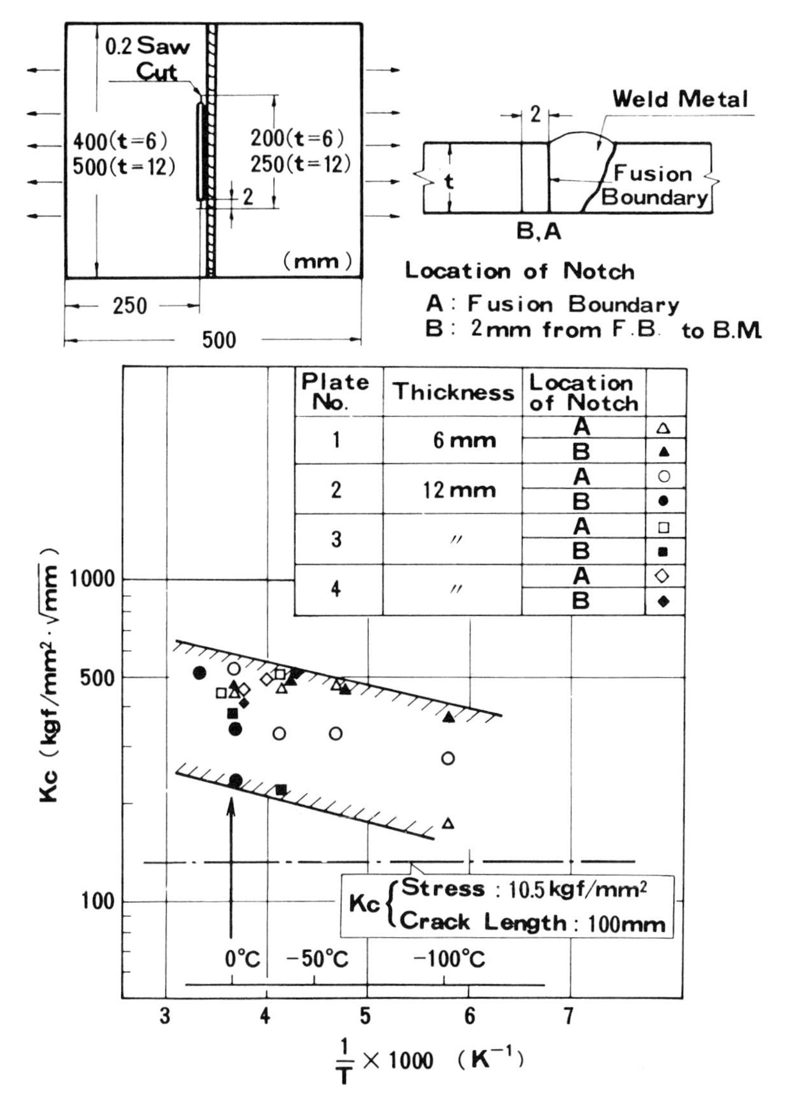

FIG. 11—K_c-*value of welded joint obtained by Deep-Notch test.*

chromium containing steels such as the 18Cr and 18Cr-2Mo steels. According to Grobner, in the case of low-chromium steels the precipitation of the α'-phase takes place on dislocations rather than by spinodal decomposition. It was also reported by Grobner that embrittlement can also be casued by precipitation of carbonitrides on dislocations at 538°C.

Figure 16 shows the TEM's of specimens aged at 475 and 550°C. Compared with the annealed condition (Fig. 2), the TEM's show no definite change in their microstructure. The precipitation on the dislocations reported by Grobner or fine precipitation in the matrix are not observed. Such differences resulted from the following facts. These specimens were annealed at 950°C, a temperature at which sufficient recrystallization takes place such that they had a considerably lower dislocation density which

Source: ASTM STP 706, 1980, 99-122

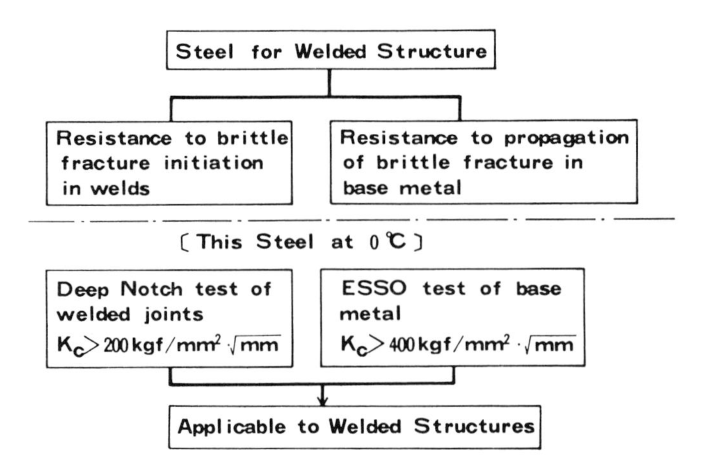

FIG. 12—*Application of steel plates for welded structures.*

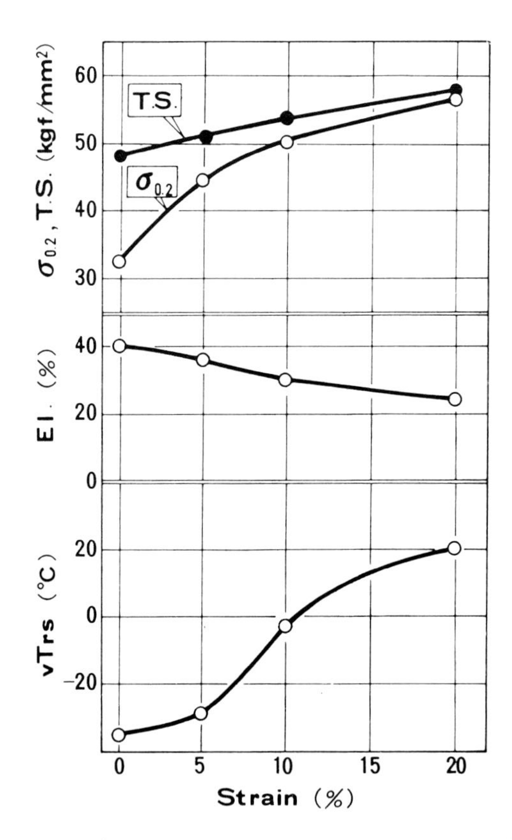

FIG. 13—*Effects of cold-working and room-temperature aging on the tensile and V-notch Charpy impact properties of base metal* (vTrs: *V-notch Charpy 50 percent shear fracture transition temperature*).

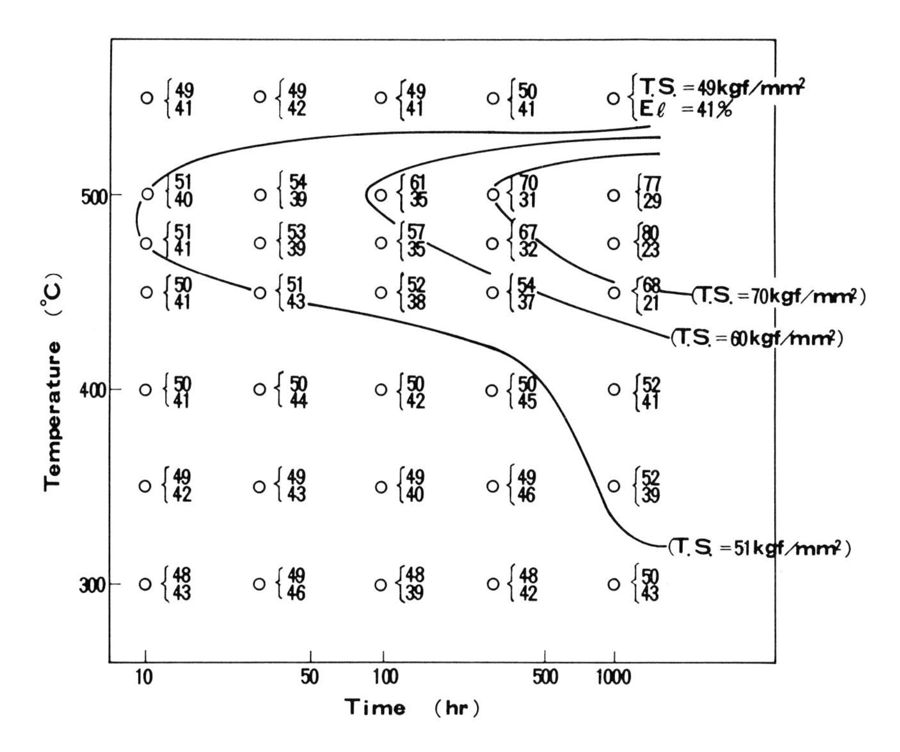

FIG. 14—*Effects of aging time and temperature on the tensile properties of base metal.*

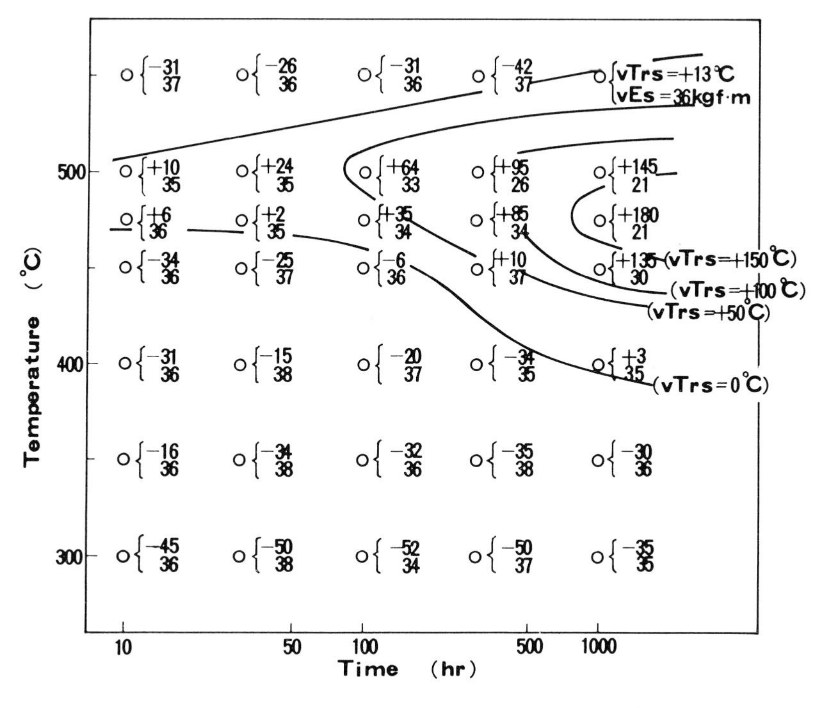

FIG. 15—*Effects of aging time and temperature on the V-notch Charpy impact properties of base metal.*

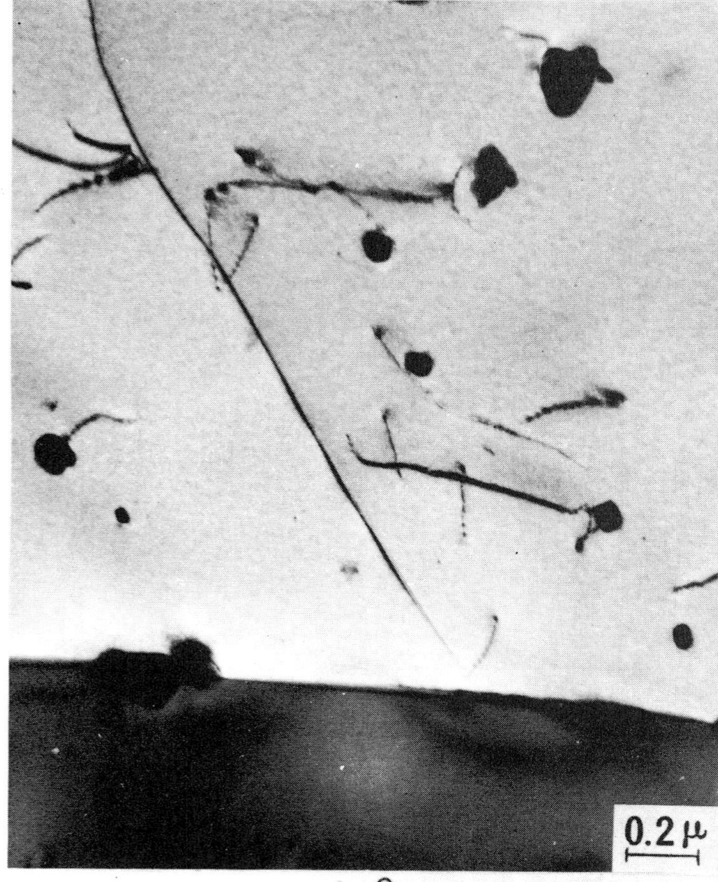

(a) aged at 475°C for 1000hr

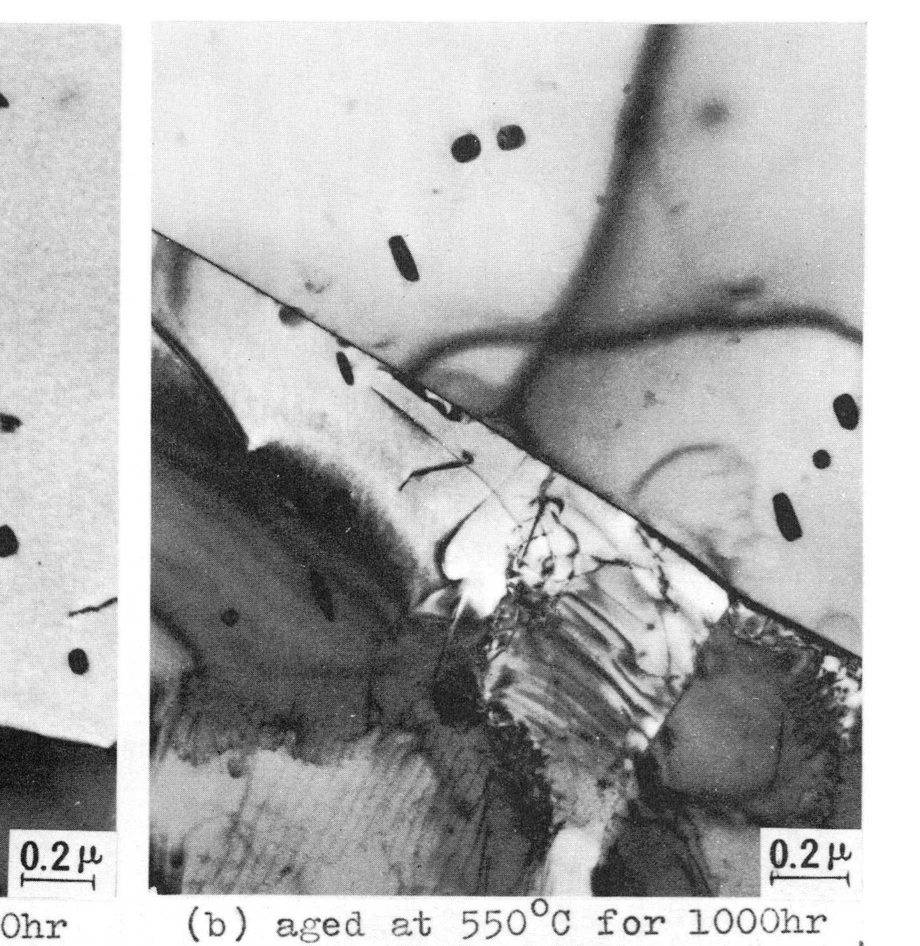

(b) aged at 550°C for 1000hr

FIG. 16—*Transmission electron micrographs of aged specimens.*

would serve as precipitate nuclei. The carbon and nitrogen which are said to promote the precipitation of the α'-phase [11] were fixed by niobium. Also, we could not detect clearly the fine α'-phase by the transmission electron microscope because the difference in the lattice parameter between the α-phase and the α'-phase is so small and the scattering factor of iron and that of chromium are so similar that the contrast between the two phases is very low. In other words, since this steel, being a niobium-stabilized steel, it has a very low initial dislocation density. In addition, the α'-phase has not developed to a size detectable by the transmission electron microscope at the aging time of 1000 h.

Effect of Strain Aging—The test results of strain-aged specimens are given in Fig. 17 in comparison with the data obtained by aging alone. The increase in tensile strength and 0.2 percent proof strength produced by strain aging is higher than that produced by aging alone. The difference becomes more pronounced at higher aging temperatures. The toughness of the steel is decreased by strain aging, and the extent of deterioration is greater than that produced by aging alone. The higher the aging temperature, the more pro-

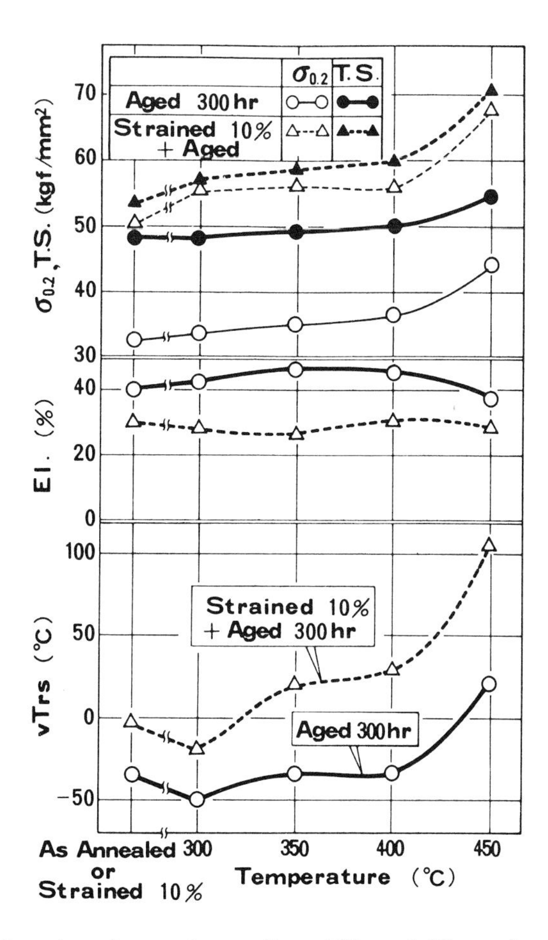

FIG. 17—*Effect of strain-aging on the tensile and V-notch Charpy impact properties of base metal.*

nounced this trend became. Thus, this steel is susceptible to strain-aging, a phenomenon explained by Grobner [*11*], who postulated that α' precipitates on dislocations.

Our data are, however, different from the results obtained by Grobner and Steigerwald [*12*], since they reported that up to 30 percent cold-work does not accelerate the 475°C embrittlement of 18Cr-2Mo steels stabilized with either niobium or titanium.

Effect of Precipitation Treatment—The results of hardness measurement and Charpy test of precipitation-treated specimens are shown in Fig. 18. The hardness increases substantially to about Hv 250 at temperatures between 550 and 600°C. The DBTT shows two peaks, at 550 and 950°C, while between these two it shows the best DBTT value of −35°C after a 750°C aging. Thus, at 550°C both hardening and embrittlement take place, while at 950°C the DBTT increases and the hardness decreases to a level lower than that of the solid solution state of the steel.

The effect of the heating time was studied at these three characteristic temperatures, that is, 550°C, at which both hardening and embrittlement occur; 750°C, at which toughness alone improves; and 950°C, at which only embrittlement occurs. Figure 19 shows the results of heating the specimens up to 100 h. Hardness slightly rose with time at 550°C, but toughness underwent little change. At 750°C, the change in hardness with holding time was negligible, while the toughness decreased substantially with longer time. At 950°C the hardness showed only a little change, but toughness deteriorated further with time. The analysis of electrolytically extracted residues was conducted for these test specimens. The results are shown in Fig. 20. Precipita-

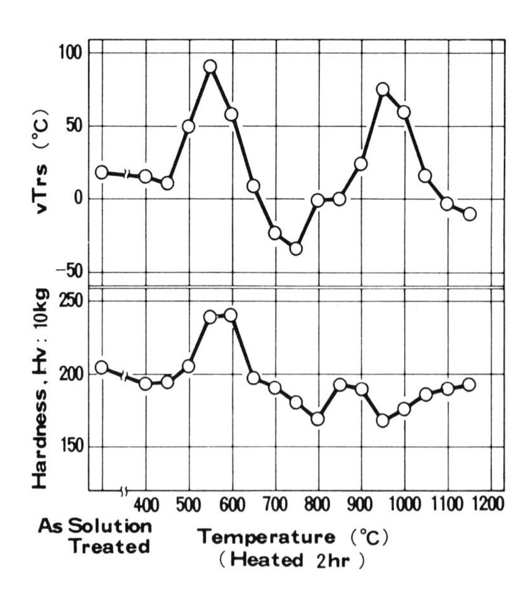

FIG. 18—*Effect of precipitation treatment on the V-notch Charpy impact property and hardness of base metal.*

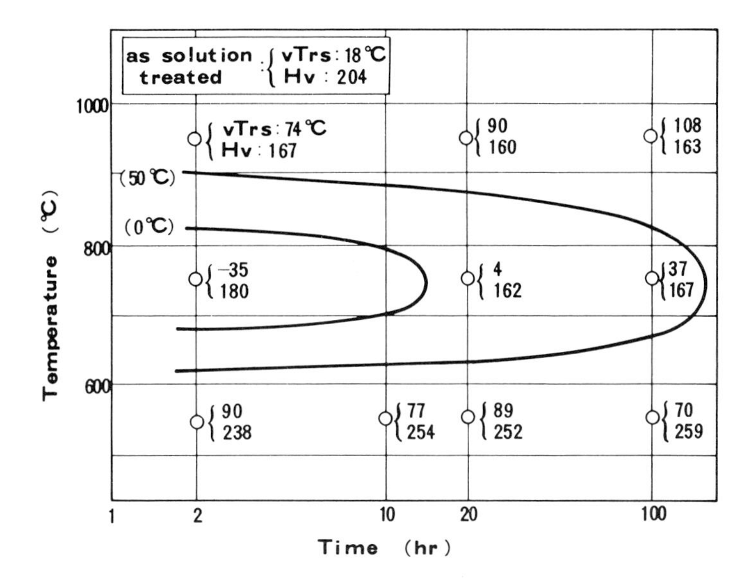

FIG. 19—*Effects of precipitation treatment on the V-notch Charpy impact property and hardness of base metal.*

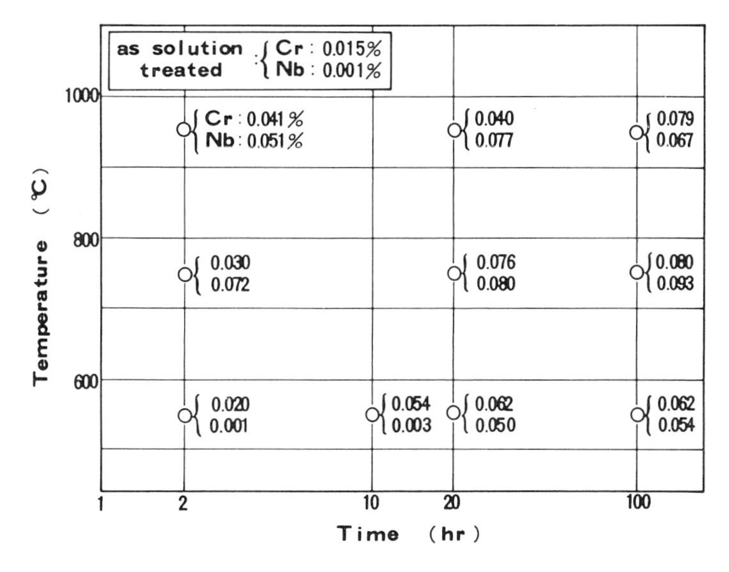

FIG. 20—*Effects of precipitation treatment on the amount of chromium and niobium in the electrolitically extracted residues base metal.*

tion of elements, mostly chromium, took place at the shorter time at 550°C, but in the longer times niobium began to precipitate. At 750 and 950°C, on the other hand, both niobium and chromium precipitated even in a short time. The *Z*-phase (Cr_2-Nb_2-N_2) and niobium carbonitride were identified by X-ray diffraction analysis of the residues precipitated at 750 and 950°C.

Figure 21 shows the TEM's of the steel under different heating conditions. Highly dense and homogeneous precipitation of very fine particles was seen

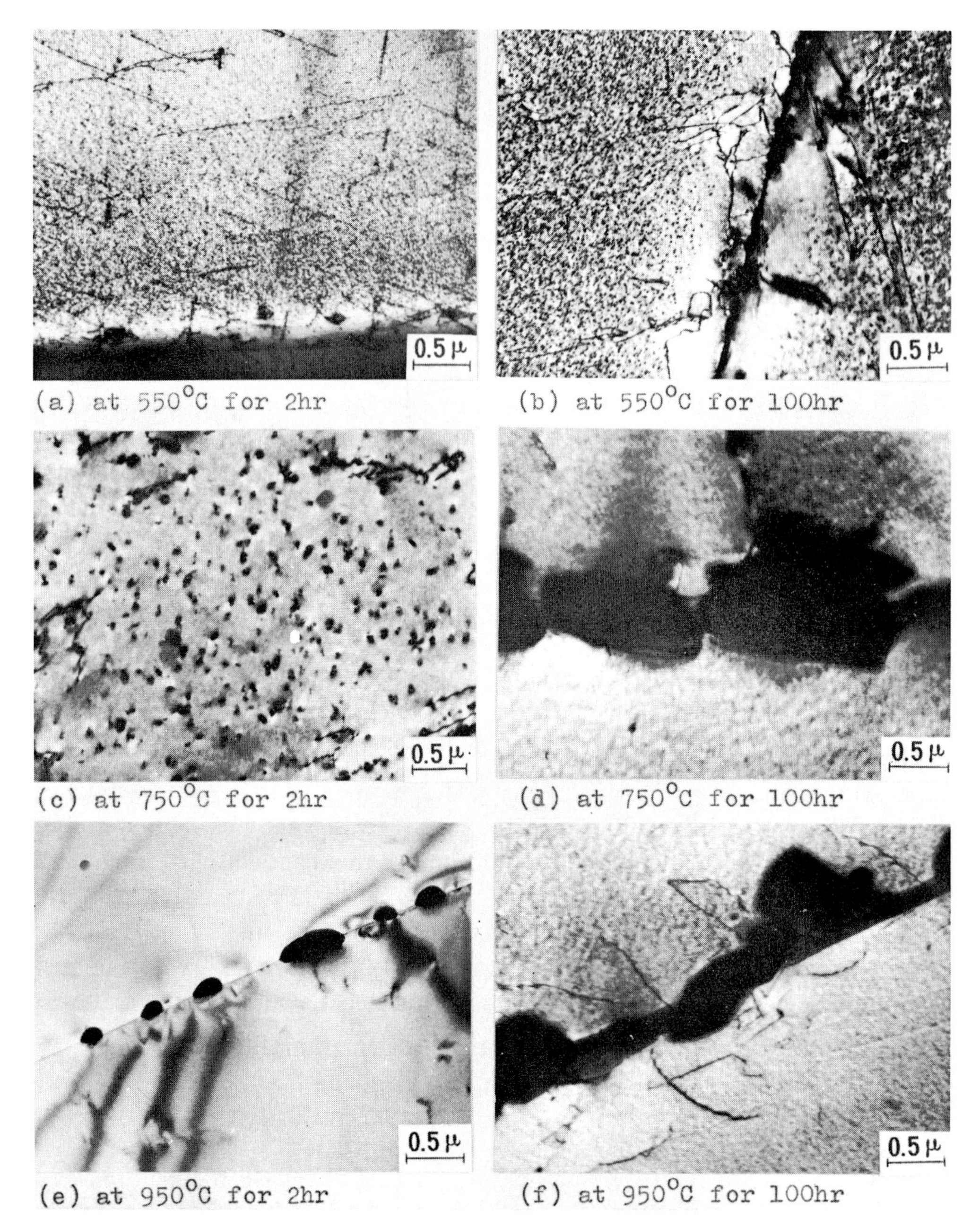

FIG. 21—*Transmission electron micrographs of specimens precipitation-treated after solution-treating at 1250°C for 1 h.*

at 550°C after 2 h of heating. When the heating time was extended to 100 h, the precipitated particles coalesced, and some coarse precipitates were found on the grain boundaries and on dislocations. No fine precipitates were found near these grain boundaries and dislocations. Such a precipitation-free zone is caused by solute atom depletion or vacancy depletion [*13,14*] near the grain boundary and dislocations. In this case it is not clear which factor is dominant. While precipitates of approximately 200 Å in diameter were scattered homogeneously within the grains after 2 h at 750°C, coarse precipitates started to appear side by side on the grain boundaries in 100 h. The coarser precipitates were observed at the grain boundaries and within the grains at 950°C for 2 h. The grain boundary precipitates were closely aligned and after 100 h were further coarsened.

Based on the preceding observations, the changes in the mechanical properties of the solution-treated and aged alloy can be explained as follows. The hardening and embrittlement occurring at 550°C is caused by a fine precipitation which is not chromium-rich α' phase, since this temperature is well over the precipitation temperature range of the chromium-rich α'-phase [*11*]. It can also be assumed that the niobium and chromium carbonitrides precipitated at 750°C in short time (2 h) were too isolated to bring about precipitation hardening or to decrease the toughness. Therefore we assume that the improvement in toughness was caused by the decrease in the amount of carbon and nitrogen in solid solution. In contrast, either at 750°C in longer time or at 950°C, the toughness was deteriorated because the carbonitrides were considerably coarsened and were almost continuous at the grain boundaries, thus acting akin to a notch.

Fabrication of a Model Tank

The applicability of this steel to welded structures was confirmed by fabricating a model tank. The 200-mm-diameter nozzles were made out of welded 6-mm-thick plate. The spherical heads were made by cold-spinning 8-mm-thick plate, and the actual 1000-mm-diameter, 2000-mm-long tank also used 8-mm plate welded with Type 316L ($C < 0.02$ percent) wire. Figure 22 is an outside view of the model tank.

No problem was found by the X-ray test and by the hydraulic test (7 kgf/cm^2), and thus the applicability of this steel to welded structures was demonstrated.

Conclusions

1. Based on the fracture toughness values of the base metal and weld zone, steel plates up to 12 mm thick can be used safely for welded structures at 0°C or higher.
2. The toughness of this steel decreased by cold-working. Some care therefore should be exercised when forming it by more than 10 percent.
3. This steel is susceptible to the 475°C embrittlement by aging. The increase in strength and the decrease in toughness concur around 475°C. Hence, the upper temperature limit for use of this steel should be about 300°C. In view of the strain-aging sensitivity of this steel strained plates should not be heated above 300°C.
4. After a solid-solution treatment and aging, this steel undergoes the

FIG. 22—*View of the model tank.*

following three property changes: (1) At an early stage of reheating, that is, at 550°C, hardening and embrittlement occur due to a dense precipitation of fine particles. (2) Heating at 750°C for a short time improves the toughness since both carbon and nitrogen are tied up by niobium. (3) Longer time exposure at 750 or 950°C produces precipitation of coarse particles and embrittlement. Because of these embrittling phenomena, any welding or fabrication which requires excessively large heat input should be avoided.

5. A model tank was successfully fabricated with this steel plate by welding with Type 316L (C < 0.02 percent) wire. Thus, the applicability of the steel to welded structures was well demonstrated.

References

[1] Binder, W. O. and Spendelow, H. R., *Transactions*, American Society for Metals, Vol. 43, 1951, p. 759.

[2] Baerlecken, E., Fisher, W.A., and Lorenz, K., *Stahl und Eisen*, Vol. 81, No. 12, 1961, p. 768.

[3] Semchyshen, M., Bond, A. P., and Dundas, H. J. in *Proceedings*, Symposium Toward Im-

proved Ductility and Toughness, Kyoto, Japan; Climax Molybdenum Co., Greenwich, Conn., 1971, p. 239.
[4] Jarleborg, O. H., Sawhill, J. M., and Steigerwald, R. F., *Stahl und Eisen*, Vol. 97, No. 1, 1977, p. 29.
[5] Abo, H., Nakazawa, T., Takemura, S., Onoyama, M., Ogawa, H., and Okada, H. in *Proceedings*, Stainless Steel '77, A Global Forum; Climax Molybdenum Co., Greenwich, Conn., 1977, p. 35.
[6] Ikeda, K., Akita, Y., and Kihara, H., *Welding Journal*, Vol. 46, No. 3, 1967, p. 133-s.
[7] Hall, W. J., Kihara, H., Soete, W., and Wells, A. A., *Brittle Fracture of Welded Plate*, Prentice-Hall, Englewood Cliffs, N.J., 1967, p. 290.
[8] Jack, D. H. and Jack, K. H., *Journal of the Iron and Steel Institute*, Vol. 210, 1972, p. 790.
[9] Marcinkowski, M. J., Fisher, R. M., and Szimae, A., *Transactions*, American Institute of Mining, Metallurgical, and Petroleum Engineers, Vol. 230, 1964, p. 676.
[10] Williams, R. O., *Transactions*, American Institute of Mining, Metallurgical, and Petroleum Engineers, Vol. 212, 1958, p. 497.
[11] Grobner, P. J., *Metallurgical Transactions*, Vol. 4, 1973, p. 251.
[12] Grobner, P. J. and Steigerwald, R. F., *Journal of Metals*, Vol. 29, No. 7, 1977, p. 17.
[13] Rosenbaum, H. S. and Turnbull, D., *Acta Metallurgica*, Vol. 6, 1958, p. 653.
[14] Rosenbaum, H. S. and Turnbull, D., *Acta Metallurgica*, Vol. 7, 1959, p. 664.

Source: ASTM STP 706, 1980, 99-122

N. Ohashi,[1] *Y. Ono,*[1] *N. Kinoshita,*[1] *and K. Yoshioka*[1]

Effects of Metallurgical and Mechanical Factors on Charpy Impact Toughness of Extra-Low Interstitial Ferritic Stainless Steels

---

REFERENCE: Ohashi, N., Ono, Y., Kinoshita, N., and Yoshioka, K., **"Effects of Metallurgical and Mechanical Factors on Charpy Impact Toughness of Extra-Low Interstitial Ferritic Stainless Steels,"** *Toughness of Ferritic Stainless Steels, ASTM STP 706,* R. A. Lula, Ed., American Society for Testing and Materials, 1980, pp. 202-220.

ABSTRACT: The effects of precipitates, grain size, plate thickness, and notch sharpness on the Charpy impact toughness of extra-low interstitial 18Cr-2Mo, 26Cr-1Mo, and 29Cr-2Mo steels were investigated.

The steels exhibit good toughness in the solution-treated condition at 1000 to 1200°C and subsequently in the 2-mm V-notched condition. If the steels are subjected to fatigue cracks or brittle weld cracks, their transition temperatures increase and their shelf energies decrease.

When the steels are slowly cooled from the solution temperatures or reheated at intermediate temperatures, 700 to 900°C, after solution treatments, precipitation of second phases occurs, such as Laves phase in 18Cr-2Mo steel and sigma phase in 26Cr-1Mo and 29Cr-2Mo steels. The steels are embrittled by these precipitates and behave very similarly to the solution-treated specimens, having sharp notches such as fatigue cracks or brittle weld cracks.

A decrease in grain size or thickness is beneficial in lowering the transition temperatures of solution-treated and 2-mm V-notched specimens. This effect, however, decreases, when the specimens have brittle weld cracks.

These test results suggest that the solution-treated steels have good resistance to brittle fracture initiation and poor resistance to brittle crack propagation. Precipitation of second phases assists mainly crack initiation but also crack propagation to a lesser extent.

KEY WORDS: ferritic stainless steels, Laves phase, sigma phase, grain size, specimen size, notch sharpness, V-notch, fatigue notch, brittle weld crack notch, fracture toughness

[1] Assistant director and senior researchers, respectively, Research Laboratories, Kawasaki Steel Corp., Chiba 260, Japan.

In many studies devoted to improving the toughness of ferritic stainless steels, the effects of composition [*1-9*],[2] heat treatment [*9-11*], precipitates [*2*], 475°C holding [*12,13*], grain size [*11*], plate thickness [*11,14*], etc. have been extensively reviewed. From a practical point of view, studies on optimizing the manufacture of the plates [*11*], including welding [*14*], are of special interest. The general conclusion of these studies is that the toughness of the steels can be improved by reducing the contents of interstitial elements and oxygen to an extremely low level [*1,9*], by additions of proper contents of niobium [*5,6*], nickel [*3,8*] and aluminum [*4*], and by solution treatments at high temperatures [*9*].

Recent developments in steelmaking techniques [*15*] have made it possible for extra-low interstitial steels to be produced commercially without special facilities or refined raw materials. The practical application of these steels will be expanded in the near future as more data on their toughness and corrosion properties are accumulated.

The present investigation deals with the effect of some metallurgical and mechanical factors on the toughness of three kinds of superferritic stainless steels. These factors are precipitation of second phases, grain size, plate thickness, and notch sharpness.

Experimental Procedure

The steels used in the study were 18Cr-2Mo, 26Cr-1Mo, and 29Cr-2Mo. Their chemical compositions are given in Table 1. These steels are characterized by very low interstitials and the addition of niobium to avoid intergranular corrosion at the welds. They were melted in a modified vacuum oxygen decarburization (VOD) furnace of 50 metric tons capacity (strongly stirring-VOD process [*15*]). The 26Cr-1Mo steel was hot-rolled to 10-mm-thick and 4-mm-thick plates while the 18Cr-2Mo and 29Cr-2Mo steels were hot-rolled to 4-mm-thick plates by commercial production processes.

Experiments were divided into roughly two groups in order to find the effects of the metallurgical and mechanical factors. First, 2-mm V-notched Charpy impact tests were conducted using subsize specimens, 4 by 10 by 55 mm^3, taken from solution-treated plate and also from embrittled plate; the longitudinal axis of the specimens was parallel to the rolling direction of the plates throughout the study. Solution heat treatments were done by heating the specimens at 1000, 1100 and 1200°C for 10 min and subsequent quenching in water. Heat treatments for embrittlement were performed by (1) reheating the solution-treated specimens at 700, 800, and 900°C and quenching in water; and (2) slow cooling the specimens from 1200°C to 700, 800, and 900°C and quenching in water. These heat treatment schedules are shown in Fig. 1. Changes in the microstructures

[2]The italic numbers in brackets refer to the list of references appended to this paper.

TABLE 1—*Chemical compositions of steels used (weight percent).*

Steel	C	Si	Mn	P	S	Cu	Nb	Cr	Mo	Ni	N	Al	O
18Cr-2Mo	0.0023	0.26	0.28	0.029	0.006	0.03	0.21	18.72	1.79	0.27	0.0039	0.008	0.0045
26Cr-1Mo	0.0022	0.33	0.08	0.018	0.004	0.02	0.17	25.88	1.20	0.16	0.0055	0.059	0.0027
29Cr-2Mo	0.0027	0.29	0.06	0.018	0.006	0.02	0.16	28.61	1.95	0.16	0.0070	0.098	0.0018

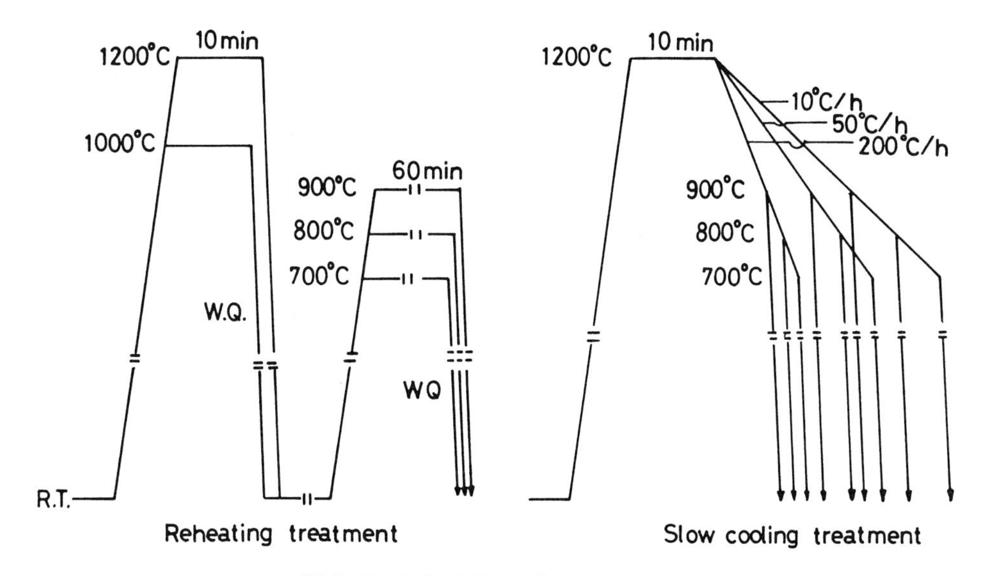

FIG. 1—*Schedules of heat treatments.*

of the specimens were observed through electron microscopy (Type JEPL-200, 200 kV), and the precipitates were identified by selected area transmission electron diffraction patterns. The foil specimens for the microscopic observation were prepared by "double jet" electrochemical polishing in a 10 percent perchloric acid plus 90 percent acetic acid solution at room temperature.

Subsequently, the effects of notch sharpness on Charpy impact toughness were evaluated using full-size specimens of 26Cr-1Mo steel, 10 by 10 by 55 mm^3, in the solution-treated (1020°C, 10 min, water-quenched) and embrittled (800°C, 60 min, water-quenched) conditions. Three types of notches were made in the cross-sections of the specimens: machined 2-mm V-notch, fatigue crack notch, and brittle weld crack notch. Fatigue cracks were introduced about 1.5 mm deep from the bottom of machined 2-mm V-notch under cyclic loads of 30 to 270 kgf for about 1.8×10^5 cycles. Brittle weld beads were made by electron beam welding after setting thin titanium and mild steel strips on the central part of the specimens, where natural brittle cracks were expected to be formed. The method of preparing specimens is illustrated in Fig. 2 and a typical cross section of a brittle weld bead is shown in Fig. 3. Both fatigue and brittle weld cracks were red-dyed prior to the impact tests to measure their depths on the fractured surfaces of the specimens.

The effect of grain size on the Charpy impact toughness was investigated using full-size and subsize 26Cr-1Mo steel specimens, provided with the usual 2-mm V-notch and brittle weld cracks. The grain size of the specimens was changed by heating at various temperatures between 1000 and 1200°C for 10 min and water-quenching. Before testing, all specimens were reheated at 1000°C for 10 min and quenched in water.

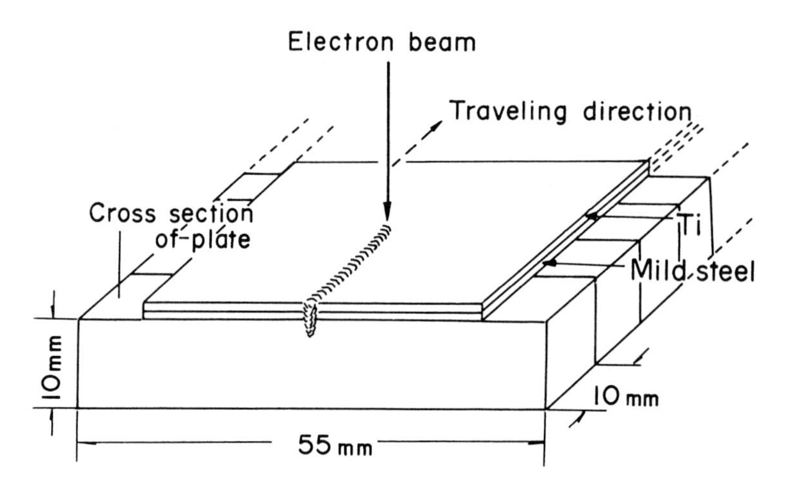

FIG. 2—*Preparation of the specimen having brittle weld crack.*

Results

Effect of Heat Treatments

The results of 2-mm V-notched Charpy impact tests on the solution-treated specimens are shown in Fig. 4. All specimens exhibit good toughness above −20°C. The 18Cr-2Mo steel treated at 1000°C is fully ductile even at −80°C. With increasing solution-treatment temperature, all steels tend to be brittle at low temperatures and this tendency is slightly stronger in the higher-chromium steels. It should be noted that the absorbed energies at low test temperatures are scattered at roughly two extremes, the entirely ductile and entirely brittle levels. Typical examples of the fracture surfaces of the two cases are shown in Fig. 5.

The results on the specimens reheated at 700 to 900°C after solution treatment at 1000°C are shown in Fig. 6. The 18Cr-2Mo steel is embrittled when reheated at 800 and 900°C, while the 29Cr-2Mo steel is only slightly embrittled by reheating at 800°C. The 26Cr-1Mo steel, however, shows no sign of embrittlement in the present reheating conditions. Results of the test on the specimens reheated at 700 to 900°C after solution treatment at 1200°C are also shown in Fig. 6. All specimens are brittle even at higher test temperatures and the scatter of absorbed energies at each test temperature becomes large.

Figure 7 shows the results for the specimens which were cooled from 1200°C to 900, 800, and 700°C with controlled cooling rates. All steels are clearly brittle even at high test temperatures and there is a tendency for embrittlement to be enhanced by lower cooling rates from the 1200°C anneal.

Figure 8 shows the transmission electron micrographs (TEM's) of the specimens solution-treated at 1200°C. No precipitates can be observed

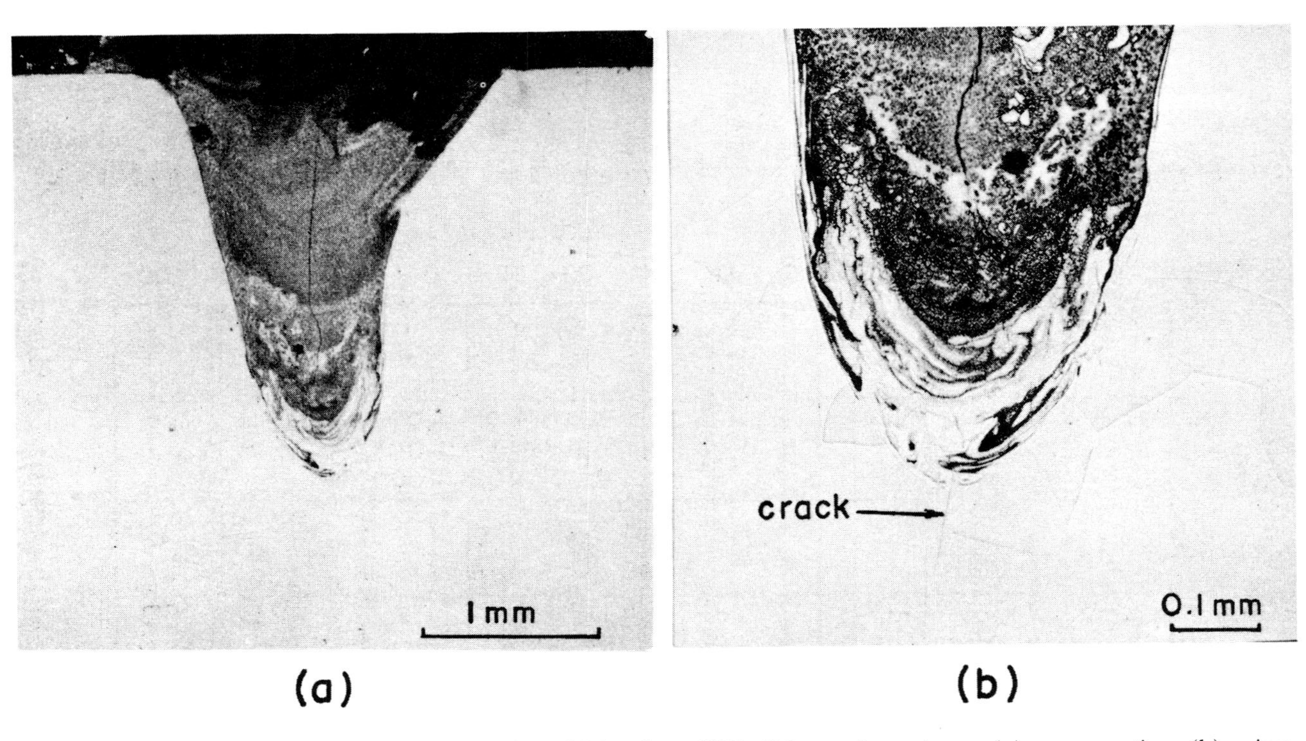

FIG. 3—*Example of cross section of brittle weld bead on 26Cr-1Mo steel specimen.* (a) *cross section;* (b) *microstructure in the root of the bead,* (a).

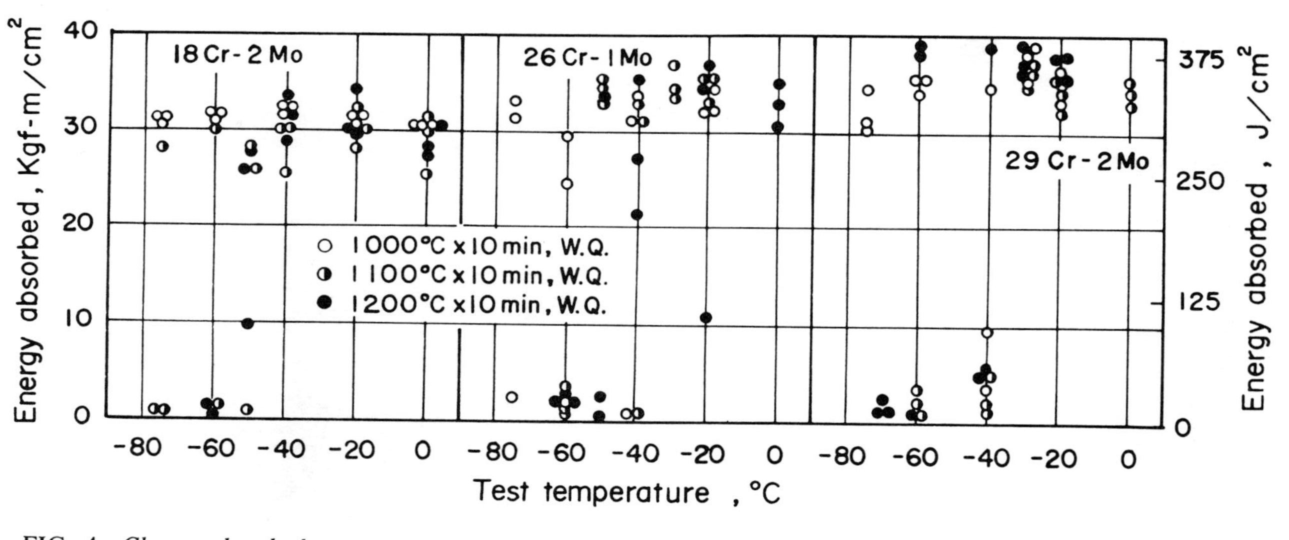

FIG. 4—*Charpy absorbed energy versus test temperature plots of solution-treated specimens (4 mm thick, 2-mm V-notch).*

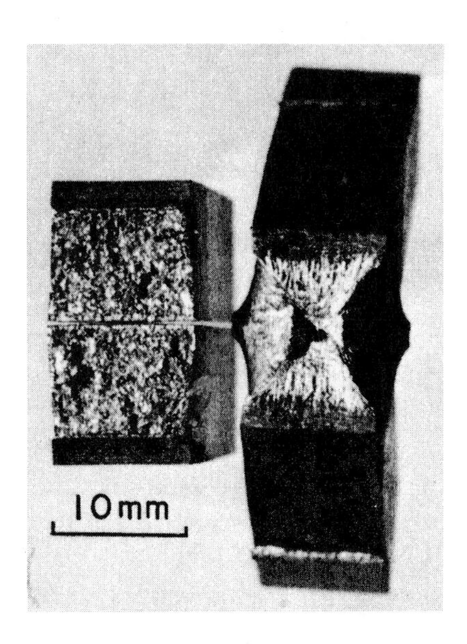

FIG. 5—*Typical examples of fracture surfaces of solution-treated and 2-mm V-notched 26Cr-1Mo steel specimens broken at −40°C.*

either at grain boundaries or within the grains. On the other hand, Fig. 9 shows the structures of the specimens embrittled by reheating at intermediate temperatures as shown in Fig. 6. Many precipitates are observed at the grain boundaries and within the grains. Electron diffraction patterns taken from the precipitates show the crystallographic structures characteristic in each steel: Laves phase [*16*] in 18Cr-2Mo steel and sigma phase [*17*] in 26Cr-1Mo and 29Cr-2Mo steels. Besides these phases, NbCrN [*18*] was detected by X-ray diffraction of extracted residues of 26Cr-1Mo and 29Cr-2Mo steels. The precipitates such as Laves and sigma phases are larger in size and are observed continually at grain boundaries in clearly embrittled specimens; for example, in 18Cr-2Mo and 29Cr-2Mo steels reheated at 800°C. In all other specimens embrittled by heat treatments, including slow cooling from 1200°C as shown in Fig. 7, similar precipitates are observed. No carbide can be detected by electron diffraction analysis in any steels, although it is suspected it precipitates as very fine particles in the specimens heated at intermediate temperatures.

Effect of Notch Sharpness

As indicated earlier, large scatter of Charpy absorbed energies was experienced, especially in the transition temperature range, for solution-treated "ductile" specimens. In order to clarify the effect of mechanical factors on this problem, three types of notches were made in full-size specimens of 26Cr-1Mo steel as described in the section on experimental procedure.

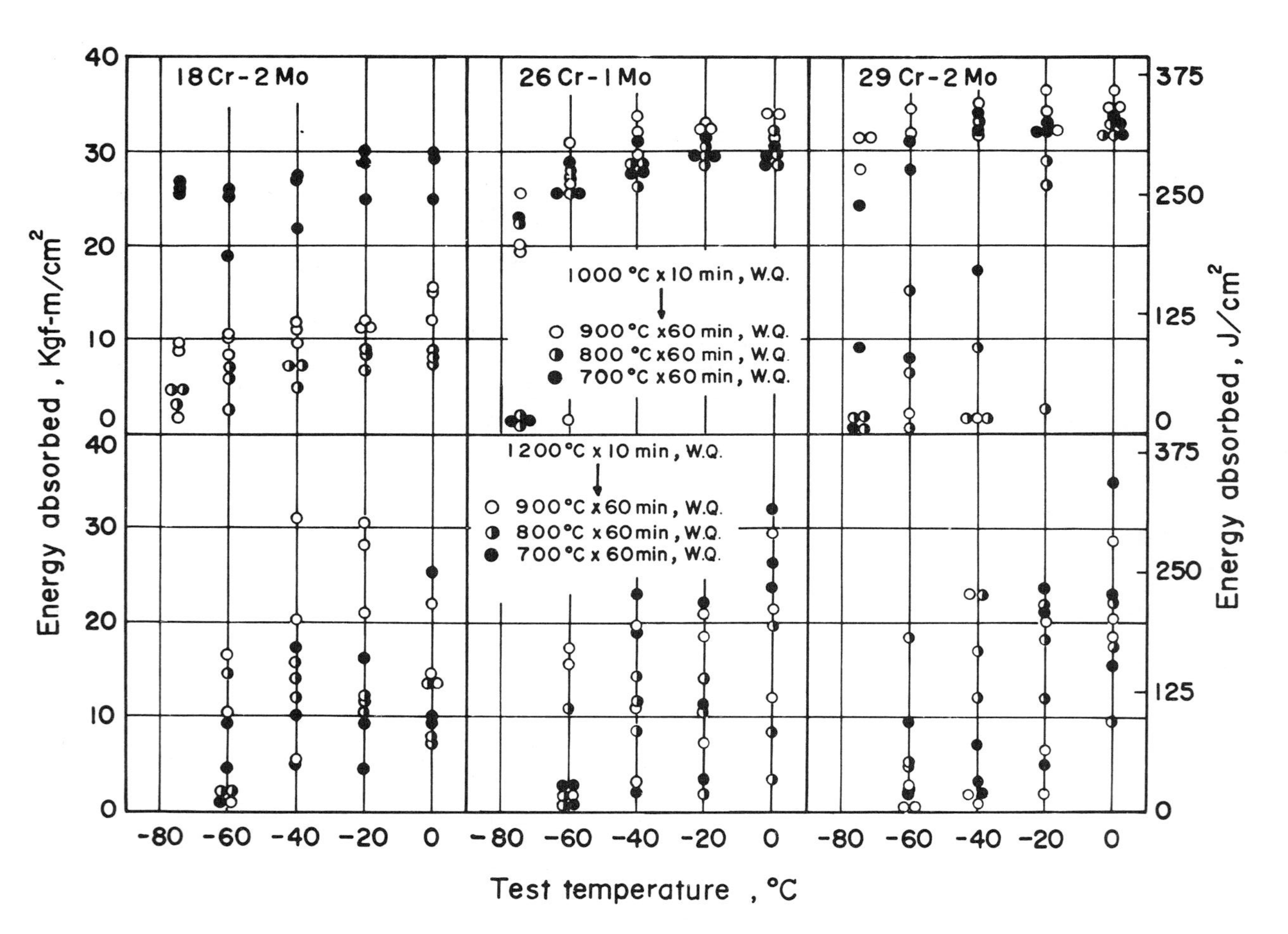

FIG. 6—*Charpy absorbed energy versus test temperature plots of specimens reheated after solution treatment at 1000 and 1200°C (4 mm thick, 2-mm V-notch).*

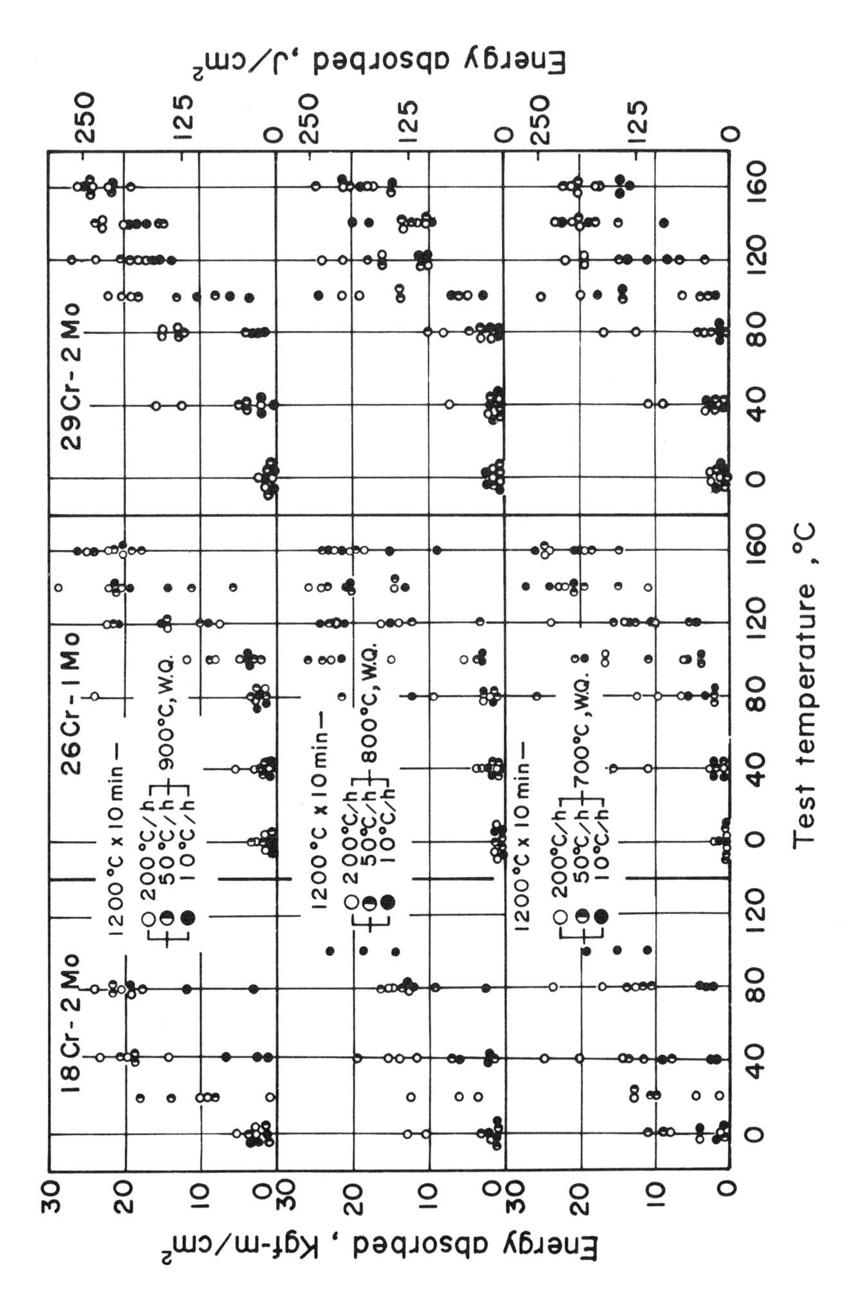

FIG. 7—*Charpy absorbed energy versus test temperature plots of specimens slowly cooled from 1200°C (4 mm thick, 2-mm V-notch).*

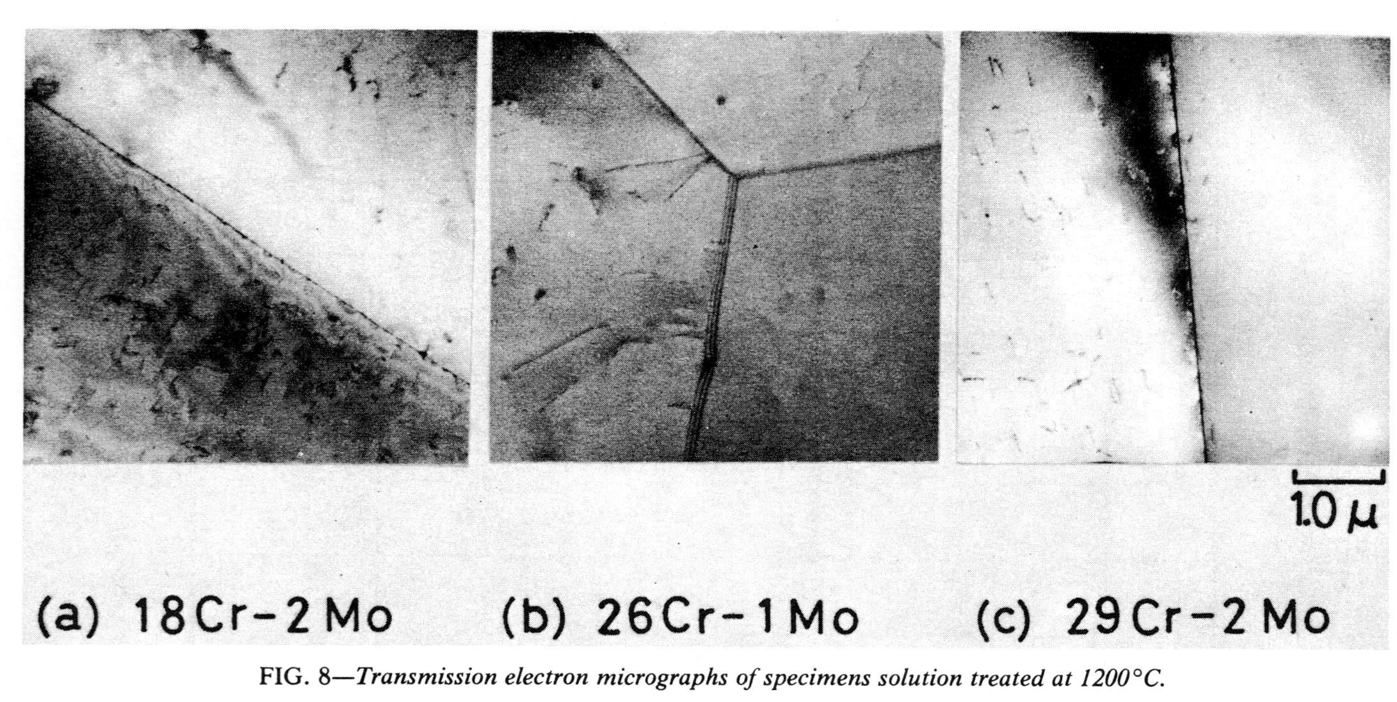

FIG. 8—*Transmission electron micrographs of specimens solution treated at 1200°C.*

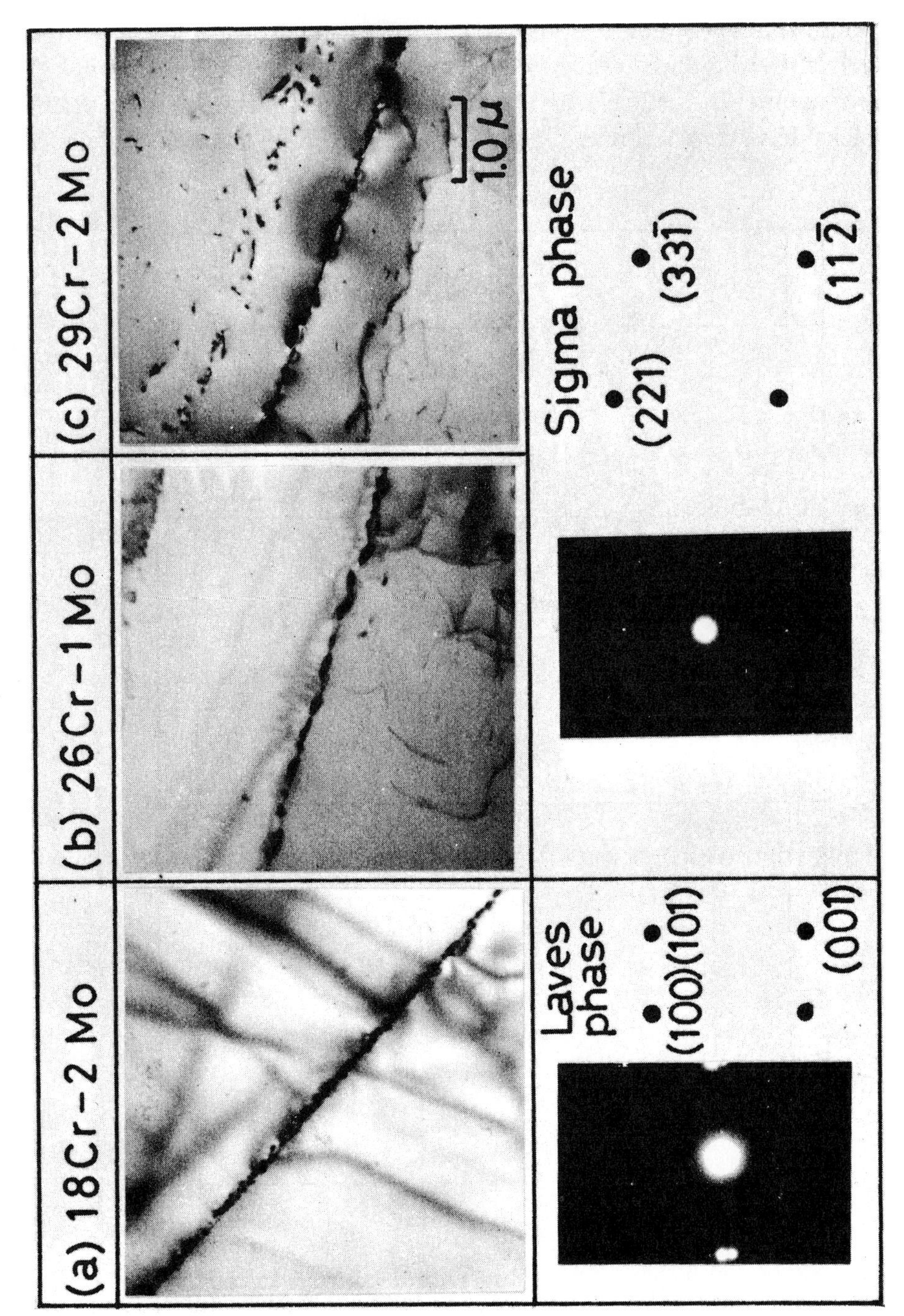

FIG. 9—*Transmission electron micrographs of embrittled specimens and electron diffraction patterns of precipitates.*

The effects of notch sharpness are shown in Fig. 10. In the solution-treated specimens, very large scatter of absorbed energies is observed in the 2-mm V-notched specimens. In the fatigue-cracked specimens, the absorbed energy versus temperature shows a wide scatter, but is slightly smaller than that of 2-mm V-notched specimens. The test results are also characterized by the appearance of intermediate values between the values of entirely ductile and entirely brittle fractures. The specimens with brittle weld cracks have much lower absorbed energies and smaller scatter of

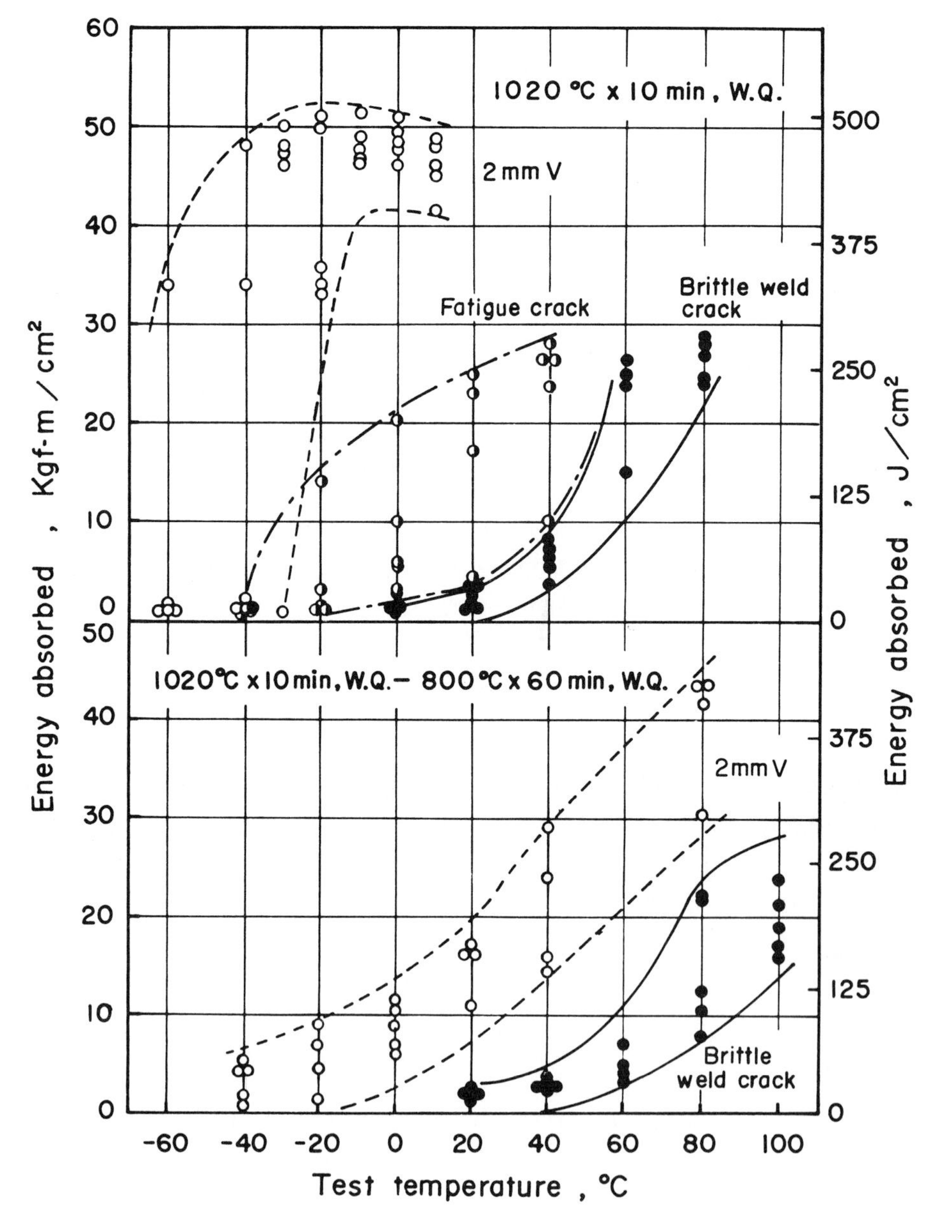

FIG. 10—*Effects of notch sharpness on Charpy impact toughness of 26Cr-1Mo steel specimens solution treated at 1020°C or embrittled at 800°C for 1 h (10 mm thick).*

data than fatigue-notched specimens, and the transition temperature shifts to a higher temperature, about 50°C.

The absorbed energy transitions of the specimens which were embrittled at 800°C and 2-mm V-notched or brittle welded are more sluggish than those of the as-solution-treated specimens. The embrittled and 2-mm V-notched specimens behave like as-solution-treated and fatigue-notched specimens. The embrittled and brittle welded specimens exhibit the smallest absorbed energies and the highest transition temperature among all the specimens tested.

Effect of Grain Size

Figures 11 and 12 show, respectively, the results of Charpy tests on solution-treated 4-mm-thick and 10-mm-thick 26Cr-1Mo steel plates whose grain sizes were varied in the range 0.03 to 1.6 mm by changing the heating temperature as described previously.

In both subsize and full-size 2-mm V-notched specimens, the energy transition temperature is raised with increasing grain size, while the shelf energy is almost independent of grain size in each specimen size. Comparing specimens of nearly the same grain size as shown in Figs. 11 and 12, the transition temperature is lower in the subsize specimens than in full-size ones. A very large scatter of data is observed, especially in the coarse-grained specimens.

On the other hand, in both subsize and full-size specimens having brittle welded cracks, the transition temperatures are higher than those of the 2-mm V-notched specimens. The grain size dependency of shelf energy and transition temperature in brittle welded specimens is very small, but is still observed. The transition temperatures of brittle welded specimens are about 30°C for subsize specimens and about 50°C for full-size ones. This again shows the effect of specimen size on the toughness of embrittled specimens. Examples of fracture surfaces of full-size specimens having brittle weld cracks are shown in Fig. 13. From the latent natural crack in brittle weld bead, brittle fracture is generated and changed into ductile fracture, which continues to the edge of the specimens. Scatter of the absorbed energies at each test temperature in Figs. 11 and 12 is caused mainly by the variation of the depth of the propagating brittle fracture.

Discussion

The results of this study of the effect of metallurgical factors on the Charpy impact toughness of ferritic stainless steels agree with the work of other investigators [*9–11*].

In a solution-treated condition, low-interstitial, high chromium and molybdenum bearing steels exhibit good toughness even at low tempera-

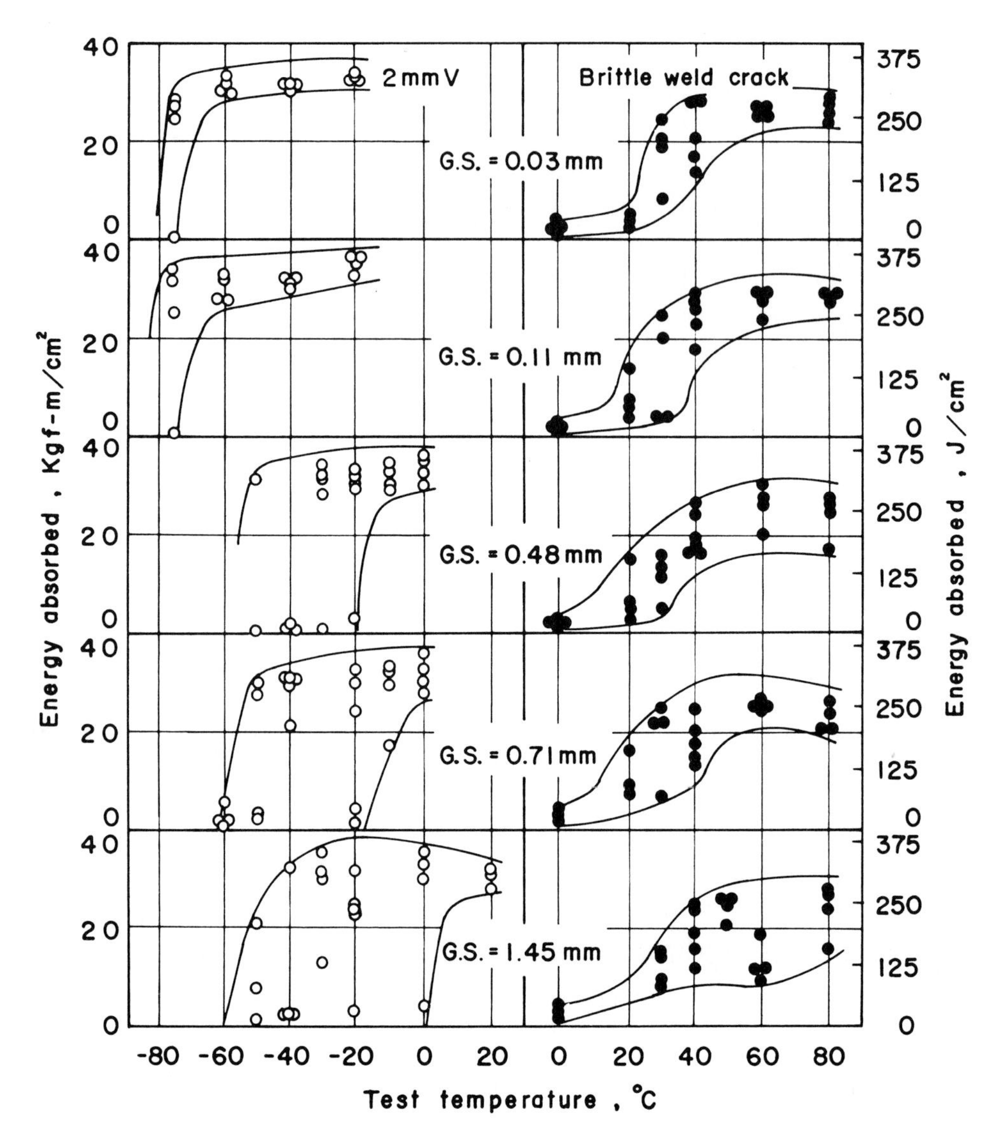

FIG. 11—*Effects of notch sharpness and grain size on Charpy impact toughness of the 4-mm-thick plate of 26Cr-1Mo steel solution treated at 1000°C.*

tures. A comparison of data in Figs. 4 and 10 showing the results of Charpy impact tests of the 26Cr-1Mo steel indicates a difference in toughness due to the difference in thickness of specimens. This well-known fact [*9*] is attributed to the difference in a mechanical factor—restriction of strain—during impact deformation of the specimens. In these figures, the solution-treated and V-notched specimens show the bimodal distribution of absorbed energies over a wide transition temperature range. At these temperatures fracture occurs either by entirely ductile or entirely brittle means as seen in Fig. 5. This phenomenon makes the transition range very broad. As shown in Fig. 10, some fatigue-cracked specimens exhibit intermediate absorbed energies in the transition temperature

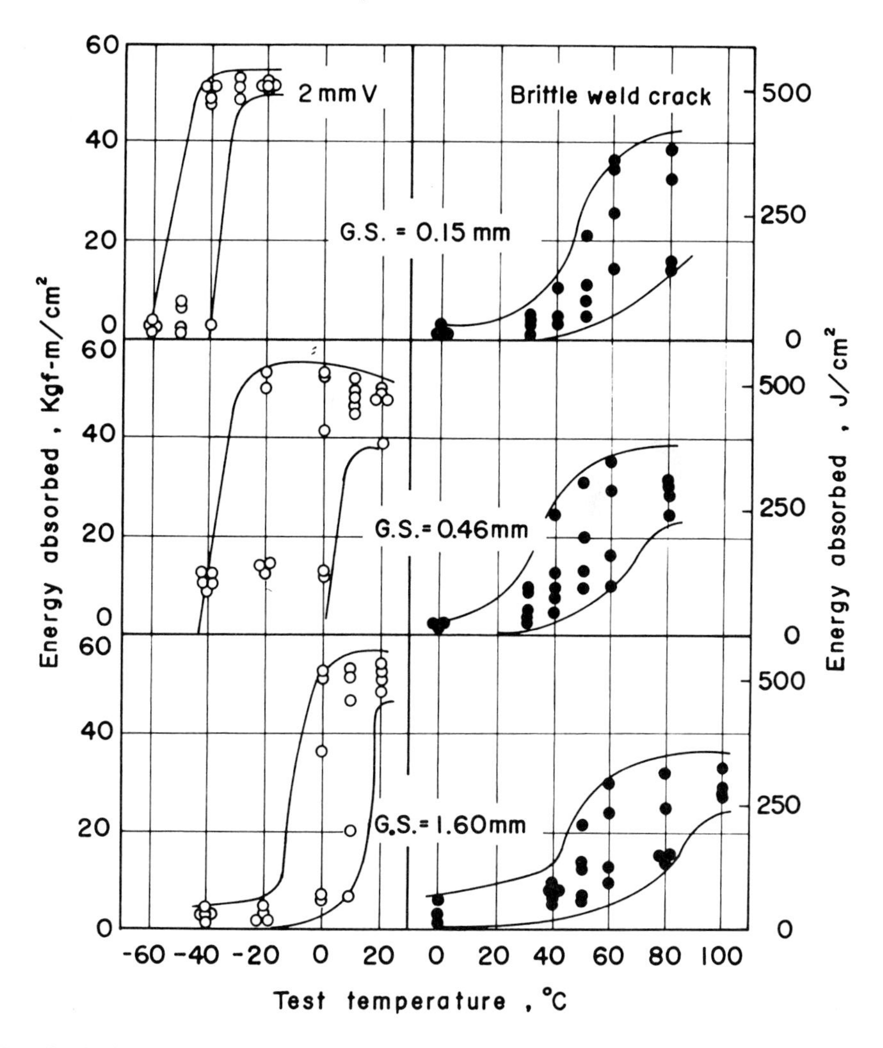

FIG. 12—*Effects of notch sharpness and grain size on Charpy impact toughness of the 10-mm-thick plate of 26Cr-1Mo steel solution treated at 1000°C.*

range and a fracture surface of one of these specimens is shown in the center of Fig. 14. The fracture surface of the specimen is divided clearly into two areas, one entirely ductile and one entirely brittle.

On the other hand, fatigued notch or brittle weld cracks bring out lower absorbed energies and higher transition temperatures. In these cases, fracture starts in a brittle manner, consuming small amounts of absorbed energy, and subsequently changes into a ductile fracture. The path length of brittle fracture is dependent on the temperature and specimen size as well as the grain size, and consequently this determines the level of total absorbed energy.

The facts just mentioned suggest that resistance to fracture initiation during dynamic loading is quite high in these types of steels in the solution-treated condition and in the presence of blunt notches. If a brittle crack

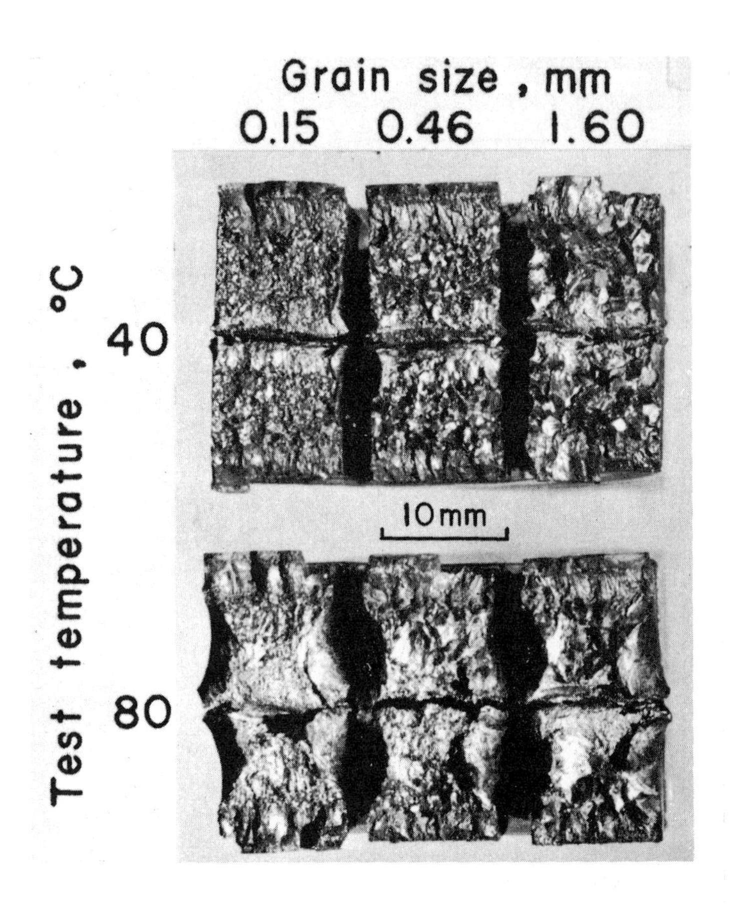

FIG. 13—*Examples of fracture surfaces of the specimens having brittle weld cracks of 26Cr-1Mo steel whose grain sizes are 0.15 mm to 1.6 mm (test temperatures: 40 and 80°C).*

FIG. 14—*Example of particular fracture surface of the specimen which was solution-treated, fatigue cracked, and tested at the transition temperature range (26Cr-1Mo). The particular fracture surface is on the specimen in the center.*

is present, however, it can easily propagate in a brittle fashion and produces low absorbed energy. Although the factors which influence the incidental initiation of brittle fracture in the solution-treated and bluntly notched specimens have not been clarified yet, it is suspected that the crystallographic orientation of the grains at the root of the notch is important, especially in coarse-grained specimens.

V-notched specimens embrittled by the precipitation of Laves or sigma phase show toughness behavior very similar to that of the solution-treated and sharply notched specimens. This shows that the precipitates assist greatly in the initiation of brittle cracks at the time of dynamic loading. In Fig. 10, however, the embrittled and weld-cracked specimens show a lower toughness than the solution-treated and weld-cracked ones, and this indicates the adverse effect of precipitates on the resistance to brittle crack propagation.

In Figs. 11 and 12, which show the brittle-to-ductile transition of solution-treated 26Cr-1Mo steel, the effect of grain size is noticeable in the V-notched specimens but slight in the specimens with brittle weld cracks. This indicates that coarse grains tend to promote crack initiation even in the blunt-notched specimens; in other words, the grain size effect contributes mainly in the resistance to initiation of brittle fracture and only slightly to the propagation of brittle fracture.

The results obtained in this investigation indicate the importance of preventing brittle crack initiation and brittle fracture in these types of ferritic stainless steels. From this point of view a fracture mechanics approach is suggested to evaluate the practical application of these steels, especially those involving welding. The addition of nickel is proposed as a method of improving the toughness of the steels [*3,8*]. To apply this method, however, more information on the corrosion properties, especially on the stress corrosion behavior under various environmental conditions, should be acquired.

Conclusion

The effects of some metallurgical and mechanical factors on the Charpy impact toughness of extra-low interstitial 18Cr-2Mo, 26Cr-1Mo, and 29Cr-2Mo steels have been investigated. The results obtained are as follows.

1. The steels have good toughness in the as-solution-treated and bluntly notched conditions. There is, however, a large scatter of the data resulting in a wide range of transition temperature.

2. When the steels are embrittled by precipitation of second phases—for example, Laves or sigma phases in the present case—they behave very similarly to the solution-treated specimens having sharp notches such as fatigue cracks or brittle weld cracks. In these cases, Charpy impact

energies decrease and the transition temperatures increase with little scatter in the data points.

3. Small grain size or thin gage lowers the transition temperature of solution-treated and bluntly notched specimens. This beneficial effect decreases in the specimens having brittle weld cracks.

4. These test results suggest that the solution-treated steels exhibit good resistance to the initiation of brittle cracks, but relatively low resistance to crack propagation. Precipitation of second phase assists crack initiation mainly, but also crack propagation to a lesser extent.

References

[*1*] Semchyshen, M., Bond, A. P., and Dundas, H. J. in *Proceedings,* Symposium Toward Improved Ductility and Toughness, sponsored by Climax Molybdenum Co., Kyoto, Japan, 25–26 Oct. 1971, p. 239.

[*2*] Brandis, H., Kiesheyer, H., and Lennartz, G., *Archiv für das Eisenhüttenwesen,* Vol. 46, 1975, p. 799.

[*3*] Brandis, H., Kiesheyer, H., Küppers, W., and Oppenheim, R., *TEW-Technische Berichten,* Vol. 2, No. 1, 1976, p. 3.

[*4*] Colombie, M., Condylis, A., Desestret, A., Grand, R., and Mayoud, R., *Revue de Metallurgie,* Vol. 70, 1973, p. 947.

[*5*] Jarleborg, O. H., Sawhill, J. M., and Steigerwald, R. F., *Stahl und Eisen,* Vol. 97, 1977, p. 29.

[*6*] Steigerwald, R. F., Dundas, H. J., Redmond, J. D., and Davison, R. M. in *Proceedings,* Stainless Steel '77, sponsored by Climax Molybdenum Co., London, England, 26–27 Sept. 1977, p. 57.

[*7*] Pollard, B., *Metals Technology,* Vol. 1, 1974, p. 32.

[*8*] Nakano, K., Kanao, M., and Hoshino, A., *Tetsu to Hagane,* Vol. 62, 1976, p. 1219.

[*9*] Abo, H., Nakazawa, T., Takemura, S., Onoyama, M., Ogawa, H., and Okada, H. in *Proceedings,* Stainless Steel '77, sponsored by Climax Molybdenum Co., London, England, 26–27 Sept. 1977, p. 35.

[*10*] Oppenheim, R. in *Proceedings,* Stainless Steel '77, sponsored by Climax Molybdenum Co., London, England, 26–27 Sept. 1977, p. 121.

[*11*] Nichol, T. J., *Metallurgical Transactions,* Vol. 8A, 1977, p. 229.

[*12*] Jacobsson, P., Bergström, Y., and Aronsson, B., *Metallurgical Transactions,* Vol. 6A, 1975, p. 1577.

[*13*] Grobner, P. J., *Metallurgical Transactions,* Vol. 4, 1973, p. 251.

[*14*] Matejka, W. A. and Knoth, J., *Journal of Testing and Evaluation,* Vol. 3, No. 3, 1975, p. 199.

[*15*] Iwaoka, S., Kaito, H., Ohtani, T., Ohashi, N., Takeda, M., and Kinoshita, N. in *Proceedings,* Stainless Steel '77, sponsored by Climax Molybdenum Co., London, England, 26–27, Sept. 1977, p. 139.

[*16*] Vowles, M. D. J. and West, D. R. F., *Journal of the Iron and Steel Institute,* Vol. 211, 1973, p. 147.

[*17*] Duwez, P. and Baen, S. R. in *Symposium on the Nature, Occurrence, and Effects of Sigma Phase, ASTM STP 110,* American Society for Testing and Materials, 1951, p. 48.

[*18*] Jack, D. H. and Jack, K. H., *Journal of the Iron and Steel Institute,* Vol. 210, 1972, p. 790.

APPLICATIONS OF NEW HIGH CHROMIUM FERRITIC STAINLESS STEELS IN THE CHEMICAL PROCESS INDUSTRIES

T. J. Nichol
Allegheny Ludlum Steel Corporation
Tubular Products Division
80 Valley Street
Wallingford, Ct. 06492

I. A. Franson
Allegheny Ludlum Steel Corporation
Research Center
Brackenridge, Pa. 15014

and

G. E. Moller
Consultant
2224 Chelsea Road
Palos Verdes, Ca. 90274

ABSTRACT

Conventional high chromium ferritic stainless steels have good resistance to corrosion including resistance to stress corrosion cracking. However, poor weldability and restricted ductility limit their use in the process industries. The new, high purity E-BRITE® (XM-27) and 29-4-2 ferritic alloys offer considerably improved fabricability as well as improved corrosion resistance. These steels resist stress corrosion cracking. General corrosion resistance is better than that of standard austenitic stainless steels in many environments and in some cases, equals or betters the resistance of some high nickel alloys. For instance, these ferritic stainless steels offer outstanding resistance to chlorides, alkalis, nitric acid, urea/ammonium carbamate, amines and organic acids in addition to environments normally handled by stainless steels. The 29-4-2 alloy also provides exceptional resistance to dilute reducing acids. The E-BRITE alloy with its high chromium content provides excellent resistance to high temperature oxidation and

® Registered Trade Mark, Allegheny Ludlum Industries, Inc.

sulfidation. Case histories covering the use of the E-BRITE alloy for as long as eight years of service are presented. Potential applications for the newer 29-4-2 alloy are also discussed.

Introduction

The conventional ferritic stainless steels, e.g. AISI Types 430 and 446, have good corrosion resistance because of their high chromium content and ferritic structure. For instance, they are superior to the AISI 300 Series austenitic stainless steels in resisting chloride stress corrosion cracking. However, these steels have not been utilized in the chemical process industries to the extent that their corrosion resistance might dictate because of poor weldability and limited ductility.

In the late 1960's , the high purity, 26 Cr-1 Mo ferritic E-BRITE alloy (ASTM XM-27) was introduced to the process industries. Later, the 29 Cr-4 Mo-2 Ni ferritic alloy also became available for use. These steels are vacuum-processed to ultra-low carbon and nitrogen content. The result is improved corrosion resistance, ductility and weldability compared to AISI Type 430 and Type 446 stainless steels and others made by argon-oxygen decarburization.

This paper reviews the properties of the high purity XM-27 and 29-4-2 ferritic alloys. Corrosion resistance is discussed in terms of typical applications. Details of nearly ten years of field experience with the E-BRITE alloy, along with potential applications for the new 29-4-2 alloy, are also discussed.

Properties of Ferritic Alloys

Composition and Physical Properties

Chemical compositions of standard and high purity ferritic stainless steels are given in Table 1. The ultralow carbon and nitrogen content of the E-BRITE and 29-4-2 alloys aids weldability and ductility and contributes to corrosion resistance. The E-BRITE alloy contains a columbium addition for intergranular corrosion resistance. A stabilizing addition is not necessary for the 29-4-2 alloy. Physical properties are given in Table 2. The ferritic alloys have body-centered cubic structure in the annealed condition and are ferromagnetic at room temperature.

Physical Metallurgy and Mechanical Properties

The physical metallurgy and mechanical properties of the E-BRITE and 29-4-2 alloys have been discussed(1-6). Minimum tensile properties and maximum hardness values for tubing, as defined by ASTM A268-77 specification, are compared in Table 3. Allowable stresses for XM-27 are given in Section VIII, Division 1 and Section III, Division 1 (Class 1, 2 and 3) of the ASME Boiler and Pressure Vessel Code. Design stresses for the 29-4-2 alloy are currently being established.

All high chromium ferritic stainless steels become embrittled on extended exposure to temperatures in the range 343 C to 538 C (650 F to 1000 F)[1, 6, 7]. Although the alloys may be ductile at these temperatures, they can be quite notch-brittle at room temperature. For this reason, use of the ferritic alloys covered by the ASME Code is limited to 343 C (650 F). The 29-4-2 alloy is also subject to formation of embrittling sigma and chi phases on exposure to 704 C to 927 C (1300 F to 1700 F)[2, 3, 5].

Types 430 and 446 stainless can be notch brittle in sections as light as 1.6 mm. (0.063 inch) causing problems on bending of tube, roller expansion into tubesheets, and other fabrication operations on plate. In addition, welds and heat-affected zones of these steels may be brittle in any section thickness in the as-welded condition because of formation of martensite in the microstructure.

The ultralow carbon and nitrogen alloys are ferritic at all temperatures and are not hardenable. Section thickness to 12.7 mm. (0.50 in.) exhibit good notch ductility in base plate and welds[1, 4]. Careful inert-gas (pure argon or helium only) shielded welding techniques must be utilized to maintain ductility of welds[8].

Corrosion Resistance and Applications

The E-BRITE and 29-4-2 alloys offer excellent corrosion resistance[5, 6, 9-11]. They are highly resistant to stress corrosion cracking. Molybdenum, combined with chromium, provides resistance to pitting and crevice corrosion that ranges from superior to vastly superior, when compared to Type 316 stainless. These elements also provide exceptional resistance to general corrosion by a wide range of corrodents. Included in the list are hot alkaline solutions, oxidizing acids, organic acids, amines, urea/ammonium carbamate, high temperature oxidation and sulfidation, as well as those environments generally considered acceptable to stainless steels. The nickel content of the 29-4-2 alloy provides exceptional resistance to hot, dilute reducing acids. Both the E-BRITE and 29-4-2 alloys provide excellent resistance to intergranular corrosion.

Experience with the ferritic alloys in petroleum refinery heat exchangers has been discussed recently[12]. The following sections describe the usefulness of the E-BRITE and 29-4-2 alloys for chemical plant equipment in terms of laboratory and in-plant corrosion data. Typical applications are given along with illustrative plant experience.

Chloride Environments

Stress Corrosion Cracking (SCC)

The ferritic alloys are highly resistant to chloride SCC as shown by the data in Table 4. The nickel-containing 29-4-2 alloy fails in the boiling magnesium chloride test, but has high resistance to sodium chloride environments which are felt to be more representative of actual plant service conditions[5]. The resistance of the 29-4-2 alloy, although not as good as that of XM-27 alloy, is equivalent to the 35 percent nickel alloy.

Pitting and Crevice Corrosion

Laboratory potentiostatic tests (Table 5) show the E-BRITE alloy to be considerably more resistant to pitting by chlorides than Type 304 or Type 316 austenitic stainless steels. The 29-4-2 alloy is shown to be even more resistant as high pitting potential values correlate with better resistance to pitting.

Additional laboratory data in Table 6 compare the behavior of various stainless steels and high nickel alloys in a 2% $KMnO_4$-2% NaCl solution. This test simulates contact of chlorinated water with manganese dioxide deposits which occur on heat exchanger tubing in many plants which utilize river water[10]. The results illustrate that conventional stainless steels and even more highly alloyed materials may suffer pitting corrosion even at room temperature when these conditions exist. They also illustrate the high resistance to pitting which the high purity ferritic alloys provide. The data also illustrate the strong influence of temperature in promoting pitting. For resistance at the higher temperatures, high chromium and molybdenum content, such as is present in the 29-4-2 ferritic alloy, is required.

Other studies have demonstrated the excellent resistance of the new ferritic alloys to pitting and crevice corrosion by 10% ferric chloride[5], and to saturated sodium chloride brines[15]. Table 7 presents some laboratory data in sodium hypochlorite solutions.

The 29-4-2 alloy, with higher chromium and molybdenum than XM-27, provides pitting and crevice corrosion resistance that is often comparable to titanium or Ni-Cr-Mo alloys. This alloy was not corroded in a five-year atmospheric exposure 244 m (800 ft.) from the ocean, or by a nine-month exposure to seawater[2].

Chloride Applications

The E-BRITE alloy is being used in many heat exchangers because of its resistance to aggressive cooling waters. Examples are given in Table 8. Process fluids vary from simply air (Case No. 20) to sour gases (Case No. 1), other gases (Case Nos. 10, 11, 12 and 18) to more complex liquids. A heat exchanger tube bundle is shown in Figure 1. Bi-metallic (copper and steel) tubes were replaced by more economical XM-27 ferritic tubes when it was time to rebuild the bundle. The ferritic tubes were roller-expanded into Type 316 tubesheets. Roller expansion of high purity ferritic stainless tubes into austenitic stainless steel tubesheets is common practice. In a few cases, e.g., Cases 8, 9 and 21, Table 8, seal welded tube-to-tubesheet joints have been employed.

The E-BRITE alloy has better resistance to pitting by chlorides than Type 316 or Type 317 stainless steel. However, if temperature or chloride concentration are high enough, or pH is low enough, it may also suffer pitting or crevice corrosion. While it is difficult to establish specific limits, two cases have been documented in which the XM-27 alloy did show evidence of pitting by chlorides. The first case involved a seawater-cooled condenser at relatively low temperature[14]. The second case involved a heat exchanger used to cool hot [149-171 C (300-340 F)] chlorinated organics with river water containing from 50 to 1700 PPM chloride. Perforation occurred under deposits from the water side. Under severe conditions of high chloride content, high temperature or acid pH, the 29-4-2 alloy is preferable. More than five years of trouble-free service have been accumulated on trial tubes of this alloy in power plant surface condensers utilizing seawater cooling.

The E-BRITE alloy is being used to contain a number of corrosive chloride-containing process streams. For instance, this alloy has provided a cost-effective solution to a large polymer dryer SCC problem. The polymer contained 100-500 PPM chlorides which were sufficient to cause SCC of Type 316L stainless steel pipe. A total of 1220 m (4000 ft.) of 4" Schedule 10 XM-27 pipe was utilized to fabricate the dryer which has functioned continuously for four years with no corrosion problems reported.

In another application tubing is being used in gas-to-gas exchangers and well-stream coolers to cool natural gas at a well site before liquefaction for transport. The E-BRITE alloy was selected because of resistance to corrosion by carbonic acid condensate combined with resistance to SCC by chlorides. Each bank of coolers has 40 Km (25 miles) of 19.05 x 1.82 mm (3/4" x .072") and 45 Km (28 miles) of 25.4 x 2.1 mm (1" x .083") E-BRITE tubing. The smaller tubes were seal welded to Type 316L overlaid tubesheets. The larger air cooler tubes were wrapped with aluminum fins and were seal welded into Type 410S stainless steel tubesheets with nickel alloy 82 weld wire. These units were put into service three years ago and additional systems will soon be added.

E-BRITE alloy tubing has also been used in a double-effect evaporator crystallizer circuit installed as part of an effluent control system for a tungsten mine and mill[16]. Sodium sulfate waste is concentrated from 12% to 30% at 99 C (210 F) and a crystallized product is obtained. Corrosion tests indicated that the heating surfaces should be constructed of either the E-BRITE alloy or Ni-Cr-Mo alloy 825. E-BRITE alloy tubing was selected on the basis of lower cost. Type 316L tubesheets were used. The exchangers have been in service for over four years.

Caustic Solutions

One of the outstanding features of the new high chromium ferritic stainless steels is their excellent resistance to hot caustic environments. Historically, utilization of stainless steel in hot caustic solutions has been limited either because of inadequate resistance to general corrosion or susceptibility to stress corrosion cracking. Nickel or high-nickel alloys have been the standard materials of construction.

In contract to conventional austenitic stainless steels, the ferritic alloys are highly resistant to stress corrosion cracking in hot caustic solutions. Laboratory testing of stressed (U-bent) E-BRITE alloy samples in boiling 149 C (300 F) , 45% NaOH, 5% NaCl solution produced no evidence of cracking in 800 hours. Several years experience with tubing under similar conditions in caustic evaporator systems verifies the resistance of the E-BRITE alloy to caustic cracking. Similarly, bent samples of the 29-4-2 alloy, which contained welds, were exposed to hot 200 C (392 F) 50% NaOH solution. No evidence of cracking was observed in 134 hours of exposure whereas Type 304 austenitic stainless steel samples suffered severe stress corrosion cracks[2].

The new ferritic alloys also show excellent resistance to general corrosion by hot caustic solutions[11]. Laboratory test results show typical corrosion rates of less than 0.025 mm/y (1MPY) for the E-BRITE and 29-4-2 alloy in caustic of up to 50% concentration to the boiling point, as shown in Table 9. Increase in caustic concentration to 70% and temperature to 177 C (350 F) results in higher corrosion rates of about 0.1-0.4 mm/y (4-15 MPY).

Caustic Applications

Chlor-Alkali Plants

The E-BRITE alloy tubing has become a standard material of construction for caustic evaporator steam chests and other associated heat exchangers[11]. In these applications, the liquor contains sodium chloride and a small amount of sodium chlorate in addition to caustic. The E-BRITE alloy is not adversely affected by the presence of chloride or chlorate in caustic as shown in Table 10. Chlorate is known to accelerate corrosion of nickel[17]. This metal is also corroded by noncondensible gases (CO_2, O_2) in steam condensate[18] to which stainless steels are resistant.

Steamchests and Preheaters

Conditions typical of multiple-effect caustic evaporator system steamchests and preheaters are given in Table 11. In most cases, E-BRITE alloy tubing has been installed into solid nickel or nickel-clad tubesheets. One unit under construction will utilize seal welds between E-BRITE tubing and nickel tubesheets. More than 20 chlor-alkali plants are now successfully using E-BRITE tubing for one or more effects of their caustic evaporation systems. The E-BRITE alloy tubing has given good service for more than six years in three-effect systems and for four years in the newer four-effect systems. The hottest stage (first-effect) in a four-effect system is generally designed for liquor (44% NaOH, 7% NaCl) temperature of about 160 C (320 F) while in three-effect systems the hottest liquor (45% NaOH, 5% NaCl) is about 143 C (290 F). A steamchest in which E-BRITE tubes are being installed is shown in Figure 2.

The initial rationale for selecting E-BRITE tubing for cautic evaporators was corrosion resistance and lower cost than nickel. Now, after more than six years of experience, it appears that the E-BRITE alloy has significantly better resistance to corrosion/erosion by hot caustic-salt solutions typical of multiple-effect evaporation systems, in the presence of chlorates. In side-by-side tests of tubes for two years, in the first-effect of a triple-effect system, the E-BRITE alloy tubes showed no preferential erosion attack and very little corrosion. The nickel tubes showed considerable wall thinning, particularly on the ends where liquor turbulence was greatest. In another test E-BRITE and nickel tubes from the first-effect of a three-effect system were examined after two and one-half years of exposure. Corrosion rates on the E-BRITE tubes were 0.07-0.12 mm/y (2.7-4.8 MPY). Rates on the nickel tubes were 3 to 5 times higher. In another system process changes resulted in increased chlorate levels and higher velocity causing severe erosion/corrosion on nickel surfaces. Some nickel tubes were replaced with E-BRITE tubes and others were fitted with 0.3 m (1 ft.) long E-BRITE 2" Schedule 10 pipe "safe-ends" which were tack welded to the existing nickel tubesheets. After more than one year of service, the E-BRITE "safe-ends" and tubes have demonstrated excellent resistance to this environment which continues to corrode nickel.

Pulp and Paper Mills

The E-BRITE alloy has also demonstrated usefulness for handling a variety of alkaline cooking liquors used in kraft paper pulp processes. Compared to Type 316 Ti and 317 austenitic stainless steels, the E-BRITE alloy demonstrates low corrosion rate and freedom from SCC (Table 12)[19]. Other tests demonstrate good resistance of the E-BRITE alloy to white liquor under conditions of heat flux (Table 13).

Liquor Heaters

Typical pulp mill liquor heater conditions are given in Table 11. Type 304 stainless steel tubing often shows excessive corrosion and/or SCC in this environment. The E-BRITE alloy has been used as a replacement for Type 304 for several years, demonstrating low corrosion rate and freedom from SCC

(Table 14). Type 304 or 304 clad tubesheets have generally been used in conjunction with E-BRITE tubes in liquor heaters.

Evaporator Crystallizer

E-BRITE alloy tubing has given good performance for more than two years in the first-effect of a three-effect evaporator crystallizer in a pulp mill salt recovery plant under approximate operating conditions given in Table 11. The ferritic XM-27 alloy replaced Ni-Cr Alloy 600 tubing which suffered excessive corrosion due to overheating and sodium sulfide attack on plugging. Recently, 29-4-2 tubing replaced Ni-Cr Alloy 600 tubes which were pitted from the shell side of the crystallizer due to presence of about 100 PPM chloride. Selection of the 29-4-2 alloy was based on its excellent ability to resist chlorides in addition to hot caustic.

Nitric Acid

For usefulness in hot nitric acid a stainless steel must provide high resistance to general corrosion and freedom from intergranular attack. Chromium has the strongest positive alloying effect with regard to resisting nitric acid[21]. Resistance to intergranular corrosion is provided by low carbon content or by stabilization by titanium or columbium. Unfortunately, Ti or Cb in amounts needed to stabilize austenitic steels are themselves deleterious to resistance to nitric acid, leading to end-grain attack or increased general corrosion[21].

The E-BRITE and 29-4-2 alloys have high chromium content and excellent resistance to intergranular corrosion. They offer excellent resistance to nitric acid. Huey test data are given in Table 15. The 29-4-2 alloy exhibits the lowest rates, i.e., 0.08-0.10 mm/y (3-4 MPY), followed by the E-BRITE alloy with 0.10-0.12 mm/y (4-5 MPY), in this boiling 65% nitric acid environment. Types 329, 304L, 347 and 430 exhibit higher rates. Autoclave tests in 10, 20 and 30% nitric acid at 149 C (300 F), Figure 3, again demonstrate the excellent ability of the high chromium E-BRITE alloy to resist nitric acid.

Huey test data on as-welded versus non-welded 29-4-2 and E-BRITE samples are given in Table 16. No intergranular corrosion is observed on either alloy following this test.

The presence of halides in nitric acid accelerate corrosion of stainless steel[22, 23]. The E-BRITE alloy demonstrates low corrosion rates in 50% nitric acid containing chlorides and/or fluorides at 79 C (175 F) compared to Type 304 stainless or titanium (Table 17). The E-BRITE alloy has also been shown to provide resistance to nitric-hydrofluoric acids[24].

Nitric Acid Applications

The E-BRITE alloy is being used in heat exchangers in nitric acid plants. Examples are given in Table 18.

Weak Acid Condensers

E-BRITE alloy tubes and tubesheets have been used for more than five years in weak acid condensers. This ferritic alloy has given better service in one condenser than Type 304, Type 430 or Type 329 which were tried previously. In a cooler-condenser of vertical design, Type 347 stainless steel tubes suffered chloride stress corrosion cracking in the vapor space beneath the upper tubesheet. E-BRITE alloy "safe-ends" were welded to the Type 347 tubes, providing protection from the waterside chlorides as well as the hot, condensing nitric acid. The composite tubes served well for at least 15 months at which time the condenser was taken out of service because of severe corrosion on the Type 304 tubesheets.

Tail Gas Heaters

E-BRITE tubes and tubesheets have been used in tail gas heaters under conditions given in Table 18. In a unit which has operated for three years, Types 304L, 430 and 329 tubing had corroded in less than six months on the upper half of the tube bundle.

Turbine Gas Heater

An E-BRITE U-tube bundle has been in service for eighteen months in a turbine gas heater under conditions given in Table 18. E-BRITE tubes were selected following inadequate performance of Type 304 stainless. The E-BRITE tubes might be partially subject to "475 C (885 F) embrittlement" in this application. However, this is not of great concern to the end user who has experience with sigma phase embrittlement of Type 310 stainless in this application. The 475 C (885 F) embrittlement on ferritic alloys can be reversed by a reanneal at 760 C (1400 F) or higher.

Although the 29-4-2 alloy has not yet been applied to nitric acid service, its superior resistance to nitric acid and chlorides makes it an excellent candidate for conditions too servere for the E-BRITE alloy and other stainless steels.

Amines

The corrosion rate of the E-BRITE alloy has been described as distinctly better than Type 316L for alkylene and aromatic amines[14]. The corrosion rate of the ferritic steel was one-fifth to one-fiftieth the rate of Type 316L stainless in aromatic amines. Other data in monoethanolamine solutions, with and without CO_2 , show the E-BRITE alloy to offer lower corrosion rates than carbon steel, Type 304 or Type 317L stainless steel[12, 25].

Amine Applications

Gas Treating Plants

Operating conditions for a monoethanolamine regenerator reboiler are given in Table 18. E-BRITE tubes showed essentially no corrosion in this environment following two years service whereas Type 316 and Type 304 tubes exhibited 0.12 mm/y (5 MPY) and 0.79 mm/y (31 MPY) respectively, by pitting(12). Other experience with tubing in gas plants handling MEA and DEA has also been reported(12). The E-BRITE alloy is a candidate for reboilers, rich-lean exchangers, overhead condensers and reclaimers in gas treating plants.

Ethanolamine Manufacture

E-BRITE tubes seal welded to E-BRITE explosively clad tubesheets are used in a reboiler used in the manufacture of mixed ethanolamines as described in Table 18. An ethanolamine absorption column has also been retrofitted with E-BRITE alloy support plate and pall rings. Type 316L which was replaced, suffered severe selective de-nickelification at these temperatures, 200 C (392 F), similar to that associated with ammonium carbamate corrosion. In-plant corrosion studies and other data reported in the literature(26) indicated the E-BRITE alloy to be very resistant under these conditions. No corrosion was evident on the column packing and support plate after ten months service.

Alkyl Amine Reboiler

Another plant has utilized an E-BRITE alloy tube bundle in alkyl amine reboiler service as given in Table 18 for more than five years. Amine decomposition products caused severe corrosion on a variety of alloys before the E-BRITE tubing was installed.

Urea/Ammonium Carbamate

Corrosion of Type 316 or 317 stainless steel can be severe in the presence of ammonium carbamate in urea synthesis equipment, particularly if oxygen for passivation is not present(27). Increasing nickel content in austenitic stainless steel appears to be detrimental(28, 29). The nickel-free ferritic E-BRITE alloy has been shown to be resistant under urea plant conditions that severely corrode Type 316 and 316L stainless steel(26, 29). Such favorable experience has led a European urea plant designer to specify the E-BRITE alloy to be acceptable for critical areas where low oxygen conditions may prevail such as in a stripping unit. Data from an industrial urea reactor are given in Table 19. Low corrosion rate and freedom from crevice corrosion characterize the performance of the E-BRITE alloy.

Urea/Ammonium Carbamate Applications

The resistance of the E-BRITE alloy to urea/ammonium carbamate has led to its use in process equipment handling these environments. Included in the list are pumps, valves (trim and plugs), gaskets, fasteners, component parts of reactors, strippers and heat exchangers. Urea stripper distributor tubes fabricated from E-BRITE alloy have provided satisfactory service for a number of years in Europe[29] as well as in the United States.

Carbamate Reboiler

E-BRITE alloy tubes, lined tubesheets, and bonnets, have been in use for over sixteen months in carbamate reboiler service under conditions given in Table 18. Tube-to-tubesheet joints are fusion welds. The bottom section of an adjacent scrubber, also exposed to the hot carbamate environment is also fabricated of the E-BRITE alloy. Type 316L stainless suffered severe corrosion in about four months in this service.

High Pressure Carbamate Condenser

E-BRITE alloy tubes and tubesheets have given good service for five years in a high pressure carbamate condenser under conditions given in Table 18. This performance has been considerably better than that of Type 316L stainless, previously used in this service.

Although the 29-4-2 alloy has no ammonium carbamate service, it is anticipated this alloy should perform well because of its chromium and low nickel content[29].

Organic Acids

The E-BRITE and 29-4-2 alloys are very resistant to a variety of organic acids. They are used instead of austenitic stainless steel because they offer better resistance to chlorides, or they are cost-effective alternatives to nickel-base alloys.

Corrosion rates for the high purity ferritic and standard austenitic stainless steels in organic acids are given in Table 20. These and other test data given in Tables 21 and 22 illustrate that the high purity ferritic steels offer excellent resistance to organic acids compared to austenitic stainless steels[31] and to high-nickel alloys[32].

Organic Acid Applications

Laboratory and in-plant corrosion tests performed by several major chemical companies have verified the outstanding resistance of the E-BRITE alloy to several organic acids. The test results have led to use of this alloy in heat exchangers and vessels used in processing organic acids.

Organic Acid Reboilers, Calandrias, Evaporators

The E-BRITE alloy tubing is being used in reboilers handling acetic-formic acids. Details are given in Table 24. It replaced Type 316L stainless steel. Field tests demonstrated that the ferritic alloy was as resistant as more highly alloyed Ni-Cr-Mo alloys.

The E-BRITE alloy has also been specified for several heat exchangers which handle hot, dilute mixtures of formic and oxalic acids as indicated in Table 23. In-plant corrosion studies showed that the XM-27 alloy corroded at a rate of 0.03-0.05 mm/y (1-2 MPY) as compared to more than 1.27 mm/y (50 MPY) for Type 316 stainless steel. The E-BRITE alloy was cost-effective alternative to Ni-Cr-Mo alloys. One of the heat exchangers is shown in Figure 4. E-BRITE alloy tubing and explosively clad tubesheets are utilized (Figure 5). Details of the E-BRITE clad Type 316 tubesheet and tubes-to-tubesheet seal welds are shown in Figures 6a and 6b, respectively. Four heat exchangers similar to the one shown in Figure 4 are currently in service, one for more than three years. Because of continued good performance, the E-BRITE alloy has been specified for two additional units.

High Temperature Oxidation/Sulfidation

The high chromium content of the E-BRITE alloy provides resistance to oxidation that equals or exceeds that of Type 446 or Type 310 stainless steel and Ni-Cr-Fe Alloy 800 at 982 C (1800 F) under cyclic or semicontinuous conditions[6]. The high chromium, nickel-free composition also provides substantial resistance to sulfur-and vanadium-bearing environments[6] (Table 24). Other tests illustrate high resistance to wet air/SO_2 mixtures and combustion gases at 816 C (1500 F)[6].

When considering the E-BRITE alloy for elevated temperature applications, two factors must be considered:

(1) it is subject to 475 C (885 F) embrittlement on extended exposure to 350-570 C (700-1060 F)[7], and

(2) strength falls off rapidly above 540 C (1000 F)[6].

High Temperature Applications

The E-BRITE alloy is being used in a number of high temperature applications because of its resistance to oxidation and sulfidation. In most of these applications, it was selected over Type 446 because of better weldability and fabricability. In another application, the E-BRITE alloy has found use as electrodes for salt baths (sodium and barium chlorides) at 800-975 C (1470-1790 F), providing more than 50 percent improvement in life over Type 310 stainless steel.

Air Preheaters

E-BRITE elements have been in service for more than five years in a Ljungstrom wheel air preheater to a pulp mill recovery boiler[33]. On inspection, no sign of accelerated corrosion or thermal cracking has been detected. The elements, arranged in a large rotating wheel, are alternately heated by hot flue gases containing SO_3 and chlorides, and cooled by combustion air.

E-BRITE alloy tubes and wrapped-on and resistance welded fins are utilized in a shell-and-tube air preheater. The shell and tubesheets are also fabricated of this alloy (Figures 7 and 8). Flue gas containing SO_2 and SO_3 at about 400 C (750 F) flows through the shell side of this preheater. Air passes through the tubes.

Heat Recuperators

The E-BRITE alloy is being utilized in heat recuperators associated with steel mill soaking pits, glass furnaces, and in at least one case, a coke gassification plant. These recuperators recover heat from hot [up to 1150 C (2100 F)] flue gases and preheat air from ambient temperature to 760 C (1400 F). In all cases, the E-BRITE alloy was selected over Type 446 stainless steel.

Miscellaneous High Temperature Applications

In gas turbine systems, E-BRITE alloy "tiles" are used to line the combustor section, contacted by combustion gases from a variety of fuels ranging from distillates to residuals. Skin temperatures range from as low as 480 C (900 F) to as high as 1090 C (2000 F). The E-BRITE alloy replaced Type 446 and Type 442 stainless steel in this application.

E-BRITE alloy nozzles and refractory lining anchors have also been used for at least four years in a gassifier system where petroleum coke is reacted with air and steam to form a mixture of H_2, CO, CO_2, N_2, steam, H_2S and a small amount of carbonyl sulfide. Temperatures range from 815-980 C (1500-1800 F). The E-BRITE alloy was selected because of its resistance to sulfur, sodium and vanadium in the coke gases.

Reducing Acids

The E-BRITE alloy offers little resistance to reducing acids such as hydrochloric and sulfuric. Laboratory tests indicate acceptable corrosion rates for this material in up to 0.25 Wt. % HCl to 66 C (150 F), and up to 1.0 Wt. % H_2SO_4 to 38 C (100 F). The 29-4-2 alloy, on the other hand, is capable of handling 1.5% HCl and 12.5 H_2SO_4 solutions to their boiling points.

Data in Table 25 illustrates excellent resistance of the 29-4-2 alloy to boiling 1%, 5% and 10% sulfuric acid. Substantially higher corrosion rates are observed on the ferritic alloys in reducing acids when samples are activated at the start of test. This phenomenon is not observed with the austenitic Type 316 alloy. Users of ferritic alloys should be aware of this active-passive behavior before applying these alloys to reducing acids.

The 29-4-2 alloy has been applied to a few initial hot process streams where dilute reducing acids are present. One application involves an acid hydrolysis process for making alcohol from starch.

The 29-4-2 reactor tubes are exposed to HCl of less than 1% concentration at temperatures as high as 177 C (350 F). No problems were evident on the 29-4-2 alloy on inspection following three months of service.

In another test application, 29-4-2 U-bent tubes were inserted into a power plant flue gas stream. Cold water is circulated through the tubes. Hot flue gas, high in sulfur and also containing chlorides, fluorides and fly ash solids were on the outside of the tubes. Acid condensation, with chlorides and fluorides under fly ash deposits, makes the test environment extremely aggressive. After one year the tubes evidence little corrosion. Laboratory tests in an acid chloride environment (7% H_2SO_4, 3% HCl, plus $CuCl_2$ and $FeCl_3$) with crevices in place, predicted that the 29-4-2 alloy would do well (Table 26). The 29-4-2 alloy is expected to find application in flue gas scrubbers and reheater tubing where acid chlorides are present.

Summary

Ten years of laboratory and in-plant corrosion tests followed by plant installations have produced a number of applications for the ferritic E-BRITE alloy. The newer 29-4-2 ferritic alloy is just beginning to make its mark. Both of these alloys offer excellent resistance to stress corrosion cracking, pitting and crevice corrosion in the presence of chlorides. This makes these alloys suitable for aggressive water-side conditions. In the case of the 29-4-2 alloy, this includes seawater. The high chromium content of these alloys confers exceptional resistance to a number of environments including hot caustic, nitric acid, amines, urea/ammonium carbamate, organic acids, and high temperature oxidation and sulfidation. Although the E-BRITE alloy has little to offer by way of resistance reducing acids, the more highly alloyed (nickel-containing) 29-4-2 alloy has outstanding resistance to dilute reducing acids. A number of case histories are presented which give details as to how these versatile ferritic alloys are being used in heat exchangers and other applications in the chemical process industries. As the excellent resistance to corrosion which these alloys possess becomes more widely recognized, their acceptance will increase. Recognition by the CPI that these high chromium ferritics are in most cases more cost effective than many alternative alloys, should help acceptance.

REFERENCES

1. I. A. Franson, "Mechanical Properties of High Purity Fe-26 Cr-1 Mo Ferritic Stainless Steel", Met. Trans., Vol 5, p. 2257 (1974).

2. M. A. Streicher, "Microstructures and Some Properties of Fe-28% Cr-4% Mo Alloys", Corrosion, Vol. 30 (4), p. 115 (1974).

3. T. J. Nichol, "Mechanical Properties of a 29 Pct. Cr-4 Pct. Mo-2 Pct. Ni Ferritic Stainless Steel", Met. Trans. A, Vol. 8A, p. 229 (1977).

4. H. E. Deverell, "Toughness Properties of Vacuum Induction Melted High-Chromium Ferritic Stainless Steels", in ASTM STP706-Toughness of Ferritic Stainless Steels, ed. by R. A. Lula, p. 184 (1980).

5. M. A. Streicher, "Stainless Steels: Past, Present and Future", in Stainless Steel '77, Ed. by R. Q. Barr, Climax Molybdenum Company, p. 1 (1978).

6. F. K. Kies and C. D. Schwartz, "High Temperature Properties of a High Purity Ferritic Stainless Steel", J. Testing and Evaluation, Vol. 2 (2), p. 118 (1974).

7. T. J. Nichol, A. Datta and G. Aggen, "Embrittlement of Ferritic Stainless Steels", Met. Trans. A, Vol. 11A, p. 573 (1980).

8. J. M. Beigay and H. E. Deverell, "Welding the Ultrahigh-Purity Steels", Am. Mach., Vol. 123 (5), p. 112 (1979).

9. R. J. Hodges, C. D. Schwartz and E. Gregory, "Corrosion Resistance of an Electron Beam Refined 26% Cr-1% Mo Ferritic Stainless Steel", Br. Corros. J., Vol. 7 (3), p. 69 (1972).

10. M. A. Streicher, "Development of Pitting Resistant Fe-Cr-Mo Alloys", Corrosion, Vol. 30 (3), pp. 77-91 (1974).

11. A. B. Misercola, R. P. Tracy, I. A. Franson and R. J. Knoth, "The Use of E-BRITE 26-1 Ferritic Stainless Steel in Production of Caustic Soda", paper presented at the Electrochemical Society Meeting, May 4, 1976, Washington, D.C.

12. G. E. Moller, I. A. Franson, and T. J. Nichol, "Experience With Ferritic Stainless Steel in Petroleum Refinery Heat Exchangers", paper presented at CORROSION/80, NACE March 3-7, 1980, Chicago, Illinois.

13. R. J. Knoth and W. A. Matejka, "E-BRITE 26-1, A New Stainless Steel for Chemical Process Equipment", paper presented at CORROSION/74, NACE, March 4-8, 1974, Chicago, Illinois.

14. C. P. Dillon, "Use of Low Interstitial 26 Cr-1 Mo Stainless Steel in Chemical Plants", Mater. Perf., Vol. 14 (8), p. 36 (1975).

15. M. J. Johnson and I. A. Franson, "Corrosion of Stainless Steel in Saturated NaCl Brine Solutions", paper presented at CORROSION/80, NACE, March 3-7, 1980, Chicago, Illinois.

16. O. J. Malacarne and A. M. Washburn, III, "Crystallization of Sodium Sulfate For Effluent Abatement", AICHE Symp. Series, Vol. 72 (153), p. 74 (1976).

17. B. M. Barkel, "Accelerated Corrosion of Nickel Tubes in Caustic Evaporator Service", paper presented at CORROSION/79, NACE, March 12-16, 1979, Atlanta, Georgia.

18. W. Z. Friend, Corrosion of Nickel and Nickel-Base Alloys, J. Wiley & Sons, New York (1970), p. 43, 45.

19. J. P. Audouard, A. Desestret, G. Vallier, J. Chevassut, J. P. Mader, "Study and Development of Special Austenitic-Ferritic Stanless-Steel Linings For Kraft Pulp Batch Digesters", paper presented at Third International Symposium on Pulp and Paper Industry Corrosion Problems, CPPA, NACE, TAPPI, May 5-8, 1980, Atlanta, Georgia.

20. G. E. Moller, "Use and Misuse of Stainless Steels", paper presented at CPPA 1980 Maintenance Conference, Sept. 10, 1980, Vancouver, B.C.

21. U. Blom and B. Kvarnbäck, "The Importance of High Purity in Stainless Steels in Nitric Acid Service-Experience From Plant Service", Mater. Perf., Vol. 14 (7), p. 43 (1975).

22. J. B. Lowe, "Influence of Acid and Chloride Concentrations on Corrosion in a Nitric Acid Concentrator", Corrosion, Vol. 17, p. 26 (1961).

23. I. I. Tingley, "Corrosion Resistance of Five Stainless Alloys in Nitric Acid Containing Chloride", Corrosion, Vol. 14, p. 273t (1958).

24. R. Mah, K. Terada and D. L. Cash, "Corrosion of Distillation Equipment by HNO_3-HF Solutions During HNO_3 Recovery", Mater. Perf., Vol. 14 (11), p. 28 (1975).

25. K. Z. Slavoljub and M. R. Svetlana, "Corrosion Resistance of Materials of Construction in Monoethanolamine Solutions-Inhibition with Sodium Metavanadate", Proc. 4th European Symposium on Corrosion Inhibitors, p. 420 (1975).

26. J. M. A. Van der Horst, "Weld Corrosion in Urea Synthesis", Corros. Sci., Vol. 14, p. 631 (1974).

27. D. W. McDowell, Jr., "Corrosion in Urea-Synthesis Reactors", Chem. Eng., Vol. 8 (10), p. 118 (1974).

28. J. M. A. Van der Horst, "Urea Synthesis Corrosion of Stainless Steel", Ammonia Plant Safety, Vol. 14, p. 98 (1972).

29. D. Droin, "Problems of Corrosion in the Manufacture of Urea", Information Chimie, (No. 161), p. 121, Dec. (1976).

30. J. M. A. Van der Horst, "Grain Boundary Attack of Stainless in Ammoniacal Atmospheres", Werkstoff U. Korr., Vol. 26 (2), p. 128 (1975).

31. G. B. Elder, "Corrosion by Organic Acid", in Process Industries Corrosion, NACE, Houston, p. 247 (1975).

32. A. I. Asphahani, P. E. Manning, W. L. Silence, F. G. Hodge, "Highly Alloyed Stainless Materials for Seawater Applications", paper presented at CORROSION/80, NACE, March 3-7, 1980, Chicago, Illinois.

33. C. R. Morin and J. E. Slater, "Corrosion Problems in Power Generation Equipment - Individual Case Histories", paper presented to Third Int. Symp. on Corr. in the Pulp & Paper Ind., CPPA, TAPPI, and NACE, May 5-8, 1980, Atlanta, Ga.

TABLE 1

Typical Chemical Composition-Ferritic Stainless Steel

Element	Typical Analysis, Weight %			
	Type 430 (S43000)	Type 446 (S44600)	E-BRITE®(1) (S44627)	29-4-2 (S44800)
Carbon	0.07	0.10	0.002	0.005
Nitrogen	0.025	0.10	0.010	0.013
Chromium	17	27	26	29
Molybdenum	--	--	1	4
Nickel	--	--	0.10	2
Manganese	0.45	0.45	0.05	0.05
Silicon	0.45	0.45	0.25	0.10
Phosphorus	0.020	0.020	0.010	0.015
Sulfur	0.010	0.010	0.010	0.010
Columbium	--	--	0.10	--

® Registered Trade Mark, Allegheny Ludlum Industries, Inc.

(1) ASTM XM-27

TABLE 2

Physical Properties of Ferritic Stainless Steels

Property	Type 430	Type 446	E-BRITE®(1)	29-4-2
Elastic Modulus, Tension, MPa	20000	20000	20000	20000
(10^6 psi)	(29)	(29)	(29)	(29)
Density, g/cc	7.6	7.5	7.6	7.6
(lb./cu. in.)	(0.28)	(0.27)	(0.28)	(0.28)
Thermal Conductivity 100°C, W/mK	0.182	0.146	0.124	0.114
(212°F), (Btu/ft.^{2}.h.F/ft.)	(15.1)	(12.1)	(10.3)	(9.5)
Expansion Coefficient, 21-100°C, 10^{-6}/°C	10.4	10.4	10.6	9.4
(70-212°F), 10^{-6}/°F	(5.8)	(5.8)	(5.9)	(5.2)

® Registered Trade Mark, Allegheny Ludlum Industries, Inc.

(1). ASTM XM-27

TABLE 3

Mechanical Properties of Annealed Ferritic Stainless Steel Tubing[1]

Alloy	Tensile Strength MPa	Tensile Strength (ksi)	0.2% Yield Strength MPa	0.2% Yield Strength (ksi)	Elongation, % 50 mm. (2 in.)	Rockwell B Hardness
Type 430	414	(60)	241	(35)	20	90
Type 446	483	(70)	276	(40)	18	95
E-BRITE®(2)	448	(65)	276	(40)	20	90
29-4-2	483	(70)	379	(55)	20	95

® Registered Trade Mark, Allegheny Ludlum Industries, Inc.

TABLE 4

Resistance of Alloys to Stress Corrosion Cracking in Various Laboratory Tests

Alloy	Time to Failure, (Hours)[1] 42% $MgCl_2$ Boiling	26% NaCl[3] Aerated 102°C (215°F)	26% NaCl Autoclave 155°C (310°F)	26% NaCl Autoclave 200°C (390°F)
Type 304	F (8)	F (72)	F (250)	F (48)
Type 316	F (24)	--	--	--
Alloy 20Cb3	F (40)[3]	NF (2544)	--	NF (655)
29-4-2	F (19)	NF (2528)	NF (487)	NF (655)
E-BRITE®(2)	NF (200)	NF (1000)[4]	--	--

® Registered Trade Mark, Allegheny Ludlum Industries, Inc.

(1). F - Stress Corrosion Failure in Annealed U-Bend specimen.

(2). NF - No Failure in time indicated.

(3). Source: M. A. Streicher, Ref. 5.

(4) Source: M. J. Johnson and I. A. Franson, Ref. 15.

TABLE 5

Critical Pitting Potentials of Stainless Alloys In Saturated NaCl Brine at 38°C (100°F)

Alloy	mV Versus SCE		
	pH 10	pH 6	pH 2
29-4-2	+990	+990	+860
E-BRITE®(1)	+400	+420	+430
Type 316	+120	+ 10	- 20
Type 304	+ 40	- 50	- 50

® Registered Trade Mark, Allegheny Ludlum Industries, Inc.

(1). ASTM XM-27.

TABLE 6

$KMnO_4$-NaCl Pitting Tests[1]

Alloy	Pitting Resistance[2] 20°C (70°F)	50°C (120°F)	75°C (165°F)	90°C (195°F)
Type 430	F	-	-	-
Type 304	F	-	-	-
Type 316L	F	-	-	-
Alloy 20Cb3	F	-	-	-
Alloy 600	R	F	-	-
Alloy 825	R	R	F	-
E-BRITE®	R	R	F	-
Alloy "C"	-	R	R	R
Alloy 625	R	R	R	R
Titanium	R	R	R	R
AL 29-4-2	R	R	R	R

® Registered Trade Mark, Allegheny Ludlum Industries, Inc.

(1). 2% $KMnO_4$-2% NaCl, no crevices (pH 7.5)

(2). F = Failure by pitting

R = Resistant, no pitting

Source: M. A. Streicher, Ref. 5.

TABLE 7

Crevice Corrosion Tests in
Sodium Hypochlorite Solutions
96 Hour Tests at 71°C (160°F)

Alloy	Weight Loss, g/cm^2 200 PPM NaOCl	5.25% NaOCl Bleach Solution
Type 304	0.0001, C[(1)]	0.0013, C[(1)]
Type 316	0.0000, C	0.0010, C
E-BRITE®[(2)]	0.0000, N	0.0001, N

® Registered Trade Mark, Allegheny Ludlum Industries, Inc.

(1). Crevice corrosion tests per ASTM G-48 procedures:

C - Crevice corrosion present

N - No corrosion evident

(2). ASTM XM-27.

TABLE 8

EXAMPLES OF APPLICATION OF E-BRITE®[1] ALLOY TUBING TO RESIST CHLORIDES IN PROCESS INDUSTRIES HEAT EXCHANGERS

Case No.	Service	New or Retube	Previous Alloy, Replacement Reason[2]	Tube Side			Shell Side			Tubesheet Material	E-BRITE Installed
				Temperature °C	(°F)	Environment	Temperature °C	(°F)	Environment		
1.	Vent Gas Recovery Condenser	New	—	71-60	(160-140)	H_2O, approx. 500 ppm Cl^-	82	(180)	Steam, CO_2, H_2S, Hydrocarbons	T316L	1979
2.	Tetralin Oxidizer	Retube	T304, SCC	124	(255)	Tetralin/air mixture	Ambient		River water	T304	1972
3.	Oxidizer Overhead Condenser	New	—	160	(320)	Tetralin, Tetralone, residues, water	Ambient		River water	T304	1977
4.	Refining Still Condenser	Retube	T304, T316, SCC	Ambient		River water	—		Dimethylacetamide (DMA) H_2O, NaCl, 0.5-3% NaOH, trace amine hydrochloride and HCl	T304	1975
5.	Crystallizer Condenser	Retube	T304, SCC	60	(140)	Toluene, trace amounts - H_2O	Ambient		River water	T304L	1976
6.	Caustic Cooler	Retube	Nickel, pitted	49	(120)	45% NaOH, 5% NaCl	Ambient		Cooler tower H_2O, 150 PPM Cl^-, pH 7.0	Nickel	1979
7.	Condensate Cooler	New	—	121	(250)	Steam condensate	Ambient		River water	T304	1978
8.	Vent Scrubber Recycle Cooler	New	—	46	(115)	H_2O, NaOH, salts	Ambient		River water	Carbon Steel	1978
9.	Reactor Cooler	New, finned E-BRITE tubes	—	60	(140)	Tempered river H_2O	154	(310)	Propylene oxide, ethylene oxide, and catalyst	T304	1978

® Registered Trade Mark, Allegheny Ludlum Industries, Inc.

(1). ASTM XM-27

(2). SCC - Stress Corrosion Cracking

TABLE 8 (Cont.)

Case No.	Service	New or Retube	Previous Alloy, Replacement Reason (2)	Tube Side Temperature °C	(°F)	Tube Side Environment	Shell Side Temperature °C	(°F)	Shell Side Environment	Tubesheet Material	E-BRITE Installed
10.	CO Compressor Intercooler	New	—	182	(360)	CO, N_2	Ambient		River water	Carbon Steel	1978
11.	CO Compressor Aftercooler	New	—	171	(340)	CO, N_2	Ambient		River water	Carbon Steel	1978
12.	Natural Gas Recycle Cooler	New (U-tubes)	—	54	(130)	Natural gas	Ambient		River water	Carbon Steel	1978
13.	Residues Evap. Tails Cooler	Retube	T304, SCC	160	(320)	Toluene diisocyanate, residues	35	(95)	Tempered H_2O steam sparged occasionally	T304L	1974
14.	Tempered Water Cooler	Retube	T304, SCC	274	(525)	Napthol, tetralone, tetralol, napthalene, tetralin, dihydronapthalene, Dowtherm	77	(170)	Tempered H_2O	T304	1977
15.	Reactor Cooler	Retube	T304, SCC	154	(310)	Polymer polyols	71	(160)	Tempered H_2O	T304	1974
16.	Tergitol Reactor	Retube	T304, SCC	Ambient		River water	121	(250)	Reaction of oxide, glycol with caustic catalyst	T304	1975
17.	Dowtherm Cooler	New	--	349	(660)	Dowtherm	Ambient		River water	Carbon Steel	1978
18.	Feedwater Heater	New	—	129	(265)	H_2, CO, methane	182	(360)	Treated water	T304	1978
19.	Evaporator Cond.	New	—	Ambient		River water	160	(320)	Toluene diisocyanate, residues	T304L	1974
20.	Air Compressor Aftercooler	New	—	Ambient		River water			Hot air	Carbon Steel	1974
21.	Waste Heat Steam Generator	Retube	T316, SCC	139	(281)	Boiler Feedwater	192	(375)	Pthalic Anhydride Vapor	T316	1978

Source: *Corrosion/81*, Paper No. 117, NACE

TABLE 9

Resistance of Ferritic Stainless Steels To Hot Caustic Solutions

% NaOH	Temperature °C	(°F)	Duration Days	Corrosion Rate, mm/y. (MPY) E-BRITE®(1)	29-4-2
15	104	(220)	5	0.001 (0.04)	-
25	110	(230)	7	0.000 (0.01)	-
30	116	(240)	5	0.001 (0.05)	-
50	143	(290)	5	0.003 (0.11)	0.003 (0.12)
60	157	(315)	4	0.084 (3.30)	0.020 (0.78)
70	177	(350)	4	0.15-0.38 (6-15)	0.097 (3.8)

® Registered Trade Mark, Allegheny Ludlum Industries, Inc.

(1). ASTM XM-27.

TABLE 10

Resistance of E-BRITE®[1] Alloy to Caustic Solutions Containing NaCl and $NaClO_3$

Concentration, %			Temperature		Corrosion Rate	
NaOH	NaCl	$NaClO_3$	°C	(°F)	mm/y.	(MPY)
20	10	-	104	(220)	0.015	(0.6)
45	5	-	143	(290)	0.041	(1.6)
50	-	-	135	(275)	0.003	(0.1)
50	5	-	152	(305)	0.076	(3.0)
50	5	0.1	152	(305)	0.069	(2.7)
50	5	0.2	152	(305)	0.028	(1.1)
50	5	0.4	152	(305)	0.028	(1.1)

® Registered Trade Mark, Allegheny Ludlum Industries, Inc.

(1). ASTM XM-27

TABLE 11

Examples of Use of E-BRITE®(1) Alloy Tubing in Heat Exchangers Handling Caustic Liquors

Service	New or Retube	Previous Alloy Replacement Reason	Tube Side			Shell Side			Tubesheet Material
			Temperature °C	(°F)	Environment	Temperature °C	(°F)	Environment	
Caustic Evaporator Steamchests, Preheaters	New	Nickel, Cost Performance	52	(125)	15% NaOH, 15% NaCl (3rd or 4th effect)			Steam	Nickel
			163	(325)	50% NaOH, 5% NaCl (1st effect)			Steam	Nickel
Pulp Mill Liquor Heaters	New	T304, SCC	149	(300)	7% NaOH, 3% Na_2S, 1.6% Na_2CO_3	185	(365)	Steam	T304
Pulp Mill First Effect Evaporator/Crystallizer	Retube	Alloy 600, Excessive Corrosion	118	(245)	21% NaOH, 5% Na_2S, 0.3% Na_2SO_4, 1.5% Na_2CO_3, 7.8% NaCl	149	(300)	Steam	Alloy 600

® Registered Trade Mark, Allegheny Ludlum Industries, Inc.

(1). ASTM XM-27

TABLE 12

Corrosion of E-BRITE®[1] Vs. Austenitic Alloys In Pulp Liquors

Alloy	Corrosion Rate, mm/y (MPY)					
	White Liquor[1]		2 Parts White Liquor to 1 Part Black Liquor (by Vol.)[1]			
	170°C	(338°F)	130°C	(266°F)	170°C	(338°F)
E-BRITE[2]	0.02	(0.8)	0.02	(0.8)	0.02	(0.8)
Type 316 Ti	0.10	(3.9)	0.10	(3.9)	0.03	(1.2)
Type 317L	0.10	(3.9)	0.10	(3.9)	0.03	(1.2)

(1)

Liquor Composition		White Liquor	Black Liquor
NaOH	g/l	133.6	6.2
Na_2S	g/l	15.0	0.9
Sulfidity	%	11.2	-
$Na_2S_2O_3$	g/l	11.9	9.2
Na_2CO_3	g/l	37.5	36.3
NaCl	g/l	1.6	0.9

Source: J. P. Audouard et.al., Ref. 19.

® Registered Trade Mark, Allegheny Ludlum Industries, Inc.

(2) ASTM XM-27

TABLE 13

Corrosion of E-BRITE®[1] Vs. Austenitic Alloys In White Liquor Under Conditions of Heat Flux

Alloy	Corrosion Rate in White Liquor, mm/y (MPY)			
	Heat Flux[2]		Jet Impingement[3]	
E-BRITE[1]	.005-.005	(0.2)	.06	(2.4)
Type 316 Ti	.10 -.20	(3.9-7.9)	.35	(13.8)
Type 317L	.05 -.10	(2.0-3.9)	.20	(7.9)

® Registered Trade Mark, Allegheny Ludlum Industries, Inc.

(1). ASTM XM-27

(2). Sample kept at 160°C (320°F) immersed in white liquor at 80°C (176°F).

(3). Sample temperature 160°C (320°F), white liquor jet at 80°C (176°F) projected against sample.

Source: J. P. Audouard et.al., Ref. 19.

TABLE 14

In-Plant Corrosion Tests of Alloys In White Liquor[1]

Alloy	Corrosion Rate mm/y	(MPY)
E-BRITE®[2]	0.0	(0.0)
Alloy 600	0.005	(0.2)
Type 329	0.008	(0.3)
Type 310	0.010	(0.4)
Alloy 800	0.020	(0.8)
Alloy 400	0.023	(0.9)
Alloy 825	0.041	(1.6)
Type 304	0.168	(6.6), SCC[3]
Alloy 625	0.173	(6.8)
Type 316	0.516	(20.3), SCC[3]
Carbon Steel	0.886	(34.9)

® Registered Trade Mark, Allegheny Ludlum Industries, Inc.

(1) White Liquor: 28% ($NaOH+Na_2S$), 7.8% NaCl 1.5% Na_2CO_3, 3% Na_2SO_4

Temperature: 127°C (261°F)

Duration: 154 Days

(2) ASTM XM-27

(3) SCC - Stress Corrosion Cracking

Source: G. E. Moller, Ref. 20.

Source: *Corrosion/81*, Paper No. 117, NACE

TABLE 15

Corrosion of Stainless Steel in the Huey Test[1]

Alloy	Corrosion Rate mm/y (MPY)
29-4-2	0.08-0.10 (3-4)
E-BRITE®[2]	0.10-0.12 (4-5)
Type 329	0.15-0.18 (6-7)
Type 304L	0.20-0.25 (8-10)
Type 347	0.25-0.41 (10-16)
Type 430	0.69-0.91 (27-36)

® Registered Trade Mark, Allegheny Ludlum Industries, Inc.

(1). ASTM A262, Practice C, Boiling 65% nitric acid, five 48-hour periods. Annealed samples.

(2). ASTM XM-27.

TABLE 16

Huey Tests - Welded Vs. Non-Welded
E-BRITE® and 29-4-2 Ferritic Alloys

Alloy	Huey Test(1) Corrosion Rate, mm/y (MPY)	
	As-Welded	Non-Welded
E-BRITE®(2)	0.11 (4.4)(3)	0.11 (4.3)
	0.11 (4.5)(4)	
29-4-2	0.07 (2.9)(3)	0.07 (2.9)

® Registered Trade Mark, Allegheny Ludlum Industries, Inc.

(1). Huey Test: ASTM A262, Practice C. Boiling 65% nitric acid, five 48-hour periods.

(2). ASTM XM-27.

(3). Autogenous welds.

(4). Filler metal added.

Source: *Corrosion/81*, Paper No. 117, NACE

TABLE 17

Influence of Chloride And/Or Fluoride On Corrosion of Stainless Steel and Titanium In 50% Nitric Acid at 79°C (175°F)

Solution[2]			Corrosion Rate, mm/y (MPY)[1]					
HNO_3	PPM Cl^-	PPM F^-	E-BRITE®[3]		Type 304		Titanium	
50%	--	--	0.005	(0.2)	0.015	(0.6)	0.091	(3.6)
50%	300	--	0.010	(0.4)	0.025	(1.0)	0.117	(4.6)
50%	--	20	0.025	(1.0)	0.079	(3.1)	0.305	(12.0)
50%	300	20	0.041	(1.6)	0.102	(4.0)	0.914	(36.0)
50%	1000	100	0.089	(3.5)	0.241	(9.5)	1.115	(43.9)

® Registered Trade Mark, Allegheny Ludlum Industries, Inc.

(1). Corrosion rate is average of two 48-hour periods.

(2). Chloride added as HCl, fluoride as HF.

(3). ASTM XM-27.

TABLE 18

Examples of Use of E-BRITE®(1) Alloy Tubing in Heat Exchangers Handling Nitric Acid, Amines and Urea/Ammonium Carbamate Environments

Service	New or Retube	Previous Alloy, Why Replaced	Tube Side			Shell Side			Tubesheet Material
			Temperature °C	(°F)	Environment	Temperature °C	(°F)	Environment	
Weak Acid Condenser	Retube, New	T304, T430, T329	232	(450)	NO, NO_2, O_2, H_2O Condensing to form 40% HNO_3	21	(70)	Cooling water	E-BRITE, T304L
Tail Gas Heater	New	T304L, T430, T329	399 230	(750) in, (450) out	Process gas: NO, NO_2, O_2, N_2, H_2O	66 230	(150) in, (450) out	Tail gas: N_2, H_2O, residual NO_X	E-BRITE
Turbine Gas Heater	Retube U-bundle	T304L, inadequate life	230 590	(450) in, (1100) out	Tail gas: N_2, H_2O residual NO_X	910 590	(1670) in, (1100) out	Process gas: NO, NO_2, O_2, N_2, H_2O	T304
MEA Regenerator Reboiler	Retube	T304, T316, show pitting	182 126	(360) in, (260) out	Shift gas	116	(240)	MEA + CO_2, 23-308 PPM Cl^-	
Mixed Ethanolamines Manufacture, Reboiler	New	T316L showed de-nickelification	200	(392)	Mixed ethanolamines			Steam	E-BRITE Clad
Alkyl Amine Reboiler	Retube	Various, high corrosion	177	(350)	Alkyl amines			Steam	E-BRITE
Carbamate Reboiler	New	T316L, short life	127	(260)	NH_3, CO_2, H_2O	163	(325)	Steam	E-BRITE lined
High Pressure Carbamate Condenser	New	T316L severely corroded	149	(300)	Ammonium carbamate, NH_3, CO_2	100	(212)	Steam generated from water	E-BRITE

® Registered Trade Mark, Allegheny Ludlum Industries, Inc.

(1). ASTM XM-27

Source: *Corrosion/81*, Paper No. 117, NACE

TABLE 19

Corrosion Study in Urea/Ammonium Carbamate Industrial Reactor Environment[1]

Alloy	Corrosion Rate Mg/dm^2/d (MPY)		Appearance
E-BRITE®[2]	12.4	(2.3)	No Corrosion
20 Cr-25 Ni-4.5 Mo-1.5 Cu	26.5	(4.8)	Crevice Corrosion under PTFE Washer

® Registered Trade Mark, Allegheny Ludlum Industries, Inc.

(1). Environment: 30% Urea, 17.5% CO_2, 33% NH_3, 19.5% H_2O

Temperature: 185°C (365°F)

Test Duration: 62.5 Days
Oxygen added for passivation during test.

(2). ASTM XM-27

Source: R. Droin, Ref. 29

TABLE 20

Corrosion by Boiling Organic Acids

Alloy	Corrosion Rate, mm/y (MPY)[2]					
	45% Formic		10% Oxalic		20% Acetic	
29-4-2	0.018	(0.7)	0.018	(0.7)	0.003	(0.1)
E-BRITE®(1)	0.067	(3)	0.072	(3)	0.000	(0)
Type 316	0.277	(11)	1.018	(40)	0.003	(0.1)
Type 304	1.225	(48)	1.271	(50)	0.024	(1)

® Registered Trade Mark, Allegheny Ludlum Industries, Inc.

(1). ASTM XM-27

(2). Average of five 48-hour periods.

Source: *Corrosion/81*, Paper No. 117, NACE

TABLE 21

Corrosion in Boiling Formic Acid(1)

% Formic Acid	Corrosion Rate, mm/y (MPY)					
	E-BRITE®(2)		Type 304		Type 316	
5	<0.03	(<1.0)	0.79	(31.1)	0.04	(1.5)
70	<0.03	(<1.0)	4.04	(159.0)	0.50	(19.5)

® Registered Trade Mark, Allegheny Ludlum Industries, Inc.

(1). 96-hour tests

(2). ASTM XM-27

Source: G. B. Elder, Ref. 31.

TABLE 22

Comparison of E-BRITE and Other Alloys In Boiling 88% Formic and 99% Acetic Acids

Alloys	Corrosion Rate, mm/y (MPY)[1]			
	88% Formic Acid		99% Acetic Acid	
E-BRITE®[2]	<0.003	(<0.1)	0.013	(0.5)
Alloy C276	0.046	(1.8)	0.010	(0.4)
Alloy G	0.101	(4)	0.041	(1.6)
Alloy 825	0.076	(3)	0.051	(2)
Alloy 625	0.229	(9)	0.010	(0.4)
Alloy 200	0.330	(13)	0.101	(4)
Alloy 600	0.381	(15)	0.203	(8)
Type 304	2.438	(96)	0.457	(18)
Type 316	0.229	(9)	0.051	(2)

® Registered Trade Mark, Allegheny Ludlum Industries, Inc.

(1). Average of four 24-hour exposures

(2). ASTM XM-27

Source: A. I. Asphahani et.al., Ref. 32.

TABLE 23

Examples of Heat Exchangers Used in Processing Organic Acids Which Utilize E-BRITE®(1) Alloy Tubing

Service	Tube Side		Shell Side		Tubesheet Material
	Temperature °C (°F)	Environment	Temperature °C (°F)	Environment	
Organic Acid Reboilers	99-110 (210-230)	6-8% Acetic Acid, 2-3% Formic Acid, Water		Steam	Alloy C-276
Organic Acid Calandrias, Evaporators	149-166 (300-330)	2-5% Formic Acid, 0.1-1% Oxalic Acid, 70% Water		Steam	E-BRITE Clad Type 316

® Registered Trade Mark, Allegheny Ludlum Industries, Inc.

(1). ASTM XM-27.

TABLE 24

Resistance of E-BRITE and Other Alloys To High Temperature Environments

Alloy	Weight Change, mg/cm^2 Cyclic Oxidation[1]	Sulfidation[2]	Fuel Ash[3]
E-BRITE®[4]	+ 2.2	- 2.3	-13.5
Type 446	+10.0	- 6.1	-13.6
Type 310	-90.3	- 5.9	-22.0
Alloy 800	-83.2	-96.7	-46.7

® Registered Trade Mark, Allegheny Ludlum Industries, Inc.

(1). Specimens alternately exposed to 1800°F (982°C) for 15. min. and room temperature for 5 min., total 1,000 hours.

(2). Specimens exposed to 90% NA_2SO_4/10% NaCl mixture for 6 hours at 1380°F (750°C).

(3). Specimens exposed to 80% V_2O_5/20% Na_2SO_4 mixture for 16 hours at 1292°F (700°C).

(4). ASTM XM-27.

Source: Kies and Schwartz, Ref. 6.

Source: *Corrosion/81*, Paper No. 117, NACE

TABLE 25

Corrosion of Stainless Steel in Boiling, Dilute Sulfuric Acid

Alloy	Corrosion Rate, mm/y (MPY)[1]					
	1% H_2SO_4		5% H_2SO_4		10% H_2SO_4	
	Non-Activated	Activated[2]	Non-Activated	Activated[2]	Non-Activated	Activated[2]
E-BRITE®[3]	0.018 (0.7)	13.74 (541)	0.356 (14)	76.71 (3020)[4]	879.0 (34,630)[4]	2580 (101,400)[4]
Type 316	0.551 (21.7)	0.660 (26)	2.489 (98)	2.72 (107)	8.61 (339)	8.74 (344)
29-4-2	0.005 (0.2)	0.076 (3)	0.025 (1)	0.279 (11)	0.025 (1)	0.457 (18)

® Registered Trade Mark, Allegheny Ludlum Industries, Inc.

(1). Corrosion Rates: Average of five 48-hour test periods.

(2). Sample activated at beginning of each test period.

(3). ASTM XM-27.

(4). One period or less because of high corrosion rate.

TABLE 26

Crevice Corrosion of Stainless Steel and Nickel Alloy in Simulated Flue Gas Scrubber Environment(1)

Alloy	Weight Loss, g/cm^2 (2)		
	24°C (75°F)	50°C (122°F)	70°C (158°F)
Type 316	.0002	.0081	.0206
Nickel Alloy 625	.0000	.0001	.0042
29-4-2	.0000	.0000	.0002

(1). 7 vol. % H_2SO_4, 3 vol. % HCl, 1 wt. % $CuCl_2$, 1 wt. % $FeCl_3$

(2). Samples 25.4 x 50.8 x 1.6 mm. (1 x 2 x .062 in.) with PTFE crevices.

Source: *Corrosion/81*, Paper No. 117, NACE

Figure 1 Heat Exchanger Tube Bundle With E-BRITE® Alloy Tubes and Type 316 Stainless Steel Tubesheets

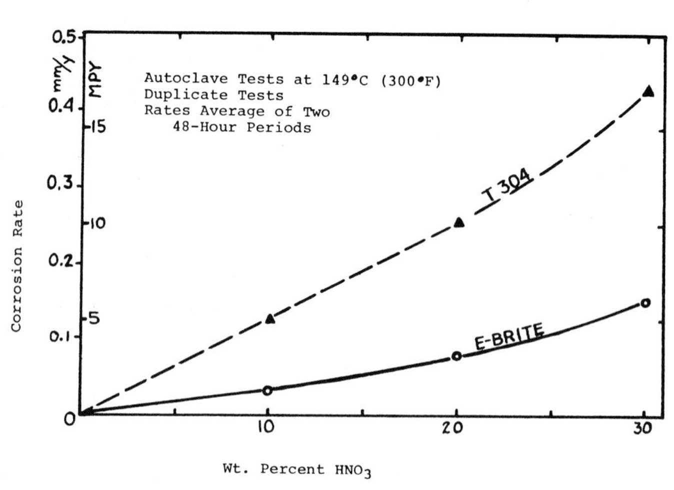

Figure 3 Corrosion of E-BRITE® Alloy and Type 304 Stainless Steel in Nitric Acid at 149°C (300°F)

Figure 2 E-BRITE® Alloy Tubes Being Installed in Steamchest for Caustic Evaporation System

Figure 4 Evaporator With E-BRITE® Alloy Tubing in Service Handling Formic and Oxalic Acids

Figure 5 E-BRITE® Alloy Tubes and Clad Tubesheet

Figure 6a Cross Section of E-BRITE® Explosively Clad Type 316 Tubesheet

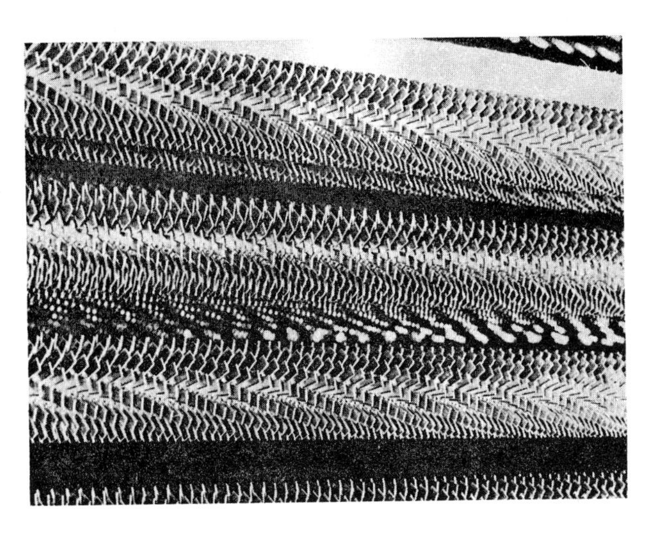

Figure 7 E-BRITE® Alloy Tubing With Wrapped and Resistance Welded E-BRITE® Alloy Fins

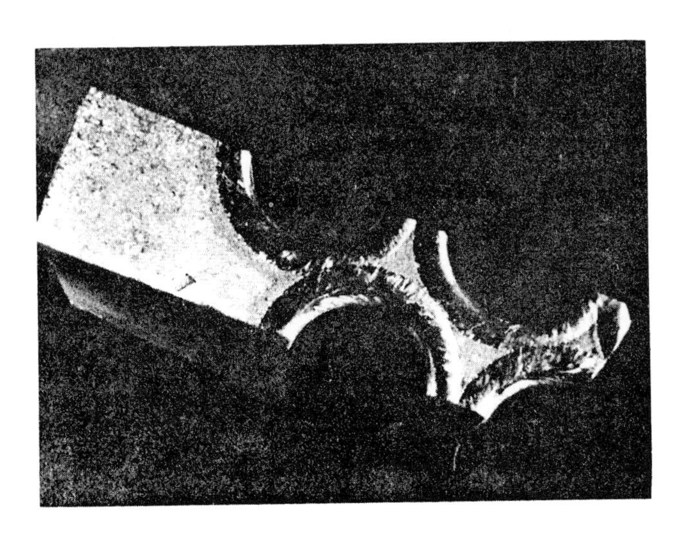

6b Seal Welds on E-BRITE® Clad Tubesheet

Figure 8 Shell and Finned Tube Air Preheater Fabricated Entirely of E-BRITE® Alloy Components

Experience With Ferritic Stainless Steel In Petroleum Refinery Heat Exchangers

G. E. Moller
Consultant
2224 Chelsea Road
Palos Verdes, Ca. 90274

I. A. Franson
Allegheny Ludlum Steel Corporation
Research Center
Brackenridge, Pa. 15014

and

T. J. Nichol
Allegheny Ludlum Steel Corporation
Tubular Products Division
80 Valley Street
Wallingford, Ct. 06492

INTRODUCTION

The ferritic E-BRITE® alloy (ASTM XM-27), and AISI Type 430 ferritic stainless steel have drawn increasing attention by petroleum refiners for use in heat exchangers. To a limited extent, the 26-1S (ASTM XM-33) and Type 444 (18 Cr-2 Mo) ferritic alloys have also found application. The choice of these ferritic alloys is made because of their resistance to stress corrosion cracking (SCC) in addition to their generally good corrosion resistance to most refinery streams. Current development of new, highly alloyed, cost effective ferritic stainless steels, offer the refining industry new alternatives for combating severely corrosive situations.

This paper will first examine the properties of the new and conventional ferritic stainless steels as background for

® Registered Trade Mark, Allegheny Ludlum Industries, Inc.

evaluation of a large number of actual refinery experiences from twelve refineries.

Historical

AISI Types 405, 409 and 410 ferritic stainless steels have been used extensively in petroleum refinery equipment[1-4]. These alloys are routinely used for fractionation column trays, vessel cladding, catalyst screens, heat exchanger tubing, valve trim, pump shafts, pump cases, impellers and wear rings. AISI 405 and 410 are also occasionally used for high temperature piping and less frequently for fired heater tubing.

In the late 1950's, AISI 430 started being applied as a high temperature sulfidation-resistant heat exchanger tubing material. Type 430 was chosen instead of Type 304 stainless steel because of its resistance to chloride stress corrosion cracking (SCC) and its compatibility with ferritic alloy tube sheets.

In the late 1960's the more corrosion-resistant, E-BRITE (ASTM XM-27) ferritic alloy was introduced and was broadly applied in the chemical, petrochemical and oil refining industries. The E-BRITE alloy, which is vacuum-induction melted (VIM), has low levels of the interstitial elements, carbon and nitrogen. This high level of purity improves corrosion resistance, ductility and weldability as compared to the conventional ferritic stainless steels.

Recently, VIM is also being utilized to produce more highly corrosion-resistant ferritic alloys such as 29 Cr-4 Mo and 29 Cr-4 Mo-2 Ni[5]. Other ferritic stainless steels such as Type 439 (18 Cr), Type 444 (18 Cr-2 Mo) and 26-1S (ASTM XM-33), have been produced by argon-oyxgen-decarburization (AOD). These steels, because of their higher carbon and nitrogen content, have inferior ductility, compared to the VIM alloys.

The ferritic stainless steels are most commonly used as tubing and bar products in the petroleum refining industry. Welded tubing, the least costly tubular product, is welded under mill conditions requiring strict process control. Vessels and tanks have been fabricated successfully from the E-BRITE alloy using proper welding procedures. However, the ferritic stainless steels, as a group, are not as easily welded as the conventional austenitic stainless steels.

PROPERTIES OF FERRITIC STAINLESS STEELS

Available Alloys

Typical analyses for ferritic stainless steels which are being used or considered for petroleum refinery heat exchanger service are given in Table 1. It should be noted that AOD processing typically yields carbon and nitrogen levels of about 250 PPM each as illustrated by the compositions of Type 439 (ASTM XM-8), 26-1S (ASTM XM-33), and Type 444 (18-2) alloys. Further reductions in carbon and nitrogen to 20-50 PPM and 100-130 PPM, respectively, are achieved by VIM as illustrated by the E-BRITE (ASTM XM-27), 29-4 and 29-4-2 alloys. Also to be noted from Table 1 is that the new ferritic alloys, except 29-4 and 29-4-2, contain titanium and/or columbium additions for stabilization against the possibility of intergranular corrosion in spite of their low carbon and nitrogen content.

Table 1 also shows that several of the new ferritic alloys contain molybdenum in addition to chromium for added resistance to pitting and crevice corrosion. The nickel addition to the 29-4-2 alloy significantly improves resistance to reducing acids at the expense of some resistance to stress corrosion cracking. Streicher has reviewed the development of these and other new ferritic alloys in detail(5).

Physical Properties

The physical properties of the conventional and new ferritic stainless steels are quite similar for all the steels as shown in Table 2. Thermal conductivity and coefficient of expansion decrease with increase in alloy content. Compared to austenitic stainless steels, the ferritic alloys as a group have similar elastic modulus and density and considerably lower coefficient of expansion. The thermal conductivity of the ferritics with 17-18 percent chromium is about 50 percent higher than that of Type 304 stainless steel. The difference is less in the more highly alloyed ferritic alloys such as 29-4-2. All of the ferritics have a body-centered-cubic crystal structure and are ferromagnetic at room temperature.

Mechanical Properties

Minimum tensile properties and maximum hardness values for ferritic stainless steel tubing, as defined by ASTM A268-77, are given in Table 3. It can be seen that reasonable strength and ductility are available in tubing made from these alloys. Design stresses for Type 430, E-BRITE (XM-27) and 26-1S (XM-33) are given in Table UHA-23 of Section VIII, Division 1, of the ASME Boiler and Pressure Vessel Code. The Type 439 (XM-8) alloy and the Type 444 (18 Cr-2 Mo) alloy are covered by Code Cases 1653 and 1825, respectively. Code coverage for the 29-4 and 29-4-2 alloys has not yet been granted.

All of these steels experience an embrittlement phenomenon called "475 C (885 F) embrittlement" between 343 C (650 F) and 510 C (950 F). Some are also subject to sigma phase precipitation between 540 C (1000 F) and 880 C (1600 F)(5). Although the alloys are ductile at the operating temperature, they may be brittle at room temperature. For these reasons, use of these steels in Code applications is limited to 343 C (650 F).

The conventional ferritic stainless steels, such as Type 430, may lack notch ductility at room temperature in sections as light as 1.6 mm (.063 in.). Thus, bending and roller expansion of tubing into tubesheets have sometimes been problems. In addition, welds and heat-affected zones on Type 430 may be quite brittle in the as-welded condition if martensite is present in the microstructure.

The new alloys, because of their low carbon and nitrogen contents, are ferritic at all temperatures and are not hardenable. The VIM alloys, E-BRITE, 29-4 and 29-4-2, because of ultralow carbon and nitrogen content, exhibit excellent notch ductility in sections up to about 12.7 mm (0.50 in.) thickness. It should be clear that careful inert-gas (argon or helium) shielded welding techniques, which preserve the low carbon and nitrogen levels, must be utilized in order to maintain ductility in welds made on these alloys[6].

One other important mechanical characteristic of the ferritic stainless steels is their good fatigue properties. The ratio of fatigue limit (10^7 cycles) to tensile strength for these alloys is typically 60 percent as compared to about 35 percent for the austenitic stainless steels[7].

Corrosion Resistance

The ferritic stainless steels, ranging from the straight chromium Type 430 or 439 alloys to the high purity, 29-4-2 alloy, are moderately to highly resistant to corrosion by petroleum hydrocarbons, water and contaminants. This will be discussed in the following sections.

Hydrocarbons

Pure hydrocarbons, whether they are straight chain or cyclic, saturated or unsaturated, light or heavy molecules, are not corrosive to steel. Corrosion results from contaminants present in the hydrocarbons such as sulfur, salt, organic acids, water, oxygen, carbon dioxide, and products of nitrogen.

Sulfidation

Most crude oil contains sulfur as hydrogen sulfide or organic sulfur compounds. The attack of alloys at high temperature by sulfur compounds is called sulfidation. The mechanisms of attack have been described by Couper[8], and Sharp and Haycock[9]. The corrosion rate of iron based alloys increases substantially with temperature above 250 C (482 F). Resistance to sulfidation is conferred by chromium[10, 11], as shown in Figure 1. The ferritic stainless steels, with nominally 12 percent chromium up to 29 percent chromium, thus provide modest to extremely high resistance to sulfidation, depending on the alloy.

To be resistant to high temperature hydrogen sulfide in the presence of hydrogen, alloys also require chromium. A minimum of 12 percent chromium is generally recognized as the lower limit for satisfactory resistance[12]. The corrosion rates of several alloys from a long-term hydrotreater exposure are given in Table 4[13]. The beneficial effect of high chromium content is obvious.

High Temperature Oxidation

Although the ferritic stainless alloys have low strength at temperatures above 649 C (1200 F), their high chromium content provides excellent resistance to oxidation at these temperatures. The E-BRITE alloy, with 26 percent chromium, has been shown to exhibit better cyclic oxidation resistance than Type 310 stainless steel and iron-nickel-chromium Alloy 800[14]. In this respect, the E-BRITE alloy is similar in performance to conventional Type 446 stainless steel.

Naphthenic and Other Organic Acids

The new ferritic alloys with high chromium content and molybdenum have excellent resistance to organic acids[5]. Because of this value, the E-BRITE alloy is replacing Type 316L stainless steel in petrochemical processes. Likewise, the E-BRITE, 29-4 and 29-4-2 alloys should perform well in naphthenic and/or carbolic acid environments.

Chloride Pitting and Crevice Corrosion

Pitting and crevice corrosion by chlorides can occur on either process side or water side of oil refinery heat exchanger tubing if alloy content is inadequate. Type 430, for instance, has sometimes been found to be inadequate from the water side[1]. High chromium content and molybdenum additions to the new ferritic stainless steels improve their pitting and crevice corrosion resistance. The critical pitting potential of the Type 439, E-BRITE, 29-4 and 29-4-2 ferritic alloys in a saturated salt brine at 38 C (100 F) are compared to potentials for Types 304 and 316 stainless in Table 5. These data illustrate the improvement to pitting corrosion resistance offered by the new ferritic alloys. The data in Table 6 show the maximum temperature to which the ferritic alloys, nickel-base alloys "625" and "C" and titanium can be heated in 10% ferric chloride with a crevice present and not be corroded. High temperature resistance in this test is indicative of excellent resistance to crevice corrosion[15].

By way of comparison, the Types 430 and 439 ferritic steels resist pitting to roughly the same extent as Type 304 austenitic stainless steel and the 18 Cr-2 Mo alloy approaches the pitting resistance of Type 316. The E-BRITE alloy has demonstrated superior pitting resistance than Type 316[5] and is widely used in chemical industry heat exchangers to handle aggressive river waters. Types 29-4 and 29-4-2 ferritic alloys are considerable improvements over conventional stainless steels and are comparable to nickel-chromium alloy 625, nickel-base alloy "C", and to titanium, which are resistant to seawater. The 29-4 alloy, for instance, has performed well for four years in a power plant surface condenser which utilizes seawater for cooling purposes.

Stress Corrosion Cracking by Chlorides or Caustic

The ferritic stainless steels are highly resistant to chloride stress corrosion cracking. Nickel and copper content of these steels must be controlled to low levels in order to be resistant in laboratory magnesium chloride stress corrosion tests. The 2 percent of nickel in the 29-4-2 alloy, for instance, causes failure in the magnesium chloride test. However, as shown by the data in Table 7, cracking does not occur in sodium chloride environments[5]. Chlorides are present in small quantities in refinery process streams as well as various water streams. Stress corrosion cracking potential is widespread. The ferritic stainless steels provide a good solution to this problem.

In addition to resisting stress corrosion cracking by hot chloride environments, the ferritic stainless steels have also demonstrated resistance to stress corrosion cracking by hot caustic environments[16, 17]. The E-BRITE alloy tubing, for instance, is used widely in caustic evaporator systems.

Polythionic Acid

Polythionic and sulfurous acids form from iron sulfide scale upon exposure to air and moisture[18]. These dilute, cold acids cause intergranular corrosion and intergranular cracking of sensitized and stressed austenitic stainless steels. Little information is available concerning the resistance of the ferritic stainless steels to polythionic acid cracking. However, the ferritic stainless steels, if improperly stabilized, can become sensitized[19]. Intergranular corrosion has been experienced in non-stabilized and sensitized E-BRITE produced prior to 1972. It might then be expected that sensitized ferritic stainless steels could be attacked by polythionic acid. However, currently produced ferritic steels are stabilized with columbium and/or titanium to prevent sensitization. Therefore, these alloys should be resistant to polythionic acid attack. Most heat exchangers using ferritic stainless steels do not operate in the sensitizing range because the use of these alloys above 344 C (650 F) is not recommended due to "475 C (885 F) embrittlement".

Mineral Acids

The high chromium content of the new ferritic stainless steels provides excellent resistance to oxidizing acids. This is shown by the Huey Test (boiling 65 percent nitric acid) data in Table 8. On the other hand, as shown by the data in Table 9, the stainless steels are heavily corroded by reducing acids such as boiling 1% HCl or 10% H_2SO_4. A standout exception is the 29-4-2 ferritic steel which remains passive and exhibits a low corrosion rate in these environments. The 29-4-2 alloy will resist up to 1.5% HCl and 12.5% H_2SO_4 from room temperature to their boiling points. As shown by the data in Table 10, the 29-4-2 alloy exhibits extraordinary resistance to an acid chloride crevice corrosion test. The combination of stress corrosion resistance, resistance to chlorides and to reducing acids makes the 29-4-2 alloy an excellent candidate for severe heat exchanger service such as found in crude distillation overhead systems.

Wet Hydrogen Sulfide

Wet hydrogen sulfide, in the absence of oxygen, is not very corrosive to carbon steel. Corrosion rates vary from 0.10-0.30 mm/year (4-12 mpy). The addition of oxygen, which forms sulfurous acid, makes H_2S corrosive. Other contaminants such as chlorides, make water-bearing H_2S much more corrosive. As such, ferritic stainless steels are extremely resistant to wet H_2S. Hence, their choice as the standard API valve trim. However, Types 410 and 410S, when hardened above Rc 22, will display cracking in moist sulfide environments.

Carbonic Acid

Stainless steels, both austenitic and ferritic, are essentially immune to corrosion by carbon dioxide and water at all refinery operating temperatures. Illustrative data are presented in Table 11.

Miscellaneous Environments

In addition to the environments already discussed, the ferritic stainless steels are also useful for other refinery corrodents such as ammonia[21]. The E-BRITE alloy has also been shown to exhibit excellent resistance to caustic[16]. This alloy has also been shown to resist monoethanolamine (MEA) environments[22]. Data for 20 and 70 percent MEA, with and without carbon dioxide, are given in Table 12.

APPLICATIONS OF FERRITIC STAINLESS STEELS

Seven companies contributed information for this study about the use of Type 430 and E-BRITE (XM-27) in petroleum refining heat exchanger service. A few case histories for the 26-1S (XM-33) alloy were also provided. Data concerning the Type 430 installations are given in Table 13. The E-BRITE and 26-1S data are given in Table 14.

Tables 13 and 14 are arranged by refining process, and by individual type of service within the process. If the ferritic stainless steel was chosen to retube a heat exchanger, the alloy used previously and its life to replacement is provided. Important construction information such as size and number of tubes and tubesheet material is given as are approximate process conditions. The length of time the ferritic stainless steel tubing has been in service or the time to failure, in months, is given in the "History" column. For instance, in Case Number 1 in Table 13, the 120 indicates 120 months service with no failure. This tubing would be expected to last much longer and, possibly, for the life of the plant. In Case Number 9 of Table 13, the tubing lasted only 10 months and was replaced with Alloy 800.

Applications of Type 430; Table 13

Type 430 stainless steel tubing is being used successfully in crude distillation, hydrodesulfurizing and hydrocracking services. This alloy is most widely used for hot services such as crude preheat and feed preheat, exchanging with hot, heavy oil, and feed-effluent service in hydrotreaters.

In two gas plant reboiler applications, Cases 20 and 21 in Table 13, Type 430 tubing perforated after 68 to 73 months service. The Type 430 alloy has a low level of resistance to wet chloride conditions such as may be found in wet naphtha streams. A photograph of Type 430 tubing removed from an unsaturated gas plant reboiler is shown in Figure 2. The tubing shows deep pitting on the tube side. The bundle was retubed with the E-BRITE (XM-27) alloy which has a greater resistance to chloride environments than 18 Cr steels as illustrated by the data of Tables 5 and 6.

In Case 22, Table 13, Type 430 tubes in a prefractionator feed heater, heavily fouled with iron oxide and sulfide from upstream, perforated from the shell side in the center of the bundle. This condition was attributed to corrosion which occurred during shutdown periods after water washing and hydrotesting.

Case 9, Table 13, is an example of a reported pitting failure of Type 430 tubing in hydrotreater effluent coolers under iron sulfide and ammonium hydrosulfide deposits. The reason for this performance has not been thoroughly explained. The attack may be attributable to crevice corrosion by ammonium chloride under heavy deposits of iron sulfide.

In Case 10 and 11, Type 430 tubes were found to be embrittled from 475 C (885 F) embrittlement within the roller expanded area of the tubesheet. Both heat exchangers were in the same refining unit and were being heated with 390 C (750 F) reactor effluent. The tube ends apparently were heated to temperatures in excess of 343 C (650 F) in the tubesheet with the hot effluent on one side versus 260-395 C (500-560 F) hydrocarbon on the opposite side. Performance of the Type 430 alloy was satisfactory until rerolling was attempted. Nevertheless, tube life was excellent and the exchangers were retubed with Type 430.

Type 430 ferritic stainless steel is a stress corrosion cracking resistant material which is successfully being used in petroleum refining service. Type 439 alloy, which is a new, higher purity version of Type 430, should find equal success in refineries. These alloys remain free of voluminous self-generated corrosion product compared to carbon steel and, therefore, maintain excellent heat transfer characteristics. Type 430, and therefore Type 439, should not be applied to wet hydrocarbon services such as in reboilers and in locations where tubes might remain wet during lengthy shutdowns. To extend life of Type 430 or 439 tubes, bundles should be dried after hydroblast and hydrotest.

Applications of the E-BRITE (XM-27) Alloy and 26-1S (XM-33); Table 14

The E-BRITE (XM-27) alloy is being used successfully in crude distillation, gas separation, hydrodesulfurization, hydrogen production, fuel gas sweetening, lube oil refining and sour water stripping. To a limited extent the 26-1S (XM-33) alloy has also been applied to some of these services. These alloys are chosen because they couple resistance to high temperature sulfidation, chloride pitting, acid gases such as CO_2, H_2S and HCN, organic acids and ammoniacal compounds with immunity to chloride stress corrosion cracking. They are limited by 475 C (885 F) embrittlement, attack by chlorides under heating and boiling conditions, and by brackish and saltwater. The E-BRITE (XM-27) and 26-1S (XM-33) alloys are not applicable to every heat exchange condition in the processes mentioned above.

Case 19 in Table 14 covers the successful use of E-BRITE (XM-27) alloy tubing for an unsaturated gas plant debutanizer reboiler. This alloy has been in this service for 73 months to date. Figure 2 shows Type 430 tubing which had previously perforated by pitting in this service. An extremely deteriorated carbon steel tube bundle from a saturated gas plant debutanizer reboiler is shown in Figure 3. This bundle, retubed with E-BRITE (XM-27) and 26-1S (XM-33) is performing well with 58 months service life to date. Figures 4 and 5 are photographs of E-BRITE (XM-27) tubes from Case 27, Table 14, which replaced Admiralty and Type 430 tubing previously used. The Admiralty suffered deposit attack under moist sulfidic sludge. The Type 430 tubes pitted on the water side. The high chromium-molybdenum ferritic stainless steel is working satisfactorily as a stripper overhead condenser in this case.

Figure 6 shows comparative performance of Types 304, 316 and E-BRITE (XM-27) alloy tubing from a hydrogen plant monoethanolamine regenerator reboiler, Case 35, Table 14. MEA polymerizes at temperatures above 200 C (390 F) to form very corrosive organic complexes. E-BRITE (XM-27) alloy tubing, which is resistant to organic acids, urea and ammonium carbamate, shows superior resistance in this application also.

However, in Case 39, Table 14, E-BRITE alloy tubing corroded rapidly in an MEA reclaimer. The MEA, in this case, was in a fuel gas treating plant, so H_2S, CO_2 and chlorides are present in addition to sludge and the heat stable salts and degradation products of MEA. Corrosion was smooth and occurred close to the hot tube sheet only. Reclaimers are, by intention, traps for solids and degraded, high boiling polymers.

Case 5, Table 14, a rerun reboiler, illustrates that the E-BRITE (XM-27) alloy can suffer localized deposit attack or crevice corrosion under hot evaporative conditions in the presence of water. The extract flash column condenser in Case 44, Table 14, also suffered pitting from saltwater intrusion. This plant is serviced by saltwater for cooling duty.

Cases 1 and 47, Table 14, illustrate that some E-BRITE (XM-27) tubing was produced prior to 1977 with welds contaminated with carbon and/or nitrogen. The result was that the weld bead and heat affected zone of the tubing lacked ductility and corrosion resistance. Case 48, Table 14, indicates that some tubes were produced without correct annealing and intergranular corrosion cracking was encountered. Present production of the E-BRITE (XM-27) alloy by an integrated steel producer is not subject to these early problems. Quality control for ductility is maintained through flattening, flaring and bend tests on finished tubing as dictated by ASTM A268-77. Resistance to intergranular corrosion is checked via ASTM A-262, Practice B, the ferric sulfate, sulfuric acid test.

The case histories outlined in Table 14 illustrate that the E-BRITE (XM-27) alloy, and to some extent 26-1S (XM-33) has provided petroleum refiners with excellent performance in a number of services. These steels combine resistance to chloride stress corrosion cracking with a level of corrosion resistance substantially greater than that of Type 430 stainless steel. Care should be exerted in applying these steels where they might be corroded such as under wet reboiling or strong, hot acid salt conditions. The more highly alloyed 29-4-2 ferritic alloy, with better resistance to acid and hot chlorides, may be an appropriate material for these more severe refinery applications.

SUMMARY

Type 430 ferritic stainless steel tubing has been used for a variety of petroleum refinery heat exchanger applications for some time. For the most part this alloy has performed well. Exceptions have been where hot chlorides have been present, causing failures due to pitting or under-deposit crevice corrosion, or where operating temperatures in excess of 343 C (650 F) have caused 475 C (885 F) embrittlement. The newer, more highly alloyed E-BRITE (XM-27) alloy has provided the refiner even broader application. In addition to providing excellent resistance to chloride stress corrosion cracking as Type 430 and other ferritic stainless steels do, the E-BRITE (XM-27) alloy offers better pitting and general corrosion resistance. Now, an even more highly alloyed ferritic stainless tubing alloy, 29-4-2, is available to the petroleum refiner for combatting corrosion. This alloy will handle aggressive cooling waters including seawater, in addition to a wide variety of refinery process stream corrodents. Because of this broader corrosion resistance now offered by the ferritic stainless steels, in addition to stress corrosion cracking resistance, these alloys are expected to find ever increasing usage in petroleum refineries.

ACKNOWLEDGEMENTS

The authors gratefully acknowledge and appreciate the contributions of the Union Oil Company, Mobil Oil Corporation, Chevron U.S.A., Exxon Company, U.S.A., Suntech, Inc., Lion Oil Company, and Getty Refining Company which made this presentation possible.

References

1. E. L. Bereczky, "Stainless Steels in the Petroleum Industry", in Handbook of Stainless Steels, Ed. by D. Peckner and I. M. Bernstein, McGraw-Hill, 44-1 (1977).

2. "The Role of Stainless Steels in Petroleum Refining", American Iron and Steel Institute, Washington, D.C., (1977).

3. R. Q. Barr, "A Review of Factors Affecting the Selection of Steels for Refining and Petrochemical Applications", Climax Molybdenum Company, Greenwich, Conn., (1971).

4. J. F. Lancaster, "Materials For the Petrochemical Industry", International Metals Reviews, Vol. 23, No. 3, p. 101 (1978).

5. M. A. Streicher, "Stainless Steels: Past, Present and Future", in Stainless Steel '77, Ed. by R. Q. Barr, Climax Molybdenum Company, p. 1 (1978).

6. J. M. Beigay and H. E. Deverell, "Welding the Ultrahigh-Purity Steels", American Machinist, Vol. , No. 5, p. 112 (1979).

7. I. A. Franson, "Mechanical Properties of High Purity Fe-26 Cr-1 Mo Ferritic Stainless Steel", Met. Trans., Vol. 5, p. 2257 (1974).

8. A. S. Couper, "Laboratory Tests to Evaluate Corrosion Resistance of Sulfidic Media", API, Refining Division, Houston, (1961).

9. W. H. Sharp and W. E. Haycock, "Sulfide Scaling Under Hydrorefining Conditions", API, Refining Division, New York, May (1959).

10. N. F. McConomy, "High Temperature Sulfidic Corrosion in Hydrogen Free Environment", Proc. Am. Pet. Inst., Vol. 43, Sec. 3, p. 79 (1963).

11. G. A. Nelson, Corrosion Data Survey, NACE, (1967).

12. A. S. Couper and J. W. Gorman, "Computer Correlations to Estimate High Temperature H_2S Corrosion in Refinery Streams", Mater. Prot., Vol. 10, No. 1, p. 31, (1971).

13. G. E. Moller, Unpublished data.

14. F. K. Kies and C. D. Schwartz, "High-Temperature Properties of a High-Purity Ferritic Stainless Steel", J. Testing and Evaluation, Am. Soc. for Testing and Materials, Vol. 2, No. 2, p. 118 (1974).

15. R. J. Brigham, "Temperature as a Pitting Criterion", Corrosion, Vol. 29, No. 1, p. 33 (1973).

16. A. B. Misercola, et.al., "The Use of E-BRITE 26-1 Ferritic Stainless Steel in Production of Caustic Soda" Paper presented at the Electrochemical Society Meeting, May 4, 1976, Washington, D.C.

17. M. A. Streicher, "Microstructures and Some Properties of Fe-28% Cr-4% Mo Alloys", Corrosion, Vol. 30, No. 4, 115 (1974).

18. R. L. Piehl, "Stress Corrosion Cracking by Sulfur Acids, Proc. API, Div. 3, Vol. 44, p. 189 (1964).

19. R. L. Cowan and C. S. Tedmon, "Intergranular Corrosion of Iron-Nickel-Chromium Alloys", Advances in Corrosion Science and Technology, Vol. 3, p. 332, Plenum Press, (1973).

20. L. T. Overstreet, Unpublished data.

21. H. H. Uhlig, The Corrosion Handbook, John Wiley & Sons, New York, 799 (1948).

22. K. Z. Slavoljub, and M. R. Svetlana, "Corrosion Resistance of Materials of Construction in Monoethanolamine Solutions-Inhibition with Sodium Metavanadate", Proc. 4th European Symposium on Corrosion Inhibitors, p. 420 (1975).

TABLE 1

Typical Chemical Composition - Ferritic Stainless Steel

Element	Type 430 (S43000)	Type 439[1] (S43025)	Type 444[2] (S44400)	26-1S[3] (S44626)	E-BRITE® [4] (S44627)	29-4 (S44700)	29-4-2 (S44800)
	Typical Analysis, Weight %						
Carbon	0.07	0.03	0.02	0.02	0.002	0.005	0.005
Nitrogen	0.025	0.020	0.025	0.025	0.010	0.013	0.013
Chromium	17	18	18	26	26	29	29
Molybdenum	-	-	2	1	1	4	4
Nickel	-	0.20	0.20	0.25	0.10	0.10	2
Manganese	0.45	0.35	0.35	0.30	0.05	0.05	0.05
Silicon	0.45	0.30	0.15	0.30	0.25	0.10	0.10
Phosphorus	0.040	0.020	0.020	0.030	0.010	0.015	0.015
Sulfur	0.010	0.010	0.010	0.010	0.010	0.010	0.010
Titanium	-	0.80	0.30	0.50	-	-	-
Columbium	-	-	0.35	-	0.10	-	-

® Registered Trade Mark, Allegheny Ludlum Industries, Inc.

(1) ASTM XM-8

(2) 18-2

(3) ASTM XM-33

(4) ASTM XM-27

TABLE 2

Physical Properties of Ferritic Stainless Steels

Alloy	Elastic Modulus Tension MPa	(10^6 psi)	Density g/cc	(lb./cu.in.)	Thermal Conductivity 100 C (212 F) W/m.K	(Btu/ft.^{2}hr.F/ft.)	Expansion Coefficient x 10^6 21-100 C °C^{-1}	(70-212 F) (°F^{-1})
Type 430	20000	(29)	7.6	(0.28)	26.1	(15.1)	10.4	(5.8)
Type 439[1]	20000	(29)	7.6	(0.28)	24.2	(14.0)	10.1	(5.6)
Type 444[2]	20000	(29)	7.6	(0.28)	24.2	(14.0)	10.4	(5.8)
26-1S[3]	20000	(29)	7.6	(0.28)	17.8	(10.3)	10.1	(5.6)
E-BRITE® [4]	20000	(29)	7.6	(0.28)	17.8	(10.3)	10.6	(5.9)
29-4	20000	(29)	7.6	(0.28)	17.3	(10.0)	9.4	(5.2)
29-4-2	20000	(29)	7.6	(0.28)	16.4	(9.5)	9.4	(5.2)

® Registered Trade Mark, Allegheny Ludlum Industries, Inc.

(1) ASTM XM-8

(2) 18-2

(3) ASTM XM-33

(4) ASTM XM-27

TABLE 3

Mechanical Properties of Annealed
Ferritic Stainless Steel Tubing(1)

Alloy	Tensile Strength MPa	(ksi)	0.2% Yield Strength MPa	(ksi)	Elongation 50 mm (2 in.)	Rockwell B Hardness
Type 430	414	(60)	241	(35)	20	90
Type 439(2)	414	(60)	207	(30)	20	90
Type 444(3)	414	(60)	310	(45)	20	95
26-1S(4)	469	(68)	310	(45)	20	100
E-BRITE®(5)	448	(65)	276	(40)	20	90
29-4	483	(70)	379	(55)	20	95
29-4-2	483	(70)	379	(55)	20	95

(1) Minimum tensile properties and maximum hardness per ASTM A268-77

(2) ASTM XM-8

(3) 18-2

(4) ASTM XM-33

(5) ASTM XM-27

TABLE 4

Sulfidation (Corrosion) Rates of Metals and Alloys
In High Temperature Hydrotreating Environments(1)

Alloy	Chromium Content, %	Corrosion Rate mm/Yr.	(mpy)
Carbon Steel	0	0.76	(30)
2 1/4 Cr-1/2 Mo	2.2	1.24	(49)
9 Cr-1 Mo	9	0.39	(15.2)
Type 410	12	0.14	(5.5)
Type 304	18	0.05	(2.1)
Type 310	25	0.01	(0.4)
Cast HF Modified	23	0.01	(0.4)

Source: Ref. 13

(1) Temperature: 421 C (790 F)
Partial Pressure of H_2S: 158.6 kN/m^2 (23 psia)
Time of Exposure: 16000 hours

TABLE 5

Critical Pitting Potentials for Stainless Steel In Saturated NaCl Solution at 38 C (100 F)

Alloy	Potential, mV vs. SCE		
	pH 10	pH 6	pH 2
Type 304	+ 40	- 50	- 50
Type 316	+ 120	+ 10	- 20
E-BRITE®(1)	+ 400	+420	+430
29-4	+1020	+940	+880
29-4-2	+ 990	+990	+860

®Registered Trade Mark, Allegheny Ludlum Industries, Inc.

(1)ASTM XM-27

TABLE 6

Crevice Corrosion Resistance of Stainless Steel
10% Ferric Chloride With Crevices*

Alloy	Maximum Temperature of No Crevice Corrosion, °C	(°F)
Type 304	<-2.5	(<27.5)
Type 316	2.5	(27.5)
Type 444[(1)]	2.5	(27.5)
E-BRITE®[(2)]	20-25	(68-77)
Alloy 625	<50	(<120)
29-4	50	(120)
29-4-2	50	(120)
Alloy "C"	65-74	(150-165)
Titanium	77	(170)

*Artificial PTFE crevices in contact with metal specimen during test.

(1) 18-2

(2) ASTM XM-27

TABLE 7

Stress Corrosion Cracking Resistance of Stainless Steels (U-Bend Samples, Boiling Solutions)

Alloy	Time to Failure, Hours	
	42% $MgCl_2$	26% NaCl
Type 304	<3	<72
Type 439[1]	No Failures (200 Hrs)	No Failures (200 Hrs.)
E-BRITE®[2]	No Failures (200 Hrs.)	No Failures (1000 Hrs.)
29-4	No Failures (2136 Hrs.)	No Failures (200 Hrs.)
29-4-2	5-19	No Failures (8760 Hrs.)[3]

®Registered Trade Mark, Allegheny Ludlum Industries, Inc.

(1) ASTM XM-8

(2) ASTM XM-27

(3) Other exposures of 2528 hours produced no failures at pH 7.3 or pH 4.0.

TABLE 8

Corrosion Rates in Boiling 65% Nitric Acid (Huey Test - ASTM A262-Practice C)

Alloy	Chromium Content, %	Corrosion Rate mm/yr.	(mpy)
Type 430	17	.5-.9	(20-36)
Type 304/304L	18	.2-.6	(8-24)
Type 347	18 (Cb)[1]	.25-.4	(10-16)
Type 309 Cb	22 (Cb)[1]	.13	(5)
E-BRITE®(2)	26 (Cb)[1]	.11	(4.3)
29-4-2	29	.06	(2.4)

®Registered Trade Mark, Allegheny Ludlum Industries, Inc.

(1)Stabilized against intergranular corrosion with columbium.

(2)ASTM XM-27.

TABLE 9

Corrosion Rate of Stainless Steel In Boiling Hydrochloric and Sulfuric Acid

Alloy	Corrosion Rate, mm/yr. (mpy)			
	1% HCl		10% H_2SO_4	
Type 304	81	(3,200)	400	(15,750)
Type 316	71	(2,800)	22	(870)
Type 430	1500	(59,000)	6400	(252,000)
Type 444[1]	850	(33,500)	2400	(94,500)
E-BRITE®[2]	Active 2000 Passive 0.7	(78,700) (28)	3400	(133,900)
29-4	Active 500 Passive 0.2	(21,600) (8)	1300	(51,000)
29-4-2	0.2	(8)	0.2	(8)

Source: Streicher, Reference 5

® Registered Trade Mark, Allegheny Ludlum Industries, Inc.

(1) 18-2

(2) ASTM XM-27

Source: *Corrosion/80*, Paper No. 53, NACE

TABLE 10

Crevice Corrosion of Stainless Steel In Scrubber Reheater Environment(1)

Alloy	Weight Loss, g/cm^2(2) 24 C (75 F)	50 C (122 F)	70 C (158 F)
Type 316	.0002	.0081	.0206
Nickel Alloy 625	.0000	.0001	.0042
29-4-2	.0000	.0000	.0002

(1) 7 vol. % H_2SO_4, 3% vol. % HCl, 1 wt. % $CuCl_2$, 1 wt. % $FeCl_3$

(2) 25.4 x 50.8 x 1.6 mm (1 x 2 x .062 in.) coupons with PTFE crevices

TABLE 11

Corrosion Data From Hydrogen Production Illustrating Relative Resistance of Iron-Base Alloys With Increasing Chromium Content Vs. Carbonic Acid

Temperature: 150 C (300 F)

Environment: 56 mol. % H_2O
6.8 mol. % CO_2
35.6 mol. % H_2
1.6 mol. % $CO+CH_4$

Alloy	Nominal Chromium Content, %	Corrosion Rate mm/yr.	(mpy)
Carbon Steel	0	>2	(>79)
Type 505 (5 Cr)	5	.4	(16.1)
9 Cr	9	.2	(6.8)
Type 410	12	.06	(2.5)
Type 430	17	.005	(0.2)
Type 304	18	NIL	(.02)

Source: Reference 20

TABLE 12

Corrosion of Alloys In Boiling Monoethanolamine (MEA) Solutions (504 Hour Tests)(1)

Concentration MEA, %	Corrosion Rate, mm/yr. (mpy) Carbon Steel	Type 304	Type 317L	E-BRITE®(2)
20	0.0 (0.0)	0.0 (0.0)	0.0 (0.0)	0.0 (0.0)
20 + CO_2	1.3 (49.7)	0.0 (0.0)	0.0 (0.0)	0.0 (0.0)
70	0.07 (2.7)	0.02 (1.0)	0.01 (0.3)	0.0 (0.0)
70 + CO_2	3.3 (128.6)	0.24 (9.4)	0.01 (0.4)	0.01 (0.2)

®Registered Trade Mark, Allegheny Ludlum Industries, Inc.

(1) Source: Slavoljub and Svetlana, Reference 22

(2) ASTM XM-27

TABLE 13

PETROLEUM REFINING

AISI TYPE 430 STAINLESS STEEL HEAT EXCHANGER TUBING EXPERIENCE

ASTM A268-78 TP430

Case No.	Process	Service	New or Retube	Previous Alloys	Life of Previous Alloy	No. of Tubes	Size (2)	BWG	Shellside (3)(4) Temp. Range	Shellside Press	Shellside Stream	Tubeside Temp. Range	Tubeside Press	Tubeside Stream	Tube Sheet Mat'l	History (5)
1	Crude Dist.	Crude vs Hvy G.O. - 4 units	New	-	-	1100 each	1	16	345/415		Desalted Crude	570/400		Hvy G.O.		120; light pitting to 1mm deep.
2		Flashed crude vs. Lower P.A. 2 units	New	-	-	670 each	1	16	565/465		Lower P.A.	415/445		Flashed Crude		120; very fine pitting only.
3		Flashed crude vs. Diesel	New	-	-	494	1	16	640/520		Diesel Dist.	445/465		Flashed Crude		120; very sharp pits 0.7mm deep on shell side.
4		Lt. Dist. Reboiler	New	-	-	632	1	16	515/520		Lt. Dist.	665/600		Crude Col. Botts.		120; no corrosion.
5		Sweetner Feed Coolers. (Air Cooler)	New	-	-	38	1					410/120		Lt. Dist. 435F cut		120; no corrosion. Remainder of bundle is carbon steel.
6		Waste Heat Boiler 2 units	New	-	-	404 each	1	16	270		Boiler F.W.	600/465		Vac. Tower Botts.		120; no corrosion.
7	FCC	Steam Generator	New	-	-	1250	1	16	270		Boiler F.W.	570/490		G.O.		120; no corrosion. Carbon steel top half.
8	HDS.	Feed/Effluent	Retube	9 Cr.	18	300 U	1	14	775/700	700	HC, H_2 1.5% S	550/610	700	HC, H_2, H_2S, NH_4	304 Cl.	62; Pitting..01mm deep from hydro-test water. 9Cr sulfidized.
9	HDS.	Effluent Cooler (Air cooler)	New	-	-	1232	1	10	-	-	-	310/120	1270	HC, H_2, H_2S, NH_4	Steel	10 months only; heavy corr. under FeS deposits. Repl. with Inc. 800.
10		Feed/Effluent	Retube	9Cr	14	554	1	14	506/560	750	HC, H_2 H_2S, NH_4	750/600	750	HC, H_2, H_2S, NH_4	9Cr	132 months life; retubed with 430; tubes suffered 475C embritt.
11		Dehex. Reboiler	Retube	9Cr	14	250	1	12	380/500	30	HC	750/600	750	HC, H_2, H_2S, NH_4	9Cr.	98; 89; 475C Em-britt in tubesheet zone.
12		Prefract. Reboiler	Retube	9Cr	36	600	3/4	14	480/550	50	HC	650/500	50	HC+H_2S	C.S.	82; no corr.

TABLE 13 (continued)

Case No.	Process	Service	New or Retube	Previous Alloys	Life of Previous Alloy	No. of Tubes	Size (2)	BWG	Shellside (3)(4) Temp. Range	Shellside Press	Shellside Stream	Tubeside Temp. Range	Tubeside Press	Tubeside Stream	Tube Sheet Mat'l	History (5)
13		Feed/Effluent	New	-	-	513 U	1	16	390/475	650	React. Feed.	600/435	650	React. Eff.	304 Cl	86; no reported corr.
14		Stripper Feed	Retube	C.S.	12	244	3/4	16	630/494	145	Stripper Botts.	383/490	490	Stripper Feed	410 and C.S.	67; no corrosion.
15		Stripper Pre-heater	New	-	-	544 U	1	11	95/325	90	Stripper Feed	430/331	2400	Stripper Effluent	Inconel clad.	84; pitting occured after 48 mos; 0.5 mm deep. Now arrested.
16		Feed/Effluent	Retube	410	132	256 U	3/4	12	475	680	React. Feed	650	650	React. Eff.	410	48 then plant re-vamp. new 430 in-stalled in 1978; 430 shows no corr. 410 scaled and coked.
17	HDS.	Stripper Feed vs. React. Effluent (2 units)	New	-	-	705 each	3/4	16	280/550		Stripper Feed.	650/400	-	React. Eff.		132; pits I.D. & O.D. 0.5mm deep.
18		Feed/Effluent	New	-	-	344	1	12	500/530		React. Feed	570/535	-	React. Eff.	304 Cl	96; very fine pit-ting 0.05mm.
19		Stripper Feed vs Reactor Effluent	New	-	-	176 U	3/4	16	296/500	-	Stripper Feed	560/400	-	React. Eff.	304 Cl	132; sharp pits about .12mm deep.
20	Sat. Gas Plant	Debutanizer Reboiler	New	-	-	1792	3/4	16	369/402	166	HC + S + Cl + H_2O	541/426	135	HC + S + Cl	C.S.	68 with perfora-tions; replaced with 1351 XM-27 & 441 XM-33 tubes. 2mm loss under de-posits in storage.
21	Unsat. Gas Plant	Debutanizer Reboiler	New	-	-	1350	3/4	16	573/341	160	Hvy.Cycle Oil, H_2S, S	335/356	90	He	C.S.	73 with perfor-ations. Retubed with XM-27.
22	HC	Prefraction-ator Feed Heater	New	-	-	1218	3/4	16	550/350	155	FCC Charge Stock	320/400	170	G.O. Feed	C.S.	96 but perforated from shellside corrosion in center of bundle. Moist sludge.
23		Debutanizer Reboiler	New	-	-	728 U	3/4	14	480/526	240	Debut. Botts.	660/563	1660	1 stage effluent H_2S 22 psia	410 overlay	96; no corrosion.

Source: *Corrosion/80*, Paper No. 53, NACE

TABLE 13 (continued)

Case No.	Process	Service	New or Retube	Previous Alloys	Life of Previous Alloy	No. of Tubes	Size (2)	BWG	Shellside (3)(4) Temp. Range	Shellside Press	Shellside Stream	Tubeside Temp. Range	Tubeside Press	Tubeside Stream	Tube Sheet Mat'l	History (5)
24		Heavy Hydro-crackate Reboiler.	New	-	-	343	1	10 & 12	430/443	31	Heavy Crackate	663/532	1650	1 Stage Effluent	410 overlay	96; no corrosion.
25		Medium Hydro-crackate Reboiler	New	-	-	270	1	10 & 12	287/291	29	Medium Crackate	532/515	1650	Reactor Effl.	C.S.	96; no corrosion.
26		Recycle Gas Heater	New	-	-	300	1	14 & 16	130/399	1840	Recycle Gas H_2S 8.7 psia	515/440	1642	1 Stage React. Eff.	C-Mo	96; no corrosion.
27		Feed Heater 2 units	New	-	-	425 U each	3/4	14 & 16	150/521	1900	G.O. Feed.	602/315	1660	2nd stage React. Eff. H_2S 8.9 psia	C-Mo	96; no corrosion.

Notes:
(1) Life is in months
(2) Size is in inches; 3/4 inch equals 19.1mm; 1" = 25.4mm
(3) Temperatures are in ^oF; $^oC = (^oF-32) \times 5/9$
(4) Press is in PSIG; X 6.89 = kN/m^2
(5) Numbers are months of life to 1980 with no problems or corrosion as noted.
(6) Abbreviations:
CTW - Cooling Tower Water
HC - Hydrocarbon; Hydrocracker
ADM - Admir - Admiralty
U - U bent tubes
HDS - Hydrodesulfurizer
FCC - Fluid catalytic cracker.

TABLE 14

PETROLEUM REFINING

E-BRITE - FERRITIC STAINLESS STEEL HEAT EXCHANGER TUBING EXPERIENCE

ASTM A 268-76 GRADES XM 27 & XM 33 (5)

									Shell Side (3)(4)			Tubeside				
Process	Case No.	Service	New or Retube	Previous Alloys	Life(1) Previous Alloy	No. of Tubes	Size (2)	BWG	Temp. Range	Press	Stream	Temp. Range	Press	Stream	Tube Sheet Mat'l	History(5)(6)
Crude Distill	1	Atmos. O.H. Condenser	Retube	C.S.	—	2200	3/4	16	240/150	35	O.H. Vapor	80/150		CTW	C.S.	53; 7 faulty tubes plugged; no prob. since
	2	Atmos. O.H. vs. Crude	Retube	CA 715 70-30	144	288	1	16	304/170	28	O.H. Vapor	150/300	304	Crude	Monel Cl.	52.
	3	Atmos. O.H. vs. Crude	Retube	CA 715 70/30 Cu Ni	180	576	1	16	304/170	35	O.H. Vapor	230	220	Crude	NRB	54; 14 trial tubes with no corrosion.
	4	Atmos. O.H. vs. Crude	Retube	C.S.	22	6	1	16	325/225	35	O.H. Vapor	200/300	50	Crude	C.S.	57
	5	Rerun Reboiler	Retube	5 Cr 321	17 4	224	3/4	16	340/330	58	Naphth + H_2O + Cl-	450	400	Steam	C.S.	41 on XM-27 to failure. 5Cr and E-Brite had broad pits under deposits
	6	Crude Charge	Retube	C.S.	41	1140	3/4	16	590/455	120	Bottoms	425/475	357	Crude	C.S.	40 } Test tubes 48 with no loss. Corr. rate on C.S. was 1mm/year.
	7	Crude Charge	Retube	C.S.	31	1140	3/4	16	590/455	120	Bottoms	425/475	357	Crude	C.S.	34 }
	8	Depentanizer Feed Preheat	Retube	C.S. 5 Cr Monel	6 7 93	580	3/4	16	250/300	98	Naphtha	485/400	130	Vac. Tower Pump Around	C.S.	41
Crude Distill	9	Atmos. O.H. vs. Crude	Retube	C.S.	18	274U	1	18	180/250	280	Crude	460/340	90	Kerosene	C.S.	23, XM-33; Carbon steel suffered under deposit attack.
	10	Vacuum Col. O.H. vs Crude. Two crude units	Retube	C.S.	24	1010 each	1	16	410/480		Crude	720 in	Vac	Steam non-cond.	C.S.	58 for Unit 1 44 for Unit 2
	11	Crude vs. Vac. Botts	Retube	5 CR	60	300	1	16	675/595	150	Vac. Botts.	475/495	200	Crude	5 Cr.	34

TABLE 14 (continued)

Process	Case No.	Service	New or Retube	Previous Alloys	Life(1) Previous Alloy	No. of Tubes	Size (2)	BWG	Shell Side (3)(4) Temp. Range	Shell Side Press	Shell Side Stream	Tubeside Temp. Range	Tubeside Press	Tubeside Stream	Tube Sheet Mat'l	History(5)(6)
Delayed Coker	12	Waste Heat Steam Generator	Retube	Steel	24	950	3/4	18	240/377	175	Boiler Feed Water	600/425	175	Hvy G.O.	C.S.	30; Partial retube; steel corroded due to level control prob.
	13	Interstage Cooler	Retube	Admir.	24	238U	3/4	18	233/100	75	Lt. HC + NH_4, H_2S	80/115	50	CTW(1)	Adm.	42; Admir. corroded shellside.
	14	After cooler	Retube	Admir.	12	188U	3/4	18	194/100	230	Lt. HC + NH_4 + H_2S	85/120	55	CTW	Adm.	54; Admir. corroded shellside.
	15	Absorber Inter Cooler	Retube	Admir.	12	159U	3/4	18	111/100	260	Lt. HC + NH_4 + H_2S	85/100	55	CTW		54; Admir. corroded shellside.
FCC	16	Feed Preheat (2)	Retube	5 Cr.	47	1132 each	3/4	16	365/405	150	Gas Oil Feed.	560/415	65	Hvy. Cycle Oil	5 Cr.	41; 5 Cr failed by TS pitting at rate pf 0.5mm/year.
FCC	17	Debutanizer Reboiler	Retube	5 Cr.	47	1920	3/4	16	402/424	260	Naphtha	560/460	90	Hvy. Cycle Oil	5 Cr.	41; tube side pitting of 5 Cr.
Gas Plants	18	Unsat - Hi Press. Compr. after cooler.	New	-	-	541	3/4	16	180/100	455	HC + H_2S + NH_3 + CN	90/110	70	C.T.W.	C.S.	38; no insp.; no failures.
	19	Unsat - Debutanizer Reboiler	Retube	430	73	1350	3/4	16	573/341	160	Hvy. Cycle Oil; H_2S + S + HC	335/396	90	Naphtha	5 Cr	46; no corrosion; TP-430 perforated in 73 mos.
	20	Sat-Debut. Feed Heater	Rebuilt	C.S.	36	1056	3/4	16	207/269	175	Naphtha + S + CL	387/294	160	Naphtha + S + Cl	C.S.	58; XM-27 and XM-33 are both excellent.
	21	Sat-Debut. Reboiler	Retube	430	68 with perf.	1792	3/4	16/	369/402	166	Naphtha + S + Cl	541/426	135	Naphtha + S + Cl	C.S.	34 mos; 430 lasted 68 mos; sample XM-27 tubes corroded 2mm in storage after 17 mos. good service; retube with XM-27 in 6/77; no corr. in service since.

TABLE 14 (continued)

Process	Case No.	Service	New or Retube	Previous Alloys	Life(1) Previous Alloy	No. of Tubes	Size (2)	BWG	Shell Side (3)(4) Temp. Range	Shell Side Press	Shell Side Stream	Tubeside Temp. Range	Tubeside Press	Tubeside Stream	Tube Sheet Mat'l	History(5)(6)
	22	Sat-Debut. Reboiler	Retube	C.S.	no history	1792	3/4	16	369/402	166	Naphtha + S + Cl	541/426	135	Naphtha + S + Cl.	C.S.	47 mos. with perforation due to 9 mo. storage uncleaned; partial retube with XM-33.
	23	Unsat-Debut Reboiler	Retube	5 Cr.	18	1847	1	18	374/414	176	Deb. Botts.	680/470	105	FCC Main Col. Botts.	5 Cr.	23; C.S. corroded and fouled, could not pull bundle easily.
Gas Plants	24 25 26	Sat-De-ethanizer - top, middle & bottom Inter-coolers	Retube	Admir.	24	115U each	3/4	18	112/100	115	NC; H_2S; NH_3	85/100	58	Water	C.S.	36 on 3 indiv. units. One unit showed pitting while out of service.
HDS	27	H_2S stripper O.H. Cond.	Retube	Adm. 430 S.S.	38 24	440	3/4	18	190/110	80	HC + H_2O + NH_3 + H_2S	90/110	70	CTW	316	73; Adm. suffered deposit attack. 430 pitted waterside.
	28	Compressor After cooler (double pipe)	New	-	-	8	1¼	12	90/110	70	CTW	240/115	35	Recycle Gas	None	28; some pitting at welds; 1 mm deep on waterside.
	29	Feed/ Effluent	Retube	410	156 then 11 mos.	196U	1	16	675/500	482	React. Feed	250/400	347	React. Effl.	410	96; 410 good service for 144 mos.; then constant perforation.
	30	Feed/ Effluent	Retube	410	168	196U	1	16	500/360	482	React. Feed	120/250	347	React. Effl.	410	77; 410-168 mos. of service originally.
	31	Fractionator Reboiler	Retube	C.S. 304 430	38 88 29	244U	1	16	500	225	Fract. Botts.	615/520	225	React. Effl.	304 Cl	54; fine pits in 24 months; none since.
	32	Diesel Stripper Feed vs. React. Efflu.	Retube	C.S.	12 to 19 mos.	518U	3/4	16	301/281	480	Reactor Effluent	134/300	135	Stripper Feed	C.S.	71; minute sharp pits on O.D.
	33	Effluent Cooler	Retube	CA 715 70 30 Cu Ni	36	800	3/4	16	260/106		HC + H_2S + NH_4	45/70	50	CWT	C.S.	64; 4 leaking tubes from SW corr.; now using CTW.
HDS	34	Hi Press. Sep. Cond. (air cooler)	Retube	C.S.	24	279	1	18			Air	315/110	750	HC + H_2S + NH_4 + NH_4Cl	C.S.	37

TABLE 14 (continued)

Process	Case No.	Service	New or Retube	Previous Alloys	Life(1) Previous Alloy	No. of Tubes	Size (2)	BWG	Shell Side (3)(4) Temp. Range	Press	Stream	Tubeside Temp. Range	Press	Stream	Tube Sheet Mat'l	History(5)(6)
Hydrogen Plant	35	MEA Regenerator Reboiler	Retube	304	24	721U	3/4	16	240	25	MEA + CO_2 + Polymers	360/260	320	Shift Gas	316	50; 304 pitted on top half of bundle.
	36	MEA Reclaimer	Retube	304	24	195	3/4	16	240/280	25	Foul MEA	365	150	Steam	Monel Cl	50; 304 had hvy. corr. next to tube sheet.
Gas Treating	37	MEA Rich - Lean	Retube	C.S.	42	21	3/4	16	260/199	18	Lean MEA	135/199	66	Rich MEA	C.S.	53; XM-27 at top; 329 remainder; good service with XM-27 and 329.
	38	DEA heat recovery exchanger	Retube	C.S.	36 Deep pitting	152U	3/4	18	214	1	Steam Condensate	85/115	65	CTW	C.S.	36
	39	MEA Reclaimer	Retube				3/4	16	300/400	20	Degraded MEA			Steam		2 only; E-Brite corroded in crevice next to tube sheet.
Lube	40	Phenolic Water Cooler	Retube	70-30	84	64	1	16	230/150	VAC	Phenolic water	90/115	25	CTW	304 Cl	47
	41	Waste Heat Steam Gen.	Retube	C.S 304 Monel 70-30	7,37 20,24 22,22,29 35,44	29	3/4	16	100/335	95	Process condensate	585/430		Raffinate	Monel	68
Lube	42	Waste Heat Steam Gen.	Retube	Monel C.S.	31,33 14,14	38	3/4	16	340	135	Feedwater	574	100	Raffinate	Monel	66; Monel & steel failed from CO_2
	43	Waste Heat Steam Gen.	Retube	Monel C.S.	33,32, 19,26 4,20	38	3/4	16	340	135	Water + Cresylic Acid	625/585	200	Extract	Monel	66; Monel & steel failed from CO_2
	44	Extract Flash O.H. Cond. (Phenol Treater)	Retube	C.S.	19	754	3/4	16	364/244	15	Phenol 90%; oil 7%; Cresol 3%	160/264	40	Phenol 66%; Extr. 25%; water	C.S.	30 mos. life only; salt water intrusion in plant caused corrosion.
Sour Water Stripper	45	Phenolic Feed/Efflu.	Retube	C.S.	24	84U	1	14	252/130	16	H_2O + traces of contam.	100/220	150	H_2O + NH_3 + COOH + H_2S + 5	C.S.	94; tubes in good cond.

TABLE 14 (continued)

									Shell Side (3)(4)			Tubeside				
Process	Case No.	Service	New or Retube	Previous Alloys	Life(1) Previous Alloy	No. of Tubes	Size (2)	BWG	Temp. Range	Press	Stream	Temp. Range	Press	Stream	Tube Sheet Mat'l	History (5)(6)
	46	Non-Phenol Feed/Efflu.	Retube	C.S.	24	84U	1	14	252/130	16	H_2O + traces of contam.	100/220	130	H_2O + NH_3 + H_2S + S	C.S.	94; tubes in good cond.
	47	Phenolic Reboiler	Retube	C.S.	24	167U	3/4	14	252/275	16	H_2O + trace of contam.	280/200	50	Steam	316 Cl	94; small cracks at tube ends - rolling; contaminated seam welds.
	48	Non-Phenol Reboiler	Retube	C.S.	24	167U	3/4	14	252/300	16	H_2O + trace of contam.	358/200	150	steam	316 Cl	33 Mos.; slight cracks in U bends; faulty tubes.
Sour Water Stripper	49	Desalter Eff. vs. Feed.	Retube	C.S.	12	270	3/4	18	50/180	53	Sour feed	260/175	120	Desalted waste water	C.S.	33; carbon steel displayed heavy pitting on feed side.
Tail Gas	50	Stretford cooler	New	C.S.	—	440	3/4	16	125/100	80	Stret-ford	85/120	75	CTW	C.S.	60 mos. on one unit.
	51	Two Units														5 mos. to failure on second unit. Pitted during S/D Carbon steel corroded under deposits.
Misc.	52	Ammonia Cooler	Retube	C.S. Adm. Bi. Met.	-	224	2	16	230	200	NH_3	90/110	0	CTW	Adm. clad	53; leak at interface caused corr. of Adm. & clad.

Notes:
(1) Life is in months
(2) Size is in inches; 3/4 inch equals 19.1mm; 1" = 25.4mm
(3) Temperatures are in °F; °C = (°F-32) X 5/9
(4) Press is in PSIG; X 6.89 = kN/m^2
(5) E-Brite (XM-27) used in all cases except as noted in cases 9, 20, & 22.
(6) Numbers are months of life to present with no problems or no corrosion. Problems and corrosion for XM-27, XM-33 or the original alloy are specifically noted.
(7) Abbreviations:
CTW - Cooling Tower Water
HC - Hydrocarbon; Hydrocracker
ADM - Admir - Admiralty
U - U bent tubes
HDS - Hydrodesulfurizer
FCC - Fluid catalytic cracker

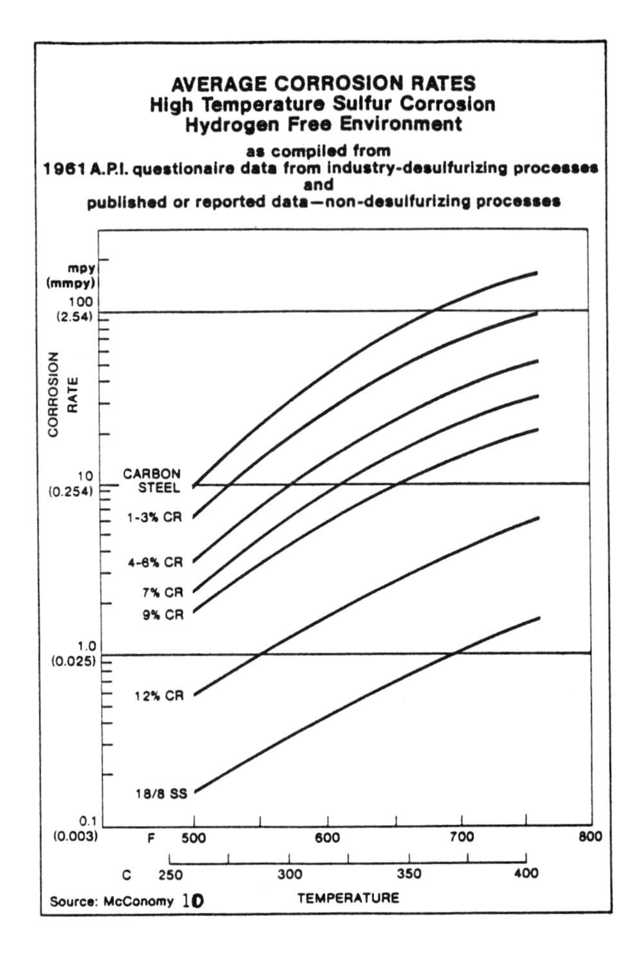

1. Average Corrosion Rates of Iron Base Alloys in High Temperature H_2S.

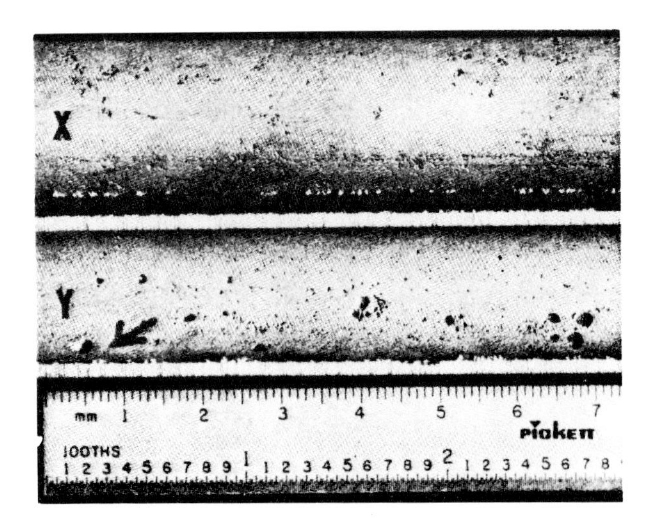

2. TP-430 Tubing from Debutanizer Reboiler - 73 Months Service.

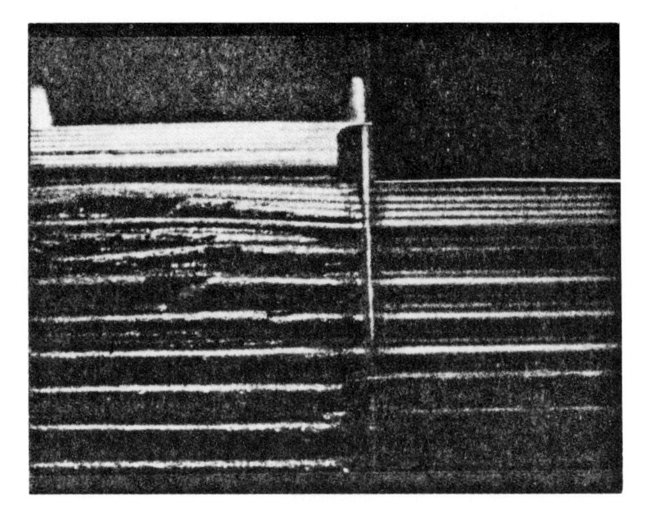

3. Carbon Steel After 36 Months Debutanizer Reboiler Service.

4. XM-27 (E-BRITE) After 24 Months in an H_2S Stripper Condenser-Shellside.

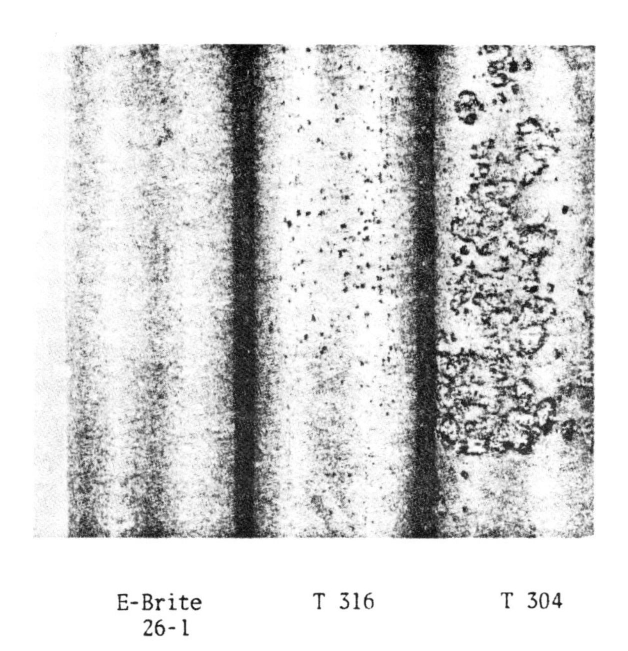

6. Sample Tubes From an MEA Regenerator Reboiler After 24 Months.

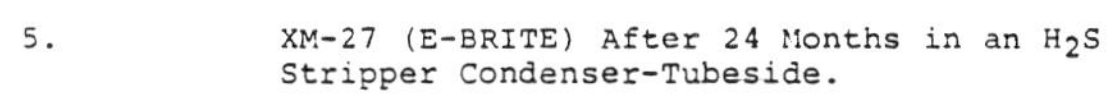

5. XM-27 (E-BRITE) After 24 Months in an H_2S Stripper Condenser-Tubeside.

Appendix

Pessall and Nurminen have correlated the pitting potential, E_c^{scr}, determined by polarization tests in deaerated seawater, with resistance to crevice attack, determined in ferric chloride immersion tests. The contours of equal degree of resistance to crevice corrosion do not occur at constant critical potential values, indicating that the resistance to crevice attack is a function not only of the critical potential but also of the sum of Cr plus Mo. Comparative corrosion resistance of several commercial alloys is shown, and the graph can also be used to determine the expected properties of newly developed alloys.

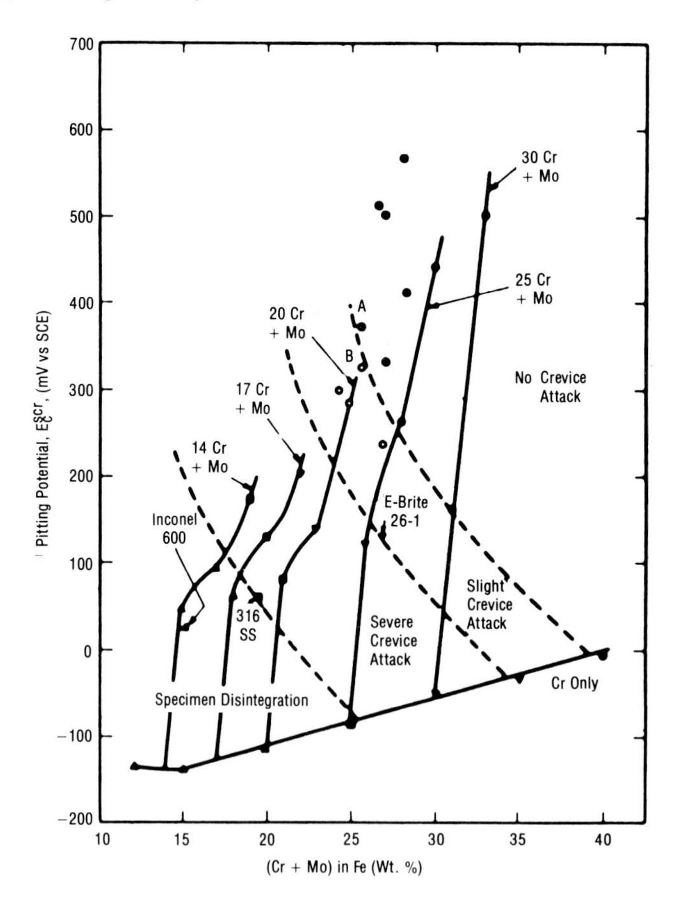

Reprinted with permission from N. Pessall and J. I. Nurminen, "Development of Ferritic Stainless Steels for Use in Desalination Plants," Paper No. 65, Corrosion/73, National Association of Corrosion Engineers, © 1973 NACE.

INDEX

D

E

G

T

U

W

X